3rd International Conference on Global Sustainability and Chemical Engineering (ICGSCE 2017)

AA006765

IOP Conference Series: Materials Science and Engineering
Volume 358

Putrajaya, Malaysia
15 – 16 February 2017

ISBN: 978-1-5108-6340-8
ISSN: 1757-8981

Printed from e-media with permission by:

Curran Associates, Inc.
57 Morehouse Lane
Red Hook, NY 12571

Some format issues inherent in the e-media version may also appear in this print version.

Copyright© (2017) by the Institute of Physics
All rights reserved. The material featured in this book is subject to
IOP copyright protection, unless otherwise indicated.

Printed by Curran Associates, Inc. (2018)

For permission requests, please contact the Institute of Physics
at the address below.

Institute of Physics
Dirac House, Temple Back
Bristol BS1 6BE UK

Phone: 44 1 17 929 7481
Fax: 44 1 17 920 0979

techtracking@iop.org

Additional copies of this publication are available from:

Curran Associates, Inc.
57 Morehouse Lane
Red Hook, NY 12571 USA
Phone: 845-758-0400
Fax: 845-758-2633
Email: curran@proceedings.com
Web: www.proceedings.com

Table of contents

Volume 358

3rd International Conference on Global Sustainability and Chemical Engineering (ICGSCE)

15–16 February 2017, Putrajaya, Malaysia

Accepted papers received: 30 April 2018
Published online: 22 May 2018

Preface

Papers

Renewable and Sustainable Energy Technology

Biofuels

Carbon/Water Foot Printing

Green Technology

Environmental Technology

IOP Conf. Series: Materials Science and Engineering **358** (2018) 011001 doi:10.1088/1757-899X/358/1/011001

IOP Conference Series: Materials Science and Engineering

3rd International Conference on Global Sustainability and Chemical Engineering (ICGSCE 2017)

15th to 16th February 2017

Putrajaya, Malaysia

Content from this work may be used under the terms of the Creative Commons Attribution 3.0 licence. Any further distribution of this work must maintain attribution to the author(s) and the title of the work, journal citation and DOI.
Published under licence by IOP Publishing Ltd

IOP Conf. Series: Materials Science and Engineering **358** (2018) 011001 doi:10.1088/1757-899X/358/1/011001

PREFACE

It is our great pleasure to welcome you to 2017 The 3rd International Conference on Global Sustainability and Chemical Engineering (ICGSCE) which has been held in Marriot Hotel, Putrajaya, Malaysia on 15th to 16th February 2017. ICGSCE is an international event organized by the Faculty of Chemical Engineering, Universiti Teknologi MARA (UiTM), Malaysia, with the purpose to highlight the roles and responsibilities that modern chemical engineering play in preserving the environment whilst satisfying the continually increasing demands of consumers and industry.

ICGSCE endeavors to serve as a platform that allows the exploration of new ideas and developments through the various spectrums of chemical engineering emphasizing on global sustainability. Building on the success of the previous conferences, ICGSCE 2017 with the theme "Catalysing Technologies for Sustainable Solutions" is therefore aiming to showcase the advances in research and technologies through a variety of state-of-the-art sessions which include keynote lectures, oral and poster presentations, and discussions. ICGSCE 2017 provides the medium for scientists and academicians from universities and industries all around the globe to promote and share various new issues and sustainable developments in the identified scopes – renewable and sustainable energy technology, sustainable policy and education, global warming, biofuels, carbon sequestration, carbon/water footprinting, green technology, environmental technology, fuel cells, biotechnology, food technology, separation technology, advanced materials, nanotechnology, process modeling and simulation, process control, oil and gas exploration and production technology, oleochemical technology, process safety and risk assessment. In line with the 2017 theme, the conference itself is an opportunity for the researchers to initiate networking and establish new research collaborations to deliver sustainable solutions.

This proceeding present a selection from papers submitted to the conference from universities, research institute and industries. All of the papers were subjected to peer-review by conference committee members and international reviewers. The papers selected depended on their quality and their relevancy to the conference.

We would like to thank all the authors who have contributed to this volume and also to the organizing committee, reviewers, speakers, chairpersons, sponsors and all the conference participants for their support to ICGSCE2017.

Assoc Prof Dr Noor Fitrah Abu Bakar
Conference Chair
ICGSCE2017

IOP Conf. Series: Materials Science and Engineering **358** (2018) 011001 doi:10.1088/1757-899X/358/1/011001

Conference Committee

Partron
Prof. Emeritus Dato'Hassan Said

Advisor:
Associate Prof. Dr Norazah Abd Rahman

Conference Chair:
Dr Noor Fitrah Abu Bakar

Deputy Chairperson:
Associate Prof. Dr Ruzitah Mohd Salleh

Editorial and Publication Board
Dr Safari Zainal (Head)
Dr Fazlena Hamzah
Dr Abdul Hadi

International Committee
Professor Dr. Jakie Ying, Institute of Bioengineering and Nanotechnology, Singapore
Professor Dr. Hidehiro Kamiya, Tokyo University of Agriculture and Technology, Japan
Assoc. Prof. Dr. Nayef Mohamed Ghasem, UAE University, United Arab Emirates
Professor Dr. J.C Diniz da Costa, University of Queensland, Australia
Professor Dr. Kang Li, Imperial College London, United Kingdom
Assoc. Prof. Dr. Wuled Lenggoro, Tokyo University of Agriculture and Technology, Japan
Dr. Motoyuki Iijima, Yokohama National University, Japan
Professor Dr. Jirí Jaromír Klemeš, University of Pannonia, Hungary
Professor Mohammed Farid, The University of Auckland, New Zealand
Dr. Mahidin A. Taleb, University of Syiah Kuala, Indonesia
Dr. Zuchra Helwani, University of Riau, Indonesia
Professor Dr. Pattarapan Prasassarakich, Chulalongkorn University, Thailand
Professor Dr. Masayuki Horio, Tokyo University of Agriculture and Technology, Japan
Mr. Clive Walter, Walter Energy Limited
Assoc. Prof. Yoichi Tominaga, Tokyo University of Agriculture and Technology, Japan
Assoc. Prof. Susumu Inasawa, Tokyo University of Agriculture and Technology, Japan

IOP Conf. Series: Materials Science and Engineering **358** (2018) 011001 doi:10.1088/1757-899X/358/1/011001

IOP Conf. Series: Materials Science and Engineering **358** (2018) 011001 doi:10.1088/1757-899X/358/1/011001

IOP Conf. Series: Materials Science and Engineering **358** (2018) 011002 doi:10.1088/1757-899X/358/1/011002

Peer review statement

All papers published in this volume of *IOP Conference Series: Materials Science and Engineering* have been peer reviewed through processes administered by the proceedings Editors. Reviews were conducted by expert referees to the professional and scientific standards expected of a proceedings journal published by IOP Publishing.

Content from this work may be used under the terms of the Creative Commons Attribution 3.0 licence. Any further distribution of this work must maintain attribution to the author(s) and the title of the work, journal citation and DOI.
Published under licence by IOP Publishing Ltd

Biohydrogen Production from Pineapple Waste: Effect of Substrate Concentration and Acid Pretreatment

K Cahyari[1], A M Putri[2], E D Oktaviani[2], M A Hidayat[2] and J D Norajsha[2]

[1] Dept. of Chemical Engineering, Faculty of Industrial Technology, Universitas Islam Indonesia, Jl. Kaliurang km 14.5, Yogyakarta 55584 Indonesia
[2] Undergraduate Student, Dept. of Chemical Engineering, Faculty of Industrial Technology, Universitas Islam Indonesia, Jl. Kaliurang km 14.5, Yogyakarta 55584 Indonesia

khamdan.cahyari@uii.ac.id

Abstract. Biohydrogen is the ultimate choice of energy carrier in future due to its superior qualities such as fewer greenhouse gases emission, high energy density (142 kJ/gram), and high energy conversion using a fuel cell. Production of biohydrogen from organic waste e.g. pineapple waste offers a simultaneous solution for renewable energy production and waste management. It is estimated that pineapple cultivation in Indonesia generated more than 1 million ton/year comprising of rotten pineapple fruit, leaves, and stems. Majority of this waste is dumped into landfill area without any treatments which lead to many environmental problems. This research was meant to investigate the utilization of pineapple waste i.e. peel and the core of pineapple fruit and leaves to produce biohydrogen through mesophilic dark fermentation (30°C, 1 atm, pH 5.0). Effect of dilute acid treatment and substrate concentration was particularly investigated in these experiments. Peel and core of pineapple waste were subjected to fermentation at 3 various substrate concentration i.e. 8.8, 17.6 and 26.4-gram VS/liter. Meanwhile, pineapple leaves were pretreated using dilute acid (H_2SO_4) at 0.2, 0.3 and 0.4 N and followed by dark fermentation. Results show that the highest yield of biohydrogen was obtained at a substrate concentration of 26.4-gram VS/liter both for peel and core of the waste. Pretreatment using dilute acid (H_2SO_4) 0.3 N might improve fermentation process with a higher yield at 0.8 ml/gram VS. Hydrogen percentage in biogas produced during fermentation process was in the range between 5 – 32% of volume ratio. In summary, it is possible to utilize pineapple waste for production of biohydrogen at an optimum substrate concentration of 26.4-gram VS/liter and acid pretreatment (H_2SO_4) of 0.3 N.
Keywords: hydrogen, fermentation, pineapple waste, core, peel, leaf

1. Introduction
Hydrogen is the cleanest energy carrier to substitute conventional fossil fuel for the betterment of the environment. It emits fewer greenhouses gases, more energy efficient conversion up to 49.7% through fuel cell platform, higher energy densities, and it is renewable [1]. The most feasible and common method to generate hydrogen is through water electrolysis, only if there is sufficient free excess electricity supply generated from solar [2], hydropower, wind, geothermal or else [3, 4]. However, recent studies show that the ratio of energy input-to-energy output through the electrolysis method is still high indicating that this method has negative energy balance. On the other hand, there is a more promising method to produce hydrogen from organic waste with least amount of energy input i.e. dark fermentation [5]. This production of hydrogen via dark fermentation process exploits microbial community to metabolize organic matter into volatile fatty acids along with electron transfer to hydrogenase to convert a proton into hydrogen (H_2). This method is influenced by many factors e.g.

Content from this work may be used under the terms of the Creative Commons Attribution 3.0 licence. Any further distribution of this work must maintain attribution to the author(s) and the title of the work, journal citation and DOI.
Published under licence by IOP Publishing Ltd

type and concentration of substrate, temperature, pressure, pH, indigenous chemical inhibitors, organic loading rate, hydraulic retention time and reactor configuration [6, 7].

Pineapple waste comprises of peel and core of the fruit, leaves, and the stem is produced from cultivation and industrial processing of pineapple fruit. It leaves behind some portion of waste comprising of those components in huge quantities. This waste is commonly dumped in open area without any proper treatment creating environmental problems such as bad odors, health-related problems, contamination, and global warming. Pineapple fruit is produced more than 1.8 million tons/year in Indonesia of which some portion turned into waste as rotten fruit [8]. Core, peel, and crown of pineapple fruit become waste during processing of pineapple fruit in many fruit industries. Moreover, leaves and pseudo-stem of pineapple plant are considered as waste since they are commonly treated through open-air combustion to reduce its volumetric quantity. Therefore, it is estimated that more than 1 million tons of pineapple waste are generated every year.

Physio-chemical characteristics of pineapple waste have been reported in many previous types of research. Peel of the pineapple fruit has contents of 75 – 80% moisture, 35% cellulose, 19,7 hemicellulose, 16% lignin, 4.7% total ash, 0.46% total fat, 23.71% total crude fiber, and 0.33% total proteins [9]. It also contains of 27.08% total carbohydrate, 1.9 mg/kg magnesium, 26.096 mg/kg potassium, and 298.184 mg/kg zinc [10]. Leaf of pineapple plantation which has been decorticated contained 70.98% alpha cellulose, 15.34% hemicellulose, 4.9% lignin, 0.96% fats and waxes, 3.0% pectin matter and 0.95% total ash [11]. Extraction of a crude waste mixture of pineapple parts namely crown, peel and core using two-stage ultrafiltration (75 kiloDaltons (kDa) or 7.5 nm of pore size and 10 kDa membrane filters) showed bromelain enzyme activity of 129.4±23.0 Casein Digestion Unit (CDU) per ml extract or 76.1±9.5 CDU/milligram protein [12]. Saccharification of core and peel of pineapple fruit using microwave pretreatment (270 W, 15 minutes) and boiling pretreatment (100°C, 10 minutes) could deliver sugars yield of 35 gram/kilogram fresh waste of peel and 30 gram/kilogram fresh waste of core, respectively. Moreover, pretreatment heat at high pressure with an autoclave (121°C, 205.24 kPa, 30 minutes) could generate higher sugars yield of 60 and 50 gram/kilogram fresh waste of peel and core, respectively [13]. This research was meant to investigate the utilization of pineapple waste namely peel and core of the fruit as well as leaves of pineapple trees as raw material for biohydrogen production. Particularly, it investigated the effect of substrate concentration and concentration of dilute acid pretreatment.

2. Materials and Method

2.1. Feedstock

Peel and core of pineapple fruit were collect from local fruit market in Yogyakarta Province, Indonesia. Leaves of pineapple tree were collect from local farming in the province. Those materials were cut and ground individually to reduce the size in the range of 1-5 mm in diameter, thereafter stored at cold room (5°C, 1 atm) prior to the experimental run, called as feedstock.

2.2. Inoculums

Hydrogen-producing microorganisms were originally from the residual sludge of cows dung biogas plant followed with acidic pretreatment at pH 5.0 for 24 hours and heat-shock treatment (95°C, 45 min).

2.3. Experimental procedures

The experiment was conduct using 120 ml serum vial bottles as a bioreactor. Peel and core of pineapple were subject into dilute acid pretreatment using 0.3 N H_2SO_4 solution at 60°C for 30 minutes. The solid fraction of pretreated feedstock of peel and core was individually weighted and introduced into bioreactor at 4, 8, and 12 grams, added with 20 ml of inoculums,10 ml buffer solution, and distilled water. The bioreactor was then closed and sealed using butyl rubber stopper and aluminum cap, followed by N_2 gas flushing for 3 minutes. Fermentation was carried out at pH 5.0, 30°C and 1 atm. Dilute acid pretreatment of pineapple leaves was conducted through submerging of the feedstock at 0.2, 0.3 and 0.4 N of sulfuric acid (H_2SO_4) at 60°C and 30 minute's exposure time. Thereafter, the solid fraction of the feedstock was subject to the fermentation experiment procedure. Each experiment was conduct in duplicate.

2.4. Analysis
Hydrogen content was analyzed using gas chromatography, which equipped with molecular sieve 5A column and thermal conductivity detector with injection, oven and detector temperatures at 50°C, 80°C, and 80°C, respectively. Helium gas was used as a gas carrier at volumetric flowrate 40 ml/min. Meanwhile, physio-chemical properties namely moisture content, total solid (TS), volatile solid (VS) and ash measured based on the standard method of waste and wastewater analysis according to American Public Health Association (APHA) [14].

3. Results and Discussion
Dark fermentation of organic waste occurs through complex metabolic pathways according to the simplified equation in (3.1) and (3.2) [15]. When glucose is the main substrate, the highest hydrogen yield is 4 mol H_2/mol glucose or equivalent to 498 ml STP H_2/gram glucose theoretically. However, when pineapple waste (peel, core, and leaves) is utilized as a substrate, those materials contain a non-readily fermentable substrate for hydrogen-producing microorganisms to digest and produce hydrogen. It has to be pre-treated to unleashed organo-polymeric structures into fermentable sugars, called as saccharification process. Method to conduct this process was through dilute acid pre-treatment using sulphuric acid (H_2SO_4) at various concentration of 0.2, 0.3, and 0.4 N.

$$C_6H_{12}O_6 + 2H_2O \rightarrow 2CH_3COOH + 4H_2 + 2CO_2 \quad (3.1)$$

$$C_6H_{12}O_6 + 2H_2O \rightarrow 2(CH_2)_3COOH + 2H_2 + 2CO_2 \quad (3.2)$$

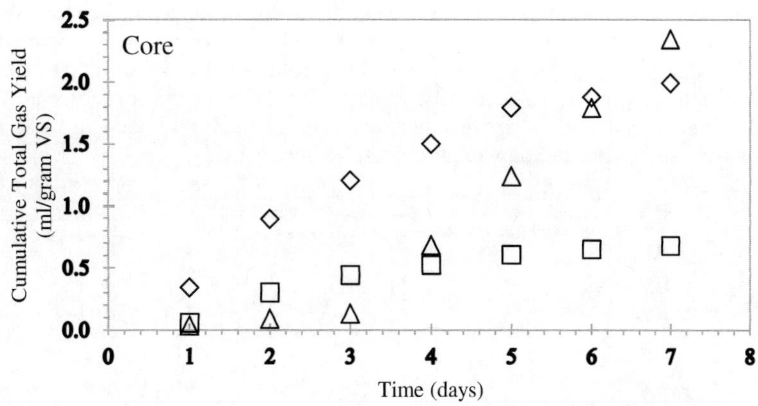

Figure 1. Total gas yield of pineapple fruit core at substrate concentration (□ 8.8 gram VS/liter, ◊ 17.6, Δ 26.4)

Profile of cumulative total gas yield during dark fermentation of pineapple fruit core at a substrate concentration of 8.8, 17.6 and 26.4-gram VS/liter was illustrated in Figure 1. It was noticed that the highest yield was achieved at a substrate concentration of 26.4-gram VS/liter corresponding to a yield of 2.4 ml STP/gram VS. It indicated that the higher concentration of substrate, the higher quantity of the organic compound which can be metabolized to produce more hydrogen. Meanwhile, a lower concentration of substrate of 8.8-gram VS/liter culture produced lower yield due to the limited quantity of carbon source for the microorganism to grow and metabolite.

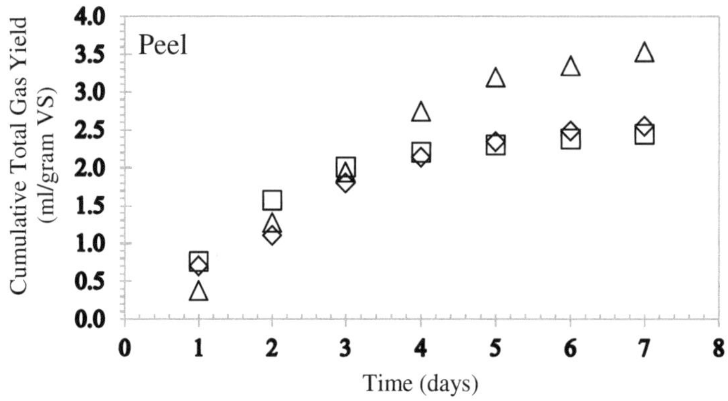

Figure 2. Total gas yield of pineapple fruit peel at substrate concentration (□ 8.8 gram VS/liter, ◊ 17.6, Δ 26.4)

Profile of total gas yield production from fermentation of pineapple peel at substrate concentration 8.8, 17.6 and 26.4-gram VS/liter culture was depicted in figure 2. A similar pattern of the gas production between peel and core was noticed in this experiment. It was demonstrated that the higher substrate concentration, the higher total gas production. Furthermore, the highest yield was also achieved at a substrate concentration of 26.4-gram VS/liter culture with a yield of 3.5 ml STP/gram VS. When higher substrate concentration introduced into fermentation, more bromelain enzymes was also added. It was reported that crude waste mixture of pineapple peel and core contain 129 - 153 CDU/ml extract [16]. Bromelain is proteolytic enzymes which can act as antimicrobial agents toward the majority of the microbial community of bacteria in dark fermentation [17]. In this experiment, the effect of bromelain as antimicrobial agents was not demonstrated as inhibiting agent toward biohydrogen production up to substrate concentration of 26.4-gram VS/liter culture.

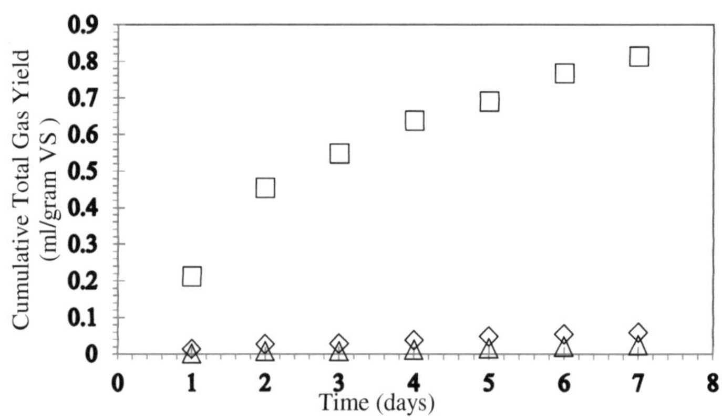

Figure 3. Profile of total gas yield of pineapple leaves fermentation at acid concentration of pre-treatment (□ 0.2 N, ◊ 0.3 N, Δ 0.4 N)

Figure 3 demonstrates profile of total gas yield from the fermentation of pineapple leaves with dilute acid pre-treatment at a various acid (H_2SO_4) concentration of 0.2, 0.3 and 0.4 N. The highest yield was achieved at pre-treatment condition of 0.3 N H_2SO_4 with total gas yield 0.8 ml STP/gram VS. This yield is considered lower compare to the yield from the fermentation of core and peel due to its higher content of cellulose and hemicellulose. It indicated that dilute acid pre-treatment was not able to provide a significant quantity of fermentable sugars from cellulose and hemicellulose to be digested by hydrogen-producing microorganisms. When limited carbon source is introduced into dark fermentation, it affects the effectiveness of microbial metabolism to grow and produce hydrogen properly. It is suggested to improve acid pre-treatment condition for increasing the concentration of fermentable sugars. Analysis of hydrogen percentage of the total gas produced was conducted with results value in the range between 5 to 32% volume ratio of hydrogen.

4. Conclusion

It is possible to produce biohydrogen from pineapple waste (peel, core, and leaf). This method can solve both environmental problems of the waste and generation of renewable energy source. Substrate concentration and acid pre-treatment affect biohydrogen production with the optimum condition at 26.4-gram VS/liter substrate concentration and 0.3 N H_2SO_4 acid pre-treatment. Total gas yield of peel fermentation was higher than that of core corresponding to 3.5 ml STP/gram VS with hydrogen content in the range between 5 to 32 % volume ratio.

Acknowledgment

Authors acknowledge the Department of Chemical Engineering, Faculty of Industrial Technology, Universitas Islam Indonesia for providing financial support.

References

[1] Purnima P and Jayanti S 2016 *Int. J. Hydrogen Energy* **41** 13800
[2] Bhattacharyya R, Misra A, and Sandeep K C 2017 *Energy Convers. Manag.* **133** 1
[3] Basile A and Lulianelli A 2014 *Advances in Hydrogen Production, Storage and Distribution* (Cambridge: Woodhead Publishing) p 159
[4] Yilmaz C and Kanoglu M 2014 *Energy* **69** 592
[5] Ren N, Guo W, Liu B, Cao, G and Ding J 2011 *Curr. Opin. Biotechnol.* **22** 365
[6] Heyndrickx M, De Vos P, Hibau B, Stevensm P and De Ley J 1987 *Syst. Appl. Microbiol.* **9** 163
[7] Wang J and Wan W 2009 *Int. J. Hydrogen Energy* **34** 799
[8] FAOUN 2015 *Statistical Pocketbook World Food and Agriculture* (Rome: FAO Publishing)
[9] Khedkar M A, Nimbalkar P R, Gaikwad S G, Chavan P V and Bankar S B 2017 *Bioresour. Technol.* **225** 359
[10] Vega-Castro O, Contreras-Calderon J, León E, Segura A, Arias M, Pérez L and Sobral P J A 2016 *J. Biotechnol.* **231** 232
[11] Hazarika D, Gogoi N, Jose S, Das R and Basu G 2017 *J. Clean Prod.* **141** 580
[12] Nor M Z M, Ramchandran L, Duke M, Vasiljevic T 2016 *Food Bioprod. Process* **98** 142
[13] Roda A, De Faveri DM, Giacosa S, Dordoni R, and Lambri M 2016 *J. Clean Prod.* **112** 4477
[14] APHA 2005 *Standard Methods for the Examination Water and Wastewater* (Washington: APHA Publication)
[15] Cahyari K, Syamsiah S, and Prasetya A 2016 *IOP Conf. Ser.: Mater. Sci. Eng.* 162 012013
[16] Chaurasiya R S and Umesh H H 2013 *Se.p Purif. Technol.* **111** 90
[17] Praveen N C, Rajesh A, Madan M, Chaurasia V R, Hiremath N V and Sharma A M 2014 *J. Int. oral. Heal.* **6** 96

IOP Conf. Series: Materials Science and Engineering **358** (2018) 012002 doi:10.1088/1757-899X/358/1/012002

Integrated Assessment of Palm Oil Mill Residues to Sustainable Electricity System (POMR-SES): A Case Study from Peninsular Malaysia

I F Md Jaye, J Sadhukhan and R J. Murphy*

Centre of Environment and Sustainability, University of Surrey, UK

i.mdjaye@surrey.ac.uk, j.sadhukhan@surrey.ac.uk, *rj.murphy@surrey.ac.uk

Abstract. Generating electricity from biomass are undeniably gives huge advantages to the energy security, environmental protection and the social development. Nevertheless, it always been negatively claimed as not economically competitive as compared to the conventional electricity generation system using fossil fuel. Due to the unfair subsidies given to renewable energy based fuel and the maturity of conventional electricity generation system, the commercialization of this system is rather discouraging. The uniqueness of the chemical and physical properties of the biomass and the functionality of the system are fully depending on the availability of the biomass resources, the capital expenditure of the system is relatively expensive. To remain competitive, biomass based system must be developed in their most economical form. Therefore the justification of the economies of scale of such system is become essential. This study will provide a comprehensive review of process to select an appropriate size for electricity generation plant from palm oil mill (POM) residues through the combustion of an empty fruit bunch (EFB) and biogas from the anaerobic digestion of palm oil mill effluent (POME) in Peninsular Malaysia using a mathematical model and simulation using ASPEN Plus software package. The system operated at 4 MW capacity is expected to provide a return on investment (ROI) of 20% with a payback period of 6.5 years. It is notably agreed that the correct selection of generation plant size will have a significant impact on overall economic and environmental feasibility of the system.

1. Introduction

The transition towards non-fossil electricity has been an inspiration for many countries of the world for many years. Factors such as the finite nature of fossil fuels, volatility of the global oil market prices, electricity supply security and limiting global warming to no more than 2°C are among the main drivers for this transition. Lignocellulosic plantation residues (e.g palm oil empty fruit bunches, rubber, sugar cane baggase, forestry and timber processing residues) and their availability, indicate a high potential for use as an alternative fuels for sustainable electricity generation [1]. This potential encompasses several other advantages e.g carbon footprint reduction from electricity generation, facilitating 'waste to resources' transition, supporting integrated waste management system and contributing to local social-economic development [2]. In Malaysia, palm oil reported to be the major commodity harvested between 2000-2014 compared to rice, cocoa and rubber. According to the Malaysia Economic Statistics Time Series 2015 report, 72.7 million tonnes of palm oil fresh fruit bunch (FFB) was harvested from 4.6 million hectares of palm oil cultivation in 2014 [3]. This makes the oil palm industries the largest contributor of plantation residues with more than 95% of the country's total accumulated value. In 2015, it has been further estimated that 175.6 million tonnes wet of field and processing residues was produced from 95.38 million tonnes of FFB [4]. The potential electricity generation capacity from empty fruit bunch (EFB) and biogas from the anaerobic digestion of palm oil mill effluent (POME) in 2015 has been estimated to be 9.7 million MWh and 3.3 million MWh respectively. However, only a very small fraction of these amounts (approx. 13,600 MWh) were

Content from this work may be used under the terms of the Creative Commons Attribution 3.0 licence. Any further distribution of this work must maintain attribution to the author(s) and the title of the work, journal citation and DOI.
Published under licence by IOP Publishing Ltd

actually generated, showing that a substantial gap exists in delivering the full potential of this biomass feedstock for electricity production. To date, relatively few attempts have been made to provide suitable business cases to identify the economies of scale of the generation plant to optimize delivery of both economic and environmental benefits. Such business cases can serve as an appropriate technological and investment solutions to reduce the gap between aspiration for, and implementation of, renewable, sustainable electricity generation from palm oil residues.

2. Objective

The research undertakes a detailed resource and system analysis to identify the opportunities and constraints in developing the maximum potential for electricity generation from the combustion of palm oil EFBs and biogas from POME.

3. Methodology

A comprehensive technical, economic and environmental evaluation for a combined heat and power (CHP) system has been conducted based on the process flowchart in Figure 1. The works has been conducted in five different stages begins with the resource scoping activities to establish the full range of potential generation plant size according to the amount of available residues. The scaling process is based on a combination of top-down and bottom-up resource assessment of the current availability of EFB and POME residues from palm oil mill operations in Peninsular Malaysia. A well-established CHP system was selected for this work due to its capability to co-generate heat and power in the same generation plant and having better overall efficiency as compared to only generating electricity [5,6]. A conventional CHP configurations using multiple combination of input fuels as listed below are used in this work to extend the versatility of the technology.

- Configuration 1 – Conventional CHP configuration using EFB as a single input fuel to the boiler.
- Configuration 2 – Conventional CHP configuration using biogas as a single input fuel to the boiler.
- Configuration 3 – Conventional CHP configuration using EFB and biogas as an input fuels to the boiler.
- Configuration 4 – Conventional CHP configuration using EFB as an input fuel to the boiler and biogas as an input to a gas turbine.

The amount of potential electricity generation from EFB and biogas are obtained from theoretical calculation and simulation process using Aspen Plus software taking consideration of stoichiometric condition of full combustion process as suggested in reference [7]. Such techniques are also been adopted in other CHP technical performance evaluation studies [8,9]. In order to determine the amount of potential electricity generation, the extractable heat from the biomass boiler is first to be quantified using Eq. (1).

$$H_{i,n}^{Out} = M_i^{In} \, CV_i \, n_n^{boiler} \tag{1}$$

where $H_{i,c}^{Out}$ is the total extractable heat from biomass i via biomass boiler in configuration n, M_i^{In} is the fed rate of the biomass i , CV_i is the calorific value of dried biomass i and n_n^{boiler} is the efficiency of the boiler in configuration n. The efficiency of biomass based boiler can reach 80% - 90% when it is installed in a CHP system [10]. For this work, an average boiler efficiency of 80% is used throughout the analysis. The extractable heat from the biomass boiler is then used as the input to produce high pressure superheated steam (50 bar, 500°C) to full fill the mill's routine operation steam demand and for electricity generation purposes.

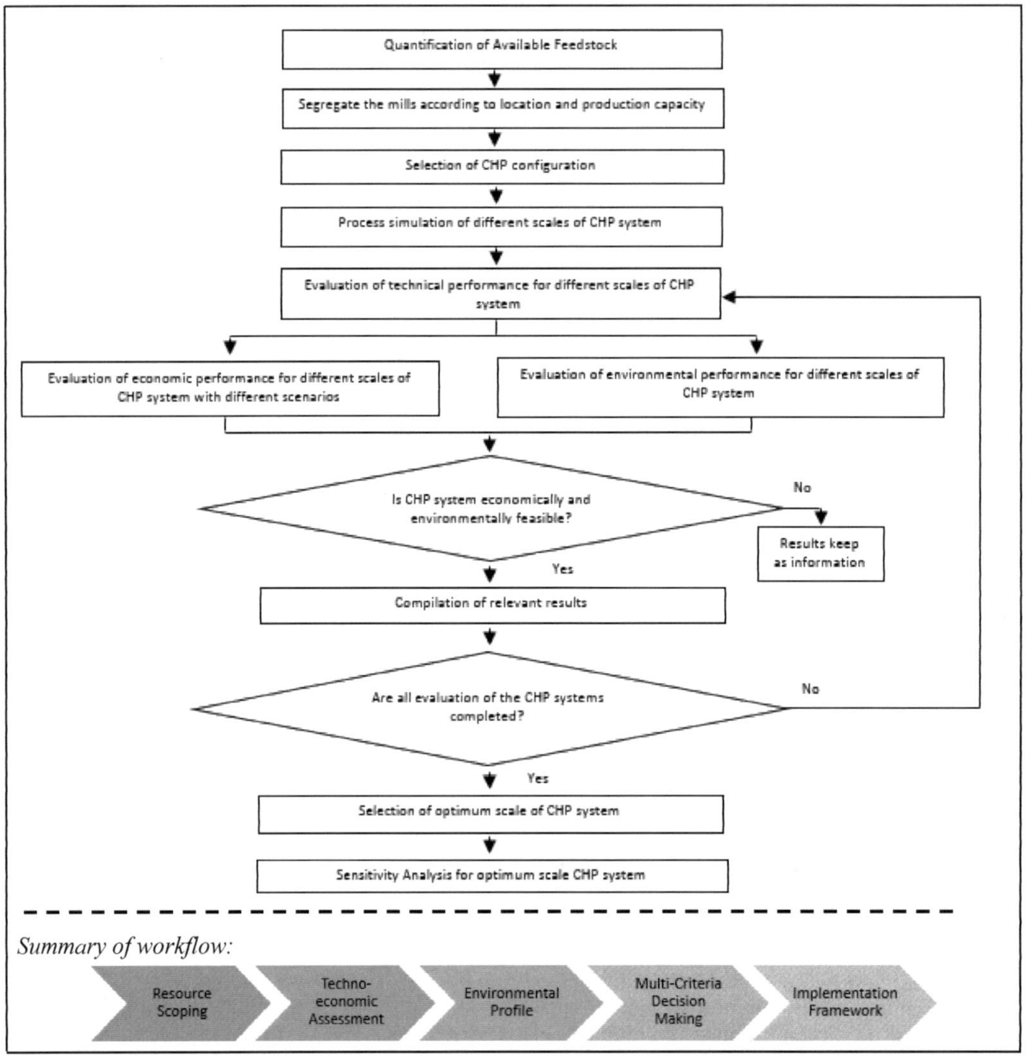

Figure 1: Technological assessment framework for optimizing CHP for oil palm residues

From the amount of extractable heat quantified from Eq. (1), the total mass flow rate of the steam generated (m_{steam}) can be calculated using Eq. (2).

$$H_{i,n}^{Out} = m_{steam}\left[C_p(T_{sat} - T_{BWF}) + \Delta h_{vap} + \left(h_{sup} - h_v\right)\right] \quad (2)$$

The operating condition of steam generation given as:

- Specific enthalpy of superheated steam $\left(h_{sup}\right)$ = 3433.7 kJ/kg.
- Specific enthalpy of saturated steam (h_v) = 2794.2 kJ/kg at 50 bar.
- Saturation temperature of steam (T_{sat}) = 264°C at 50 bar.
- Temperature of BFW (T_{BFW}) = 105°C.
- Heat vaporization of water $\left(\Delta h_{vap}\right)$ = 1639.6 kJ/kg.
- Heat capacity of water $\left(C_p\right)$ = 4.187 kJ/kg/°C.

The available steam flow rate for electricity generation (m_{elec}) can be derived from Eq. (3) where m_{mill} is the steam demand from the mill.

$$m_{elec} = m_{steam} - m_{mill} \tag{3}$$

Note that, the mass flow rate of the steam fed back to the mill must be not more than 50% of the total flow rate of the steam generated to adhere the condensate steam turbine requirement for electricity generation. In due course to maximize the electricity generation potential for this study, the heat from the combustion flue gas is used to dry the EFB from 60% to 11% moisture content [11,12]. Drying the EFB is needed to maximize the recoverable energy from the fuel, improve the efficiency of the boiler and to reduce the air emissions.

A techno-economic assessment spreadsheet is used as a tool to measure the economic and investment feasibility of the CHP system as described in detail in Ng & Sadhukhan and Patel et al. [13,14]. Two different scenarios pertaining the existence of grid connection from the proposed generation location to the national grid is used to represent the actual implementation conditions of such system in Peninsular Malaysia. This scenario will evaluate the volatility of capital and operational expenditure of the CHP system. The assessment method suggested in Wan et al. is adopted for plant lifetime of 15 years and interest rate of 12% [7]. The total capital expenditure (CAPEX) for the CHP is the summation of direct capital cost, indirect capital cost, working capital and cost of installation of the new electricity distribution inter-connection cost. For solid-liquid based electricity generation plant, the summation of direct capital cost, indirect capital cost and working capital can be specified as 5.03 of the total delivered cost of equipment (DCoE). In order to estimate the DCoE, list of required equipment with the designated size is prepared and the concept of economy of scale and capital cost index is used to calibrate the cost to the present DCoE. The operational expenditure (OPEX) for the CHP can be define as the gross value of direct production cost (fixed OPEX + variable OPEX) and miscellaneous cost. The development of POMR-SES will be within the mill compound thus the feedstock will be supplied by the mill with no cost. The number of workers employed is based on the generation plant scale. The workers worked 10 hours per day and paid on the basis of USD 10 per hour. ROI and simple payback period (SPP) and discounted payback period (DPP) are used as an economic and investment indicators to measure the economic and investment competitiveness of the system. The revenue counted for this analysis are coming from selling the surplus electricity to the grid, claiming the eligible FiT rates and the saving from the mill's steam and electricity self-sufficiency. The CHP system was then co-evaluated using life cycle assessment method to establish the CHP individual environmental profile by quantifying the carbon emission saving (CES) per year as compared to the conventional fossil generation [15]. The value of carbon emission saving from the electricity generated in the CHP ($Elec$) can be identified using Eq. (4):

$$CES = Elec\ x\ EF_{grid}\ x\ OH \tag{4}$$

Given the carbon emission factor of electricity from the grid (EF_{grid}) = 0.899 kg CO_2 / kWh and operating hours of the CHP is 8000 hours/year [7]. Multi criteria decision making analysis is employed to justify the appropriate scale of generation for various cases to meet combined/optimized economic and, environment (and, less formally social) constraints to foster future decision making and planning on potential implementations of POMR-SES [16,17]. This equitable assessment results is then compiled and proposed as an effective operational and/or business plan to assist in the implementation of the POMR-SES schemes considering business risk and an equalable balance between human development and sustainable technology.

4. Results & Discussion

4.1 Resource Scoping

The amount of potential electricity generation from EFB and biogas are accounted based on the residues production report obtained from series of visit to selected mills across Peninsular Malaysia and from a full residue production report of 236 POM operated in Peninsular Malaysia obtained from a reliable data base. This allows the full-range of potential electricity generation been translated more accurately. The statistics are then further anticipated using a standard waste-to-energy conversion method demonstrated in [15,16]. These references treated EFB and biogas as a stand-alone fuel for combustion (similar to Configuration 1 and Configuration 2 in this study). The full-range of potential electricity generation is classified into small (< 3 MW), medium (3-10 MW) and large (>10MW). Table 1 summarized the result for the resource scoping phase. From the tabulated results, it was found that Configuration 3 which is combusting EFB and biogas in the same boiler giving the largest potential of electricity generation (25 MW) followed by Configuration 4 (20 MW), Configuration 1 (15 MW) and Configuration 2 (5 MW). However, any mill intended to employ Configuration 1 and/or Configuration 2 to generate electricity must ensure the availability of 4500 tonne EFB and 20000 tonne POME every year to obligate Eq.(3) and the note that following the equation about the mass flow rate of the steam for electricity generation. Therefore, the minimum size for Configuration 1 and Configuration 2 is 0.28 MW and 0.68 MW respectively.

Table 1: Full-range of potential electricity generation from EFB and biogas in Peninsular Malaysia

Items	Standard (MW)	Theoretical Calculation (MW)	Aspen Plus Simulation (MW)
Configuration 1	0.08 – 16.00	0.07 – 15.00	0.08 – 15.00
Configuration 2	0.00 – 4.00	0.00 – 5.00	0.00 – 5.00
Configuration 3	-	1.00 – 20.00	1.00 – 25.00
Configuration 4	-	1.00 – 18.00	1.00 – 20.00

4.2 Techno-economic Assessment

In this work, the techno-economic assessment was conducted for two scenarios to find the optimal scale of the POMR-SES that provide at least 20% ROI, reasonable SPP and DPP. The scenario was selected because the POMR-SES will required a new electricity distribution inter-connection to supply the surplus electricity to the national grid. This will allow an evaluation of the significant effect of new electricity distribution inter-connection cost on the economic performance of the POMR-SES. It was quoted that, the installation cost of new electricity distribution inter-connection is USD 312,500 per kilometer. The distance of the new electricity inter-connection is kept at 5 km. Table 2 illustrates the result of the selected indicator for both scenarios. In order to operate above reasonable chance of being economically viable, the plant should be operated at the optimum size of 4 MW for Scenario 1 and 4.5 MW for Scenario 2. In favor of coping with the ROI and payback period of palm oil mill cultivation, the plant should be running at a scale of 7.5 MW. It was found that, there are a significant effect of new electricity distribution inter-connection cost on the overall economic performance of the system.

Table 2: POMR-SES optimal size economic performance

		Scenario 1			**Scenario 2**	
Plant Size (MW)	ROI (%)	SPP (years)	DPP (years)	ROI (%)	SPP (years)	DPP (years)
0.28	-193.00	5.50	>15.00	-249.00	5.50	>15.00
4.00	20.00	3.00	6.50	3.20	3.00	7.00
4.50	31.00	3.00	5.50	21.00	3.00	6.50
7.50	90.00	2.50	4.50	80.00	2.75	4.50

4.3 Environmental Profile

The environmental profile of the POMR-SES is the measure of the amount of carbon dioxide (CO_2) saving as compared to the conventional central generation that been dominated by fossil fuel. The amount of CO_2 saving based on the optimal size of POMR-SES is shown in Table 3.

Table 3: CO_2 saving from electricity generation from EFB and biogas in Peninsular Malaysia

Plant Size (MW)	Standard (tonne CO_2/y)
0.28	2014
4.00	28800
4.50	32400
7.50	54000

4.4 Multi-Criteria Decision Making

Multi criteria decision making technique is applied to identify the most convincible optimize solution(s) for POMR-SES based on multiple criteria set from the economic and environmental perspectives. In the interest of promoting an effective operational and/or business plan and to attract the stakeholder engagement, the main criteria to be met is providing a ROI of 20% within a reasonable timeframe with admissible amount of CO_2 saving. A dedicated multi criteria decision tree from the two scenarios assessed before can be internalized through Figure 2.

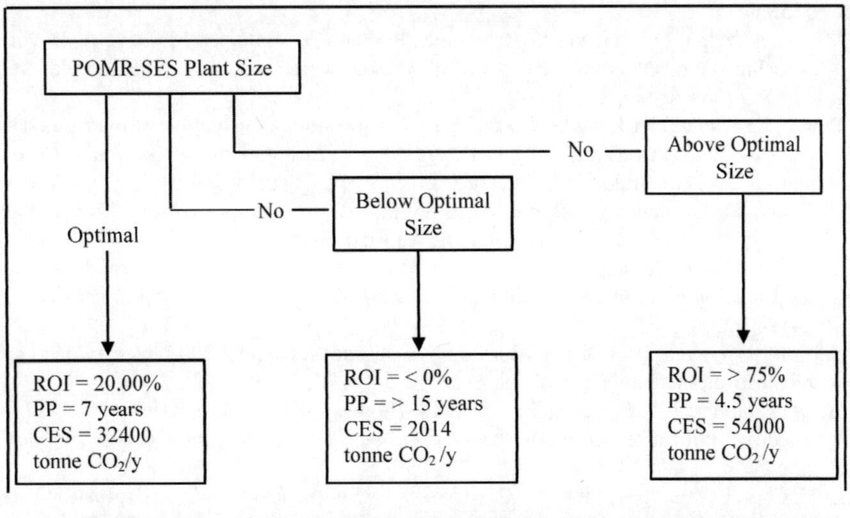

Figure 2: Multi-criteria decision tree

4.5 Implementation Framework

A compilation result of an extensive techno-economic and environmental assessment for site specific optimal size POMR-SES can use as a standard implementation framework to stimulate the good business environment for the investor and other stakeholders to get in to the renewable energy electricity generation based business.

5. Conclusion

The relatively slow pace of implementation of electricity generation from PO mills in Peninsular Malaysia does not appear to be due to a lack of available technologies, specific scientific data or even some supportive policies. Rather is seems to result from shortage of systematic, techno-economic and risk modelling and demonstration of specific benefits and uncertainties on a case-by-case basis and on inertia by the industry to enter the space due to understandable priority focus on meeting demand in robust primary markets for crude palm oil. From the perspective of economic viability, only medium size and large size POMR-SES are eligible to become the independent power producers (IPP) in Peninsular Malaysia. To qualify as one of the IPP, the mill should produce at least 65,000 tonnes of EFB and about 136,000 tonnes of POME every year. The research that described in this paper to identify the optimal business, environmental and societal cases for POMR-SES scheme should be consider as the necessary steps in encouraging the PO industry and government institutions to develop investment and support mechanism that will enables this important potential resources of sustainable, low carbon electricity to be tapped effectively.

References

[1] Shafie S, Mahlia T, Masjuki H 2012 A review on electricity generation based on biomass residue in Malaysia *Renewable and Sustainable Energy Reviews*, **16(8)**, pp.5879-5889

[2] Sadhukhan J, Martinez-Hernandez E, Ng KS 2016 Biorefinery value chain creation, *Chemical Engineering Research and Design*, **107**, pp. 1-3

[3] Department of Statistics Malaysia 2015 *Malaysia Economic Time Series*, pp. 1-277

[4] Loh SK 2016 The potential of the Malaysian oil palm biomass as a renewable energy source, *Energy Conversion Management*

[5] Wood S R, Rowley P N 2011 A technoeconomic analysis of small scale, biomas-fueled combined heat and power for community housing, *Biomass and Bioenergy*, **35(9)**, pp.3849-3858

[6] McIlveen-Wright D R, Huang Y, Rezvani S, Redpath D, Anderson M, Dave A, Hewit N J 2013 A technical and economic analysis of three large scale biomass combustion plants in the UK, *Applied Energy*

[7] Wan Y K, Sadhukhan J, Ng D K S 2015 Techno-economic evaluation of intergrated bioethanol plant for sago starch processing facility, *Chemical Engineering Research and Design*

[8] Hunpinyo P, Narataruksa P, Tungkamani S, Pana-Suppamassadu K, Chollacoop N 2013 Evaluation of techno-economic feasibility biomass-to-energy by using ASPEN Plus: A case study of Thailand, *Energy Procedia*, **42(August)**, 640-649

[9] Ying T, Lam G, Zhang Y, Wai C 2014 Chemical engineering research and design an optimization of biomass power plant, *Chemical Engineering Research and Design*, **92(8)**, 1412-1427

[10] Thornley P, Upham P, Huang Y, Rezvani S, Brammer J, Roger J 2009 Intergrated assessment of bioelectricity technology options, *Energy Policy*, **37(3)**, 890-903

[11] Gebreegziabher T, Olajire A, Yu Z 2014 Biomass drying for an intergrated power plant: effective utilization of waste heat, *European Symposium on Computer Aided Process Engineering*, **Vol 33** gbfg

[12] Li H, Chen Q, Zhang X, Finney K N, Shari V N, Swithenbank J 2012, Evaluation of a biomass drying process using waste heat from process industries: a case study, **35**, 71-80

[13] Ng K S, Sadhukhan J 2011, Process intergration and economic analysis of bio-oil platform for the production of methanol and combined heat and power. *Biomass and Bioenergy*, **35(3)**, 1153-1169

[14] Patel C, Lettieri P, Simons S J R, Germana A 2011, Techno-economic performance analysis of energy production from biomass at different scales in the UK context, *Chemical Engineering Journal*, **171(3)**,986-996

[15] Rahim K A 2014, Towards low carbon economy via carbon intensity reduction in Malaysia, **5(16)**, 123-133

[16] Catrinu M D 2006, Decision aid for planning local energy system: application of multi-criteria decision analysis, *Electrical Power Engineering*

[17] Scott J A, Ho W, Dey P K 2012, A review of multi-criteria decision-making methods for bioenergy system, *Energy*, **42(1)**, 146-156

Synergistic Effect of Co-utilization of Coal and Biomass Char: An Overview

M E S Paiman[1], N S Hamzah[1], S S Idris[1], N A Rahman[1] and K Ismail[2]

[1] Faculty of Chemical Engineering, Block 5, Tower 2, Sultan Aziz Mu'adzam Shah Engineering Complex, University Technology of MARA (UiTM), Shah Alam, Selangor, Malaysia

[2] Faculty of Applied Sciences, University Technology of MARA (UiTM), 40450 Shah Alam, Selangor, Malaysia

Email: shawal075@salam.uitm.edu.my

Abstract: Global concerns on impact of greenhouse gases emission, mostly released from coal-fired power plant, and the depletion of fossil fuel particularly coal, has led the production of electricity from alternatives resources such as co-utilization technologies. Previous studies proved that the co-utilization of coal and biomass/biomass chars has significantly reduced the emission of greenhouse gases either during the pyrolysis, combustion or gasification process in laboratories, pilots as well as in the industrial scales. Interestingly, most of the studies reported the presence of synergistic effect during the co-utilization processes particularly between coal and biomass char while some are not. Biomass chars were found to have porous and highly disorder carbon structure and belong to the class of most reactive carbon material, resulting to be more reactive than those hard coal and lignite. Up to date, microwave assisted pyrolysis is one of the best and latest techniques employed to produce better quality of biomass chars and it is also reduce the processing cost. Lot of works has been done regarding on the existence of synergistic effects during its co-utilization. However, the knowledge is limited to thermal and product characteristics so far. Even so, the specific reasons behind its existence are yet to understand well. Therefore, in this paper, the emphasis will be given on the synergistic effects on emission characteristics of co-utilization of coal and biomass chars so that it can be apply in energy-based industries to help in reduction of the greenhouse gases emission.

Keywords: Coal, Biomass chars, Synergistic effects, Microwave assisted pyrolysis.

1. Introduction

Energy demand increases over the year especially in developed and developing countries. Between 1971 and 2014, world total primary energy supply (TPES) was multiplied by almost 2.5 times. In which, oil remained the dominant fuel in 2014; nonetheless fell from 44% to 31%. Meanwhile, coal natural gas and nuclear was increased from 26–29%, 16–21%, and 1–5%, respectively. As for Asia, coal remains by far the main energy source, supplying more than half (54%) of its main energy demand compared to 29% globally in 2014 followed by oil (21%), biofuels (13%), and natural gas (8%). According to the International Energy Agency (IEA) 2016, the significant utilisation of coal in Asia is partly explained by the use of coal in power generation where in 2014, coal represented 67% of the electricity mix versus 41% globally [1].

Coal is black or dark brown solid type of fossil fuel. It is combustible, sedimentary, organic rock, which is composed mainly of carbon (C), hydrogen (H), oxygen (O), and volatile matters. As stated above, it was a dominant fuel in global power generation where it full power generation cost are below

Content from this work may be used under the terms of the Creative Commons Attribution 3.0 licence. Any further distribution of this work must maintain attribution to the author(s) and the title of the work, journal citation and DOI.
Published under licence by IOP Publishing Ltd

than oil, gas or renewables [2]. In comparison to other type of fossil fuels, coal has more than 100 years reserves globally based on 2015 production [3]. This is the main reason for coal being widely used as dominant fuel in power generation plants. However, coal is the dirtiest among fossil fuels since it emitting excessive greenhouse gases such as SOx, NOx and CO_2 during its utilization for power generation.Increasing concerns on the adverse effect of the emissions arising from coal conversion technologies on the environmental and gradually depletion of the fossil fuel reserves had led to global initiatives on using renewable and other opportunity resources to meet the future energy demands in a sustainable manner. Biomass has been reported to be the most promising alternative resources that is renewable which promoted the significant reduction of greenhouse gases emission during its co-utilization with coal as well as potentially to the replacement of fossil fuels.

2. Co-utilization of coal and biomass

Co-utilization is a promising thermochemical conversion way to utilize the abundant biomass residues and wastes into cleaner energy products for reducing the consumption of the fossil fuel particularly coal. The most highlighted one is the impact on the emission of greenhouse gases i.e. SOx, NOx, CO_2, volatile organic compound (VOC) and poly-aromatic hydrocarbon (PAH) which largely released from their formation during incomplete combustion. As compared to coal, there was proved that co-firing or gasification of coal and biomass results in a reduction in the emission of the latter. It is believe that the competitive char burnout contributed in the reduction of NOx, while the reduction of SO_2 can be explained by sulphur fixation in the ash due to the increase in potassium and calcium from biomass. Also, biomass contains virtually has no sulphur, therefore, SO_2 emissions are reduced in direct proportion to the coal replacement [4,5]. Co-firing may also reduce fuel costs, minimize waste and reduce soil and water pollution, depending upon chemical composition of the biomass used [6].

2.1. Environmental impact of co-utilization of coal and biomass

According to Tillman, 2000 [7], almost all demonstrations of co-firing shown a significant reduction on emission of SOx, NOx and CO_2. In his study, he found that the NOx reduction is typically disproportionate to the co-firing percentage on a heat-input basis, and result from the fuel volatility. In addition, co-firing also provides for significant reduction in fossil CO_2 emission. Demirbas, 2003 [6] stated that the benefits of co-firing include the reduction in CO_2 emission from the combustion of fossil fuels, the reduction of SO_2 formation through a reduction in fuel. To reduce greenhouse gas emission, the pressure is on conventional coal fired utilities to burn renewable fuels such a waste product or energy crop-derived biomass fuels, as a lowest cost option for reducing greenhouse emission. The impacts of co-firing of coal with biomass residues could substantially contribute to reduction in CO_2 emission from the power generation sector. Co-firing has a larger potential contribution to CO_2 reduction than the most other options for the power sector. In addition, Dayton, 2002 [8] stated biomass usually has lower sulphur content than coal. Therefore, the co-firing results in a reduction of SOx emission because of a displacement of sulphur in the fuel blend. Similar reduction also observed for NOx emission because the nitrogen content of the co-fired fuel generally lower than the nitrogen content of coal. Based on the previous studies, similar findings reported the positive reduction of greenhouse gases emission done by co-utilization of coal and biomass [9–11] as well as the most effective ways in reduction of the excessive biomass wastes i.e. oil palm biomass and municipal waste in Malaysia.

The co-utilization of coal and biomass is an interesting way to solve the environmental problems particularly on greenhouse gases emission. As a consequence from this field of study, a number of studies has been carried out to understand the combustion characteristics and its behaviour during co-utilization processes under thermogravimetric analyser (TGA). Most interestingly, there are few number of studies has been reported the existence of synergistic effect in the co-utilization of coal and biomass [12–20] while others are not [21–27].

3. Synergistic effect of co-utilization of coal and biomass char

The studies on co-utilization of biomass char and coal are relatively new field all over the world. Co-utilization of biomass char in existing coal-fired power plants may result to environmental, technical, and economical benefits. Combustion reactivity assessment by thermogravimetric analysis (TGA) technique has been reported in literature for either coal [28–30] or biomass fuels [31–33], but there exist a few comparison studies [34,35] and even less concerning about coal-biomass char blends [36,37].

According to Kastanaki & Vamvuka 2006 [38], biomass char were generally more reactive than those of hard coal and lignite based on their finding. In addition, biomass chars were found to have porous and highly disorder structure and belong to the class of most reactive carbon materials. The porosity within the chars causes more accessible of the reactive gas to active site resulting in the very good combustion reactivity.Biomass chars production is derived from pyrolysis process where the biomass is decomposed by heating in oxygen free or oxygen-limited environment. Pyrolysis has been used commercially for the wide production including bio-gas and bio-oil. Lot of technologies also has been invented to meet the desired i.e. muffle furnace [39], fixed bed furnace [40], and drop tube furnace (DTG) [41] in effort to investigate the biomass chars characteristics. However, microwave assisted pyrolysis reaction is one of the latest technique employed for pyrolysis reaction nowadays. Its benefits will reduce processing cost, produce better product quality and reduce hazard to human, environment and enhance quality of life [42].

3.1. Synergistic effects

The unexpected finding was discovered by previous studies claimed that there was the existence of synergistic effects during co-utilization of coal and biomass chars. However, not all studies discovered the same synergistic effects output as previous studies i.e. combustion reactivity, weight lost, thermal stability, and de-volatilization. Meanwhile, some of studies claimed there was lack and even no synergistic effect reported. By far, basically, the observed synergistic effects of co-utilization of primary and secondary fuels can be divided into three categories; thermal characteristics, product characteristics and emission characteristic. Summary of previous studies on synergistic effects by co-utilization of coal and biomass chars are shown in Table 1.

Table 1. Summary of previous studies on synergistic effects by co-utilization of coal and biomass char

Author(s)	Feedstock	Method & Conditions	Remarks
Lu, Lee, Chen, & Lin, 2013 [19]	Biomass – Torrefied *C. japonica* wood (char). Coal – Australian's anthracite coal.	Thermogravimetric analysis (co-pyrolysis reaction) – Heating rate (20 °C/min), Temperature range (25–800 °C), inert air condition (N_2 influenced, 100 cc/min)	Synergies observed – weight percentage
Sahu, Sarkar, Chakraborty, & Adak, 2010 [43]	Biomass – Pyrolyzed Saw dust & rice husk (char). Coal – Indian's coal	Thermogravimetric analysis (co-combustion reaction) – Heating rate (10 °C/min), Temperature range (ambient up to 750 °C), air flow rate (50 ml/min)	Synergies observed (lack) – improvement of reactivity in major combustion zone. – Blends containing less than 50% are better performing.
Gil et al., 2012 [44]	Biomass – Torrefied pine sawdust (char). Coal – Anthracite (AC), semi anthracite (HVN), high-volatile bituminous (DAB, M6N and NZ)	Thermo-balance (oxy-fuel combustion reaction) – Heating rate (2, 3, & 5 K/min), Temperature range (room to 1273 K), air flow rate (50 NmL/min, 30 %O_2 and 70 %CO_2)	No synergies observed
Kastanaki & Vamvuka,	Biomass – Cotton,	Thermogravimetric	Synergies observed –

2006 [38]	forest residues, olive kernel, & wood chars. Coal – Lignite	analysis (co-combustion reaction) – Heating rate (10 °C/min), Temperature range (25–850 °C), air flow rate (45 mL/min)	combustion performance of the blends showed some deviation from the expected weighted average of the constituents chars.
Idris, 2014 [42]	Biomass – EFB, PKS, PMF chars (*microwave irradiation*). Coal – Mukah Balingian	Thermogravimetric analysis (co-combustion reaction) – Heating rate (20 °C /min).	Synergies observed – improved coal reactivity. – Low burnout temperature.

4. Conclusion

It can be seen that, a lot of efforts have been done in order to understand the existence of the synergistic effects and behaviour of the combustion characteristics during co-utilization of coal and biomass/biomass char so that it can be apply beneficially in energy-based industry. By far, the works on synergistic effect during co-utilization is limited only to the investigation on thermal and product characteristics. Details works focusing on emission characteristics which might be related with the existence of the synergistic effect is yet to be explore. Therefore, further studies are very much needed to be conducted.

5. Acknowledgement

The author also would like to express his gratitude to the Research Management Institute (RMI) for the financial support (600-RMI/FRGS 5/3 (127/2015)), committee of Faculty of Chemical Engineering (FKK) and ICGSCE for publishing this article.

6. References

[1] IEA 2016 Key World Energy Trends *Iea* 19
[2] IEA Coal Industry Advisory Board 2014 The Impact of Global Coal Supply on Worldwide Electricity Prices 57
[3] Global B 2016 Coal Reserves *BP Glob.*
[4] Nussbaumer T 2003 Combustion and Co-combustion of Biomass: Fundamentals, Technologies, and Primary Measures for Emission Reduction *Energy and Fuels* **17** p 1510–21
[5] Kumar M, Kumar P and Kumari A 2015 Co-Utilization of Coal and Biomass – A Review **2** p 4–8
[6] Demirba A 2003 Biomass Co-firing for Coal-Fired Boilers **21** p 269–78
[7] Tillman D A 2000 Cofiring benefits for coal and biomass *Biomass and Bioenergy* **19** p 363–4
[8] Dayton D 2002 A Summary of NO x Emissions Reduction from Biomass Cofiring 7
[9] Kazagic A and Smajevic I 2007 Experimental investigation of ash behavior and emissions during combustion of Bosnian coal and biomass *Energy* **32** p 2006–16
[10] Spliethoff H and Hein K R . 1998 Effect of co-combustion of biomass on emissions in pulverized fuel furnaces *Fuel Process. Technol.* **54** p 189–205
[11] Sahu S G, Chakraborty N and Sarkar P 2014 Coal-biomass co-combustion: An overview *Renew. Sustain. Energy Rev.* **39** p 575–86
[12] Jones J M, Kubacki M, Kubica K, Ross A B and Williams A 2005 Devolatilisation characteristics of coal and biomass blends *J. Anal. Appl. Pyrolysis* **74** p 502–11
[13] Haykiri-Acma H and Yaman S 2007 Synergy in devolatilization characteristics of lignite and hazelnut shell during co-pyrolysis *Fuel* **86** p 373–80
[14] Cai J, Wang Y, Zhou L and Huang Q 2008 Thermogravimetric analysis and kinetics of coal/plastic blends during co-pyrolysis in nitrogen atmosphere *Fuel Process. Technol.* **89** p 21–7
[15] Haykiri-Acma H and Yaman S 2010 Interaction between biomass and different rank coals during co-pyrolysis *Renew. Energy* **35** p 288–92

[16] Park D K, Kim S D, Lee S H and Lee J G 2010 Co-pyrolysis characteristics of sawdust and coal blend in TGA and a fixed bed reactor *Bioresour. Technol.* **101** p 6151–6
[17] Moon C, Sung Y, Ahn S, Kim T, Choi G and Kim D 2013 Effect of blending ratio on combustion performance in blends of biomass and coals of different ranks *Exp. Therm. Fluid Sci.* **47** p 232–40
[18] Krerkkaiwan S, Fushimi C, Tsutsumi A and Kuchonthara P 2013 Synergetic effect during co-pyrolysis/gasification of biomass and sub-bituminous coal *Fuel Process. Technol.* **115** p 11–8
[19] Lu K M, Lee W J, Chen W H and Lin T C 2013 Thermogravimetric analysis and kinetics of co-pyrolysis of raw/torrefied wood and coal blends *Appl. Energy* **105** p 57–65
[20] Wu Z, Wang S, Zhao J, Chen L and Meng H 2014 Synergistic effect on thermal behavior during co-pyrolysis of lignocellulosic biomass model components blend with bituminous coal *Bioresour. Technol.* **169** p 220–8
[21] Meesri C and Moghtaderi B 2002 Lack of synergetic e ects in the pyrolytic characteristics of woody biomass = coal blends under low and high heating rate regimes *Meesri, C., Moghtaderi, B. (2002). Lack Synerg. e ects pyrolytic Charact. woody biomass = coal blends under low high Heat. rate regimes, 23, 55–66.* **23** p 55–66
[22] Vuthaluru H B 2004 Thermal behaviour of coal/biomass blends during co-pyrolysis *Fuel Process. Technol.* **85** p 141–55
[23] Sadhukhan A K, Gupta P, Goyal T and Saha R K 2008 Modelling of pyrolysis of coal-biomass blends using thermogravimetric analysis *Bioresour. Technol.* **99** p 8022–6
[24] Idris S S, Rahman N A, Ismail K, Alias A B, Rashid Z A and Aris M J 2010 Investigation on thermochemical behaviour of low rank Malaysian coal, oil palm biomass and their blends during pyrolysis via thermogravimetric analysis (TGA) *Bioresour. Technol.* **101** p 4584–92
[25] Gil M V., Casal D, Pevida C, Pis J J and Rubiera F 2010 Thermal behaviour and kinetics of coal/biomass blends during co-combustion *Bioresour. Technol.* **101** p 5601–8
[26] Muthuraman M, Namioka T and Yoshikawa K 2010 Characteristics of co-combustion and kinetic study on hydrothermally treated municipal solid waste with different rank coals: A thermogravimetric analysis *Appl. Energy* **87** p 141–8
[27] Wang Z, Li Q, Lin Z, Whiddon R, Qiu K, Kuang M and Cen K 2016 Transformation of nitrogen and sulphur impurities during hydrothermal upgrading of low quality coals *Fuel* **164** p 254–61
[28] Varhegyi G, Szabo P, Jakab E, Till F and Richard J-R 1996 Mathematical Modeling of Char Reactivity in Ar-O2 and CO2-O2 Mixtures *Energy & Fuels* **10** p 1208–14
[29] Russell N V, Beeley T J, Man C-K, Gibbins J R and Williamson J 1998 Development of TG measurements of intrinsic char combustion reactivity for industrial and research purposes *Fuel Process. Technol.* **57** p 113–30
[30] Ciuryla V T, Weimer R F, Bivans D A and Motika S A 1979 Ambient-pressure thermogravimetric characterization of four different coals and their chars *Fuel* **58** p 748–54
[31] Adánez J, de Diego L F, García-Labiano F, Abad A, Abanades J C and Luis F 2001 Determination of biomass char combustion reactivities for FBC applications by a combined method *Ind. Eng. Chem. Res.* **40** p 4317–23
[32] Di Blasi C, Buonanno F and Branca C 1999 Reactivities of some biomass chars in air *Carbon N. Y.* **37** p 1227–38
[33] Henrich E, Bürkle S, Meza-Renken Z I and Rumpel S 1999 Combustion and gasification kinetics of pyrolysis chars from waste and biomass *J. Anal. Appl. Pyrolysis* **49** p 221–41
[34] Stenseng M, Zolin a and Cenni R 2001 Thermal Analysis in Combustion Research *J. Therm. Anal. Calorim.* **64** p 1325–34
[35] Zolin A, Jensen A, Jensen P A, Frandsen F and Dam-Johansen K 2001 The influence of inorganic materials on the thermal deactivation of fuel chars *Energy and Fuels* **15** p 1110–22
[36] Backreedy R I, Fletcher L M, Jones J M, Ma L, Pourkashanian M and Williams A 2005 Co-firing pulverised coal and biomass: A modeling approach *Proc. Combust. Inst.* **30 II** p 2955–64
[37] Backreedy R I, Jones J M, Pourkashanian M and Williams A 2003 Burn-out of pulverised coal

and biomass chars q **82** p 2097–105

[38] Kastanaki E and Vamvuka D 2006 A comparative reactivity and kinetic study on the combustion of coal-biomass char blends *Fuel* **85** p 1186–93

[39] Zhang Y, Zhai M, Wang X, Sun J, Dong P, Liu P and Zhu Q 2015 Preparation and characteristics of biomass char *BioResources* **10** p 3017–26

[40] Avila C, Pang C H, Wu T and Lester E 2011 Morphology and reactivity characteristics of char biomass particles *Bioresour. Technol.* **102** p 5237–43

[41] Borrego A G, Garavaglia L and Kalkreuth W D 2009 Characteristics of high heating rate biomass chars prepared under N2 and CO2 atmospheres *Int. J. Coal Geol.* **77** p 409–15

[42] Idris S S 2014 *Production of solid fuel from oil palm biomass via microwave irradiation pyrolysis technique for co-combustion with coal* (University Technology of MARA)

[43] Sahu S G, Sarkar P, Chakraborty N and Adak A K 2010 Thermogravimetric assessment of combustion characteristics of blends of a coal with different biomass chars *Fuel Process. Technol.* **91** p 369–78

[44] Gil M V, Riaza J, Álvarez L, Pevida C, Pis J J and Rubiera F 2012 Kinetic models for the oxy-fuel combustion of coal and coal/biomass blend chars obtained in N2 and CO2 atmospheres *Energy* **48** p 510–8

IOP Conf. Series: Materials Science and Engineering **358** (2018) 012004 doi:10.1088/1757-899X/358/1/012004

Bioethanol Production From Banana Stem By Using Simultaneous Saccharification and Fermentation (SSF)

Kusmiyati[1], A Mustofa[2] and Jumarmi[3]

[1] Renewable Energy Research Center, Department of Chemical Engineering, Faculty of Engineering, Universitas Muhammadiyah Surakarta, Surakarta, Indonesia
[2,3] Department of Chemical Engineering, Faculty of Engineering, Universitas Muhammadiyah Surakarta, Surakarta, Indonesia

[1]kusmiyati@ums.ac.id.

Abstract. The rapid growth and development of industries in the world result in a greater energy needs. Some studies show that ethanol can be used as an alternative energy. However, bioethanol production from food raw materials such as sugar and starch has drawback that cause the food crisis. This aim of this study was to convert banana stem into bioethanol. Banana stem contained of 44.6% cellulose, 36.0% hemicellulose and 19.4% lignin. After banana stems were pretreated with acid (H_2SO_4) and alkaline (NaOH) at a concentration of 2% w/v at 121 °C for 30 minutes, then subsequently the simultaneous saccharification and fermentation (SSF) were carried out by using mixed cultures of *Aspergillus niger, Trichoderma reesei* and *Zymomonas mobilis* at various enzymes ratios of (1:1:1), (1:2:1), (1:2:2), (1:1:2) and various pH (4, 5 and 6) with SSF time for 144 hours and temperature of 30°C. The results show that acid pretreatment showed better results than the alkali pretreatment. After acid pretreatment and alkali pretreatment, lignin content of pretreted banana stem reduced to 15.92% and 16.34%, respectively, cellulose increased to 52.11% and 50.6% respectively, hemicellulose reduced to 28.45% and 28.83%, respectively The SSF showed that pH 5 gave the highest bioethanol. The highest concentration of bioethanol (8.51 g /L) was achieved at the SSF process at pH 5 with a ratio *Aspergillus niger, Trichoderma reesei* and *Zymomonas mobilis* enzymes of (1:1:2).

1. Introduction

Energy is one of the most basic need in human life. The growing of transportation sector has contributed the rise of petroleum consumption that accounts for 60% of fossil fuels consumption which also caused the pollution generation to the environment. [1]. However, the limitation the fossil oil feedstock and the negative impact of the fossil oil environmentally, ecologically or economically have led to a search for renewable sources of energy globally in recent years [2]. Several studies have emphasis on the use of foresty and agricultural residues for the production of lignocellulosic material. Banana pseudo-stem is one of lignocellulosic material which has higher holocellulose content than straw and lower lignin content than wood and straw. Banana pseudo-stem has potential as bioethanol feeedstocks because of its acceptable content of cellulose and low lignin content. Moreover, in Indonesia, banana pseudo–stems were found abundantly as natural resource like other subtropical and tropical regions [3]. In Indonesia, the planting area of banana is large which reached 88.728 ha in 2015 [4]. The banana pseudo-stem has the high cellulosic content 42.2 - 63 % that can be utilized for the production of fermentable sugars for bioethanol production [6].

Prior to the conversion of cellulose and hemicelluloses into monomeric sugars and the sugar into ethanol using a microorganism, the pretreatment step is required. The goal of the pretreatment is to remove the hemicellulose and lignin barrier around the cellulose fibres and to open the crystalline

Content from this work may be used under the terms of the Creative Commons Attribution 3.0 licence. Any further distribution of this work must maintain attribution to the author(s) and the title of the work, journal citation and DOI.
Published under licence by IOP Publishing Ltd

structure of the cellulose, resulting the cellulosic fibres more accessible to enzymatic attack[7]. Cellulose, the major component of pretreated biomass is then undergo enzymatically saccharification using cellulolytic enzymes. The cellulose saccharification is further undergo enzymatic fermentation to produce ethanol. Cellulose saccharification and the hexose fermentation can be carried out either separately (SHF process) or simultaneously (SSF process). The simultaneous saccharification and fermentation processes combines enzymatic saccharification of cellulose with simultaneous fermentation of its main derived sugar (glucose) to ethanol [8]. For the complete hydrolysis of cellulose, cellulases enzymes is employed to breakdown of cellulose to glucose. *T. reesei* and *Aspergillus* strains are known to have ability to produce β-glucosidase, the cellulolytic enzyme. The β-glucosidase production from mixed culture of *A. Niger* BKMF 1305 with *T. reesei* RUT C30 has been examined using waste paper lignocellulosic material as carbon source [9]. *Z. mobilis* was reported had good results for ethanol synthesis in varied environmental conditions, including temperature and phase of growth, concentration or type of carbon source [10]. The aims of the current study were to investigate the effect of acidic (H_2SO_4) or alkaline (NaOH) agents on banana stem pretreatment and ethanol concentration produced by simultaneous saccharification and fermentation (SSF). The effect of mixed culture ratios of *A. niger, T. reesei* and *Z. Mobilis* enzymes in SSF process on ethanol produced from acid pretreated banana stem were also studied.

2. Materıal and method

2.1. Preparation of raw materials and Pretreatment

Banana stem was obtained from local farmer in Central Java. The banana stems were chopped to small pieces approximately 3cm x 3 cm x 3cm x 0.5 cm and dried until the water content less than 10%. The dried banana stem was milled and sieved to 40 mesh size. Material was placed in sealed container and stored in a dry place at room temperature. The acid (H_2SO_4) and alkaline (NaOH) treatments were carried out by an autoclaving the 10% (w/v) banana stem dried solution at 121°C for 30 minutes using 2% (w/v) H_2SO_4 (p.a) or NaOH (p.a) solution. After pretreatment, the slurry was filtered to obtain the solid of pretreated banana stem contained a lignocellulosic material. Then the solid was repeated washed with distilled water until pH 7 and the solid was dried at 65°C until its constant weight.

2.2. Preparation ef enzymes

Aspergillus niger and Trichoderma reesei, Zymomonas mobilis microbes were used in this study. Pure *A. Niger* 6018 and *T. reesei* 6012 were cultured on the Potato Dextrose Agar (PDA) at temperature of 30°C. Pure cultures of *Z. mobilis*-0056 ENCC was incubated on RM medium containing nutriens: 20 g/L glucose, 10 g/L yeast extract, 2 g/L KH_2PO_4, 15 g/L agar. The culture time for *A. niger* 6018 and *T. reesei* 6012 was 3 days and 7 days, respectively then stored at 4°C. For enzymes production, pre-culture and main-culture of *A. niger* 6018 and *T. reesei* 6012 were prepared, respectively. The pre-culture medium (10 ml) for both *A. niger* 6018 and *T. reesei* 6012 contained similar composition of nutrients: 3 g/L urea, 10 g/L $(NH_4)_2SO_4$, 3 g/L KH_2PO_4, 0.5 g/L $MgSO_4.7H_2O$, 0.5 g/L $CaCl_2$. The pre-culture mediums were sterilized by autoclaving at 121°C and a pressure of 0.18 MPa for 20 minutes. The pre-cultures of *Z. mobilis* was incubated for 72 hours at 30°C using a shaker at speed of 120 rpm. The pre-culture media of *Z. mobilis* (10 ml) contains: 100 g/L glucose, 10 g/L yeast extract, 1 g/L $(NH_4)_2SO_4$, 1 g/L KH_2PO_4, and 0.5 g/L $MgSO_4.7H_2O$ was sterilized at a temperature of 121°C and a pressure of 0.18 MPa for 20 minutes. The main cullture (100 ml) of both *A. niger* 6018 and *T. reesei* 6012 had similar composition of nutriens with pre-culture and it was adding with pretreated banana stem (dry) substrat: 60 g/L substrat, 3 g/L of urea, 10 g/L $(NH_4)_2SO_4$, 3 g/L KH_2PO_4, 0.5 g/L $MgSO_4.7H_2O$, 0.5 g/L $CaCl_2$. The main culture (100 mL) medium of *Z. mobilis* was containing: 60 g/L substrate of pretreated banana stem (dry), 10 g/L yeast extract, 1 g/L $(NH_4)_2SO_4$, 1 g/L KH_2PO_4, and 0.5 g/L $MgSO_4.7H_2O$. Prior to incubation, the main culture mediums were autoclaved at 121°C and a pressure of 0.18 MPa for 20 minutes. The main cultures of *A. niger* 6018 , *T. reesei* , *Z. mobilis* were incubated for 72 hours at 30°C with shaker speed of 120 rpm.

2.3. Simultaneous Saccharification and Fermentation (SSF)

The Simutaneous Saccharification and Fermentation (SSF) of banana stem pretreated was carried out in 600 ml fermentation medium with a pretreated banana stem concentration of 10 % (w/v). A 0.05 M sodium citrate buffer solution was used to maintain the pH (4; 5; 6). The pretreatments were carried out using autoclave at 121°C and pressure 0.18 MPa for 20 minutes. The sterilized SSF samples were inoculated with mixed cultures of *A. niger, T. reesei, Z. mobilis* using a total loading of enzymes volume was 300 m. The effect of microbes ratios of *A. niger: T. reesei: Z. mobilis* were (1: 1: 1), (1: 2: 1), (1: 2: 2); (1: 1: 2) in SSF were carried out using a substrate of acid pretreated banana steam. SSF was conducted for 6 days at a temperature of 30°C with a shaker speed of 150 rpm.

2.4. Analysis of Lignin

The content of lignin in biomass was determined by using the method of "Klason lignin" (TAPPI T222 om-88) [11]. 1 gram of biomass was hydrolyzed using a 72% H_2SO_4 solution of 15 cm^3 for 2 hours at room temperature. Then the sample was diluted until the concentration of sulfuric acid up to 3 wt% by adding water as much as 560 cm^3 water. Samples boiled for 4 hours. Then the sample was separated between solid and liquid by filtration. Lignin was washed using hot water until it reaches a neutral pH. The content of lignin was obtained from the amount of dried residue.

2.5. Analysis of Holocellulose

Milled wood samples (4 g) was added with acid solution (Sodium acetate solution 160 cm^3) at 75 °C for 5 hours. Then sodium chlorite solution (4 cm^3) was added every hour for 4 hours. The solution was then cooled and the residue was filtered. The residue was washed with the first liquid (1 dm^3) and with acetone (15 cm^3). The residue was finally dried at room temperature. For the determination of holoselulose aliquots were dried at 105 ° C and it was weighed .

2.6. Analysis of Cellulose

5 g holocellulose was added 100 cm^3 of NaOH solution (17.5 wt%) at room temperature for 30 minutes. The residue was filtered and washed clean twice with water (with 200 cm^3) and then filtered again. Then, a acetic acid solution (10% by weight of 15 cm^3) was added to hydrolyze cellulose and degraded hemicellulose. Hemicellulose is more easily hydrolyzed than cellulose. The residue was filtered and washed with hot water (500 cm^3), and dried at 105 °C. Cellulose was counted ultimately determined by gravimetry.

2.7. Analysis of Bioethanol

The ethanol was obtained by distillation of fermentation broth at 78 °C until the impurities removed. Ethanol concentration was analyzed by Gas Chromatograpy (GC Shimadzu GC-8A, Japan), equipped with a Flame Ionisation Detector and and a polar factor four capillary colomn (GP 106 SP 1200). Detector temperature was set at a temperature of 250°C. The carrier gas N_2 was kept at 40 ml/min. Velocity gases flow rate of H_2 was set at 40 ml/min and O_2 at 300 ml/min. Each sample was analyzed in duplicate.

3. Results and dıscussıon

Banana stem was analyzed its chemical composition as shown in Table 1. The result shows that the lignin, sellulose and hemicellulose contents of dry banana steam are 19.39%, 47.74%, 36.01%, respectively.

Table 1. Compositition of Banana Stem.

Component	Percentage
Lignin	19.4%
Cellulose	44.6%
Hemicellulose	36.0%

This result seem to be suitable with the previous studies reported in the literature. It has beeen reported that banana stem contains considerable amount of cellulose 47%, hemicellulose 13%, lignin 13 %, ash 8.2% and extactives 3.05% [5]. Due to its high lignocellulosic material contents therefore banana stems has been considered as raw material for bioethanol. Cellulose and hemicellulose content in

IOP Conf. Series: Materials Science and Engineering **358** (2018) 012004 doi:10.1088/1757-899X/358/1/012004

banana stem can be converted to be fermentable sugar by enzymatic hydrolysis. However lignin content was a barrier for the succesful of hydrolysis. Pretreatment was used to remove the lignin content.

3.1. The effect of delignification

In this study, the pretreatment process was carried out using acid and alkali pretreatment in an autoclave at 121°C for 30 minutes with 2% H_2SO_4 and 2 % NaOH. Based on the composition of pretreated banana stem in figure 1, the lignin and hemicellulose of pretreated banana stem was significantly reduced as compared to untreated banana stem. The lignin content of pretreated banana stem using acid and alkali solution were 15.92% and 16.34%, respectively. Figure 1 shows the slightly different of lignin for the banana stem pretreated using autoclaving of 2%(w/v) acid and alkali solution. The result showed that all pretreatments were successful in degradating the lignin of banana stem. The results show that cellulose contents of the treated banana stem was increased as compared to untreated stem.

The acid pretreatment has a higher efficiency in degrading lignin than alkaline pretreatment which produced the cellulose of 52.11% and 50.6%, respectively and the hemicellulose of 28.45% and 28.83%, respectively. This results are in accordance with results found in the literature [7], [12],[13]. Pretreatment of cactus pear biomass using hydrogen peroxide, NaOH and H_2SO_4-alkaline delignification at 121 °C for 30 min was reported. The pretreatment resulted an increased of cellulose content, however an increase in lignin and hemicellulose content of treated catus pear biomass was found using all chemical treatments. The greater delignification values were obtained with alkaline peroxide pretreatment[7]. Another work studied on the Kans *(S. Spontaneum)* grass pretreated with 2% sulfuric acid for 90 min at 120 °C followed by saccharification of the pretreated kans grass. The treated biomass with dilute acid (2% H_2SO_4, 90 min, and 120 °C) removed 27.63% of the total lignin along with 77.05% of tthe total hemicellulose and 19.7% of total cellulose content [12]. In this work the simultaneous saccharification and fermentation (SSF) was employed where the hydrolysis and fermentation process taken place in one process simultaneously. In this work the SSF process was using mixed cultures of *Aspergillus niger, Trichoderma Reseei,* and *Zymomonas Mobilis* with the microbes ratios 1:1:1. The SSF of treated and untreated banana stem lignocellulose to ethanol production was studied at different pH (pH 4 and 5) and temperature of 30 °C and the result was shown in figure 2.

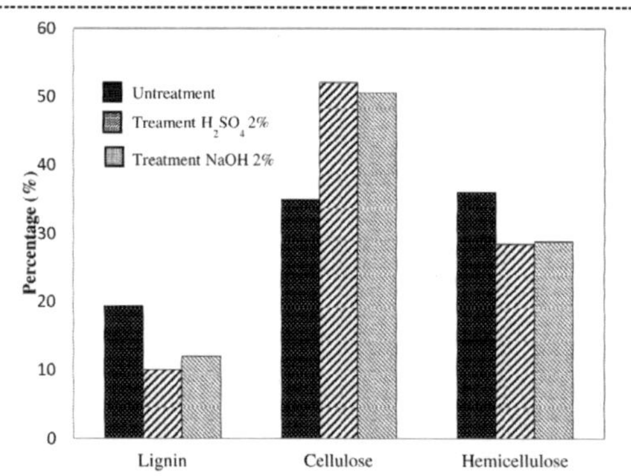

Figure 1. The results of the analysis of lignin, cellulose and hemicellulose before and after pretreatment

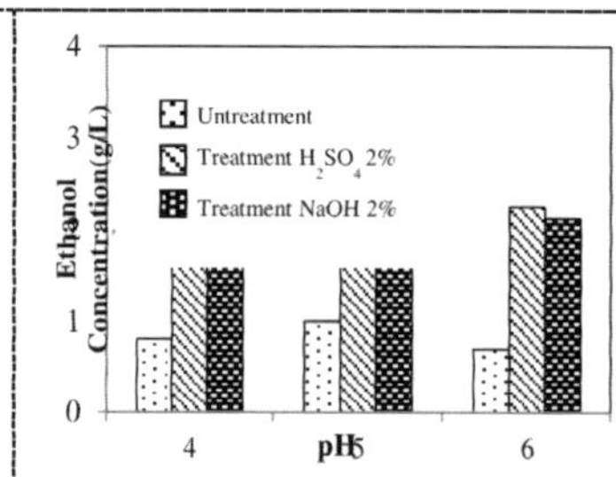

Figure 2. Effect of acid and alkaline pretreatment and pH on ethanol production

The highest ethanol was observed at pH 5 (3.493 g/L) for the treated and untreated banana stem raw materials. While at pH 4 and 6, a decreased in ethanol was observed. Previous studied reported the effect of initial pH value of food waste hydrolysate fermentation on ethanol production *by Z.mobilis* at pH range 3.5-6. The highest ethanol was found at pH 4 a decreased in ethanol was observed when pH increased up to 6. This was due to food waste raw material contained lactic acid bacteria that produced high concentration of contaminants such as acetic acid and lactic acid which inhibited ethanol formation [14]. Figure 2 shows the acid and alkaline pretreatment banana stem resulted the higher ethanol concentration than untreatment banana stem. Previous work reported SSF of delignification of bagasse steam pretreated sweet sorghum (SSB) as a substrate indicated that delignification of SSB could increase ethanol concentration obviously [15]. Steam-pretreatments were employed for deliginification of sweet sorghum bagasse (SSB) and Douglas-fir (DF). In SSF, the delignified SSB and DF showed an increase final ethanol concentration to 2.76 g/L (SSB) and 3.43 g/L (DF) with an increase of 26% and 157% compared with controls [15]. A co-fermentation process of hydrothermally pretreated wheat straw as substrate was studied in which *F. oxysporum culture (submerged or solid-state)* was added at the SSF stage along with *S. cerevisiae*. The addition of solid *F. oxysporum* culture increased ethanol production by 19%, leading to a final ethanol concentration of 58 g L^{-1}[16].

3.2. The effect of mixed culture ratios in Simultaneous Saccarification and Fermentation of acid pretreated banana stem

Simultaneous saccarification and fermentation of acid pretreated banana stem was studied at temperature 30°C using pH 4 and 5. The influence of mixed culture ratios of Aspergillus niger, Trichoderma Reseei, and Zymomnas Mobilis (1:1:1; 1:2:1, 1:2:2, 1:1:2) were studied. The lowest ethanol was observed at ratios microbes A. Niger : T. Reesei : Z. Mobilis 1: 1: 1. Figure 3 shows the highest ethanol concentration at pH 5 (8.51 g/L) was observed at ratio of microbial A. Niger : T. Reseei : Z. Mobilis (1: 1: 2). The higher ethanol concentrations were obtained on pH 5 than that on pH 4. The ethanol concentration for the ratios of A. Niger : T. Reseei : Z. Mobilis of (1:2:1), (1:2:2), and (1:1:2) at pH 4 were 2.78 g/L, 6.25 g/L, and 6.67 g/L while at pH 5 were 3.93 g/L, 7.41 g/L, 8.51 g/L, respectively. Based on this result mixed cultures of A. Niger : T. Reesei : Z. Mobilis produced higher concentration of ethanol from pretreated banana stem.This research is consistent with the results of research conducted by [16] that the higher the levels of concentration Z. mobilis ethanol produced will be higher levels of ethanol. Mixed cultures of A. Niger and T. Reseei was examined for the production of enzime complex of β-glucosidase and cellulase enzymes for the conversion of lignocellulosic material to ethanol. The higher β-glucosidase activity was reported for the cocultures as compare to monoculture A. Niger and monoculture T. Reseei [9]. The simultaneous saccharification and fermentation to ethanol of microcrystalline cellulose using co-cultures of Klebsiella oxytoca and strains of Saccharomyces pastorianus, Kluyveromyces marxianus and Z. mobilis showed that more ethanol and less time fermentation was observed for co-cultures than pure cultures of the above organisms. This was due to the higher thermotolerance of Klebsiella oxytoca, which permitted the use of a temperature (37 °C) more favourable to cellulase activity in the yeast SSFs [17].

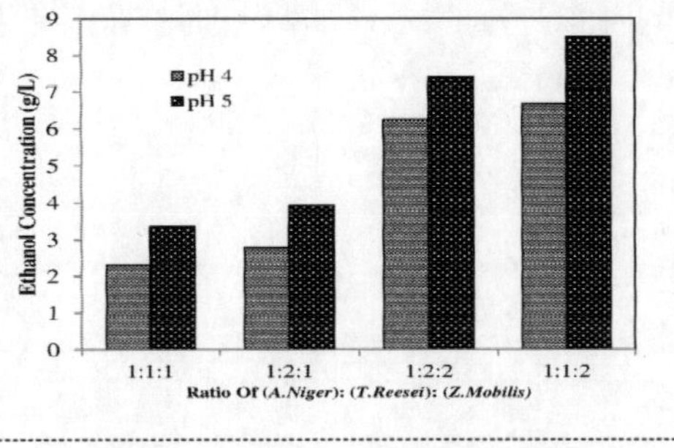

Figure 3. Effect of the concentration of microbes on ethanol production

4. Conclusıon

Simultaneous saccharification and fermentation (SSF) of acid pretreated banana stem was carried out using different ratios of mixed cultures of *A. niger, T. reesei* and *Z. Mobilis* enzymes.At the culture ratios *A. Niger*: *T. Reesei* : *Z. mobilis* of 1:1:2 yielded the highest ethanol concentration. Prior to SSF, banana stem pretreatments were carried out using acid (2 % w/v H_2SO_4) and alkaline (2% w/v NaOH) solution and heated in autoclaved 121°C for 30 minutes. The highest ethanol was obtained at pH 5 as compared to pH 4 and pH 6. The highest ethanol (8.51 g/L) was obtained at SSF condition of pH 5 with a ratio mixed cultures of 1:1:2.

Acknowledgments

The author express her gratitude to Directorate of Research and Higher Education of Indonesia (Kemenristekdikti) for Finacial Assistance under the Competitve Research Grant by year 2016 (No Contract HIKOM 006/SP2H/LT/DRPM/II/2016).

References

[1] Aditiya H B, Mahlia T M I, Chong W T, Nur H and Sebayang A H 2016 *Renew. Sustain. Energy Rev* **66** 631–653
[2] Adekunle A, Orsat V and RaghavanV 2016 *Renew. Sustain. Energy Rev* **64** 518–530
[3] Li K, Fu S, Zhan H, Zhan Y and Lucia L A 2010 *BioResources* **5-2** 576–585
[4] BPS 2016 "Luas Panen Pisang Menurut Provinsi , 2011-2015"
[5] Idrees M,Adnan A, Malik F,Qureshi F A, Road K, Control D, Division T M, Shahzad C, Shahzad C and Road P 2013 *J. Technol* 269–281
[6] Saraiva A B,Pacheco E B A V, Visconte L L Y, Pereira E, Bispo, Escócio V A, de Sousa A M F, Soares A G, Junior M F, do C. Motta L C and da C. Brito G F 2012 *Int. J. Environ. Bioenergy* **4- 2** 101–119
[7] Ferreira P, Filho D S, Ribeiro V T and Silvino E 2016 *Ind. Crop. Prod.* **89** 425–433
[8] Ferreira V, Faber M D O and Mesquita S 2010 *Electronic J of Biotech* **13-2** 1–7
[9] Juhász T, Kozma K, Szengyel Z and Réczey K 2003 *Food Technol Biotechnol* **41-1** 49–53
[10] Shakirova L, Grube M, Goodacre R, Gavare M, Auzina L and Zikmanis P 2013 *Vib. Spectrosc* **64** 51–57
[11] TAPPI Test Method T222 om-88, Acid-insoluble lignin in wood and pulp. In: Tappi test methods. Atlanta, GA: Technical Association of the Pulp and Paper Industry
[12] Kataria R and Ghosh S 2011 *Bioresour. Technol* **102-21** 9970–9975
[13] Zaafouri K, Ziadi M, Ben R, Farid M, Hamdi M and Regaya I 2016 *Biomass and Bioenergy*, **94** 66–77
[14] Ma K, Ruan Z, Shui Z, Wang Y, Hu G and He M 2016 *Bioresour. Technol* **203** 295–302
[15] Shen F, Kumar L, Hu J and Saddler J N 2011 *Bioresour. Technol* **102-19** 8945–8951
[16] Paschos T, Xiros C and Christakopoulos P 2015 *Ind. Crop. Prod* **76** 793–802
[17] Golias H, Dumsday G J, Stanley G A and Pamment N B 2002 *J of Biothec* **96** 155–168

IOP Conf. Series: Materials Science and Engineering **358** (2018) 012005 doi:10.1088/1757-899X/358/1/012005

Production of Biodiesel by Esterification of Free Fatty Acid over Solid Catalyst from Biomass Waste

N. I. F. Mukti, B. Sutrisno and A. Hidayat

Chemical Engineering Department, Universitas Islam Indonesia
Jalan Kaliurang km 14,5 Yogyakarta 55584, Indonesia

nurindah_fm@uii.ac.id

Abstract. Recently, low cost feedstocks have been utilized to replace vegetable oils in order to improve the economic feasibility of biodiesel. The esterification of free fatty acid (FFA) on Palm Fatty Acid Distillate (PFAD) with methanol using solid catalyst generated from bagasse fly ash is a promising method to convert FFA into biodiesel. In this research, the esterification of FFA on PFAD using the sulfonated bagasse fly ash catalyst was studied. The performances of the catalysts were evaluated in terms of the reaction temperatures, the molar ratios of methanol to PFAD, and the catalyst loading. The effects of the mass ratio of catalyst to oil (1-10%), the molar ratio of methanol to oil (6:1-12:1), and the reaction temperature (40-60°C) were studied for the conversion of PFAD to optimize the reaction conditions. The results showed that the optimum conditions were methanol to PFAD molar ratio of 12:1, the amount of catalyst of 10%wt. of PFAD, and reaction temperature of 60°C. The reusability of the solid acid carbon catalysts was also studied in this work. The catalytic activity decreased up to 38% after third cycle. The significant decline in catalyst esterification activity was due to acid site leaching. The physico-characteristics and acid site densities were analyzed by Nitrogen gas adsorption, surface functional groups by Fourier transform infrared spectroscopy (FT-IR), elemental analysis using X-ray fluorescent (XRF), and acid-base back titration methods for determination of acid density.

1. Introduction

Increased demand for energy, global warming due to emission of greenhouse gases, environmental pollution, and fast diminishing supply of fossil fuels are the major key factors leading to search for alternative sources of energy [1, 2]. Biodiesel is considered a promising alternative substitute for petroleum diesel fuel. Biodiesel, defined as monoalkyl fatty acid ester preferentially methyl and ethyl esters, has several advantages over petroleum diesel, such as it is biodegradable, non-toxic, originates from renewable sources and contains no sulphur or aromatic compounds and would therefore, contribute to the reduction of emissions from diesel engines [3]. Biodiesel is generally synthesized by the transesterification of vegetable oils or animal fats, which is carried out by acid or base catalysis. Biodiesel can be also produced by esterification of free fatty acids, which are present in animal fats (ex. lard or tallow), with methanol over acid catalysts [4].

The problems related to biodiesel production are high cost feedstock and the conflict of interest between energy crisis and food security. The use of low cost feedstocks such as waste cooking oils and non-conventional seed oils can reduce biodiesel production cost and increase supply while avoiding the food versus fuel problem. Therefore, it is necessary to search for non-conventional feedstocks for biodiesel production. Palm Fatty Acid Distillate (PFAD), which has high content of free fatty acids, is by-product of palm oil refinery. Compared to the expensive refined palm oil (1 USD per Liter), PFAD costs only 0.3 USD per liter. Thus, the utilization of PFAD for biodiesel production via esterification

Content from this work may be used under the terms of the Creative Commons Attribution 3.0 licence. Any further distribution of this work must maintain attribution to the author(s) and the title of the work, journal citation and DOI.
Published under licence by IOP Publishing Ltd

reaction would reduce the production cost of biodiesel, and would consequently enable it to compete economically with petroleum-based fuel. Commercial biodiesel production employs homogeneous alkali catalysts. The homogeneous catalysts for biodiesel production are alkaline catalysts (such as sodium hydroxide, potassium hydroxide, and alkali metal hydroxides) and acids catalysts (such as sulfuric acids or hydrochloric acid) [5]. Conventional homogeneous catalysts are no longer feasible for biodiesel production. The homogeneous catalysts are generally corrosive to equipment, have to be usually neutralised after reaction, cannot be regenerated and recycled, also give rise to serious environmentally problems. These problems can be overcome by the use of solid acid catalysts. A literature survey shows that esterifications of free fatty acids to produce biodiesel involve use of different heterogeneous catalysts such as, clays [6], zeolite [7], ion exchange resin [8], carbon based material [9], metal oxides as well as sulphated metal oxides [10], WO_3/ZrO_2 [11], propylsulphonic acid functionalized mesoporous silica [12], heteropolyacids [13] and heteropolyacids supported on to different supports such as zirconia [14], activated carbon, silica [15] and tantalum pentoxide [16].

A novel catalyst which has not been fully explored is carbon based catalysts. Carbon supported catalysts have several distinct advantages over alumina or silica supported systems; they are stable under acidic basic conditions, can have very high surface area (200–1500 m^2/g), renewable biomass sources can be used to generate the carbon, and the non-polar nature of the support matrix may reduce adsorption of polar molecules (e.g., water or glycerol) that can deactivate the catalyst [17]. Sulfonated carbons are the most promising solid acids and their use have been demonstrated in many acid catalyzed reactions including that of FFAs esterification as well as cellulose hydrolysis [15,16]. Such catalysts are generally produced by sulfonation of incompletely pyrolyzed biomass, such as sucrose[17], glucose [18], starch [19], glycerol [20] or biochar [21]. Carbon precursors are directly pyrolyzed and sulfonated in concentrated H_2SO_4 at high temperatures. However, this simple process usually produces sulfonated carbons with low specific surface area, low acid density and poor reusability. In this research, the esterification of PFAD using the sulfonated bagasse fly ash catalyst was studied. The performances of the catalysts were evaluated in terms of the reaction temperatures, the molar ratios of methanol to PFAD, the catalyst loading and the reaction times. The reusability of the solid acid carbon catalysts was also studied in this work.

2. Experiment

2.1. Catalysts Preparation
Bagasse fly ash used as raw materials were obtained from Bantul regency, D.I. Yogyakarta province. The starting materials were manually chosen, cleaned with deionized water, and dried at 110°C for 48 h. Prior to use, bagasse fly ash were crushed and sieved to obtain particles with sizes less than 1 μm. Sulfonation was carried out at 15°C for 12 h by immersed and stirred 10 g activated carbon in 100 ml concentrated sulfuric acid (98%). The sulfonated sample was filtered water washed with hot distilled water (>80°C) until the sulphate ions were no longer detected in the wash water. Then the sulfonated catalyst was dried at 110°C for overnight in a oven to remove water.

2.2. Catalyst Characterization
The specific surface area and pore structure characteristic of prepared activated carbons were determined by nitrogen adsorption at −196°C. The BET surface area was calculated using the BET equation; total pore volumes were estimated to be the liquid volumes of N_2 at a high relative pressure (~0.99). The surface functional groups and structure were studied by Fourier transform infrared spectroscopy (FTIR). The FTIR spectra of the raw material and the resulting activated carbons were recorded from a wavenumber of 4000–400 cm^{-1}. Elemental analysis was used to determine the composition of the catalysts. The sulphur contents of the catalyst were determined by elemental analysis. The total acid density of the catalysts was determined using the standard acid-base back-titration.

2.3. Catalyst testing
Esterification of PFAD using the prepared catalysts was performed in batch runs on a hot plate with a magnetic stirrer at different process conditions. The samples were analyzed by titration procedure for the evaluation of free residual acidity. Once the sample was taken from the reactor, it was washed with

water in order to stop the reaction and separates the catalyst and the alcohol from the oil phase. A weighted amount of the sample was dissolved in ethanol and sulfuric ether in order to be able to make the titration analyses, and some drops of phenolphthalein as indicator were added to the solution to be able to measure the conversion. The titration was done with a 0.05 mol/L KOH solution. The acid value was calculated using the following equation:

$$AV = \frac{56.1 \times C \times V}{M} \quad (1)$$

where $AV$is the acid value of the sample (mg KOH.g^{-1}), V is volume of KOH consumed for titration (mL), c is concentration of the KOH used for titration (mol/L), m is the weight of the sample (g), and 56.1 is the molecular weight of KOH. The conversion of free fatty acid was calculated using the following equation:

$$X_F = \left(1 - \frac{AV_2}{AV_1}\right) \times 100 \quad (2)$$

where X_F is the FFA conversion, AV_1 is the initial acid value of the mixture and AV_2 is the acid value of mixture after reaction. To study the effect of the catalyst amount on FFA conversion, the reaction was carried out at four different catalyst loadings of 1, 2.5, 5, and 10%wt. of PFAD. Each reaction was carried out at 60°C for 4 h using molar ratio of methanol to oil of 12:1. In this study the molar ratio of methanol to PFAD was varied from 6:1 to 12:1. At the end of each experiment, the catalyst was filtered from the reaction mixture, washed with fresh methanol, dried at 100°C for 2 h, and then reused in the next cycle of experiments. The esterification reaction was carried out at temperature of 60°C with a 6:1 molar ratio of methanol to oil and a 10 wt.% PFAD of catalyst amount.

3. Results And Discussions

3.1. Catalysts Characterization

Nitrogen isotherms of biochar and catalyst at −196°C are presented in Figure 1.

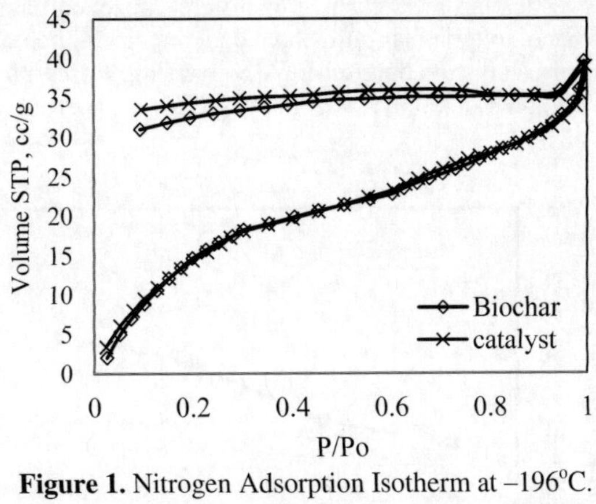

Figure 1. Nitrogen Adsorption Isotherm at −196°C.

The shape of adsorption isotherm can provide preliminary qualitative information about the adsorption mechanism as well as the pore structure of activated carbons. These isotherms show that the biochar from bagasse fly ash has type IV isotherm which is an indication of mesoporous carbon. Catalyst, on the other hand, also has type IV isotherm. The characterization of biochar and catalyst presented in Table 1. It can be seen that sulfonation process change the carbon characterization. Moreover, the pore volumes and surface area of mesopore were increased by sulfonation process. Pore size distribution of the biochar and catalyst obtained from BJH model is presented in Figure 2. As expected from N_2 adsorption isotherms (Figure 1), all samples showed that has type IV isotherm which is an indication of mesoporous carbon. However, the PSDs calculated using BJH model which is related to the fundamental assumptions on describing the porous structure and/or mechanism of adsorption for mesoporous structure.

Table 1. Characterization of Carbons.

Sample	S, m^2/g	S_{mes}, m^2/g	% S_{mes}	V, cm^3/g	V_{mes}, cm^3/g	% V_{mes}	D_{avg}, nm
Biochar	65.96	25.02	37.93	0.060	0.04	66.00	3.46
Catalyst	70.60	35.19	49.84	0.061	0.04	68.76	3.64

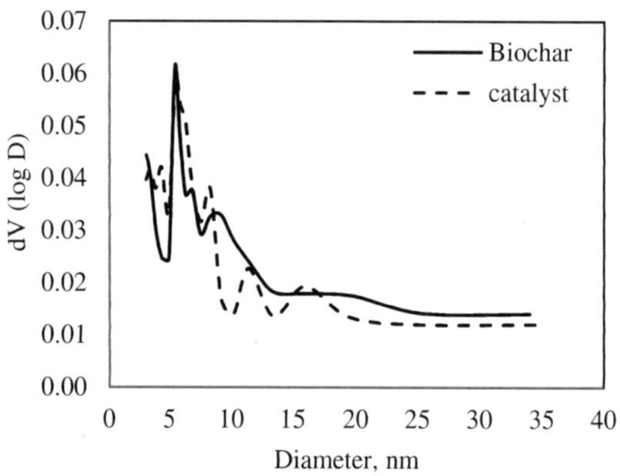

Figure 2. Pore Size Distribution using BJH Model.

3.2. Catalyst Testing

The esterification reaction is affected by various factors depending upon the reaction conditions used. Further experiments for investigating the influence of different reaction parameters such as catalyst loading, reaction time, methanol to oil molar ratio and reaction temperature on the conversion of FFA. Figure 3 shows the effect of catalyst loading on FFA conversion. Based on these results, it can be clearly seen that, when the amount of catalyst was increased from 1 to 10%, conversion of FFA was found to be increased.

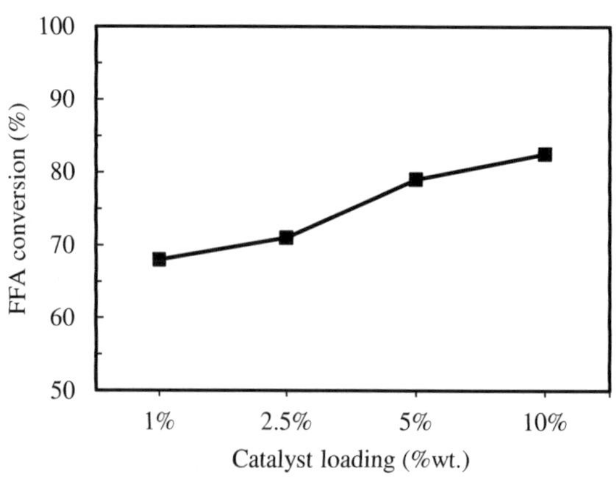

Figure 3. Effect of catalyst loading on FFA conversion (reaction conditions: reaction temperature 60°C, stirring speed 800 rpm, and molar ratio of PFAD to methanol 1:12).

The highest conversion of 82.5% was obtained at catalyst loading of 10%wt. of PFAD. When the catalyst amount was increased, the total number of available active sites increased resulted in faster reaction rate to reach reaction equilibrium, with more catalyst addition [17]. Molar ratio is one of the important factors affecting the conversion of FFA. Theoretically, the esterification of FFA requires one moles of methanol per mole of FFA. Since the esterification of FFA is a reversible reaction, the excess of methanol shifts the equilibrium towards the direction of fatty acid methyl ester formation. In addition to this, usage of excess methanol is one of the better options to improve the rate of the esterification reaction catalyzed by heterogeneous catalysts. In this study the molar ratio of methanol to PFAD was varied from 6:1 to 12:1. Figure 4 shows the effect of methanol to oil molar ratio on the conversion at the end of 4 h reaction time at 60°C in the amount catalyst of 10%wt. of PFAD. It can be clearly seen from Figure 4, when the molar ratio was increased from 6:1 to 12:1, the conversion was found to increase from 68% to 82.5%. The highest conversion was found at molar ratio of 12. Further increase in the methanol amount has not showed any significant improvement in the conversion.

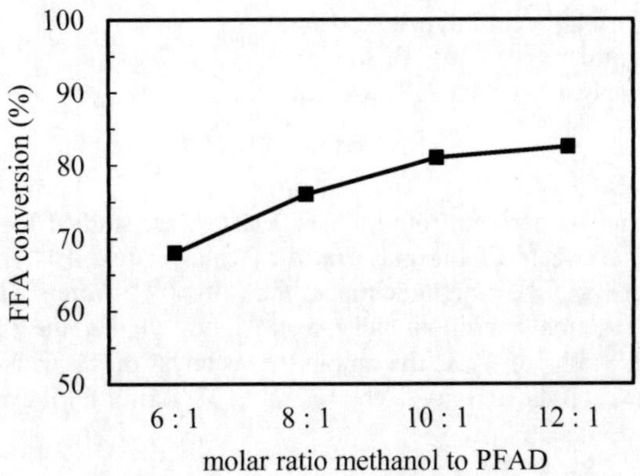

Figure 4. Effect of oil to methanol molar ratio on FFA conversion (reaction conditions: reaction temperature 60°C, stirring speed 800 rpm and catalyst loading 10% wt. of PFAD).

The influence of temperature was studied in the esterification of FFA applying methanol to oil molar ratio of 12:1 and using amount catalyst of 10 %.wt PFAD by varying the reaction temperature from 40 to 60°C. Figure 5 illustrates the influence of temperature on the conversion of FFA. Figure 5 shows that the FFA conversion increases from 61 to 83% as the temperature increases from 40 to 60°C.
The increased of FFA conversion due to the effect of increase of the reaction rate by increasing temperatures. Some improvement of the mass transfer limitation between reactant and catalyst when the temperature increase also effect on the FFA conversion. As the reaction temperature is increased, all reactant molecules will gain more kinetic energy that will eventually accelerate the mass transfer rate among the oil–methanol–catalyst phases that resulted in the formation of more FAME [17]. Reusability of the catalyst in the esterification of FFA, under the optimum condition, was studied. The catalyst recovered after the reaction by filtration and directly used for another reaction. The results are presented in Figure 6. It can be seen that the catalyst's esterification activity was reduced to 38% at the end of 3rd cycle. This decay could be due to the leaching of active sites to the reaction media.

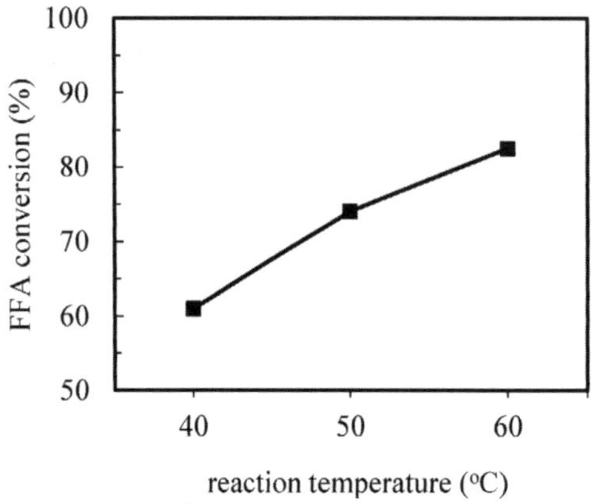

Figure 5. Effect of reaction temperature on FFA conversion (reaction conditions: stirring speed 800 rpm, molar ratio of PFAD to methanol 1:12, catalyst loading 10%wt. of PFAD).

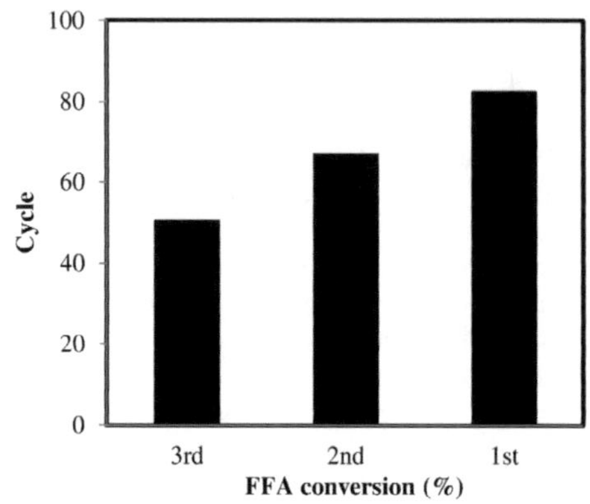

Figure 6. The reusability of the catalyst.

4. Conclusions

The sulfonated carbon catalyst derived from biomass waste were studied for esterification of the free fatty acid on PFAD. The effects of the mass ratio of catalyst to oil (1–10%), the molar ratio of methanol to oil (6:1–12:1), and the reaction temperature (40–60°C) were studied for the conversion of PFAD to optimize the reaction conditions. The results showed that the optimum conditions were methanol to PFAD molar ratio of 12:1, the amount of catalyst of 10 wt.% of PFAD, and reaction temperature of 60°C. The catalytic activity decreased up to 38% after third cycle.

Acknowledgments

This work is fully financially supported by the Direktorat Riset dan Pengabdian Masyarakat, Direktorat Jenderal Pendidikan Tinggi, Kemenristekdikti, Republik Indonesia.

References

[1] Kawentar W A and Budiman A 2013 *Energy Proc.* **32** 190-199.
[2]. Fatimah, Hidayat A and Setiawan K H 2010 *Asian J. Chem.* **22** 5 3793-3801.
[3] Budiman A and Ishida M 1997 *J. Chem. Eng. Japan* **30** 5 966-969.
[4] Budiman A, Sutijan and Sawitri D R 2011 *Int. J. Exergy* **8** 4 477-493.
[5] Kusumaningtyas R D, Rochmadi, Purwono S and Budiman A 2014 *Int. J. Exergy* **1** 4 447-467.
[6] Do Nascimento L S, Tito L M Z, Angelica R S, Da Costa C E F, Zamian J R and Da Rocha G N Filho 2011 *Appl. Catal. B: Environ.* **101** 3-4 495–503.
[7] Chung K H and Park B G J 2009 *Ind. Eng. Chem.* **15** 388–392.
[8] Kiss A A, Dimian A C and Rothenberg G 2006 *Adv. Synth. Catal.* **348** 1-2 75–81.
[9] Okuhara T 2002 *Catal. Today* **73** 167–176.
[10] Wang Y, Liu Y and Liu C 2008 *Energy Fuels* **22** 4 2203–2206.
[11] Park Y M, Lee J Y, Chung S H, Park I S, Lee S Y, Kim D K, Lee J S and Lee K Y 2010 *Bioresour. Technol.* **101** 1 S59–S61.
[12] Melero J A, Bautista L F, Morales G, Iglesias J and Briones 2009 *Energy Fuels* **23** 1 539–547.
[13] Silva V W G, Laier L O and Da Silva M J 2010 *Catal. Lett.* **135**, 207–211.
[14] Oliveira C F, Dezaneti L M, Garcia F A C, De Macedo J L, Dias Silvia J A, Dias C L and Alvim K S P 2010 *Appl. Catal. A: Gen.* **372** 153–161.
[15] Caetano C S, Fonseca I M, Ramos A M, Vital J and Castanheiro J E 2008 *Catal. Commun* **9** 1996–1999.

IOP Conf. Series: Materials Science and Engineering **358** (2018) 012005 doi:10.1088/1757-899X/358/1/012005

[16] Xu L, Wang Y, Yang X, Yu X, Guo Y and Clark J H 2008 *Green Chem.* **10** 746–755.
[17] Hidayat A, Rochmadi, Wijaya K, Hinode H and Budiman A 2013 *Asian J. Chem.* **25** 3 1569-1575.
[18] Toda M, Takagaki A, Okamura M, Kondo J N, Hayashi S and Domen K 2005 *Nature* **438** 178.
[19] Okamura M, Takagaki A, Toda M, Kondo J N, Domen K and Tatsumi T 2006 *Chem. Mater.* **18** 3039–3045.
[20] Guo F, Xiu Z L and Liang Z X 2012 *Appl. Energ.* **98** 47–52.
[21] Hidayat A, Rochmadi, Wijaya K, Nurdiawati A, Kurniawan W, Hinode H, Yoshikawa K and Budiman A 2015 *Energy Proc.* **75** 969-974.

IOP Conf. Series: Materials Science and Engineering **358** (2018) 012006 doi:10.1088/1757-899X/358/1/012006

Pretreatment of Cellulose By Electron Beam Irradiation Method

N A A Jusri[1,a], A Azizan[1,b*], N Ibrahim[1,c], R Mohd Salleh[1,d] and M F Abd Rahman[2]

[1] Faculty of Chemical Engineering, Universiti Teknologi MARA, Shah Alam, Malaysia
[2] Agensi Nuklear Malaysia, Bangi, Malaysia

[1,a]aiedajusri@yahoo.com, [1,b*]amizon@salam.uitm.edu.my, [1,c]norli816@salam.uitm.edu.my, [1,d]ruzitah@salam.uitm.edu.my, [2]faizal@nuclearmalaysia.gov.my

Abstract. Pretreatment process of lignocellulosic biomass (LCB) to produce biofuel has been conducted by using various methods including physical, chemical, physicochemical as well as biological. The conversion of bioethanol process typically involves several steps which consist of pretreatment, hydrolysis, fermentation and separation. In this project, microcrystalline cellulose (MCC) was used in replacement of LCB since cellulose has the highest content of LCB for the purpose of investigating the effectiveness of new pretreatment method using radiation technology. Irradiation with different doses (100 kGy to 1000 kGy) was conducted by using electron beam accelerator equipment at Agensi Nuklear Malaysia. Fourier Transform Infrared Spectroscopy (FTIR) and X-Ray Diffraction (XRD) analyses were studied to further understand the effect of the suggested pretreatment step to the content of MCC. Through this method namely IRR-LCB, an ideal and optimal condition for pretreatment prior to the production of biofuel by using LCB may be introduced.

1. Introduction

The production of bioethanol through the pretreatment of LCB is one of the alternatives to produce environmental friendly and safer fuel as well as increasing the reduction of CO_2 emissions [1]. Waste that is bio-based material can be applied as the raw material for bioethanol production [2]. Biomass such as rice husk showed great potential to be one of the second generation bioethanol as it contained high MCC as well as nanocrystalline cellulose (NCC) [3]. There are several of methods that can be applied for the pretreatment of the LCB. Some of the pretreatment technologies showed impressive outcome as well as cost-effective pretreatment for LCB for the conversion to fuels and chemicals. These LCB pretreatment strategies are focusing on the reduction of lignin content and the crystalline structure of cellulose without damaging the fermentable sugar content [4]. Irradiation method for the pretreatment of LCB does not require any chemicals for the process and it reduces the crystallinity of the cellulose [5]. The conversion of sugar monomer to ethanol is obtained from the hydrolization of cellulose [6]. Cellulose is a natural polymer that can be easily obtained from LCB material such as oil palm frond and other agricultural residue [7].

Cellulose is the main structural components in the plant cells walls and located at the fibrous structure of the walls. D-glucose subunits comprised in the cellulose are linked to one another by β-(1,4) –glycosidic bond [8]. According to previous studies, the conversion yield of cellulose to glucose is increasing by 4% after 24 hours and 6% after 48 hours with 20 kGy of irradiation dose [9, 10, 11]. However, irradiation method alone is highly energy intensive and highly inapplicable. The indirect

Content from this work may be used under the terms of the Creative Commons Attribution 3.0 licence. Any further distribution of this work must maintain attribution to the author(s) and the title of the work, journal citation and DOI.
Published under licence by IOP Publishing Ltd

measurements of cellulose were studied via the analyses by using various methods such as Fourier Transform Infrared Spectroscopy (FTIR) and X-Ray Diffraction (XRD).

2. Material and Method

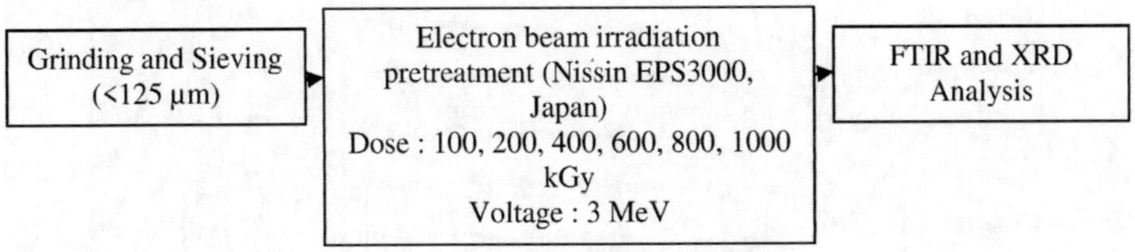

The irradiated MCC was then analyzed via FTIR and XRD for the Lateral order index (LOI) and Crystallinity Index (CrI), respectively. The LOI for FTIR sample was calculated by using the absorbance (A) ratio A_{1430}/A_{898}. From the raw data of FTIR, the transmittance (T) value was converted to absorbance by using Equation 1 from Beer's Law [12]. Crystallinity Index (CrI) of samples were measured by using a High-Resolution X-Ray Diffactrometer (Multi-Purpose X-Ray Diffractrometer, Ultima IV, Rigaku, USA) with Cu Kα at 50 kV and 300 mA. Samples were scanned over a scattering angle (2θ) from 10° to 50° at a rate of 2θ per min. The CrI was calculated by using Segal's Equation (Equation 2) in order to determine the crystalline structure of each treated samples [13] where CrI, I_{am} and I_{002} are defined as the crystallinity, the lowest intensity of the peak at 2θ =18.0° and the maximum intensity about 2θ = 22.6°, respectively.

$$A = 2 - \log_{10}(\%T) \quad (1) \qquad CrI\,(\%) = \frac{I_{002} - I_{am}}{I_{002}}\, x\, 100 \quad (2)$$

3. Result and Discussion

3.1 FTIR Analysis

FTIR spectroscopy was frequently used to investigate the structure of constituents and the chemical changes occurring during the pretreatment of lignocellulosic biomass [14]. Table 1 shows the characteristics and variations of wavenumber in FTIR spectra possible for MCC. The FTIR spectra of MCC samples irradiated with different doses are shown in Figure 1.

Table 1. Characteristic and variations of wavenumbers in FTIR spectra possible for MCC [15, 16, 17].

Wavenumber	Possible Functional Group	Wavenumber	Possible Functional Group
3350	O-H	1158	C-O asymmetric stretching
3175	-OH stretching	1050	C-O stretching in cellulose I and cellulose II
2918	C-H stretching	850	β-glycosidic bonds between sugar
1430	C-H_2 bending	550	β-glycosidic linkages
1315	C-H_2 wagging		

In FTIR spectra, due to the absorption of certain amount of energy produced from the radiation, the covalent bonds of each functional group are stretched, vibrated, bended, rocked or twisted [14]. According to the structure of MCC, all the wavenumbers obtained from FTIR spectra show almost similar characteristic as explained in Table 1. Functional bond of O-H is shown at the range of 3600-3200 cm^{-1} while C-H stretching as well as C-H_2 bending and wagging are observed at wavelength bonds between 2900-2800 cm^{-1} and 1500-1300 cm^{-1}, respectively. Other than that, C-O-C stretching and bonds between sugar are defined at the wavenumber of 1200-800 cm^{-1}. Table 2 below shows the LOI value for the effect of irradiation dose on MCC pretreatment.

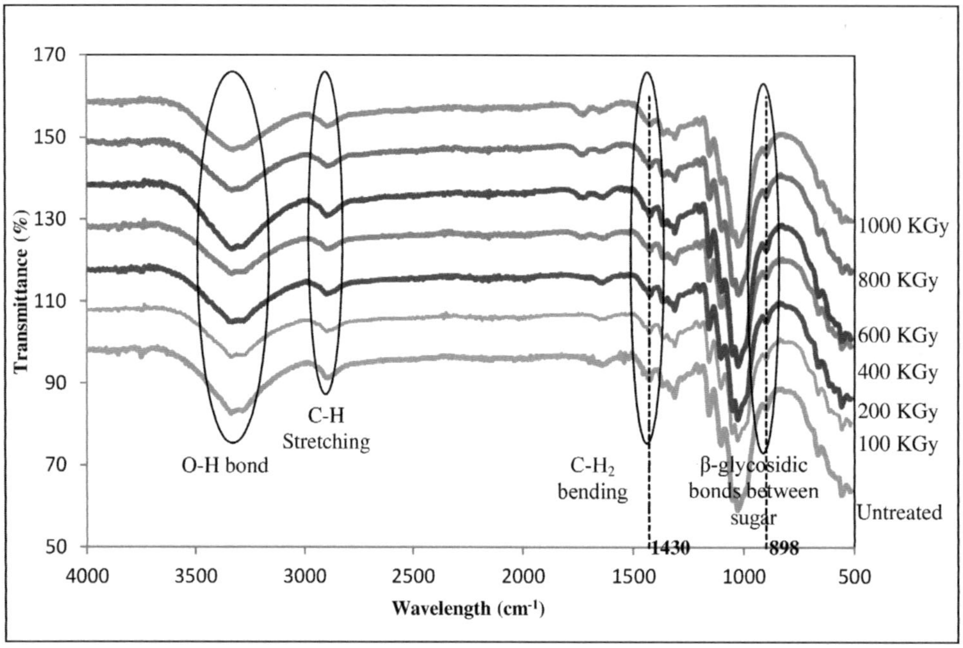

Figure 1. FTIR band for microcrystalline cellulose (MCC) with different irradiation dose (100 to 1000 kGy).

Table 2. Lateral order index (LOI) value for the effect of irradiation dose on MCC pretreatment.

Irradiation Dose (kGy)	LOI	Irradiation Dose (kGy)	LOI
Untreated	0.551	600	0.486
100	0.536	800	0.476
200	0.518	1000	0.464
400	0.508		

The absorbance values at ~1430 cm^{-1} and ~898 cm^{-1} are quite sensitive in the crystal structure of cellulose in cellulosic material [18].Therefore, the absorbance ratio of A_{1430}/A_{898} which is known as crystallinity index known as LOI was used in determining the effect of pretreatment to the crystallinity in cellulose. The lower the value of LOI reflects that the material has a lower crystallinity. High dosage of irradiation resulted in lower LOI values indicating that the MCC structure has changed to amorphous structure. Thus, with higher amorphous structures present, it enables better hydrolysis via suitable enzyme in comparison to with the presence of ordered arrangement of crystalline structures resulting in a higher effectiveness in biofuel production. The initial investigation of irradiation opens to possible different prospective in pretreatment on Malaysian LCB without using any additional chemicals.

3.2 XRD Analysis

XRD was used to identify the crystallinity index of the MCC (after pre-treatment). From Table 3, it is shown that CrI of pretreated MCC decreases with the increment of irradiation dose. XRD pattern for untreated and irradiated MCC are shown in Figure 2. Due to cross linking which may take place in the reaction and the removal of the crystalline fraction, there was a slight decrease in the degree of

IOP Conf. Series: Materials Science and Engineering **358** (2018) 012006 doi:10.1088/1757-899X/358/1/012006

crystallinity after pretreatment by the irradiation. The untreated MCC gave the highest crystallinity index as compared to the irradiated MCC which was 85.9%. The lowest crystallinity index was shown by the irradiated MCC at 1000 kGy. Table 3 was constructed to show the effectiveness of pretreatment by crystallinity index (CrI) for each irradiation dose in percentage (%) form. The lowest CrI values represent the most efficient dose for pretreatment due to the highest value in effectiveness which is 14.32% at 1000 kGy irradiation dose as compared to the untreated one.

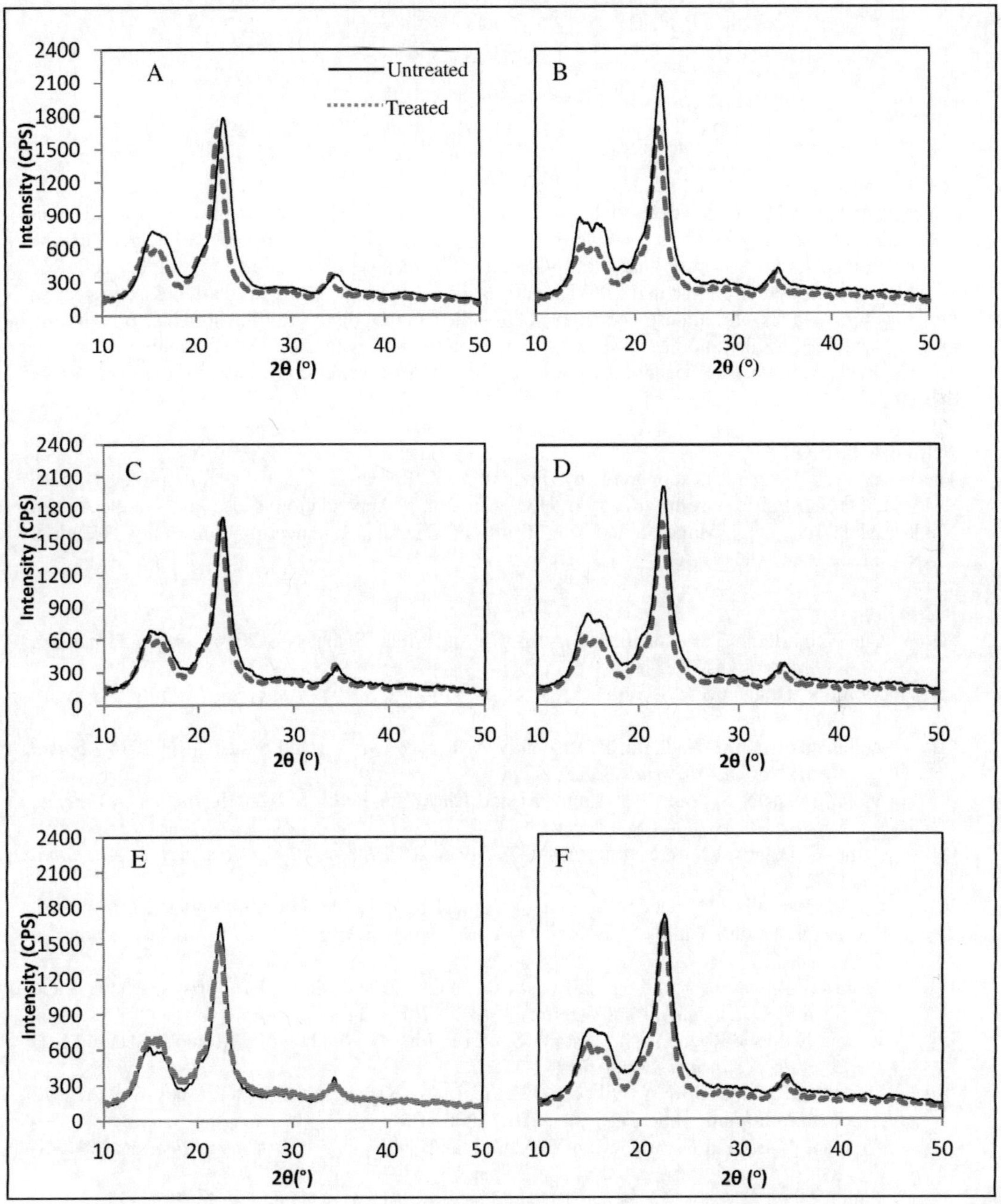

Figure 2. XRD pattern of untreated (unirradiated) and treated (irradiated) MCC [A: 100 kGy, B: 200 kGy, C: 400 kGy, D: 600 kGy, E: 800 kGy and F: 1000 kGy].

Table 3. Values of crystallinity index and effectiveness of pretreatment.

Irradiation Dose (kGy)	Crystallinity Index (%)	Effectiveness (%)
0	85.9	-
100	84.2	1.98
200	80.9	5.82
400	80.0	6.87
600	77.4	9.90
800	76.1	11.41
1000	73.6	14.32

4. Conclusion

Microcrystalline cellulose powder which was treated by electron beam irradiation process at different dose indicates that the method is possible for further improvement From the FTIR spectra, it can be concluded that the lower transmittance at O-H bonds wavelength may indicate that higher cellulose could be recovered after pretreatment. XRD analysis also shows that as the irradiation dose increases, the crystallinity index of cellulose decreases. This indicates that the crystalline structure of cellulose may be leading towards amorphous states. If this is consistent with the lower crystallinity index it becomes with the irradiation technology introduced on MCC, a better conversion into biofuel can be expected.

Acknowledgements

The work was financially supported by Fundamental Research Grant Scheme (FRGS) (600-RMI/FRGS 5/3 (12/2015)) from Ministry of Higher Learning. Appreciation is also given to the Agensi Nuklear Malaysia, Bangi, Malaysia and The Faculty of Chemical Engineering, University Teknologi MARA, Shah Alam, Malaysia.

References

[1] Che Kamarludin S N, Jainal M S, Azizan A and Mohd Safaai N S 2014 *Applied Mechanics and Material* **625** 838-841
[2] Mohd N, Draman S F S, Salleh M S N and Yusof N B 2017 *American Institute of Physics* **1809**
[3] Zuliahani A, Nurul Nadhirah R, Rozyanty A R, Nawawi W I and Seman A N 2016 *Applied Mechanics and Materials* **835** 272-276
[4] Mohd Safaai N S, Azizan A, Ramli M and Che Kamarludin S N 2015 *Advanced Material Research* **1125** 260-265
[5] Saini A, Aggarwal N K, Sharma A and Yadav A 2015 *Biotechnology Research International* 157139
[6] Che Kamarludin S N, Ubong S, Idris N, Azmi I S, Jainal M S, Jalil R, Wan Omer W S A, Tengku Zainal Mulok T E, Azizan A and Mohd Safaai N S 2014 *Advanced Material Research* **911** 307-313
[7] Azmi I S, Azizan A, Mohd Salleh R, Jalil R, Tengku Zainal Mulok T E, Idris N, Ubong S and Sihab A L 2013 *Advanced Material Research* **701** 243-248
[8] Bujang N, Mohd Rodhi M N, Musa M, Subari F, Idris N, Makhtar N S M and Ku Hamid K H 2013 *Procedia Engineering* **68** 372-378
[9] Duarte C L, Ribeiro M A, Oikawa H, Mori M N, Napolitano C M and Galvao C A 2012 *Radiation Physics and Chemistry* **81**(8) 1008-1011
[10] Kristiani A, Effendi N, Aristiawan Y, Aulia F and Sudiyani Y 2015 *Energy Procedia* **68** 195-204
[11] Jeun J P, Lee B M, Lee J Y, Kang P H and Park J K 2015 *Renewable Energy* **79** 51-55
[12] Hurtubise F G and Krassig H 1962 *Analytical Chemistry* **32** 177-181
[13] Swatloski R P, Spear S K, Holbrey J D and Roger R D 2002 Dissolution of cellulose with ionic liquids, *J. Am. Chemistry. Society* **124** 4974–4975

IOP Conf. Series: Materials Science and Engineering **358** (2018) 012006 doi:10.1088/1757-899X/358/1/012006

[14] Li H, Qu Y, Yang Y, Chang S and Xu J 2016 *Bioresource Technology* **199** 34-41
[15] Azizan A, Mohd Shafaei N S, Sidek N S, Hanafi F, Mokhti N and Zaharudin S 2016 *International Journal of Applied Engineering Research* **11** 10048-10050
[16] Popescu C M, Popescu C M, Singurel G, Vasile C, Argyropoulos D S and Willfor S 2007 *Applied Spectroscopy* **61**
[17] Hu Q, Su X, Tan L, Liu X, Wu A, Su D, Tian K and Xiong X 2013 *Bioscience Biotechnology Biochemistry* **77**(11) 2181-2187
[18] Ciolacu D, Kovac J and Kokol V 2010 *Carbohydrate Research* **345**(5) 621-630

IOP Conf. Series: Materials Science and Engineering **358** (2018) 012007 doi:10.1088/1757-899X/358/1/012007

Study of Crystallinity Index (CrI) of Oil Palm Frond Pretreatment using Aqueous [EMIM][OAc] in a Closed System

R. Abu Darim[1], **A. Azizan**[2] **and J. Salihon**[2]
[1]Brackishwater Aquaculture Research Division, Department of Fisheries Malaysia.
[2]Chemical Engineering Faculty, Universiti Teknologi MARA (Shah Alam), Malaysia

rosmaria.abudarim@gmail.com

Abstract. The objective of this preliminary study is to identify the Crystalinity Index (CrI) of Oil Palm Frond (OPF) pretreated with 40% concentration of 1-ethyl-3-methylimidazolium acetate ionic liquid ([EMIM][OAc]) in a closed system. The morphology and structural changes of the pretreated OPF were examined by using Fourier Transform Infrared Spectrometer (FTIR) and X-Ray Diffraction (XRD). The pretreatment process was carried out in triplicates by loading 40% of [EMIM][OAc] concentration with 10 wt% of OPF loading in the Bio-ionic liquid-reactor. The pretreatment process was conducted for 3 hours with working volume of 70 mL and temperature of 110°C. A Bio-ionic liquid reactor was purposely designed for the lignocellulosic pretreatment by using aqueous ionic liquid at high temperature (higher than boiling point of water). The CrI of OPF pretreated with 40% concentration of [EMM][OAc] in a closed system observed was 9% lower from the untreated OPF and the result showed significant difference with 95% confidence level. Further examination of the untreated and pretreated OPF by using XRD showed that the diffraction pattern of the pretreated OPF samples was decreasing compared to the untreated OPF. The characteristic of the FTIR spectra of the pretreated OPF showed the presence of the cellulose I and occurrence of amorphous cellulosic in the samples. The findings from this study are expected to improve knowledge on pretreatment of OPF by using aqueous [EMIM][OAc] as a green economically viable process for future renewable energy.

1. Introduction

Bioethanol is one of the alternative renewable energy with low net Greenhouse Gases (GHG) emissions [1] produced from lignocellulosic biomass such as plant mass, grasses, wood and agricultural waste [2]. The cellulose obtained through lignocellulosic biomass pretreatment will be converted into sugars via enzymatic hydrolysis/saccharification process and sugars will be converted into bioethanol via fermentation process by bacteria/yeast. There are different kinds of chemical and physicochemical pretreatment methods, for instance alkaline delignification, diluted acid hydrolysis, steam explosion, alkaline peroxide, aqueous ammonia soaking, ultrasound irradiation, ionic liquid and ionic liquid-co-solvent were used for dissolution of lignocellulosic biomass [3, 4, 5, 6]. Deconstruction and fractionation of lignocellulosic biomass or called as pretreatment of lignocellulosic biomass [7] is an extremely important step in commercial biorefinery [8]. Towards low environmental impact on process, ionic liquid (IL) has provided a platform on the application of pretreatment of lignocellulosic materials in a green way [7]. Previous study indicated that IL pretreatment is a promising alternative in pretreatment of switch grass in terms of total process time to produce high yield of sugars from the recovered product, as compared to the dilute acid pretreatment [8]. Pretreatment of lignocellulosic biomass by using IL are resulted by a low inhibitor, low cost of equipment and energy and easier

Content from this work may be used under the terms of the Creative Commons Attribution 3.0 licence. Any further distribution of this work must maintain attribution to the author(s) and the title of the work, journal citation and DOI.
Published under licence by IOP Publishing Ltd

chemical recovery as compared to other methods [9]. However, the cost of IL is expensive [9]. Therefore, by considering the cost of IL, ionic liquid-water mixture is considered to optimize the IL utilization. However, a study showed that the solubility of cellulose significantly decreased in the presence of water in the ionic liquid, presumably through competitive of hydrogen bonding to the cellulose microfibrils which inhibits the solubilisation [10]. Previous studies have demonstrated that 50–80% 1-ethyl-3-methylimidazolium acetate ionic liquid ([EMIM][OAc]) in water was efficient for cellulose dissolution which can match the performance of 100% [EMIM][OAc] in terms of glucose yields [11].

Interaction of 1-butyl-3-methylimidazolium acetate ionic liquid [BMIM][OAc] with native crystalline cellulose in plant cell wall shows that IL molecules progressively swell the biomass samples and the fraction of biomass samples swollen with IL transformed into a disordered structure [12]. [EMIM] [OAc] was indicated to be the most efficient for dissolving cellulose among ionic liquids tested [13]. Cellulose crystallinity, can be measured as Crystallinity Index (CrI) through X-Ray Diffraction (XRD) and Fourier-transform infrared spectroscopy (FTIR) [14, 15, 16, 17, 18, 19]. CrI were found to be significantly correlated with hydrolysis yield potential (HYP) after 24 hours of hydrolysis [20]. CrI were obtained through Lateral Order Index (LOI) in the FTIR spectra as a ratio of A_{1430} to A_{898} [16]. Absorption band, between 1420 and 1430 cm^{-1} in the FTIR spectra is assigned to a symmetric CH_2 bending vibration, which is known as "crystallinity band' and a band between 893 and 898 cm^{-1} (A_{898}) is assigned to C-O-C stretching at β-(1→4)-glicosidic linkage, which is known as the amorphous band [17]. Malaysia produces about 47% of the world's supply of palm oil. 12.9 metric tonne (mT) of Oil Palm Frond (OPF) and 15.8 mT of Empty Fruit Bunch (EFB) of biomass were produced in Malaysia [21, 22]. The OPF are available daily throughout the year when the palms are pruned during the harvesting of fresh fruit bunch for the production of oil [15]. OPF was identified with a higher cellulose content composition with 45.2% as compared to EFB of only 38.2% [23]. The objective of this preliminary study is to identify the Crystalinity Index (CrI) of (OPF) pretreated with 40% concentration of 1-ethyl-3-methylimidazolium ionic liquid in a closed system. The morphology and structural changes of the pretreated OPF were examined by using FTIR and XRD. The findings from this study are expected to improve knowledge on pretreatment of OPF by using aqueous 1-ethyl-3-methylimidazolium ionic liquid as a green and economically viable process for future renewable energy.

2. Materials and Methods

2.1. Materials and Preparation
Oil palm frond (OPF) stem pieces were obtained from Shah Alam were thoroughly washed with water, dried and stored in a closed tight container at room temperature. The OPF samples were then grounded by a grinder (A11, IKA, Malaysia) and sieved using a 120 mesh siever (OCTAGON 2000, Endecotts, United Kingdom) then stored in a sealed plastic bag before being used in the experiments. The ground OPF was named as 'untreated OPF'. Ionic liquid, 1-ethyl-3-methilimidazolium acetate ([EMIM][OAc]) was purchased from Sigma-Aldrich (Germany). The OPF was weighted to 10 wt% OPF loading in 70 mL of the pretreatment working volume. The 40% of ([EMIM][OAc]) was prepared by mixing the ([EMIM][OAc]) with deionized water.

2.2. Bio-ionic liquid reactor
A Bio-ionic liquid reactor was designed for lignocellulosic pretreatment by using aqueous ionic liquid at high temperature pretreatment process (higher than boiling point of water) and fabricated up to 200 mL of working volume, 0-130°C working temperature and 150 psi maximum pressure.

2.3. Aqueous ionic liquid pretreatment
The pretreatment process was carried out in triplicates by loading 40% of [EMIM][OAc] concentration with 10 wt% of OPF loading in the Bio-ionic liquid reactor. The pretreatment process was conducted for 3 hours with a working volume of 70 mL and temperature at 110°C. The heat-up

time was ~10-15 minutes. After 3 hours, warm water was added to the slurry by ratio 1:1. Cellulose could be precipitated from the IL solution by adding water [10]. The water was used as an anti-solvent for cellulose regeneration and for recovering any solubilized biomass [11]. The samples were centrifuged at 10,000g for 20 minutes in order to separate the solid and the liquid. The supernatant containing IL was removed, and the precipitate was washed 4 to 5 times [8, 11, 12] until there is no residue of ionic liquid.

2.4. Crystallinity and chemical changes measurements

X-Ray diffraction samples were collected on a RIGAKU RINT 2500 and was used to determine the CrI of both untreated and pretreated OPF. The OPF crystallinity, as expressed by CrI was determined from XRD data and calculated by using Segal formula in the equation (1) below. Based on the equation (1), the I_{002} is the maximum intensity of the 002-lattice diffraction and I_{am} is the intensity of diffraction at 2θ. The I_{002} and I_{am} were in arbitrary units (a.u.) [14]. The XRD data were collected using CuKα radiation, which generated at 40 kV and 26 mA. The scans were collected from 5°–35° 2θ in 2.5 minutes per 2θ intervals.

$$CrI = \frac{I_{002} - I_{am}}{I_{002}} \times 100 \quad (1)$$

Fourier Transform Infrared spectra of approximately 2 mg of untreated and pretreated OPF samples were measured by using PerkinElmer Spectrum One Fourier Transform Infrared Spectrometer. The sample spectra were obtained with a resolution of 4 cm^{-1} and the frequency range between 4000–500 cm^{-1} in the transmission mode.

3. Result and Discussion

3.1. Bio-ionic liquid reactor

A Bio-ionic liquid reactor for aqueous ionic liquid pretreatment was fabricated as shown in figure 2(a) and 2(b).

3.2. Aqueous ionic liquid pretreatment

Normality test shown in table 1 demonstrates that the data was significantly normal (Shapiro-Wilk). The CrI of OPF pretreated with 40% concentration of [EMIM][OAc] in a closed system was compared to the untreated OPF.The result showed significant difference with 95% confidence level as shown in table 2 and 3.

Table 1. Normality of CrI data of the pretreated OPF

	Kolmogorov-Smirnov			Shapiro-Wilk		
	Statistic	Degree of Freedom	Significant	Statistic	Degree of freedom	Significant
CrI	0.204	6	0.200	0.883	6	0.281

Table 2. CrI of untreated and pretreated OPF

Oil Palm Frond (OPF)	N	Mean	Standard Deviation	Standard Error	95 % Confidence Interval for Mean		Minimum	Maximum
					Lower Bound	Upper Bound		
Untreated	3	62.5745	1.4378	0.8301	59.0028	66.1461	60.99	63.81
Pretreated	3	54.1319	2.2408	1.2937	48.5655	59.6982	52.62	56.71
Total	6	58.3532	4.9212	2.0091	53.1886	63.5177	52.62	63.81

IOP Conf. Series: Materials Science and Engineering **358** (2018) 012007 doi:10.1088/1757-899X/358/1/012007

Table 3. Anova analysis on CrI of untreated and pretreated OPF between and within groups.

CrI	Sum of Squares	Degree of freedom	Mean Square	F	Significant
Between Groups	106.916	1	106.916	30.167	0.005
Within Groups	14.177	4	3.544		
Total	121.093	5			

Figure 2 (a). Bio-ionic liquid reactor for biofuel production©.

Figure 2 (b). The reactor without lid

3.3. Crystallinity and chemical changes measurements

An X-ray diffractogram of untreated and pretreated OPF is shown in figure 3. The diffraction pattern of the pretreated OPF samples is lower as compared to the untreated OPF, showed by the two main peaks of the cellulose and amorphous background band. I_{002} at the Bragg angle (2θ) in the x axis represents the maximum intensity at 2θ = 22°. I_{am} shows the minimum intensity at 2θ =18. The CrI of untreated OPF and pretreated OPF are 63% and 54%, respectively. Figure 4 shows the FTIR spectra in the fingerprint region between 4000 and 500 cm^{-1} for untreated and pretreated OPF. The characteristic of hydrogen bonds or O-H stretching vibration are shown in the 3600-3100 cm^{-1} region of the OPF pretreated peak (3320 cm^{-1}) where the peak became sharper as compared to the untreated OPF. This is correlated with the scission of the intra and intermolecular hydrogen bonds. Figure 4 shows the intensity of the adsorption band of 2908 cm^{-1} (C-H stretching vibration) which strongly decreased as correlated to the presence of amorphous or cellulose II. The adsorption band 1165 cm^{-1} assigned to C-O-C asymmetric (bridge oxygen stretching) of the cellulose. Adsorption band 1425- 1430 cm^{-1} showed CH_2 bending as correlated to the presence of cellulose (crystallized I and amorphous) or also known as "crystallinity band". Decrease in the intensity of the band reflects a reduction in the degree of crystallinity of the samples (CrI).

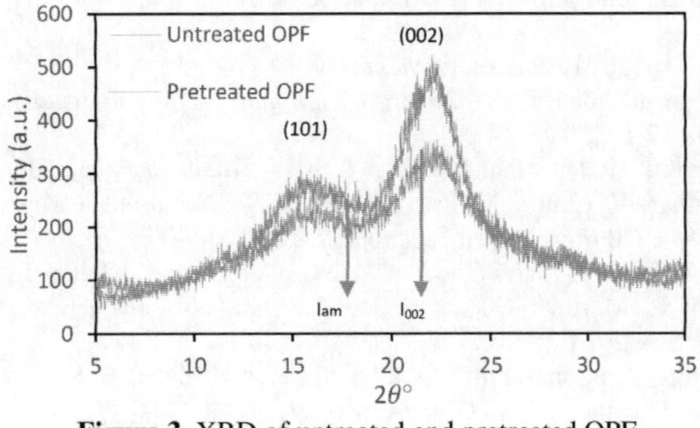

Figure 3. XRD of untreated and pretreated OPF

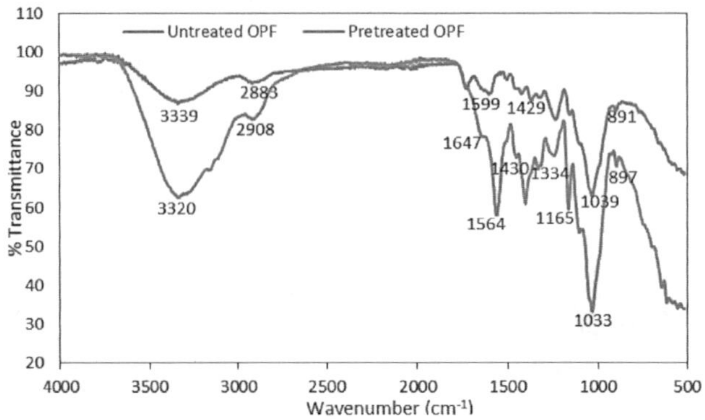

Figure 4. FTIR spectra of untreated and pretreated OPF

4. Conclusion

The CrI of pretreated OPF with 40% concentration of 1-ethyl-3-methylimidazolium ionic liquid in a closed system obtained from XRD analysis was 14% lower from the untreated OPF and the result shows significant difference with 95% confidence level. The diffraction pattern of the pretreated OPF samples was lower compared to the untreated OPF. The characteristic of the FTIR spectra of the pre-treated OPF shows a presence of cellulose I and occurrence of amorphous cellulosic in the samples. The findings from this study are expected to improve knowledge on pretreatment of OPF by using aqueous 1-ethyl-3-methylimidazolium acetate ionic liquid as a green and economically viable process for future renewable energy.

Acknowledgements

The authors gratefully express their sincere appreciation to the Ministry of Higher Education (MOHE) for the research grant (FRGS/1/2014/TK06/UITM/03/5) and Department of Bioprocess Engineering, Faculty of Chemical Engineering, Universiti Teknologi MARA (Shah Alam) for their support and provision in conducting this study

References

[1] Soccol C R, Faraco V, Karp S, Vandenbergh L P S, Thomaz-Soccol V, Woiciechowski A and Pandey A. 2011. *Biofuels: Alternative Feedstocks and Conversion Processes* vol 1, ed A Panday *et al* (USA: Academic Press) chapter 5 pp 101-122

[2] Balat M, Balat H and Cahide O 2008 Progress in bioethanol processing *Progress in Energy and Combustion Science* **34** 551-73

[3] Eliana C, Jorge R, Juan P and Luis R 2014 *Fuel.* **118** 41-47

[4] Karimi M, Jenkins B and Stroeve P 2014 *Renewable and Sustainable Energy Reviews* **40** 400–21

[5] Lan N, Ho S, Koo Y 2014 *Process Biochemistry.* **49** 1144–51

[6] Godson A and Sokan–Adeaga A A. 2014 *International Journal of Sustainable Energy and Environment.* **2** 1–22

[7] Brandt A, Gräsvik J, Hallett J P and Welton T. 2013. *Green Chemistry*, **15** 550–83

[8] Chenlin L, Bernhard K, Chithra M, Rohit A, Henrik V S, Manfred A, Kenneth P V, Blake A S and Seema S. 2010 *Bioresource Technology.* **101** 4900–06

[9] Liu C Z, Wang F, Stiles A R and Guo C. 2012 Applied *Energy.* **92** 406–14

[10] Gude V, Patil P, Martinez-Guerra E, Deng S and Nirmalakhandan N. 2013 *Sustainable Chemical Processes.* **1** 5

[11] Shi J, Balamurugan K, Parthasarathi R, Sathitsuksanoh N, Zhang S, Stavila V, Venkatesan S, Blake A S and Seema S. 2014 *Green Chemistry.* **16** 3830–40

[12] Zhang J, Wang Y, Zhang L, Zhang R, Liu G and Cheng G. 2014 *Bioresource Technology.* **151** 402–5

[13] Zavrel M, Bross D, Funke M, Büchs J and Spiess A C. 2009 *Bioresource Technology*. **100** 2580–87.
[14] Segal L, Creely J J, Martin A E and Conrad C M. 1959 *Textile Research Journal*. **29** 786–94
[15] Li L, Zhou W, Wu H, Yu Y, Liu F and Zhu D. 2014 *BioResources* **9** 3993-4005
[16] O'Connor R T, DuPre E F and Mitcham D. 1958 *Textile Res. J.* **28** 382-92
[17] Nelson, M. L. and O'Connor 1964 J. Appl. Polym. Sci. **8** 1325–1341
[18] Karimi K and Taherzadeh M J. 2016 *Bioresource Technology* **200** 1008-18
[19] Karimi K and Taherzadeh M J. 2016 *Bioresource Technology* **203** 348-56
[20] Vandenbrink J P, Hilten R N, Das K C, Andrew H P and Frank A F. 2012 *Bioenergy Research*, **5** 387–97
[21] Sumathi S, Chai S P and Mohamed A R. 2008 *Renewable and Sustainable Energy Reviews*. **12** 2404–21.
[22] Aainaa N, Ramli S, Aishah N and Amin S. 2014 *Fuel Processing Technology*. **128** 490–8.

IOP Conf. Series: Materials Science and Engineering **358** (2018) 012008 doi:10.1088/1757-899X/358/1/012008

Solid Catalyst Nanoparticles derived from Oil-Palm Empty Fruit Bunches (OP-EFB) as a Renewable Catalyst for Biodiesel Production

H Husin[1,2*], T M Asnawi[1], A Firdaus[1], H Husaini[3], I Ibrahim[4], and F Hasfita[4]

[1]Department of Chemical Engineering, Syiah Kuala University, Darussalam, 23111, Banda Aceh, Indonesia
[2]Graduate Program in Chemical Engineering, Syiah Kuala University, Darussalam, 23111, Banda Aceh, Indonesia
[3]Department of Mechanical Engineering, Syiah Kuala University, Darussalam, 23111, Banda Aceh, Indonesia
[4]Department of Chemical Engineering, Malikussaleh University, Lhoekseumawe, 24300, Aceh Utara, Indonesia

husni_husin@che.unsyiah.ac.id*

Abstract. Solid nanocatalyst derived from oil-palm empty fruit bunches (OP-EFB) fiber was successfully synthesized and its application for biodiesel production was investigated. The OP-EFB was treated by burning, milling and heating methods to generate ashes in a nanoparticle size. The nanoparticle palm-bunch ash was characterized by scanning electron microscopy (SEM) and x-ray diffraction (XRD). The effects of the calcination temperature and catalyst amounts for transesterification reactions were investigated. XRD analysis of palm bunch ash exhibited that the highest composition of peaks characteristic were potassium oxide (K_2O). SEM analysis showed that the nano palm bunch ash have a particle size ranging of 150-400 nm. The highest conversion of palm-oil to biodiesel reach to 97.90% was observed by using of palm bunch ash nanocatalyst which heated at 600°C, 3 h reaction time and 1% catalyst amount. Reusability of palm bunch ash catalysts was also examined. It was found that of its high active sites, reusable solid catalyst was obtained by just heating of palm bunch ash. It has a capability to reduce not only the amount of catalyst consumption but also reduce the reaction time of transesterification process.

1. Introduction

Recently, alternative energies, for example, hydrogen energy [1-3], solar energy [4, 5], and biodiesel [6, 7], have been focused worldwide because of recent energy. Biodiesel is one of popular alternative energy in the world due to biodegradable and renewable fuel [8]. Biodiesel is mono-alkyl esters of long chain fatty acids. Biodiesel has physical and chemical properties that resemble diesel fuel. In addition, biodiesel is an environmentally friendly fuel due to its nontoxic and lower emissions of CO emission [9]. Currently, most commercial processes used for biodiesel synthesis employ a homogeneous base catalyst, such as NaOH or KOH. However, it suffers several drawbacks, such as equipment corrosion and deal with H_2SO_4 waste produced from neutralization process [10, 11]. This situation has deficiency not only due to high catalyst consumption but also biodiesel purification process would be difficult. This will certainly lead to increase production costs[12]. On the other hand, heterogeneous catalysts could replace homogeneous ones to eliminate the problems associated with homogeneous catalysts. Solid based catalysts are easily regenerated and less corrosive, safer, cheaper

Content from this work may be used under the terms of the Creative Commons Attribution 3.0 licence. Any further distribution of this work must maintain attribution to the author(s) and the title of the work, journal citation and DOI.
Published under licence by IOP Publishing Ltd

and more environmentally friendly process, and therefore, attracted considerable attention recently [13, 14].

Oil-palm empty fruit bunch (OP-EFB) is a solid lignocellulosic waste, abundantly generated in Southeast Asia (SEA), particularly Indonesia, Malaysia, and Thailand, with total production almost 88.5% of world's palm oil (48.99 million metric tons per year) [15, 16]. Pyrolysis of OP-EFB will produce volatile matters, carbonaceous substances and ash residue. Ash residue contains organic and inorganic materials. Inorganic constituent of OP-EFB ash is mainly potassium in crystalline forms. This metal rich constituent is easily converted into potassium oxide (K_2O) by burning and calcination in air atmosphere at moderate temperature [17]. So far, studies on the preparation and application of biomass catalyst are very limited in open literatures. These results have motivated the present study involving the conversion of EFB biomass to catalyst for transesterification reaction of palm oil into biodiesel. Therefore, it is possible to utilize potassium oxide from palm bunch ash as a low-cost solid catalyst for biodiesel production. The aim of the present work is to obtain an active and cheap solid catalyst from empty fruit bunch ash for transesterification of vegetable oils into biodiesel. For this purpose, the EFB was treated by burning, milling and heating methods to generate ashes in a nanoparticle size. The catalysts were characterized by analytical techniques. In this work, the performance of non-nanocatalyst and nanocatalyst on transesterification of palm oil into biodiesel were investigated.

2. Materials and methods

2.1. Preparation of Nanocatalyst Palm Bunch Ash

Oil-palm empty fruit bunch was cleaned up from impurities, washed and dried for 1 to 2 days, respectively. The dried materials then burned in open air to get ashes and put into siever screening 100 mesh. OP-EFB ash pass through a 100 mesh Siever can be directly used for nanocatalyst preparation stage. While the ash retained on Siever recycled into milling and sieved again. K_2O-riched ash put into a porcelain mortar and calcined at temperature of 600 °C under air atmosphere for 4 hours to vaporize remaining volatile substances and impurities to its enhance active sites.

2.2. Catalyst Characterization

The catalysts were characterized by powder X-Ray Diffraction (XRD) and scanning electron microscopy (SEM). In the XRD method, the samples were recorded on Bruker diffractometer using Cu-Kα radiation (λ = 0.154 nm) operating at 40 kV and 100 mA. The diffractograms were recorded in the 2θ ranges of 10 – 60° at a scanning speed of 5° min^{-1}. Diffraction peaks of crystalline phases were compared with those of standard compounds reported in the JCPDS (Joint Committee of the Powder Diffraction Standards) database (JCPDS-International Center for Diffraction Data (ICCD), 2000). The morphology and structure of OP-EFB ash nanocatalyst powder was recorded by Scanning Electron Microscopy (SEM). Fatty acid composition of biodiesel was determined by gas chromatography with Mass Spectrometry (GC-MS).

2.3. Evaluation of transesterification reaction

Testing of the catalysts was conducted by transesterification reaction in a set of three-neck flask 500 ml as a reactor and equipped with impeller, temperature controller, and reflux cooling system. The catalytic reaction was carried out at 65°C for 3 h. Catalyst loading of 0.1- 5.0 wt.% and methanol to palm-oil mole ratio of 6:1 were used in the catalyst testing. After the reaction finished, the biodiesel product (FAME) was separated after aged for 24 h forming 3 layers (residual methanol, FAME, and glycerol as well as used solid catalyst). The resulted FAME biodiesel was characterized and identified using GC-MS (SHIMADZU) equipped with Agilent DB-1 column and helium as a gas carrier.

3. Results and discussion

3.1. Characterization Palm Bunch Ash Catalyst by XRD

Figure 1.a shows the XRD pattern OP-EFB ash non-calcined catalyst. K_2O formation was confirmed by XRD patterns obtained from the study. According to XRD pattern, OP-EFB ash composes of K_2O as a major component. XRD spectra show that K_2O have identical crystal structures to cubic type

(JCPDS card no. 26-1327, space group Fm3m). The characteristic peaks of K_2O appear at 2θ = 25.121°, 28.607°, 31.843°, 40.625°, and 50.101°, respectively. It also demonstrated that the highest intensity of characteristic peaks of K_2O of non-calcined catalyst at 2θ= 28.607° reaches up to 674 cps, as can be seen in Figure 1.a.

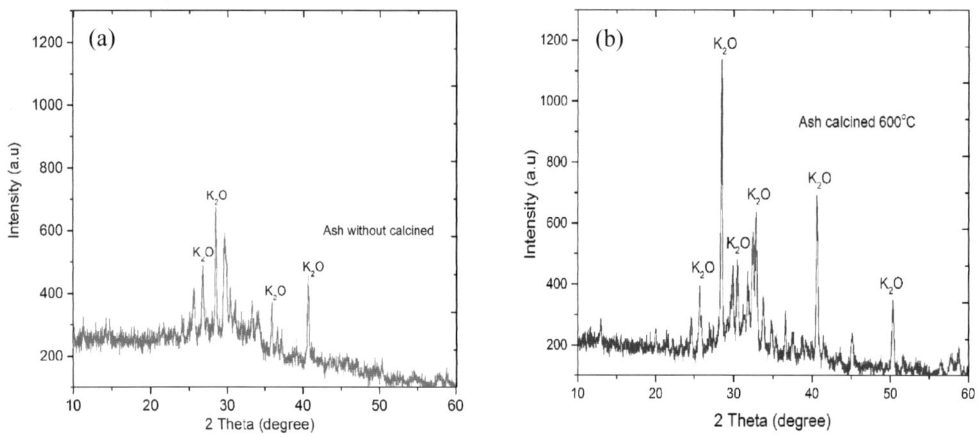

Figure 1. XRD patterns of (a) OP-EFB ash without calcined, (b) palm bunch ash calcined at 600 °C.

Figure 1.b depicts the XRD pattern of the sample after annealed at 600 °C. The OP-EFB ash nanocatalyst powder showed a significant increase of the intensity of K_2O main peak up to 1135 cps, as depicted in Figure 1.b. The K_2O as major compound in OP-EFB ash possess basic surface sites, which make them highly efficient in catalytic process applications, although it can be seen some small peaks due to phases of silica compound. Comparing the XRD pattern of non-calcined and calcined catalyst demonstrated that the intensity of K_2O increased from 674 to 1135 cps (at 2θ = 28.607°). This findings indicated that calcination temperature at 600 °C led to increase the degree of crystallinity of K_2O. Calcination can enhance crystals growth of a compound, wherein the crystals growth will increase with increasing temperature of heating to form a complete crystalline structure at optimum temperature. It can vaporize impurities or volatile matters that cover the pores of the catalyst.

3.2. Characterization of Palm Bunch Ash by SEM

In order to elucidate the effect of milling on particle sizes of palm bunch ash, the SEM measurement of ash catalysts was also performed. The SEM morphologies of non-nano and nanocatalyst are shown in Figure 2. The palm bunch ash catalyst was previously calcined in a muffle furnace for 4 h at 600 °C. The non-nanocatalyst sample is composed by large particles, exposing irregular particles, which sizes ranged between 2000-5000 nm, as illustrated in Figure 2.a. The SEM image of non-nanocatalyst showed a compact agglomeration. After milling process the particle sizes of catalyst become smaller, as can be seen in Figure 2.b. The nanoparticles with average size around 150 to 400 nm are observed. The catalyst in nanosizes is able to increase the catalyst surface area significantly.

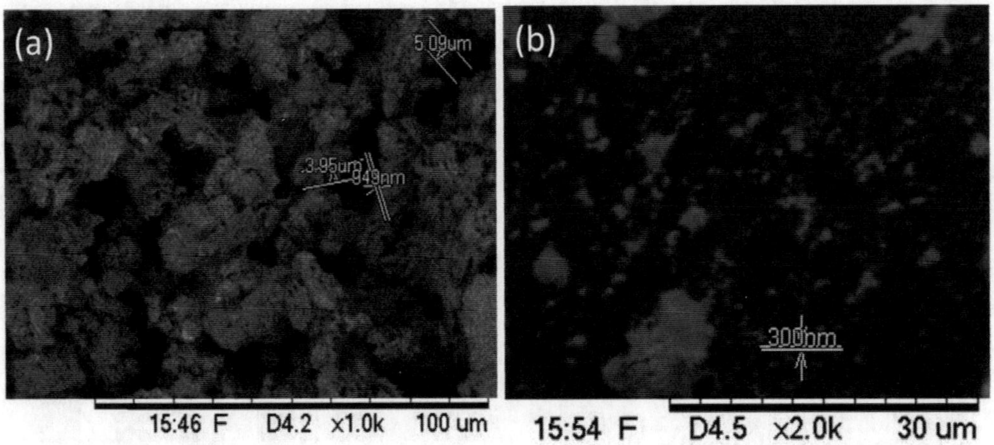

Figure 2. SEM Images of palm bunch ash: a) non-nano and b) nanosizes

3.3. Effect of Catalyst Loading on Biodiesel Production

Catalyst weight is one of parameters determining the performance of a catalyst. Reactivity of the catalyst can be known from its ability to accelerate the reactions in which reactants are converted into products obtaining optimum yield with less consumption of the catalyst.In this study, the weight of catalyst is used as one of the parameters to determine catalyst performance. The effect of the catalyst loading on the biodiesel yield was investigated by varying the catalyst mass ratio from 0.1-5.0 wt.%. Then biodiesel yield from transesterification reaction compared between nanocatalyst and non-nanocatalyst loadings were also determined. Nanocatalyst palm bunch ash denoted as N-catalyst, whiles the non nanocatalyst denoted as S-catalyst.

The activity of N-catalyst and S-catalyst was evaluated in terms of ester conversion during transesterification of palm-oil. The biodiesel yield exhibited a gradually increasing trend as the catalyst loading increased. However, further increase in the amount of catalyst, the biodiesel yield decreased. In detail, the highest biodiesel yield gains (optimum), i.e 97.13% results from N-catalyst loading with ratio catalyst 1% by weight of oil and earned the highest yield (optimum), i.e 93.32% resulting from the use of S-catalyst by weight of the catalyst 3% by weight oil. This evidence demonstrated that the use of N-catalyst requires less amount of catalyst compared to S-catalyst to produce optimum yield of biodiesel. This is resumed due to the catalyst in nanometer size has a specific surface area greater than the mass or volume of the same of non-nanocatalyst. The active compound potassium oxide (K_2O) contained in the ash palm bunches catalyst distributed not only on the surface of the catalyst but also inside of the catalyst particles. During high speed grinding treatment bulk catalyst was initially broken up and reduced in nanometer size to form new higher surface area, as well evidenced by SEM images.

Potassium oxide compounds play an active role in the formation of methoxide anion in the catalyst activation process. Methoxide anion formation becomes very important because the anion active role in the termination of the carbonyl carbon bond of triglycerides to form methyl esters (biodiesel). If the distribution of potassium oxide on the surface of the catalyst increases, the possibility of methoxide anions will be increased. Increasing the amount of methoxide anion in the catalyst activation process will accelerate the reaction time required by anion methoxide break the chains of carbon carbonyl compounds of triglycerides. By utilizing nanocatalyst which has a high reactivity then the transesterification process would be more effective due to the process only requires less amount of catalyst loading. Conversely, biodiesel yield decreased after optimum yield is reached with a higher amount of catalyst loading. This is because by increasing the amount of the catalyst in the transesterification process, the density of the particles of the catalyst will increase. If the densities of catalyst particles are high, it will trigger agglomeration of catalyst particles. Catalyst agglomeration will affect the active surface area of the catalyst decreases which subsequently decreasing the reactivity. That situation will complicate the process of activation of the catalyst. Where the active compound will be difficult ionized of potassium oxide resulting in a decrease in the amount of methoxide anions in the process.

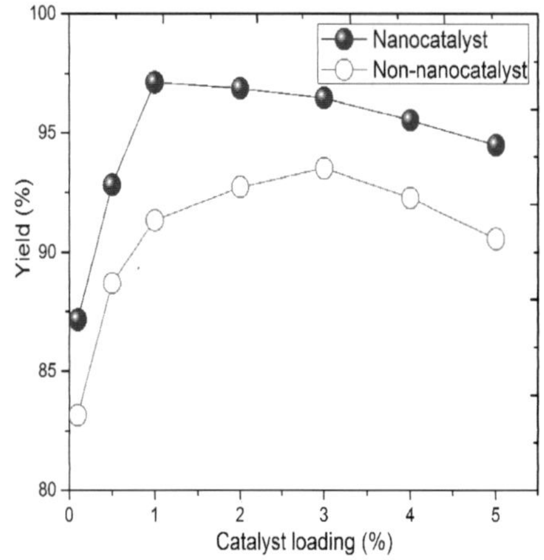

Figure 3. Effect of catalyst loading on biodiesel production

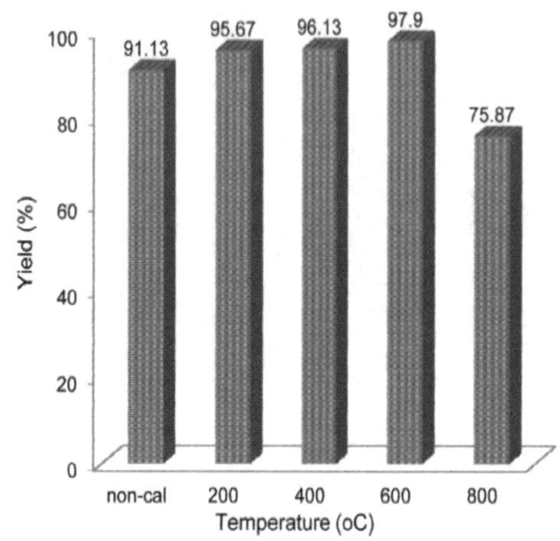

Figure 4. Effect of calcination temperature of catalyst on biodiesel production

In addition agglomeration of catalysts may also affect the process of mass transfer of the reactants that inhibit diffusion onto the surface of the catalyst after being covered by the other agglomerated particles. This is because most of the active compound which is located on the surface of the catalyst particles is covered by other particles. Where the active compound will be difficult ionized of potassium oxide resulting in a decrease in the amount of methoxide anions in the process. Decreased amount of methoxide anions in the catalyst activation and mass transfer process would lead to suppressed biodiesel yield. It is caused due to insufficient of methoxide anion availability to break carbonyl carbon bonds so that the conversion of triglycerides into methyl ester decreased, consequently lowering the yield of biodiesel. There is slight reduction of conversion of reactants due to reaction system being as viscous slurry and emulsified [18].

3.4. Effect of calcination temperature of ash catalyst

In this study, the calcination temperature of ash catalyst is used as one of the parameters to determine catalyst performance. In order to study the influence of the calcination temperature on the catalyst performance, N-catalysts have been calcined from 200 to 800°C. As presented in Figure 4, it can be seen that the increase of biodiesel yield is in accordance with increasing of calcination temperature which lead to enhance crystal growth. The improved crystal growth affected by increasing heating at appropriate temperature would lead to form a complete crystal structure of the compound. In addition, from Figure 4 can also be seen that the yield of biodiesel at temperature 600°C and 800°C drastically decreased from 97.89 to 75.87%. The decrease was due to reduced active compound in crystal phase of potassium oxide (K_2O) contained in nanocatalyst. The active phase on the catalyst was sintered at higher temperature and total basicity [19, 20]. As evidenced by the catalyst XRD characterization, the ash catalyst comprises of K_2O. The active species of the solid catalysts were K_2O. The K_2O is incorporated on the surface of empty fruit bunch ash support, leading to the improved activity of the catalyst. The calcination process for the solid ash catalyst is favourable for the interaction between K_2O and other metal oxides, which may lead to the enhanced stability of catalyst [21]. However, the declining of the crystal phase of the compound would be caused by calcination temperature at 800°C is above K_2O melting point 740 °C. Heating at excessive high temperatures would developed crystals agglomeration which decreasing catalyst surface area and subsequently suppressed nanocatalyst reactivity and lead to decline biodiesel yield drastically [22].

3.5. Composition of palm oil

Figure 5 displays biodiesel chromatogram spectra that identify the presence of methyl esters contained in biodiesel. Methyl myristate and methyl palmitate are identified at peak 1 and 2. Methyl stearate, methyl oleate and methyl linoleate are demonstrated at peak 3, 4 and 5. While, methyl linolenate and methyl arachidate are identified at retention time of 25.167 and 25.31 seconds, as recorded at peak 6 and peak 7, respectively. All data of the biodiesel composition are tabulated in Table 1. From Table 1, it is reported that methyl oleate is the highest concentration of 38.98%. The lowest concentration of methyl ester biodiesel constituents are methyl myristate and methyl linolenate.

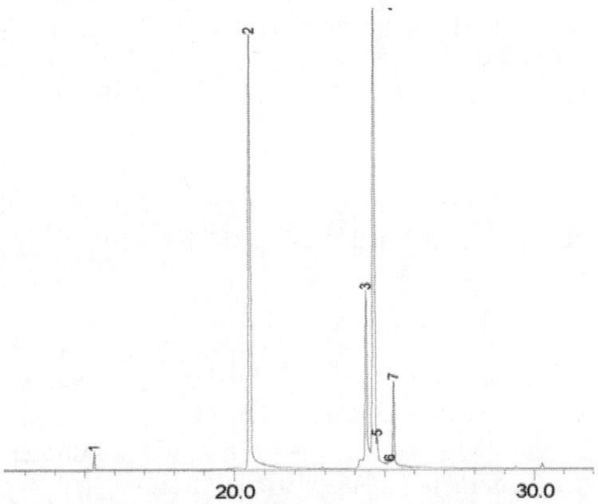

Figure 5. GC-MS chromatogram of biodiesel product

Table 1. Fatty Acid Methyl Ester (biodiesel)

Peaks No.	Component	Composition (%)
1	Methyl Myristate (C16:0)	0,76
2	Methyl Palmitate (C16:1)	35,52
3	Methyl Stearate (C18:0)	11,88
4	Methyl Oleate (C18:1)	38,98
5	Methyl Linoleate (C18:2)	3,16
6	Methyl Linolenate (C18:3)	0,76
7	Methyl Arachidate (C20:0)	8,94

3.6. Characteristic of biodiesel

To study the physicochemical properties, the transesterification reaction was carried out in the presence of the nanocatalyst under calcined at 600 °C. The characteristics of the biodiesel sample were shown in Table 2.

Table 2. Characteristic of biodiesel of this work, standard ASTM, prEN & SNI

Parameter	This work	ASTM	prEN	SNI
Kinematic viskosity (cSt)	3,59 – 3,69	1,90 – 6,00	3,50 – 5,00	2,30 – 6,00
Density (gr/cm^3)	0,8717 – 0,8741	0,80 – 0,88	0,86 – 0,90	0,85 – 0,89
Acid number (mg KOH/gr)	0,11 – 0,2805	max 0,80	max 0,3	max 0,80

As listed in Table 2, the fuel properties of the prepared biodiesel have been found to comply with the ASTM, prEN and SNI standards. Refer to Table 2, there can be seen that the kinematic viscosity of biodiesel is ranging from 3.59 - 3.69 cSt. It can be seen that the density of biodiesel is ranging from 0.8717 - 0.8741 g/cm^3 and the acid number is ranging from 0.1122 - 0.2805 mg KOH/g. These values are also in compliance with ASTM standards, prEN and SNI [23-25].

4. Conclusions

A palm bunch ash solid catalyst was successfully synthesized through calcination method. The catalyst was calcined at temperature of 600 °C. The major component of nanocatalyst derived from palm bunch ash is potassium oxide (K_2O). The particle sizes of palm bunch ash ≤ 400 nm. Utilization of as-produced nanocatalyst derived palm bunch ash resulting the highest conversion of palm oil into biodiesel reached up to 97.90%, reactions for 1 h and catalyst loading 1.0 wt.%..

Acknowledgment

We acknowledge the financial support through the Research Grants of of Scopus h-index of Syiah Kuala University, Republic of Indonesia (No: 1712/UN11/SP/PNBP/2016).

References

[1] Husin H, Mahidin Zuhra and Hafita F 2014 H_2 evolution on Lanthanum and Carbon co-doped $NaTaO_3$ Photocatalyst *Bulletin of Chemical Reaction Engineering & Catalysis* **9** p 81-86.

[2] Husin H, Pontas K, Sy Y, Syawaliah and Saisa 2014 Synthesis of Nanocrystalline of Lanthanum Doped $NaTaO_3$ and Photocatalytic Activity for Hydrogen Production *Journal of Engineering and Technological Sciences* **46** p 318-327

[3] Husin H, Adisalamun A, Sy Y, Asnawi TM and Hasfita F 2017 Pt nanoparticle on $La_{0.02}Na_{0.98}TaO_3$ catalyst for hydrogen evolution from glycerol aqueous solution in *AIP Conference Proceedings* p 030073

[4] Susanti D, Nafi M, Purwaningsih H, Fajarin R, and Kusuma G E 2014 The Preparation of Dye Sensitized Solar Cell (DSSC) from TiO_2 and Tamarillo Extract *Procedia Chemistry*. **9** p 3-10

[5] Husin H, Mahidin Yunardi and Hafita F 2015 Visible Light Driven Photocatalytic Hydrogen Evolution by Lanthanum and Carbon-co-Doped $NaTaO_3$ Photocatalyst *Key Engineering Materials*. **659** p 231-236

[6] Taufiq YH, Lee HV, Hussein MZ and Yunus R 2011 Calcium-based mixed oxide catalysts for methanolysis of Jatropha curcas oil to biodiesel *Biomass and Bioenergy* **35** p 827-834

[7] Husin H, Mahidin and Marwan 2011 Studi penggunaan katalis abu sabut kelapa, abu tandan sawit dan K_2CO_3 untuk konversi minyak jarak menjadi biodiesel *Reaktor* **13** p 254-261

[8] Fidalgo W R R, Ceron A, Freitas L, Santos J C and de Castro H F 2016 A fluidized bed reactor as an approach to enzymatic biodiesel production in a process with simultaneous glycerol removal *Journal of Industrial and Engineering Chemistry* **38** p 217-223

[9] Dai YM, Chen KT, Wang PH and Chen CC 2016 Solid-base catalysts for biodiesel production by using silica in agricultural wastes and lithium carbonate *Advanced Powder Technology* **27** p 2432-2438

[10] Istadi I, Anggoro D D, Buchori L, Rahmawati D A and Intaningrum D 2015 Active Acid Catalyst of Sulphated Zinc Oxide for Transesterification of Soybean Oil with Methanol to Biodiesel *Procedia Environmental Sciences* **23** p 385-393

[11] Abomohra AEF, Jin W, Tu R, Han SF, Eid M, and Eladel H 2015 Microalgal biomass production as a sustainable feedstock for biodiesel: Current status and perspectives, *Renewable and Sustainable Energy Reviews* **64** p 596-606

[12] Zuhra, Husin H, Hasfita F and Rinaldi W 2015 Preparasi Katalis Abu Kulit Kerang Untuk Transesterifikasi Minyak Nyamplung Menjadi Biodiesel *Agritech* **35** p 69-77

[13] Alonso D M, Granados M L, Mariscal R and Douhal A 2009 Surface chemical promotion of Ca oxide catalysts in biodiesel production reaction by the addition of monoglycerides, diglycerides and glycerol *J Catal.* **276** p 229-236

[14] Feng Y, Zhang A, Li J and He B 2011 A continuous process for biodiesel production in a fixed bed reactor packed with cation-exchange resin as heterogeneous catalyst, *Bioresource Technology* **102** p 3607-3609

[15] Abdullah N and Gerhauser H 2008 Bio-oil derived from empty fruit bunches. *Fuel* **87** p 2606-2613

[16] Idris J, Shirai Y, Ando Y, Ali A A M, Othman M R, Ibrahim I 2014 Production of Biochar With High Mineral Content from Oil Palm Biomass *Malaysian Journal of Analytical Sciences* **18** p 700-704.

[17] Purwandari F A, Sanjaya A P, Millati R, Cahyanto M N, Horvath I S and Niklasson C 2013 Pretreatment of oil palm empty fruit bunch (OPEFB) by N- methylmorpholine-N-oxide (NMMO) for biogas production: structural changes and digestion improvement *Bioresour Technol* **128** p 461-667

[18] Boro J, Deka D, and Thakur A J 2012 A Review on Solid Oxide Derived from Waste Shell as Catalyst for Biodiesel Production *Renewable and Sustainable Energy Reviews* **16** p 904-910

[19] Chen G, Shan R, Shi J, Liu C, and Yan B 2015 Biodiesel production from palm oil using active and stable K doped hydroxyapatite catalysts *Energy Conversion and Management* **98** p 463-469

[20] Husin H, Zuhra, Hasfita F and Rinaldi W 2013 Waste shells of cockle (Clinocardium nuttalli) as solid catalysts for transesterification of calophyllum inophyllum L. oil to biodiesel production in *The Annual International Conference Syiah Kuala University* AAC Hall Syiah Kuala University Darussalam Banda Aceh Indonesia **3** p 14-18

[21] Xie W and Zhao L 2013 Production of biodiesel by transesterification of soybean oil using calcium supported tin oxides as heterogeneous catalysts, *Energy Conversion and Management* **76** p 55-62

[22] Bet-Moushoul E, Farhadi K, Mansourpanah Y, Nikbakht A, Molaei R, and Forough M 2016 Application of CaO-based/Au Nanoparticles as Heterogeneous Nanocatalysts in Biodiesel Production *Fuel* 164 p 119-127

[23] *ASTM D6751-15ce1 2015 S*tandard Specification for Biodiesel Fuel Blend Stock (B100) for Middle Distillate Fuels ASTM International West Conshohocken, PA, *www.astm.org*.

[24] *ACEA* 2009 "Biodiesel Guidelines" European Automobile Manufacturers Association, Brussels, Belgium March 2009 http://www.acea.be/uploads/publications/ 20090423_B100_ Guideline.pdf

[25] *Standar* Nasional Indonesia *SNI* 04-7182-2006 *Biodiesel* ICS 75.160. Badan Standardisasi Nasional. *bsn.go.id/index.php?/sni_main/sni/detail_sni/7240*

Simulation of CO_2 Solubility in Polystyrene-b-Polybutadiene-b-Polystyrene (SEBS) by artificial intelligence network (ANN) method

R W Sharudin[1,2], S AbdulBari Ali[1], M Zulkarnain[2] and M A Shukri[1]
[1]Faculty of Chemical Engineering, Universiti Teknologi MARA, 40450 Shah Alam, Malaysia
[2]Institute for Infrastructure Engineering & Sustainable Management (IIESM), Faculty of Civil Engineering, Universiti Teknologi MARA, 40450 Shah Alam, Malaysia

rahida@salam.uitm.edu.my*, abdulbari@salam.uitm.edu.my, zulkarnain9902@salam.uitm.edu.my, muhd.aziz.azyzy@gmail.com

Abstract. This study reports on the integration of Artificial Neural Network (ANNs) with experimental data in predicting the solubility of carbon dioxide (CO_2) blowing agent in SEBS by generating highest possible value for Regression coefficient (R^2). Basically, foaming of thermoplastic elastomer with CO_2 is highly affected by the CO_2 solubility. The ability of ANN in predicting interpolated data of CO_2 solubility was investigated by comparing training results via different method of network training. Regards to the final prediction result for CO_2 solubility by ANN, the prediction trend (output generate) was corroborated with the experimental results. The obtained result of different method of training showed the trend of output generated by Gradient Descent with Momentum & Adaptive LR (traingdx) required longer training time and required more accurate input to produce better output with final Regression Value of 0.88. However, it goes vice versa with Levenberg-Marquardt (trainlm) technique as it produced better output in quick detention time with final Regression Value of 0.91.

1. Introduction

Rubber and elastomer foams have a wide variety of applications like in transportations, packaging, construction and building insulation. Thermoplastic elastomers, often called thermoplastic rubbers are consist of a soft midblock and hard end-block which behaves both thermoplastic and elastic properties. SEBS is one type of thermoplastic elastomer that has captured attention in recent years due to its unique characteristics compared to natural rubber. The advantage of synthetic rubber is it has good resistance in temperature and oil. Nowadays, thermoplastic elastomer is being foamed to prepare light weight polymeric material which possesses many applications such as thermal insulation, shock absorber, automotive, aerospace and biomedical. However, there are many factors to be controlled like processing conditions and its physical properties in order to prepare fine cellular structure of thermoplastic elastomer foam. According to Rahida et al. [1] stable dimension of SEBS foam can be prepared when optimum rigidity, CO_2 solubility and diffusivity were achieved. They claimed that controlling diffusion of CO_2 through SEBS during foaming is of great importance. In their study they found that the shrinkage problem in elastomer foaming can be controlled by increasing the hard-polystyrene end-blocks or blending with thermoplastic polystyrene. This is because greater styrene contents increased the storage modulus and reduced the CO_2 solubility and diffusivity which then led to stabilized elastomer cell foam. In recent years, most sophisticated magnetic suspension balance machine was used to measure the CO_2 solubility and diffusivity in polymers. This experimental

Content from this work may be used under the terms of the Creative Commons Attribution 3.0 licence. Any further distribution of this work must maintain attribution to the author(s) and the title of the work, journal citation and DOI.
Published under licence by IOP Publishing Ltd

technique has made possible to measure CO_2 solubility and to estimate diffusion coefficient in any kind of polymers at various temperature and pressure. Some studies have been done on solubility measurement of different gases in polymers. However, this method is highly expensive, time consuming, and highly affected by pressure-volume-temperature (PVT) data of polymers. Due to these limitations, there are some efforts to predict the CO_2 solubility and diffusivity in polymers by intelligence methods as reported elsewhere. Golzar et. al. [2] for example, predicted the solubility of CO_2 and nitrogen (N_2) in common polymers via linear and non-linear models. They reported on structure-properties relationship via intelligence methods like ANN, ANFIS and GA by using set of solubility data for CO_2 and N_2 in different polymers from literatures.

This study reported on the ability of ANN to predict combination of experimental and interpolated data of CO_2 solubility and diffusivity in SEBS in the temperature range of 61.76 to 101.9°C and pressure range of 5 to 20 MPa without repeating such costly and time consuming experimental measurement by MSB. In addition, to the best authors' knowledge, no study has been done on verification of interpolated experimental data of CO_2 solubility in SEBS by ANN. ANN has become one of the most promising techniques in dealing with complex and non-linear problem relations that cannot be solved by simple electronic calculator. The operation of the ANN is likely the operation of the nervous system containing neurons that functional to solve engineering- based problem, scientific or no-scientific based problem. For example Taskinen et al.[3] predicted physicochemical properties of organic compounds which focusing on pharmaceutical research and Fatehi et al. [4] studied on estimation of viscosities of pure ionic liquids using an artificial neural network based on only structural characteristics. In addition, the ability of ANN in processing the complex system even in critical condition makes the ANN as one of the extremely powerful methods for prediction process. Therefore, the intention of this study is to predict and further validate the interpolated data that might be useful for reducing the need of repeating MSB measurement as well as facilitating the control of SEBS cell stabilization. The idea behind is to predict the optimum CO_2 solubility at varying temperature and pressure using ANN that could possibly control the foam shrinkage in SEBS.

2. Methodology

2.1 Data input

Solubility of CO_2 in SEBS data was collected from previous MSB measurement in the temperature range of 61.76 to 101.9°C and the pressure range of 5 to 20 MPa. The selected data were divided into three sections which are the most concerned part in making a perfect ANN model. Randomly, those data were separated according to the following ratio 70% used for training and learning session, and 30% will equally divided into two parts which are validation purpose and for test sets. According to Huang et al.[5], the appropriate distribution of data set is very important in order to obtain the accurate results. Besides, the percentage for test set was in the range of 5 to 35%.

2.2 Neural Network Structure Used

There is various type of neural network structures can be used for prediction. In this study, Multilayer Perceptrons (MLP) with a basic backpropagation training and universal theorem was used as it could work with limited sources of inputs. This network is the most recommended ANNs to handle a simple task. It is classified as feedforward network that could approximate generic functions (continuous and integrable). In this study, feedforward backpropagation (FFBP) was chosen to be used in predicting the CO_2 solubility in SEBS. In feed forward neural network, the process happens in forward direction and the neuron is connected from input to output. The three layers must be connected with each other in order for the process to be done. For feed forward neural network, the connection of the layer is from input to hidden and lastly to output. The direction cannot be reversed to the opposite direction and it is so called as feedforward backpropagation.

After the training, new output data is obtained and further been compared with the actual output. The output data that was keyed in at the starting of the training is known as actual output while the new output is known as expected output data. The adjustment on the data from output to hidden followed by input was made by utilizing the backward propagation together with feedforward for obtaining better result. In this study, the most effective and suitable method of feedforward backpropagation method was used because the variables in this study; temperature, pressure and CO_2 solubility have no linear relationship between one another.

2.3 Artificial Neural Network Training

The most important part in using ANNs is to find the almost perfect set of weight, *w* parameters. To achieve that, abundant training is required. In the training process, theoretically the output from neural model is compared with the desire outputs for all of the training samples using the following formula where the error could be defined.

$$E = \frac{1}{2}\sum_{k\in T_r}\sum_{j=1}^{m}(y_j(x_k, w) - d_{jk})^2 \quad (1)$$

Where,

$d_{jk} = j^{th}$ element of d_k
$y_j(x_k, w) = j^{th}$ neural network output for input x_k
T_r = Index set of training data

Foremost, the value of weight is initialized. During training session, *w* is updated with the following formula,

$$= w - n\frac{\partial E}{\partial w},$$

until the value of E is small enough. Value of n is called learning rate. Per-sample of E_k is,

$$E_k = \frac{1}{2}\sum_{j=1}^{m}(y_j(x_k, w) - d_{jk})^2 \quad (2)$$

The E_k value is used and been updated as

$$= w - n\frac{\partial E_k}{\partial w}.$$

Next, the value of $\frac{\partial E_k}{\partial w}$ is computed by using basic differentiation,

$$\frac{\partial E_k}{\partial w_{ij}^l} = \frac{\partial E_k}{\partial z_i^l}\cdot\frac{\partial z_i^l}{\partial w_{ij}^l} \quad (3)$$

And,

$$\frac{\partial z_i^l}{\partial w_{ij}^l} = \frac{\partial \sigma}{\partial \gamma_i^l}\cdot z_j^{l-1} \quad (4)$$

The gradient $\frac{\partial E_k}{\partial z_i^l}$ can be initialized at output layer as,

$$\frac{\partial E_k}{\partial z_i^L} = (y_i(x_k, w) - d_{ik}) \quad (5)$$

The error is then calculated between both outputs from ANNs and the desired outputs. For a network that using sigmoid equation, the back propagation process is then give by the following formula:

$$\delta_i^L = (y_i(x_k, w) - d_{ik}) \quad (6)$$

$$\delta_i^l = \sum_{j=1}^{N_{l+1}}(\delta_j^{l+1}w_{ji}^{l+1})z_i^l(1 - z_i^l), l = L-1, L-2, \dots 2 \quad (7)$$

And the derivative with respect to weights is,

$$\frac{\partial E_k}{\partial w_{ij}^l} = \delta_i^l z_j^{l-1} l = L, L-1, \ldots, 2 \tag{8}$$

3. Results and Discussion

3.1. Comparison of CO_2 solubility interpolated data between trainlm and traingdx methods

After interpolating solubity result at various ranges of temperature and pressure, a set of CO_2 solubity data was divided into three sections for training and learning by ANN model. The input data was then normalized in the range between 0 and 1. The biggest value of solubility from experimental data is marked as 1 while the smallest value is noted as 0. For the purpose of verifying the ability of ANN to predict the interpolated data, two types of training functions; trainlm and traingdx are employed. The data was imported to ANN database to run the simulation process. Figure 1(a) and (b) shows the training results from trainlm and traingdx networks, respectively. The prediction values are clearly corroborated with experimental values for both networks where it can be seen that the predicted values of training set, validation set and testing set are closed to the experimental value line. However, the predicted value for trainlm is more intense and closer than traingdx. This result is further supported by the values of squared correlation coefficients (R^2) for trainlm and traingdx of 0.911 and 0.886, respectively. The R^2 values for both networks verified are very close to the unity and it shows that ANN model can be used for predicting interpolated CO_2 solubility data. However, trainlm has higher ability to predict CO_2 solubility because the value of R^2 is higher than traingdx.

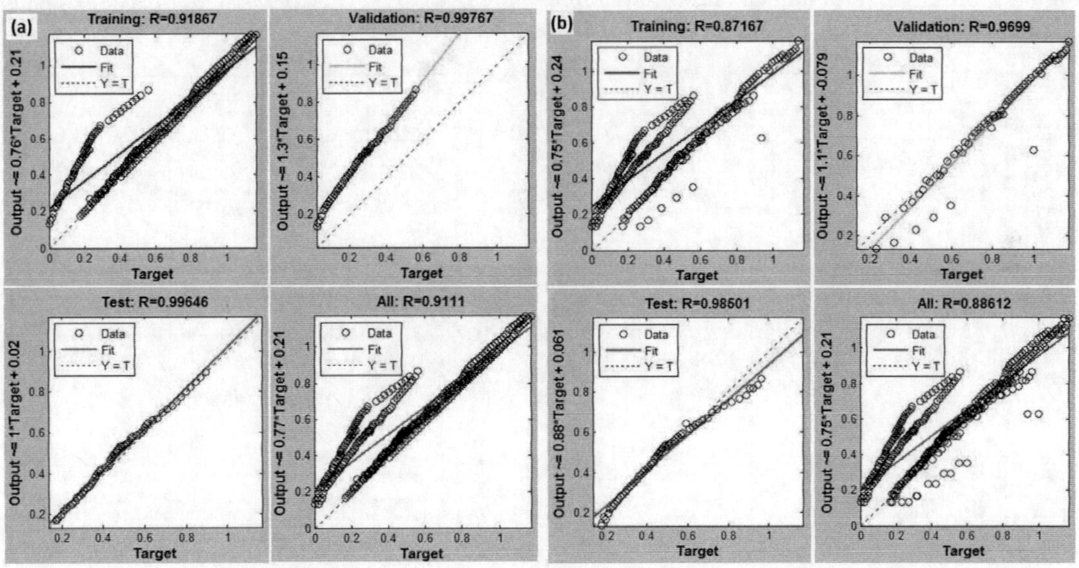

Figure 1. Training results; (a) trainlm technique and (b) traingdx technique

3.2. Error Yield between Input Data and Output Generated

The reliability of ANN model with trainlm and traingdx transfer functions was further examined by measuring the variance between predicted and experimental data. The CO_2 solublity data was trained at fix temperature and varying the pressure to show the trend of trained output (predicted) with the input (combination result of experimental and interpolated data) as a reference in a plot form. Figure 2 and 3 show the error plot obtained from the differences of solubility between input and output by trainlm and traingdx, respectively. For traingdx, at temperature of 61.76 and 62.18°C, the trend was appeared unsmoothed due to some data fluctuated and deviated from the input data.

IOP Conf. Series: Materials Science and Engineering **358** (2018) 012009 doi:10.1088/1757-899X/358/1/012009

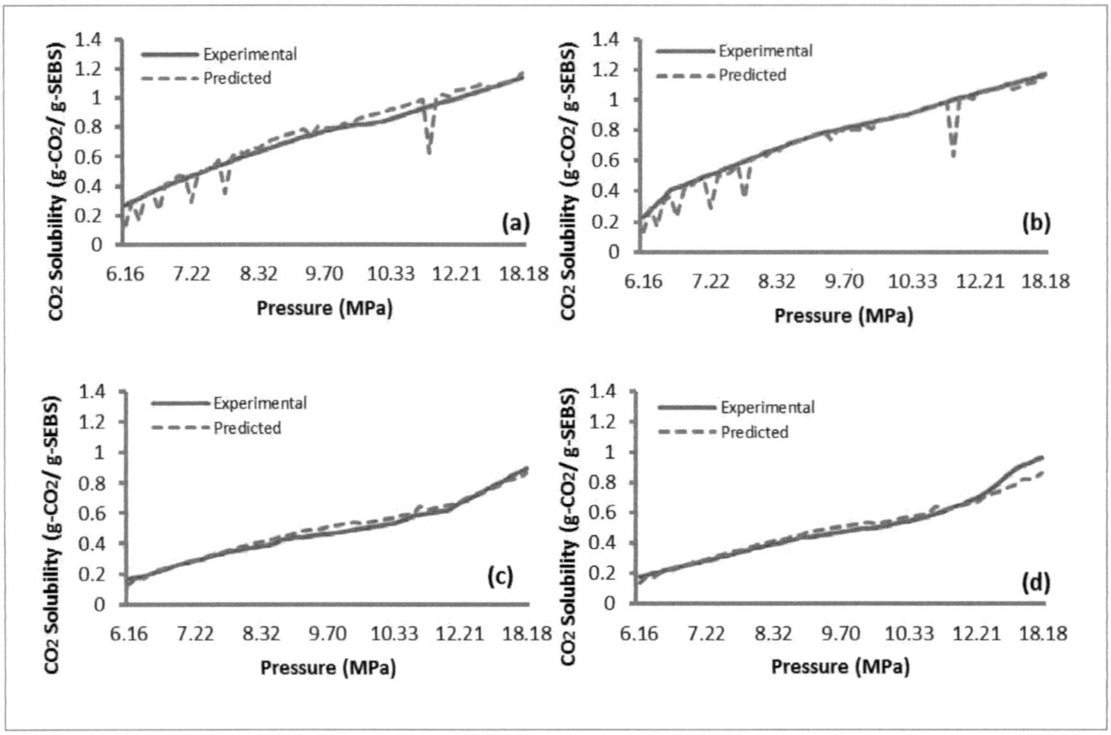

Figure 2. Error yield for traingdx at various temperatures; (a) 61.76°C, (b) 62.18°C, (c) 101.5°C, and (d) 101.9°C

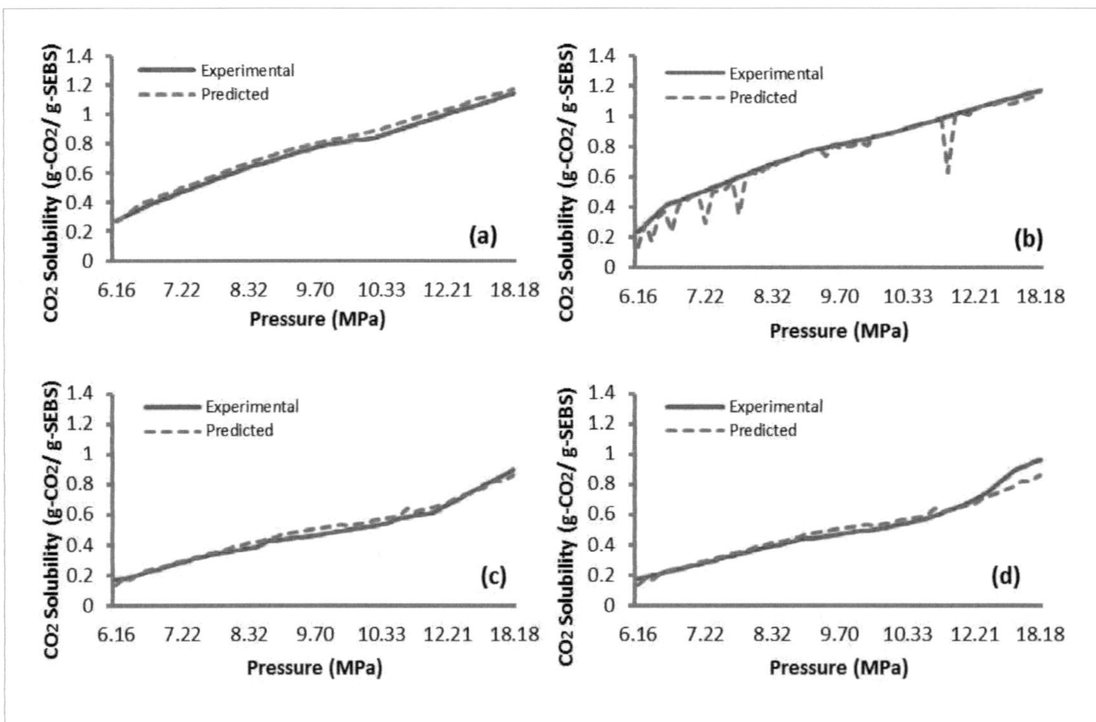

Figure 3. Error yield for trainlm at various temperatures; (a) 61.76°C, (b) 62.18°C, (c) 101.5°C, and (d) 101.9°C

However, at 101.5°C the trend became smooth and corroborated with the input. It shows that the model is reliable and able to train the interpolated data at 101.9°C by traingdx. The similar trend was observed for the predicted data at temperature of 101.9°C which is indicated the reliability of network to predict. For trainlm prediction at similar temperature and pressure, as expected the trends look better at 61.76°C compared to traingdx. It was observed that data fluctuation was reduced and the predicted data line was near to the experimental one, which related to the decreased of the error yield. The smaller gap between predicted and experimental lines by trainlm indicated higher accuracy of prediction is achieved, and thus confirm the ability of ANN model to predict a set of interpolated data. As the function of trainlm algorithm is usually considered for optimizing data, the trend obtained therefore smoother than that of traingdx. Furthermore, as the traingdx algorithm is very sensitive to the learning rate settings, it can oscillate during training and thus the output data became unstable.

3.3. Renormalized Output Data

The presence of large errors in both transfer functions at high temperatures, each of the solubility values were renormalized and tabulated in standard solubility range. By using the value of error generated from the training, two graphs were plotted to show the true pattern of CO_2 solubility predicted by ANNs as depicted in Figure 4. It was observed that the data generated by traingdx was not stable at some temperatures as it might be due to traingdx transfer function required more training as compared to trainlm. In addition, traingdx is quite sensitive to the learning rate setting where when the learning rate is too high, the algorithm might oscillate and became unstable. Whereas the trend obtained by trainlm algorithm appeared smooth and stable as it is usually serves for optimization process. Obviously, this method cannot be used in order to generate output with limited parameter and input data. In contrast, Figure 4(b) clearly shown that the generated pattern by trainlm was smooth and in agreement with the experimental data flow pattern. However, more training is required to produce better R^2 hence, give the best solubility prediction data with lower error.

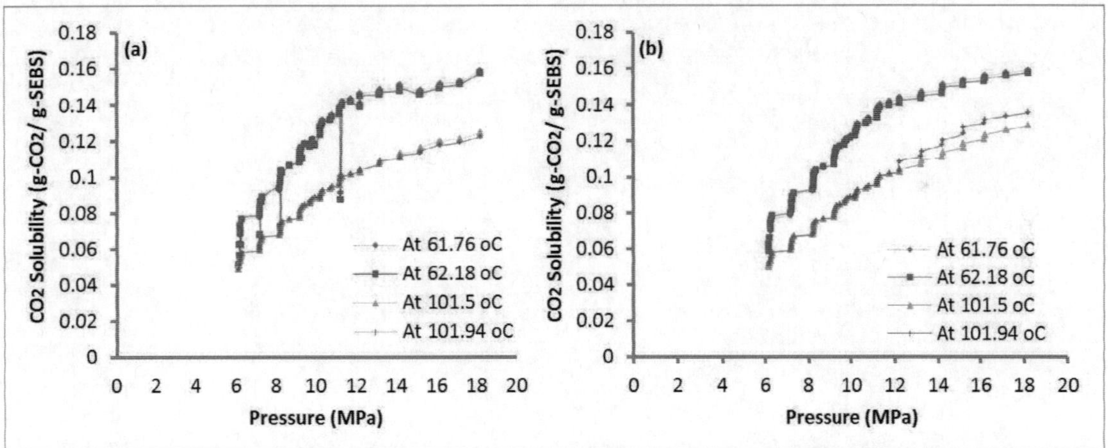

Figure 4. Error yield at various temperatures for technique; (a) traingdx and (b) trainlm.

4. Conclusion

CO_2 solubility control in elastomer foam is vital in producing high performance insulator. Stable dimension with uniformly distributed pores in microcellular elastomer foam promises good heat transfer and perhaps reduce heat loss from insulated system. Thus, the appropriate level of CO_2 solubility needs to be control to help in retaining the stable-shape foam dimension. This can be achieved by measuring the optimum CO_2 solubility prior to foaming process. In this study, the prediction of optimum CO_2 solubility in SEBS by ANN method is proposed as an alternative for direct high cost and time consuming MSB measurement. The simulation results showed the reliability of ANN model with trainlm transfer function to simulate the interpolated CO_2 solubility data from the experimental one. The validated simulation results of R^2 close to the unity reveals the trainlm transfer function is applicable for prediction of interpolated solubility data at temperature range of 61.76 to 101.9°C at various pressure compared to traingdx transfer function.

References

[1] Sharudin R W, Ohshima M, 2012 *J.of Applied Polym. Sci.* **128(4)** 2245
[2] Golzar K, Iranagh S A, Modarres H, 2013 *Measurement, 46* **4206**
[3] Taskinen J, Yliruusi J, 2003 *Advanced Drug Delivery Reviews* **55** 1163
[4] Fatehi M R, Raeissi S, Mowla D, 2016 *J. Of Molecular Liquids* **227** 309
[5] Huang C, Moraga C, 2004 *International Journal of Approximate Reasoning.* **35** 137

Thermodynamic Equilibrium Solubility of Diethanolamine – N-Butyl-1-Methylpyrrolidinium Dicyanamide [DEA-BMPYRR DCA] Mixtures for Carbon Dioxide Capture

R M Salleh and S N Jamaludin

Chemical Engineering Department, Faculty of Chemical Engineering, Universiti Teknologi MARA, 40450 Shah Alam, Selangor Malaysia

ruzitah@salam.uitm.edu.my

Abstract. Solubility data of carbon dioxide (CO_2) in aqueous Diethanolamine (DEA) blended with pyrrolidinium-based ionic liquid: N-Butyl-1-Methylpyrrolidinium Dıcyanamıde [Bmpyrr][DCA] are presented at various temperatures (313.15K-333.15K) and pressure up to about 700 psi. The concentration of [Bmpyrr][DCA] ranges from 0-10wt% and 30-40wt% for DEA. The solubility of CO_2 was evaluated by measuring the pressure drop in high pressure stirred absorption cell reactor. The CO_2 loading in all studied mixtures increases with an increase in CO_2 partial pressure and decreases with temperature. It was also found that the CO_2 loading capacity decrease as the concentration of [Bmpyrr][DCA] increases. The experimental data were correlated as a function of temperature and CO_2 partial pressure to predict the solubility of CO_2 in the mixtures. It was found that the model predicted results in a good agreement with experimental value.

1. Introduction

Antropogenic CO_2 emissions is a big threat to the environment and has led to global climate change since CO_2 concentration has been increasing over the year due to rapid usage of natural gas, coal, petroleum and etc. One of the alternatives to mitigate the excessive CO_2 amount in the atmosphere is by applying post combustion capture technology onto the power plants. Chemical absorption technique using alkanolamine is the most extensively used benchmark technology in the industries. DEA (secondary amine) is one of the common alkanolamine used in the industries due to their high reactivity and low cost however, low loading of this amine developed researchers interest to explore new amines/solvents which may have a better loading capacity of CO_2 gas [1]. Other than that, owing to volatility and corrosive nature of the amine with high energy required for regeneration make this solvent unattractive thus efficient alternative routes urgently needed [2][3].

Nowadays, Ionic liquids (ILs) have received much attention and considered as one of the promosing absorbents for capturing CO_2 since its have advantages over basic aqueous alkanolamine such as high thermal stability, low vapor pressure with added benefit of minimal solvent loss and corrosion [4]. These great advantages of ionic liquids make it suitable to be mix with amine solvent. Nevertheless, most ILs are not always cost effective, thus developing the most efficient hybrid solvent for CO_2 capture technologies is urgently needed. The objective of this work is to measure the total CO_2 loading in aqueous Diethanolamine (DEA) solutions mixed with N-Butyl-1-Methylpyrrolidinium Dicyanamide [Bmpyrr][DCA] over wide range of CO_2 partial pressure, temperature and concentration. A simple model of solubility was performed to predict the CO_2 solubility in the mixtures.

Content from this work may be used under the terms of the Creative Commons Attribution 3.0 licence. Any further distribution of this work must maintain attribution to the author(s) and the title of the work, journal citation and DOI.
Published under licence by IOP Publishing Ltd

IOP Conf. Series: Materials Science and Engineering **358** (2018) 012010 doi:10.1088/1757-899X/358/1/012010

2. Experimental

2.1. Materials

Purified CO_2 (purity ≥ 99.95%) was obtained from Linde Malaysia Sdn Bhd. N-Butyl-1-Methylpyrrolidinium Dicyanamide [Bmpyrr][DCA] with minimum purity of 98% and Diethanolamine, DEA (purity ≥ 99.5%) were purchased from Merck and used as received. The structure of DEA and [BmPyrr][DCA] is shown in Figure 1. Aqueous DEA solutions were prepared by adding distilled water to the desired concentration. Then [BmPyrr][DCA] was added to the aqueous DEA at different concentration from 0 to 10wt%. The mixtures were listed in Table 1.

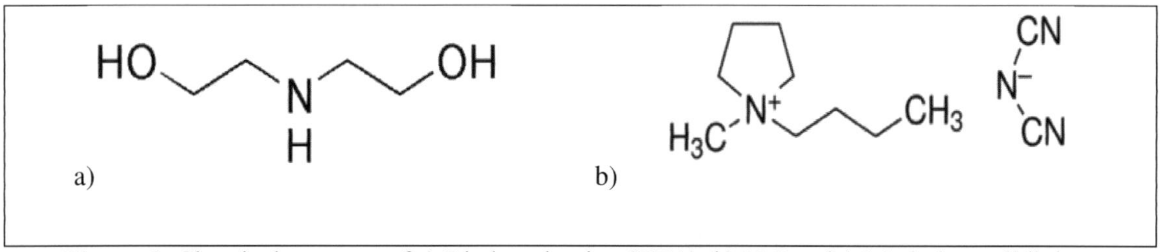

Figure 1. Chemical structure of a) Diethanolamine (DEA), b) N-Butyl-1-Methylpyrrolidinium Dicyanamide [BmPyrr][DCA].

Table 1. Mixtures of DEA with different weight percentage of [BmPyrr][DCA].

Sample	DEA Concentration (wt %)	[BmPyrr][DCA] Concentration (wt %)
	40	0
Aqueous DEA + [BmPyrr][DCA]	35	5
	30	10

2.2. Methods

The CO_2 solubility was measured using high-pressure jacketed reactor which equipped with magnetic stirrer, glass liner and thermocouple. For this work, temperature were set in the ranges of 313.15K-333.15K while pressure up to 700 psi. In each run, 20ml of solvent mixtures was placed into the reactor. Operation was started by heating up the CO_2 gas reservoir and the reactor to the desired temperature. Once temperature was stable, CO_2 gas was released into the reactor and constantly stirred throughout the experiments. Equilibrum was achieved indicated by a constant pressure. Finally, the CO_2 loading for equilibrium cell was calculated based on pressure drop using equation (1).

$$\alpha = \frac{\frac{(PTi-Pf)\times Vvessel}{zRT} - \frac{(Peq-Pvi)\times Vreactor}{zRT}}{n_{total}} \quad (1)$$

Where α is mol CO_2 loading/ total amine, P_{Ti} is initial total pressure on the vessel, P_f is final pressure in the vessel, V_{vessel} is 1 litre, $V_{reactor}$ is gas volume in reactor, Z is gas compressibility factor, R is gas constant, P_{eq} is equilibrium pressure in reactor, P_{vi} is vapor pressure.

3. Results and Discussion

In order to verify the reliability of equipment and the deviation of equilibrium experimental data, the CO_2 solubility experiment was measured at 313.15K using 1M concentration of AMP and compared with the literature [5]. Figure 2 demonstrated that the measured data exhibit good agreement with the literature where the average absolute deviation is only 4.28%.

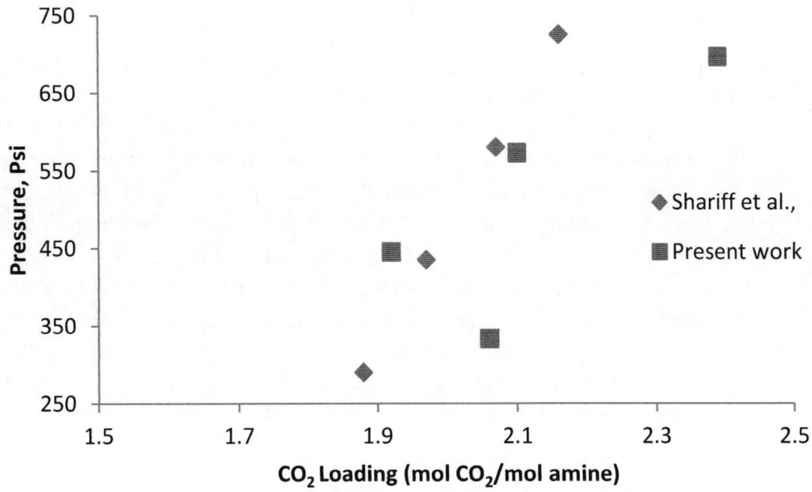

Figure 2. Comparison between this work and literature values.

3.1. Effect of [BmPyrr][DCA] concentration on CO_2 loading

In this work, the CO_2 loading were measured at various weight percent of ILs [BmPyrr][DCA] in aqueous DEA solution. The composition of DEA+ [BmPyrr][DCA] mixtures are 40wt% DEA, 35wt% DEA + 5wt% ILs and 30wt% DEA + 10wt% ILs respectively.

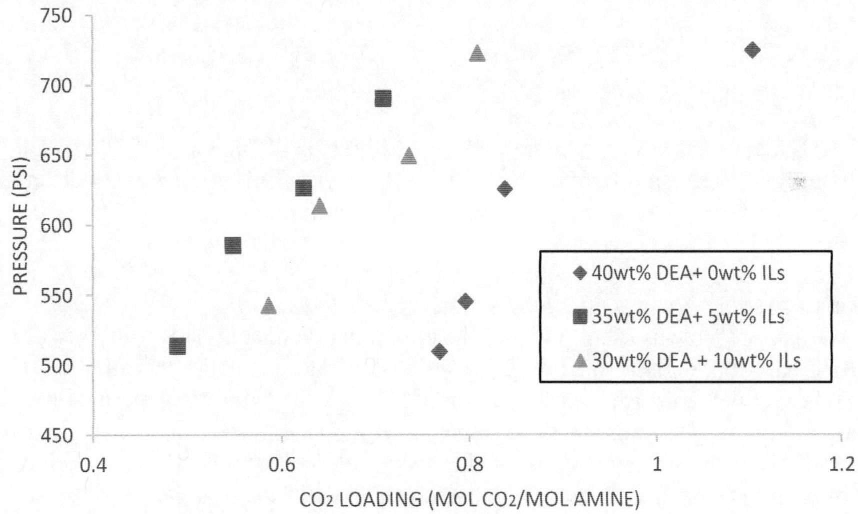

Figure 3. CO_2 loading for aqueous mixtures of DEA + [BmPyrr][DCA] with different composition at 333.15K.

The results indicated that the addition of [BmPyrr][DCA] into aqueous DEA reduced the absorption capacity of CO_2. This can be seen from the data on Figure 3 since the uptake of CO_2 into aqueous mixtures of [DEA][BmPyrr][DCA] was decreased compared to aqueous DEA alone. This behavior can be explained the main reason of reducing CO_2 loading due to reduced water which replaced by ILs. Since the water content on the mixture was reduced, the CO_2 cannot react directly with the alkanolamine for absorbing CO_2. Other than that, this is due to the acidity of [BmPyrr][DCA] which reduced the basicity of aqueous DEA in the mixture, as a result the driving force for the mass transfer to occur is decreasing.

3.2. Effect of temperature and pressure

The temperature and pressure effect toward CO_2 loading were conducted within temperature range of 313.15K- 333.15K and pressure range up to 700psi. From Figure 4 and Figure 5, there is a clear trend of decreasing CO_2 absorption performance in amine-ILs mixures when the temperature increases. This CO_2 solubility pattern is in agreement with the findings of other researchers [6][7][8][9]. The results show that the curve shift to the left side when the temperature increases from 313.15K to 333.15K indicating that less CO_2 absorbed by the solutions. This behaviour demonstrated that more gas is present in the solution with lower temperature, besides that the increasing of kinetic energy at high temperature has led to vigorous dynamic motion of CO_2 molecules causing the breaking of the intermolecular bond between CO_2 and the solution [10]. As a results, less CO_2 will be absorbed since more CO_2 was escaped from the solutions. In terms effects of pressure toward CO_2 absorption, CO_2 solubility increased with increasing of CO_2 partial pressure as shown in Figure 4 and Figure 5. This is because the gas pressure increases the number of collision between the gas molecules and the solution surface, therefore more gas is dissolved as the pressure increased [8][11]. The similar trends with respect to temperature and pressure was also observed for 35wt% DEA + 5wt% [BmPyrr][DCA].

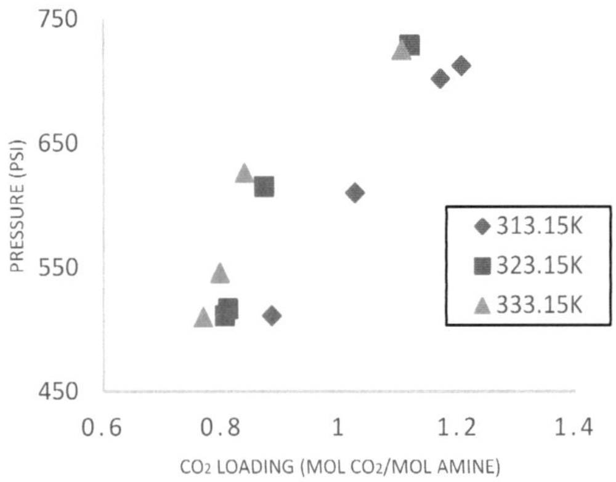

Figure 4. CO_2 loading of 40wt% of DEA + 0wt% of [BmPyrr][DCA] at different temperature.

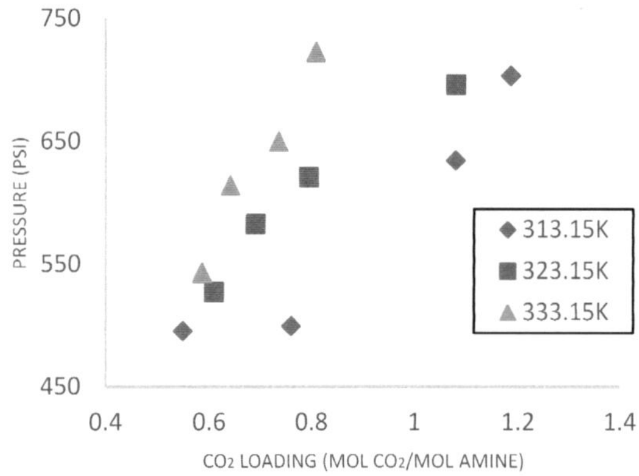

Figure 5. CO_2 loading of 30wt% of DEA + 10wt% of [BmPyrr][DCA] at different temperature.

3.3. Correlations of solubility as a function of pressure and temperature

Linear regression approach was used to model the experimental data of CO_2 solubility as a function of temperature and partial pressure. Equation 2 is used to correlate the data according to Jou and Mather model [12]. Where P is the pressure in Psi, α is mol CO_2 loading/ mol total amine and A and B were determined by plotting ln P against ln α at temperature of 313.15K, 323.15K and 333.15K. Once the values of A and B were obtained these values were plotted against the temperature to get the coefficients of A and B. The coefficients were listed in Table 2.

$$\ln P = A \ln \alpha + B \quad (2)$$

Table 2. Jou and Mather correlations.

System	Correlations A	Correlations B
40 wt% DEA	-0.0103T + 4.3473	0.0070T + 4.1978
35 wt% DEA + 5 wt% [BmPyrr][DCA]	0.0015T + 0.4177	0.0184T + 0.7509
30 wt% DEA + 10 wt% [BmPyrr][DCA]	0.0177T + 5.1241	0.0160T + 1.3955

The solubility correlations as a function of temperature at various pressures were conducted to determine the CO_2 loading measurement. Comparison between the predicted and experimental values of CO_2 solubility is shown in Figure 6. The model was further quantified by determining the average absolute deviation (AAD) using (3) as follows:

$$AAD = \frac{\sum_{i-1}^{N} \left| (\alpha_{\exp} - \alpha_{cal}) / \alpha_{\exp} \right|}{N} \quad (3)$$

It was found that the correlation of CO_2 solubility has average absolute deviation (AAD) percentages of 5.427%. Thus, it shows good agreement between the experimental and predicted data.

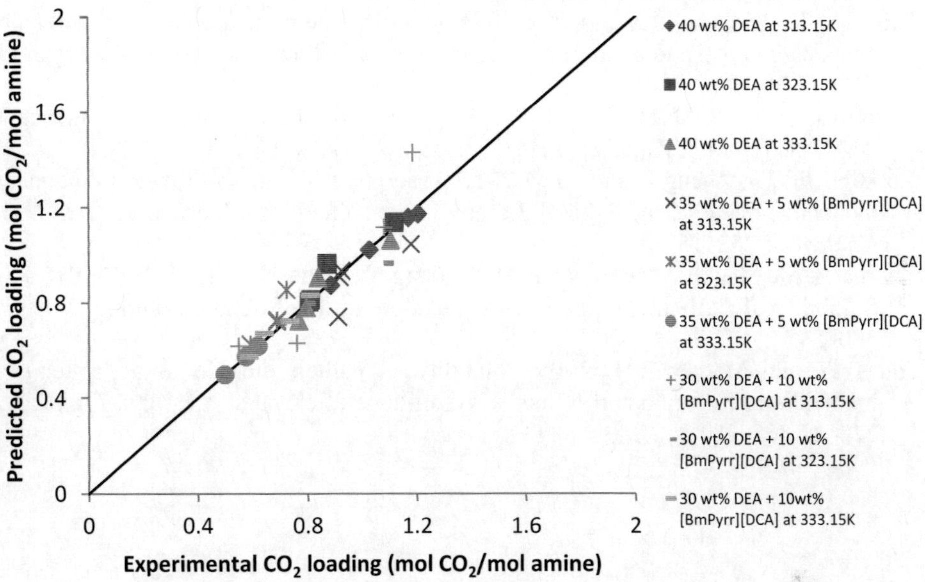

Figure 6. Comparison between correlated and experimental data for CO_2 loading.

4. Conclusions

In this work, the absorption of CO_2 into aqueous DEA blended with [BmPyrr][DCA] was measured. The results exhibit that addition of [BmPyrr][DCA] into aqueous DEA reduced the CO_2 loading capacity. On the other hand, it was found that increasing the pressure has led to increasing of CO_2 loading and increasing of temperature reduces the CO_2 solubility performances. Predicted CO_2 solubility using Jou and Mather model was found to be in good agreement with the experimental data.

Acknowledgement

The authors would like to thank the Ministry of Higher Education (MOHE) and Universiti Teknologi MARA (UiTM) for financial support given through the Fundamental Research Grant Scheme (FRGS) (600-RMI/FRGS 5/3 (91/2014)) for completing this research project.

References

[1] Abu-Zahra M R M, Abbas Z, Singh P and Feron P 2013 Carbon Dioxide Post-Combustion Capture : Solvent Technologies Overview, Status and Future Directions *Mater. Process. energy Commun. Curr. Res. Technol. Dev.* 923–934

[2] Dutcher B, Fan M and Russell A G 2015 Amine-Based CO2 Capture Technology Development from the Beginning of 2013—A Review *ACS Appl. Mater. Interfaces*, **7** 2137–2148

[3] Zhou L, Fan J, Shang X and Wang J 2013 Solubilities of CO2 , H2 , N2 and O2 in ionic liquid

1- n -butyl-3-methylimidazolium heptafluorobutyrate *Thermodyn. J Chem*, **59** 28–34
[4] Althuluth M, Mota-Martinez M T, Kroon M C and Peters C J 2012 Solubility of Carbon Dioxide in the Ionic Liquid 1-Ethyl-3-methylimidazolium Tris(pentafluoroethyl)trifluorophosphate *Fluid Phase Equilib.* **332** 35–39
[5] Shariff A M, Murshid G, Lau K K, Bustam M A and Ahamd F 2011 Solubility of CO2 in Aqueous Solutions of 2- Amino-2-Methyl-1-Propanol at High Pressure *Int. J. Environ. Chem. Ecol. Geol. Geophys. Eng.* **5**, 825–828
[6] Feng Z, Cheng-Gang F, You-Ting W, Yuan-Tao W, Ai-Min L and Zhi-Bing Z 2010 Absorption of CO2 in the aqueous solutions of functionalized ionic liquids and MDEA *Chem. Eng. J.* **160** 691–697
[7] Xu F 2014 Solubility of CO2 in aqueous mixtures of monoethanolamine and dicyanamide-based ionic liquids *Fluid Phase Equilib.* **365** 80–87
[8] Wong M K, Bustam M A and Shariff A M 2015 Chemical speciation of CO2 absorption in aqueous monoethanolamine investigated by in situ Raman spectroscopy *Int. J. Greenh. Gas Control* **39** 139–147
[9] Nordin S, Salleh R M and Ismail N 2014 Separation of CO2 by Using Aqueous Mixtures of AMP and [B3MPYR][BF4] at High Pressure *J. Teknol.* **67** 1–3
[10] Lu B H, Jin J J, Zhang L and Li W 2012 Absorption of carbon dioxide into aqueous blend of monoethanolamine and 1-butyl-3-methylimidazolium tetrafluoroborate *Int. J. Greenh. Gas Control* **11** 152–157
[11] Shariff A M, Shaikh M S, Bustam M A, Garg S, Faiqa N and Aftab A 2016 High-pressure Solubility of Carbon Dioxide in Aqueous Sodium L- Prolinate Solution *Procedia Eng.* **148** 580–587
[12] Jou F Y and Mather A E 2005 Solubility of carbon dioxide in an aqueous mixture of methyldiethanolamine and N-methylpyrrolidone at elevated pressures *Fluid Phase Equilib.* **228–22** 465–469

Absorption of Carbon Dioxide in Aqueous Solutions of *N*-methyldiethanolamine Mixtures

S Ma'mun[1, 2] and H F Svendsen[2]

[1]Department of Chemical Engineering, Universitas Islam Indonesia
Jl. Kaliurang Km. 14.5, Yogyakarta 55501, Indonesia
[2]Department of Chemical Engineering, Norwegian University of Science and Technology, N–7491 Trondheim, Norway

sholeh.mamun@uii.ac.id

Abstract. Carbon dioxide (CO_2) is one of the greenhouse gases (GHG) that has contributed to the global warming problem. Carbon dioxide is produced in large quantity from coal-fired power plants, iron and steel production, cement production, chemical and petrochemical industries, natural gas purification, and transportation. Some efforts to reduce the CO_2 emissions to the atmosphere are then required. Amine-based absorption may be an option for post-combustion capture. The objective of this study is to measure the effect of promoter addition as well as MDEA concentration for the CO_2 absorption into the aqueous solutions of MDEA to improve its performances, i.e. increasing the absorption rate and the absorption capacity. Absorption of CO_2 in aqueous solutions of MDEA mixtures were measured at 40 °C in a bubble tank reactor. The systems tested were the mixtures of 30 wt% MDEA with 5 and 10 wt% BEA and the mixtures of 40 and 50 wt% MDEA with 6 wt% AEEA. It was found that for MDEA-BEA-H_2O mixtures, the higher the promoter concentraation the higher the CO_2 absorption rate, while for the MDEA-AEEA-H_2O mixtures, the higher the MDEA concentration the lower the CO_2 absorption rate.

1. Introduction

Carbon dioxide is known as one of the major contributors to the global warming problem. Carbon dioxide is produced in large quantities from coal-fired power plants, iron and steel industries, petrochemical industries, cement production, and natural gas purification as well as the exhaust gas from the transportation sectors [1, 2]. CO_2 emissions seem to rise from year to year. In 2005, 26.3 Gt of CO_2 were emitted globally and reached 32 Gt in 2010 [3]. Aqueous solutions of alkanolamines are the most commonly used absorbents for the removal of acidic gases from natural, refinery, and synthesis gas streams, such as monoethanolamine [MEA – $H_2N(CH_2)_2OH$], diethanolamine [DEA – $HN(CH_2CH_2OH)_2$], and N-methyldiethanolamine [MDEA – $CH_3N(CH_2CH_2OH)_2$]. A chemical that is used as a commercial absorbent should meet some criteria such as high net cyclic capacity, high absorption rate, low regeneration energy requirement, reasonable thermal stability, etc. In addition, other criteria that need to consider in selecting CO_2 capturing solvents can also be found in [4].

MDEA is a tertiary amine. This solvent has been widely used for high-pressure CO_2 removal and for selective removal of hydrogen sulfide (H_2S) from gas streams containing both CO_2 and H_2S. The advantages of MDEA over primary and secondary amines are, besides the selectivity for H_2S, a lower enthalpy of reaction with the acidic gases which leads to lower energy requirements for regeneration, a lower corrosiveness, better thermal and chemical stability, and a lower vapor pressure. Because of its low vapor pressure, the concentrations of aqueous solutions of MDEA can be used up to 60 wt% without appreciable evaporation losses [5]. The limitations of MDEA include a slower reaction rate with CO_2 and a lower absorption capacity at low CO_2 concentrations. To encounter the MDEA

Content from this work may be used under the terms of the Creative Commons Attribution 3.0 licence. Any further distribution of this work must maintain attribution to the author(s) and the title of the work, journal citation and DOI.
Published under licence by IOP Publishing Ltd

limitations, the addition of promoter may, therefore, be required. The objective of this work is to measure the effect of promoter addition and MDEA concentration for the CO_2 absorption into the aqueous solution of MDEA to improve its performances, e.g. increasing the absorption rate and the absorption capacity.

2. Materials and Method

2.1. Materials

The CO_2 (min. 99.99 mol %) and N_2 (min. 99.6 mol %) gases were obtained from AGA Gas GmbH. The alkanolamines used were MDEA min. 98.5%, BEA min. 98% [2-(butylamino)ethanol – $CH_3(CH_2)_3NH(CH_2)_2OH$], and AEEA min. 97% [2-(2-aminoethyl-amino)ethanol – $H_2N(CH_2)_2NH(CH_2)_2OH$]. They were obtained from Acros Organics and were used without further purification.

2.2. Apparatus and Method

The experiments were conducted using a similar screening apparatus used by [6] as shown in figure 1. According to [6], the apparatus was designed to operate at atmospheric pressure and temperatures up to 80 °C. It consists of six 1–L glass absorbers including six water saturators, six K–type thermocouples, a HETO circulating heater (Type 02 T 623), a Hartmann & Braun Uras 3G IR CO_2 analyzer, and a BRONKHORST HI–TEC $N_2/CO_2/H_2S$ mass flow controller (Type E–7100). The data acquisition system uses LabVIEW. Before starting the experiment, a CO_2–N_2 gas mixture containing 10 vol% CO_2 with flowrate of 5 NL/min was circulated through a by–pass valve to calibrate the analyzer. Once the process started, the by–pass valve closed automatically. The gas mixture was passed through a water saturator and then to the absorber containing 750 mL of the absorbent. The gas phase leaving the absorber was cooled and the CO_2 concentration was directly measured by the IR CO_2 analyzer. The temperature of the water bath was maintained at 40 ± 0.1 °C. The gas CO_2 content, the temperatures, and the gas flowrates were collected by the LabVIEW data acquisition system. The process automatically terminated when the concentration of CO_2 in the outlet reached 9.5 vol% (9.5 kPa CO_2 partial pressure).

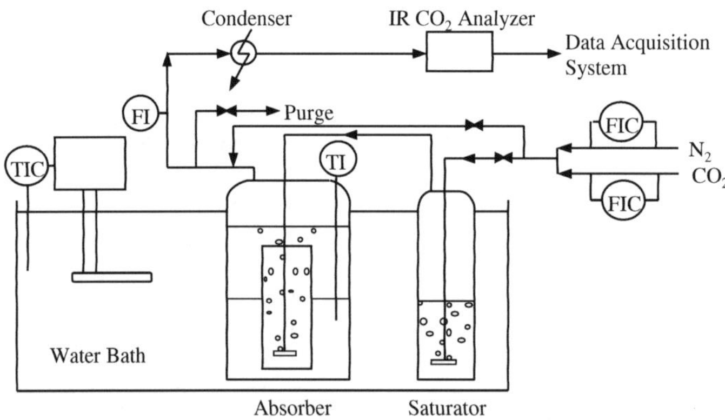

Figure 1. Experimental setup for CO_2 absorption.

2.3. Gas and Liquid Analyses

The vapor bleed extracted for CO_2 concentration measurement was cooled to about 10 °C to condense water, MDEA, BEA, and AEEA. The CO_2 content was directly determined by IR analysis. The vapor phase in the IR analyzer, therefore, consists of N_2, CO_2, and little amounts of H_2O and alkanolamines (MDEA, BEA, and AEEA). The concentration of CO_2 in the IR analyzer is calculated as follows:

$$y_{CO_2} = \frac{n_{CO_2}}{n_{CO_2} + n_{N_2} + n_{H_2O} + n_{MDEA} + n_{BEA/AEEA}} \quad (1)$$

where the *n*'s denote molar flow (mol/s). Due to the low vapor pressures of MDEA, BEA and AEEA at 10 °C, i.e. P^{o}_{MDEA}= 0.007 Pa, P^{o}_{BEA}= 3.395 Pa and P^{o}_{AEEA}= 0.020 Pa [7, 8], thus MDEA, BEA and AEEA contents in the vapor phase through the analyzer can then be disregarded. After terminating the experiment, a liquid sample was then analyzed by the barium chloride ($BaCl_2$) method [9]. The amount of HCl that was not used to dissolve $BaCO_3$ was titrated by 0.1 M NaOH carried out with an automatic titrator (Metrohm 702 SM Titrino) with an endpoint at pH 5.2.

3. Results and Discussion

Absorption rates of CO_2 in aqueous solutions of MDEA mixtures were measured at 40 °C. The objective of this work is to evaluate the effect of promoter addition, i.e. BEA and AEEA, and MDEA concentration to improve MDEA performances such as increasing the absorption rate and the absorption capacity. BEA and AEEA were selected in this study because they give, according to [6], better CO_2 absorption rates and capacities in coparision to those of MDEA. To calculate the CO_2 absorption rate, equation (2) below was used.

$$r_{CO_2} = \frac{1}{V}\left[n^{in}_{CO_2} - \frac{y^{out}_{CO_2}\, n_{N_2}}{\left(1-y^{out}_{CO_2}\right)}\right] \tag{2}$$

where *V* is the volume of the solution (L), *n* is molar flow (mol/s), and *y* is mole fraction of CO_2 in the gas phase measured by the CO_2 IR analyzer. Ma'mun *et al.* [6] tested 30 wt% MDEA solution and the result showed poor performnace in terms of CO_2 absorption rate. Therefore, addition a promoter may be required to eliminate its drawback. To evaluate the effect of BEA as a promoter, aqueous solutions of 30 wt% MDEA were mixed by 5 and 10 wt% BEA. In addition, mixtures of 40 wt% MDEA + 6 wt% AEEA and 50 wt% MDEA + 6 wt% AEEA were tested. These additional data were run to evaluate the effect of MDEA concentration and AEEA addition as a promoter. However, direct comparison can only be evaluated between 30 wt% MDEA + 5 wt% BEA and 30 wt% MDEA + 10 wt% BEA mixtures, and between 40 wt% MDEA + 6 wt% AEEA and 50 wt% MDEA + 6 wt% AEEA mixtures. The last mixtures were tested to evaluate the effect of MDEA concentration at constant AEEA concentration.

Table 1 shows the experimental data of CO_2 absorption in the aqueous mixture of 30 wt% MDEA + 5 wt% BEA at 40 °C. The CO_2–N_2 gas mixture that was bubbled into the reactor contains 9.74 vol% CO_2 with the flowrate of 5.11 NL/min. It can be seen from table 1 that the CO_2 absorption rate at the beginning reached 2.717×10^{-4} kmol/m^3/s. This value is much higher compared to the absorption rate of CO_2 into 30 wt% MDEA solution which is only about 1.87×10^{-4} kmol/m^3/s [6]. This indicates that the addition of BEA as a promoter was able to improve the CO_2 absorption rate into the MDEA solution, i.e. ~45% increase). Moreover, the accumulated CO_2 that can be absorbed was found to significantly increase from 0.77 to 0.99 mol CO_2/L. The concentration of CO_2 in the outlet reached 9.5 vol% after the experiment was run for 159 minutes. Figure 2 shows the results of this study. It can be seen that for 30 wt% MDEA solutions, the higher the promoter concentraation, the higher the CO_2 absorption rate. According to Ma'mun *et al.* [6], the CO_2 absorption rate into BEA solution are higher than that into MDEA. That is why addition of BEA to MDEA solution will improve MDEA performance, i.e. CO_2 absorption rate. In addition, the effect of MDEA concentration was studied. It can be seen from figure 2 that the CO_2 absorption rate into 40 wt% MDEA + 6 wt% AEEA mixture is relatively higher than that into 50 wt% MDEA + 6 wt% AEEA mixture for the first two hours, but it becomes lower afterwards. Since the CO_2 absorption into the MDEA solution is relatively low, therefore, increasing MDEA concentration in a mixture may lead to decreasing the overall absorption rate. It should be noted that the comparison is semi–quantitative, in the sense that there is, according to [6], no guarantee that the bubble structure, and therefore the gas–liquid interfacial area was exactly the same during all experiments. However, the superficial gas velocity was the same, so differences would arise mainly due to variations in interfacial tension, bubble coalescence properties, and viscosity.

Table 1. Experimental data of CO_2 absorption in aqueous mixture of 30 wt% MDEA + 5 wt% BEA at 40 °C.

Time, min	$y_{CO2\text{-}out}$, vol%	$[CO_2]_{absorbed}$, mol/L	CO_2 loading, mol CO_2/mol amines	r_{CO2} x 10^5, kmol/m³/s
1	4.59	0.008	0.003	27.17
5	5.21	0.069	0.022	24.19
10	5.48	0.139	0.045	22.90
15	5.65	0.206	0.066	21.91
20	5.87	0.270	0.087	20.82
25	6.09	0.331	0.106	19.63
30	6.34	0.376	0.121	18.34
35	6.61	0.429	0.138	17.05
40	6.87	0.478	0.153	15.57
45	7.11	0.523	0.168	14.28
50	7.32	0.564	0.181	13.09
55	7.49	0.602	0.193	12.29
60	7.66	0.638	0.205	11.40
65	7.81	0.671	0.215	10.51
70	7.97	0.702	0.225	9.82
75	8.10	0.730	0.234	9.12
80	8.22	0.756	0.243	8.53
85	8.34	0.780	0.250	7.63
90	8.46	0.802	0.257	7.04
95	8.53	0.819	0.263	6.74
100	8.62	0.838	0.269	6.35
105	8.70	0.856	0.275	5.75
110	8.77	0.872	0.280	5.25
115	8.84	0.888	0.285	5.16
120	8.90	0.902	0.290	4.56
125	8.96	0.916	0.294	4.36
130	9.01	0.928	0.298	4.06
135	9.06	0.940	0.302	3.67
140	9.10	0.951	0.305	3.67
145	9.15	0.962	0.309	3.27
150	9.19	0.971	0.312	3.17
155	9.23	0.980	0.315	2.78
159	9.25	0.987	0.317	2.68

Figure 3 shows the CO_2 absorption rates as a function of CO_2 loading. It can be seen that all the mixtures give higher absorption rates up to a CO_2 loading of 0.30, except for the mixture of 50 wt% MDEA + 6 wt% AEEA. The maximum CO_2 loading, i.e. at equilibrium, that could be achieved by the mixtures of 30 wt% MDEA + 5 wt% BEA, 30 wt% MDEA + 10 wt% BEA and 40 wt% MDEA + 6 wt% AEEA is about 0.34, while the 50 wt% MDEA + 6 wt% AEEA mixture could achieve the maximum CO_2 loading of 0.28. It can also be seen that for the MDEA-BEA-H_2O mixtures, the higher the promoter concentration the higher the CO_2 absorption rate, while for the MDEA-AEEA-H_2O mixtures, increasing MDEA concentration leads to lower absorption rate. In general, the addition of promoters has improved the MDEA performances.

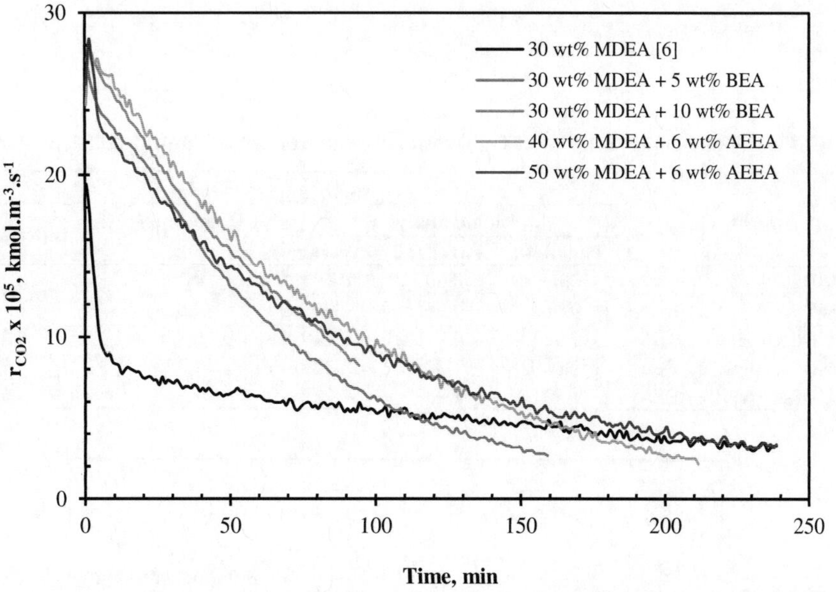

Figure 2. Absorption rates of CO_2 in aqueous solutions of MDEA at 40 °C.

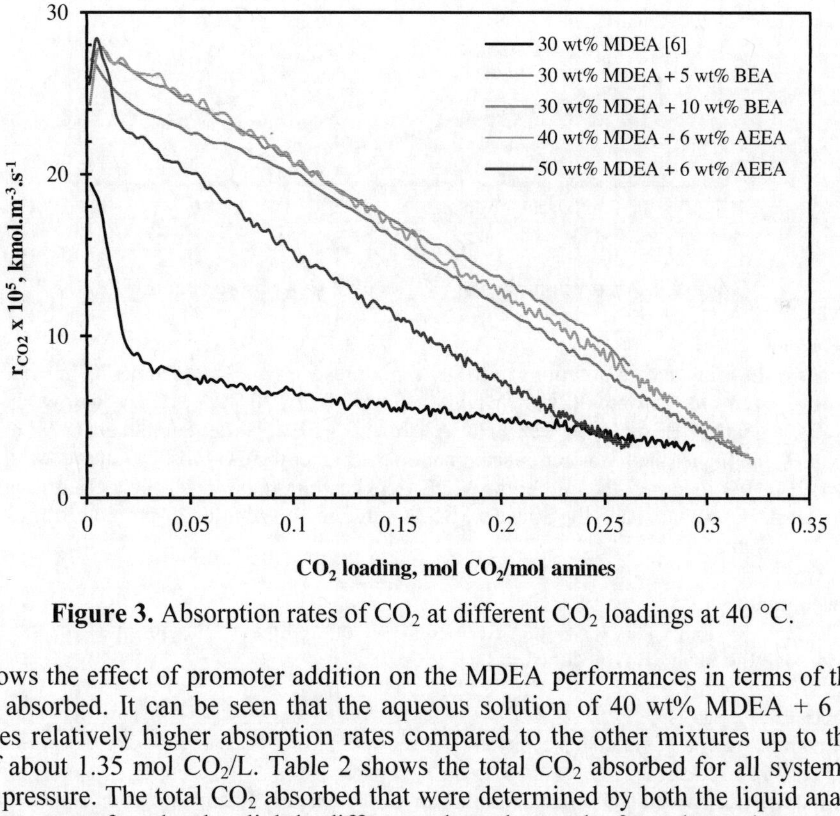

Figure 3. Absorption rates of CO_2 at different CO_2 loadings at 40 °C.

Figure 4 shows the effect of promoter addition on the MDEA performances in terms of the total CO_2 that can be absorbed. It can be seen that the aqueous solution of 40 wt% MDEA + 6 wt% AEEA mixture gives relatively higher absorption rates compared to the other mixtures up to the total CO_2 absorbed of about 1.35 mol CO_2/L. Table 2 shows the total CO_2 absorbed for all systems at 9.5 kPa CO_2 partial pressure. The total CO_2 absorbed that were determined by both the liquid analysis and by the IR analyzer were found to be slightly different where the results from the analyzer were relatively higher than those determined by the liquid analysis. This might occur due to dilution of the solution in the reactor by water condensate from the cooler. The CO_2 concentration in the solution that was determined by the analyzer was calculated from the CO_2 mass balance.

Table 2. Gas and liquid analyses for the CO_2 absorption in aqueous solutions of MDEA mixtures.

System	$[CO_2]_{absorbed}$, mol/L				Difference, %
	Liquid analysis			CO_2 gas analyzer	
	Parallel 1	Parallel 2	Average		
30 wt% MDEA	0.693	0.695	0.694	0.777	10.58
30 wt% MDEA + 5 wt% BEA	0.967	0.955	0.961	0.987	2.60
30 wt% MDEA + 10 wt% BEA	0.921	0.918	0.919	0.930	1.15
40 wt% MDEA + 6 wt% AEEA	1.330	1.324	1.327	1.379	3.76
50 wt% MDEA + 6 wt% AEEA	1.231	1.233	1.232	1.365	9.76

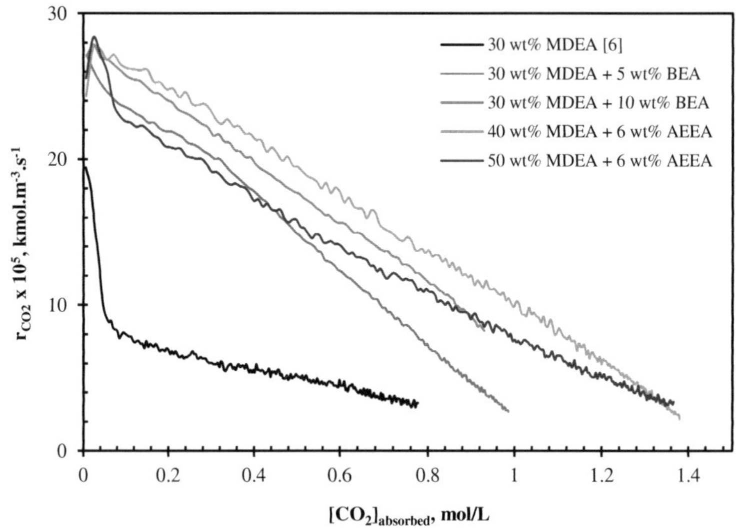

Figure 4. Absorption rates of CO_2 vs total CO_2 absorbed at 40 °C.

4. Conclusion

Absorptions of CO_2 in aqueous solutions of MDEA mixtures were measured at 40 °C in a bubble tank reactor. The systems tested were 30 wt% MDEA + 5 wt% BEA, 30 wt% MDEA + 10 wt% BEA, 40 wt% MDEA + 6 wt% AEEA and 50 wt% MDEA + 6 wt% AEEA. It was found that for MDEA-BEA-H_2O mixtures, the higher the promoter concentraation the higher the CO_2 absorption rate, while for the MDEA-AEEA-H_2O mixtures, the higher the MDEA concentration the lower the CO_2 absorption rate. For all systems, in comparison to the 30 wt% MDEA solution, the addition of promoters has improved the MDEA performances.

Acknowledgment

The authors acknowledge the Norwegian Research Council "Klimatek Programme" for the financial support through the SINTEF Project 661292.01.

References

[1] Ma'mun S, Amelia E, Rahmat V, Alwani DR and Kurniawan D 2017 *IOP Conf. Ser.: Mater. Sci. Eng.* **162** 012003
[2] Ma'mun S, Kamariah, Sukirman, Kurniawan D, Amelia E, Rahmat V and Alwani DR 2016 *MATEC Web Conf.* **101** 02001
[3] van der Hoeven 2012, *World Energy Outlook 2012*, International Energy Agency, Paris, France
[4] Hoff KA, da Silva EF, Kim I, Grimstvedt A, Ma'mun S 2013 *Energy Procedia* **37** 292
[5] Kohl A, Nielsen R 1997 *Gas Purification* Gulf Publishing Company, Houston, USA
[6] Ma'mun S, Svendsen HF, Hoff KA and Juliussen O 2007 *Energy Convers. Manage.* **48** 251
[7] Yaws CL 2015 *The Yaws Handbook of Vapor Pressure: Antoine Coefficients*, 2nd ed., Elsevier

Inc., Oxford, UK
[8] Wilson GM, VonNiederhausern DM, Giles NF 2002 *J. Chem. Eng. Data* **47** 761
[9] Ma'mun S, Nilsen R., Svendsen HF and Juliussen O 2005 *J. Chem. Eng. Data* **50** 630

IOP Conf. Series: Materials Science and Engineering **358** (2018) 012012 doi:10.1088/1757-899X/358/1/012012

Building Electricity Consumption as an Indicator of Indirect Carbon Dioxide Emissions

S Ma'mun[1], Sukirman, A E Alel and M Hasanah

Department of Chemical Engineering, Universitas Islam Indonesia
Jl. Kaliurang Km. 14.5, Yogyakarta 55501, Indonesia

sholeh.mamun@uii.ac.id

Abstract. The global CO_2 emissions have continually increased from year to year and reached 32 Gt in 2010. The increased CO_2 emissions may lead to a higher temperature and cause climate change on a global scale. Building energy-using equipment in Indonesia continuously increases annually leading to increasing indirect CO_2 emissions from the buildings. The objective of this study is to measure the indirect CO_2 emissions from the Faculty of Industrial Technology (FIT), Universitas Islam Indonesia (UII) Yogyakarta, Indonesia. The research data were taken from the electricity consumption by reading the electric meter at specified time intervals for 7 weeks from 26 September to 13 November 2016. The amount of electricity consumption indirectly indicates the amount of CO_2 emission in the FIT where the FIT has consumed the electricity of 18.6 kWh/day corresponding to the average indirect CO_2 emission of 15.9 kg CO_2-eq/day. The results obtained would, therefore, give some recommendations to the FIT to take some policy actions related to the indirect CO_2 emission by improving energy management system to minimize the indirect CO_2 emission in the FIT.

1. Introduction

Greenhouse gases such as carbon dioxide (CO_2) absorb heat emitted from Earth's surface. Increases in concentrations of these gases in the atmosphere cause Earth to warm by trapping more of this heat. Continued emissions of these gases will cause global warming which in turn cause climate change. Various climate models estimate that the global average temperature may rise by about 1.4 to 5.8 °C by the year 2100 [1, 2]. However, long-term climate change over many decades will depend mainly on the total amount of CO_2 and other greenhouse gases emitted from human activities. The CO_2 emissions tend to rise from year to year. In 2005, CO_2 emissions reached 26.3 Gt and rose to 32 Gt in 2010 [3] where building sector was responsible for 40% of the global energy consumption and one third of the global GHG emissions [4-6]. As the global CO_2 emissions continue to increase, the total CO_2 emissions in Indonesia also increased from 356 Mt in 2004 to 453 Mt in 2014 [7].

A previous study by [8] calculated the total emissions of CO_2 caused by energy consumption in four types of hotels in Taiwan. As the result, CO_2 emissions are produced more by the hotel with highest facilities and services, and the electricity use is the highest and main contributor of the CO_2 emissions in hotel building.Indonesia is one of developing countries that has big populations where the amount of inhabitant increase gradually year by year and most of the people in Indonesia prefer living in urban area. There were significant differences in household energy use between household in urban and nonurban areas, where such differences are, according to [9], driven part by contextual factors such as population density that indirectly affects the energy use. Building energy-using equipment and energy consumption in Indonesia continuously increase annually. This, therefore, increases the indirect CO_2 emissions from the buildings due to using electrical energy generated by the fossil-fuel-based power plants [10]. In 2014, Perusahaan Listrik Negara (PLN), the Indonesian state-owned electricity company, provides national electricity with the total installed capacity of 51.62 GW and

Content from this work may be used under the terms of the Creative Commons Attribution 3.0 licence. Any further distribution of this work must maintain attribution to the author(s) and the title of the work, journal citation and DOI.
Published under licence by IOP Publishing Ltd

93.2% of which are the fossil-fuel-based power plants such as steam, gas, and diesel power plants. The total CO_2 emissions from these electricity generation activities were 128 Mt of CO_2-eq [11]. This present study aims to measure the indirect CO_2 emissions from Faculty of Industrial Technology (FIT), Universitas Islam Indonesia (UII) Yogyakarta, Indonesia by measuring the electricity use for a period of time. The results could be used for electricity utilization management in the faculty as an effort to decrease a number of CO_2 emissions in the FIT UII Yogyakarta, Indonesia.

2. Methods

The methodology used in this research was by reading the electricity meters as an indicator of energy consumption from the electricity consumed by the FIT at specified time intervals, i.e. every three hours, for 12 hours from 06:00 am to 06.00 pm in seven weeks (49 days). The electricity meters were monitored both during the busy days and also during the weekends or holidays. The amount of electricity consumed indirectly indicates the total amount of CO_2 emission caused by the activities in the FIT UII Yogyakarta

3. Results and Discussion

The objective of this research is to measure the indirect CO_2 emission from the electricity consumption in FIT buildings in the period of time. The data were taken from three electricity meters for seven weeks (49 days) starting from 26 September 2016 at 09.00 am to 13 November 2016 at 06.00 pm. The 1st electricity meter is used to calculate the electricity consumed for teaching activities including the laboratory works, the 2nd meter is for the sports center, and the 3rd meter is for street lighting. The electricity meters were read every 3 hours from 06:00 am to 06.00 pm and divided into two periods of time to compare the electricity consumption during both normal day and exam term as described in table 1. The total data collected for 49 days were 238 data and those were presented in figure 1.

Table 1. The electricity meter reading activity from 26 September to 13 November 2016.

Date	Reading Interval	Comment
26-09-2016 to 23-10-2016	3 hours	Teaching period: classrooms, laboratories, offices, corridors, sport centers and street lighting
24-10-2016 to 04-11-2016	3 hours	mid-term exam period: classrooms, offices, corridors, sport centers and street lighting
05-11-2016 to 13-11-2016	3 hours	teaching period: classrooms, laboratories, offices, corridors, sport centers and street lighting

CO_2 emissions may come from power plants from the combustion process of fuels which are used in, such as coal, natural gas, LNG, HSD (High-Speed Diesel), and MFO (Marine Fuel Oil). The CO_2 emission is calculated by the following equation [12]:

$$CO_2 \text{ emitted (kg } CO_2) = (\text{Grid Emission Factor}) \times (\text{kWh}) \quad (1)$$

Grid emission factor (GEF) is defined as the intensity of CO_2 emissions per electricity energy unit in grid/transmission system (kg CO_2-eq/kWh). With the total amount of national CO_2 emissions of 201 Mt in 2015, it was obtained the national average GEF of 0.867 kg CO_2-eq/kWh. However, the GEF for Java-Bali region is a bit lower than the national GEF, i.e. 0.857 kg CO_2-eq/kWh. Since UII is located in Yogyakarta, a part of Java Island, thus the CO_2 emitted from the fossil-fuel-based power plants for the Java-Bali interconnection can be calculated by using equation (1) with the GEF for the Java-Bali transmission system. By use of equation (1), the amount of CO_2 emissions contributed indirectly by the FIT-UII from the electricity use can then be measured. Figure 1 shows the total electricity use measured by the three electricity meters in the FIT. Due to the difference in electricity need for each day, the data seem to be fluctuating. During the meter reading periods, it was found the electrical shut down several times, thereby it caused the electricity use decreased significantly. Moreover, the student's activities in the evening would also contribute the large amount of kWh. In general, the electricity consumption reached the minimum during weekends because of no teaching and laboratory work activities.

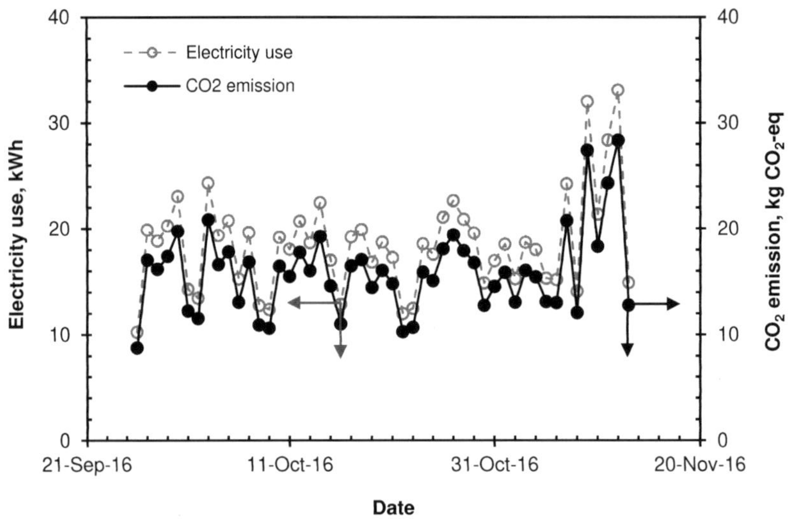

Figure 1. Daily electricity consumptions and corresponding CO_2 emissions at the FIT UII Yogyakarta, Indonesia from 26 September to 13 November 2016.

The electricity consumption, as well as the corresponding CO_2 emissions during 5 working days and weekends for 7 weeks, are depicted in figure 2. It can be seen that the increase of electricity use in the building would also escalate the CO_2 emissions. According to the data, the highest CO_2 emissions happened in the working days, i.e. Monday to Friday. The teaching and laboratory activities had already begun in September and finished in the beginning of December 2016. Moreover, some sport competition events were held in the sports center. Those activities have significantly increased the electricity consumption in the FIT. It can also be seen from the figure that the use of electricity during the mid-term exam (weeks 5 and 6) are almost similar to those during the teaching periods. This indicates that the use of electricity in the classrooms for mid-term exam purpose were very significant because there were no laboratory work activities

The accumulative electricity consumption and corresponding CO_2 emissions at the FIT UII Yogyakarta are shown in figure 3. It can be seen that the electricity consumption, as well as the CO_2 emissions during the day (06:00 am – 06:00 pm), accounts for about 56.4% where the peak occurred at 09:00 am till 03.00 pm, i.e. 30.9%, while the lowest occurred in the early morning (06:00 pm – 09:00 pm) and in the late afternoon (03:00 pm – 06:00 pm) where the activities begin and finish. The accumulative electricity consumption for 7 weeks from 26 September to 13 November 2016 at the FIT UII Yogyakarta were 912 kWh (18.6 kWh/day) with the indirect CO_2 emissions of 781 kg CO_2-eq (15.9 kg CO_2-eq/day). The current results were a bit lower compared to those reported by [10] where they measured the electricity consumption at the FIT in April 2016 for 30 days with the average daily electricity usage of 20.3 kWh and the indirect CO_2 emissions of 17.2 kg CO_2-eq/day. To minimize the energy consumption and CO_2 emissions, the FIT is planning to install a solar panel with a design capacity of 10 kW. Moreover, all old lamps have been replaced by the LED lamps in all rooms at the FIT as well as installing air conditioners (ACs) with low energy consumption. These actions have indeed slightly reduced the electricity consumption and CO_2 emissions in the FIT UII Yogyakarta and thus contribute to achieving the national target to reduce 26% of the CO_2 emissions by 2020 [13].

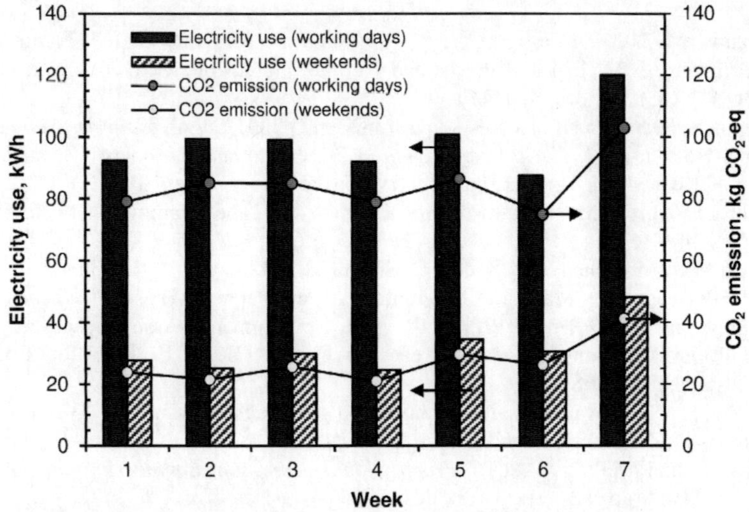

Figure 2. The electricity consumptions and corresponding CO_2 emissions during working days and weekends at the FIT UII Yogyakarta, Indonesia for 7 weeks from 26 Sept to 13 Nov 2016.

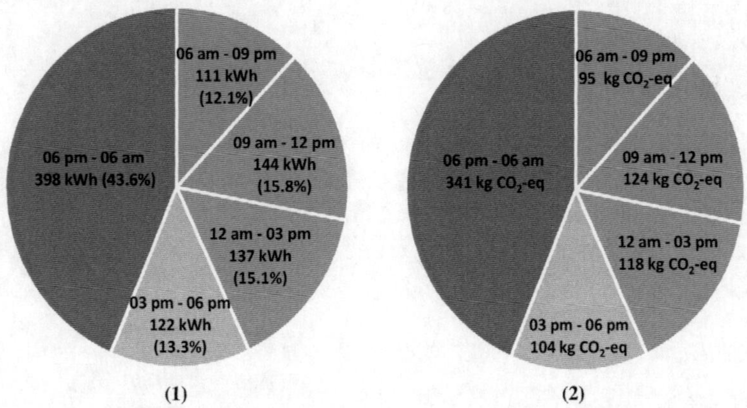

Figure 3. (1): Accumulative electricity use and (2): accumulative CO_2 emissions at the FIT UII Yogyakarta, Indonesia for 7 weeks.

4. Conclusion

The indirect CO_2 emission at the FIT UII Yogyakarta was measured by determining the electricity consumption for 7 weeks from 26 September to 13 November 2016. The results indicated that the average indirect CO_2 emission was 15.9 kg CO_2-eq/day resulted from 18.6 kWh electricity consumed per day. The FIT has taken some actions to improve energy management system to minimize the indirect CO_2 emission in the FIT.

Acknowledgment
The authors acknowledge the Universitas Islam Indonesia for the financial support through the Department of Chemical Engineering Research Grant 2016.

References

[1] Williams M 2002, *Climate Change: Information Kit*, the United Nations Environment Programme (UNEP) and the United Nations Framework Convention on Climate Change (UNFCCC), Geneva, Switzerland
[2] Ma'mun S, Svendsen HF, Hoff KA and Juliussen O 2007 *Energy Convers. Manage.* **48** 251
[3] van der Hoeven 2012, *World Energy Outlook 2012*, International Energy Agency, Paris, France
[4] Costa A, Keane MM, Torrens JI and Corry E 2013 *Appl. Energy* **101** 310
[5] Shaikh PH, Nor NBM, Nallagownden P, Elamvazuthi I and Ibrahim T 2014 *Renew. Sust. Energ. Rev.* **34** 409
[6] Ahmad MW, Mourshed M, Mundow D, Sisinni M and Rezgui Y 2016 *Energ. Buildings* **120** 85
[7] Olivier JGJ, Janssens-Maenhout G, Muntean M and Peters JAHW 2015 *Trends in Global* CO_2 *Emissions: 2015 Report*. PBL Netherlands Environmental Assessment Agency, The Hague; European Commission, Joint Research Centre (JRC), Institute for Environment and Sustainability (IES)
[8] Tsai KT, Lin TP, Hwang RL and Huang YJ 2014 *Tourism. Manage.* **42** 13
[9] Timmons D, Zirogiannis N and Lutz M 2016 *Energy Research and Social Science* **22** 137
[10] Sukirman and Ma'mu S 2016 *Proc. Global Conf. on Engineering and Applied Science* (Hokkkaido) p 193
[11] Perusahaan Listrik Negara 2014a *Sustainability Report 2014 – Towards new performance level (*Jakarta: PT PLN Persero).
[12] Perusahaan Listrik Negara 2014b *Rencana Usaha Penyediaan Tenaga Listrik (RUPTL) 2015-2024* (Jakarta: PT PLN Persero)
[13] Syahrial E, Pasarai U and Iskandar UP 2010 *Lemigas Scientific Contribution* **33** 129

Semi-Empirical Model to Estimate the Solubility of CO_2 NaCl Brine in Conditions Representative of CO_2 Sequestration

E Mohammadian*[1], H Hamidi[2] and A Azdarpour [3]
[1] Faculty of Chemical Engineering, Universiti Teknologi MARA, 40450 UiTM, Shah Alam, Selangor, Malaysia
[2]383 Fraser Noble Building, School of Engineering, King's College, University of Aberdeen, Scotland, UK
[3] Department of Petroleum Engineering, Marvdasht Branch, Islamic Azad University, Marvdasht, Iran

Erfan.723@gmail.com

Abstract. CO_2 sequestration is considered as one of the most anticipated methods to mitigate CO_2 concentration in the atmosphere. Solubility mechanism is one of the most important and sophisticated mechanisms by which CO_2 is rendered immobile while it is being injected into aquifers. A semi-empirical, easy to use model was developed to calculate the solubility of CO_2 in NaCl brines with thermodynamic conditions (pressure, temperature) and salinity gradients representative CO_2 sequestration in the Malay basin. The model was compared to the previous more sophisticated models and a good consistency was found among the data obtained using the two models. A Sensitivity analysis was also conducted on the model to test its performance beyond its limits.

1. Introduction

Burning fossil fuels has released enormous amounts of greenhouse gases in the atmosphere since the industrial revolution. The greenhouse gases are the main reason behind the global warming [1]. Carbon dioxide (CO_2) sequestration in the aquifers is one of the most anticipated methods to mitigate CO_2 concentration in the atmosphere. CO_2 sequestration potentially provides around 25 % of the required mitigation to global emissions, which can delay global warming to an acceptable extent If successfully conducted, CO2 may be safely sequestrated in depleted (or active) saline aquifers, un-minable coal beds, oil and gas reservoirs. [1-2] Due to known geological formation and existence of seal traps, CO_2 may be more safely sequestrated in depleted oil and gas reservoirs as compared to saline aquifers and coal bed methane reservoirs. On the other hand, the abundance and higher storage capacity are two major motivations for sequestration of CO_2 in saline aquifers. [3]

There are four main mechanisms of sequestration, which contribute in rendering the injected carbon dioxide immobile: structural trapping, residual phase trapping mineralization and dissolution of and carbon dioxide in formation brine. Mineral trapping is reaction of CO_2 with minerals existed in rock to form stable components i.e. carbonates and almino-silicate. The carbon dioxide secured by mineralization mechanism is proven to be the safest in terms of releasing back to atmosphere. [4] However the time scales of the reaction is known to be very long [1]. Therefore the attempt has been undertaken to accelerate the mineralization reactions.The importance of solubility mechanism is that it could be manipulated to increase the sequestration efficiency. Besides, in the newer methods such as surface mixing, solubility of CO_2 in water (or brine) is the key parameter that could increase the CO_2 mitigation efficiency. For CO_2 sequestration operations on the other hand, the importance of solubility mechanisms is that rate under which CO_2 is secured using solubility mechanism are higher than those of mineralization. [2,4] So far, there have been numerous studies that experimentally investigated

Content from this work may be used under the terms of the Creative Commons Attribution 3.0 licence. Any further distribution of this work must maintain attribution to the author(s) and the title of the work, journal citation and DOI.
Published under licence by IOP Publishing Ltd

effects of various parameters on the solubility of CO_2 in brine in a variety of conditions [5]. There are plenty of experimental and theoretical studies regarding the future implementation of CO_2 sequestration in the areas such as North Sea and the U.S. However, in case of Malay basin, there is a lack of experimental and modelling studies on CO_2 sequestration. Since Malaysia has been one of the countries that is a member of the Kyoto protocol [6] and therefore taking measures to mitigate its produced CO_2 in inevitable.

2. Model Development

For sequestration purposes, CO_2 is injected as the supercritical fluid, i.e. at pressures above 72.9 atm, so the risk of leakage is minimized [7]. There are limited modelling studies by which solubility of CO_2 in brine could be estimated in conditions in favor of CO_2 sequestration, Moreover, some of the previous models use sophisticated techniques and lots of assumptions. This study seeks to propose an empirical model to describe CO_2 solubility in conditions representative of Malay basin. The formation brine composition and pressure and temperature gradients were chosen in an order to achieve that objective. The model data were obtained from the various experiments conducted in the author's previous work [8]. The experiments were conducted in an autoclave reactor up to 300 atm of pressure. The detailed about experimental setup and procedure could be found in the authors' previous work [8]. Figure 1 shows the parameters that were studied in the experimental phase.

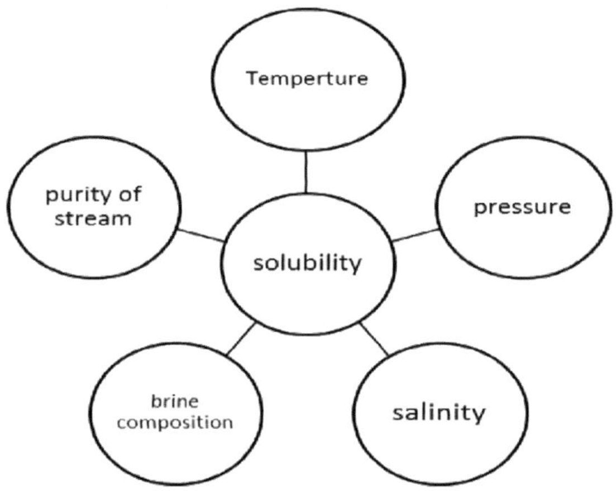

Figure 1. The parameters that affect the solubility were experimentally investigated

Once the experimental phase completed, the outliers (data with irrationally high or low values) were identified and removed using static data analysis software (such as SPSS™ 18). The latter was done using z-score method. Once the outliers were removed, the model can be developed by using various regression techniques according to the number of independent variables (pressure, temperature, salinity and purity of the stream) and the changes of dependent variable (solubility) with that independent variable. The correlation between the dependent and independent variable can be shown as follows:

$$x = f(P, T, S, F) \qquad [1]$$

In the above correlation, x is the solubility of CO_2 in brine, P is the pressure, T is the temperature, S represents the salinity of the brine and F represents the purity of stream. The range of salinity is in this study was (1000, 10,000, and 15,000 ppm). In order to develop models, nonlinear regression was used. In non-linear regression analysis, the observational (experimental) data were modelled by a function, which is the combination (nonlinear) of the model parameters and depends on a number of independent variables. In order to do the regression, a raw equation was defined together with number of constants. The data are fitted by a method of consecutive iterations to the point where the best

IOP Conf. Series: Materials Science and Engineering **358** (2018) 012013 doi:10.1088/1757-899X/358/1/012013

model was obtained. Various models have been investigated and one with the highest R2 (goodness of the fitting line or curve) was chosen as the proper model for the solubility (dependent variable). The model than checked with the experimental data of author's previous work [8] to verify its performance. The developed model is able to predict the solubility of CO_2 in water in a variety of conditions. Once the model was developed, it was tested beyond its range of P-T-s and compared to the previous models to test its performance. Figure 2 shows the flowchart of model development.

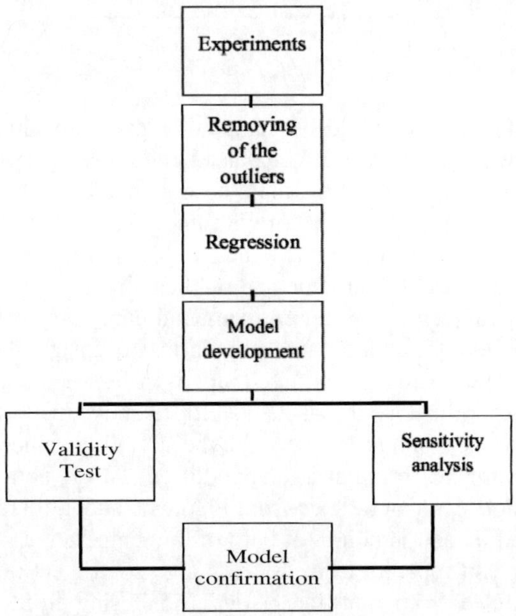

Figure 2. Model development flowchart

2.1. Model preparation

The model was developed through several attempts of regression solubility data with those of pressure, temperature, salinity and purity of stream. The purity of the stream was added to investigate the option of co-injection of other gases with CO_2 stream (Such as N_2 and Ar). Due to several technical and economic reasons co-injection of impurities with CO_2 stream could be of a great significance [9]. Once the various models were tested, the one with the highest goodness of the fit was chosen. The R_2 of the following model is 98.3%.

$$x = \left(\left[\left(c + \frac{RaT^2}{Pb} \right) + 1 \right]^{-1} - (dS) \right) \times F \,.\, g \qquad [2]$$

where x is the solubility of CO_2 in brine in mol/kg, R is the universal constant for gases (82.023 cm3.atm. K^{-1} .mol^{-1}), T is the temperature in Kelvin, P is the pressure in atm, S is the salinity of brine, in this case NaCl is the only salt present in the brine, in wt% , F is purity of the stream based on N2 content in volume percent (for a stream with 100% CO_2 the value is 1 and for 90% CO_2 +10% N_2 the value is 0.9) the value is *a, b, c, d,* and *g* are regression constants as shown in Table 1. Unlike the few available models, this model is an empirical, simple to use method as it has been solely obtained based on the statistical analysis of the experimental data. The previous models are more complicated and calculate the solubility by aid of parameters such entropy, fugacity at any pressure, temperature and salinity of interest etc [11].

Table 1. Parameters estimated from SPSS ($R^2 = 0.983$)

Parameter	Estimate
a	0.063
b	745.515
c	6.882
d	0.005

2.2. Model verification

All of the tables, images and figures should be centered. Figures and images should be numbered together (Figure 1) and figure definitions should be placed under the figure or image; as for the tables, they should also be numbered (Table 1) and the table header should be placed at the top of the chart. Table, image and figure headers should be written with upper case initial letters, bold and should be centered. References (if any) of the tables, figures and images should be presented just in the tables, figures and images in the form of author surname and publication date. In order to test the integrity of this empirical model, first it was verified using experimental data. Once verified, the outcomes were compared with previous models [10] at the same conditions to ensure the applicability of models developed in this research. Figure 3 illustrates the comparison between solubility data of 100% pure CO_2 stream obtained using experiments and models developed above in 10000 ppm NaCl brine, at 353 and 373 K. Figure 4 illustrates the Comparison between the solubility values of stream containing 95% CO_2+ 5% N2 in a 15,000 ppm NaCl brine at a temperature of 333 K obtained in the experiments to those obtained using the model.As it can be seen from Figures 3 and 4, there is a very good agreement between the experimental and model outcomes at both of the temperature series and different purities of the stream. The latter was also apparent from the very high coefficient of correlation, R squared, of 98.3% for NaCl Brine obtained from non-linear regression using SPSS software. It is useful to compare the solubility values obtained from the current model to previous models. Very few models are available in a wide range of pressure, temperature and salinity comparable to those of this study. One of the models comparable is the one developed by Duan and Sun [10] which covers solubility of CO_2 in seawater as well as brine saturated with various types of salts. Thus, it is among the most comprehensive models developed to date. Figure 5 illustrates solubility of CO_2 in brine obtained from Duan and Sun [10] and that of calculated in this research. It can be seen that at both temperature series of 353 K, there is a very good agreement with the model developed by Duan and Sun [10] and that of this study. At 353°K the values calculated using the two models are more identical, however just like the previous case, at higher pressures, 190 bar as an example, the highest difference in solubility can be observed to be 1.9 %.

2.3. Model limitations

This section presents a sensitivity analysis, that was conducted to test the model performance beyond its allowable range of function i.e. pressure range of 2-210 atm, 298-373 °K and TDS of 0- 1.5 wt%. The validity of the model beyond it limits, i.e. the solubility at higher temperature, pressure and salinity was calculated using the model developed in this study (Mohammadian's model) and was compared to the literature. Table 1 summarizes the "off-limits" calculation of the current model, in contrast with those calculated from Duan and Suns' (2003) model. In the first step, the solubility values are calculated at 333 °K, from 220 to 400 atm. The Current model works fine beyond its pressure limits with a negligible error of less than 3% at pressures up to 500 atm. It can be seen that the error is increased at higher pressures. Considering normal gradient of 0.44 psi/ft in Malay basin, pressure of 500 atm, represents a formation with Depths more than 16700 feet which is considered ultra-deep formation. Therefore the application of CO_2 sequestration method in ultra-deep formation will be challenging and would requires all-inclusive understanding from the technical and economical point of view.

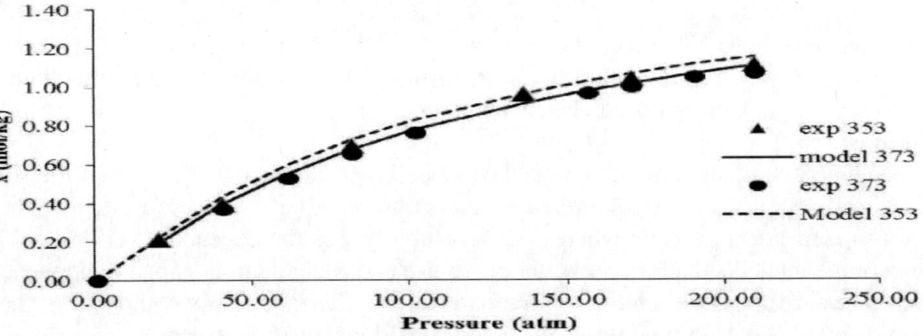

Figure 3. Model and experimental data of solubility of pure CO_2 in brine of 10,000 ppm at 353 and 373 K.

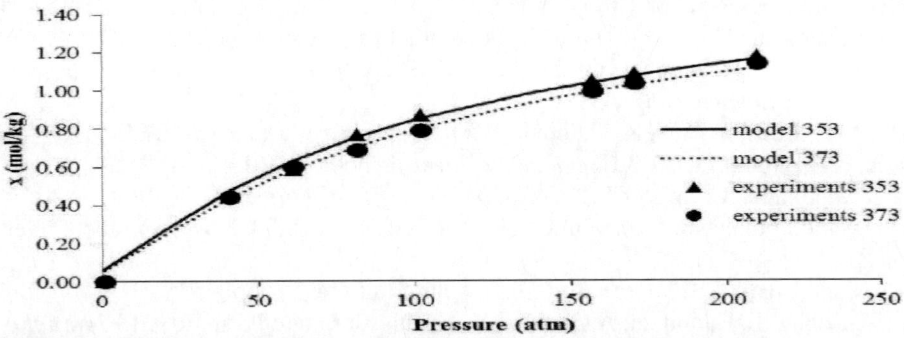

Figure 4. Model and experimental data of solubility of 95% CO_2 +5 N_2% in brine of 1000 ppm at 353 and 373 K.

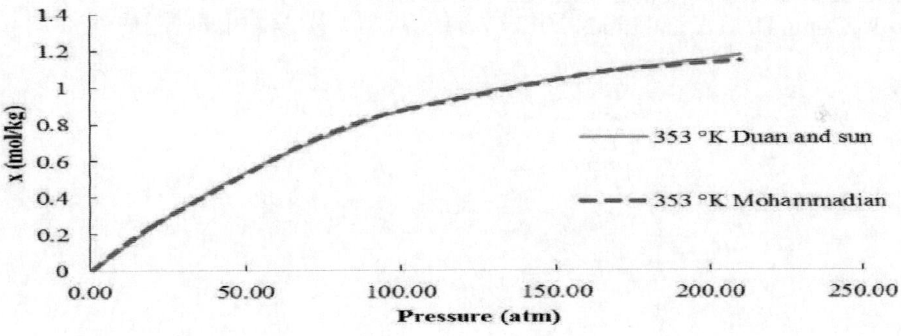

Figure 5 Comparison of solubility obtained from the current model with those of Duan and Sun (2003).

However, using the model beyond its temperature range produces rather larger errors (up to 41%). The latter is due to the inversion of temperature effect on solubility, which occurs at high pressures. This effect causes a rather unexpected increase in the solubility of CO_2 as temperature increases and has been reported by some of the previous researchers (Duan and Sun, 2003; Duan *et* al., 2006; Tong *et* al., 2013) without any explanation of the cause. The reason that model is not able to detect this effect, may be due to the fact that the model was developed based on experimental data obtained at lower temperature ranges. Therefore, the effect was not observed from experimental data gained in this research. Testing the model beyond its salinity limits generated errors up to 7%. Although the tolerance for error is usually around 5%, it is still acceptable to use the model out of its salinity range which is 0-15000 ppm. Moreover, the model could not be tested in various conditions of impurity in the stream as to date, to the authors' knowledge, there is no model or experimental data on effects of impurities in the stream and therefore there is no benchmark to test the current model with. To sum up, within an acceptable range of error, the model works fine, even at pressures and brine salinities beyond its scope. However care should be taken in using the model at temperatures higher than its limit i.e. 373 K as significant errors can be encountered.

3. Conclusions

In this research an empirical correlation was developed based on the experiments on the solubility of CO_2 in NaCl brines. The effects of pressure, temperature, salinity and purity of CO_2 stream was considered as independent variable whereas CO_2 solubility was the dependent variable in the model. The novelty of the model is that, it is very simple to use and accurate in its range of application and to some extends beyond them (depending on the parameter). The model was compared to the previous most widely-known models of literature and a good consistency was observed.

Referances

[1] Intergovernmental Panel on Climate Change 2003. Good Practice Guidance for Land Use, Land-use Change and Forestry. The IPCC good practice guidance for LULUCF.
[2] Bachu S S 2005 Special report on carbon dioxide capture and storage. *IPCC–Intergovernmental Panel on Climate Change.*
[3] Godec M L, Kuuskraa V A and Dipietro P 2013 *J.of Energy and Fuels*, **27** (8) 4183–4189
[4] Blyton, C A Bryant S L 2013 *Chemical Engineering Science* **101** 461–469
[5] Hou S, Maitland G C, Trusler J P 2013 *J. of Supercritical Fluids* **73** 87– 96
[6] United Nation Framework Convention on Climate Change 2002. Koyoto protocol reference manual
[7] Zhao, X and Liao X 2012 *Journal of Energy and Fuels* **26** (8) 5350–5354
[8] Mohammadian, E, Hamidi H, Azdarpour A, Asadualah M and Junin, R 2015 *Journal of Chemical & Engineering Data* **60** (7)
[9] Thodla, R, Francois, A, and Sridhar, N 2009. Materials Performance. In Supercritical CO_2 Environments. NACE conference. Presented in Las Vegas, USA. No. 09255
[10] Duan Z and Sun R 2003 *Chemical Geology* **193** 257– 27
[11] Mao S, Zhang, D, Li Y and Liu N 2013 *Chemical Geology* **347** Pages 43-58

IOP Conf. Series: Materials Science and Engineering **358** (2018) 012014 doi:10.1088/1757-899X/358/1/012014

Preparation of Activated Carbon/N-doped Titania Composite for Synergistic Adsorption-photocatalytic Oxidation of Batik Dye

A A Aziz[1] and S Ibrahim[2]

[1]Faculty of Engineering Technology, Universiti Malaysia Pahang, Lebuhraya Tun Razak, Gambang, 26300 Kuantan, Pahang, Malaysia
[2]Department of Civil Engineering, Faculty of Engineering, University of Malaya, 50603, Kuala Lumpur, Malaysia

[1]azrinaaziz@ump.edu.my, [2]shaliza@um.edu.my

Abstract. A synergetic improved composite TiO_2 photocatalysts was successfully synthesized by using nitrogen (N) as a dopant and activated carbon (AC) as synergetic compound. Two different types of AC prepared from Garcinia mangostana shell and commercial AC obtained from palm shell were chosen as synergetic compound. Thus synthesized composites was further characterized by Brunauer-Emmett-Teller (BET) surface analyzer and UV-visible light spectroscope. The doping of N resulted in a better solar light utilization potential. Furthermore, synergizing with AC contributed for the improved BET surface area and pore size distribution. The synergetic adsorption-photocatalytic activity was investigated by removing a commercial batik dye namely Remazol Brilliant Blue (RBB) under direct solar irradiation. The synergetic experiments showed that commercial AC synergized with N-TiO_2 resulted with a maximum removal efficiency of ~80% in 6 h.

1. Introduction

Over the decades, textile industry is gaining popularity in Malaysia especially in East Coast of Peninsular Malaysia like Kelantan and Terengganu. Among them locally well know batik industries are traditionally inherited over generations. In present days, these industries are commercialized and contributed significantly for the Malaysian Gross Domestic Product (GDP). However, this industry produces multiple dye waste that is released into the aquatic environment without proper treatment. Among the batik dyes, Remazol Brilliant Blue dye (RBB) has a major commercial contribution. Thus released dyes exert great harmfulness to the well-being, flora and fauna of the aquatic ecosystem [1]. Moreover their chemical structure makes more resistivity towards the natural photodegradation [1, 2]. Therefore, in order to protect the environment and to meet the stringent government law, it has driven many researchers to develop an effective and economical way to treat the dye wastewater [3].The developed methods include chemical coagulation, precipitation, biological techniques, adsorption, advanced oxidation processes (AOPs), membrane process, etc. [1-3]. Among them photocatalytic oxidations process, a classification of AOPs is more significant route over the others. They are highly efficient oxidation process with lower cost, zero toxicity with greater chemical and biological stability. This process employs semiconductors include TiO_2, ZnO, CeO_2, WO_3, SnO_2, ZrO_2, CdS and ZnS. They have been widely examined over the last few decades. Off these, titanium dioxide (TiO_2) has been proved to be an effective in removing organic pollutants from water bodies. Though they are efficient, their activity highly depends on recombination rate of photogenerated electron (e^-) and hole (h^+) pairs as well as the band gap energy [4-6]. The conventional TiO_2 has wider band gap energy (3.2 eV) and faster recombination rate of e^- and h^+ pairs. This wider band gap hinders its activation only to the ultraviolet (UV) region (λ <380 nm) than that of visible.

Many researchers had attempted to overcome these limitations either by doping metal ions or non-metal ions into the TiO_2 lattice. These doping could narrow the band gap energy in order to utilize visible light irradiation [7]. These photocatalysts can be employed as either as slurry, on a support,

Content from this work may be used under the terms of the Creative Commons Attribution 3.0 licence. Any further distribution of this work must maintain attribution to the author(s) and the title of the work, journal citation and DOI.
Published under licence by IOP Publishing Ltd

immobilization with glass fibers, glass beads or steel. However, they have issue due to adhesion force between the photocatalyst surface and the support. Therefore in present years researchers are moving towards a synergetic approach. Among them AC is one of the cost effective synergist with enriched adsorption capacity [4-6]. Hence in the present study N doped TiO_2 photocatalyst (N-TiO_2) was successfully synthesized and thus was synergized with low cost AC prepared from *Garcinia Mangostana* shell (N-TiO_2/GAC) and commercial activated carbon derived from palm shell (N-TiO_2/CAC). The fabricated photocatalysts was characterized and functionalized for the synergetic adsorption-photocatalytic removal of commercial batik dye viz., RBB under sunlight irradiation.

2. Materials and methods

2.1. Materials

The commercial AC (CAC) of < 1 mm, manufactured using commercial palm oil mill waste as precursor was obtained from Bravo Green Sdn Bhd, while the *Garcinia Mangostana* shell derived AC was synthesized in our laboratory. The chemicals including orthophosphoric acid, H_3PO_4 ($\geq$ 99% trace metals basis), sodium hydrogen carbonate, $NaHCO_3$ (99.5-100.5%), sodium sulphate (Na_2SO_4, $\geq$ 99% trace metals basis), barium chloride ($BaCl_2$, $\geq$ 99% trace metals basis) and titanium (III) sulphate solution ($Ti_2(SO_4)_3$, $\geq$ 99% trace metals basis) were purchased from Sigma Aldrich and used without further purification. Milli-Q water ($\geq$ 18.2 M Ω cm) was used for all experiments.

2.2. Synthesis of photocatalyst

The synthesis of the AC from *Garcinia Mangostana* shell and N doping procedure is detailed in our earlier reports [8, 9]. The synergization with AC was prepared by using 1 g of N-TiO_2 and 1 g of AC mixed into 1 L of Milli-Q water. The mixture was stirred for 24 h at room temperature (27°C±2°C) to obtain a homogeneous solution. Then the product was centrifuged and dried over night at 100°C ±2°C. The dried product was then carbonized in the furnace at 400 °C for an hour.

2.3. Characterization

Brunauer-Emmett-Teller (BET) surface area, pore volume, and Barret-Joyner-Halenda (BJH) pore size distribution (PSD) based on N adsorption-desorption isotherms were analyzed by Quantachrome Autosorb Automated Gas Sorption System. The optical absorption study was evaluated using UV-Vis (Shimadzu UV-2600) equipped with $BaSO_4$ diffuse reflectance integrating sphere (DRS) at ambient temperature in the wavelength range between 200 and 800 nm.

2.4. Adsorption-photocatalytic oxidation study

Synergetic adsorption-photocatalytic oxidation study was carried out in a batch reactor of 1 L capacity with a working volume of 250 mL. The reactor adopted is a simple glass beaker, under stirred condition using a magnetic stirrer in presence of sunlight as a light source. All the experiments were carried in a good sunny day under identical conditions. The intensity of sunlight varied between 2100 and 101200 lx. Solar luminance was measured using LT Lutron LX-101 Light Meter and the average luminance for the experimental run was calculated.

The photocatalytic experiments were performed with a catalyst loading of 1 g and 50 mg/L of initial dye concentration. The dye was vigorously stirred by a magnetic stirrer for uniform dye concentration and better mass transfer. This photocatalytic experiments were followed by a one hour dark reaction, which was sufficient to attain adsorption equilibrium. The samples were collected at regular time interval, filtered and the supernatant was subsequently analyzed for the residual dye concentration. The concentrations of RBB solution were determined by measuring the absorbance at 596 nm with the UV-Vis spectrophotometer (Shimadzu UV-2600). Figure 1 shows the experimental set-up for the photocatalytic study [8, 9].

IOP Conf. Series: Materials Science and Engineering **358** (2018) 012014 doi:10.1088/1757-899X/358/1/012014

Figure 1. Photograph of experimental set-up for the photocatalytic study.

3. Results and discussion

3.1. BET analysis

Table 1 lists the BET surface area, pore volume and average pore diameter of the both prepared photocatalyst and synergized photocatalysts.From the table, it is observed that the surface area and pore volume of GAC is higher than that of CAC. N-TiO_2 showed Type II BET isotherm which attributes to the cylindrical pores of the prepared TiO_2. On the other hand GAC synergized photocatalyst showed a Type IV isotherm, common isotherm type for AC, indicates that N-TiO_2 is condensed inside the tiny pores of the AC. Furthermore, after doping with N, the surface area and pore volume for N-TiO_2/GAC is lower compared to N-TiO_2/CAC. This proved a proper method of N doping for N-TiO_2/GAC where majority N-TiO_2 covered the pores of GAC which leads to a better photocatalytic process.

Table 1. BET surface area, pore volume and average pore diameter of the AC and prepared photocatalysts.

Samples	BET surface area [m^2 g^{-1}]	Pore volume [cm^3 g^{-1}]	Average pore diameter [nm]
N-TiO_2	106.3	0.3592	13.51
GAC	1218.0	0.7510	2.47
CAC	974.8	0.4480	1.84
N- TiO_2/GAC	343.6	0.3968	4.62
N- TiO_2/CAC	689.0	0.4357	2.53

3.2. UV-vis absorption spectra

The UV-visible absorption spectra obtained for the samples were presented in Figure 2. The figure illustrates the response of both N-TiO_2 and AC synergized N-TiO_2 between 200 – 800 nm of the spectrum. All the prepared photocatalysts exhibited an enhanced absorption in the visible light region along with a markable shift towards visible spectrum. Among them N-TiO_2/GAC emerged with slightly higher absorption and better shift towards the visible spectrum. This was further attributed to the contribution of the N and carbon support as a result of absorption under the visible light region.

IOP Conf. Series: Materials Science and Engineering **358** (2018) 012014 doi:10.1088/1757-899X/358/1/012014

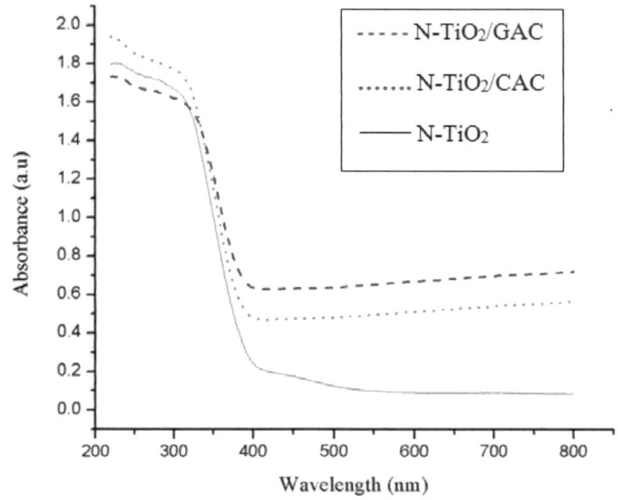

Figure 2. UV-visible absorption spectra of prepared photocatalysts.

3.3. Synergetic adsorption-photocatalytic oxidation study

The synergetic adsorption-photocatalytic degradation of RBB dye with the prepared photocatalysts were shown in Figure 3.

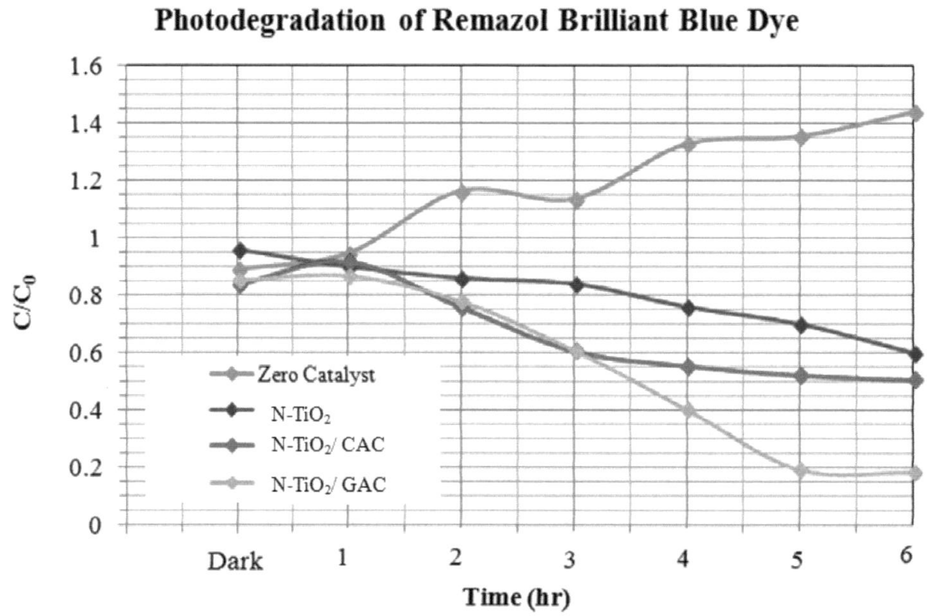

Figure 3. Synergetic adsorption-photocatalytic oxidation of the prepared photocatalysts.

The N-TiO_2/CAC resulted with a maximum removal efficiency of ~ 80%; while N-TiO_2/GAC resulted with ~45% degradation within 6 h of illumination. It is evident that both prepared synergized photocatalysts played an exclusive role in degrading the RBB dye. The carbon derived from the AC played a crucial role for the enhancement of the photocatalytic activity by decreasing the band gap energy and increased the adsorption process compared to N. A low removal efficiency of RBB dye was observed for N-TiO_2/GAC photocatalyst was attributed due to the large condensation of N-TiO_2 on the pores on AC. However for the better removal of RBB dye, the adsorption phenomenon driven by the synergization had its major contribution in comparison to the photocatalysis.

4. Conclusion

The synergetic N-TiO_2/AC composites have been successfully synthesized. The prepared photocatalysts had a better absorbance towards the visible region ensuring the utilization of the solar irradiation. The synergetic effect was also well observed in terms of adsorption and contributed for the better removal of RBB dye. Thus shows the significance of the developed synergized photocatalysts. On the whole the prepared N-TiO_2/AC exhibited its vibrant applicability towards wastewater treatment by incorporating both the sorption and photocatalytic effect.

Acknowledgement

The authors acknowledge the financial support of the RDU Grant No. RDU160317 and RDU 1603137, Faculty of Engineering Technology, University Malaysia Pahang, Malaysia which enabled this research study.

References

[1] Rahmat N A, Ali A A, Salmiati, Hussain N, Muhamad M S, Kristanti R A, Hadibarata T 2016 *Water Air Soil Pollut.* **227** 105
[2] Yonten V, Tanyol M, Yildrim N, Cikcikoglu N, Ince M 2016 *Des Water Treat* **57** 15592
[3] Ratnamala G M, Vidya S K, Srinikethan G 2016 *Des Water Treat* **57** 11361
[4] Sim L C, Leong K H, Saravanan P, Ibrahim, S 2015 *Appl. Surf. Sci.* **358** 122
[5] Leong K H, Sim L C, Bahnemann D, Jang M, Ibrahim S, Saravanan, P 2015 *APL Mater.* **3** 104503
[6] Sim L C, Ng K W, Ibrahim S, Saravanan P 2014 *J. Nanosci. Nanotech.* **14** 7001
[7] Xing J, Sun X, Qiu J 2015 Russian *J. Phys. Chem. A* **89** 1108
[8] Aziz A A, Yong K S, Ibrahim S, Saravanan P 2012 *J. Hazard. Mater.* **199** 143
[9] Kang Y L, Toh S K S, Monash P, Ibrahim S, Saravanan P 2013 *Asia Pac J. Chem. Eng.* **8** 811

IOP Conf. Series: Materials Science and Engineering **358** (2018) 012015 doi:10.1088/1757-899X/358/1/012015

Effect of Germanium on the TiO_2 Photoanode for Dye Sensitized Solar Cell Applications. A Potential Sintering Aid

M S Ahmad, AK Pandey and N A Rahim

UM Power Energy Dedicated Advanced Center (UMPEDAC) Jalan Pantai Baharu, Level 20, Wisma R&D University of Malaya, 59990 Kuala Lumpur, Malaysia

shakeelalpha@gmail.com, adarsh@um.edu.my

Abstract: Anatase titanium-germanium (TiO_2-Ge) nanocomposite has been prepared by using colloidal suspension process and investigated for photoanode to be used in dye sensitized solar cell. Ge possesses lower band gap energy compared to TiO_2 and has the capability to absorb infrared region of solar spectrum. Its remarkable absorption and good electron transfer ability due to lower band gap energy makes it a potential candidate material in the field of DSSCs to counter important disadvantages such as high probability of electron recombination reactions and absorption of small region (UV region) of solar spectrum. Another advantage is its low sintering temperature which proved to be an added advantage to increase inter-particle contact which in turn leads to improved electron transfer. Scanning electron microscopy (SEM), uv-vis spectroscopy and electron impedance spectroscopy (EIS) have been employed to evaluate the effect of Ge on TiO_2photoanode.

1. Introduction:

Today, the increasing energy demand leads to more CO_2 emission and other pollutants in our environment which is the basic problem in sustainable development[1]. To cope with the energy demand and to save the environment, a concept of renewable or green energy is getting pace and the major challenge for scientists working in the field of technology is to develop an efficient and cost effective renewable technology to harvest energy from natural and renewable resources[2]. Photovoltaic cells in this regard are the strong candidates to effectively harvest solar energy directly into electricity. Silicon based solar cells achieve 25% efficiency [3]and 41% efficiency [2]has been claimed with three junction photovoltaic cells. Still this efficiency is well below their theoretical limits and these technologies require expensive processes and fabrication equipments. In this regards Dye sensitized solar cells (DSSCs) are cost effective and easy to fabricate with moderate efficiency.

DSSCs work on the principle of photosynthesis [4]. In brief, there are three main components of DSSC (1) photoanode (TiO_2 nanostructure), (2) dye sensitizer (ruthenium based dye to absorb sun light) and (3) electrolyte (Iodide/tri-iodide redox couple). Photons are absorbed by the dye sensitizer and an electron has been ejected from dye to conduction band of photoanode material. This electron travels from photoanode towards external circuit. Meanwhile, the positively charged dye pick electron from electrolyte and become stable. The charged electrolyte pick electron from the outer circuit and hence the process continues [5]. Various approaches have been trialed to improve the incident photon conversion efficiency (IPCE) of DSSCs such as size quantization, development of different nano-architectures, surface Plasmon effect and doping with different cations & anions [5-6]. With improvement in light absorption capacity, recombination reactions and dye loading characteristics, the maximum claimed IPEC reached to 12-14% [7-8].

Noval titania/germanium (TiO_2-Ge)nanocomposites and sandwich structures have beendeveloped recently for the removal of environmental pollutants [9]and for high performance thin film

Content from this work may be used under the terms of the Creative Commons Attribution 3.0 licence. Any further distribution of this work must maintain attribution to the author(s) and the title of the work, journal citation and DOI.
Published under licence by IOP Publishing Ltd

photovoltaic devices[10-12].Bulk Ge has direct band gap of 0.8eV(band gap of TiO_2 is 3.2ev). And an indirect band gap of 0.66eVat room temperature[10]. These characteristics make it a potential candidate in the field of DSSCs.In this study, we report for the first time the effect of Ge on TiO_2photoanode and support our conclusion in the light of results obtained through SEM, uv-vis spectroscopy and EIS techniques.

2. Experimental:

TiO_2(sigma Aldrich 100 to 200 nm, irregular shape) and Ge (sigma Aldrich 100 to 200nm size range, irregular shape) nanoparticles were weighed and stirred using in acetone separately for two hrs with subsequent sonication (model XA2600) for 1hr and then stirred over nigh. The two suspensions were mixed together and again sonicated for 1hr followed by stirring for 12hrs. The mixture was then filtered using filter paper and dried at 80°C for 4hrs. Three specimens i.e. (1) TiO_2 nanoparticles, (2) Ge nanoparticles and (3) TiO_2-2wt%Ge nanocomposite have been investigated. SEM was performed to examine and investigate the surface morphology of particles. Field emission scanning electron microscope (FE–SEM) (QUANTA FEG 450) was used in secondary electron imaging (SEI) mode and back scattered electron mode with an accelerating voltage of 1-10kV. Uv-vis spectroscopy (model: uv 2600 spectrometer) and EIS studies have been conducted to evaluate Ge nanoparticles as potential nanomaterial to improve the performance of DSSCs.

3. Results and discussion:

3.1. Surface morphology: scanning electron micrographs (Figure 1) were taken to examine the surface morphology, shape and size of as received nanoparticles. Figure 1(a&b) is showing irregular shaped TiO_2 nanoparticles, the size of these particles ranges between 90 to 150nm. Some TiO_2 nanoparticles larger than 150nm can also be seen in the SEM micrographs. Figure 1(c&d), is showing as received Ge nanoparticles. The shape of Ge particles was observed to be irregular. Ge particles range between 100 to 200 nm.

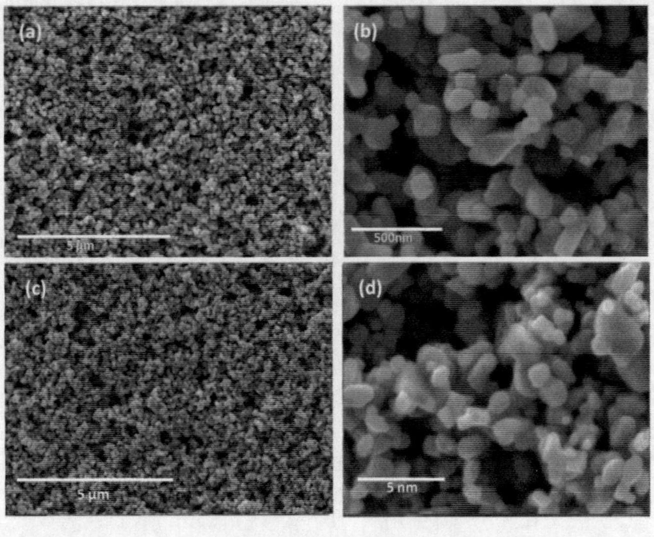

Figure 1. SEM micrographs of raw powders (a,b) TiO_2 nanoparticles and (c,d) Ge submicron particles.

Figure 2 is showing SEM micrographs of un-annealed and annealed (annealed at 450°C to 30 min at the rate of 5°C/min) nanocomposite specimen. Homogeneous and agglomeration free mixing of individual particles can be seen in these micrographs. Excellent mixing has been achieved by employing vigorous stirring and ultrasonication for extended period of time. Figure 2(a,a1) is illustrating un-annealed specimen i.e. pure TiO_2 and TiO_2-2wt%Ge nanocomposite, it can be observed that there is no surface diffusion among nanoaprticles which is the major reason of increase in charge recombination among the semiconductor, dye and electrolyte interface. Semiconductor nanoparticles due to lack of good contact cannot efficiently transfer electrons from sensitizer to the transparent conducting electrode. To facilitate efficient charge transfer from the semiconductor, the contact between the surfaces of semiconductor nanoparticles is improved with the aid of thermal energy which reduces the surface tension of the particles and facilitates surface diffusion. Heating to the temperature range 450-500°C is expected to give optimum results [12]. Annealed composite specimen can be seen in Figure 2(b,b1). Round red circles are indicating surface diffusion of nanoparticles. At this stage complete sintering is not favorable as complete sintering will render a highly compact surface layer with less surface area. High surface area porous layers are highly desirable for photoanode to improve pickup of dye sensitizer molecules along with sufficient inter-particle contact.

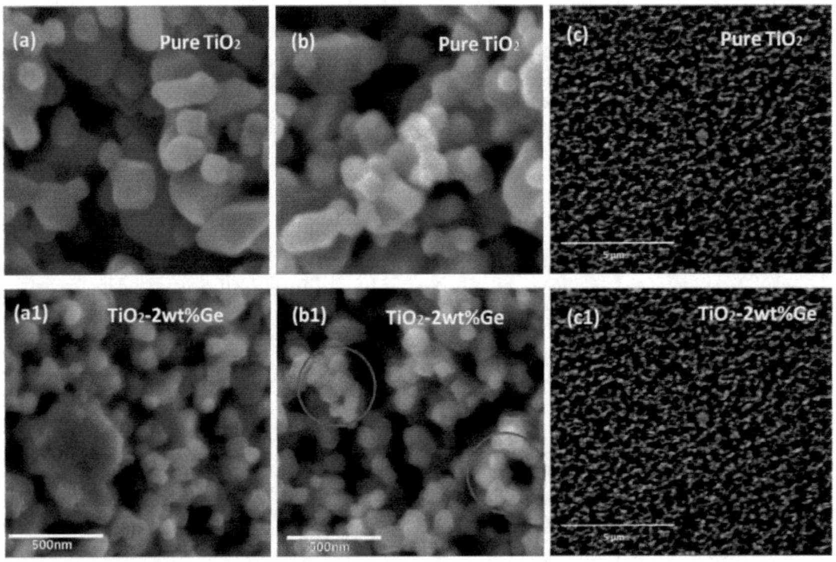

Figure 2 SEM micrographs of annealed and un-annealed composite specimen (a,a1) un-annealed TiO_2&TiO_2-2wt%Ge nanocomposite (b,b1) annealed TiO_2 & TiO_2-2wt%Ge nanocomposite and (c) backscattered SEM images of TiO_2 & TiO_2-2wt%Ge nanocomposite.

It has been observed that with the incorporation of Ge particles, the surface tendency of nanoparticles to diffusion increased. Higher surface diffusion has been observed in TiO_2-2wt%Ge nanocomposite specimen. This effect can be co-related with the specific heat and melting temperature of Ge (938°C) and TiO_2(1843°C). Ge require less energy to activate surface diffusion compared to that of TiO_2 that is why with the addition of Ge, nanocomposite underwent higher degree of surface diffusion which is required to improve the contact between nanoparticles to pave the way for uninterrupted transport of electrons from dye sensitizer to the external circuit and to reduce recombination reactions at the capacitive interface.

3.2. Light absorption

Solar spectrum mainly consists of ultraviolet (Uv), visible (vis) and infrared light (IR). For efficient working of photoanode it is necessary to absorb as much incident spectrum as possible to facilitate the dye sensitizer to eject electron from its valance band. TiO_2 can only absorb Uv spectrum. As Ge has the ability to absorb near infrared region of solar spectrum, combining these two semiconductor elements (i.e. TiO_2 and Ge) will absorb more incident energy and in-turn increase the ability of photoanode to harvest more energy. Figure 3 is showing the optical characteristics of TiO_2 and Ge under different regions of solar spectrum. It is evident from the graph that TiO_2 can only absorb UV region of solar spectrum while Ge is showing absorption in near infrared region. Iso-propyle alcohol has been used to prepare specimens for Uv-vis spectroscopy. Nanocomposite prepared by using these two elements will be able to absorb light from UV and near-infrared region of solar spectrum and in-turn improve the light absorption characteristics of photoanode to facilitate dye sensitizer. This improved light absorption capability improves the incident photo conversion efficiency of solar device.

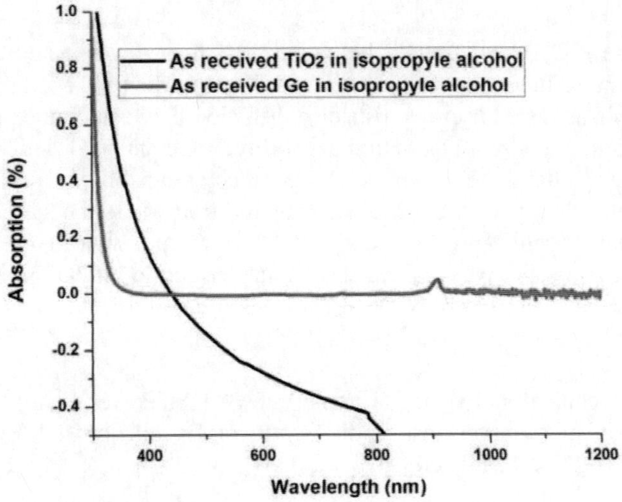

Figure 3. Uv-vis spectra of Ge and TiO_2

3.3. EIS response

The EIS study has been conducted (frequency range was 0.01Hz to 1MHz and ac voltage of 10mV have been chosen for experiments) to examine the capacitance i.e charge transfer ability of Ge and its nanocomposite with TiO_2. It is a powerful tool to study the kinetic behavior of electron transfer from photoanode. The charge transfer properties of TiO_2 and TiO_2 – 2wt% Ge nanocomposite is shown in Figure 4.

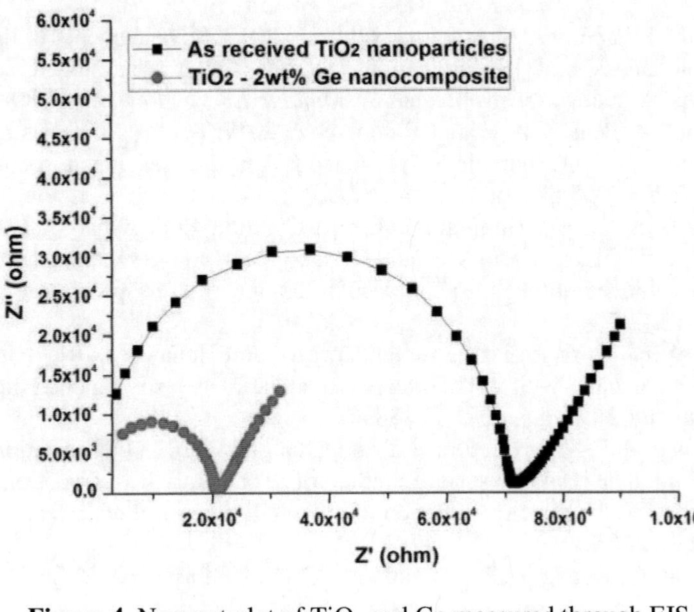

Figure 4. Nyquest plot of TiO_2 and Ge measured through EIS

The TiO_2 - 2wt% Ge nanoparticles show good electron transfer ability compared to TiO_2 nanoparticles. It is evident from the nyquest plot that with the addition of Ge in TiO_2 matrix, the charge transfer ability of photoanode has been increased and electron will have the opportunity to be first picked up by the Ge nanoparticles from the dye efficiently and then transfer to TiO_2 semiconductor matrix. This hetro-junction so called step approach effectively controls the recombination reactions and will improve the photo conversion efficiency of the device.

4. Conclusion:

The effect of Ge on the TiO_2 nanoparticle photoanode has been investigated as potential candidate material to improve the efficiency of dye sensitized solar device. Morphological observations conducted through SEM suggested improved surface diffusion between nanoparticles. Uv-vis analysis has been performed to study the light absorption capability of Ge and TiO_2 individually and EIS study has been conducted to understand the electronic charge transfer capability of TiO_2 with the addition of Ge nanoparticles. Results clearly showed that with the addition of Ge the surface diffusion and light harvesting capability have been improved. Also with the incorporation of Ge, the electron transfer capability has been increased. From the above results and discussion it can be concluded that Ge has the potential to improve the performance of DSSCs and should be further explored for optimization.

Acknowledgements
The authors thank the technical and financial assistance of UM Power Energy Dedicated Advanced Centre (UMPEDAC) and the Higher Institution Centre of Excellence (HICoE) Program Research Grant, UMPEDAC - 2016 (MOHE HICOE - UMPEDAC).

References:

[1] Mehmood U, Hussein I A, Harrabi K, Mekki M, Ahmed S and Tabet N 2015 Hybrid TiO_2–multiwall carbon nanotube (MWCNTs) photoanodes for efficient dye sensitized solar cells (DSSCs) *Solar Energy Materials and Solar Cells* **140** 174-9

[2] Barberio M, Grosso D, Imbrogno A and Xu F 2016 Preparation and photovoltaic properties of layered TiO 2/carbon nanotube/TiO 2 photoanodes for dye-sensitized solar cells *Superlattices and Microstructures* **91** 158-64

[3] Green M A 2009 The path to 25% silicon solar cell efficiency: History of silicon cell evolution *Progress in Photovoltaics: Research and Applications* **17** 183-9

[4] Sengupta D, Das P, Mondal B and Mukherjee K 2016 Effects of doping, morphology and film-thickness of photo-anode materials for dye sensitized solar cell application–A review *Renewable and Sustainable Energy Reviews* **60** 356-76

[5] Shakeel Ahmad M, Pandey A K and Abd Rahim N 2017 Advancements in the development of TiO2 photoanodes and its fabrication methods for dye sensitized solar cell (DSSC) applications. A review *Renewable and Sustainable Energy Reviews* **77** 89-108

[6] Shakeel Ahmad M, Pandey A K and Rahim N A 2017 Towards the plasmonic effect of Zn nanoparticles on TiO2 monolayer photoanode for dye sensitized solar cell applications *Materials Letters* **195** 62-5

[7] Mathew S, Yella A, Gao P, Humphry-Baker R, Curchod B F, Ashari-Astani N, Tavernelli I, Rothlisberger U, Nazeeruddin M K and Grätzel M 2014 Dye-sensitized solar cells with 13% efficiency achieved through the molecular engineering of porphyrin sensitizers *Nature chemistry* **6** 242-7

[8] Kakiage K, Aoyama Y, Yano T, Oya K, Fujisawa J-i and Hanaya M 2015 Highly-efficient dye-sensitized solar cells with collaborative sensitization by silyl-anchor and carboxy-anchor dyes *Chemical Communications* **51** 15894-7

[9] Chun S Y, Chung W J, Kim S S, Kim J T and Chang S W 2015 Optimization of the TiO 2/Ge composition by the response surface method of photocatalytic degradation under ultraviolet-A irradiation and the toxicity reduction of amoxicillin *Journal of Industrial and Engineering Chemistry* **27** 291-6

[10] Li X, He F, Liu G, Huang Y, Pan C and Guo C 2012 Fabrication of Ge quantum dots doped TiO 2 films with high optical absorption properties via layer-by-layer ion-beam sputtering *Materials Letters* **67** 369-72

[11] Abe S, Adachi Y, Matsuda K and Nose M 2014 Single-phase anatase structure and dominant metallic Ge in Ge/TiO 2 multi-layer films using a differential pumping co-sputtering system *Thin Solid Films* **562** 104-8
[12] Khan A F, Mehmood M, Ali T and Fayaz H 2013 Structural and optical studies of nanostructured TiO 2–Ge multi-layer thin films *Thin Solid Films* **536** 220-8

Techno-economic Analysis of Evacuated Tube Solar Water Heater using F-chart Method

H Fayaz[1], N A Rahim[1], R Saidur[2] and M Hasanuzzaman[1]

[1]UM Power Energy Dedicated Advanced Centre (UMPEDAC), Level 4, Wisma R&D UM,
University of Malaya, Jalan Pantai Baharu, 59990 Kuala Lumpur, Malaysia
[2]Sunway University, Jalan Universiti, Bandar Sunway, 47500 Subang Jaya, Selangor, Malaysia

katperfayaz@gmail.com

Abstract. Solar thermal utilization, especially the application of solar water heater technology, has developed rapidly in recent decades. Solar water heating systems based on thermal collector alone or connected with photovoltaic called as photovoltaic-thermal (PVT) are practical applications to replace the use of electrical water heaters but weather dependent performance of these systems is not linear. Therefore on the basis of short term or average weather conditions, accurate analysis of performance is quite difficult. The objective of this paper is to show thermal and economic analysis of evacuated tube collector solar water heaters. Analysis done by F-Chart shows that evacuated tube solar water heater achieves fraction value of 1 to fulfil hot water demand of 150liters and above per day for a family without any auxiliary energy usage. Evacuated tube solar water heater show life cycle savings of RM 5200. At water set temperature of 100°C, RM 12000 is achieved and highest life cycle savings of RM 6100 at the environmental temperature of 18°C are achieved. Best thermal and economic performance is obtained which results in reduction of household greenhouse gas emissions, reduction of energy consumption and saves money on energy bills.

1. Introduction

Hot water is necessity for the basic life activies from domestic to industrial purposes. Basically, fuels used for heating water release huge amount of coarbon dioxide in the environment, which cause the pollution. It has been predicted that there will be a great shortage in the supply of conventional fuels if a suitable alternatives are not found [1, 2]. Therefore, solar energy has been the best option to mitigate the pollution and avoid the threat of depleting resources of energy. Solar water heaters are developed and are being used to heat the water with solar energy all over the world. The SWH system investigated consists of mainly three parts, namely a flat plate solar collector, a heat exchanger (storage tank) and a circulating pump[3]. Solar water heating systems have been the famous application to convert solar radiations as an energy source into thermal energy [4]. Furthermore, these days solar water heating is obtained by combination of thermal collector with photovoltaic (PV) modules called as PVT, which gives two fold benefits. Firstly, it gives improved electrical efficiency of PV module by extracting heat form PV panel which is used for water heating and some other applications. Secondly, size of overall system reduces which occupies lesser area as compared to both thermal collectors and PV panels if installed separately. In today's market, evacuated tube solar water heaters are most cost-effective, efficient and reliable as compared to others types of solar water

Content from this work may be used under the terms of the Creative Commons Attribution 3.0 licence. Any further distribution of this work must maintain attribution to the author(s) and the title of the work, journal citation and DOI.
Published under licence by IOP Publishing Ltd

heaters[5]. There are many types of evacuated tube solar collectors to choose from, making them extremely versatile in the applications they can be used for. These systems are smart in size and efficient in thermal conversion with reasonable prices [6, 7]. Fortunately, Malaysia is blessed with abundant solar energy. In Malaysia, monthly average daily sunshine duration is from 4hrs to 8hrs with monthly average daily solar radiation value of 4000-5000 Whr/m2[8]. Due to the favourable climate conditions in Malaysia, it has a high solar energy potential specially in harnessing solar thermal energy [9]. Three years monthly average data of solar irradiations was collected from Subang/KL meteorological station to analyse solar water heaters in Malaysian climate. Evacuated tube solar water heater is analysed through F Chart method to investigate the economic and thermal perfoamnce.

1.1 F-Chart Method

F–Chart which is designed by S.A. Klein and W.A Beckman is an authoritative source for design and analysis of solar system. The chart scheme provides means for predicting the thermal behaviour of the dynamic solar heating systems easily (by means of air or liquid as fluid used) and the solar domestic hot water systems [10]. The F–chart scheme is principally a relation of effects from numerous simulations of the solar heating systems. The environments of the resultant relation gives *f*, the part of scheduled heating load delivered by the solar energy as a formula of 2 dimensionless quantities (variables) involving a collector features, heating of loads, and the local climate [11]. The part of monthly heating load (for space heating and hot water) is delivered by the solar energy which is a formula of 2 dimensionless quantities, *A* and *B*. Where the ratio of collector losses to heating loads is *A* and the ratio of absorbed solar radiation to the heating loads is *B* [12].
Reference collector energy

$$X = \frac{Loss\ during\ a\ month}{Total\ heating\ load\ during\ a\ month} \tag{1}$$

$$Y = \frac{Total\ energy\ absorbed}{Total\ heating\ load} \tag{2}$$

The equations (1) and (2) for X and Y can be written as follow:

$$X = F_R.\,U_L.\frac{F'_R}{F_R} \times \left(T_{ref} - \overline{T_a}\right) \times \Delta\tau \times \frac{A_c}{L} \tag{3}$$

$$Y = F_R(\tau\alpha)_n \times \frac{F'_R}{F_R} \times \frac{\overline{(\tau\alpha)}}{(\tau\alpha)_n} \times \overline{H_T}\mathrm{N} \times \frac{A_c}{L} \tag{4}$$

where
Ac = Area of solar collector (m2 or ft2),
F'R = Collector-heat exchanger efficiency factor (%),
FR = Collector heat removal factor (%),
UL = Collector overall energy loss coefficient (W/m2-°C or Btu/hr-ft2-oF),
Δτ = Total number of seconds (SI) or hours (IP) in the month,
T a = Monthly average ambient temperature (°C or °F),
L = Monthly total heating load for space heating and hot water (GJ or MMBtu),
HT = Monthly averaged, daily radiation incident on collector surface per unit area (MJ/m2 or Btu/ft2),
N = Number of days in the month,
(τα) = Monthly average transmittance-absorptance product (%),
(τα)n = Normal transmittance-absorptance product (%), and
Tref = An empirically derived referenc$f$e temperature (100 °C or 212 oF).

The F-Chart equations for the fraction *f* of the monthly space and water heating loads supplied by solar energy are the following.

$$f = 1.04Y - 0.065X - 0.159Y^2 + 0.00187X^2 - 0.0095Y^3 \quad \text{(Air system)} \qquad (5)$$

$$f = 1.029Y - 0.065X - 0.245Y^2 + 0.0018X^2 + 0.0215Y^3 \quad \text{(Liquid system)} \qquad (6)$$

The F-Chart equations for the fraction *f* of the monthly space and water heating loads supplied by solar energy are given as equations 5 and 6 respectively.

2. Methodology

Using F-chart software, thermal and economic output of the system has been calculated by putting solar evacuated collectors data and monthly weather data from Subang/Kuala Lumpur meteorological station, active domestic hot water system data and economic data. The input data is shown in Tables 3.1, 3.2, 3.3, 3.4 respectively.

Table 1. Evacuated tube collector data (F-Chart).

Parameters	Units	Evacuated tube
Number of collector panels	-	3
Collector panel area	m^2	2.86
FR×UL (Test slope)	W/m^2-°C	1.652
FR×TAU×ALPHA	-	0.480
Collector slop	Degrees	15
Collector azimuth (South=0)	Degrees	0
Collector flow rate/area	kg/sec-m^2	0.020
Collector fluid specific heat	kJ/kg-°C	4.15

Evacuated tube solar water heating system is taken from RETScreen database, which is Solar Rating and Certification Corporation certified collector commercially used all over the world [13].

Table 2. Weather view/change input data (F-Chart).

Months	Solar Rad. [kJ/m^2]	Temp. [°C]	Humidity [kg/kg]	Mains [°C]	Reflect.	°C-days
Jan	15270	26.1	0.0173	27	0.30	44
Feb	15290	26.5	0.0170	27	0.30	32
Mar	19100	26.8	0.0182	27.1	0.30	34
Apr	17690	27.0	0.0190	27.3	0.30	29
May	21630	27.2	0.0192	27.4	0.30	28
Jun	17930	27.0	0.0187	28.0	0.30	29
Jul	20310	26.6	0.0183	27.2	0.30	36
Aug	18430	26.6	0.0182	27.0	0.30	36
Sep	15940	26.4	0.0183	27.5	0.30	37
Oct	15700	26.3	0.0184	27.1	0.30	41
Nov	19110	26.1	0.0184	27.6	0.30	42
Dec	20140	26.0	0.0187	27.3	0.30	46

Where,
Solar rad are the radiations incident on solar panel in kJ/m^2
Temp is atmospheric temperature in °C.
Humidity is the humidity ratio in kg/kg.
Mains is the temperature of main water supply in °C.

Reflect is the fraction of reflected solar radiations.
°C-days is the degree-days in temperature °C.

Table 3. Active domestic hot water system data input (F-Chart).

Active Domestic Hot water system		
Location	Subang/Kuala Lumpur	
Water volume/collector area	300	Litres/m^2
Fuel	Electricity	
Efficiency of fuel usage	90%	%
Daily hot water usage	150	Liters
Water set temperature	55	°C
Environmental temperature	27	°C

Where,
Water volume/collector is the amount of water volume in single tank system.
Fuel is auxiliary fuel used along with solar energy.
Daily hot water usage is the amount of water used per day.
Water set temperature is the required temperature of the water set by user.
Environmental temperature is the temperature of surrounding air.

Table 4. Economics Parameters inputs (F-Chart).

Parameters input	Detailed	Units
Cost per unit area	n/a	RM/m^2
Area independent cost	6000	RM
Price of electricity	0.2500	RM/kW-hr
Annual % increase in electricity	1.0	%
Period of economic analysis	20	Years
% Down payment	100	%
Ture % property tax rate	3.0	%
Commercial system?	No	%

Where,
Cost per unit area is the cost of the heating system unit wise but not as a whole price.
Area independent cost is the system's whole cost, if it is given then cost per unit area is not applicable.
Price of electricity is the unit cost of electricity used as auxiliary energy.
Annual % increase in electricity is the expected increase in the unit price of electricity.
Period of economic analysis is the total time period for which the solar water heaters are analysed for their performance.
% downpayment is the partial payment made at the time of purchase of the system with the balance paid later as a term of mortgage.
True % property tax rate is the ratio of the increment in real estate taxes due to the solar system to the cost of the solar system.
Commercial system? Set "Yes" or "No" to indicate if the solar system is on private homes or used as commercial purpose.

IOP Conf. Series: Materials Science and Engineering **358** (2018) 012016 doi:10.1088/1757-899X/358/1/012016

3. Results and discussions

F-chart is an analysis tool to design solar water heating systems. Many types of solar system can be analysed and designed for long term average performance. It is carefully constructed correlation which based on simulations with a streamlined version of TRANSYS program. F-chart has been used as verification tool for TRANSYS simulations, experiments and other programs for comparing accuracies. The average availability of solar energy on monthly basis based on last 3 years data from Subang/KL meteorological station is used. From the F-chart calculations solar fraction of value 1.00 is obtained through solar water heater throughout the year. This shows that evacuated tube solar water heater with can fulfil 100% domestic hot water demand of about 150 litres minimum on daily basis.
Figure 1 draws the picture of long run economic benefits (in terms of yearly savings) of shifting to solar energy source. As per this analysis, in 20 years, RM5200 can be saved in the case of evacuated collector solar water heater. While solar fraction is constant as value of 1.0 along the time period of 20 years, which results in total life time saving of amount RM5200 as mentioned earlier. The figure shows the progressive improvement in life cycle savings even after time of 20 years. Hence, it is a significant contribution to overall national as well as personal savings.

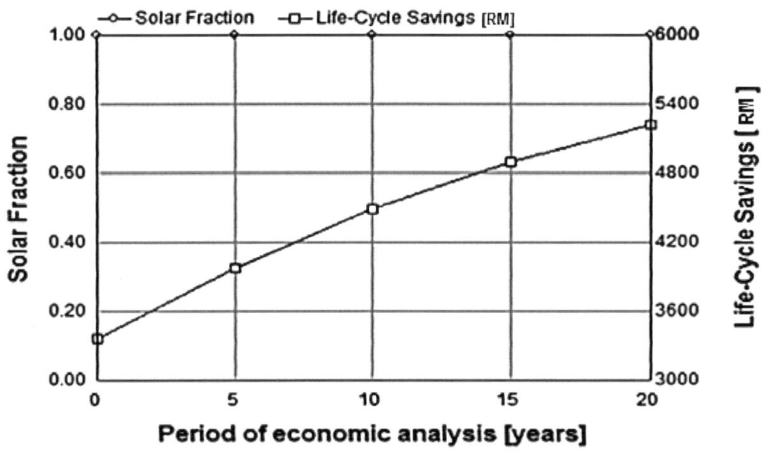

Figure 1. Solar fraction and life cycle savings versus period of economic analysis (years).

Figure 2 shows the association of daily hot water consumption (horizontal axis) with solar fraction and projected life-cycle savings (both scaled vertically). It shows that as water consumption increases, the life-cycle savings also increases, according to this, higher the consumption, higher the savings will be gained. It can, hence, be concluded that large consumers are more to be benefitted from the solar heating system. It is however important to note that the solar fraction falls sharply once the consumption reaches 275 litres a day. This decline in the solar fraction sets in because as demand of hot water increases, it requires more of non-solar more expensive heating sources. With evacuated collector installed, life-cycle savings can go higher than RM 12, 000 even with the increase in hot water usage more than 500 liters. Hence, it shows that evacuated solar water heaters are with more capability and flexible to serve the domestic needs.

The Figure 3 portrays how solar fraction and life-cycle savings behave with change in water set temperature. For evacuated collector solar water heater, the solar option works well at lower temperatures, sufficing the energy requirements and maximizing the savings at the same time. Below 55°C, there appears no need of additional source of heating technology as the solar system suffices. Life-cycle savings also are affected at higher temperatures, due to significant decline in solar fraction. It shows that higher the water set temperature, lesser will be the solar fraction, while life cycle savings tend to increase to certain level. From the results it shows that evacuated collector solar water heaters are more stable and flexible in serving hot water at higher temperatures more than 55°C with the solar fraction of 0.70. In terms of life cycle savings, if water set temperature is set at 100°C then RM12000 is achieved by evacuated tube collector solar water heaters.

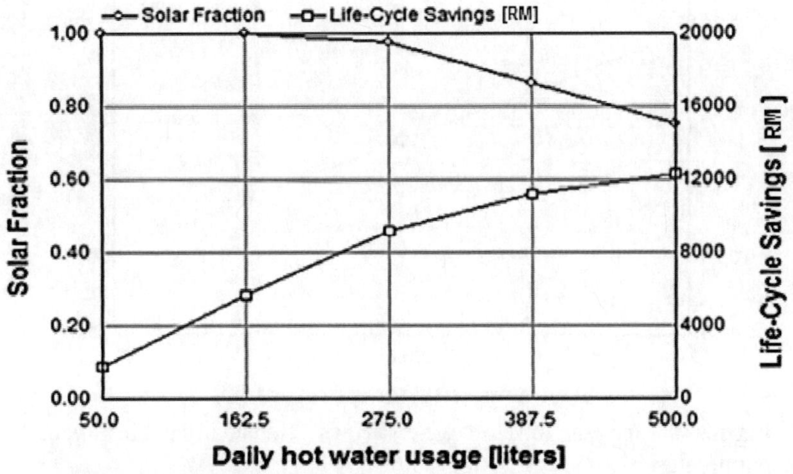

Figure 2. Solar fraction, life cycle savings and daily hot water usage.

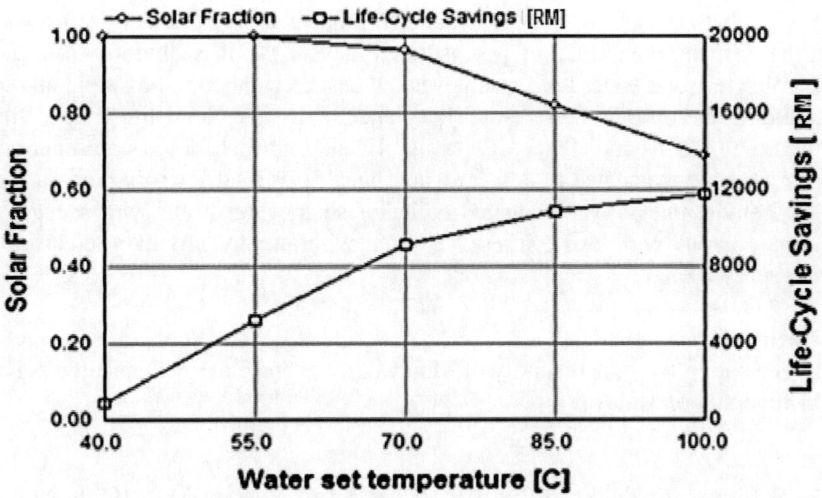

Figure 3. Solar fraction, life cycle savings versus daily water set temperature.

Figure 4 analyses the impact of environmental temperature on solar fraction and life cycle savings. Evacuated collector solar water heaters are highly efficient as they show good performance even at lower environmental temperatures of 18°C. Throughout the temperature scale the solar fraction obtained through these collectors is constant as 1.0. Where, life cycle savings decline because of the reason that cost of solar water heaters is set at RM6000 and the water set temperature at 55°C as input values. Considering these input values, life cycle savings will obviously decrease as the solar water heaters will no longer play major role to heat the water if the environmental temperature goes higher. Evacuated solar water heater results in the highest life cycle savings of RM 6100 at the environmental temperature of 18°C and lowest savings of RM3900 at the temperature of 40°C.

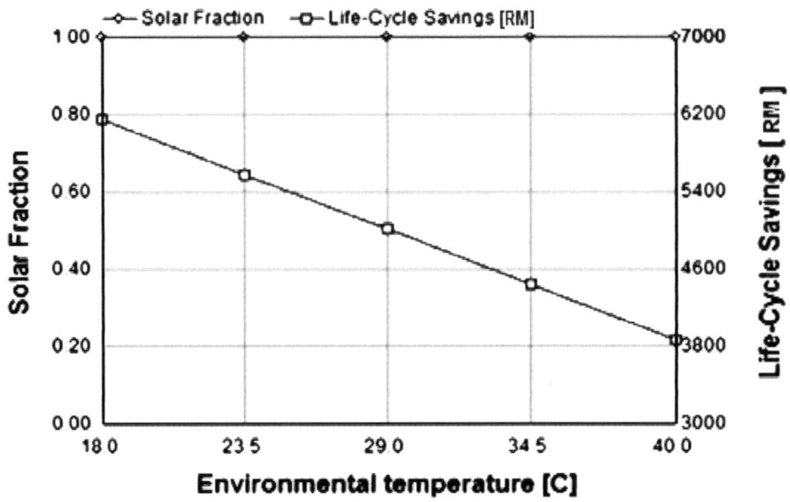

Figure 4. Solar fraction, life cycle savings versus daily environmental temperature.

4. Conclusion

It can be concluded that F-Chart analysis of evacuated tube solar water heater performance shows the best results in the climatic conditions of peninsular Malaysia. As it is obvious from the results that evacuated tube collector solar water heater achieves the best thermal output as it obtains solar fractions of 1.0 with excluding the option of using auxiliary energy. It saves total life time saving of amount RM5200 for the time of 20 years. Hence, it is a significant contribution to overall national as well as personal savings. Overall; evacuated tube solar water heater gives satisfactory performance thermally, economically and environmentally. Evacuated collector solar water heater will serve the needs with maximum thermal output with extra litres of water without any auxiliary electric water heater installation and energy use.

Acknowledgement

This work has been supported by University of Malaya Power and Energy Dedicated Advanced Center (UMPEDAC),University of Malaya.

References

[1] Fayaz H, Rahim N, Saidur R, Solangi K, Niaz H and Hossain M. 2011 Solar energy policy: Malaysia vs developed countries. Clean Energy and Technology (CET), IEEE First Conference on: IEEE; 2011. p 374-8.

[2] Ali B, Sopian K, Ghoul MA, Othman MY, Zaharim A and Razali AM. 2009 Economics of Domestic Solar Hot Water Heating Systems in Malaysia. European Journal of Scientific Research. **26**:20-8.

[3] Kreider JF and Kreith F. 2011 Solar energy handbook. OSTI: 6408304. McGraw-Hill series in modern structures: systems and management.1120.

[4] Khan SI and Obaidullah M. Fundamental of Solar Water Heaters. 2007 Centre for Energy Studies.125-32.

[5] NREL NREL 1996 Solar Water Heating. Energy Efficiency and Renewable Energy.

[6] Ahmad S, Kadir MZAA and Shafie S. 2011 Current perspective of the renewable energy development in Malaysia. Renewable and Sustainable Energy Reviews.**15**:897-904.

[7] Qazi A, Fayaz H and Raj RG. 2010 Discourse on Data Mining Applications to Design Renewable Energy Systems.

[8] Ab Kadir MZA, Rafeeu Ya and Adam NM. 1982 Prospective scenarios for the full solar energy development in Malaysia. Renewable and Sustainable Energy Reviews.14:3023-31.
[9] Chuah DGS and Lee SL. Solar radiation in peninsular Malaysiaâ€"Statistical presentations. Energy Conversion and Management.**22**:71-84.
[10] Beckman WA, Klein, S.A and Duffie, J.A. 1977 Solar Heating Design by the F-Chart Method. New York: John Wiley & Sons.;.
[11] Klein SA and Beckman WA. 2001 User's Manual, www.fchart.com. F-CHART.
[12] Okafor I and Akubue G. F-Chart Method for Designing Solar Thermal Water Heating Systems. International Journal of Scientific & International Research.3.
[13] RETScreen. 2014 Solar Water Heating. Canada.

Physicochemical properties of betaine monohydrate-carboxylic acid mixtures

I Zahrina[1,2], **M Nasikin**[1] **and K Mulia**[1*]
[1]Department of Chemical Engineering, Faculty of Engineering, Universitas Indonesia, Depok 16424, Indonesia
[2]Department of Chemical Engineering, Faculty of Engineering, University of Riau, Pekanbaru 28293, Indonesia

*E-mail: kmulia@che.ui.ac.id

Abstract. Green solvents are widely used to minimize environmental problems associated with the use of volatile organic solvents in many industries. DES are new green solvents in recent. The physicochemical properties of DES can be varied by properly combining of salts with different hydrogen bond donors. The objective of this work is to investigate the effect of varying molar ratios on the physicochemical properties of betaine monohydrate-carboxylic acid (*i.e*,. propionic or acetic acid) mixtures. Properties of mixtures were measured at 40°C. The viscosity, polarity scale (E^{NR}), density, pH, and water content tend to decrease with the decrease in a molar ratio of betaine monohydrate to acid. Conversely, the ionic conductivity was increased. The physicochemical properties of these mixtures depend on the hydrogen bonding interactions between betaine, water and acid molecules. Betaine monohydrate-carboxylic acid mixtures have wide range of polarity, low viscosity, high ionic conductivity, and density higher than 1 $g{\cdot}cm^{-3}$ that make them fit for numerous various applications. Additionally, due to these mixtures have acidic pH, it should be properly selected of metal type to minimize corrosion problems in industrial application.

1. Introduction

Deep eutectic solvents (DES) are new green solvents that have attractive physicochemical properties, such as non-toxic, non-volatility, non-flammability, thermal stability, non-flammability, high conductivity, and good solubilizing capacity [1]. DES are formed by mixing a quaternary ammonium salt (such as choline chloride, betaine) with hydrogen bond donor (HBD) compound, such as an alcohol, organic acid, sugar or amino acid [2]. DES were applied in catalysis, CO_2 removal, and metal ion adsorption, and extraction of organic and inorganic compounds [3-7].

The physicochemical properties of DES depend on the molar ratio of salt to HBD, and type of HBD. For instance, the density of choline chloride-fructose based DES, and choline chloride-glucose based DES tend to decrease with increase in concentration of salt in the mixture (in the range of choline chloride to sugar molar ratio of 1:1 to 2.5:1). At the same temperature and molar ratio of salt to sugar, the densities of choline chloride-fructose based DES are higher than choline chloride-glucose based DES [8,9]. Zahrina, et al. [10] reported that the mixture of betaine monohydrate and organic acid (propionic or acetic acid) form eutectic at a mole ratio of 1:2. The thermodynamic study revealed that interactions between betaine monohydrate and propionic acid mixture were greater than between betaine monohydrate and acetic acid mixture at eutectic. In addition, the interactions between betaine monohydrate and acid at the eutectic composition were higher than those at the hyper-eutectic.

The use of choline chloride-alcohol based DES as the solvents for extraction process shown that the extraction efficiency increase with the addition of alcohol concentration in the mixture [7,11]. The reasons for this phenomenon were probably related to physicochemical property of DES (viscosity,

Content from this work may be used under the terms of the Creative Commons Attribution 3.0 licence. Any further distribution of this work must maintain attribution to the author(s) and the title of the work, journal citation and DOI.
Published under licence by IOP Publishing Ltd

surface tension, and basicity) and the interactions of hydrogen bonding between DES and solutes. The effect of diffusion and mass transfer improved with the increase in the polyol ratio in mixture due to the viscosity and surface tension decrease. However, a further decrease in the ratio of choline chloride can reduce the interactions between solutes and chloride anion [11].

The physicochemical properties of mixture of betaine monohydrate and carboxylic acid (propionic or acetic acid) at eutectic and hyper-eutectic should be studied in order to be utilized for several possible industrial applications. In this work, the effect of varying mole ratios on physicochemical properties, and the relationship between physicochemical properties and interaction between molecules in betaine monohydrate-carboxylic acid mixtures were investigated.

2. Materials and Methods

2.1. Materials

The synthesis grade of propionic acid and acetic acid (glacial) were purchased from Merck. Betaine monohydrate and nile red (dye) were purchased from Sigma Aldrich.

2.2. Preparation of betaine monohydrate-carboxylic acid mixtures

The betaine monohydrate and carboxylic acid (propionic or acetic acid) was mixed at different molar ratios. The mixture was placed in a bottle with a screw-cap and stirring bar. The mixture was heated in a hotplate stirrer (IKA C-MAG HS7) at 50±1°C with agitation of 150 rpm for 90 min.

2.3. Determination of physicochemical properties of mixtures

The viscosity and density of betaine monohydrate-carboxylic acid mixtures were measured at 40°C by triplicate runs. The viscosity was performed according to the procedure in ASTM D445 using a viscometer type U-tube reverse (Cannon-Fenske) in a water-bath with controlled temperature. The density was measured using a calibrated density bottle [IP Method 733].

The procedure in polarity scale (E_{NR}) is following the previous work [2]. Polarity scale measurements were carried out using a Spectroquant Pharo 300 model spectrophotometer. The pH of each mixture was measured by EcoTest pH2. The water content in mixtures was measured by titration method using MKS-1s Karl-Fischer moisture meter.

2.4. NMR analysis

The NMR spectra of NOESY (^1H-^1H-nuclear overhauser enhancement spectroscopy correlation) and COSY (spectroscopy correlation) of betaine monohydrate-acid mixtures were recorded with a JEOL JNMECA 500 NMR spectrometer using D_2O as solvent.

3. Results and discussion

The polarity and viscosity are important properties for solvent which corresponding with their solubilizing ability for solutes [12]. As can be seen in Figure 1a, the polarity scale (E^{NR}) of betaine monohydrate-propionic acid mixture is higher than betaine monohydrate-acetic acid mixture at the same molar ratio. Methanol has the polarity scale in the range of betaine monohydrate-propionic acid mixtures. The betaine monohydrate-acetic acid mixtures have higher polarity scale than water, and lower than methanol. The betaine monohydrate-acetic acid mixture inferred has higher hydrogen bonding ability and thus lower polarity scale (or higher polarity) than the betaine monohydrate-propionic acid mixture. Generally, the polarity of DES rises from the hydrogen bonding ability of the involved hydrogen bond donor. While upon the salt interact with the hydrogen bond donor, partially offset by a loss of hydrogen-donor ability due to some of the hydrogen bond donor groups interact with the ions [13]. In this work, if the molar ratio of salt to acid decrease thus the hydrogen bonding ability also increased and thus leading to increase in the polarity (or decrease in polarity scale). Similar result also reported by Abbot et al. [14] that the polarity scale $E_T(30)$ (using Reichardt's dye) of choline chloride-glycerol DES decrease with an increase in glycerol concentration.

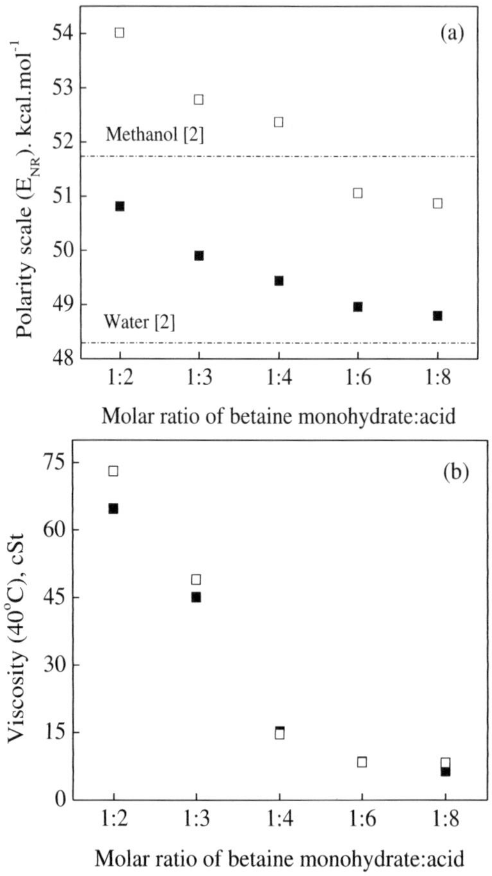

Figure 1. Polarity scale, E^{NR} (a) and viscosity (b) of betaine monohydrate-propionic acid mixtures (□) and betaine monohydrate-acetic acid mixtures (■)

The viscosity of betaine monohydrate-carboxylic acid mixtures in the range of 6.4–73 cSt at a temperature of 40°C (Figure 1b). The low viscosity value makes these mixtures fit for numerous various applications including as the solvents for the extraction processes. The viscosities of betaine monohydrate-propionic acid mixtures were higher than betaine monohydrate-acetic acid mixtures. However, viscosity of the mixture tends to similar with decrease in the mole ratio of salt to acid. The high viscosity of DES is often attributed to the presence of an extensive hydrogen bond network between salt and hydrogen bond donor which causing the lack of free volume within the DES [14]. The betaine monohydrate-propionic acid mixture has greater stability than betaine monohydrate-acetic acid mixture at the eutectic composition [10]. Therefore, the betaine monohydrate-propionic acid mixture has higher viscosity if compared with the betaine monohydrate-acetic acid mixture at a molar ratio of 1:2. At a molar ratio of 1:3, the excess Gibbs free energy (G^E) of betaine monohydrate-acid mixtures was higher than those at a molar ratio of 1:2 [10]. The presence fewer interactions between components in addition of acid concentration in the mixture, and it cause the viscosity decrease. The NMR spectra of betaine monohydrate-propionic acid mixtures are depicted in Figures 2 and 3.

In the NOESY spectra of betaine monohydrate-propionic acid mixtures, the interactions between methyl group proton of betaine monohydrate and both of the protons of the methyl and methylene groups of propionic acid were present. Also, there are intra-molecular interactions in the acid molecules. However, the interactions between betaine monohydrate and propionic acid are similar at the molar ratios of 1:2 and 1:3 (Figures 2a and 3a). In the COSY spectra of betaine monohydrate-propionic acid mixture at a molar ratio of 1:2 (Figure 2b) revealed that the proton of methyl group interact with H_2O of the betaine monohydrate molecule. However, this interaction is loss at a molar ratio of 1:3 (Figure 3b). The water content in the mixtures tends to significantly decrease with the increase in acid concentration in the mixture (Figure 4), and the hydrogen bonding interactions can be formed between hydroxyl group of water and carbonyl group of acid. The carbonyl group of betaine monohydrate can also interact with the hydroxyl group of acid. In addition, the carbonyl group of acid can also interact with hydroxyl group of acid if the acid concentration increases. Therefore, the hydrogen bonding interactions between salt and acid also tend to decrease and lead to decrease in the viscosity of mixtures.

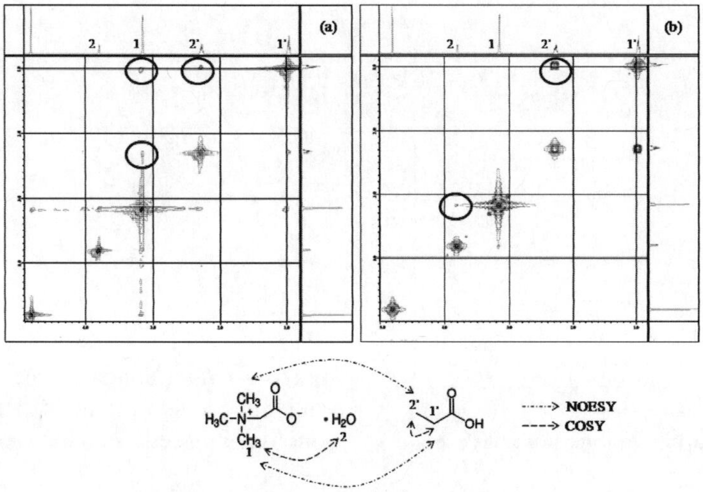

Figure 2. NOESY (a) and COSY (b) spectra of betaine monohydrate-propionic acid at a molar ratio of 1:2

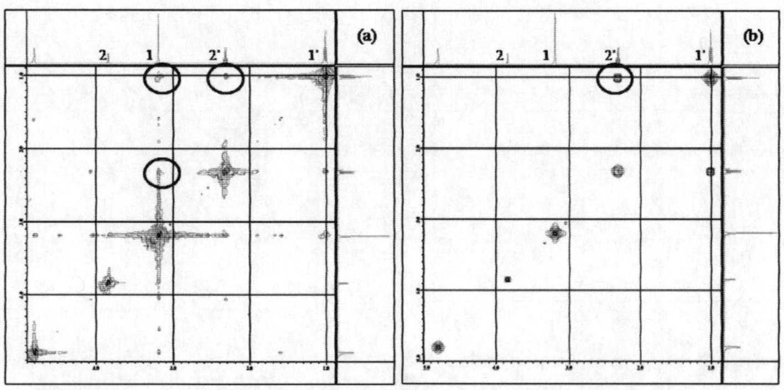

Figure 3. NOESY (a) and COSY (b) spectra of betaine monohydrate-propionic acid at molar ratio of 1:3

The viscosity and conductivity of DES are strongly correlated with the ability of the involved molecules of the salt and hydrogen bond donor to move [15]. Therefore, the relationship between viscosity and ionic conductivity of NADES were analyzed using Walden plots. The ionic conductivities of the betaine monohydrate-acetic acid mixtures were higher than betaine monohydrate-propionic acid mixtures (Figure 5a). The ionic conductivity of betaine monohydrate-acetic acid mixtures tends to linearly increase with the addition in acid concentration. However, the ionic conductivity of betaine monohydrate-propionic acid mixture has slightly increase if mole ratio of salt to acid decrease. The ion mobility and conductivity are determined by not only the availability of suitable holes but also the type and strength of interactions between ion of salt and HBD [15]. Because the betaine monohydrate-propionic acid mixture has stronger interactions between components than betaine monohydrate-acetic acid mixture in the previous study [10], thus the betaine monohydrate-acetic acid mixture has higher ionicity and ability to move. According to the NMR spectroscopic results, the hydrogen bonding interactions between salt and acid decrease induce the ionic conductivity tend to increase with the decrease in molar ratio of salt to acid.

The betaine monohydrate-propionic acid mixtures present larger deviations from the ideal reference line (obtained for 0.01M KCl aqueous solution) than betaine monohydrate-acetic acid mixtures (Figure 5b). This indicates that the betaine monohydrate-propionic acid mixtures have lower ionicity, and thus their viscosity is higher than the betaine monohydrate-acetic acid mixtures. The lower ionicity was also obtained with decrease in a molar ratio of salt to acid whereas the viscosity of mixtures decreases.

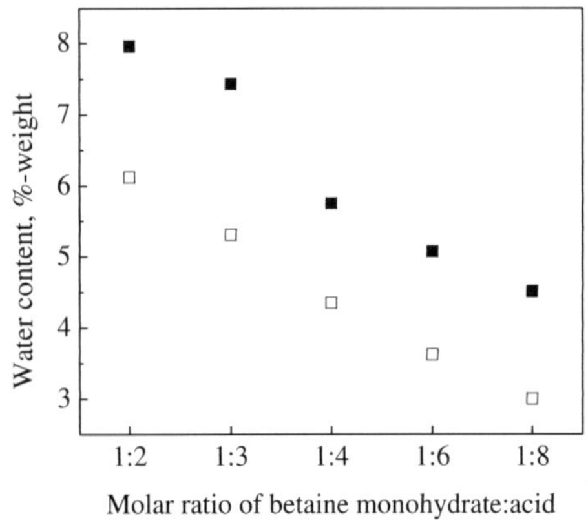

Figure 4. Water content in betaine monohydrate-propionic acid mixtures (□) and betaine monohydrate-acetic acid mixtures (■)

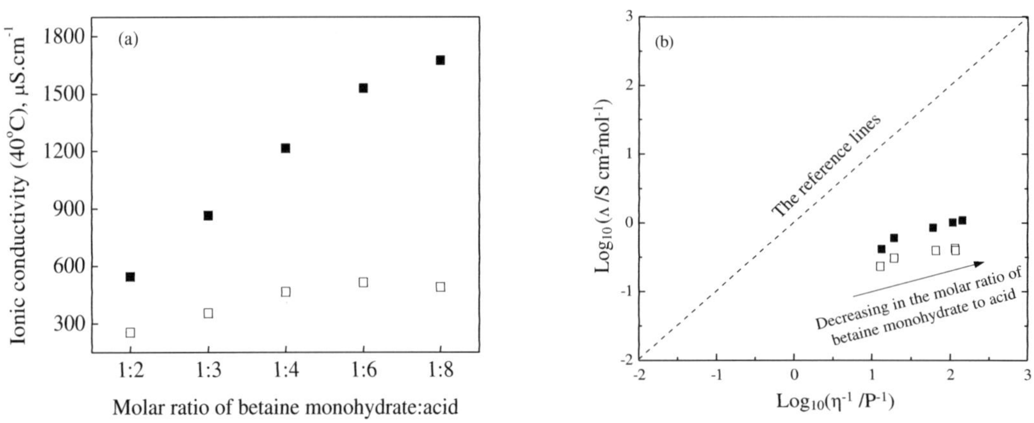

Figure 5. Ionic conductivity (a), and Walden plots (b) of betaine monohydrate-propionic acid mixtures (□) and betaine monohydrate-acetic acid mixtures (■)

The densities of betaine monohydrate-propionic acid mixtures were lower than betaine monohydrate-acetic acid mixtures at the same molar ratio (Figure 6a). This result is indicating that the increase in chain length of acid induce the density decrease. Moreover, the density of these mixtures tends to decrease gradually as the acid concentration increases (or molar ratio of salt/acid decrease). Likewise, the density of DES from choline chloride-diacid mixture decreases with increasing chain length (oxalic acid > malonic acid > glutaric acid) at a molar ratio of salt to acid of 1:1 [16]. The molecular characteristics of the HBD and the mole ratio of salt to HBD have a large effect on the density of DES [9]. According to the hole theory (hard sphere model), the density of DES is affected by free volume [14]. The NMR spectroscopic result revealed that the hydrogen bonding interactions between salt and acid decrease with the decrease in mole ratio of salt to acid which the free volume can be increased in this mixtures and the density decrease. As presented in Figure 6b, the pH of betaine monohydrate-propionic acid mixtures was higher than the betaine monohydrate-acetic acid mixtures at the same molar ratio of salt to acid. Additionally, the decrease in mole ratio of salt to acid was gradually increase acidity of mixtures.

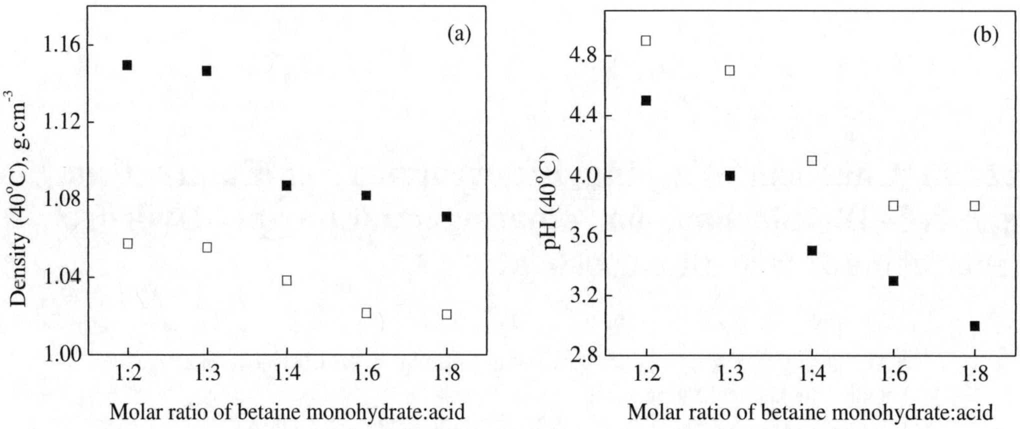

Figure 6. Density (a) and pH (b) of betaine monohydrate-propionic acid mixtures (□) and betaine monohydrate-acetic acid mixtures (■)

4. Conclusions

The physicochemical properties of betaine monohydrate-carboxylic acid mixtures depend on molar ratio of salt to acid and type of acid. The hydrogen bonding interactions between components in the mixtures affect their physicochemical properties. The betaine monohydrate-carboxylic acid mixtures have wide range of polarity, low of viscosity, high ionic conductivity, and density higher than water. It makes these mixtures fit for numerous various applications including for the extraction processes. These mixtures have acidic pH (3-4.9), it should be properly selected of metal type to minimize corrosion problems in industrial application.

5. References

[1] Zhang Q, Vigier KDO, Royer S and Jerome F 2012 *Chem. Soc. Rev.* **41** 7108-46
[2] Dai Y, van Spronsen J, Witkamp GJ, Verpoorte R and Choi YH 2013 *Anal. Chim. Acta* **766** 61-68
[3] Gu CD, You YH, Wang XL and Tu JP 2012 *Surf. Coat. Technol.* **209** 117-123
[4] Francisco M, van den Bruinhorst A, Zubeir LF, Peters CJ and Kroon MC 2013 *Fluid Phase Equilibria* **340** 77-84
[5] Rimsza JM and Corrales LR 2012 *Computational and Theoretical Chemistry* **987** 57-61
[6] Bi W, Tian M, Row KH 2013 *J. Chromatogr. A* **1285** 22-30
[7] Cui Q, Peng X, Yao XH, Wei ZF, Luo M, Wang W, Zhao CJ, Fu YJ and Zu YG 2015 *Sep. Purif. Technol.*150 63-72
[8] Hayyan A, Mjalli FS, AlNashef IM, Al-Wahaibi T, Al-Wahaibi YM and Hashim MA 2012 *Thermochim. Acta* **541** 70-75
[9] Hayyan A, Mjalli FS, AlNashef IM, Al-Wahaibi YM, Al-Wahaibi T and Hashim MA 2013 *J. Mol. Liq.* **178** 137-141
[10] Zahrina I, Nasikin M and Mulia K 2017 *J. Mol. Liq.* **225** 446-450
[11] Peng X, Duan MH, Yao XH, Zhang YH, Zhao CJ and Zu YG 2016 *Sep. Purif. Technol.* **157** 249-257
[12] Dai Y, van Spronsen J, Witkamp GJ, Verpoorte R and Choi YH 2013 *Journal of Natural Products* **76** 2162-73
[13] Pandey R, Rai R, Pal M and Pandey S 2014 *Phys Chem Chem Phys.* **16** 1559-68
[14] Abbot AP, Harris RC, Ryder K, D'Agostino C, Gladden LF and Mantle MD 2011 *Green Chemistry* **13** 82-90
[15] Garcia G, Aparicio S, Ullah R and Atilhan M 2015 *Energy & Fuels* **29** 2616-44
[16] Florindo C, Oliveira FS, Rebelo LPN, Fernandes AM and Marrucho IM 2014 *ACS Sustainable Chemistry & Engineering* **2** 2416-25

IOP Conf. Series: Materials Science and Engineering **358** (2018) 012018 doi:10.1088/1757-899X/358/1/012018

Effect of Chemical Washing Pre-treatment of Empty Fruit Bunch (EFB) biochar on Characterization of Hydrogel Biochar composite as Bioadsorbent

N H Meri, A B Alias, N Talib, Z A Rashid and W A Wan Ab Karim Ghani
Faculty of Chemical Engineering
Universiti Teknologi MARA, 40450 Shah Alam, Selangor, Malaysia

norhidayahmeri@yahoo.com

Abstract. Hydrogel biochar composite (HBC) is a recent interest among researchers because of the hydrophilic characteristic which can adsorb chemical fluid and showed a versatile potential as adsorbent in removing hazardous material in wastewater and gas stream. In this study, the effect of chemical washing pre-treatment by using two different type of chemical agent Hydrochloric Acid (HCL) and Hydrogen Peroxide (H_2O_2) was analysed and investigated. The raw EFB biochar was prepared using microwave assisted pyrolysis under 1000W for 30 min under N2 flow with 150 mL/min. To improve the adsoprtion ability, the EFB biochar has been chemical washed pre-treatment with Hydrochloric Acid (HCl) and Hydrogen Peroxide (H_2O_2) before polymerization process with acrylamide (AAm) as monomer, N,N'-methylenebisacrylamide (MBA) as crosslinker and ammonium persulfate (APS) as initiator. The characterization has studied by using Fourier transform infrared spectroscopy (FTIR) and Differential Scanning Calorimetry (DSC). FTIR result shows that, the formation of Raw EFB to Hydrogel Biochar Composite (Raw EFB > EFB Biochar > Treated Biochars (HCl & H_2O_2) > Hydrogel Biochar Composite) have changed in functional group. For DSC result it shows that the thermal behaviour of all samples is endothermic process and have high thermal resistance.

1. Introduction

Palm oil industry is the largest agricultural industry in Malaysia aligned with wastes and residues produced and became the major disposal problems [1]. Palm fibre (PF), palm kernel (PK), Empty Fruit Bunch (EFB), and palm oil mill effluent (POME) are the biomass produced from oil palm industry [2]. It is a good oportunity to produce more beneficial product from the agriculture waste such as biochar which is high potential in adsorption sector. Some methods were used to produced biochar they are included the fixed bed pyrolyser, carbonizer and furnace [3]. But in recent years, researchers found a new alternative technique applied for pyrolysis in producing biochar with high efficiency, energy saving, selective, no pollution and also friendly used which is by using microwave assisted pyrolysis [4]. Microwave is very effective equipment in producing chars product from biomass such as Empty Fruit Bunch or materials have high moisture content because microwave mechanism in particle heating where the electromagnetic field will penetrates the solid and interacts directly with dipoles in the chemical structure to vaporize the moisture [5]. On the other hands, Hydrogel Biochar Composite (HBC) is recent findings combination of biochar and hydrogel polymer. Hydrogel is a unique material in terms of physical and chemical properties because it is hydrophilic, swellable and modifiable composite which is good in adsorption process [6]. Then the aim of this study is to enhance the performance of biochar as bioadsorbent by chemical washing pre-treatment with Hydrochloride Acid (HCl) and Hydrogen Peroxide (H_2O_2) in order to remove the impurities and enlarge the porosity. Uses

Content from this work may be used under the terms of the Creative Commons Attribution 3.0 licence. Any further distribution of this work must maintain attribution to the author(s) and the title of the work, journal citation and DOI.
Published under licence by IOP Publishing Ltd

of dilute acid such as 0.1M HCl as chemical agent because the dilute HCl will not damage the lignocellulose [7].

Meanwhile, H_2O_2 is an oxidizing agent which have potential to bleach and in the same time will remove impurities and toxical element [8]. Then two types of Hydrogel Biochar Composite (HBC) were produced, which are the biochar were pre-treated with HCl (EFB-HBC H100) and the other one is biochar were pre-treated with H_2O_2 (EFB-HBC P100). This paper are going to compare the characteristic of EFB-HBC H100 and EFB_HBC P100. As previous sturdy, experiment by Sanyang et al. [9] was successful with 99% and 35.75 mg/g sorption capacity of zinc removal in wastewater by modified RH-HBC (hydrogel biochar composite form rice husk). Earlier researcher, Karakoryun et al. [10] has conclude that, hydrogel biochar composites has great application in removing of toxic organic species such as phenols from wastewater. As expected, the HBC will apply in industrial waste treatment for example in removing hazardous gas with high temperature. This study will compare and investigate in terms of their functional group and thermal behaviour in production of HBC.

2. Material and Method

2.1. Raw Material & Chemical

Empty Fruit Bunch (EFB) has chosen as raw material and obtained from one of the palm oil industry located in Banting, Selangor. The chemical washing agent for this study are 0.1M of dilute Hydrochloric acid (HCl) and 0.1M Hydrogen Peroxide (H_2O_2) for biochar washing pre-treatment and distilled water is used as diluting agent. Other chemical for HBC synthesize such as Acrylamide (AAm) as monomer, N,N′-methylenebisacrylamide (MBA) as cross-linker and ammonium persulfate (APS) as initiator are supplied from R&M chemical.

2.2. Biochar Production

Microwave assisted pyrolysis technique has applied in producing EFB biochar in this study. 200g of raw Empty Fruit Bunch (EFB) were placed in quartz reactor in the microwave pyrolyzer at 1000 W of microwave power level 30 minutes under nitrogen flow at 150 mL/min. The method to prepared EFB char by using microwave assisted pyrolysis is referring to Zakiuddin Januri et al. [5] with some modification.

2.3. Acid Washing Pre-treatment

The acid washing experiments were carried out using Hydrochloric Acid (HCl) and Hydrogen Peroxide H_2O_2 [10][11]. 2 L of diluted 0.1 M HCl and 0.1 M H_2O_2 solutions were prepared in 2 L volumetric flask for stock chemical solution. Then 10g of EFB biochar were pre-treated with 200 mL of prepared acid solutions in a closed beaker for 6 hrs. Finally, the biochar were washed with distilled water until a neutral pH is obtained and oven dried at 80 ˚C overnights. Then pre-treated biochar named as biochar H100 and biochar P100 for HCl and H_2O_2 solutions respectively.

2.4. Hydrogel Biochar Production

EFB-HBC were synthesized using 1.0 g of AAm was desolved in 1.0 mL of distilled water. Then, 0.6 g of Empty Fruit Bunch (EFB) biochar and 0.001 g of MBA were added to the AAm solution. After thorough mixing, 0.2 mL of 0.1 g aqueous solution of APS was added to initiate the polymerization. The hydrogel biochar precursor solution was immediately placed into plastic mould and placed in an oven at 40 ºC for 30 min. The plastic mould containing hydrogel biochar composites were removed from the oven after 30 min and left for 24 hrs at room temperature to ensure complete polymerization and crosslinking. EFB-HB was taken from the plastic mould, cut to desired sizes and washed several times with distilled water to remove all unreacted monomers and low molecular weight polymeric matter from the hydrogel. The washed EFB-HB was first dried in air before drying in a vacuum oven at 40 ˚C for 24 hrs. Then, hydrogel biochar named as EFB-HB H100 and EFB-HB P100. This hydrogel formation method are following to procedure used by [10].

2.5. Characterization

Fourier transform infrared spectroscopy (FTIR) will be measuring and analysing the functional group of samples. This instrument also can be applied to investigate the structural changes of biochars as a function of different sample preparation (pre-treatment process). In this research, FTIR will be performed by using Perkin – Elmer Spectrum 2000 FTIR with the wavelength 4000 – 600 cm^{-1}[12]. DSC is a thermal analysis technique that shows the relation between material's heat capacity changes by temperature or in other word is to determine the heat required to increase temperature of sample. This instrument operated with sample weight ranging 5 to 10 mg in powder form heating from 30°C to 300°C at a heating rate of 10°C/min [13].

3. Result and Discussion

3.1. Differential Scanning Calorimetry Analysis (DSC)

Figure 1 shows the DSC curves for all samples (Raw EFB, EFB biochar, Biochar H100, Biochar P100, EFB-HBC H100 and EFB-HBC P100).

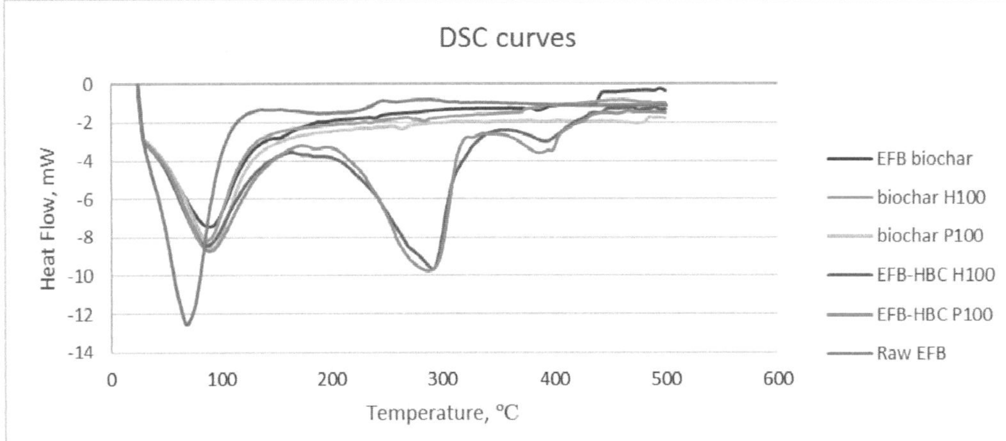

Figure 1. DSC Curves for Raw EFB, EFB biochar, biochar H100, biochar P100, EFB-HBC H100 and EFB-HBC P100

As shown in Figure 1, DSC curve for raw EFB showed a high endothermic peak that represented as the glass transition which exhibits enthalpy relaxation [14]. Because of the raw EFB only undergo a glass transition, the raw EFB were classified as amorphous substances [15]. The same result has shown for biochars (Biochar, Biochar H100 and Biochar P100) which also undergo the glass transition. Besides, comparing in glass transition temperature, from raw EFB pyrolysed became biochars, it shows that the peaks are shifted to the right from T_g = 68 °C to T_g = 89 °C which mean it experienced some changes after pyrolysis (EFB biochar) and increasing the thermal transition. As seen from DSC curves for treated biochars (biochar H100 and biochar P100) are maintained at the same peak as biochar. This can be concluded that, in thermal reaction, the pretreated biochars (biochar H100 and biochar P100) has nothing much different from its biochar. This is because of, high amount of volatile matter was released during pyrolysis which make the material less reactive. Then the washing pre-treatment step is just to eliminate the remaining of volatile matter inside the biochar.

Moreover, three consecutive endothermic curves which means substances absorbed heat are observed for hydrogel biochars. Both of the hydrogel biochar composites (EHB-HBC H100 and EFB-HBC P100) have a quite similar thermal behaviour, which means both of HBCs have approximately same glass transition temperature, T_g around 86°C to 89°C and they start to degrade around 200°C. Another two curves are represented as the decomposition of substances. Michalak et al. [16] stated on their study if there is two peak existing in DSC curves after transition phase means, they found that it lose crystallization water in two steps. Then, in this study, DSC curves shows that both of HBCs lose crystallization water in two steps at T_{m1}, 298°C and T_{m2}, 400°C. In addition, the present of double endothermic peak conclude that HBCs is enantiotropic and polymorphic compound [17].

Polymorphism is a single substances that exhibits multiple different crystal structures or in other words they have various melting phase transitions [18].

3.2. Fourier transform infrared spectroscopy (FTIR)

3.2.1. FTIR Spectrum (From the Top: Raw EFB, EFB-HBC H100, Biochar H100, EFB biochar)
FTIR were used to identify the changes of functional groups before and after chemical treatment and also in Hydrogel Biochar Composites (HBCs) form [19]. Figure 2 shows FTIR spectrum for the first part of the research.

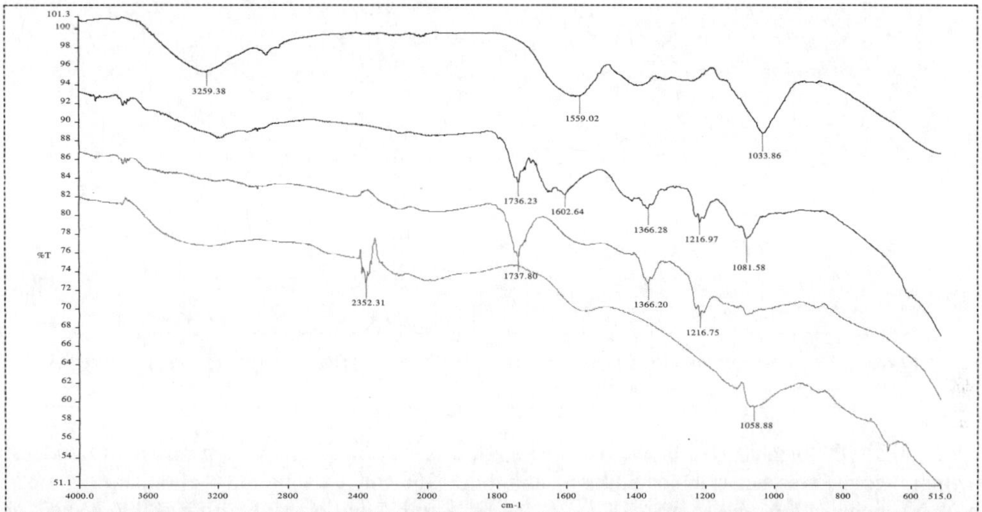

Figure 2. FTIR Spectrum (From the Top: Raw EFB, EFB-HBC H100, Biochar H100, EFB biochar)

For the first part of this research are producing EFB-HBC H100 which is from Raw EFB > EFB Biochar > Biochar H100 > EFB-HBC H100. The spectrum of raw EFB shows a broad characteristic peak at 3259.38 cm^{-1}, which indicates the presence of O–H stretching. Same result shows in Claoston et al., [20] which recognized the stretching functional group of hydrogen-bonded hydroxyl group (O-H), indicative presence as phenols and alcohols. After pyrolysis process in producing EFB biochar, the O-H groups were diminished because of the mass loss during thermal decomposition. The same result are shown for biochar that was treated by HCl (Biochar H100) which is the O-H stretching were more diminished because some of volatile matter are released during washing pre-treatment. Stretching vibration of O=C=O (CO_2) with wavelength 2352.1 cm^{-1} was observed. The presence of CO_2 in EFB biochar is from the pyrolysis process which is generally produced gaseous product rich in CO_2. In addition, CO_2 was removed and diminished after treated with HCl (Biochar H100) and shown on the FTIR spectrum for Biochar H100. The diminished of CO_2 can be conclude that, the pre-treatment is able to remove the toxical material such as CO_2 as mention by Samy Sadaka [7].

According to Asim et al., [19], peaks are around 1700 cm^{-1} indicated the presence of holocellulose and lignin which has shown for Biochar H100 and EFB-HBC H100 with wavelength 1737.80 cm^{-1} and 1736.23 cm^{-1} the respectively. This means is, by formation of hydrogel biochar composite, does not destroy the lignin which is important to structure the HBC. For EFB biochar and EFB-HBC H100 have a same bending vibration on 1366 cm^{-1} represent as existing the C-H (alkane) group and a same stretching vibration on 1216 cm^{-1} shows the existing of C-O group. After the polymerization process become hydrogel biochar, have some of changes in functional group with addition of C=C stretching and amine C-N stretching on 1602 cm^{-1} and 1081 cm^{-1} respectively. This is because of the addition Acrylamide (AAm) as monomer, N,N′-methylenebisacrylamide (MBA) as cross-linker in polymerization process.

3.2.2 FTIR Spectrum (From the Top: Raw EFB, EFB-HBC P100, Biochar P100, EFB Biochar)
Then, Figure 3 shows the FTIR spectrum for Raw EFB, EFB-HBC P100, Biochar P100 and EFB biochar. Based on FTIR spectrum in Figure 3, it shows that, the peaks of EFB-HBC P100 show that there is low transmittance, and hence large absorption ability compared to EFB-HBC H100.

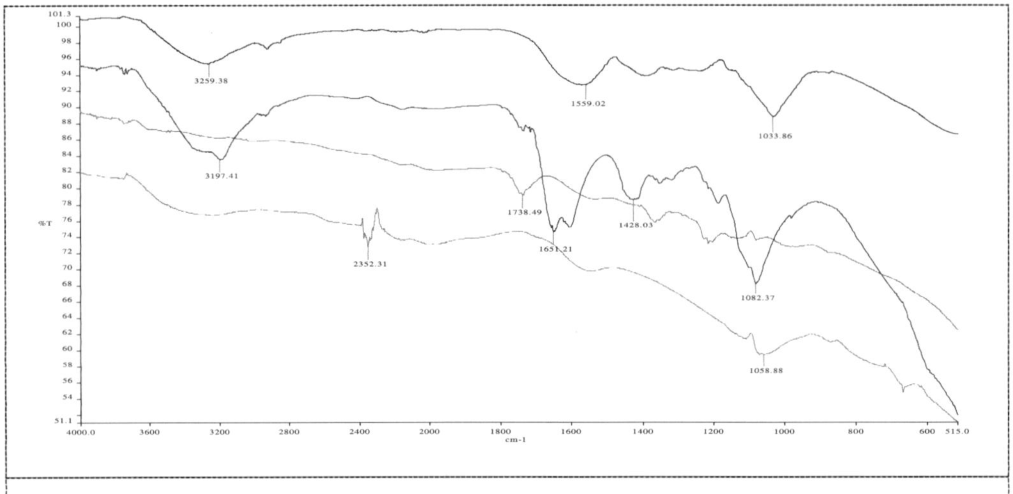

Figure 3. FTIR Spectrum (From the Top: Raw EFB, EFB-HBC P100, Biochar P100, EFB Biochar)

As we can see the biochar H100 have a same patern or spectrum which have peaks are around 1700 cm^{-1} indicated the presence of holocellulose and lignin, 1366 cm^{-1} existing the C-H (alkane) group and a peak are around 1216 cm^{-1} shows existing of C-O group. In addition, CO_2 also was diminished after treated with H_2O_2 (Biochar P100). After polymerization become EFB-HBC P100, peak around 3197.41 represented as existing phenols and alcohols in the structure. Same as EFB-HBC H100, after the polymerization process, have some of changes in functional group with addition of C=C stretching and primary amine C-N stretching on 1651 cm^{-1} and 1082 cm^{-1} respectively. The FTIR spectrum of all sample from raw EFB to HBCs are summarized in Table 1.

Table 1. Functional Group Observed in FTIR spectra for Raw EFB, EFB biochar, Biochar H100, Biochar P100, EFB-HBC H100 and EFB-HBC P100

Wave number range (cm^{-1})	Raw EFB (cm^{-1})	EFB Biochar (cm^{-1})	Biochar H100 (cm^{-1})	Biochar P100 (cm^{-1})	EFB-HBC H100 (cm^{-1})	EFB-HBC P100 (cm^{-1})	Vibration Characteristic
3550-3200	3259.38	3236	-	-	3199	3197.41	O-H stretching (alcohol& phenols)
2349	-	2352. 10	-	-	-	-	O=C=O stretching (Carbon dioxide)
1750-1735	-	-	1737.80	1738.49	1736.23	1739	C=O stretching (ester)
1650-1600	1559.02	-	-	-	1602.64	1651.21	C=C stretching (conjugate

							alkene)
1440-1395	-	-	-	-	-	1428.03	O-H bending (carboxylic acid)
1370-1365	-	-	1366.28	1366	1366.20	-	C-H bending (alkane)
1275-1200	-	-	1216.75	1216.75	1216.97	1210	C-O stretching (alkyl aryl ether)
1250-1020	-	1058.88	-	-	1081.58	1082.37	C-N stretching (amine)

4. Conclusion

Hydrogel Biochar Composites (HBCs) is a new finding in adsorption technology. By treating the EFB biochar using HCl and H_2O_2 shows the changes in functional group. However, nothing to compare between biochar H100 and biochar P100 because the role of chemical agents is same which is to removed toxical material. The functional group was changed because of the diminished of CO_2 from its biochar in both washing pre-treatment step HCl. The Differential Scanning Calorimetry (DSC) shows that Hydrogel biochar are enantiotropic and polymorphic compound with various melting transitions. In addition, the HBCs can be exposed to high temperature to the high degradation temperature. However, the DSC results for treated biochar (biochar H100 and biochar P100) has not much different compared to biochar. This is because the biochar itself has loss reactivity from pyrolysis process. Finally, from the FTIR spectra, the peaks of EFB-HBC P100 shows there is low transmittance, and hence large absorption ability compared to EFB-HBC H100.

Acknowledgement

This research was supported by a grant 600-RMI/FRGS 5/3 (96/2014). We would like to thank to Ministry of Higher Education (MOHE), Universiti Teknologi Mara (UiTM) and Faculty of Chemical Engineering for the grant sponsorship, facilities and motivation.

References

[1] Shariff A, Syairah N, Aziz M and Abdullah N 2014 vol. **25**, no. 2, pp 97–112, 2014
[2] Abdullah N and Sulaiman F 2013 *Biomass Now – Sustain. Growth Use*, pp 75–100
[3] Ivb I 2013 pp. 0–13
[4] Abas F Z and Nasir Ani F 2014 *J. Teknol.*, vol. **3**, pp 33–37
[5] Januri Z, Abd Rahman N, Idris S S, Matali S and Manaf S F A 2014 *J. Teknol.*, vol. **5**, no. 8, pp. 103–109
[6] Okay O 2010 *Hydrogel Sensors and Actuators*, vol. **6**, pp 1–15
[7] Samy Sadaka, P.E. 2008 Pyrolysis, Iowa State University, 2008
[8] Diaz A, Le Toullec J, Blandino A, De Ory I and Caro I 2013 *Chem. Eng. Trans.* vol. **32**, pp 949–954
[9] Sanyang L, Ghani W A K, Idris A and Ahmad M 2014 *Appl. Mech. Mater.* vol. **625** pp 842–846
[10] Karakoyun N, Kubilay S, Aktas N, Turhan O, Kasimoglu M, Yilmaz S and Sahiner N 2011 *Desalination*, vol. **280**, no. 1–3, pp 319–325
[11] Yakout S M, El A, Daifullah H M and A El-reefy S 2015 vol. **14**, no. 2, pp 473–480
[12] Liu Y, He Z and Uchimiya M 2015 vol. **9**, no. 4, pp 246–253.
[13] Novotny E H 2013 pp 227248
[14] Toledo M 2018 *Mettler Toledo* pp 1–40
[15] Shawe J, Riesen R, Widmann J and Schubnell M 2000 *Mettler Toledo*, pp 1–28
[16] Michalak O, Łaszcz M, Jatczak K, Witkowska A, Bujak I, Groman A and Cybulski M 2015 *Molecules*, vol. **20**, no. 8 pp 13814–13829

[17] Moriwaki C, Costa G L, Ferracini C N, De Moraes F F, Zanin G M, Pineda E A G and Matioli G 2008 *Brazilian J. Chem. Eng.* vol. **25**, no. 2, pp 255–267
[18] Namatame Y and Sato H 2013 *Rigaku J.*, vol. **29**, no. 3, pp 8–15
[19] Asim M, Abdan K, Jawaid M, Nasir M, Dashtizadeh Z, Ishak M R and Hoque M E 2015 *Int. J. Polym. Sci.*, vol. **2015**, pp 1–17
[20] Claoston N, Samsuri A, Ahmad Husni M and Mohd Amran M 2014 *Waste Manag. Res.*, vol. **32**, no. 4, pp 331–9

IOP Conf. Series: Materials Science and Engineering **358** (2018) 012019 doi:10.1088/1757-899X/358/1/012019

Design and Analysis of Photovoltaic (PV) Power Plant at Different Locations in Malaysia

M A Islam*[1,2], M Hasanuzzaman[1] and N A Rahim [1,3]

[1] UM Power Energy Dedicated Advanced Center, University of Malaya, MALAYSIA.
[2]Institute of Graduate Studies, University of Malaya, MALAYSIA.
[3]Renewable Energy Research Group, King Abdulaziz University, SAUDI ARABIA

aminulmse@gmail.com, hasan@um.edu.my, nasrudin@um.edu.my

Abstract. Power generation from sun oriented vitality through a photovoltaic (PV) system is ended up prevalent over the world due to clean innovation. Geographical location of Malaysia is very favorable for PV power generation system. The Malaysian government has also taken different steps to increase the use of solar energy especially by emphasizing on building integrated PV (BIPV) system. Comparative study on the feasibility of BIPV installation at the different location of Malaysia is rarely found. On the other hand, solar cell temperature has a negative impact on the electricity generation. So in this study cost effectiveness and initial investment cost of building integrated grid connected solar PV power plant in different regions of Malaysia have been carried. The effect of PV solar cell temperature on the payback period (PBP) is also investigated. Highest PBP is 12.38 years at Selangor and lowest PBP is 9.70 years at Sabah (Kota Kinabalu). Solar cell temperature significantly increases the PBP of PV plant and highest 14.64% and lowest 13.20% raise of PBP are encountered at Penang and Sarawak respectively.

1. Introduction

Energy is the basic needs of various types of industrial facilities development, economic growth, automotive and modernization in the social development around the world. So the development of a country directly depends on its capacity of energy harvesting. Currently, most extensively used power generation system is based on fossil fuel burning, which produces a huge amount of carbon dioxide emission. Moreover, the nonrenewable energy sources are limited and the amount of stock is declining day by day. So for sustainable energy harvesting, people's propensity is going to the renewable sources of lowest environmental pollution. Among the renewable sources of energy, solar energy is one of most abundant and largest carbon-free energy source and hourly near about 6.3×1020J energy strikes the Earth as sunlight. Electricity generation from the sunlight by utilizing photovoltaic (PV) system is becoming an encouraging over the world due to clean technology. Likewise, other developed country, the Malaysian government has also started a Feed in tariff (FIT) program since 2004. Geographical location of Malaysia is very favorable for PV power generation system. That's why different international companies are invested to build up both solar cell and module manufacturing industries in Malaysia. The authority of the Malaysian government name as Sustainable Energy Development Authority (SEDA) has also taken on many strategies to increase the use of solar energy and they have a target solar will be the main source energy supply by 2050 as shown in Figure 1. However, it has been reported that the total PV electricity production as a % of total electricity consumption was 0.04 % in a year of 2013. Table 1 shows a scenario of PV installation with a year of 2014. So the growth rate of the PV power plant either in small scale kW range (roof top building integrated) or in a large scale MW range is not sufficient in respect of Malaysia due to lack of proper designing and economic analysis. Several reports have been found in literature, such as Sopian et

Content from this work may be used under the terms of the Creative Commons Attribution 3.0 licence. Any further distribution of this work must maintain attribution to the author(s) and the title of the work, journal citation and DOI.
Published under licence by IOP Publishing Ltd

al.[1] have investigated only the performance 5.76 kWp grid-connected PV at UKM, Bangi, Malaysia. Kassim et al. [2] have reported performance and economic effects of 5 MWp solar PV plant installed at Perlis, Malaysia. However authors do not include the effect of temperature in the analysis. Therefore, a detail investigation on the feasibility, cost effectiveness and initial investment cost solar PV power plant at different sectors of Malaysia is necessary. So the objectives of this study are (i) to compare the performance and cost effectiveness of the PV plant located in different places in Malaysia (ii) to investigate the effect of temperature on the PV plant payback period.

Table 1. PV power installed during calendar year 2014 [3].

Type of Connection	Category	MWdc installed in 2014
Grid-connected	Individuals	13.46
	Non-individuals	73.27
Off-grid	Residential	0.15
	Hybrid system	1.587
Total		**88.467**

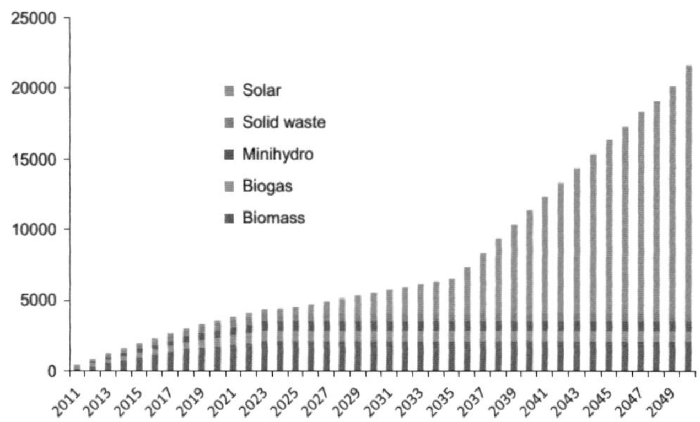

Figure 1. National renewable energy Goals of the Sustainable Energy Development Authority (SEDA) [4].

2. Research methodology

2.1 PV plant design and output estimation

The study has been carried out for the grid connected PV systems having rooftop building integrated (12kW range). Daily average global solar radiation and ambient temperature data has been taken from the Malaysian Meteorological Department and NASA Surface meteorology and Solar Energy [5]. Table 2 Shows average global solar radiation and ambient temperature of different zones of Malaysia. Electricity generation has been calculated according to average global solar radiation and average ambient temperature. 3% losses due to partial shading or dust are encountered in the analysis. PV module cell temperature is calculated by using nominal operating cell temperature (NOCT), solar irradiation and ambient temperature according to the following equation.

$$T_C = T_{amb} + (NOCT - 20) \times \frac{E}{800}$$

Where T_c is cell temperature, T_{amb} is ambient temperature, E is the solar radiation. The above expression is widely used to estimate in a for open rack systems. The roof integrated modules operate at temperatures higher than those in open rack, having NOCT values from 10 to 20 °C above depending on the convective cooling [6]. The power loss due to increase of cell temperature has been calculated based on a polycrystalline type PV module, locally manufactured, brand name of "LB250QM-60". The detail module specification is shown in the Table 3. The NOCT has taken 10°C

IOP Conf. Series: Materials Science and Engineering **358** (2018) 012019 doi:10.1088/1757-899X/358/1/012019

higher than that of the nominal value due to considering a roof top installation. The number of the required PV module is calculated as:

$$\text{No. of module required} = \frac{Total\ watt\ peak\ rating}{PV\ module\ peak\ rated\ out\ put}$$

The inverter is necessary for the grid connected PV power system. Normally the inverter size should be 25-30% higher than that of the watt required [7].

Table 2. Average meteorological data of different are of Malaysia [8].

Area	kWh/kWp/Year	Average daily ambient temperature (°C)
Johor	1171	27
Kelantan (Kota Baharu)	1229	27.3
Kuala Lumpur	1132	27.7
Melaka	1215	27.5
Pahang (Kuantan)	1154	26.9
Penang	1286	27.8
Perak (Ipoh)	1253	27.7
Sabah (Kota Kinabalu)	1369	27.4
Sarawak (Kuching)	1157	26.4
Selangor	1072	27.7
Terengganu	1235	26.8

Table 3. Specifications of LB250QM-60 polycrystalline module.

Place of origin	Malaysia
Brand name	EPV
Materials	Polycrystalline silicon
Number of solar cells	6 x 10 = 60
Maximum power	250W
Open circuit voltage (V_{oc})	37.8V
Short circuit current (I_{sc})	8.73A
Voltage at Pmax. (V_{mpp})	30.6V
Current at Pmax. (I_{mpp})	8.17A
Maximum system voltage	1000V
Maximum reverse current	15A
NOCT	46°C
Temperature coefficient Pmax	-0.39%/°C

2.2 Installation cost and revenue estimation

Sustainable Energy Development Authority (SEDA), Malaysia has reported the PV module and system prices in at different years. The average module and system prices in Malaysia have decreased per year from 2005 to 2015 as shown in the figure 2. At 2005 the module price was 21.39 MR per watt and at 2015 the price reduced to 3.07 MR per watt. The grid connected PV plant system cost includes different sorts such as PV module cost, inverter cost, wiring and racking cost and installation labor cost, etc. An average system cost for a PV plant having capacity up to 12kW have been reported by SEDA as 31.41 and 7.79 MR/W for 2005 and 2015 respectively. The Feed in Tariff (Fit)) rate for solar PV system is shown in table 4 which is effective from 1st January 2016. Activation year of this Fit rate is 21 years from the commencement date. SEDA also has installation capacity boundary limitation as: for individual the installation capacity of solar PV system must be up to and including 12kW and for non-individual the installation capacity is up to and including 30MW [9].

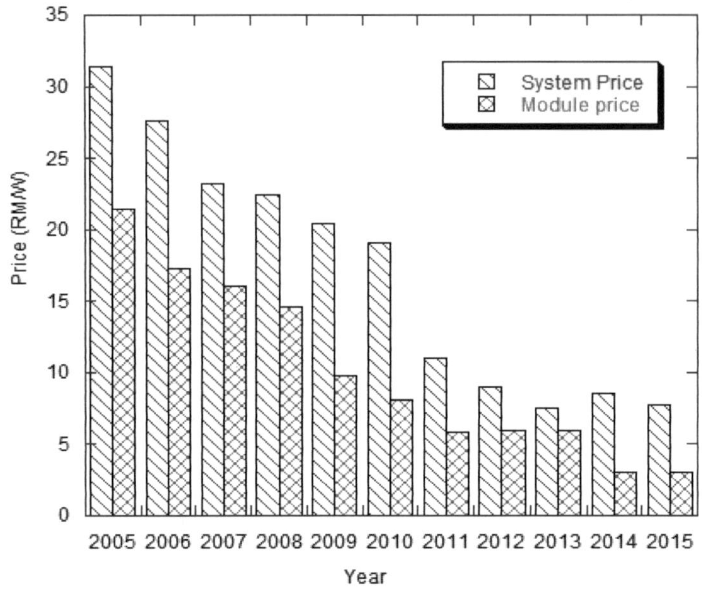

Figure 2. Module and system costs of residential PV system (up to12kW) per watt at different years in Malaysia [3].

Table 4. Fit rate of PV power system provided by Sustainable Energy Development Authority (SEDA) effective from 1st January 2016 [10].

Installed capacity	**Fit rates (RM per kWh)**
Up to and including 4kW	0.8249
Above 4kW and up to and including 24kW	0.8048
Above 24kW and up to and including 72kW	0.6139
Above 72kW and up to and including 1MW	0.5930
Above 1MW and up to and including 10MW	0.4651
Above 10MW and up to and including 30MW	0.4162
Bonus Fit rates having following criteria	
Use as installation in building and building structures	+0.1550
Use as building materials	+0.1325
Use of locally manufactured or assembled solar PV module	+0.0500
Use of locally manufactured or assembled solar inverters	+0.0500

3. Result and discussion

The installation cost of different size PV plants is shown in Figure 3. The total cost for 12 kW BIPV grid connected PV plant installation is near about RM 93480. Electricity production has been varied according to the solar radiation at different places in Malaysia as shown in Figure 4. The ambient temperature reduces the PV module performances and consequently electricity production is hampered. The highest electricity production has been found near about 10960 kWh per year at Sabah and followed by Penang where the value is 10282 kWh/year. The lowest value of electricity production has been observed at Selangor 8587kWh/year. Payback periods (PBP) of the PV power plant at different locations have shown in Figure 5. The temperature has a significant effect on the PBP. The power generation is decreased due to an increase of PV cell temperature as a consequence the PBP is also increased. The highest amount of effect is found at Penang with a figure of 14.64% addition of PBP and the lowest effect is observed at Sarawak with a figure of 13.20% raise of PBP.

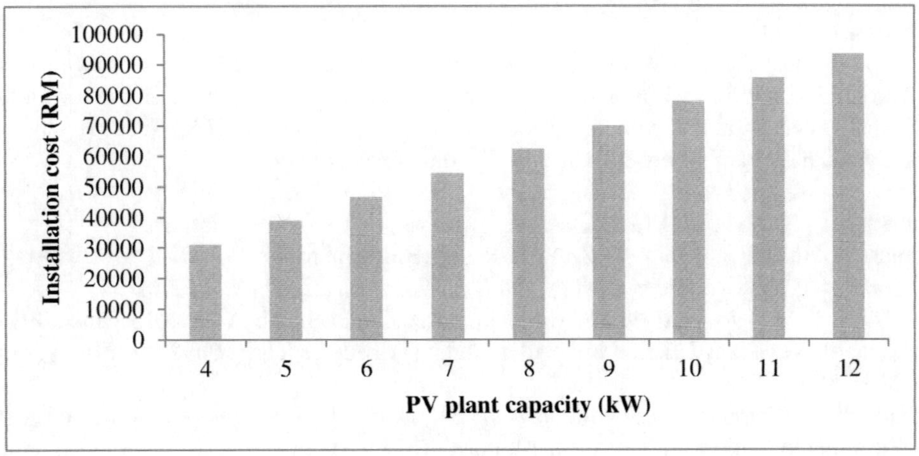

Figure 3. Total installation cost of a grid connected BIPV system up to 12kW in Malaysia

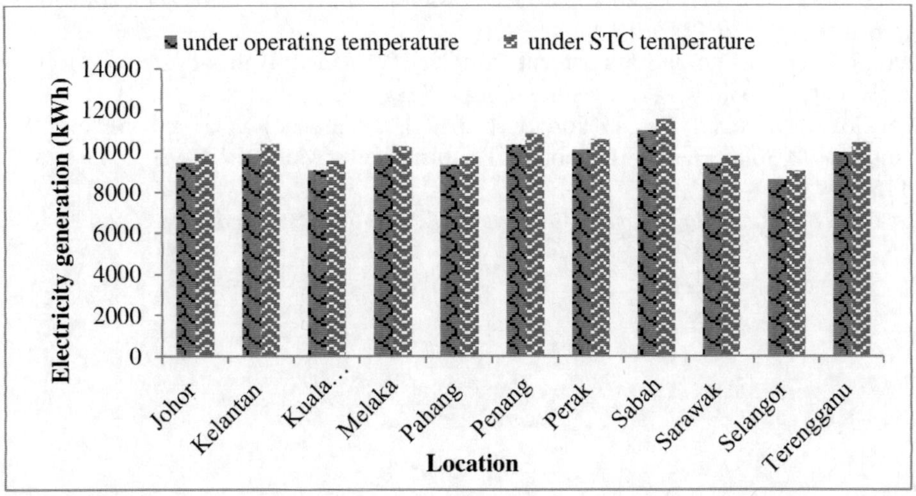

Figure 4. Electricity generation per year from 12 kW PV system at different places in Malaysia

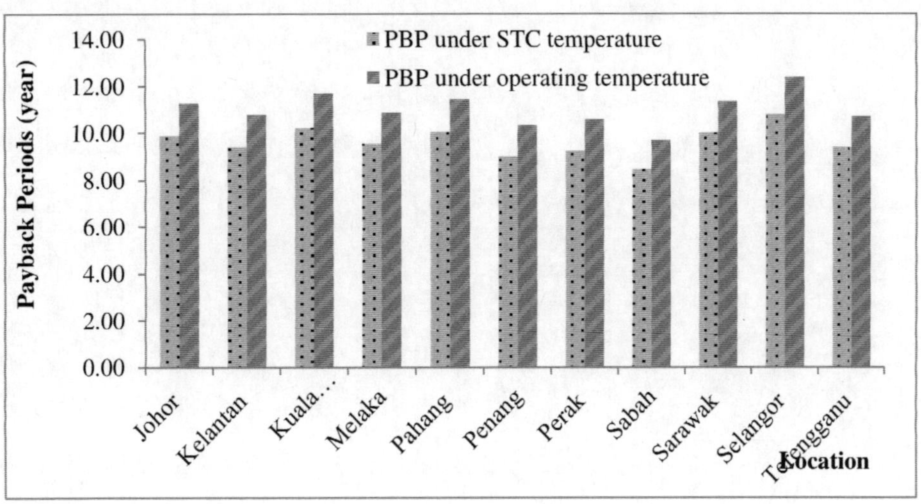

Figure 5. Payback periods of 12kW BIPV system at different places in Malaysia

4. Conclusion

Total installation cost for 12 kWp building integrated grid connected PV system is near about RM 93,480. The highest and lowest amounts of electricity generated per year are in the location of Sabah (Kota Kinabalu) and Selangor with the values of 10,961 kWh and 8587 kWh respectively. Payback periods at Sabah and Selangor are 9.70 and 12.38 years respectively.

References

[1] Sopian K, Shaari S, Amin N, Zulkifli R and Rahman M N A 2007 Performance of a grid-connected photovoltaic system in Malaysia *Int. J. Eng.Technol.* **4** 57-65
[2] Kassim M P M, Al-Obaidi K M, Munaaim M A C and Salleh A M 2015 Feasibility Study on Solar Power Plant Utility Grid under Malaysia Feed-in Tariff *Amer. J. Eng. & Appl. Sci.* **8** 210-22
[3] SEDA 2014 *National Survey Report of PV Power Applications in Malaysia* (Malaysia: Sustainable Energy Development Authority (SEDA)) pp 1-28
[4] Chen W N 2012 *Renewable energy status in Malaysia* (Malaysia: SEDA)
[5] NASA 2016 *Surface meteorology and Solar Energy* (Atmospheric Science Data Center, NASA)
[6] Alonso García M C and Balenzategui J L 2004 Estimation of photovoltaic module yearly temperature and performance based on Nominal Operation Cell Temperature calculations *Renew. Energy* **29** 1997-2010
[7] Khatri R 2016 Design and assessment of solar PV plant for girls hostel (GARGI) of MNIT University, Jaipur city: A case study *Energy Reports* **2** 89-98
[8] Engel-Cox J A, Nair N L and Ford J L 2012 Evaluation of Solar and Meteorological Data Relevant to Solar Energy Technology Performance in Malaysia *J. Sust. Energ. & Environ.* **3** 115
[9] SEDA 2016 *FiT Rates for Solar PV* (Malaysia: SEDA) http://seda.gov.my

IOP Conf. Series: Materials Science and Engineering **358** (2018) 012020 doi:10.1088/1757-899X/358/1/012020

Optimization of Acid Orange 7 Degradation in Heterogeneous Fenton-like Reaction Using Fe_3-xCoxO_4 Catalyst

M Z Ibrahim[1], R Alrozi[1], N A Zubir[1], N A Bashah[1], S A Md Ali[1] and N Ibrahim[2]

[1]Faculty of Chemical Engineering, Universiti Teknologi MARA, Cawangan Pulau Pinang, Kampus Permatang Pauh, 13500 Permatang Pauh, Pulau Pinang, Malaysia
[2]Faculty of Chemical Engineering, Universiti Teknologi MARA, Shah Alam, 40450 Shah Alam, Selangor, Malaysia

rasyidah.alrozi@ppinang.uitm.edu.my

Abstract. The oxidation process such as heterogeneous Fenton and/or Fenton-like reactions is considered as an effective and efficient method for treatment of dye degradation. In this study, the degradation of Acid Orange 7 (AO7) was investigated by using Fe_3-xCoxO_4 as a heterogeneous Fenton-like catalyst. Response surface methodology (RSM) was used to optimize the operational parameters condition and the interaction of two or more parameters. The parameter studies were catalyst dosage (X_1), pH (X_2) and H_2O_2 concentration (X_3) towards AO7 degradation. Based on analysis of variance (ANOVA), the derived quadratic polynomial model was significant whereby the predicted values matched the experimental values with regression coefficient of R^2 = 0.9399. The optimum condition for AO7 degradation was obtained at catalyst dosage of 0.84 g/L, pH of 3 and H_2O_2 concentration of 46.70 mM which resulted in 86.30% removal of AO7 dye. These findings present new insights into the influence of operational parameters in the heterogeneous Fenton-like oxidation of AO7 using Fe_3-xCoxO_4 catalyst.

1. Introduction

Water is important for human daily lives. The main sources of water came from groundwater, river and reservoir. There are many uses of water in industries that give benefit to human lives. Wastewater is any water that has been adversely affected in quality by anthropogenic influence [1]. It needs to be treated to minimize the effect on human's health and environment. The causes of pollution in the water are virtually endless. The major of water pollution is originally from textile industries. Textile manufacturing sector has been considered an important and dynamic component in the economic growth in recent years. Like any other industries, this industry also generates industrial wastes namely liquids, solids and gases. Due contaminated wastewater, dyes are difficult to be decolorized and decomposed biologically due to their complex structure [2]. The discharge of this wastewater into receiving streams causes damage not only to aquatic life but also to human beings. In facts, the dye contaminated wastewater is also considered as toxic or carcinogenic to human being. They stimulate mutagenic effects which causes the errors in DNA replications [3].

According to Su et al. [4], the oxidation process such as heterogeneous Fenton and / or Fenton-like reactions were previously found as an effective and efficient method for treatment of dye degradation. This process used iron oxide (magnetite) as the catalyst and hydrogen peroxide (H_2O_2) being used as an oxidant. One of the major drawbacks of this treatment however is due to its inability to recycle magnetite since this catalyst's activity continuously diminishes after numerous reaction cycles, which leads to unnecessary fresh catalyst required. Therefore, isomorphic substitutions by using a few

Content from this work may be used under the terms of the Creative Commons Attribution 3.0 licence. Any further distribution of this work must maintain attribution to the author(s) and the title of the work, journal citation and DOI.
Published under licence by IOP Publishing Ltd

IOP Conf. Series: Materials Science and Engineering **358** (2018) 012020 doi:10.1088/1757-899X/358/1/012020

transition metal cations have been added to the magnetite structure in order to prolong its catalytic activity for longer periods besides making it more stable [5].

The optimization of analytical procedures has been carried out by using multivariate statistic techniques. Among the most relevant multivariate techniques used in analytical optimization is response surface methodology (RSM). RSM is a collection of mathematical and statistical techniques based on the fit of a polynomial equation to the experimental data, which describe the behaviour of a data set with the objective of making statistical previsions. It can be well applied when a set of interest responses are influenced by several variables. Hence, the objective of this work is to optimize the operational parameters (i.e catalyst dosage, pH and concentration of H_2O_2) on the oxidative degradation of AO7 using Fe_3-xCoxO_4 catalyst.

2. Materials and Method

2.1. Acid Orange 7

An anionic dye, Acid Orange 7 (AO7) with molecular weight of 350.33 g/mol, was selected as model pollutant in order to evaluate the degradation efficiency of the Fe_3-xCoxO_4 catalyst in heterogeneous Fenton-like reaction. The molecular structure of AO7 is shown in Figure 1. All chemical reagents used in this study were of analytical grade.

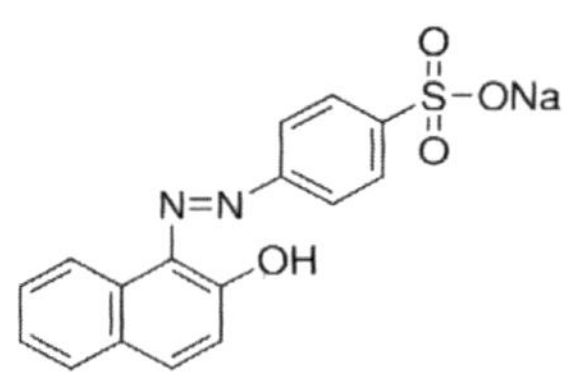

Figure 1. Structure of Acid Orange 7

2.2. Catalyst Synthesis

$Fe_{3-x}Co_xO_4$ catalyst has been synthesized according to the previous works by Yang et al. [6] with some modification. 5 mL of 2 mol L−1 $CoCl_2$•$6H_2O$ in HCl 7.4% solution and 40 mL of 0.5 mol L−1 $FeCl_3$•$6H_2O$ in double distilled water were pre-heat to 50 °C. The solutions then were mixed and decanted into a boiling solution of 200 mL of 1 mol L−1 NaOH under vigorous stirring for 30 min. After was cooled to room temperature, the dark brown precipitate was separated from the dispersion medium by using a permanent magnet. The resulting suspension was dried in oven at 100 °C for 5 h.

2.3. Catalytic Test

Batching process was applied to AO7 solution that has been prepared in the series of Erlenmeyer flask. A total of 20 Erlenmeyer flasks of AO7 solution with different catalyst dosage, pH and H_2O_2 concentration were batched by using top orbital shaker (Model SK-300, Malaysia). This process was carried out in room temperature (27-30 °C) with agitating speed of 150 rpm until it reached the desired contact time. The percentage dye degradation at equilibrium was calculated based on common equilibrium equation.

2.4. Design of Experiment

RSM was employed in this work because it is capable for modelling vast numbers of numerical experiments. RSM can evaluate the relationship of the controlled experimental factors and measured responses. CCD was applied to study the variables for AO7 degradation involve in three categories of runs which are 2n factorial runs, 2(n) axial runs and 6 center runs where n is the number of the process variables. Independent process parameters in qualitative form were calculated as follows:

$$Y = f(X_1, X_2, X_3, \dots X_n) \pm \varepsilon \quad (1)$$

where Y is the response, f is the response functions, ε is the experimental error and X_1, X_2, X_3,..., X_n are the independent parameter that involved in the experiment [7]. Experimental error and the reproducibility of the data can be obtained by using the center point [8]. The axial point is located at $\pm \alpha$ for each of variables where $\pm \alpha$ is the distance of the axial point from the center point. Hence, the response with linear function model can be written as:

$$Y = C_o + C_1X_1 + C_2X_2 + \cdots + C_nX_n \pm \varepsilon \tag{2}$$

where Y is the response, C is the coefficient, X is the coded values of the process variables and ε is the experimental error. However, when the curvature occurs, the higher order polynomial such as quadratic model may be used. The quadratic equation is written by equation:

$$Y = C_o + \sum C_iX_i + \sum C_{ii}X_i^2 + \sum C_{ij}X_iX_j \tag{3}$$

where Y is predicted response, C_o is the constant coefficient, C_i is the linear effect, C_{ii} is the squared effect, and C_{ij} is the interaction effect [9]. In this experiment, the process variables that were investigated were catalyst dosage (g/L, X_1), pH (pH, X_2) and H_2O_2 concentration (mM, X_3). The parameters were coded at three levels of -1, 0, 1, where the center point of the experiment was at 0.55 g/L catalyst dosage, pH 5.75 and 26.25 mM of H_2O_2 concentration. Table 1 shows the complete design matrix that has been developed with their responses values.

Table 1. Experimental design matrix for AO7 degradation

Run	Catalyst dosage, X_1 (g/L)	pH, X_2	H_2O_2 concentration, X_3 (mM)	AO7 degradation (%)
1	-1.00	-1.00	-1.00	42.41
2	1.00	-1.00	-1.00	99.75
3	-1.00	1.00	-1.00	9.03
4	1.00	1.00	-1.00	32.43
5	-1.00	-1.00	1.00	57.80
6	1.00	-1.00	1.00	99.59
7	-1.00	1.00	1.00	14.98
8	1.00	1.00	1.00	42.46
9	-1.00	0.00	0.00	39.80
10	1.00	0.00	0.00	48.10
11	0.00	-1.00	0.00	98.98
12	0.00	1.00	0.00	35.31
13	0.00	0.00	-1.00	42.91
14	0.00	0.00	1.00	45.68
15	0.00	0.00	0.00	42.20
16	0.00	0.00	0.00	40.78
17	0.00	0.00	0.00	43.16
18	0.00	0.00	0.00	41.38
19	0.00	0.00	0.00	42.28
20	0.00	0.00	0.00	43.11

2.5. Model Fitting and Statistical Analysis

The experimental data was analyzed by using Design Expert 7.0.0 software (Stat-Ease, Inc.) for regression analysis and statistical significance of the equation derived.

3. Result and Discussions

3.1. Development of Regression Model Equation

CCD was used as a polynomial regression in order to determine the interaction between the parameters variables and responses. The quadratic models of AO7 degradation was suggested by the software. The selection of the quadratic model is based on the higher order polynomial. Eq. 4 is the final empirical formula model for AO7 degradation:

$$Y = 45.25 + 15.83X_1 - 26.43X_2 + 3.4X_3 - 6.03X_1X_2 - 1.43X_1X_3 + 0.095X_2X_3 - 0.594X_1^2 + 17.26X_2^2 - 5.6X_3^2 \quad (4)$$

In this study, the R^2 values for Eq. 4 was 0.9399 (as shown in Figure 2) indicated that 93.99% of total variation in the AO7 discoloration was much closer to the actual value. In other words, the models developed were successful in capturing the correlation between the operational parameters variables to the response of AO7 degradation.

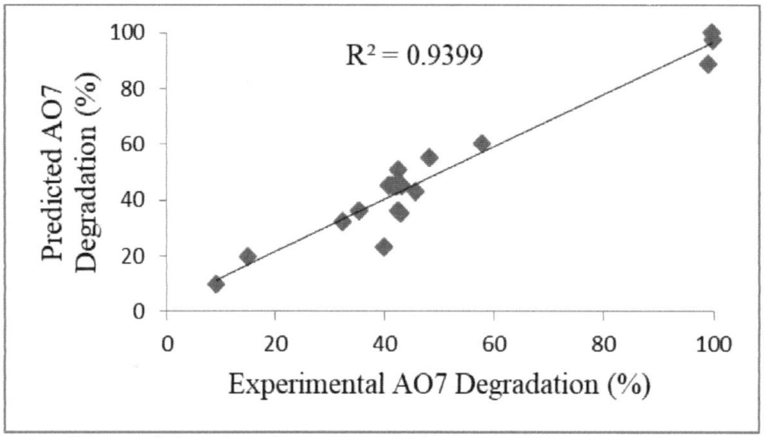

Figure 2. Predicted versus experimental AO7 degradation

3.2. *Statistical Analysis*
In this study, analysis of variance (ANOVA) was used to analyze and justify the significance and adequacy of the models. The results of the surface quadratic model of ANOVA are given in Table 2. The model terms are considered as significant if the F-value is relative high and the value Prob.>F is less than 0.05. The F-value was 17.69 and the prob.>F was less than 0.0001 indicated that the model was relevant and significant. Meanwhile, the significant model terms obtained were X_1, X_2 and X_2^2. In contrast, X_3, X_1X_2, X_1X_3, X_2X_3, X_1^2 and X_3^2 were insignificant model terms to the model. The model developed was successful because the predicted values obtained were closer to the actual experimental data.

3.3. *AO7 Degradation*
From Table 2, the catalyst dosage and solution pH were observed as the most significant factors that contributed to AO7 degradation. Meanwhile, the quadratic effect of solution pH was considered least significant. Figure 3 shows the effect of solution pH and catalyst dosage on the AO7 degradation when H_2O_2 concentration was fixed. From Figure 3, when the catalyst dosage and solution pH were decreased, the AO7 degradation increased. The results obtained was supported by previous study that was conducted by Xu et al. [10] which reported that at high catalyst dosage, more active sites were presence on the catalyst surface in accelerating the decomposition of H_2O_2. On the other hand, when the solution pH decrease, the number of negatively charged sites decrease. This can attributed to reducing of the degradation of the dye due to electrostatic repulsion. Similar observation was obtained by Xu et al. [11] where RhB decoloration was increased with increasing catalyst dosage and decreasing pH solution. In addition, Sun et al. [12] found that increasing the catalyst dosage gave significant effect on degradation of Orange G (OG) by fenton oxidation but increasing of H_2O_2 resulted to negative impact on the degaration of OG.

IOP Conf. Series: Materials Science and Engineering **358** (2018) 012020 doi:10.1088/1757-899X/358/1/012020

Table 2. Analysis of variance (ANOVA) for AO7 discoloration

Source	Sum of squares	Degree of freedom	Mean square	F value	*p*-value (Prob > *F*)	
Model	10742.96	9	1193.66	17.69	< 0.0001	Significant
X_1	2506.09	1	2506.09	37.15	0.0001	
X_2	6986.91	1	6986.91	103.57	< 0.001	
X_3	115.50	1	115.50	1.71	0.2200	
X_1X_2	291.13	1	291.13	4.32	0.0645	
X_1X_3	16.42	1	16.42	0.24	0.6324	
X_2X_3	0.072	1	0.072	0.001061	0.9747	
X_1^2	97.09	1	97.09	1.44	0.2579	
X_2^2	818.83	1	818.83	12.14	0.0059	
X_3^2	86.12	1	86.12	1.28	0.2849	
Residual	674.58	10	67.46	-	-	

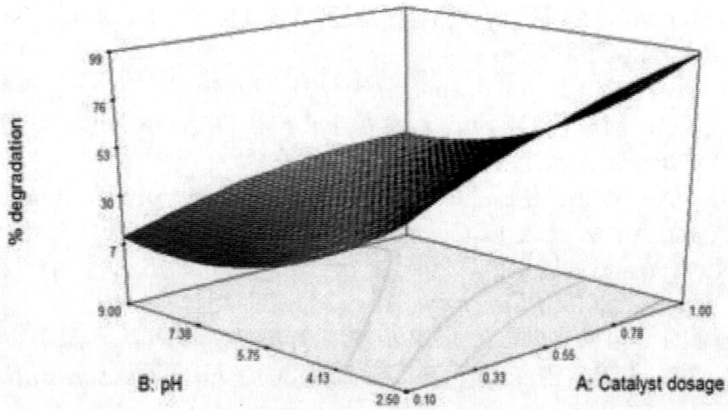

Figure 3. Three-dimensional response surface plot of AO7 discoloration (effect of pH and catalyst dosage, H_2O_2 concentration = 46.70 mM)

3.4. Process Optimization

RSM was used to determine the optimum parameters condition for the maximum AO7 degradation. During analysis process, the response was set as maximum value whereas all the parameters were fixed in the range. Table 3 shows the experimental results and predicted results of AO7 degradation at the optimum condition. The optimum condition for AO7 degradation was selected based on the highest value of model desirability. The optimum condition was obtained at catalyst dosage of 0.84 g/L, pH solution of pH 3.14 and H_2O_2 concentration of 46.70 mM which resulted in 86.30% of AO7 degradation.

Table 3. Model validation for AO7 degradation

Model of desirability	Catalyst dosage, X_1 (g/L)	pH, X_2	H_2O_2 concentration, X_3 (mM)	AO7 degradation (%)		
				Predicted	Experimental	Error (%)
1.00	0.84	3.14	46.70	86.30	85.43	0.87

4. Conclusion

In conclusion, AO7 can be efficiently removed by heterogeneous Fenton-like reaction using $Fe_{3-x}Co_xO_4$ catalyst. RSM was successfully used to investigate the parameters that influenced the AO7 discoloration. Based on the parameters studied, catalyst dosage and solution pH were found as significant factors that contribute to the AO7 discoloration. By using RSM, the optimum condition for AO7 degradation was obtained at catalyst dosage of 0.84 g/L, pH solution of pH 3 and H_2O_2 concentration of 46.70 mM which resulted in 86.30% of AO7 degradation.

Acknowledgement
This research is fully funded projects associated with Internal Research Acculturation Grant Scheme (iRAGS) Grant No. 600-RMI/DANA 5/3/IRAGS (44/2015) administered by Research Management Centre, Institute of Research Management and Innovation (IRMI) of Universiti Teknologi MARA. Authors gratefully acknowledge contribution in this paper whom directly or indirectly towards the outcome of the present research.

References

[1] Duarte F, Maldonado-Hódar F, Pérez-Cadenas A and Madeira L M 2009 *Appl. Catal., B* **85** 139-147
[2] Stoyanova M, Slavova I and Ivanova V 2014 *Appl. Catal., A* **476** 121-132
[3] Umar A, Akhtar M S, Al-Hajry A, Al-Assiri M S, Dar G N and Saif Islam M 2015 *Chem. Eng. J.* **262** 588-596
[4] Su Y, Wu Z, Wu Y, Yu J, Sun L and Lin C 2015 *J. Mater. Chem. A* **3** 8537-8544
[5] Costa R C C, Lelis M F F, Oliveira L C A, Fabris J D, Ardisson J D and Rios R R V A 2006 *J. Hazard. Mater.* **129** 171-178
[6] Yang W, Zhou X, Zheng N, Li X and Yuan Z 2011 *Electrochim. Acta* **56** 6588–6592
[7] Ahmad M A and Alrozi R 2011 *Chem. Eng. J.* **168** 280-285
[8] Can M Y, Kaya Y and Algur O F 2006 *Bioresour. Technol.* **97** 1761-1765
[9] Wang J and Chen C 2004 *Biotechnol. Adv.* **24**, 427-451
[10] Xu Y H, Shi N T and Wu C L 2013 *Water, Air, and Soil Pollut.* **224** 1740
[11] Xu H Y, Qi S Y, Li Y, Zhao Y, and Li J W 2013 *Environ. Sci. Pollut. Res.* **20** 5764–5772
[12] Sun S P, Li C J, Sun J H, Shi S H, Fan M H and Zhou Q 2009 *J. Hazard. Mater.* **161** 1052–1057

IOP Conf. Series: Materials Science and Engineering **358** (2018) 012021 doi:10.1088/1757-899X/358/1/012021

Characteristics of *Amorphophallus campanulatus* Modified Starch as Novel Adsorbent for Nickel and Cadmium Removal from Aqueous Solution

A Y D Lestari[1] and L K Dewi[2]

[1]Department of Chemical Engineering, Islamic University of Indonesia, Indonesia
2 Department of Chemical Engineering, Brawijaya University, Indonesia

aydlestari@uii.ac.id

Abstract. Novel adsorbents are succesfully manufactured from Amorphophalluscampanulatus (Porang or Suweg or Elephant Foot Yam or Foot Yam) starch. The experiment focused on modifiying the starch with citric acid and detailing those morphologies and chemical bonds. Analysis with FTIR and SEM showed that PTM (modified porous porang starch) sample is the best adsorbent which has most stable of chemical bonding and also has the most pores that influence an adsorption phenomena. Isotherm adsorption analysis showed that the adsorption mechanism of Cd and Ni ions onto the suface of PB, PT and PTM followed the Temkin and Langmuir isotherm adsorption. Adsorption ability of PTM is the best than the other PB and PT which can adsorps 256,23 mg Cd/g PTM and 87,45 mg Ni/g PTM in 500 ppm synthetic aqueous solution

1. Introduction

Adsorption is the process that can separate objects which neither impossible to be apply nor impractical by conventional techniques [1]. Zeolite, activated carbon, activated alumina are used as adsorbent. But recently there are researches that concern on making of the novel adsorbent. This novel adsorbent is made not only from unusual sources as paper waste [2] but also made from the polisaccarides.Polisaccarideswhich is choosen as adsorbents due to their easily to synthesize, have large adsorption capacity due to its large number of the –OH bond and cheap [3]. Elephant foot yam contains polisaccarides that not explore before. In the other hand, exposure of heavy metal ions on our environment especially cadmium and nickel caused some diseases, such as: liver, renal tubular disease, emphysema, dizziness, embolism, etc. So this project focused on evaluation of the characteristics of the citric acid modified the elephant foot yam starch adsorbents and their ability to adsorp both ion Cd^{2+} and Ni^{2+} in aqueous solution.

2. Materials and Methods

The main materials of this subject are foot yam tubers, pure distilled water, Merck's citric acid and Merck's ethanol. The selected fresh foot yam tubers is obtain from the farmers at Ngawi, East Java, Indonesia which has the itchy yellowish tuber.

a. Isolation and modification of starch from Foot Yam

Elephant foot yam starch (PB) is isolated from its fresh tuber by conventional process. Then, PB was porous followed by Chang's method [4]. PB powder was mixed with pure distilled water at the same time heated at 90°C for about 30 minutes. The mixture was then cooled in the 5°C freezer for 3 day to

Content from this work may be used under the terms of the Creative Commons Attribution 3.0 licence. Any further distribution of this work must maintain attribution to the author(s) and the title of the work, journal citation and DOI.
Published under licence by IOP Publishing Ltd

obtain the starch gel. The gel was cut into cubes and frozen at -10°C for another 48 hours. The frozen cubes were immersed in 90% ethanol at room temperature. The cubes were dried at 50°C for 6 hours and then heated at 110°C for 2 hours to remove the ethanol and water and to obtain white solid PT cubes. PT cubes then modified with citric acid which followed Ma's method [5]. Synthetic PTM was begun with mixing ethanol and citric acid (10:3 v/v). The PT cubes then was added with the solution of ethanol citric acid. The mixture was stirred with 500 rpm overnight at room temperature. The slurry then heated at 110°C for 4 hours. The dried slurry then soaked with 90% ethanol to remove the excess citric acid. The mixture then dried at room temperature and resulted the PTM powder.

b. Characterization of the adsorbent

There are two characterization of this project. They are suface morphology characterization and chemical bonding characterization. The adsorbents surface morphology was visualized by FEI Inspect S50Scanning Electron Microscope (SEM).The chemical bonding was determined with Thermo Nicolet Avatar 360Fourier Transform Infrared Spectroscopy (FTIR).

c. Determination adsorption properties of adsorbent

Isotherm adsorption was determined by mixing 1.5 g of sample adsorbents (PB, PT, PTM)with 25 mL differential concentration of Ni and Cd synthetic waste water (100, 200, 300, 400, 500 [ppm]). Mixture then mixed well for about 30 minutes. After the mixing the adsorbent separated from the filtrate by the Whatmann 40 ashless filter paper, the final concentration of the mixture then analized using Perkin Elmer PinAAcle 900T Atomic Adsoprtion Spectrofotoetry (AAS). Experimental data then fitted with Langmuir, Freundlich, Temkin and DubininRadushkevich isotherm model and studied which the appropriate model(s).

3. Result and Discussion

a. Effect of Citric Acid Modification

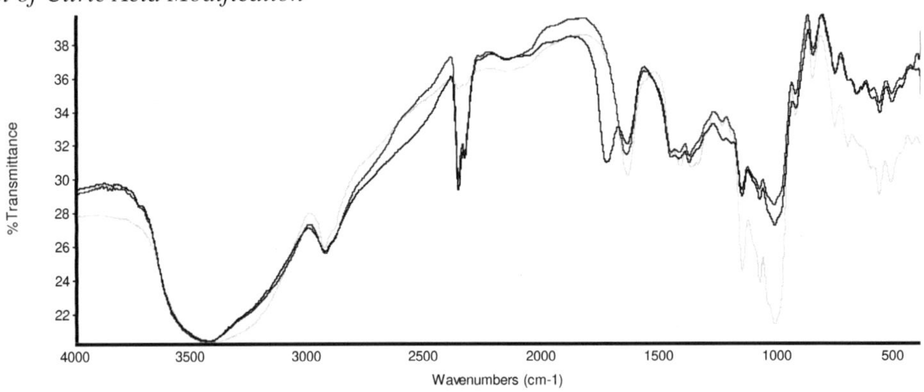

Figure 1. FTIR Spectrum of PB, PT and PTM.

Figure 1 represented FTIR spectra of the chemical bonding that occurred in the adsorbent PB, PT and PTM. The black line represented PTM spectrum, the dark blue line represented PT spectrum and the light blue line represented PB spectrum. Spectra of PB, PT, PTM showed the –OH bonds on wavelength around 3.400/cm, the C=O bonds on wavelength around 1.800/cm, the C-O bonds on wavelength 1.000/cm. Main focus on this research was the –OH bond. Initial PB –OH bond had the appearance of a little bit wide to the right. After the modification, both PT and PTM's -OH bonds was tend to be more symmetrical. It can be said that the process of modification with citric acid may alter-OH bond.

b. Characterization of Adsorbents

SEM photograph showed in Figure 2 below with 1.000x magnification of PB, PT and PTM. Figure PB that represent the pure foot yam starch showed that sphere with few pores. There was some staple surrounding the granules, it may the impurities of the starch as reported by Zhu [6] that starch

IOP Conf. Series: Materials Science and Engineering **358** (2018) 012021 doi:10.1088/1757-899X/358/1/012021

impurities could be ash, protein, lipid, fosfor and fibres. Figure PT represent the porous foot yam starch showed that the spheres are smoother and have more pores than figure PB. Figure PTM represents the modified porous foot yam starch that has more amorph, porous and clean structure.Modification process influences the morpfology and chemical bonding of adsorbents. Based on those characteristic, the PTM sample has more compromising adsorbent for adsorpstion process.

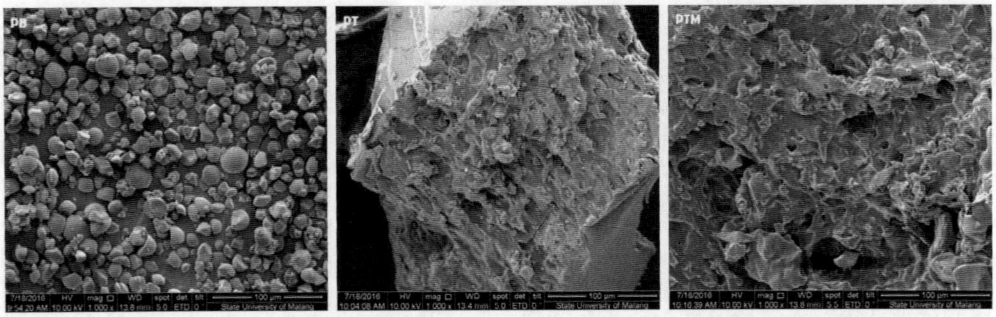

Figure 2. SEM Photograph of PB, PT and PTM.

c. Effect of the initial Cd and Ni concentration

The relationship between the initial Cd and Ni ion concentration and the adsorption capacities of PB, PT, PTM for them was studied. As shown in Figure 3 and Figure 4, the adsorption capacities of PB, PT and also PTM for Cd and Ni were correlated with the initial Cd and Ni ion concentration because the process was depent on the concentration. When the concentration rose from 100 ppm to 500 ppm, the adsorption capacities of PB, PT and PTM increased from 13.8 to 68 mg/g, 24.7 to 199 mg/g, 49.7 to 256 mg/g for Cd ion adsorption and 9.2 to 25.3 mg/g, 29.66 to 61.88 mg/g, 45.51 to 87.45 mg/g for Ni ion adsorption.

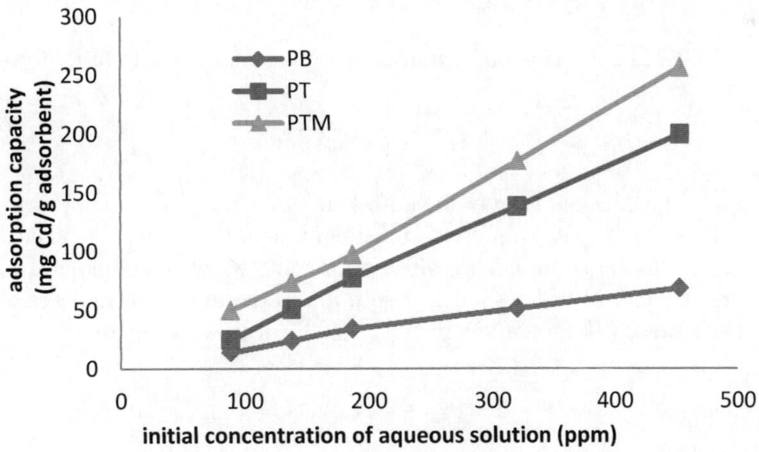

Figure 3. Effect of the initial Cd concentration onto PB, PT and PTM.

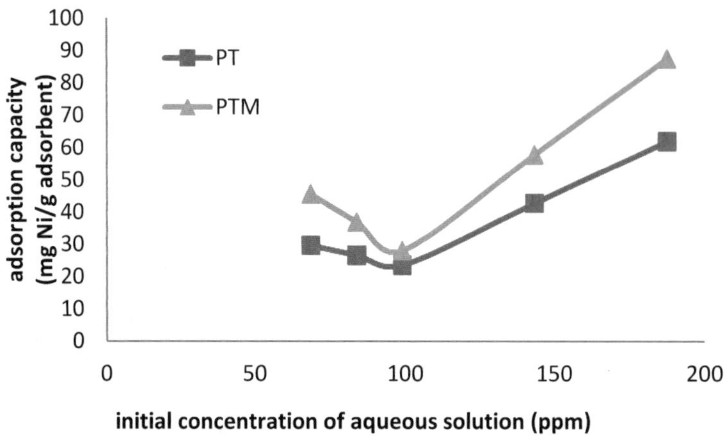

Figure 4. Effect of the initial Ni concentration onto PB, PT, PTM

d. Isoterm Adsorption of Ni and Cd onto Adsorbents

The adsorption isotherms showed the relations between the concentration of adsorbate and its degree of accumulation of Ni and Cd onto surface of adsorbent at room temperature. Several models of adsorption isotherm have been used to fit to the experimental data. Fitting model used to evaluate isotherm performances for water adsorption. These isotherm models are the Freundlich model, Langmuir model, Temkin model and DubninRadushkevich model.Langmuir adsorption isotherm describes the performance of a monolayer adsorbate on the outer surface of the adsorbent quantitatively. The Langmuir isotherm is valid for monolayer adsorption onto a surface containing a finite number of identical sites [7]. Langmuir model assumes that there is uniform adsorption energy on the surface and there is no adsorbate transmigration in the plane of the surface. Based upon these assumptions, Langmuir represented the following equation:

$$q_e = \frac{Q_0 K_L C_e}{1+K_L C_e} \quad (1)$$

Langmuir parameters could be determined by transforming the equation into linier form:

$$\frac{1}{q_e} = \frac{1}{Q_0} + \frac{1}{Q_0 K_L C_e} \quad (2)$$

where C_e is the equilibrium concentration of the adsorbate (g/L), q_e is the amount of water per gram of the adsorbent in equilibrium (g/g), Q_0 represents maximum monolayer coverage capacity (g/g) dan K_L for Langmuir isotherm constant (L/mg). The values of q_{max} and K_L were computed from the slope and intercept of the Langmuir plot of vs Freundlich adsorption isotherm is used to describe characteristics for the heteregeneous surface [7]. Freundlich respresented the following equation:

$$Q_e = K_f C_e^{\frac{1}{n}} \quad (3)$$

where K_f is Freundlich isotherm constant (g/L), n is adsorption intensity, C_e represents the equlibrium concentration of adsorbate (g/L) and Q_e is the amount of water adsorbed per gram of adsorbent at equilibrium (g/g). Linierizing of the Freundlich equation above we have:

$$\ln Q_0 = \ln K_f + \frac{1}{n} \ln C_e \quad (4)$$

The constant K_f is an approximate indicator of adsorption capacity, while is a function of the strength of adsorption in the adsorption process [8]. If n=1 then the partition between the two phase are independent of the concentration. If value > 1 it indicates not only a normal adsorption but also indicates coorperative adsorption [9]. Goldberg [10] reported that the linier least squares method and the linierly transformed equations have been widely applied to correlate sorption data where is a heterogeneity parameter, the smaller, the greater the expected heterogeneity. This expression reduces to linier adsorption isotherm when =1. If 1 < n < 10, this indicates a favorable sorption process. Temkin contains a factor that explicitly taking into the account of adsorbent-adsorbate interactions. By ignoring the extremely low and large value of concentrations, the model assumes that heat of adsorption of all molecules in the layer would decrease linearly rather that logarithmic with coverage [11]. As implied in the equation, its derivation is characterized by a uniform distribution of binding energies (up to some maximum binding energy) was carried out by plotting the quantity sorbedq_e against lnC_e and the constants were determined from the slope and intercept. The model is given in following equation:

$$q_e = \frac{RT}{b}\ln(A_T C_e) \quad (5)$$

$$q_e = \frac{RT}{b_T}\ln A_T + \left(\frac{RT}{b_T}\right)\ln C_e \quad (6)$$

$$B = \frac{RT}{b_T} \quad (7)$$

$$q_e = B\ln A_T + B\ln C_e \quad (8)$$

where A_T is Temkin isotherm equilibrium binding constant (L/g), bT is Temkin isotherm constant, R represents universal gas constant (8,314 J/molK), T is temperature at 298 K and B is constant related to heat of sorption (J/mol).DubininRadushkevich isotherm is generally applied to express the adsorption mechanism with a Gaussian energy distribution onto a heterogeneous surface [12]. The model has often successfully fitted high solute activities and the intermediate range of concentrations data well.

$$q_e = (q_s)exp(-K_{ad}\varepsilon^2) \quad (9)$$

$$\ln q_e = (q_s) - (K_{ad}\varepsilon^2) \quad (10)$$

where q_e is amount of adsorbate in the adsorbent at equilibrium (g/g), q_s is theoretical isotherm saturation capacity (g/g), K_{ad} is DubininRadushkevich isotherm constant (mol^2/kJ^2) and ε is Dubinin Radushkevich isotherm constant. This model applied to determine the physical and chemical adsorption of metal ions with its mean free energy (E) per molecule of adsorbate.

$$E = \left[\frac{1}{\sqrt{2B_{DR}}}\right] \quad (11)$$

$$\varepsilon = RT\ln\left[1 + \frac{1}{C_e}\right] \quad (12)$$

where R, T, Ce represent the ideal gas constant (8,314 J/molK), absolute temperature (K) and adsorbate equilibrium concentration (g/L). Foo and Hameed [13] said that one of the unique of this isotherm model lies on the fact that it is temperature dependent which when adsorption data at different temperatures are plotted as a function of logarithm of amount adsorbed (ln q_e) versus the square of potential energy (ε^2), all suitable data will lie on the same curve named as the characteristic curve.

e. Cadmium soprion isotherm

Figure 3, Figure 4 and Figure 5 showed the plot of data isoterm of Cd adsorption onto the PB, PT and PTM adsorbents. They showed that the Cd adsorption onto PB and PTM are suitable with Temkin isotherm while Cd adsorption onto PTM is suitable with Langmuir isotherm. It said that Cd adsorption onto PB and PTM are heterogenous whereas Cd adsorption onto PTM is occured in monolayer adsorption.

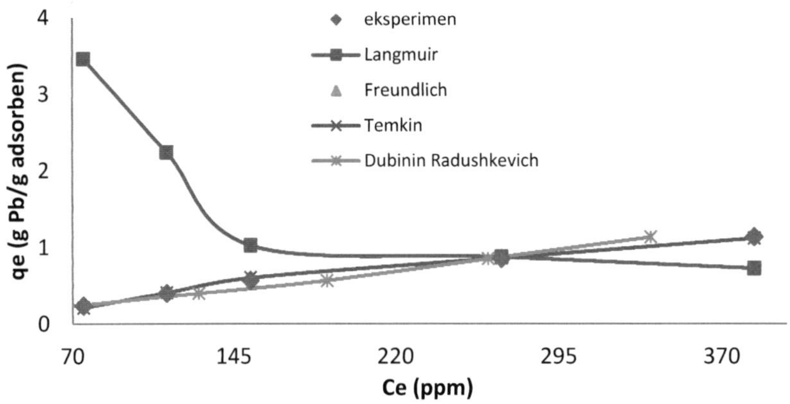

Figure 5. Cd onto PB.

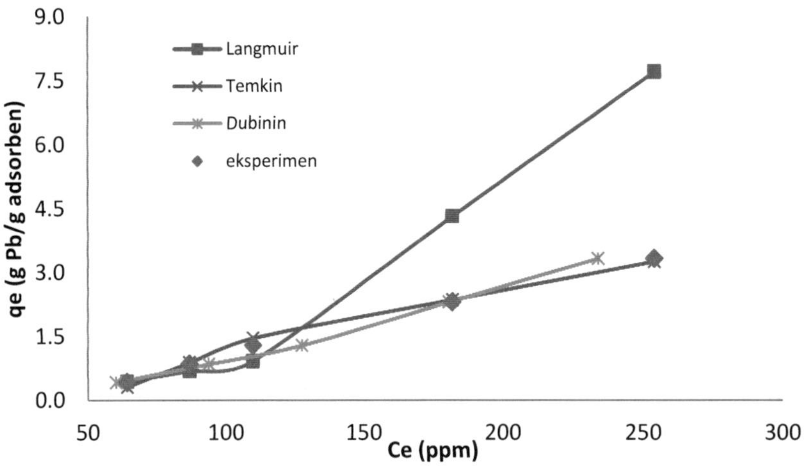

Figure 6. Cd onto PT.

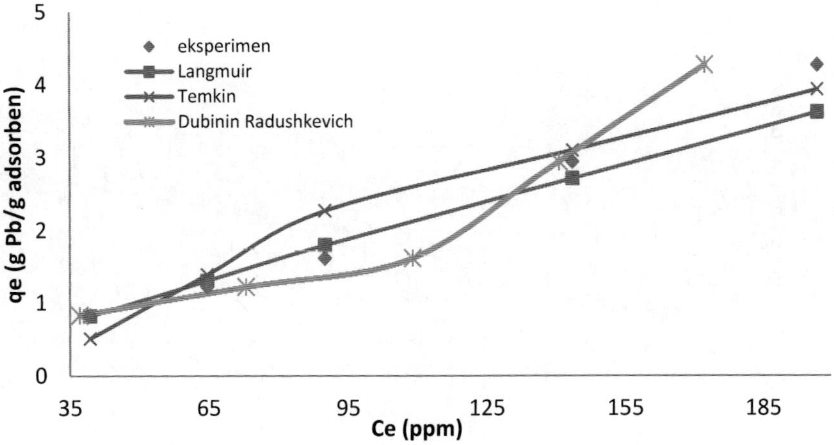

Figure 7. Cd onto PTM.

Table 1 showed the details of Cd adsorption isotherm parameters onto PB, PT and PTM adsorbents.

Table 1. Cadmium Isothem Parameters onto PB, PT and PTM

	Langmuir			Freundlich		
	K_L	q_0	R^2	K_F	n	R^2
PB	0,017	0,568	0,99	0,215	1,727	0,271
PT	0,003	3,788	0,975	0,112	0,829	0,975
PTM	0,001	26,316	0,99	0,254	1,089	0,999
	Temkin			DubininRadushkevich		
	A_t	B	R^2	K_{ad}	q_s	R^2
PB	0,559	0,158	0,078	5,354	480,583	0,941
PT	0,031	1,982	0,985	3,169	264,807	0,949
PTM	0,056	1,930	0,951	2,050	184,380	0,94

f. Nickel Sorption Isotherm

Figure 6, Figure 7 and Figure 8 showed the plot of data isoterm of Ni adsorption onto the PB, PT and PTM adsorbents. They showed that the Ni adsorption onto PB and PTM are suitable with Langmuir isotherm while Ni adsorption onto PT is suitable with Temkin isotherm. It said that Ni adsorption onto PB and PTM are occured in monolayer whereas Ni adsorption onto PT is occured in heterogeneous layer.

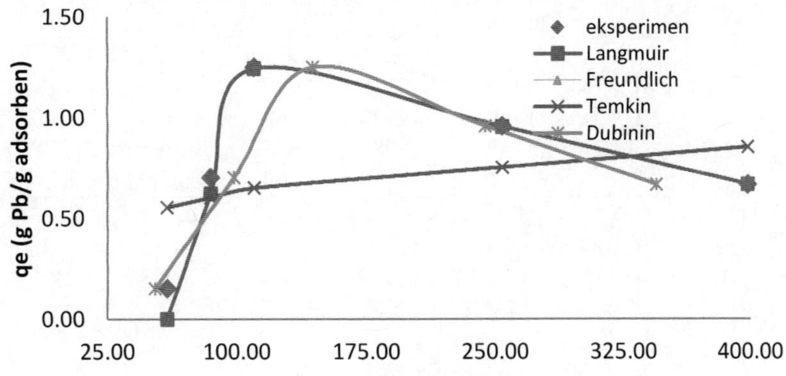

Figure 8. Ni onto PB.

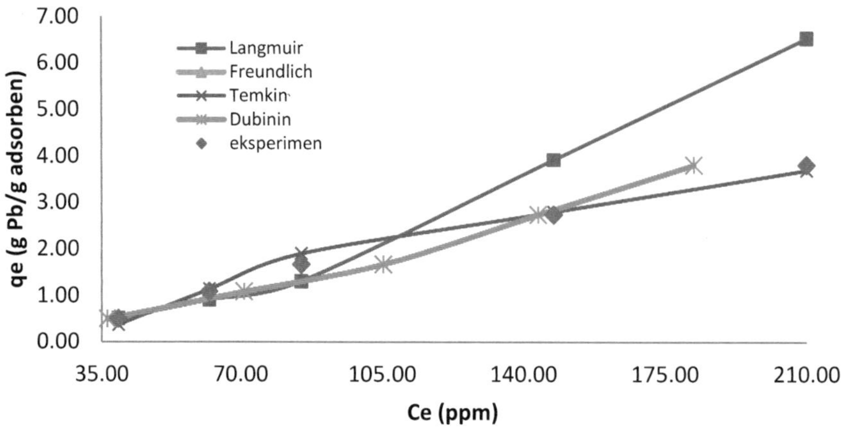

Figure 9. Ni onto PT.

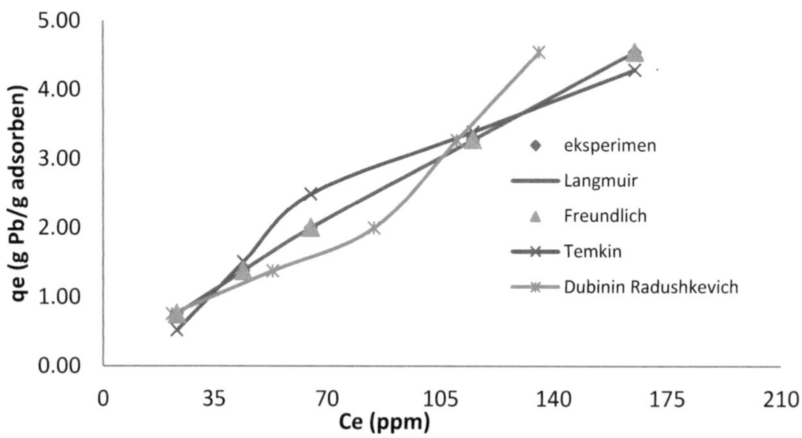

Figure 10. Ni onto PTM.

Table 2 showed the details of Ni adsorption isotherm parameters onto PB, PT and PTM adsorbents.

Table 2. NickelIsothem Parameters onto PB, PT and PTM.

	Langmuir			Freundlich		
	K_L	q_0	R^2	K_F	n	R^2
PB	0,016	0,609	0,39	0,977	1,036	0,089
PT	0,003	1,658	0,942	0,968	0,673	0,049
PTM	0,001	25,000	0,987	0,985	0,986	0,178
	Temkin			DubininRadushkevich		
	A_t	B	R^2	K_{ad}	q_s	R^2
PB	0,019	0,556	0,995	5,977	496,210	0,95
PT	0,018	2,142	0,989	4,743	371,668	0,964
PTM	0,032	2,121	0,901	3,075	253,661	0,955

g. Comparison PB, PT and PTM with other Cd and Ni adsorbents

Table 3 showed the list of adsorption capacity of Cd and Ni by some synthetic adsorbents not only conventional adsorbent but also low cost natural based adsorbent. This study said that the PTM adsorbent had more compromising adsorbent among PB and PT for the Cd and Ni adsorption processes. Another advantage of PTM especially compared to adsorbent conventional is the availability of raw materials for the manufacture of adsorbent abundant sources of biomass, easily regenerated, adsorbent regeneration after use more environmentally.

Table 3. Maximum Adsorption Capacity of Cd and Ni by Some Synthetic Adsorbents.

Adsorbents	**Adsorption Capacity (mg/g)**		**Initial Concentration (mg/L)**		**Reference**
	Cd^{2+}	Ni^{2+}	Cd^{2+}	Ni^{2+}	
PB	68.23	23.53	500	200	This study
PT	199.23	61.88	500	200	This study
PTM	256.23	87.45	500	200	This study
Activated carbon	6.16	9.15	50	50	[14]
Coconut fiber	31.12		10-140		[15]
Crosslinked carboxymethyl KGM	9.65				[16]
Amberlite IR 120 resin	19.49				[17]
Amino modified cassava starch	23.09				[18]
Crosslinked starch		397	504		[19]
Dialdehyde o phenylenediamine starch			69.69	1,242	[20]

4. Conclusion

Novel adsorbent was succesfully synthesized from *Amorphophallus campanulatus* starch modified by citric acid. The experiment focused on investigation of the effect of modification and the adsorption ability for Cd and Ni ions in aquoeus solution. Modification caused starch more porous and the changed the infrared spectrum of –OH bonding. The adsorption behaviour is dependent on the initial concentration of Ni and Cd ion. Adsorption capacity Cd and Ni on PB, PT, PTM in 500 ppm aqueous are 68 mg/g and 25 mg/g, 199 mg/g and 61.88 mg/g, 256 mg/g and 87.45 mg/g. The adsorption follows the Langmuir and Temkin isotherm. The study also evaluates that the PTM is the more promising adsorbent among the PB and PT. So citric acid modification on the *Amorphophallus campanulatus* starch gave a potential application as another low cost natural based adsorbent.

Acknowledgement
Authors would say thank to Directorate of Research and Community Service Islamic University of Indonesia for the financial support. We also thank Dr. Rudi Syahputra, head of Integrated Laboratory Islamic University of Indonesia and Abdulloh Fuad, M.Si, head of Central Laboratory of Mineral and Advanced Materials, Malang State University for their guidance with the experimental element of this research.

References

[1] Knaebel, K.S. 2009.Adsorption Research Inc's publication: 1-21.
[2] Khalili, N., Vyas, J.,Weangkaew, W., Westfall, S., Parulekar, S. andSherwood, R. 2002. Sep.Purif.Technol.**26 (2-3)**295–304.
[3] O'Connell, D.W.,Birkinshaw, C. and O'Dwyer, T.F. 2008. Bioresour.Technol.**99** 6709–6724.
[4] Chang, P.R., Yu, J. and Ma, X. 2011. Carbohydr. Polym.**83** 1016-1019.
[5] Ma, X., Liu, X., Anderson, D.P. and Chang, P.R. 2015. Food Chem. **181** 133-139.
[6] Zhu, F. (2015). Carbohydr.Polym, **122** 456-480.
[7] Dada, A.O., Olalekan, A.P., Olatunya, A.M. and Dada, O. 2012. J. Appl. Chem. **3** 38-45.
[8] Voudrias, E., Fytianos, F.andBozani, E. 2002. Global Nest, The International Journal,**4(1)**75-83
[9] Mohan, S. andKarthikeyan, J. 1997. Environ.Pollut**. 97**183-187.
[10] Goldberg, S. 2005. Soil Science Society of America, 677.

[11] Tempkin, M.I. andPyzhev, V. 1940.Acta Phyico Chimica USSR. **12**327–356.
[12] Gunay, A., Arslankaya, E. andTosun, I. 2007.J.Hazard. Mater.**146** 362–371.
[13] Foo, K.Y. and Hameed, B.H. 2010. Chem.EngJ.**156**2–10.
[14] Naiya T.K.,Bhattacharya, A.K. andDas, S.K. 2009.J. *Colloid InterfaceSc.,* **333**14–26.
[15] Júnior B.C.,Nascimento, J.H.O., Quina, M.J. andFerreira, L.G. 2014.*J. Agric.Sci. Technol.*, **4** 291–298
[16] Niu, C., Wu, W., Wang, Z., Li, S., and Wang, J. 2007. J. Hazard. Mater.**141** 209–214.
[17] Demirbas, A., Pehlivan, E., Gode, F., Altun, T., and Arslan, G. 2005. J. Colloid.Interface Sci.**282** 20–25
[18] Xie, G., Shang, X., Liu, R., Hu, J. and Liao, S. 2011. Carbohydr.Polym.**84** 430-438.
[19] Sancey, B., Trunfio, G., Charles, J., Minary, JF., Gavoille, S., Badot, PM. and Crini, G. 2011. J. Environ. Manage.**92** 765-772.
[20] Zhao, P., Jiang, J., Zhang, F., Zhao, W., Liu, J. and Li, R. 2010. Carbohydr.Polym.**81** 751-757.

IOP Conf. Series: Materials Science and Engineering **358** (2018) 012022 doi:10.1088/1757-899X/358/1/012022

Oil Palm Empty Fruit Bunch (OPEFB) Fiber as Lost Circulation Material (LCM) in Water Based Mud (WBM)

N A Ghazali, A Sauki, N F Abu Bakar and S Mohamed

Faculty of Chemical Engineering, Universiti Teknologi MARA, 40450 Shah Alam, Selangor, Malaysia.

nurulaimi@salam.uitm.edu.my

Abstract. Lost Circulation Material (LCM) is an additive used to prevent lost of mud to the formation as a results from natural or induced fractured during drilling operation. Losses of mud could give great impact to the oil industry as it increases mud cost and rig time. The objective of this research was to investigate the effect of size and concentration of Oil Palm Empty Fruit Bunch (OPEFB) as LCM in water based mud (WBM). Several important properties of WBM rheology after adding the OPEFB namely plastic viscosity, apparent viscosity, yield point and gel strength were characterized. The sizes of OPEFB added into the WBM were 150μm, 250μm, 500μm and 1000μm while the concentration of OPEFB used were 5g, 10g, 15g and 20g in 350 mL of WBM. Results indicated that the plastic viscosity and apparent viscosity increased with increasing of the OPEFB concentrations. On the other hand, the plastic viscosity and apparent viscosity decreased with increasing sizes of OPEFB. Yield point increased as the concentration and size of OPEFB increases. This study indicated that OPEFB was effective to be used as LCM for size of 150μm and concentration of 15g whereby it produced least amount of filtrate volume as well as good control in mud rheology.

1. Introduction

Lost circulation is the most troublesome problem in drilling process. It can occur when the drilling fluid losses into the formation instead of returning to the annulus [1][2][3]. It also can occur when the size of pore opening are larger than size of the solid particles in drilling fluid [4]. Lost circulation still can occur even with the best drilling practices. However, lost circulation can be reduced by adding lost circulation material (LCM) in drilling fluid that help to accelerate the formation of filter cake and therefore control the overflow of drilling fluid into the formation pores [5]. The LCM will form of a filter cake, which bridges the formation face to stabilize the wellbore [6]. There are 4 types of LCM which are fiber, flake, granular and blend of fiber, flakes and granular [1]. This research will focus on fiber types of LCM which is OPEFB. There are a few researchers have done the research using fiber types of LCM such as lemongrass and saw dust [6,10]. The smaller particle size of lemongrass forms lower filtrate volume of drilling fluid and thin filter cake when compared to drilling fluid with larger particles [8,9]. The smaller size of sawdust give least mud cake thickness but the bigger size of sawdust gives lowest filtration volume [10].

OPEFB is a by-product of the palm oil industry and is the residual bunch after removal of the fruits [11][13-15]. It is the easiest and cheapest lignocellulose resources that can be found in oil palm biomass [16][17]. Approximately fifteen million tons of OPEFB biomass waste is generated annually throughout Malaysia by palm oil mills. The waste is being discarded in open areas or burnt in incinerators which create environmental pollution problems in nearby localities by generating large quantity of solid waste and pollutant gases [11][12]. In a way to save environment, this research will evaluate the feasibility of using OPEFB as LCM additive in water based mud and the effects of its size and concentration will also be investigated.

Content from this work may be used under the terms of the Creative Commons Attribution 3.0 licence. Any further distribution of this work must maintain attribution to the author(s) and the title of the work, journal citation and DOI.
Published under licence by IOP Publishing Ltd

2. Methodology

2.1. Preparation of Oil Palm Empty Fruit Bunch (OPEFB) Fiber as LCM

The fresh OPEFB sample was pre-treated by drying in the oven at 40°C for 5 days to achieve equilibrium moisture content of below than 5 wt%. High moisture content of fresh OPEFB (±29 wt%) will affect the rheology of mud system [20]. Prior to addition in water based mud, the dried OPEFB sample was grounded using cutting mill and sieved to obtain the size range of 150μm, 250μm, 500μm and 1000μm.

2.2. Preparation of Water Based Mud as drilling fluid.

The formulation of water based mud (WBM) is shown in Table 1 [18][19]. There were four variation of OPEFB amount added in the mud formulation i.e. 5g,10g, 15g, 20 g which are based on the common range of LCM in the mud formulation. 5-10g of LCM are typically added for slight losses while 10-20g of LCM are added for major losses in drilling operation [3]. The same formulation as in Table 1 was used for different sizes of the dried OPEFB LCM i.e. 150μm, 250μm, 500μm and 1000μm. Sufficient mixing time must be allowed to ensure complete hydration and shear of the specified additives in mud during circulation. Insufficient mixing time will result Barite to settle and reduce the mud rheology. Higher amount of Barite requires longer time to homogenously mix the mud.

Table 1. Formulation for Water Based Mud.

Component	Quantity	Mixing time (mixing speed)
Water	318.00 mL	3 min (low)
Na_2CO_3 or aka soda ash	0.5 g	3 min (low)
NaOH or aka caustic soda	0.25 g	3 min (low)
Starch	1 g	3 min (low)
Bentonite	12 g	5 min (low)
Barite	109.20 g	7 min (medium)
LCM (OPEFB)	5g,10g, 15g, 20 g	5 min (medium)

2.3. Determination of Water Based Mud Rheology.

The viscometer was used to calculate the plastic viscosity (PV), apparent viscosity (AV), yield point (YP) and gel strength of the water based mud sample and mud density of the WBM sample was determined by using mud balance. The equations to calculate the plastic viscosity and yield point are shown in equations (1), (2) and (3) respectively [7][20][21][22]. The rheology reading at 600 rpm and 300 rpm were taken to simulate the real mud circulation rate during drilling operation. This experiment was repeated for three times and the average readings were presented in the results.

$$\text{Plastic Viscosity (PV)} = \text{Reading at 600 rpm} - \text{Reading at 300 rpm} \quad (1)$$

$$\text{Apparent Viscosity (AV)} = \frac{\text{Reading at 600 rpm}}{2} \quad (2)$$

$$\text{Yield Point (YP)} = \text{reading at 300 rpm} - \text{PV} \quad (3)$$

2.4. Determination of filtrate volume and mud cake thickness.

Low Pressure Low Temperature (LPLT) filter press was used to measure the filtrate volume under static conditions. The cell was filled with the samples up to approximately 3/4 inches height. The LPLT filtration was conducted at room temperature and at 100 psi for 30 minutes [6][7]. The readings were repeated for three times and the average readings were presented in the results.

IOP Conf. Series: Materials Science and Engineering **358** (2018) 012022 doi:10.1088/1757-899X/358/1/012022

3. Results and Discussion

3.1. Influence of OPEFB on rheological properties of water based mud.

OPEFB was added into WBM at different concentrations (5, 10, 15 and 20 g) and the experimental results were shown in Figure 1. According to previous literatures, AV reflects the flowability of drilling fluids and is related to rate of penetration, and PV is part of resistance to flow caused by the friction between the suspended particles and influenced by the viscosity of the base liquid [23]. As observed in Figure 1, AV and PV values increased when the concentration of OPEFB was increasing. This is due to the increment of solid content in drilling mud that increased the frictional force between the suspended particles. In contrast, AV and PV values decreased with the increasing size of OPEFB. It is expected that the reduction of OPEFB size could increase the surface area which increase the frictional drag and viscosity of mud as shown in 150µm-sized OPEFB. The findings are consistent with N.A. Ghazali [20] whereby both concentration and size of the solid particles will affect the mechanical friction and viscosity of mud.

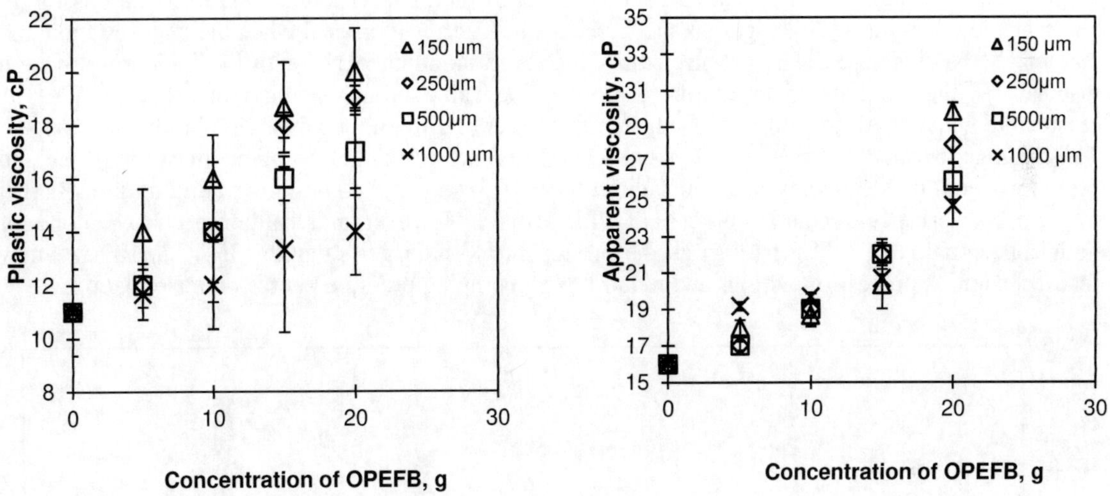

Figure 1. Plastic Viscosity (PV) and Apparent Viscosity (AV) Analysis of WBM at different concentration and size of OPEFB

Yield point is an important rheological parameter for drilling fluids and reflects the resistance to initial flow or the stress required to initiate fluid movement. Figure 2 revealed that YP goes up as the concentration and size of OPEFB increase. This is because the yield point is influenced by the concentration of solid in mud. Type and amount of solids in the mud alters the mechanical iteration and therefore influence the yield point of the mud [25]. However, the trend progresively increased at 20g concentration due to coagulation and flocculation of OPEFB particles especially for finer particles. It can be postulated that the positively charged OPEFB which is a fibrous material absorbed water around the particles and decrease the distance between the charges on the particles and increase coagulation and flocculation process.

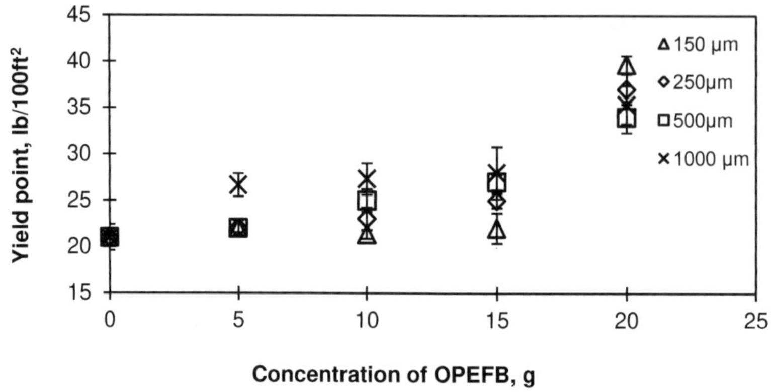

Figure 2. Yield point of WBM at different concentration and size of OPEFB

Gel strength is also an important property in drilling mud and indicates the ability of drilling fluid to suspend drill cutting. It measures the attractive forces of mud particles under static conditions for 10 seconds and 10 minutes [20]. Figure 3 shows the gel strength is increased when the concentration and size of OPEFB increased. The gel strength is increased simultaneously with the increasing of mud viscosity as the viscosity gives major relation to the value of gel strength theoretically [24]. A reasonable gel strength is vital to prevent immediate settling of solids when circulation has stopped. The findings revealed that there are no significant difference of the gel strength values for 10 sec and 10 min (overall data show less than 50% difference). This indicate no on-going build up process that might occur during re-circulation process of OPEFB-mud. However, a precaution need to be done for high concentration of OPEFB (20g) as the production of high gel strength (35-40 lb/100 ft^2) may retard separation process of cuttings and requires high pumping pressure to break the circulation.

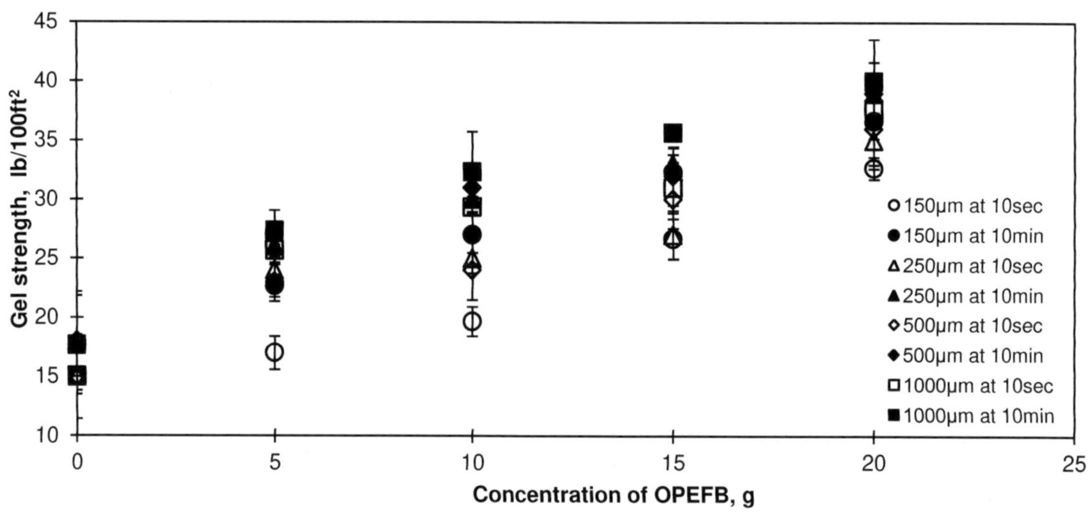

Figure 3. Gel strength of WBM at different concentration and size of OPEFB

3.2 Influence of OPEFB on filtration loss of water based mud.

The filtration loss and mud cake thickness of OPEFB-WBM are shown in Figure 4. The filtration loss is decreases when the concentration of OPEFB increased while the filter cake thickness is increased when the concentration of OPEFB increases. It shows that the OPEFB particles have filled the pore spaces and formed a thicker mud cake and prevent the fluid loss from passing through the filter paper. A good drilling mud is typically produced filter cake with a thickness of less than 3 mm. However, higher particle size of OPEFB is adversely influence the filter cake forming process, causing increased in mud cake thickness. It is advisable not to use 20g of OPEFB as it will increase the mud cake thickness more than 3mm from the based mud. This may lead to potential pipe sticking in later drilling operation. The 15g concentration of 150 μm and 250 μm sizes of OPEFB can reduce the filtrate volume up to 32% and increase the mud thickness about 3 mm from the based mud which can be considered as a fair limit.

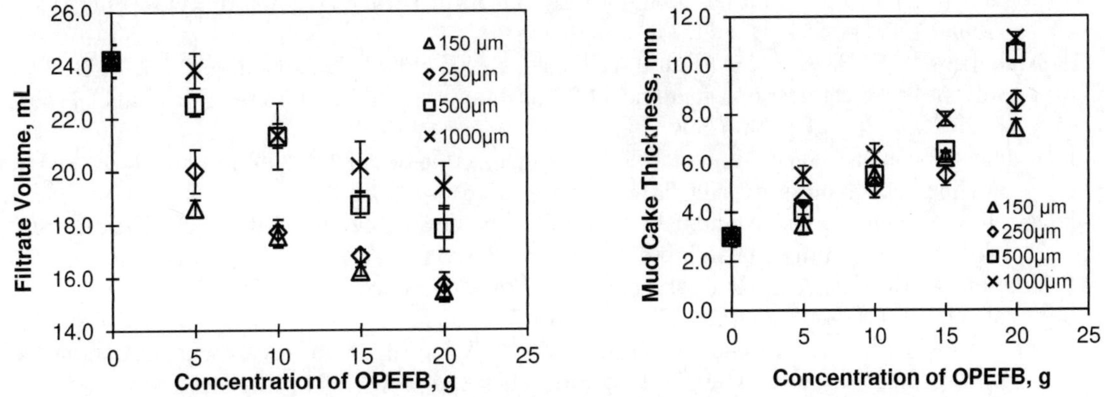

Figure 4. Filtrate analysis of WBM at different concentration and size of OPEFB

4. Conclusion

It can be generally concluded that the addition of OPEFB into the mud could improve the rheology and filtration properties of water based mud. However, proper selection of its size and concentration is vital to fit the condition of drilling operation. OPEFB is suitable to be used as LCM in WBM as it can reduce the filtrate volume and plug the the pore spaces of the formation by having thicker mud cake.

Acknowledgment

The authors would like to thank Universiti Teknologi MARA for the financial support through the ARAS Grant: 600-RMI/DANA 5/3/ARAS (37/2015), LESTARI Grant: 600-IRMI/DANA 5/3/LESTARI (0028/2016), FRGS Grant: 600-RMI/FRGS 5/3 (98/2015) and 600-RMI/RAGS 5/3 (60/2012)

References

[1] Nayberg T M and Petty B R 2013 Laboratory Study of Lost Circulation Materials for Use in Oil-Base Drilling Muds *Soc. Pet. Eng.*

[2] West G, Hall J and Seaton S 2006 Petroleum Engineering Uploaded by : *Ebooks Chemical Engineering For More Books, softwares & tutorials Related to Chemical Engineering*, ***II***

[3] Nayberg T M, Western T and America N 1987 Laboratory Study of Lost Circulation Materials for Use in Both Oil-Based and Water-Based Drilling Muds *Soc. Pet. Eng.*

[4] Fidan E, Babadagli T, and Kuru E 2013 Use of Cement as Lost-Circulation Material: Best Practices *Can. Int. Pet. Conf.*, p 1–11

[5] Egun II and Abah A M 2013 Comparative performance of Cassava Starch to PAC as Fluid Loss Control Agent in Water Based Drilling Mud *Discov. Res. Chem. Eng*. **3**(9) p 36–39

[6] Ghazali N A, Mohd T A T, Alias N H, Azizi A, and Harun A. A. 2013 The Effect of Lemongrass as Lost Circulation Material (LCM) to the Filtrate and Filter Cake Formation *J. Key Eng. Mater*. **594–595** p 68–72.

[7] Azizi A, Shahrul M, Ibrahim N, Hamid K H K, Sauki A, Ghazali N A, Mohd T A T 2013 Agarwood Waste as A New Fluid Loss Control Agent in Water-based Drilling Fluid *Int. J. Sci. Eng*. **5**(2) p 101–105

[8] Perlmutter B A 2005 Clean in place operations for thin-cake filtration technologies *Fine Technology*

[9] Kabir M A and Gamwo I K 2011 Filter cake formation on the vertical well at high temperature and high pressure : Computational fluid dynamics modeling and simulations *J. Pet. Gas Eng.* **2** p 146–164

IOP Conf. Series: Materials Science and Engineering **358** (2018) 012022 doi:10.1088/1757-899X/358/1/012022

[10] Ayotunde T and Precious C 2012 Sawdust As A Filteration Control And Density Additives in Water-Based Drilling Mud *Int. J. Sci. Eng.* **3**(7) p 7–10

[11] Rahman S H A, Choudhury J P, Ahmad A L and Kamaruddin A H 2007 Optimization studies on acid hydrolysis of oil palm empty fruit bunch fiber for production of xylose *Bioresour. Technol.* **98**(3) p 554–9

[12] Abdullah S S S, Hassan M A, Shirai Y, Funaoka M, Shinano T, and Idris A 2009 Effect of Solvent Pre-Treatment on Lignophenol Production from Oil Palm Empty Fruit Bunch Fibers *J. Oil Palm Res.* **21 p** 700–709

[13] Akhtar J, Kuang S K, and Amin N A S 2010 Liquefaction of empty palm fruit bunch (EPFB) in alkaline hot compressed water *Renew. Energy* **35**(6) p 1220–1227

[14] Ramli R, Shaler S, and Jamaludin M A 2002 Properties of Medium Density Fibreboard From Oil Palm Empty Fruit Bunch Fibre *J. Oil Palm Res.* **2** p 34–40

[15] Ibrahim O, Badri S M H, Khairiah H 2012 A Study on Glycerolysis of Oil Palm Empty Fruit Bunch Fiber *Sains Malaysiana J.* 41 (12) p 1579–1585

[16] Ibrahim M N M, Azian H, and Yusop M R M 2007 The Effects of Lignin Purification on the Performance of Iron Complex Drilling Mud Thinner *J. Teknol.* **44** p 83–94.

[17] Soom R M A T, Aziz A A B D, Hasamudin W A N, and Hassan W A N 2009 Solid-State Characteristics of Microcrystalline Cellulose from Oil Palm Empty Fruit Bunch Fibre *J. Oil Palm Res.* 21 p 613–620

[18] Onyeukwu I H, Ihuoma W I, and Agbor E S 2012 Use of Irvingia Gabonensis As A Viscosifier for Water Based Muds *Society of Petroleum Engineers*

[19] Ariyanto E Realising Potential in Oilfield Drilling Solutions Drilling Fluids Scomi Oiltools

[20] Ghazali N A, Mohd T A T, Alias N, Shahruddin M Z, Sauki A, and Maliki M B F 2014 The Effects of Temperature on Rheology Properties and Filtrate after Using Lemongrass as Lost Circulation Materials for Oil Based Drilling Mud *J. Adv. Mater. Res.* **911** p 243–247

[21] Meng X, Zhang Y, Zhou F, and Chu P K 2012 Effects of carbon ash on rheological properties of water-based drilling fluids *J. Pet. Sci. Eng.* **100** p 1–8

[22] Alias N, Tahir N F M, Mohd T A T, Ghazali N A, Yahya E, Shahruddin M Z, Ramlee N A, and Azizi A 2014 Rheological Study of Nanosilica Based Drilling Fluid *J. Appl. Mech. Mater.* **575** p 128–133

[23] Falode O A, Ehinola O A, and Nebeife P C 2008 Evaluation of local bentonitic clay as oil well drilling fluids in Nigeria *Appl. Clay Sci.* 39(1–2) p 19–27

[24] Ghazali N A, Jaih M Z M, Mohd T A T, Alias N, Azizi A, and Yahya E 2015 The Characteristic Study of Oil Palm Kernel Expeller as Lost Circulation Material in Water Based Drilling Mud (WBM) *J. Adv. Mater. Res.* **1113** p 648–653

[25] Max R A and Martin V S 1974 Drilling Fluid Technology Exxon Co. U.S.A.

IOP Conf. Series: Materials Science and Engineering **358** (2018) 012023 doi:10.1088/1757-899X/358/1/012023

Growth Kinetics for Microalgae Grown in Palm Oil Mill Effluent (POME) medium at various CO_2 Levels

S Razali, J Salihon and M A Ahmad
Faculty of Chemical Engineering, Universiti Teknologi MARA, 40450 Shah Alam, Selangor, Malaysia

jailani@salam.uitm.edu.my

Abstract. This paper sought to find the growth kinetic data of maximum specific growth rate (μ_{max}) and substrate saturation constant (K_S) for a microalgal reaction system over various dissolved CO_2 levels (0.04, 0.1, 0.3, 0.5, 0.8, 1.0, 5.0, 10.0% v/v) at a constant sparging rate of 1.2 vvm, by using logistic model and Monod kinetics. The reaction system consisted of microalgae growing in palm oil mill effluent (POME) medium in 1 L flask with constant light illumination and sparged with the specified CO_2 gas mixture. It is found from the experimental works that the values of μ_{max} and K_S to be at 0.04958 h^{-1} and 0.03523% (v/v) respectively. The results also showed that utilizing CO_2 levels (v/v) in the sparging gas mixture more than 1% (v/v) would not improve microalgae growth significantly as expressed in the values of specific growth rate μ. These data and information are critically important for bioreactor scaling up purposes, especially bioreactor system dedicated for microalgae products and CO_2 sequestration.

1. Introduction

The prospects of incorporating microalgal system into the palm oil industry for residual nutrient removal in palm oil mill effluent (POME) have been rigorously investigated, and the results are positively encouraging [1]. While the application of microalgae for bioremediation may be enticing, its potential for CO_2 sequestration is equally important to realize the zero-waste palm oil mill, where all greenhouse gases (GHG) produced are fully captured, and all the wastewater is fully recycled. POME can be used to generate renewable bioenergy such as biomethane and biohydrogen through anaerobic digestion [2]. May et al. [3] estimated that a palm oil mill with processing capacity of 60 tonnes/hr can generate a net profit of RM 3.8 mil per year from electricity generation using biogas produced from POME treatment. However, the final GHG discharged from the use of biomethane as fuel, carbon dioxide (CO_2), needs to be captured, along with all CO_2 produced elsewhere in the mill. Thus, it is envisioned that a microalgal system be incorporated into palm oil industry where bioremediation of residual nutrient in POME is coupled with sequestration of waste CO_2 gas [2,4].

Previous work by Adibah et al. [4] used factorial experimental designs to evaluate the microalgae biomass yield and the CO_2 recovery by microalgae grown in POME medium. They found that both the experimental variables of gas-mixture sparging rate (vvm) (representing mixing intensity) and the level of CO_2 (% v/v) in the CO_2-air gas mixture (approximating dissolved CO_2 – DCO_2) have significant main effects on biomass yield. Their conclusion adds to the evidence that aeration rate is proven to have influenced microalgal growth to a substantial extent [5,6] and the volume percentage of CO_2 in the sparging gas mixture can be increased to overcome the gas-liquid mass transfer limitation experienced in systems with inherently poor mass transfer capacity [6,7]. In this paper, we sought to find the overall maximum specific growth rate (μ_{max}) of the microalgae growing in POME medium over various levels of CO_2 at a constant sparging rate. The DCO_2 can be approximated from the respective levels of CO_2 content (% v/v) in the sparging gas mixture, because partial pressure of a gas

Content from this work may be used under the terms of the Creative Commons Attribution 3.0 licence. Any further distribution of this work must maintain attribution to the author(s) and the title of the work, journal citation and DOI.
Published under licence by IOP Publishing Ltd

corresponds to its solubility in a liquid as predicted by Henry's law [8]. Determining μ_{max} is of great importance in this case because it can be used for evaluating cell growth performance; which is especially useful in designing bioreactor systems. In addition, the kinetic data and information from this study might present valuable insights on the reaction system as well as imperative industrial implications.

2. Materials and methods

2.1. Microalgae culture

Chlorella sp. with identification number of UMACC324 was obtained from University of Malaya Algae Culture Collection (UMACC). The culture was maintained in 1 L flask in Bold's Basal Medium (BBM) [9] at ambient temperature with sterile air pump aeration.

2.2. Palm oil mill effluent (POME) sample

POME used in this experiment was obtained from the aeration pond of Sime Darby Bhd Mill (2°30'52.1''N101°53'34.3''E) in Sua Betong, Negri Sembilan, Malaysia. The POME was stored at 4°C until being used.

2.3. Experimental Set-up

The gas mixing system used in this study combined the air flow and gaseous CO_2 flow to make the gas mixture sparged into the culture. The pressure of each of the air flow and the CO_2 flow was regulated to be at 2 bars and the flow rate for each flow was controlled by a flow controller and measured by a flowmeter (model N082-03, Cole-Parmer, IL, USA) to give the required CO_2 content (% v/v) in the sparging gas mixture. The CO_2-air mixture entered the culture system from the bottom of the flask through a sparging tube with diameter of 1 mm. The stopper of the flask culture has one input tube for sparging and one output tube for venting. The concentration of DCO_2 in this study was approximated based on the CO_2 content (% v/v) of the sparging gas mixture, since it is recognized that the partial pressure of CO_2 gas will determine the solubility of the gas in the medium. In this regard, the composition of CO_2 in air was taken to be representative of the partial pressure of CO_2 as Dalton's law dictates that the percentage by volume of a gas in mixture corresponds with partial pressure it exerts [8]. The limitation of this approximation is in so far as that equilibrium is assumed to be achieved between DCO_2 and the CO_2 gas in the bubbles. The lighting chamber was an open-top box that was equipped with fluorescent bulbs on the inside walls that provided the light intensity of 10,000 lux. Figure 1 shows the experimental set-up for this study.

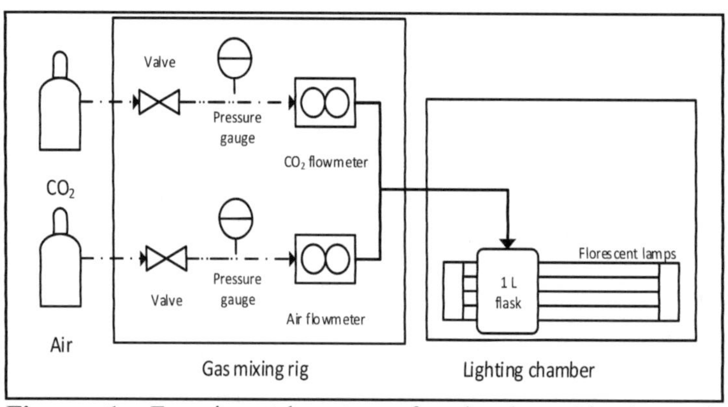

Figure 1. Experimental set-up for batch cultivation of microalgae in POME medium in 1 L flask with CO_2 supplementation. The set-up consists of a gas mixing rig which provides specified gas mixture of CO_2-air into a 1 L flask placed in a chamber with 10,000 lux illumination capacity.

2.4. Batch Cultivation of Microalgae in POME Medium in 1 L Flask

The medium was prepared by diluting 500 mL of unsterilized POME sample with 400 mL of distilled water [4]. A 10% (v/v) inoculum was then added, making the total volume of 1 L. The three-day old inoculum was prepared by the same technique as per microalgae maintenance procedure. Batch cultivation in 1 L flasks were carried out in duplicate. The flasks were placed in the lighting chamber and the cultivation proceeded up to seven (7) days. At interval of 12 hours, 30 mL sample was withdrawn, with a maximum of seven samples being drawn from a particular flask before the flask was discarded, and the sampling started on a new identical flask in the duplicate. The sample was centrifuged at 10,000 rpm for 15 minutes where the supernatant is discarded. The remaining pellet is then re-suspended in distilled water of the same volume and oven-dried in porcelain crucible at 100oC over 24 hours, and weighed to determine the cell dry weight (CDW) of the biomass.

2.5. Logistic Modeling

The logistic equation [10,11] was used to model the biomass growth profile of the microalgae in batch culture is shown in equation (1) and its integration gives equation (2); where X is actual biomass concentration (g/L), X_m is the maximum cell concentration (g/L), and μ is the specific growth rate (hr^{-1}) of the microalgae at the given level of CO_2 content. Denoting the actual microalagal biomass concentration at the beginning of the cultivation (t=0) as X_0, the logistic equation (2) was fitted to the data of cell concentration X in time t from each batch cultivation experiment using curved fitting tool in MATLAB® R2013a (Mathworks, Natick, MA). Confidence bounds of 95% were taken into consideration to find the fit.

$$\frac{dX}{dt} = \mu X(1-\frac{X}{X_m}) \tag{1}$$

$$X = \frac{X_0 X_m e^{\mu t}}{X_m - X_0 + X_0 e^{\mu} t} \tag{2}$$

2.6. Determination of Kinetic Parameters of Overall μ_{max} and K_S

The μ (hr^{-1}) of the microalgae at a given level of CO_2 content increases with increasing level of CO_2 content, until the level of CO_2 content which gives the highest value of μ has been reached. Further increase in the level of CO_2 content beyond this point will not cause any increase in the value of μ. This highest value of μ is called the overall maximum specific growth rate (hr^{-1}) or μ_{max}. The saturation constant K_S is a constant in the Monod equation (3) [11] which has a value of half the value of the concentration of DCO_2 corresponding to the level of CO_2 content first giving μ_{max}; and S in the equation is the DCO_2 as represented by the level of CO_2 content (% v/v) in the sparging gas mixture.

$$\mu = \frac{\mu_{max} S}{K_s S} \tag{3}$$

In order to study the kinetics of the batch growth of the microalgae in POME medium, eight (8) batch culture experimental runs were conducted at eight (8) different levels of CO_2 in the sparging gas mixture namely 0.04, 0.1, 0.3, 0.5, 0.8, 1.0, 5.0, and 10.0% (v/v) at 1.2 vvm constant sparging rate. The Monod equation was used to model the growth kinetics and to find the value of overall μ_{max} and the value of K_S. It was fitted to the values of μ determined from each batch cultivation experiment using the same MATLAB® curve fitting tool as per logistic equation previously. Minimum doubling time $t_{d(min)}$ is then determined using equation (4) [11].

$$t_{d(\min)} = \frac{\ln 2}{\mu_{max}} \tag{4}$$

3. Result and discussion

The growth profile of cultivation of *Chlorella* sp. in POME medium was determined at eight (8) different levels of CO_2 in the sparging gas mixture of 0.04, 0.1, 0.15, 0.2, 0.3, 0.5, 1.0, 5.0, and 10.0%

(v/v). Table 1 gives the biomass concentrations of the culture from samples taken at 12 hr intervals for seven days from the 1 L batch cultivation runs under 10,000 lux fluorescent light illumination, using constant sparging at 1.2 vvm with gas mixture different levels of CO_2 as above. The biomass growth data (table 1) was fitted into the logistic equation (2) to give the values of μ and X_m for each specific run characterized by the level of CO_2. Table 2 presents the values of μ and X_m for each batch of the cultivations. These data was then used to generate a plot of specific growth rate (μ) vs. the corresponding set point of CO_2 levels (S) by fitting it with the Monod equation (3) to give two important kinetic parameters namely maximum specific growth rate (μ_{max}) and substrate saturation constant (K_S). This was also done with aid of curve fitting tools in MATLAB® software. The plot can be seen in figure 2.

Table 1. Experimental data of growth in time (h) of *Chlorella* sp. growing in POME medium at various CO_2 levels (% v/v) in air.

CO_2%	Biomass Concentration (g/L)													
	0 h	**12**	**24**	**36**	**48**	**60**	**72**	**84**	**96**	**108**	**120**	**132**	**144**	**156**
0.04	0.20	0.30	0.22	0.30	0.36	0.44	0.52	0.56	0.52	0.54	0.58	0.56	0.58	0.64
0.10	0.30	0.35	0.37	0.43	0.47	0.48	0.47	0.53	-	0.50	0.50	0.48	0.50	0.53
0.30	0.20	0.43	0.45	0.47	0.57	0.70	0.65	0.62	0.69	0.80	0.77	0.77	0.73	0.77
0.50	0.20	0.30	0.38	0.36	0.38	0.42	0.45	0.48	0.48	-	0.48	0.54	0.46	0.48
0.80	0.20	0.33	0.37	0.40	0.40	0.43	0.50	0.67	0.50	0.53	0.50	0.43	0.50	0.48
1.00	0.30	0.38	0.44	0.46	0.50	0.53	0.56	0.60	0.60	0.56	0.53	0.53	0.50	0.52
5.00	0.20	0.23	0.30	0.33	0.36	0.37	0.37	0.40	0.40	0.43	0.40	0.39	0.40	-
10.0	0.10	0.30	0.37	0.43	0.48	0.50	0.57	0.67	0.73	0.81	0.82	0.82	0.83	0.80

Table 2. Specific growth rate μ and maximum biomass concentration X_m of *Chlorella* sp. growing in POME medium at different CO_2 level obtained from logistic modeling [equation (2)].

CO_2 (% v/v)	μ (h^{-1})	X_m (g/L)	Goodness of Fit, R^2
0.04	0.02507	0.6461	0.9338
0.10	0.03764	0.5099	0.9303
0.30	0.04361	0.7533	0.9178
0.50	0.04429	0.4866	0.8991
0.80	0.05461	0.5066	0.7092
1.00	0.05051	0.5499	0.8171
5.00	0.04396	0.4065	0.9724
10.0	0.04785	0.8079	0.8925

From the result in table 2, it can be seen that the logistic equation found to fit the experimental data reasonably well based upon the value of R^2. Most batches have near to unity R^2 values with only a few batches giving comparatively lower values. Similarly, the values of X_m also varies over different CO_2 levels with the highest is observed at 0.8079 g/L at 10% (v/v) CO_2 set point and the lowest is recorded at 0.4065 g/L at 5% (v/v) CO_2 set point. The X_m indicates the carrying capacity of the culture which can be defined as the maximum algal biomass concentration that can be supported by the culture at the particular experimental conditions [10]. Hence, the fluctuating pattern of X_m that can be observed from the lowest CO_2 set point to the highest is reasoned due to the variation in the availability of carbon source in the form of DCO_2 in these experiments. In addition, POME characteristics vary considerably with respect to variation in feedstock supply from plantation, rainfall pattern, and changing pattern of operation of the mill [12]. The aeration pond POME used in the experiment was taken from the same mill which is Sime Darby Berhad Sua Betong Mill, but obtained at different times during the course of the experiment. The resulting variation in POME characteristics actually can be visually observed while conducting the experiment as one batch is darker than the previous one and vice versa. This surely influenced the growth profile of the microalgae growing in POME, in addition to the influence of the variation in the availability of carbon sources in the form of DCO_2, which together translated in the fluctuating pattern of X_m. The same reason also applies for the lack of fit in plot of μ vs. S (figure 2) that gives R^2 of 0.8321.

The K_S for this study is observed at 0.03523% (v/v CO_2 in gas mixture). This value provides a guideline at which the CO_2 concentration should be maintained in the feed as it characterizes the CO_2 concentration required to reach half of the μ_{max}. On the other hand, minimum doubling time $t_{d(min)}$ for this study is 13.9804 h that is determined based on equation (4). These kinetic data of $t_{d(min)}$ and μ_{max} of microalgae growing in POME medium shows how fast is the microalgae growing in the system. The kinetic data gained from the system essentially gave a noticeably higher rate when compared to other Chlorella strains. Take for example, *Chlorella vulgaris* and *Chlorella kessleri* have μ_{max} of 0.01220 h^{-1} and 0.01071 h^{-1} respectively [13]. However, there are also other studies that showed their value of μ_{max} to be much lower. For instance, Kumar et al. [14] found μ_{max} for *Chlorella sorokiniana* to be at 0.1 h^{-1}. These disparities are caused by different cultivation systems being used since growth rates are very test system specific. Thus, it is reasonable for each experimentations with dissimilar cultivation medium and bioreactor configuration to give different μ_{max} values.

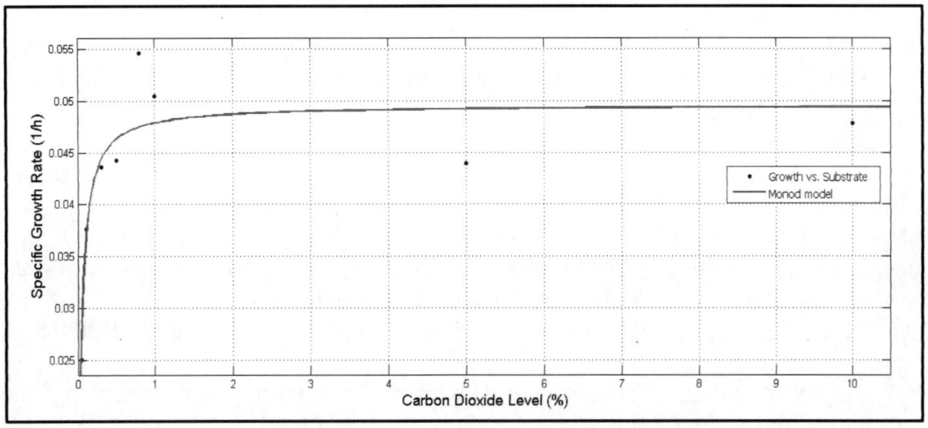

Figure 2. Plot of μ vs. CO_2 levels (S) obtained by fitting growth data against Monod equation (3). μ_{max} and K_S found at 0.04958 h^{-1} and 0.03523% (v/v) respectively, with R^2 of 0.8321. The closed symbol (•) denotes μ from logistic equation (2) (table 2) and solid blue line is the fitting curve.

Moreover, referring to plot of μ vs. S (figure 2), it is observed that the values of μ experienced a steep rise as the level of CO_2 increased from zero to 1% (v/v), followed by a very gentle slope as the level of CO_2 is increased further. So this implicates that, especially at industrial scale, for such a reaction system of microalgae growing in POME medium, it would not be economical to work at CO_2 levels higher than 1% (v/v), since it would not significantly improve the biomass performance as reflected in term of μ values.

4. Conclusion

The batch growth kinetics study conducted over various DCO_2 concentrations for a reaction system of microalgae growing in POME medium with 10,000 lux light illumination gave μ_{max} and K_S at 0.04958 h^{-1} and 0.03523% (v/v) respectively. It is found that amid variations in POME characteristics due to varying environmental factors and processing conditions, the microalgae still able to grow very well, though it affects the μ values to some extent. It is also established that utilizing CO_2 levels higher than 1% (v/v) in the sparging gas mixture would not improve biomass performance as reflected in its μ values.

Acknowledgment

This work was funded by MOSTI e-Science fund grant #02-01-01-SF0790 and MOE grant #600-RMI/RAGS 5/3 (204/2014). We are very grateful for the assistance provided by staffs of FKK UiTM and Dr Ding Gong Tao.

References

[1] Tao D G, Yaakob Z, Takriff M S, Salihon J and Abd Rahman M S 2016 Biomass production and nutrients removal by a newly-isolated microalgal strain *Chlamydomonas* sp in palm oil mill effluent (POME) *Int. J. Hydrogen Energy* **41** 4888–95

[2] Lam M K and Lee K T 2011 Renewable and sustainable bioenergies production from palm oil mill effluent (POME): Win-win strategies toward better environmental protection *Biotechnol. Adv.* **29** 124–41

[3] May C, Phaik P, Ti T, Eng C and Kit C 2013 Biogas from palm oil mill effluent (POME): Opportunities and challenges from Malaysia's perspective *Renew. Sustain. Energy Rev.* **26** 717–26

[4] Ahmad A D, Salihon J and Tao D G 2015 Evaluation of CO_2 sequestration by microalgae culture in palm oil mill effluent (POME) medium *Adv. Mater. Res.* **1113** 311–6

[5] Zhang K, Kurano N and Miyachi S 2002 Optimized aeration by carbon dioxide gas for microalgal production and mass transfer characterization in a vertical flat-plate photobioreactor *Bioprocess Biosyst. Eng.* **25** 97–101

[6] Ryu H J, Oh K K and Kim Y S 2009 Optimization of the influential factors for the improvement of CO_2 utilization efficiency and CO_2 mass transfer rate *J. Ind. Eng. Chem.* **15** 471–5

[7] Langley N M, Harrison S T L and van Hille R P 2012 A critical evaluation of CO_2 supplementation to algal systems by direct injection *Biochem. Eng. J.* **68** 70–5

[8] Peavy H S, Dowe D R and Tchobanoglous G 1985 *Environmental Engineering* (Singapore: McGraw-Hill Inc.)

[9] Andersen R A 2005 *Algal Culturing Techniques* (China: Elsevier Academic Press)

[10] Kumar K and Das D 2012 Growth characteristics of *Chlorella sorokiniana* in airlift and bubble column photobioreactors *Bioresour. Technol.* **116** 307–13

[11] Shuler M L and Kargi F 2002 *Bioprocess Engineering: Basic Concepts* (Pearson Education Inc.) p. 155

[12] Cai T, Park S Y and Li Y 2013 Nutrient recovery from wastewater streams by microalgae: Status and prospects *Renew. Sustain. Energy Rev.* **19** 360–9

[13] Zhao B and Su Y 2014 Process effect of microalgal-carbon dioxide fixation and biomass production: A review *Renew. Sustain. Energy Rev.* **31** 121–32

[14] Kumar K, Dasgupta C N and Das D 2014 Cell growth kinetics of *Chlorella sorokiniana* and nutritional values of its biomass *Bioresour. Technol.* **167** 358–66

IOP Conf. Series: Materials Science and Engineering **358** (2018) 012024 doi:10.1088/1757-899X/358/1/012024

Precious Metals Recovery from Electroplating Wastewater: A Review

A A Azmi, J Jai, N A Zamanhuri and A Yahya,

Faculty of Chemical Engineering, Universiti Teknologi MARA (UiTM), 40450 Shah Alam, Selangor, Malaysia

junejai@salam.uitm.edu.my

Abstract. Metal bearing electroplating wastewater posts great health and environmental concerns, but could also provide opportunities for precious and valuable metal recovery, which can make the treatment process more cost-effective and sustainable. Current conventional electroplating wastewater treatment and metal recovery methods include chemical precipitation, coagulation and flocculation, ion exchange, membrane filtration, adsorption, electrochemical treatment and photocatalysis. However, these physico-chemical methods have several disadvantages such as high initial capital cost, high operational cost due to expensive chemical reagents and electricity supply, generation of metal complexes sludge which requires further treatment, ineffective in diluted and/or concentrated wastewater, low precious metal selectivity, and slow recovery process. On the other hand, metal bio-reduction assisted by bioactive phytochemical compounds extracted from plants and plant parts is a new found technology explored by several researchers in recent years aiming to recover precious and valuable metals from secondary sources mainly industrial wastewater by utilizing low-cost and eco-friendly biomaterials as reagents. Extract of plants contains polyphenolic compounds which have great antioxidant properties and reducing capacities, able to reduce metal ions into zerovalent metal atoms and stabilize the metal particles formed. This green bio-recovery method has a value added in their end products since the metals are recovered in nano-sized particles which are more valuable and have high commercial demand in other fields ranging from electrochemistry to medicine.

1. Introduction

Electroplating is the application of a metal coating onto a metallic or other conducting surfaces by an electrochemical process [1]. In typical electroplating industries, large amount of wastewater are generated from rinsed water of the electroplated parts [2] which equivalent to 20% of the chemicals used in the metal salts plating baths [3]. This wastewater contains high amount of metal ions which exceeds the industrial effluent discharge limits stated by the Department of Environment and hence must be treated before being discharged due to their high toxicity [4]. However, in certain electroplating industries especially involving precious metals such as silver, gold and platinum, the wastewater contains adequate amounts of precious and valuable metals. Hence, a suitable wastewater treatment is required not only to remove the heavy metals but also to recover those precious and valuable metals for both growing economics and environmental concerns. Recently, several conventional wastewater treatment methods have been applied in electroplating industries.

Content from this work may be used under the terms of the Creative Commons Attribution 3.0 licence. Any further distribution of this work must maintain attribution to the author(s) and the title of the work, journal citation and DOI.
Published under licence by IOP Publishing Ltd

2. Conventional wastewater treatments and metal recovery methods

Currently used wastewater treatments and metal recovery methods in electroplating industries are chemical precipitation, coagulation and flocculation, ion exchange, membrane filtration, adsorption, electrochemical treatment and photocatalysis [5]–[9].

2.1. Chemical precipitation

Chemical precipitation is a process to form insoluble metal precipitates by reacting precipitant agent of hydroxides or sulfides with the dissolved metal ions [6]. This technique is the most widely used treatment methods in current electroplating industries [10], [11] which employed pH adjustment to basic conditions to reduce the metal ions solubility in the wastewater. Although sulfide precipitation could achieve high metal removal and are not amphoteric [7], but hydroxide precipitation are more preferable due to the precipitant agents (lime and limestone) availability and low-cost in most countries [12]. The metal precipitates formed are recovered using a solid separation process such as coagulation and/or sedimentation or filtration [8] and the precious metals are purified by chemical extraction process [13]. Advantages of chemical precipitation method are simple operation and low capital cost due to their availability and low-cost precipitant agents, while the disadvantages are it generate excessive amount of sludge that requires further treatment in order to recover metals, slow metal precipitation, poor settling, the aggregation of metal precipitates, and the long-term environmental impacts of sludge disposal [6]–[9].

2.2. Coagulation and flocculation

Coagulation is the destabilization of colloidal particles by neutralizing the repulsive forces that keep them apart using a coagulant [6] of ferric or alum salts [7]. On the other hand, flocculation is the action of flocculants polymer to form bridges and bind the unstable particles to form larger particles or bulky floccules [6]. The bulky metal floccules can subsequently be recovered by sedimentation or filtration [7]. However, in electroplating industries, amphoteric polyelectrolyte coagulant/flocculant such as polydiallyldimethylammonium chloride (polyDADMAC) [14], polyethyleneimine-sodium xanthogenate (PEX) [15], and mercaptoacetyl polyethyleneimine (MAPEI) [16] were commercially used to remove or recover soluble metal ions from the wastewater effectively based on their chelating ability. The advantages of coagulation–flocculation method are shorter time to settle out suspended solids, improved sludge settling, dewatering characteristics, bacterial inactivation capability and sludge stability while the disadvantages are high operational cost due to chemical consumption and treatment of the sludge generated to recover metals [6], [7], [9].

2.3. Ion exchange

In ion exchange, the ion exchanger (resin) exchanges its cations with the metal ions in the wastewater without any structural change of the resin [6]. The metal is then recovered by elution with suitable reagents after separating the loaded resin from the wastewater. Common cation exchangers are strongly acidic resins with sulfonic acid groups ($-SO_3H$) and weakly acid resins with carboxylic acid groups (–COOH) [7] where hydrogen ions in the groups serve as exchangeable ions with the metal cations in the wastewater. Instead of natural resins such as Clinoptilolite [17], synthetic resins such as Amberlite [18] and Purolite [19] were frequently used in electroplating industries since it is more effective and inexpensive [20]. The advantages of ion exchange method are no sludge generation, high treatment capacity, high removal efficiency in which certain resins ligands can selectively bond with certain metal cations and fast kinetics, while the disadvantages are required a pre-treatment systems for secondary effluent, not suitable for all types of metals and high capital and operational costs [6], [7], [21].

2.4. Membrane filtration

Membrane filtration is a method that separates particles in a fluid by applying a pressure onto the fluid to pass across a porous membrane or filter [6]. Membrane filtration generally divided into three types which are ultrafiltration, nanofiltration and reverse osmosis, depending on the size of the particle that can be retained [6], [8], [22]. Ultrafiltration utilizes permeable membrane to separate heavy metals, macromolecules and suspended solids from inorganic solution on the basis of the pore size (5-20 nm) and molecular weight of the separating compounds (1000-100,000 Da) to allow the passage of water and low-molecular weight solutes, while retaining the larger sized macromolecules [6], [8], [23]. Nanofiltration involves mechanism of steric (sieving) and electrical (Donnan) effects [24]. A Donnan potential is created between the charged anions in the nanofiltration membrane and the co-ions in the effluent to reject the latter [6]. The significance of this membrane lies in its small pore and membrane surface charge [9], which allows charged solutes smaller than the membrane pores to be rejected along with the bigger neutral solutes and salts. Reverse osmosis is a pressure-driven membrane process in which water can pass through the semi-permeable membrane (pore size down to 10^{-4} μm) [25] and retained the heavy metal inside [6], [7]. In reverse osmosis, cationic compounds were separated from water by exerting a greater hydrostatic pressure than the osmotic pressure of the feeding solution [6]. The advantages of membrane filtration method are small space requirement, low pressure, high selectivity and separation efficiency while the disadvantages are high operational cost due to membrane fouling, high capital cost of membrane sheets, and low permeate flux [6]–[8], [22].

2.5. Adsorption
Adsorption is a mass transfer process by which a substance is transferred from the liquid phase to the surface of a solid (adsorbents), and becomes bound by physical and/or chemical interactions [6]. Generally, the mechanisms involved in adsorption of metal ions onto a solid adsorbent are: (1) the transport of the metal ions from the bulk solution/wastewater to the adsorbent surface; (2) adsorption on the particle surface; and (3) transport within the adsorbent particles [8]. Widely used adsorbent types to recover metal from electroplating wastewater are activated carbons [26], carbon nanotubes (CNTs) [27], low-cost adsorbents [28] such as agricultural wastes [29] and industrial by-products [30], bioadsorbents [31] and modified biopolymers derived from chitosan [32]. The advantages of adsorption method are low-cost adsorbent, easy operating conditions, wide pH range and high metal-binding capacities while the disadvantages are low selectivity and recovery efficiency, and production of waste products [7]–[9], [33].

2.6. Electrochemical treatment
Electrochemical treatment is a method that involve plating-out of metal ions on a cathode surface using an electric current [7]. Currently applied electrochemical treatment methods in electroplating industries are electrocoagulation, electrodeposition and electrodialysis [9]. Electrocoagulation involves the generation of coagulants in situ [7] by dissolving electrically either aluminium or iron ions from aluminium [34] or iron [35] electrodes. The coagulants ions are generated at the anode, and hydrogen gas is released from the cathode to help float the flocculated particles out of the water [7]. Electrodeposition or electrowinning is an electrolytic process that involves redox (reduction–oxidation) reaction where the positively charged metal ions (cations) are reduced and deposited at the cathode while the negatively charged ions (anions) are oxidize at the anode [6]. Electrodeposition is a clean technology with no presence of the permanent residues for the recovery of metals [9]. Electrodialysis is a membrane separation in which ionized species in the solution are passed through an ion exchange membrane by applying an electric potential [6]. The membranes are thin sheets of plastic materials with either anionic or cationic characteristics. When a solution containing ionic species passes through the cell compartments, the anions migrate toward the anode and the cations toward the cathode, crossing the anion-exchange and cation-exchange membranes [8]. Electrodialysis generates two new solutions: one concentrated on ions and another consisting of almost pure water which are able to be reused back in the plating process [36]. The advantages of electrochemical treatment method are rapid and well controlled process, low chemical requirement, high selectivity, and no sludge generation while the disadvantages are high initial capital cost, and high operational cost due to energy consumption and expensive electricity supply [6]–[9].

2.7. Photocatalysis
Photocatalysis is a method that couples low-energy ultraviolet light with semiconductor catalyst particles to reduce metal ions contain in an electrolyte by the photogenerated electrons [37]. When semiconductor particles are illuminated by ultraviolet light with energy greater than the bandgap energy of the photocatalyst, electron/hole pairs are generated within the catalyst particles [37]. In photocatalytic oxidation, the electron/hole recombination are suppressed by the electron scavenger to allow the photogenerated holes undergo anodic reactions, while in photocatalytic reduction, the hole

scavenger is adopted and the photogenerated electrons are allowed to undergo cathodic reactions [8], [37], [38]. Commonly used semiconductor photocatalysts are TiO_2, ZnO, CeO_2, CdS and ZnS which derived from chalcogenides compounds [38]. The metals are deposited onto the surface of the photocatalyst and later can be extracted from the slurry by mechanical and/or chemical means. The advantages of photocatalysis method are simultaneous metal recovery and organic pollutant removal, generate less harmful by-products while the disadvantages are high capital cost of the photocatalysts, long duration time and limited applications [8], [37].

3. New alternative metal recovery method

It has long been recognized that naturally occurring phytochemicals in plants and plant parts have great antioxidant properties and reducing capacities [39]. These phenolic compounds have the abilities to act as both reducing and capping agents [40], able to reduce metallic ions into zerovalent metal atoms and stabilize the metal particles formed [41]–[43].

3.1. Bio-recovery mechanisms

The phenolic compounds extracted from plants possess hydroxyl (−OH) and carboxyl (−COOH) groups which give their significant ability to scavenge free radicals, chelate metal ions and donate hydrogen atoms or electrons [44], [45]. According to Moran *et al.* [46] this general metal chelating ability of phenolic compounds is probably related to the high nucleophilic character of the aromatic rings rather than to specific chelating groups within the molecule. According to Makarov *et al.* [47], the mechanisms of metal bio-recovery assisted by plant extracts consists of three main phases: (1) the activation phase which the reduction of metal ions occur; (2) the growth phase which the adjacent metal atoms spontaneously coalesce into larger particles (Ostwald ripening), which is accompanied by an increase in the thermodynamic stability of particles; and (3) the process termination phase by capping the metal particles formed (Figure 1). During the activation phase, the phenolic compounds liberates reactive hydrogen ions through the hydroxyl groups and get oxidized to *α,β*-unsaturated carbonyl groups (Figure 2) [48], and simultaneously reduce the metal ions [49]. On the other hand, the negatively charged oxygen in the hydroxyl groups of polysaccharides will interacts with the positively charged metal particle surfaces via electrostatic interactions and binds due to the presence of van der waals forces [50], [51]. During time, the metal particle will be fully capped and stabilized in the solution.

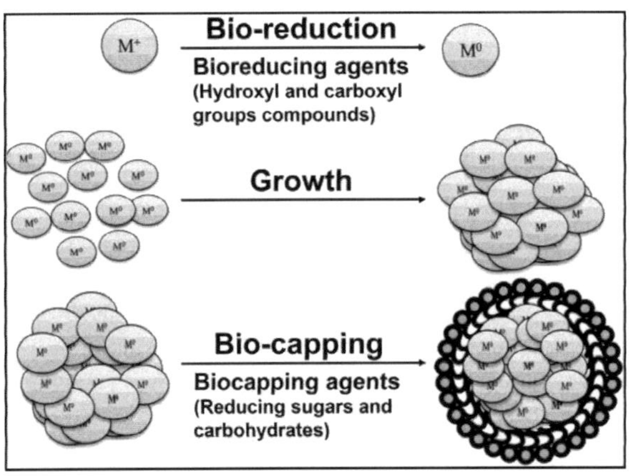

Figure 1. Mechanisms of metal bio-recovery [41].

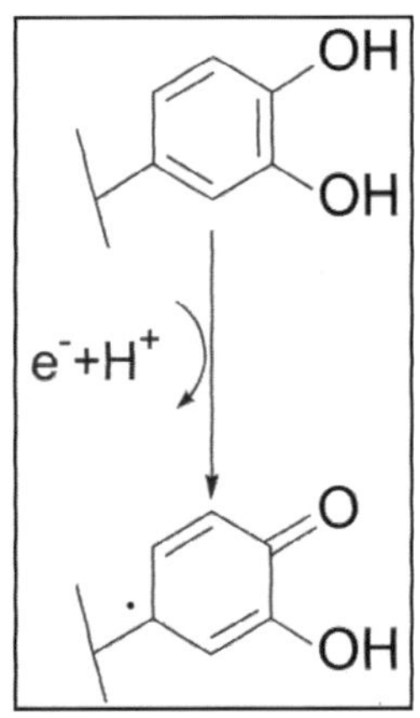

Figure 2. Hydroxyl oxidation [48].

3.2. Feasibility studies

Gatew and Mersha [52] use *Moringa stenopetala* seed powder extract to remove chromium from tannery wastewater which results 99.86% of chromium removal. Besides that, Azimi *et al.* [53] study chromium removal from tannery wastewater using acorn extract of *Quercus branti* and found 95.62% chromium removal efficiency. These results proved that the metal bio-reduction method assisted by plant extracts is very effective to remove and recover metals from wastewater. Several studies have reported that oil palm (*Elaeis guineensis*) leaves extracts are rich in phenolic compounds such as flavonoids and phenolic acids [54], [55], and showed great antioxidant activities [56], [57]. This anti-oxidative properties of palm leaves extract have been utilize by Azmi *et al.* [58] to successfully recover silver from simulated silver electroplating wastewater and in nano-sized silver nanoparticles.

4. Conclusions

To conclude, conventional physico-chemical methods have their own advantages and disadvantages in recovering those precious metals from electroplating wastewater. However, the wastewater characteristics and the metals concentration level were the two main factors that influence the recovery effectiveness. We propose this new potential bio-recovery method which utilizing eco-friendly biomaterials, palm leaves extract, to recover precious metal from electroplating wastewater and could successfully obtain the recovered metals in nano-sized metallic particles.

Acknowledgement

Authors would like to thank Ministry of Higher Education Malaysia (MOHE) for the financial sponsorship of this research under Fundamental Research Grant Scheme (FRGS), Project number FRGS/1/2014/TK05/UITM/03/2 and project leader Mdm. Asdarina Yahya to carry out this research.

References

[1] Barry D M and Kanematsu H 2016 *Corrosion Control and Surface Finishing: Environmentally Friendly Approaches* (Japan: Springer)
[2] Babu B R, Bhanu S U and Meera K S 2009 *J. Environ. Sci. Heal. C.* **27** 155–77
[3] Upadhyay K 2006 *J. Ind. Pollut. Control* **22** 59–66
[4] Adhoum N, Monser L, Bellakhal N and Belgaied J E 2004 *J. Hazard. Mater.* **112** 207–13
[5] Gupta V K, Ali I, Saleh T A, Nayak A and Agarwal S 2012 *Rsc Adv.* **2** 6380–8
[6] Kurniawan T A, Chan G Y, Lo W H and Babel S 2006 *Chem. Eng. J.* **118** 83–98
[7] Fu F and Wang Q 2011 *J. Environ. Manage.* **92** 407–18
[8] Barakat M A 2011 *Arab. J. Chem.* **4** 361–77
[9] Gunatilake S K 2015 *J. Multidiscip. Eng. Sci. Stud.* **1** 12–8
[10] Parmar M and Thakur L S 2013 *Int. J. Plant Anim. Environ. Sci*, **3** 143–57
[11] Ku Y and Jung I L 2001 *Water Res.* **35** 135–42
[12] Aziz H A, Adlan M N and Ariffin K S 2008 *Bioresour. Technol.* **99** 1578–83
[13] Babel S and del Mundo Dacera D 2006 *Waste Manag.* **26** 988–1004
[14] López-Maldonado E A, Oropeza-Guzman M T, Jurado-Baizaval J L and Ochoa-Terán A 2014 *J. Hazard. Mater.* **279** 1–0
[15] Chang Q and Wang G 2007 *Chem. Eng. Sci.* **62** 4636–43
[16] Min X, Qing C and Jinjin C 2010 *Water Environ. Res.* **82** 790–6
[17] Papadopoulos A, Fatta D, Parperis K, Mentzis A, Haralambous K J and Loizidou M 2004 *Sep. Purif. Technol.* **39** 181–8
[18] Tenorio J A and Espinosa D C 2001 *Waste Manag.* **21** 637–42
[19] Juang R S, Kao H C and Liu F Y 2006 *J. Hazard. Mater.* **128** 53–9
[20] Alyüz B and Veli S 2009 *J. Hazard. Mater.* **167** 482–8
[21] Dąbrowski A, Hubicki Z, Podkościelny P and Robens E 2004 *Chemosphere* **56** 91–106
[22] Malaviya P and Singh A 2011 *Crit. Rev. Environ. Sci. Technol.* **41** 1111–72
[23] Sablani S S, Goosen M F, Al-Belushi R and Wilf M 2001 *Desalination* **141** 269–89
[24] Muthukrishnan M and Guha B K 2008 *Desalination* **219** 171–8
[25] Lee K P, Arnot T C and Mattia D 2011 *J. Memb. Sci.* **370** 1–22
[26] Yahya M A, Al-Qodah Z and Ngah C Z 2015 *Renew. Sustain. Energy Rev.* **46** 218–35
[27] Li Y H, Ding J, Luan Z, Di Z, Zhu Y, Xu C, Wu D and Wei B 2003 *Carbon* **41** 2787–92
[28] Sousa F W, Sousa M J, Oliveira I R, Oliveira A G, Cavalcante R M, Fechine P B, Neto V O, de Keukeleire D and Nascimento R F 2009 *J. Environ. Manage.* **90** 3340–4
[29] Khorasgani F C 2013 *Int. J. Environ. Eng. Manag.* **4** 273–80
[30] Fongsatitkul P, Elefsiniotis P and Boonma C 2016 *J. Environ. Eng. Sci.* **11** 1–6
[31] Machado M D, Soares E V and Soares H M 2010 *J. Chem. Technol. Biotechnol.* **85** 1353–60
[32] Bhatnagar A and Sillanpää M 2009 *Adv. Colloid Interface Sci.* **152** 26–38
[33] Kurniawan T A, Chan G Y, Lo W H and Babel S 2006 *Sci. Total Environ.* **366** 409–26

[34] Akbal F and Camcı S 2011 *Desalination* **269** 214–22
[35] Al-Shannag M, Al-Qodah Z, Bani-Melhem K, Qtaishat M R and Alkasrawi M 2015 *Chem. Eng. J.* **260** 749–56
[36] Benvenuti T, Rodrigues M A, Bernardes A M and Zoppas-Ferreira J 2017 *J. Clean. Prod.* **155** 130–8
[37] Chen D and Ray A K 2001 *Chem. Eng. Sci.* **56** 1561–70
[38] Chong M N, Jin B, Chow C W and Saint C 2010 *Water Res.* **44** 2997–3027
[39] Mensor L L, Menezes F S, Leitão G G, Reis A S, Santos T C, Coube C S and Leitão S G 2001 *Phyther. Res.* **15** 127–30
[40] Iravani S 2011 *Green Chem.* **13** 2638–50
[41] Mittal A K, Chisti Y and Banerjee U C 2013 *Biotechnol. Adv.* **31** 346–56
[42] Shamaila S, Sajjad A K, Farooqi S A, Jabeen N, Majeed S and Farooq I 2016 *Appl. Mater. Today* **5** 150–99
[43] Singh P, Kim Y J, Zhang D and Yang D C 2016 *Trends Biotechnol.* **34** 588–99
[44] Ahmad N, Sharma S, Alam M K, Singh V N, Shamsi S F, Mehta B R and Fatma A 2010 *Colloids Surfaces B Biointerfaces* **81** 81–6
[45] Vidhu V K, Aromal S A and Philip D 2011 *Spectrochim. Acta Part A Mol. Biomol. Spectrosc.* **83** 392–7
[46] Moran J F, Klucas R V, Grayer R J, Abian J and Becana M 1997 *Free Radic. Biol. Med.* **22** 861–70
[47] Makarov V V, Love A J, Sinitsyna O V, Makarova S S, Yaminsky I V, Taliansky M E and Kalinina N O 2014 *Acta Naturae (англоязычная версия)* **6** 35–44
[48] Roy N, Alam M N, Mondal S, Sk I, Laskar R A, Das S, Mandal D and Begum N A 2012 *Process Biochem.* **47** 1371–80
[49] Nasrollahzadeh M and Sajadi S M 2016 *J. Colloid Interface Sci.* **462** 243–51
[50] Shameli K, Bin Ahmad M, Jazayeri S D, Sedaghat S, Shabanzadeh P, Jahangirian H, Mahdavi M and Abdollahi Y 2012 *Int. J. Mol. Sci.* **13** 6639–50
[51] Castillo-López D N and Pal U 2014 *J. Nanoparticle Res.* **16** 2571–86
[52] Gatew S and Mersha W 2013 *Glob. J. Environ. Sci.* **12** 29–39
[53] Azimi A A, Valizadeh B and Masoudinezhad M R 2007 *J. Environ. Stud.* **33** 5–10
[54] Han N and May C 2010 *Am. J. Appl. Sci.* **7** 1243–7
[55] Arham N A, Mohamad N A, Jai J, Krishnan J and Yusof N M 2013 *Int. J. Sci. Eng.* **5** 95–100
[56] Soundararajan V and Sreenivasan S 2012 *APCBEE Procedia* **2** 153–9
[57] Yusof Z, Gani S S, Siddiqui Y, Mokhtar F and Hasan Z A 2016 *J. Oil Palm Res.* **28** 520–30
[58] Azmi A A, Jai J and Zamanhuri N A 2017 *J. Mech. Eng. (JMechE)* **SI 2** 113–25

IOP Conf. Series: Materials Science and Engineering **358** (2018) 012025 doi:10.1088/1757-899X/358/1/012025

Coagulant from *Leucaena leucocephala* for Chromium Removal

N H Abd Razak, N Khairuddin, K N Ismail and M Musa

[1]Faculty of Chemical Engineering, Universiti Teknologi Mara

noorharliza81@gmail.com

Abstract. This research investigated the effectiveness of *leucaena leucocephala* as a natural coagulant for chromium removal. *Leucaena leucocephala* is a permanent non-climbing shrub tree which is wild and abundant in Malaysia and commonly known as petai belalang. Coagulation experiment using jar test were performed where the effect of coagulant dosage and pH were examined. The parameters investigated were suspended solid (SS), chemical oxygen demand (COD), biological oxygen demand (BOD), turbidity and chromium content. The optimum of *leucaena leucocephala* coagulant dosage for removal of suspended solid, turbidity, COD, BOD and Chromium is at range 400-600 mg/L which yielded 45, 31.4, 38.5, 27.5 and 4.05% removal respectively. While the optimum pH is at pH 2-4 (acidic) which give 33.3, 26.8, 33.75, 31.4 and 14.06% removal of suspended solid, COD, BOD, turbidity and chromium content respectively. It is concluded that the *leucaena leucocephala* showed tremendous potential for chromium removal.

1. Introduction

The present of dye in wastewater are one of the serious problems faced by many industries in all over the world. Chromium can be categorized as dangerous wastes and toxic (B3) [1].Coagulation process has been proven to be among the most effective way to remove heavy metal while aluminum salt is the main and widely used coagulant [2]. However, usage of these chemicals has several disadvantages such as relatively high costs; give harmful effects on human health, produced large volumes of sludge as well as the fact that they considerably affect pH of the treated water [3]. Thus, there is an increasing need for other environmentally acceptable and economically viable alternatives to replace the chemical-based coagulants. Plant-based coagulants could be promising alternatives as they are more cost effective, produce less voluminous sludge, highly biodegradable, non-toxic and do not alter the pH of the water under treatment [4]. Four types of plants which have been extensively used as natural coagulants are Moringa oleifera[5], nirmali seeds [6], tannin[7] and cactus [8]. This research focused on the preparation of coagulant from *leucaena leucocephala* using acid hydrolysis at 6 M of hydrochloric acid for treatment of dye wastewater. *Leucaena leucocephala* is preferable since it is abundant, wild, easy to find and naturally grown in Malaysia. The objectives of this study are to prepare coagulant from *leucaena leucocephala* using acid hydrolysis at 6 M of hydrochloric acid and determine the coagulant performance using chromium with jar test apparatus.

Content from this work may be used under the terms of the Creative Commons Attribution 3.0 licence. Any further distribution of this work must maintain attribution to the author(s) and the title of the work, journal citation and DOI.
Published under licence by IOP Publishing Ltd

2. Methodology

2.1. Material
The *leucaena leucocephala* pods were collected at Seksyen U12 Shah Alam. Samples were dried in an oven at temperature of 60 °C for 24 hours. Dried *leucaena leucocephala* were then grinded and soaked in ethanol solution to removes chlorophyll.

2.2. Acid Hydrolysis
10 g of leucaena leucocephala was added in the 100 mL of 6 M of hydrochloric acid solution. The mixture were stirred for 1 hour and being filtered using vacuum filtration equipment. The grinded LL in was then washed using distilled water until reach pH 7 and being dried in an oven until all moisture will be removed.

2.3. Preparation of standard Chromium (VI) solution
Standard Chromium (VI) solution was prepared by dissolving 0.2829 g of Potassium dichromate crystals in distilled water and making up to 100 mL. Exactly 100 mL of this solution was transferred into a 1 L volumetric flask and diluted using distilled water to get a solution containing 0.1 mg of Chromium (VI) ions per cm^3 (100 ppm or 1.923 mmol per litre).

2.4. Coagulation Jar Test Experiments
Jar Test equipment consists of six glass beakers each with a powered paddle which stirs the contents of the beaker. All tests were conducted at room temperature. 500 mL of standard chromium (VI) solution was poured into each of the six beakers. The desired amount of coagulant was added to the suspension and stirred at rapid mixing (120 rpm) for 1 minute. The speed of stirrer was reduced to slow mixing (30 rpm) for 20 minutes to keep flocs particles uniformly suspended. The paddles were then withdrawn and settling of flocs particles was observed and recorded. The mixture was left for 1 hour and then the supernatant was collected to be used in the determination of the suspended solid (SS), chemical oxygen demand (COD), biological oxygen demand (BOD), turbidity and Chromium content using the standard method.

2.5. Analytical Analysis
Suspended solid (SS) test was carried out with the aid of vacuum filtration apparatus. The initial weight of filter paper was recorded after drying in the oven at 100-105 °C for an hour. A 10 mL of the water sample was filtered through a filter paper. The residue was dried to constant weight at 100 °C - 105 °C for 1 hour. The filter paper was then cooled in dessicator before weighing. The total suspended solid content was calculated by using the equation 1;

$$\text{Suspended solid} = \frac{(A-B)\times 10^6}{C} \tag{1}$$

Where A is weight of disk + remaining solid (g), B is weight of empty disk (g), and C is volume of sample (mL)

Reactor digestion method was used to determine chemical oxygen demand (COD). 2 mL of each sample was added into COD digestion reagent vials and was inverted gently to mix the contents. It then was placed in the preheat COD reactor. The vials were heated for 2 hours at 150°C. Then, the sample vials were cooled at room temperature for about 20 minutes before the samples were analyzed by using spectrophotometer. The biological oxygen demand (BOD) test was carried out using dilution method. The dilution method is conducted by placing 50 mL of sample into bottles and filling the bottles with dilution water. The bottles are sealed and allowed to stand for 5 days at a controlled temperature of 20°C. At the end of the five-day period, the remaining dissolved oxygen is measured. The difference between the final and initial reading of dissolved oxygen was the BOD reading. Turbidity test was performed using portable turbidimeter with measurement in nephelometric turbidity unit NTU. This turbidimeter can measure turbidity from 0.001 to 1000 NTU in automatic range mode with automatic decimal point. The measurement is based on the light-transmitting properties.

Chromium (Cr) content was analyzed using Atomic Absorption Spectroscopy (AAS). 20 mL of water sample was placed into sample bottles and labeled to be analyzed by atomic absorption spectroscopy (AAS)

3. Results and discussion

3.1. The effects of coagulant dosage

Study on the effect of coagulant dosage is important for economic purpose. This is because excessive use of coagulant can be prevented during treatment of dye wastewater. The effect of coagulant dosage was studied in the range from 200 mg/L to 800 mg/L. Based on the results, the highest recorded suspended solid (SS), turbidity, chemical oxygen demand (COD), Chromium content and biological oxygen demand (BOD) removal percentages of *leucaena leucocephala* were observed at dosages range from (400-600 mg/L) at 45, 31.4, 38.5 and 27.5 % respectively. Figure 1 below shows the effect of coagulant dosage on the removal of suspended solid, turbidity, COD, BOD and Chromium content.

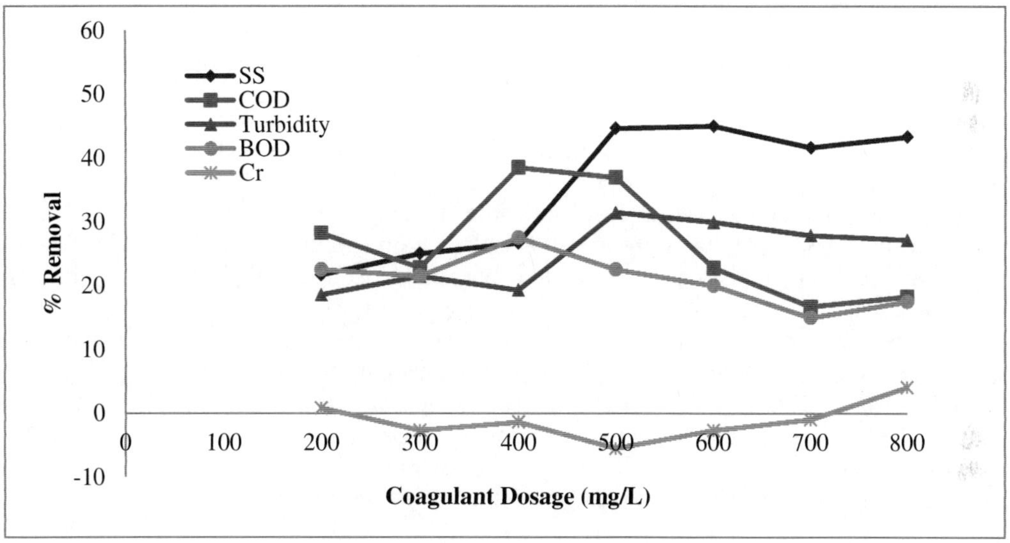

Figure 1 Effect of coagulant dosage

The percentage removal of COD and BOD was decreased in the range from 500 mg/L to 800 mg/L of coagulant dosage. This trend of graph is similar with the previous study which used *dragon fruit foliage* as the plant-based coagulant to treat concentrated latex effluent [9]. This is because the use of natural coagulants may increase the organic contents in solution resulting in increased of microbial activity. This will cause BOD concentration to increase and subsequently contributes to the increased of COD concentration [10]. The percentages of suspended solid and turbidity removal starts to decrease at coagulant dosage 600 mg/L. Similar behavior was observed when chitosan was studied to coagulate bentonite. The turbidity of the treated water started to deteriorate at increased coagulant dosage. This is because the main effect of coagulation was depending on the molecular weight of the coagulant [11]. Concentration of the chromium content in the solution was increased as the coagulant dosage increased. This may be due to the presence of heavy metal content in *leucaena leucocephala pods*. This is possible due to the geographical position of *leucaena leucocephala* [12].

3.2. The effects of pH

The effect of pH was studied in the range from 1-8. The highest recorded suspended solid, turbidity, COD, BOD and chromium removal percentages were observed at pH 2-4 (acidic) at 33.3, 31.5, 26.8, 33.75 and 14.06% respectively.

Figure 2 show that *leucaena leucocephala* have an excellent removal in the acidic solution. This is in agreement with the previous study that found *jatropha cursas seeds* coagulant is efficient under acidic conditions with turbidity removal of 99.4% at pH 3 [13]. *C. obtusifolia gum* as coagulant also showed higher coagulation activity at pH 3 to 5 with maximum total suspended solid (TTS) and COD removals of 89.9 and 33.9% respectively [14].

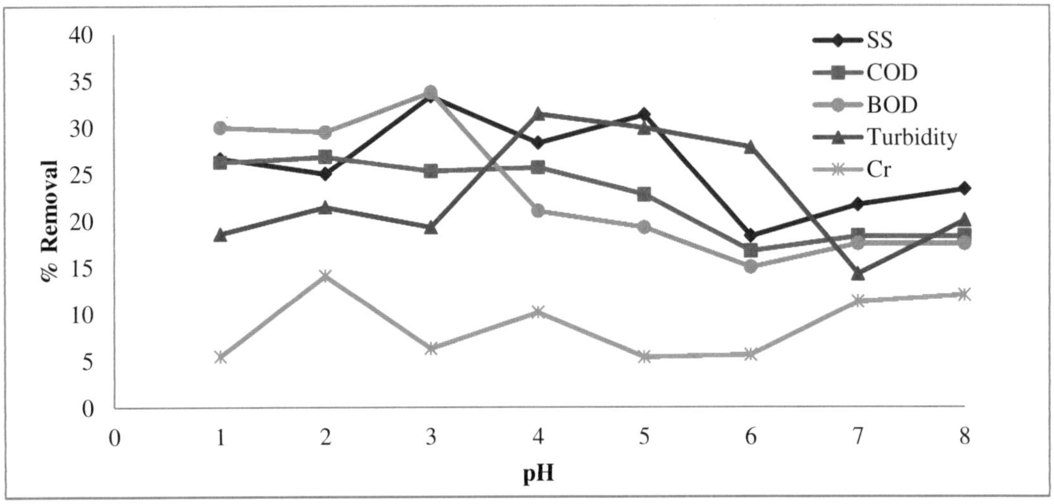

Figure 2 Effect of pH

The proposed coagulation mechanism involved could be adsorption with interparticle bridging. Thus, the charge of the coagulant in the solution did not play an important role for this mechanism. Natural polymers and organic colloids such as *leucaena leucocephala* are generally non-ionic and functioning as effective bridging coagulant in acidic solution [13],[14].Moreover, it can be seen that the percentages removal is decreasing at pH neutral. A similar observation is seen from the previous findings where the lowest turbidity, COD and suspended solid removal is in the pH range 6-8 using chitosan to treat concentrated latex effluent [10]. It appeared that SS removal percentages were consistently lower at pH neutral are caused by solution neutrality as opposed to acidic and basic where the elevated concentration of either H^+ or OH^- causes more electrostatic interactions between the coagulant and chemical species in the effluent and thus promotes coagulation [10].

4. Conclusion

The optimum coagulant dosage for removal of suspended solid, turbidity, COD, BOD and Chromium is at range 400-600 mg/L which yielded 45, 31.4, 38.5, 27.5 and 4.05% removal respectively. The optimum pH is at pH 2-4 (acidic) with 33.3, 26.8, 33.75, 31.4 and 14.06% removal of suspended solid, COD, BOD, turbidity and chromium content respectively. As such, it could be concluded that the *leucaena leucocephala* showed remarkable potential as natural coagulant for water treatment purposes.

Acknowledgment

The authors wish to acknowledge to Research Management Centre for financial support under Research Acculturation Grant Scheme (RAGS) for funding project 600-RMI/RAGS 5/3 (19/2014) by Universiti Teknologi MARA (UiTM).

References

[1] Raimon R and Said Laboratory M 2017 Effluent Treatment by Using Coagulant Alum sulphate and Poly Aluminium Chloride (PAC) *Indones. J. Fundam. Appl. Chem.* **2** 47

[2] Huang X, Bo X, Zhao Y, Gao B, Wang Y, Sun S, Yue Q and Li Q 2014 Effects of compound bioflocculant on coagulation performance and floc properties for dye removal. Bioresour *Bioresour. Technol.* **165** 116

[3] Chaibakhsh N, Ahmadi N and AliZanjanchi M 2014 Use of Plantago major L. as a natural coagulant for optimized decolorization of dye-containing wastewater *Ind. Crops Prod.* **61**.

169–175

[4] Shamsnejati S, Chaibakhsh N, Pendashteh S and Hayeripour A R 2014 Mucilaginous seed of Ocimum basilicum as a natural coagulant for textile wastewater treatment *Ind. Crops Prod.* **69** 40–47

[5] Camacho F P, Sousa V S, Bergamasco R and Teixeira M R 2017 The use of Moringa oleifera as a natural coagulant in surface water treatment *Chem. Eng. J.* 313 226–237

[6] Maruthi Y A, Dadhich A S, Hossain K and Jyothsna A 2013 Nirmali Seed as a Natural Biosorbent ; Evaluation of its Potential for Iron (II) Removal from Steel Plant Effluents and Sewage Disinfecting Capacity *Eur. J. Susatinable Dev.* **2** 77–84

[7] Thakur S S and Choubey S 2014 Use of Tannin based natural coagulants for water treatment: An alternative to inorganic chemicals *Int. J. ChemTech Res.* **6** 3628–3634

[8] Karanja A, Fengting L and Ng'ang'a W 2017 Use of Cactus Opuntia As a Natural Coagulant: Water Treatment in Developing Countries.," *Int. J. Adv. Res.*. **5** 884–894

[9] Idris J, Som A, Musa M, Ku Hamid K H, Husen R and Mohd Rodhi M N 2013 Dragon Fruit Foliage Plant-Based Coagulant for Treatment of Concentrated Latex Effluent : Comparison of Treatment with Ferric Sulfate *Journal of Chemistry*

[10] Idris J, Som A, Musa M and Ku Hamid K H 2007 Dragon fruit foliage as a low cost plant-based coagulant in latex concentrate wastewater treatment Malaysian Journal of Chemical Engineering

[11] Altaher H 2012 The use of chitosan as a coagulant in the pre-treatment of turbid sea water.," *J. Hazard. Mater.* **233–234** 97–102

[12] Kook Kwon Y, Sik Bong Y, Sik Lee K and Hwang G S 2014 An integrated analysis for determining the geographical origin of medicinal herbs using ICP-AES/ICP-MS and 1H NMR analysis *Food Chem.* **161** 168–175

[13] Subramonian W, Wu T Y and Chai S P 2014 A comprehensive study on coagulant performance and floc characterization of natural Cassia obtusifolia seed gum in treatment of raw pulp and paper mill effluent *Ind. Crops Prod.* **61** 317–324

[14] Abidin Z Z, Shamsudin M, Madehi N N S and Sobri S 2013 Optimisation of a method to extract the active coagulant agent from Jatropha curcas seeds for use in turbidity removal *Ind. Crops Prod.* **41** 319–323

Ammonium Nitrogen Removal from Urea Fertilizer Plant Wastewater via Struvite Crystal Production

I Machdar*, S D Depari, R Ulfa, S Muhammad, A B Hisbullah and W Safrul

Laboratory of Environmental Engineering, Chemical Engineering Department, Faculty of Engineering, Syiah Kuala University, Darussalam, Banda Aceh, Indonesia

machdar@unsyiah.ac.id

Abstract. Elimination of ammonium concentration from urea fertilizer plant wastewater through struvite crystal ($NH_4MgPO_4.6H_2O$) formation by adding $MgCl_2$, KH_2PO_4, and KOH were studied. This method of elimination has two benefits, namely, reducing ammonium nitrogen content in the wastewater, as well as production of a valuable material (struvite crystal). Struvite is known as a slow-release fertilizer and less soluble. This report presents the ammonium removal efficiencies during struvite formation. The growth of struvite production under different molar ratios of Mg^{2+}:NH_4^+:PO_4^{3-} and solution pH is also discussed. To find the efficiencies and measure the growth rates, lab-scale experiments were conducted in a batch crystallizer-reactor. SEM, XRD, and FTIR observation were also applied to investigate the characteristics of struvite. The reactant molar ratios of Mg^{2+}:NH_4^+:PO_4^{3-} of 1.2:1:1, 1:1:1.2, and 1:1:1 were evaluated. Each of the molar ratios was treated at the solution pH of 8, 9, and 10. It was found that, the highest ammonium removal efficiency was 94.7% at the molar ratio of 1.2:1:1 and pH of 9. Primarily, the growth rate of struvite formation complied with a first-order kinetic model. The rate constants (k_1) were calculated to be 2.6, 4.3, and 5.0 h^{-1} for solution pH of 8, 9, and 10, respectively. The findings of the study provide suggestion for an alternative sustainable recovery of ammonium nitrogen content in a urea fertilizer plant effluent.

1. Introduction

Ammonia stripping tower is commonly used in urea fertilizer plant to reduce levels of ammonium nitrogen concentration in its wastewater. Normally, the tower is equipped with a successive elevation of parallel splash-bars made of a timbered packing material to make wastewater stream into tiny liquid particles. The liquid particles are then allowed in contact with air resulting ammonia to release and reduce ammonium concentration in the liquid phase. However, this method is only efficient when a wastewater that has ammonia contents between 10 to 100 mg/L. For higher ammonia content (more than 100 mg/L), it may be more economical to use alternate techniques, such as steam stripping or biological method [1]. Several methods have been applied to remove ammonium nitrogen from industrial effluent including natural zeolite [2], membrane process [3], and electrodeionization process [4]. Exploration of a cost-effective approach to recover a valuable material from a wastewater can deliver a breakthrough idea in wastewater treatment technology. One of the novel innovations is production of struvite crystal ($NH_4MgPO_4.6H_2O$) from waste. Struvite can improve soil fertilities by providing Mg, N, and P as micro and macronutrients [5]. Struvite, as a slow-release fertilizer, has excellent qualities at certain conditions compare with standard fertilizers [6,7]. Practically, struvite can be recovered from municipal landfill leachate [8], poultry manure wastewater [9], or urine [10]. Struvite also can be produced using microbial fuel cells [11]. Recently, struvite has been produce from fertilizer wastewater [12]. This study was conducted by adding $Na_2HPO_4.12H_2O$ and $MgCl_2.6H_2O$, Mg^{2+}:NH_4^+:PO_4^{3-} molar ratio of 1:1:1, and solution pH of 9.5.

Content from this work may be used under the terms of the Creative Commons Attribution 3.0 licence. Any further distribution of this work must maintain attribution to the author(s) and the title of the work, journal citation and DOI.
Published under licence by IOP Publishing Ltd

The solution pH and molar ratio of Mg^{2+}:PO_4^{3-} have significant factor on the solubility of struvite and ammonium nitrogen removal efficiency. Struvite can be precipitated at a wide range of pH (from 7.0 to 11.5) and the effective of the ratio was 1:1 or 1.2:1 [12]. In the present study, the NH_4^+-N removal from urea fertilizer plant wastewater to produce struvite crystal was investigated in presence of $MgCl_2$, KH_2PO_4, and KOH at pH (8-10) and Mg^{2+}:PO_4^{3-} molar ratio of 1.2:1 and 1:1.2. Moreover, kinetic growth rates of struvite formation were calculated and characteristics of struvite crystal were performed by SEM, XRD, and FTIR.

2. Methodology

The experiment was conducted by using a raw urea fertilizer plant wastewater, $MgCl_2$, and KH_2PO_4. The raw wastewater was collected from a urea fertilizer plant located in Lhokseumawe City, Aceh Province, Indonesia. The wastewater as NH_4^+ contained ammonium nitrogen concentration of 824±470 mg/L and pH was in range of 8.7 to 9.2. Stock solution of $MgCl_2$ as Mg^{2+} was prepared by dissolving crystals of $MgCl_2$ in distilled water. Stock solution of KH_2PO_4 as PO_4^{3-} was made in the similar procedure. The solutions were kept individually and mixed before conducting any experiment. KOH solution (1N) was used to adjust the pH of the reactant mixture to the preferred experimental conditions (pH 8, 9, and 10).

The experiment was performed in an erlenmeyer flask (as a batch crystallizer-reactor) of a capacity of 500 ml at room and at 120 rpm. The experiment was carried out for 60 min with 5 min interval sampling time. An amount of 3 ml aliquot of reactant solution was pipetted from the flask at respective sampling period and filtered. Subsequently, the filtered samples (supernatants) were analyzed for residual ammonium nitrogen. The concentration of NH_4^+-N in the supernatant was analyzed by UV-vis spectrophotometer (Shimadzu UV-1700). The product struvite was filtered, oven dried overnight at 50°C, and weight analysis. The characteristics of struvite product were observed by SEM (JEOL JSM 6510), XRD (Shimadzu XRD 600), and FTIR (Shimadzu IR Prestige21). In order to study the mechanism of struvite formation, the first-order kinetic model was used to the experimental results as per method proposed by Rahman *et al.* [13], Nelson *et al.* [14], and Quintana *et al.* [15].

3. Results and discussions

3.1 *Effect of molar ratio and solution pH*

The effects of molar ratios of Mg^{2+}:NH_4^+:PO_4^{3-} and solution pH on the ammonium nitrogen removal are presented in figure 1. The Mg^{2+}:NH_4^+:PO_4^{3-} molar ratio of 1.2:1:1, 1:1:1.2, and 1:1:1 is given in figure 1a, figure 1b, and figure 1c, respectively. The ammonium nitrogen concentrations of 533±10 mg/L in average were applied in the experiments. The decrease of ammonium concentration in solution provides an indication of struvite formation. It was found that, the rate of ammonium reduction in bulk solution reached steady after 30 min in all experiments investigated.

The effect of the addition of excess magnesium concentration to the struvite crystallizer-reactor on the ammonium nitrogen removal is illustrated in figure 1a. During this experiment, the applied concentration of magnesium was 441±4 mg/L. The result shows that the ammonium removal efficiencies were 87.1%, 94.7%, and 90.6 % for pH of 8, 9, and 10, respectively. Similarly, the influence of the excess addition of phosphate concentration on ammonium removal is given in figure 1b. In this experiment, the added concentration of phosphate was 361±7 mg/L. It was found that the removal ammonium nitrogen efficiencies were 82.4%, 88.3%, and 86.7% at pH of 8, 9, and 10, respectively. These results demonstrated that the increase of the Mg^{2+} molar ratio from 1 to 1.2 improved the removal ammonium efficiency by 10%. Similar results have been obtained by Rahaman *et al.* [16]. The addition of chemical reactants to a wastewater to produce struvite would be needed to provide an equimolecular condition of the reactants. Figure 1c provides ammonium removal for the molar ratio of Mg^{2+}:NH_4^+:PO_4^{3-} of 1:1:1. The ammonium removal efficiencies were found of 84.6%, 93.8%, and 92.1% for pH of 8, 9, and 10, respectively. Yetilmezsoy and Zengin [9] performed a series experiment and they concluded that the Mg^{2+}:NH_4^+:PO_4^{3-} molar ratio of 1:1:1 would be sufficient for the nitrogen removal and struvite precipitation.

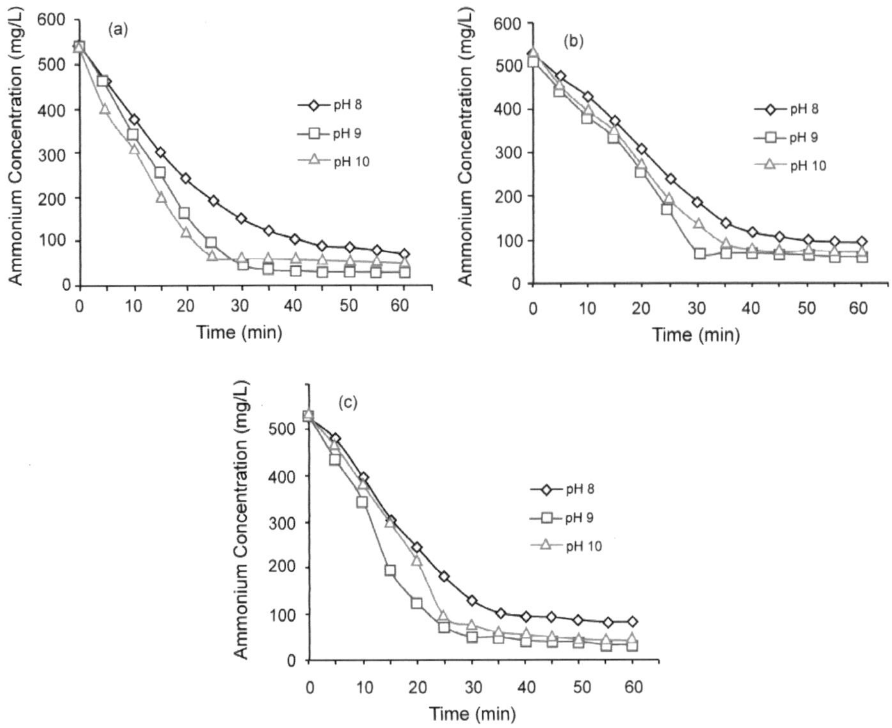

Figure 1. Effect of molar ratio and pH on the ammonium nitrogen removal. The molar ratios of Mg:NH_4:PO_4 in a) 1.2:1:1, b) 1:1:1.2, and c) 1:1:1.

Solution pH plays an important role for the removal of nutrient such as nitrogen and phosphorus from solution [17,18] by struvite precipitation. The results in figure 1 show that solution pH affects the ammonium removal or struvite production. Increasing solution pH from 8 to 9 increased ammonium removal but insignificant upgrade at pH of 10 for all ratio reactants investigated. At the ratio Mg^{2+}:NH_4^+:PO_4^{3-} of 1:1:1, for solution pH of 8, ammonium removal efficiency was 84.6%. In addition, for solution pH of 9 and 10, ammonium removal efficiency were 93.8% and 92.1%, respectively. Similar result has been found by Hao *et al.* [19] and Ariyanto *et al.* [20]. They mentioned that struvite can be precipitate at a broad range of pH, but the acceptable pH range between 7.5 to 9.0, and growth rate of struvite crystallization increased with an increase in solution pH about 8 to 9.

3.2 Kinetic model

Figure 2 presents the kinetic experimental data by plotting ln (C-Cs) versus time (t) which shows a straight line with varied slopes. The correlation coefficients (R^2) between decreased concentration and reaction time were 0.997, 0.980, and 0.970 for solution pH of 8, 9, and 10, respectively. These results demonstrated that the application of first-order kinetic model is suitable for the experimental observation. From the results, it shows that the higher ammonium nitrogen removal efficiency was achieved at higher solution pH. The first-order removal rate constants (k_1) were calculated to be 2.6, 4.3, and 5.0 h^{-1} for solution pH of 8, 9, and 10, respectively. Ariyanto *et al.* [20] investigated the influence of several of physico-chemical process parameters on kinetic of struvite crystallization using artificial wastewater. They obtained the rate constants were 3.7, 5.1, and 6.9 h^{-1} for solution pH of 8, 8.5, and 9, respectively. Comparatively lower rate constant in the present experiment is probably because of interference by other cations in the wastewater. The presence of Ca^{2+} ion in solution inhibits struvite production by the formation of calcium phosphates [12,21].

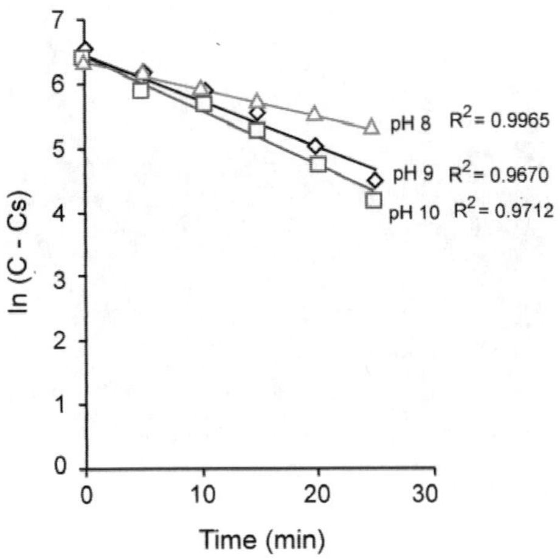

Figure 2. Plot of ln (C-Cs) versus time at various solution pH.

3.3 Characteristics of struvite crystal

The struvite product from this experiment as shown in figure 3 is white crystalline powder consisting equimolar concentration of magnesium, ammonium, and phosphate combined with six water molecules.

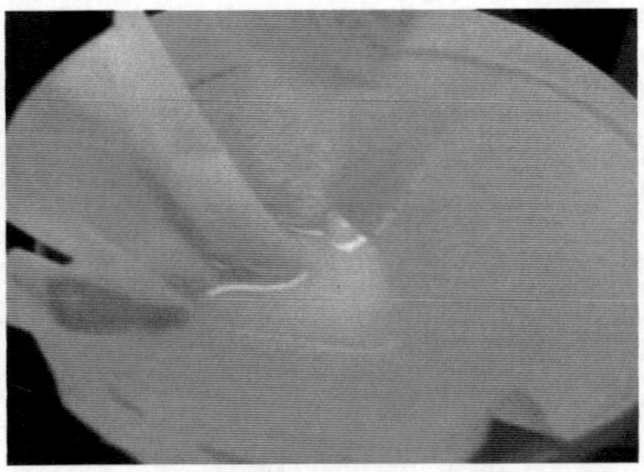

Figure 3. Struvite product from the experiment.

SEM analysis on struvite product (solution pH of 9 and Mg^{2+}:NH_4^+:PO_4^{3-} molar ratio of 1:1:1) is presented in figure 4. The struvite product has a rod like irregular structure. Similar struvite crystal formation has been reported by Le Corre *et al.* [22] using synthetic liquor in a pilot scale crystallization reactor. XRD was used to analyze the crystal structure of struvite product as shown in figure 5 (left). It notified that the XRD curve develop from this sample confirmed the database model for struvite particle. The FTIR was applied to examine potential charge and chemical bond of the struvite product. The FTIR spectrum of struvite product is presented in figure 5 (right). The absorption line spectrums appeared at 3589, 2949, and 2901 cm^{-1} are due to O-H, N-H, and NH_4^+ stretching vibrations. The weak bands occurring at 2360 cm^{-1} in the spectrum can be confirmed as a result of H-O-H stretching vibrations of water molecules in the struvite crystal. The spectrum emerged around 1688, 1578, 1441 cm^{-1} in have been indicated to N-H bending vibration. The stretching of P-OH (hydrophosphate) was confirmed at the spectrum of 1009 cm^{-1}. A medium absorption band at spectrum around 880 and 760 cm^{-1} reveals the oscillation patterns of vibration of the coordinated water and the metal-oxygen bond in the complex.

Similar descriptions of FTIR results of struvite crystal have been described by Chauhan *et al.* [23] Suguna *et al.* [24] and Ariyanto *et al.* [20]. The FTIR spectrum also established that scissoring vibration of PO_4 at about 571 cm^{-1}, NH_4^+ at around 1441 cm^{-1}, and scissoring vibration of water at close to 1688 cm^{-1} [25]. Therefore, it was confirmed that struvite crystal had been formed, and the result was comparable with the XRD examination.

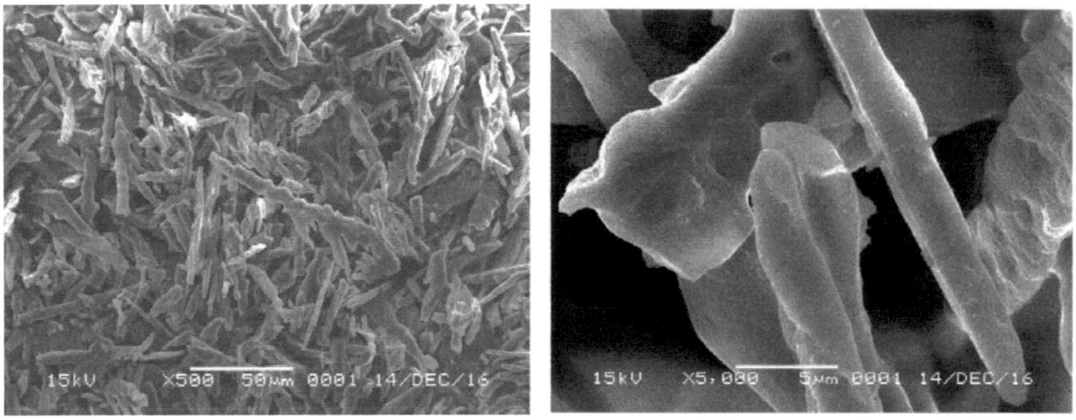

Figure 4. SEM of crystal struvite product.

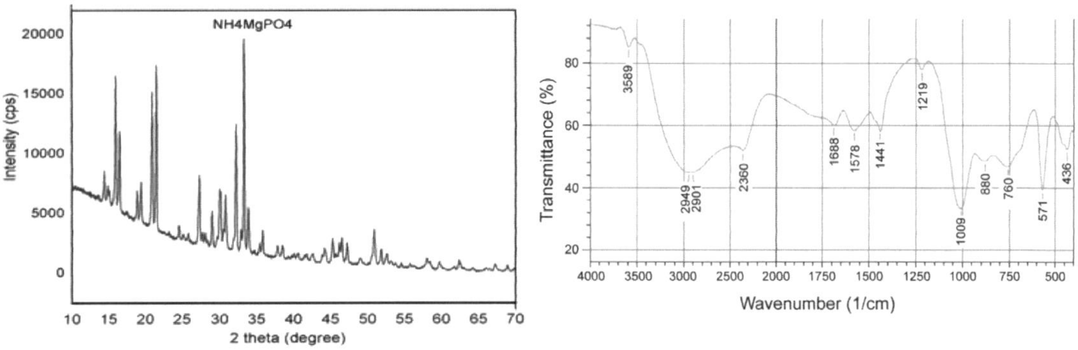

Figure 5. XRD pattern (left) and FTIR spectrum (right) of crystal struvite product.

4. Conclusions

The struvite crystal as fertilizer material was produced from a raw wastewater of urea fertilizer plant. The raw wastewater as ammonia source reacts with magnesium chloride (Mg^{2+} source) and potassium dihydrogen phosphate (PO_4^{3-} source) plus potassium hydroxide as pH adjustment to produce the crystal. Based on the result, ammonium nitrogen removal efficiency significantly depends on the molar ratio of $Mg:NH_4:PO_4$ and pH. The highest efficiency (94.7%) achieved at the molar ratio of 1.2:1:1 and pH of 9. The growth of struvite crystal fitted well with the first-order kinetic model. The lower rate constant found in this experiment is probably due to interference by other cations in the wastewater. Analysis of the powdered crystal by using XRD and FTIR confirmed the formation of struvite. This proposed method could be an alternative for the sustainability of urea fertilizer plant wastewater treatment technology through recovery of its ammonium nitrogen content.

Acknowledgment

This study was supported by the Ministry of Research, Technology, and Higher Education of the Republic of Indonesia under national competitive research scheme.

References

[1] U S EPA 1983 EPA 832-F-00-0119 Washington D C
[2] Chung Y C, Son D H and Ahn D H 2000 Wat. Sci. and Tech. **42** 127
[3] Karabelas A J, Yiantsios S G, Metaxiotou Z, Andritsos N, Akiskalos A, Vlachopoulos G and Stavroulias S 2001 Desalination **138** 93
[4] Dermentzis K, Davidis A, Chatzichristou C and Dermentzi A 2012 Global NEST Journal **4** 468

[5] El Rafie Sh, Hawash S and Shalaby M S 2013 Adv. in Applied Science Research **4** 113
[6] Ghosh G K, Mohan K S and Sarkar AK 1997 Nutr. Cycl. Agroecosyst. **46** 71
[7] Liu YH, Kwag J H, Kim J H and Ra C S 2011 Desalination **277** 364
[8] Kim D K , Ryu H D , Kim M S and Kim J, Lee S I 2006 J. Hazard. Mater **146** 81
[9] Yetilmezsoy K and Zengin Z S 2009 J. Hazard. Mater **166** 260
[10] Etter B, Tilley E, Khadka R and Udert K M 2011 Water Res. **45** 852
[11] Ichihashi O and Hirooka K 2012 Bioresour. Technol. **114** 303
[12] Yu R, Geng J, Ren H, Wang Y and Xu K 2012 Bioresour. Technol. **124** 292
[13] Rahman Md, Saleh M A M, Rashid U, Ahsan A, Hossain M M and Ra C S 2014 Arabian Journal of Chemistry **7** 139
[14] Nelson N O, Mikkelsen R L and Hesterberg D L 2003 Bioresour. Technol. **89** 229
[15] Quintana M, Colmenarejo M F , Barrera J, Sanchez E, Garcia G, Travieso L and Borja R 2008 Chem. Eng. J. **136** 204
[16] Rahaman M S, Ellis N and Mavinic D S 2008 Wat. Sci. and Tech. **57** 647
[17] Rahman M M, Liu Y H, Kwag J H, Ra C.S 2011 J. Hazard. Mater. **186** 2026
[18] Ohlinger K N, Young T M and Schroeder E D 1999 J. Environ. Eng. **125** 730
[19] Le Corre K S, Jones E V, Hobbs P and Parsons S A 2009 Crit. Rev. Environ. Sci. Tech. **39** 433
[20] Hao X D, Wang C C, Lan L, and Von Loosdrecht M C M 2008 Water Sci. Technol. **58** 1687
[21] Ariyanto E, Sen T K, and Ang H M 2014 Advanced Powder Tech. **25** 82
[22] Le Corre K S, Jones E V, Hobbs P, Jefferson B and Parsons S A 2007 Water Res. **41** 2449
[23] Chauhan C K, Joseph K C, Parekh B B and Joshi MJ 2008 Indian Journal of Pure and Applied Physics. **46** 507
[24] Suguna K, Thenmozhi M and Sekar C 2012 Bulletin of Materials Science **35** 701
[25] Doyle J D and Parsons S A 2002 Water Res. **36** 3925

IOP Conf. Series: Materials Science and Engineering **358** (2018) 012027 doi:10.1088/1757-899X/358/1/012027

Utilization of Activated Carbon Prepared from Aceh Coffee Grounds as Bio-sorbent for Treatment of Fertilizer Industrial Waste Water

M Mariana[1,2], M Mahidin[1,2], F Mulana[1,3] and F Aman[4]

[1] Department of Chemical Engineering, Faculty of Engineering, Syiah Kuala University, Darussalam, Banda Aceh 23111, Indonesia
[2] Environmental and Natural Resource Research Center, Syiah Kuala University
[3] Faculty of Chemical Engineering, Universiti Teknologi MARA (UiTM), Shah Alam 40450, Malaysia
[4] Post Graduate Program of Chemical Engineering, Syiah Kuala University Darussalam, Banda Aceh 23111, Indonesia

mariana_hasyim@yahoo.com

Abstract. The people of Aceh are well known as coffee drinkers. Therefore, a lot of coffee shops have been established in Aceh in the past decade. The growing of coffee shops resulting to large amounts of coffee waste produced in Aceh Province that will become solid waste if not wisely utilized. The high carbon content in coffee underlined as background of this research to be utilized those used coffee grounds as bio-sorbent. The preparation of activated carbon from coffee grounds by using carbonization method that was initially activated with HCl was expected to increase the absorption capacity. The prepared activated carbon with high reactivity was applied to adsorb nitrite, nitrate and ammonia in wastewater outlet of PT. PIM wastewater pond. Morphological structure of coffee waste was analyzed by using Scanning Electron Microscopy (SEM) and Fourier transform infrared spectroscopy (FTIR). The result showed that the adsorption capacity of iodine was equal to 856.578 mg/g. From the characterization results, it was concluded that the activated carbon from coffee waste complied to the permitted quality standards in accordance with the quality requirements of activated carbon SNI No. 06-3730-1995. Observed from the adsorption efficiency, the bio-sorbent showed a tendency of adsorbing more ammonia than nitrite and nitrate of PT. PIM wastewater with ammonia absorption efficiency of 56%.

1. Introduction

Used coffee grounds are inexpensive and widely available materials that can be utilized to reduce the levels of nitrite, nitrate and ammonia in wastewater. Coffee grounds are organic materials that can be converted to activated carbon to be used as adsorbent [1,2]. Activated carbon is a porous solid that produced from carbonaceous material by heating at high temperatures. The previous study mentioned that the activated carbon from coffee grounds could adsorb iron and mercury up to 99.43 and 99%, respectively [3]. In this study, the coffee grounds were used as raw material to be prepared as activated carbon. Furthermore, activated carbon was used to reduce the content of ammonia, nitrite and nitrate in the fertilizer industry wastewater effluent.

Activated carbon is an amorphous carbon material consisting mostly of free carbon and has excellent adsorption ability. Activated carbon has been generally used as an ingredient for dye relievers as well as in the application of gas adsorbent, metal adsorbent and so on [4,5].Used coffee grounds are inexpensive and widely available materials that can be utilized to reduce the content of

Content from this work may be used under the terms of the Creative Commons Attribution 3.0 licence. Any further distribution of this work must maintain attribution to the author(s) and the title of the work, journal citation and DOI.
Published under licence by IOP Publishing Ltd

ammonia, nitrite and nitrate in wastewater. Activated carbon is a porous solid that can be produced by heating up carbonaceous material at high temperatures. The larger surface area of the activated carbon caused the higher of adsorption capability [6].Even though the content of ammonia, nitrite and nitrate in the wastewater pond of outlet wastewater treatment unit of PT. PIM are still under the allowed quality standards, but with continuous discharge the levels of those wastes will accumulate and exceed the quality standards especially on the near location of this wastewater treatment unit before it is expanding. The presence of urea in certain concentrations can cause increased growth of algae. In addition to urea, excess content of nitrites and nitrates in the effluent can also cause death to the aquatic organisms. Therefore, proper treatment is required to reduce the negative impacts caused by industrial waste, particularly the urea fertilizer industry.

2. Method

Experiment was conducted as follows: (1) preparation of activated carbon bio-sorbent from coffee waste; and (2) the process of adsorption of ammonia, nitrite and nitrate using a batch process.

2.1. Preparation of activated carbon

Coffee waste was dried under the sun and then soaked in a solution of 0.1 M HCl for 48 hours. Next drained and washed with distilled water to pH neutral. Coffee waste that have been activated chemically then put in oven to reduce the moisture content, finally those materials were inserted into muffle furnace at a temperature of 350°C for 3.5 hours. The produced activated carbon was sieved using a sieve of+80-100 mesh and later activated carbon was stored in a desiccator.

2.1.1. Yield

The yield of activated carbon was calculated by comparing the weight of raw material to the weight of produced activated carbon after carbonization:

$$Yield\ (\%) = \frac{b}{a} x\ 100\% \tag{1}$$

where: a = initial weight of the carbon (g)
b = weight of dried carbon (g)

2.1.2. Water content

The water content was determined by comparing the initial weight of the carbon with the weight of dried carbon.2 grams of activated carbon was put in a porcelain cup that had been weighed beforehand. Afterwards, the cup was placed in an oven at a temperature of 105°C for 3 hours then cooled down in a desiccator and weighed.

$$Water\ content\ (\%) = \frac{(a-b)}{a} x100\% \tag{2}$$

where: a = initial weight of the carbon (g)
b = weight of dried carbon (g)

2.1.3. Ash content

The ash content was calculated by comparing the initial weight of dried carbon with the weight of the ash.2 grams of activated carbon was put in a porcelain cup that had been weighed previously. Subsequently, the cup was placed in a muffle furnace at a temperature of 600°C for 3 hours then cooled down in a desiccator and weighed.

$$\text{Ash content } (\%) = \frac{a}{b} x100\% \tag{3}$$

where: a = weight of the ash (g)
b = initial weight of dried carbon (g)

IOP Conf. Series: Materials Science and Engineering **358** (2018) 012027 doi:10.1088/1757-899X/358/1/012027

2.1.4. Iodine adsorption

A total of 2 grams activated carbon was put into a dark and sealed container. 50 mL of 0.1N iodine solution was added into the container then shaken for 15 minutes and filtered. As much as 10 mL of filtrate was titrated with sodium thiosulfate solution 0.1N. Once the yellow solution about to be disappeared, 1% of starch indicator was added. The titration was then continued right until the blue color disappeared.

$$Iodine\ Adsorption\left(\frac{mg}{g}\right)=\frac{\left\{10-\left(\frac{NxV}{0,1}\right)\right\}}{S}x\,12.69x\,5 \quad (4)$$

where: V = total sodium thiosulfate required
N = concentration of sodium thiosulfate solution
S = mass of activated carbon (g)
12.69 = amount of iodine in accordance with 1 ml of sodium thiosulfate solution 0.1

2.2. Absorption of nitrate and nitrite using batch process

Liquid waste of fertilizer industry as much as 100 mL was contacted with activated carbon as much as 0.2 grams with size of 80-100 mesh at various stirring time. The wastewater before and after contacted with adsorbents was measured the ammonia, nitrite and nitrate concentrations by using spectrophotometry.

3. Result and discussion

Characteristics of activated carbon from Aceh coffee grounds are shown in table 1.

Table 1. Characteristics of activated carbon from Aceh coffee grounds

Parameter	**Analysis Results**	**Quality requirements for activated carbon according to SNI 06-3730-1995**
Yield	18.7	-
Water content	6.38	Maximum 15 %
Ash content	1.05	Maximum 2,5 %
Absorption of iodine	856.575	Maximum 750 mg/g

Table 1 shows the analysis results of prepared activated carbon. The result of parameter was the yield value of 18.7, the water content of 6.38%, the ash content of 1.05%, and the adsorption capacity of iodine of 856.575 mg / g. From those characteristics of prepared activated carbon as shown in Table 1 it can be concluded that the prepared activated carbon meet the quality standards in accordance with the quality requirements of activated carbon refer to SNI No. 06-3730-1995.

3.1. Analysis of the functional groups and morphology of prepared activated carbon

Chemical functional groups of samples were identified by using Fourier transformation infrared spectroscopy (FTIR) with a range of 500 to 4000 cm^{-1}. Graph of infrared spectra of samples was shown in figure 1. Figure 1 shows that the results of FTIR analysis on prepared activated carbon before and after adsorption have a bit similar spectra, but there is a change and a shift in the absorption band in each region after adsorption process with the sample waste of nitrite and nitrate, in which the occurrence of absorption band shift was predicted due to the process of interaction that occurs is physically caused by the force of van der walls or hydrogen bonds. This suggests that the interaction between waste of nitrite and nitrate with functional groups that existing on the surface of the adsorbent can occur through interaction of the van der walls, hydrogen bonding, ion exchange or complex formation and this usually occurs on the surface of a solid-rich functional group [7,8].

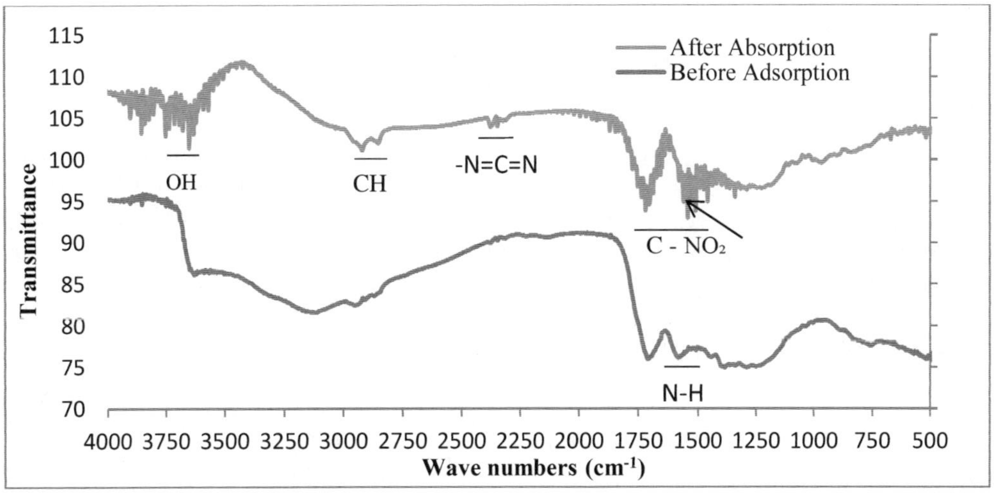

Figure 1. Spectra of prepared activated carbon.

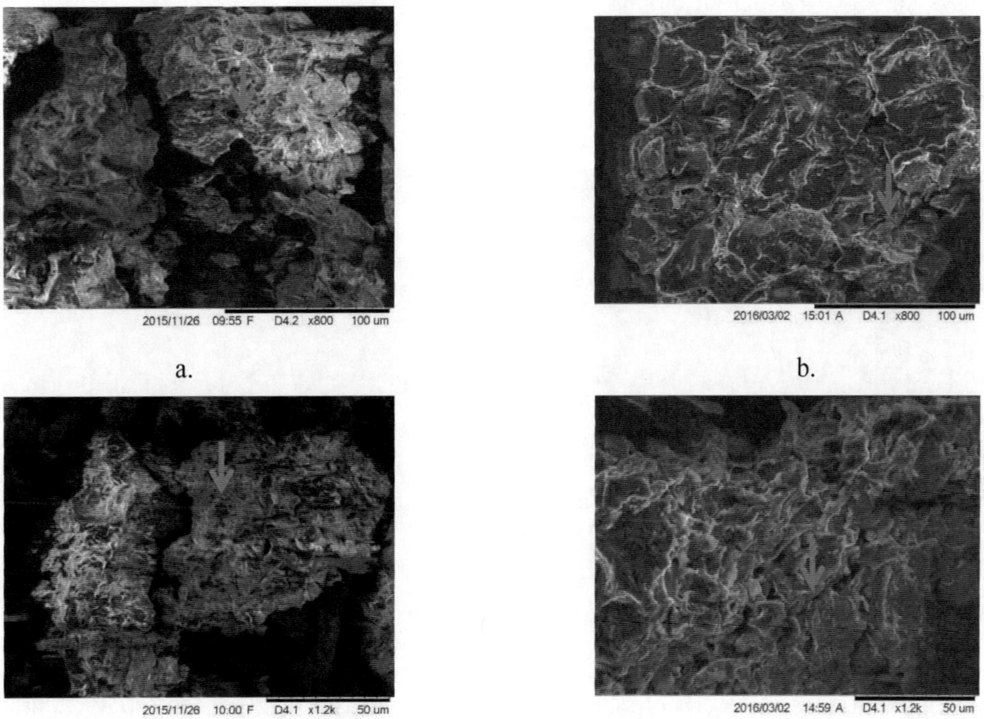

Figure 2. SEM pictures of prepared activated carbon: a. prior to adsorption process with a magnification of 800X, b. After adsorption process with 800X, c. prior to adsorption with a magnification of 1200X, and d. after adsorption with a magnification of 1200X.

The functional group obtained after analysis by FTIR is an OH group (alcohol) with a length of absorption bands (3580-3650) cm^{-1}, CH with a length of absorption bands (2900-3300) cm^{-1}, group -N=C= N (Isocyanides) with a length of absorption bands (2240-2275) cm^{-1}, group NH (amine) with a length of absorption bands (1575-1650) cm^{-1} and group NO_2 with a length of absorption bands (1550-1570) cm^{-1}. Deeply it can be seen from arrow sign inside the figure the distinguishing of the absorption bands before and after absorption that refer to the existence of nitrite and nitrate components in the liquid waste. Morphological structure of the prepared activated carbon was characterized by using Scanning Electron Microscope (SEM). SEM characterization results can be seen in figure 2 that shows the surface shape of each sample before and after the adsorption of nitrite nitrate.

Even though the results of SEM analysis are not so focus but the figure shows clearly some differences on each figure. On activated carbon sample before contacting with the adsorbate, it can be seen many visible and empty pores as shown by the arrow in figure 2. However, on sample of activated carbon after contacting with the adsorbate it seem that some visible pores were filled by a substance but not yet fully charged (as shown by the arrow). Pores contained in prepared activated carbon may enhance the ability to adsorb adsorbate concentration. In this study, it is predicted that adsorption process of the nitrites and nitrate occurs on the pores surface of the prepared activated carbon.

3.2. The adsorption efficiency of nitrite and nitrate in liquid waste of PT PIM

The adsorption efficiency stated the percentage of adsorbate adsorption on the surface of adsorbent. This adsorption efficiency is influenced by the concentration of adsorbate, where the increase of concentration of adsorbate will also increase the adsorption efficiency. The adsorption efficiency of nitrite and nitrate in liquid waste of PT PIM can be seen in figure 3.
SEM pictures of prepared activated carbon

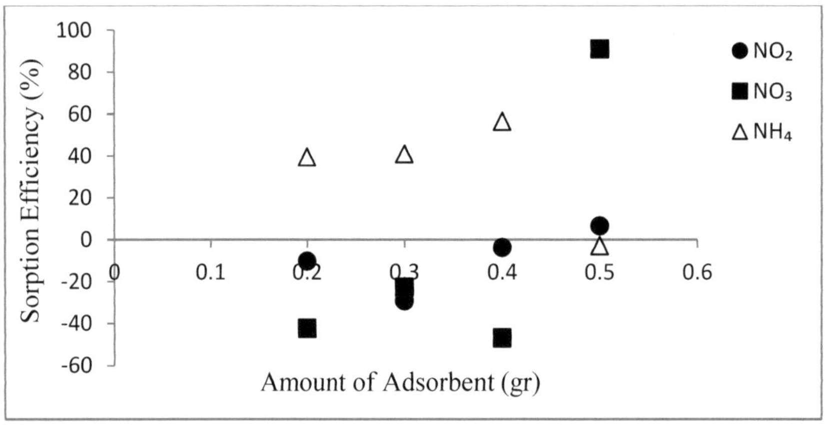

Figure 3. The adsorption efficiency of nitrite and nitrate in liquid waste of PT PIM at contact time of 40 minutes.

With increasing the used adsorbent amount, therefore the adsorption efficiency would also increase. The increasing of adsorbent weight is proportional to the increase the number of particles and the surface area of activated carbon, causing a number of binding levels of nitrite, nitrate and ammonia also increases and as a result the adsorption efficiency also increases [9]. From the above figure we can see that the adsorbent cannot adsorb all of the three components in the waste at the same time of adsorption process, but adsorbent would tend to absorb more ammonia component compare to nitrate nitrite components. Most ammonia in nature and in the wastewater is oxidized to form nitrite and nitrate that are carried out by two kinds of autotroph bacteria during nitrification process following reaction:

$NH_3 + H_2O \rightarrow NH_4 + OH^-$ (ammonia to be ammonium)

$2NH_4^+ + 3O_2 + 2OH^- \rightarrow 2NO_2^- + 2H^+ + 4H_2O$ (ammonium to be nitrite)

$NO_2^- + 1/2O_2 \rightarrow NO_3$ (nitrite to be nitrate)

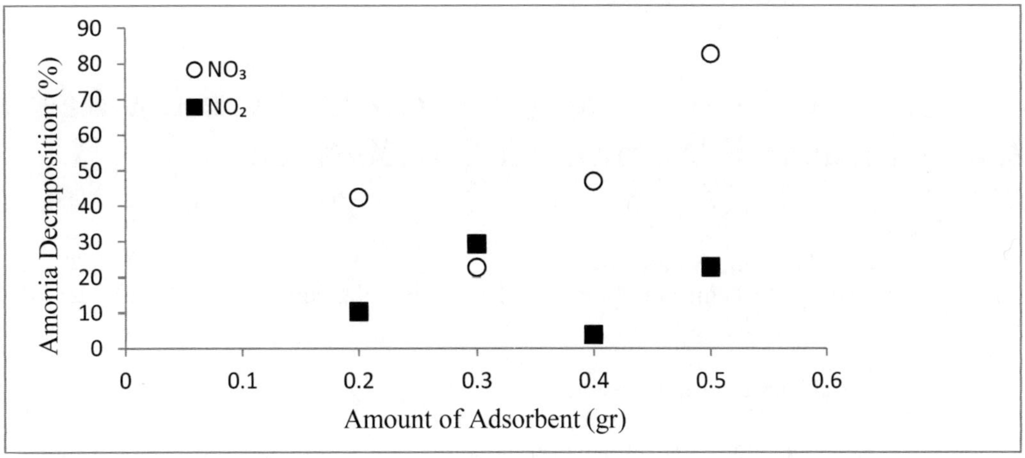

Figure 4. Percentage of ammonia decomposition after adsorption process for 40 minutes

Activated carbon from waste coffee grounds prefer absorbed components of ammonia, and ammonia remaining in the waste reacted quickly to form nitrite and nitrate and as a result the components of nitrites and nitrates in the effluent increased. From the above figure we can see that is likely to occur due to the interaction between the components in the waste cause decomposition ammonia into nitrite reached 30% and finally decompose again to be nitrate as much as 80%.

4. Conclusion

The analysis results of prepared activated carbon meet quality standards that are permitted by SNI No. 06-3730-1995. The prepared activated carbon has the capacity to absorb nitrites, nitrate and ammonia, however the adsorbent tend to adsorb more ammonia compare to sorb nitrite and nitrate. The highest ammonia adsorption efficiency was 56%. The results showed that the remaining ammonia decomposes into nitric reached 30% and breaks down into nitrates by 80%.

References

[1] Piotr N, Paulina S and Robert P 2014 *Chem. Eng. J.* **248** 208-215
[2] Nima F, Tehrani, Javiers S A and Yohannes K 2015 *J. Cleaner Prod.* **91** 64-70
[3] Lubis S and Nasution R 2002 *Jurnal Natural.* **2** 2 12-16
[4] Asbahani 2013 *J. Tek. Sipil Untan.* **13** 1
[5] Azhar A H, Mohd T L and Anuar I 2013 *Word App. Sci. J.* **22** 11-6
[6] Rizki A P and Sanjaya A S 2015 *J. Nas.* **1** 2
[7] Hasan, Nasir L, Muhammad Z, Prastawa B 2014 *Desilikasi Karbon Aktif Sekam Padi sebagai Adsorben Hg Pada Limbah Pengolahan Emas di Kabupaten Buru Provinsi Maluku Indonesia Chimica Acta* **7** 2 1-11
[8] Pujol D, Liu C, Gominho J, Olivella M.A, Fiol N, Villaescusa I and Pereira H 2013 *Industrial Crops and Products Journal.* **50** 423-429
[9] Nurhasni, Hendrawati, Nubzah S 2014 *J. Sains dan Teknol.* **4** 1

IOP Conf. Series: Materials Science and Engineering **358** (2018) 012028 doi:10.1088/1757-899X/358/1/012028

Adsorption of Used Cooking Oil (UCO) by using Raw and Modified Kapok Fibre through Esterification

N H Alias and S I Z Hasan
Faculty of Chemical Engineering, Universiti Teknologi MARA, 40450 Shah Alam, Selangor, Malaysia

norhalaliza@salam.uitm.edu.my

Abstract. UCO is one of the domestic wastes in our daily life. Normally, UCO are produced by hawkers, restaurants and household in a large quantity. The UCO usually exist in water mixture and eventually can cause water drainage problem which can lead to the environmental problem. Therefore, in order to overcome this problem, a study was conducted to test the adsorption of RKF and MKF towards the UCO. As for the MKF, the adsorption was tested by using different concentrations of Calcium Oxide (CaO) in percentage during the esterification. The oil removal percentages were calculated for RKF and MKF. Based on the results, it was found that the RKF has adsorbed 25.32g of UCO with a 50.64% of adsorption. As compared to MKF, it was able to increase the hydrophobic properties which resulted more UCO adsorption by 26.78g of UCO with the increment in the adsorption of 53.56% by using CaO of 5 wt% of RKF mass. However, when the percentage of CaO was increased, the UCO adsorption was also increased. The results showed that by using CaO of 10 wt% and 15 wt% of RKF mass, the UCO adsorption was increased to 28.50g (56.84%) and 31.73g (63.46%), respectively. Thus, MKF has higher adsorption of UCO compared to RKF. The highest amount of UCO adsorption has been achieved by using CaO of 15 wt% in the esterification, which was 31.73g corresponded to 63.46%.

1. Introduction

Nowadays, cooking oil has become one of the important element in the food preparation. There are various types of cooking oil such as olive oil, peanut oil, corn oil, coconut oil, canola oil, palm oil, grapeseed oil, avocado oil, sesame oil and etc. Malaysian use 45 million kilograms of cooking oil where each person uses 1.45 kg of cooking oil per month [1]. In Malaysia, most of the oil are derived from the palm oil. In 2008, the production of palm oil was 17.7 million tonnes which has used up about 4.5 million hectares of land in Malaysia. It stated that there was an increasing demand for the palm oil production [2]. Cooking oil usage has led to the arising problem of waste cooking oil which known as the used cooking oil (UCO). UCO could become a serious problem for home plumbing, drainage and wastewater system. Poured UCO down the sink followed by the hot water and soap or through the garbage disposal could cause the UCO to solidify into the thick layers inside the pipelines which will led to the water flow constriction and water backflow. The UCO could also eventually flow into the river and stream, causing environmental problems together with odor pollution. The treatment of the water drainage problem due to the UCO are very expensive. On the other hand, there are some operators reused the cooking oil which can cause health problem [3].

In order to overcome this problem, an alternative method has been chosen instead of the expensive treatment. Natural fiber, which is a renewable source has a good thermal properties, good insulating properties and environmentally friendly [4]. Based on the findings, these properties are present in raw kapok fiber (RKF) which makes it a good source in oil retention. However, natural fiber has low hydrophobicity and low buoyancy. Through numerous studies, the natural adsorption of RKF can be

Content from this work may be used under the terms of the Creative Commons Attribution 3.0 licence. Any further distribution of this work must maintain attribution to the author(s) and the title of the work, journal citation and DOI.
Published under licence by IOP Publishing Ltd

enhanced by chemical alterations such as acetylation [5]. Modification of the natural fiber could increase the desirable characteristic such as high hydrophobicity, good buoyancy and low cost. Acetylation alters the RKF surface characteristic from smooth into rough surface as the rougher surface enables exposure of larger surface area for adsorption which enables higher interfacial hydrophobicity. Figure 1a shows the ripe kapok pod.

(a) (b)

Figure 1: Kapok Fibre a) The ripe kapok pod. b) RKF (left) and MKF (right) submerged in UCO water mixture for UCO adsorption test.

Apart from acetylation, another method of chemical alterations is through the esterification. Esterification is a process where an ester bond is formed by the reaction between carboxylic acid and alcohol. A carboxylic acid is a group that contains –COOH group whereas in ester, the hydrogen in group is replaced by hydrocarbon group [6]. For example, the oil adsorption by using the sawdust and sugarcane as adsorbent, respectively, had been increased by using the esterification modification. The adsorption was increased by the replacement of O-H functional group in the fibers with C=O and C-O group [7]. On the other hand, sago bark (SB) had been used in the removal of oil from palm oil mill effluent (POME). The adsorption was increased by the modification SB into esterified sago bark (ESB). The results showed that ESB has higher oil removal efficiency because the modification gives rougher surface and more exposure of the internal structure with the oil entrapment sites. In addition, calcium oxide (CaO) was used as a catalyst in the esterification. The CaO was able to neutralize the negative zeta potential of oil in POME and resulted in the reduction of electrostatic repulsion that allows more oil droplets coalescence [8].

Since RKF naturally adsorps oil, conducting chemical alteration through esterification by modifying the RKF into modified kapok fiber (MKF) could enhance the adsorption of oil. Thus, it is expected that the oil adsorption is higher in the MKF compared to RKF. A few samples of MKF were prepared based on different concentration of catalyst, CaO added in the esterification reaction.

2. Experiment

2.1. Esterification of Kapok Fiber

The sample of RKF was purchased. In the esterification, the chemicals involved were stearic acid (SA), Calcium oxide (CaO) and ethyl acetate (EA). 1g of raw kapok fiber (RKF) was weighed and the ratio of SA to kapok fiber was measured at 1:1. Then, CaO of 5 wt% of RKF mass was added. The mixture was mixed in a beaker of 150 ml. Next, 100 ml of EA was added into the mixture and transferred into the round bottomed flask. Lastly, the mixture was set as in Figure 2 (reflux condensing method) with the magnetic stirrer rotating at 150 rpm and temperature of 85 °C [8]. The reflux was operated for 5 hours. The esterification reaction were repeated for CaO of 10 wt% and 15 wt% of RKF mass, respectively.

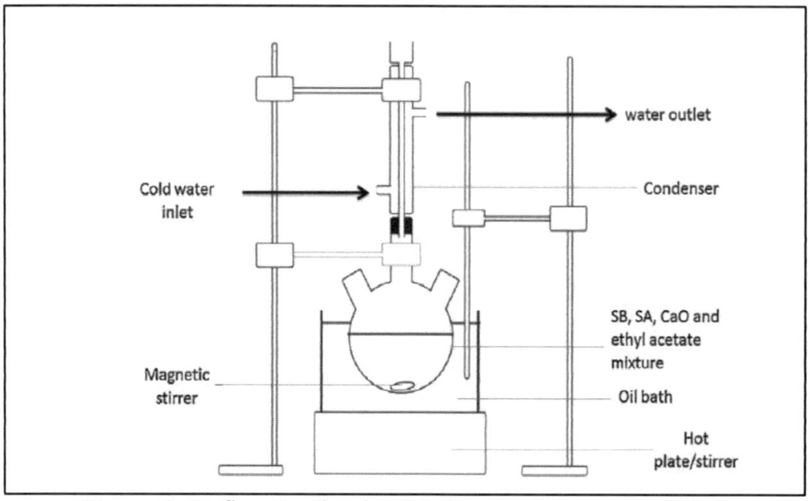

Figure 2: Reflux Condensing Method (Rafeah, et. al., 2014)

2.2. Mixture Preparation of the Used Cooking Oil (UCO) and Water

50g of UCO was mixed with 50mL of tap water. The mixture obtained was homogenized for 10 minute by using a magnetic stirrer at 150 rpm.

2.3. Measurement of Oil Adsorption Capacity

1g of dried sample of RKF was weighed to obtain W_i (weight of the initial sorbents). Then, the sample was mixed with the oil and allowed to be immersed in the oil mixture at the room temperature for 30 minutes. Next, the RKF was drained and wiped with filter paper to remove the excess oil from the bottom of the drainer. The oil adsorption capacity of the sample was determined from the calculation based on the equation below:

$$Q = \frac{W_t - W_i - W_w}{W_i} \times 100 \qquad (1)$$

Where Q is the oil sorption capacity of the sorbents, calculated as grams of oil per gram of sample. Meanwhile, W_t is the weight of the wet sorbent after draining (g) and W_w is the weight of the water adsorbed in the sorbent (g). The water content was determined by extraction separation method by using n-hexane as the solvent.

3. Results and Discussion

3.1. The Comparison of the Oil Adsorption Capacity between RKF and MKF

RKF naturally has oil adsorption ability due to it hydrophobic surface and huge hollow lumen. In the previous research, RKF was used in the treatment of oil-containing wastewaters [9]. RKF is a type of cellulosic fibre, where its structure consists of hydroxyl group which limits their oil adsorption as it behaves to more hydrophilic [10]. Thus, to enhance the adsorption of RKF, chemical modification was conducted to check whether the UCO adsorption can be enhanced by using esterification method. Figure 1b shows the comparison of the submerged RKF and MKF, respectively, in the water for UCO adsorption test.

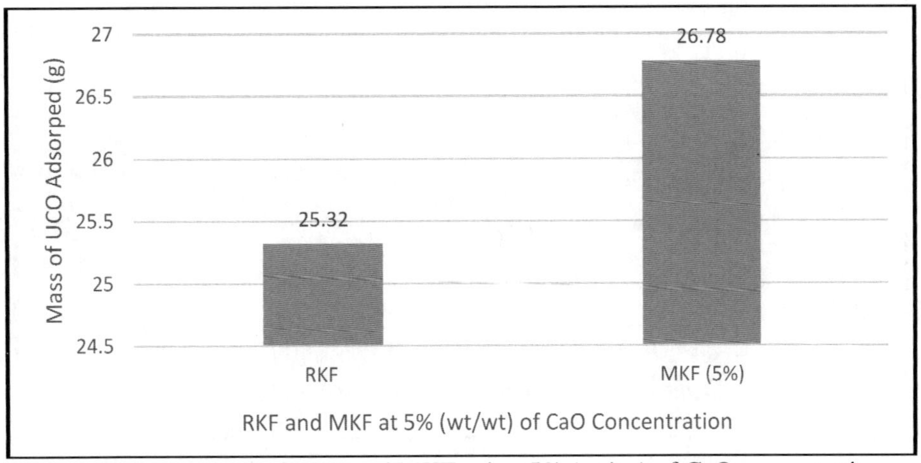

Figure 3: Comparison of RKF and MKF using 5% (wt/wt) of CaO concentration towards the UCO adsorption.

Stearic acid is a type of carboxylic acid with a chemical component of $CH_3(CH_2)_{16}COOH$ that contains carboxyl functional group which is –COOH. Stearic acid was chosen in the esterification due to the reason of having 18 carbon chain, which contribute to the higher hydrophobic properties [11]. From the results, it showed that RKF adsorped 25.32 g of UCO, gave about 50.64% of adsorption. Next, the MKF by using CaO of 5 wt% of RKF mass was tested in UCO water mixture. Based on Figure 3, the results showed that the MKF adsorped higher amount of UCO, which contributed 26.78 g (53.56%) as compared to RKF. Thus, it showed that the hydrophobic properties of RKF had been increased due to the chemical alteration by the replacing of O-H group in the RKF with the carbonyl group, C=O and carbon-oxygen bond, C-O after the esterification reaction [7].

3.2. Effect of CaO (wt%) in the UCO Adsorption by using MKF

The MKF was performed by refluxing the RKF with stearic acid (SA), calcium oxide (CaO) as a catalyst and ethyl acetate (EA). The samples were prepared at three different CaO concentrations, namely, 5 wt%, 10 wt% and 15 wt% of RKF mass, respectively. The function of CaO was to allow more chemical alteration on the RKF surface due to the high catalytic activity at mild reaction condition. The advantage of using CaO as a catalyst in this research was due to its long catalyst lifetime and low in cost [12]. Based on Figure 4, the adsorption of UCO increased as the concentration of CaO increased as compared to the RKF. This was supported by the amount of the UCO adsorption by using CaO of 5 wt%, 10 wt% and 15 wt% of RKF mass, which were 26.78g (53.56%), 28.51g (56.84%) and 31.74g (63.46%), respectively.

Based on the results of esterification reaction, the presence of O-H group in RKF was replaced by the hydrophobic group from SA onto the surface of MKF. Moreover, the increment adsorption of UCO was due to the negative zeta potential in UCO, which was neutralized by the calcium ions of calcium oxide in the MKF. The neutralization of charge resulted in the reduction of electrostatic repulsion and consequently enhanced the UCO droplets coalescence [8]. Thus, the higher concentration of CaO leads to the higher reduction of electrostatic repulsion in MKF, which allowed the adsorption of UCO to increase in the 5 hour time span and caused more available binding site for UCO adsorption.

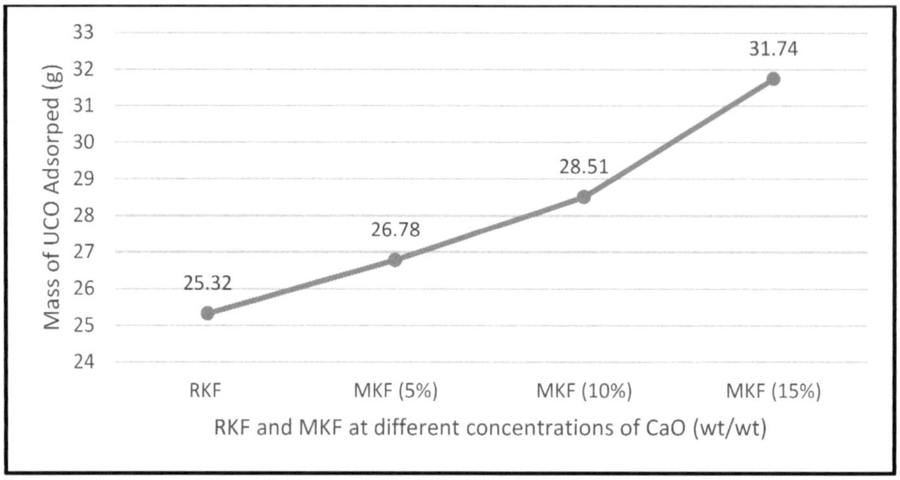

Figure 4: Comparison of the amount of UCO adsorption using RKF and MKF at 3 different concentrations of CaO

Figure 5 shows that the adsorption of the MKF towards the UCO has been increased with the increasing of amount of concentration of CaO added in the esterification reaction. Therefore, the CaO concentration can affects the conversion of RKF surface to be more hydrophobic and eventually increased the adsorption of UCO.

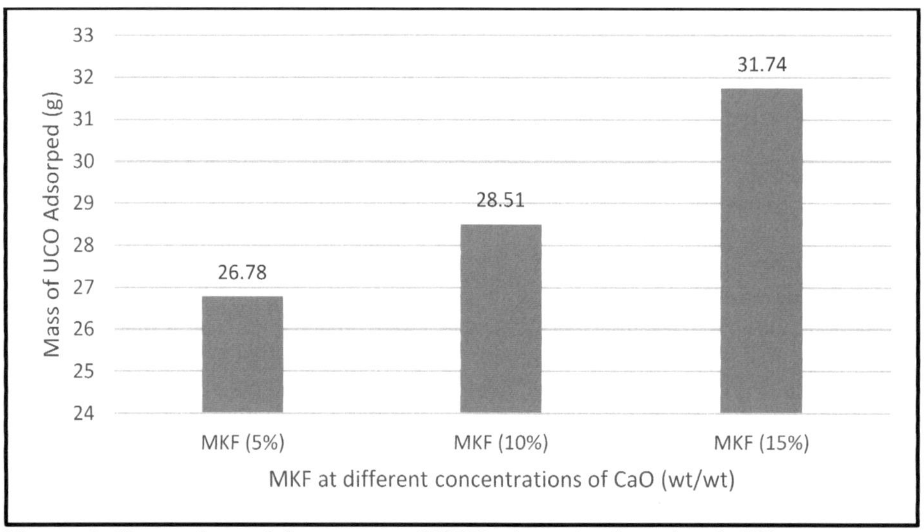

Figure 5: Comparison of the amount of UCO adsorption by using 3 different concentrations of CaO

4. Conclusion

Study on adsorption capacity shows that the RKF was able to adsorp UCO up to 25.32g with a 50.64% of adsorption. The adsorption was due to kapok fibre surface which has hydrophobic properties. However, chemical alteration by esterification of RKF to convert to MKF was able to increase the hydrophobic-oleophilic properties by replacing more O-H group with C=O and C-O group which resulted more UCO adsorption by MKF up to 26.78g with increment in the adsorption up to 53.56% using CaO of 5 wt% of RKF mass. Meanwhile, by comparing the adsorption of MKF by using different concentrations of CaO, it can be concluded that increasing the CaO concentration has reduced the electrostatic repulsion and increased the UCO adsorption. This was supported by the results obtained by using CaO of 10 wt% and 15 wt% of RKF mass, which were 28.50g (56.84%) and

31.73g (63.46%), respectively. In conclusion, MKF performed better than the RKF in the UCO adsorption. Among the amount of CaO concentrations tested, using the CaO of 15 wt% of RKF mass gave the highest amount of UCO adsorption. For future study, it is recommended to increase the CaO concentration in the process of modifying RKF into MKF in order to maximize the potential of MKF in UCO adsorption.

Acknowledgments
The author would like to acknowledge the Faculty of Chemical Engineering, Universiti Teknologi MARA, Shah Alam for the financial support and lab facilities. A special thanks to my supervisor, Madam Nor Halaliza Alias for giving the support and opportunity during the whole process of doing the research.

References

[1] Zaidi A, 18 November 2016. The Mole We Don't Care [Online]. Available: http://mole.my/malaysians-use-too-much-cooking-oil/. [Accessed on 29 November 2016].
[2] Anna B, 18 September 2014. Food Navigator [Online]. Available: http://www.foodnavigator.com/Market-Trends/Malaysia-to-double-palm-oil-production-by-2020. [Accessed on 29 November 2016]
[3] Cindy L, 26 March 2014. Borneo Post online [Online]. Available: http://www.theborneopost.com/2014/03/26/awareness-of-need-to-recycle-cooking-oil-low/. [Accessed on 29 November 2016]
[4] Chandramohan.D and Chandramohan K X, 2 August 2011. A Review of Natural Fibers [Online]. Available: www.arpapress.com/Volumes/Vol8Issue2/IJRRAS_8_2_09.pdf. [Accessed on 18 May 2016].
[5] Jintao W, Yian, Z and Aiqin W 2013 *Marine Pollution Bulletin* 91-96
[6] Jim C, October 2015. Esterification [Online]. Available: http://www.chemguide.co.uk/organicprops/alcohols/esterification.html. [Accessed on 5 December 2016].
[7] Rafeah W, Luqman A C, Thomas S Y, Zainab C N and Mohsen M N 2013 *J. Separation and Purification Technology.* **113** 51-63
[8] Rafeah W, Luqman A C, Thomas S Y, Zainab C N and Mohsen M N 2017 *J. Environmental Chemical Engineering.* **5** 170-177
[9] Jintao W, Yian Z and Aiqin W 2013 *J. Environmental Sciences.* **25(2)** 246-253
[10] Nguyen T H, Nguyen N S and Tran D H 2016 *J. Material Environment and Sciences.* **7(7)** 2498-2504
[11] Goyal R ND Carboxylic acids and their derivatives *Dept. Chemistry Indian Institute of Technology, Roorkee,* 247-667
[12] Komkanok W, Bunjerd J and Muenden P 2010 *Songklanakarin J. Science and Technology.* 627-634

Methane Synthesis from Automotive Paint Sludge via Microwave Assisted Pyrolysis

N L Rosli[1,4], N Abd Rahman[1,2,3,5] and A Kadri[1,6]
[1]Faculty of Chemical Engineering, Universiti Teknologi MARA, 40450 Shah Alam, Selangor Darul Ehsan, MALAYSIA
[2]CoRe of Frontier Materials & Industry Applications, Universiti Teknologi MARA, 40450 Shah Alam, Selangor Darul Ehsan, MALAYSIA
[3]CoRe of Green Technology & Sustainable Development, Universiti Teknologi MARA, 40450 Shah Alam, Selangor Darul Ehsan, MALAYSIA

[4]nurulliyanar@gmail.com, .,[5]noraz695@salam.uitm.edu.my,
[6]atikahkadri@salam.uitm.edu.my

Abstract. Methane gas, which has one atom of carbon and four atoms of hydrogen, is a valuable energy resource, which can be used in the energy sector. The purpose of this research work is to identify methane synthesis from Automotive Paint Sludge (APS) using microwave assisted pyrolysis. APS is known as a hazardous waste since it contains various chemicals that categorized as heavy metals and toxic substances. A modified conventional kitchen microwave was used to pyrolise the APS. The microwave was set with the power level of 600 W and 50 minutes radiation time. Through the experiment, pyrogas was collected into tedlar bag and was analysed using Gas Chromatography with Flame Ionization Detector (GC-FID). Results from the GC-FID were shown that the retention time of 3.3583, 3.2733, and 3.2267 min are proved to be methane gas. The results obtained are resembled with the results from the literature. This indicates methane gas was presented in the pyrogas of pyrolysis of APS and there is a possibility of producing methane gas. The research study suggests that it is possible to synthesize methane gas from the APS via microwave assisted pyrolysis, and in the meantime reduce the volume of APS in the landfill.

1. Introduction

The number of vehicles produced has become increasing in recent years due to the rising in market demand caused by the development in the modernization and population. Salihoglu and Salihoglu (2016) mentioned that the demand for vehicles production has been increasing at a constant rate [1]. Unfortunately, with the increasing demand for vehicle production, it will lead to the increasing of APS in the landfill. Automotive Paint Sludge (APS) is a waste paint resulting from the automotive manufacturing industry. Paint sludge is waste water containing the mixture of over sprayed paint and water. The over sprayed paint is resulted from the paint that not reaching the target in a painting by spraying method while the water in the sludge paint resulted from the circulating water that used to wash the overspray and collected in a sludge pit [2]. APS contain various chemicals and has been classified as hazardous waste. According to Gautam, Bundela and Murumkar (2010), APS contains heavy metals and other toxic substances, due to which has high pollution potential [3]. One way to overcome this problem is to use APS in order to produce valuable products. APS has been identified as one of the option of fuel production obtained from microwave assisted pyrolysis techniques [4]. According to Januri et al. (2015), APS contain calorific value about 22.6 MJ/kg, which indicate that it is a good potential for the future alternative fuel which can produce solid, liquid and gas product [5]. Table 1 shows the characteristics of APS. Rahman et al. (2015) studied on the yield of microwave

Content from this work may be used under the terms of the Creative Commons Attribution 3.0 licence. Any further distribution of this work must maintain attribution to the author(s) and the title of the work, journal citation and DOI.
Published under licence by IOP Publishing Ltd

pyrolysis of APS by using microwave absorber [6]. However, although APS has been proven very useful in producing fuel, little attention has been paid to the production of methane gas from pyrolysis of the APS using microwave.

Table 1. Characteristics of Automotive Paint Sludge [5].

Analysis Parameters		Values
Calorific value		22.6 MJ/kg
Melting point		376.56 °C
Enthalpy of melting		66.7 J/kg.K
Proximate analysis	Ash	2.9 wt.%
	Moisture	2.4 wt.%
	Volatile matter	75.9 wt.%
	Fixed carbon	3.2 wt.%
Ultimate analysis	Carbon	38 wt.%
	Hydrogen	10 wt.%
	Nitrogen	9 wt.%
	Sulfur	1 wt.%
	Oxygen	42 wt.%
Chemical composition		$C_{38}H_{10}N_9SO_{42}$

Methane is a gaseous form in its natural state with the chemical formula CH_4 and is the simplest component in alkane chemical class. Howarth (2011) stated that methane gas was the main component and is the primary components of natural gas [7]. Zhang, Liang and Harpalani (2016) produced methane from bituminous coal through biogasification process [8]. Methane also can be produced from the digestion of glucose and sludge using microbial electrolysis cell assisted anaerobic digestion systems [9]. Zou et al. (2016) conducted a research in producing methane production from pretreatment of food waste with high voltage pulse discharge using multi-needle-to-plate reactor [10]. According to Córdoba et al. (2016), methane can be produced from spent populus sawdust obtained from the cultivation of *Gymnopilus pampeanus* [11]. Many researchers doing research in the production of methane since it is important to produce methane besides depending on the natural gas because natural gas will deplete in the future.

Microwave is a heating equipment that use short waves of electromagnetic energy that increases the kinetic energy of water dipoles to its boiling point very quickly [12]. Chan et al. (2013) evaluated the anaerobic biodegradability of microwave pre-treated dairy manure in the methane production [13]. Jackowiak, Bassard, Pauss, and Ribeiro (2011) performed a research in microwave pre-treatment of wheat straw for methane production [14]. Marin, Kennedy, and Eskicioglu (2010) investigated on the methane production from a model kitchen waste using high temperature and pressure microwave irradiation [12]. Pyrolysis is the thermal decomposition process occurring in the absence of oxygen [15]. The product produced will be varied, depend on the pyrolysis method, residence time, temperature to which the biomass is exposed and the characteristics of the biomass itself. Bucko et al. (2014) also stated that pyrolysis produced four main end products include biochar, wood vinegar (from bio oils), creosote (from bio oils) and electricity (methane gas from pyrogas) [16]. The present research work proposes to identify methane synthesis from automotive paint sludge using microwave assisted pyrolysis. This method forms a novel for methane synthesis where the pyrolysation of APS significantly reduced the volume of APS in landfills and in the meantime produces valuable methane gas.

2. Experimental

2.1 Materials

The experiment involved the pyrolysis of APS via microwave and analysing the pyrogas obtained to identify the existence of methane gas via Gas Chromatography with Flame Ionization Detector (GC-FID). The sample of APS was collected from an automotive manufacturer industry in Selangor Darul Ehsan, Malaysia. The collected APS was placed in a covered container to avoid from any contamination during storage. Pure methane gas (high purity of 99.99 %) was purchased from Pure Dimension Enterprise, Malaysia and was used for comparison purpose.

2.2 Microwave assisted pyrolysis of APS

Prior to the pyrolysis process of APS, 200 g of APS and 20 g of activated carbon, which act as microwave absorber was placed in a quartz reactor before placing it inside the modified conventional kitchen microwave. The microwave was set with the power level of 600 W and 50 minutes radiation time. Nitrogen gas was purged into to microwave for 20 minutes before pyrolysis process and was continued flowing throughout the experiment to evacuate the air in the microwave and create nitrogen atmosphere. The microwave run was operated for 50 min. At 20, 30 and 40 minutes, pyrogas were collected and were stored into three different tedlar bags. After microwave had reached the radiation time of 50 minutes, the microwave was switched off.

2.3 Analysis

The collected pyrogas and pure methane gas were then analyzed using GC-FID according to the Standard practice for analysis of reformed gas by gas chromatography, American Society for Testing and Materials (ASTM) International, ASTM D1946 - 90(2015)e1 [17]. Table 2 shows the operating conditions of the GC-FID.

Table 2. Operating Conditions of Gas Chromatography with Flame Ionization Detector Used [18].

Parameter	Condition
Oven temperature	60 °C (Ramp rate 22 °C / min)
Detector temperature	250 °C
Injector temperature	150 °C
Injection volume	1 mL
Carrier and make up gas with flow rate	Nitrogen, 2 mL/min
Hydrogen flow	48 mL/min
Air flow	500 mL/min

3. Result and Discussion

In this research study, pyrogas was obtained from the pyrolysis of the APS using microwave. The results obtained from GC-FID were compared with pure methane gas. Table 3 shows the retention time of the GC-FID chromatogram of the pyrogas and pure methane gas.

Table 3. Retention Time of the Gas Chromatography with Flame Ionization Detector Chromatogram of the Pyrogas and Pure Methane Gas.

Sample	Retention time (min)	Area (Counts)
Sample 1: Pyrogas taken at 20 min of radiation time	3.3583	208684
Sample 2: Pyrogas taken at 30 min of radiation time	3.2733	1356719
Sample 3: Pyrogas taken at 40 min of radiation time	3.2267	2045462
Sample pure methane gas	2.8467	18132170

As can be seen, all three samples, which taken in 20 minutes, 30 minutes and 40 minutes of radiation time, have comparable retention time. This result is consistent with the result obtained by Wang (2010) where the retention time of methane gas obtained from the analysis of greenhouse gas standards was 3.314 minutes [18]. The results obtained of this research study also merge with the result obtained by Gedam and Regupathi (2012) in which the retention time of methane gas obtained from the analysis of syngas from municipal solid waste was 3.40 minutes [19]. To the degree that, Tipler (2010) also reported that the retention time of methane gas was 3.50 minutes, which also pursuant to other results [20]. However, the retention time of all three samples was not consistent with the retention time of pure methane gas. Pure methane gas has retention time earlier than those of the three samples, Wang (2010) and Gedam and Regupathi (2012) [18, 19]. This is due to the sample of pure methane gas used was high purity (99.99 %).

The peaks can be seen clearly in Figure 1, which shows the GC-FID chromatogram of the pyrogas and pure methane gas. The figure shows that the peak of all three samples lies under the peak of the pure methane gas. The pure methane gas has a width peak compared to those of the three sample because the pure methane gas used is a high purity gas with 99.99 %, while those of the three samples was obtained from pyrogas which has not purity as high as the pure methane gas. However, it was proved that the peaks existed in the GC-FID chromatogram of the pyrogas was methane gas whether it's taken during 20 minutes, 30 minutes or 40 minutes of radiation time. Figure 2 shows comparison of the GC-FID chromatogram of the pyrogas and pure methane gas. As can be seen, even though pure methane gas has early retention time compared to those of the three samples, pure methane gas has a width peak compared to others.

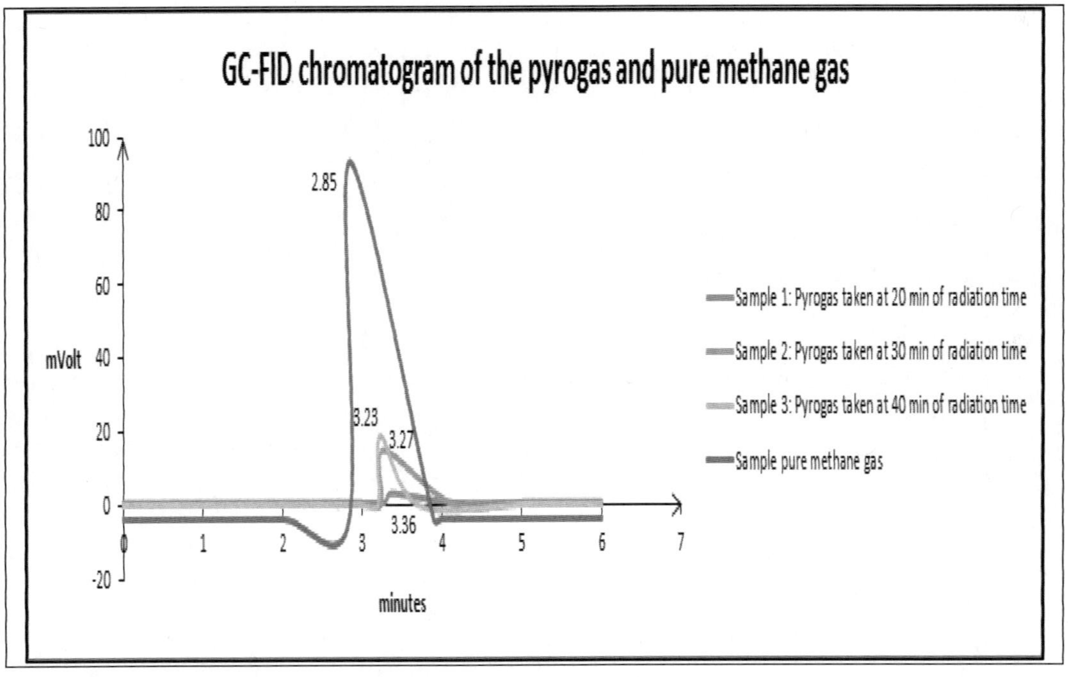

Figure 1. Gas Chromatography with Flame Ionization Detector Chromatogram of the Pyrogas and Pure Methane Gas.

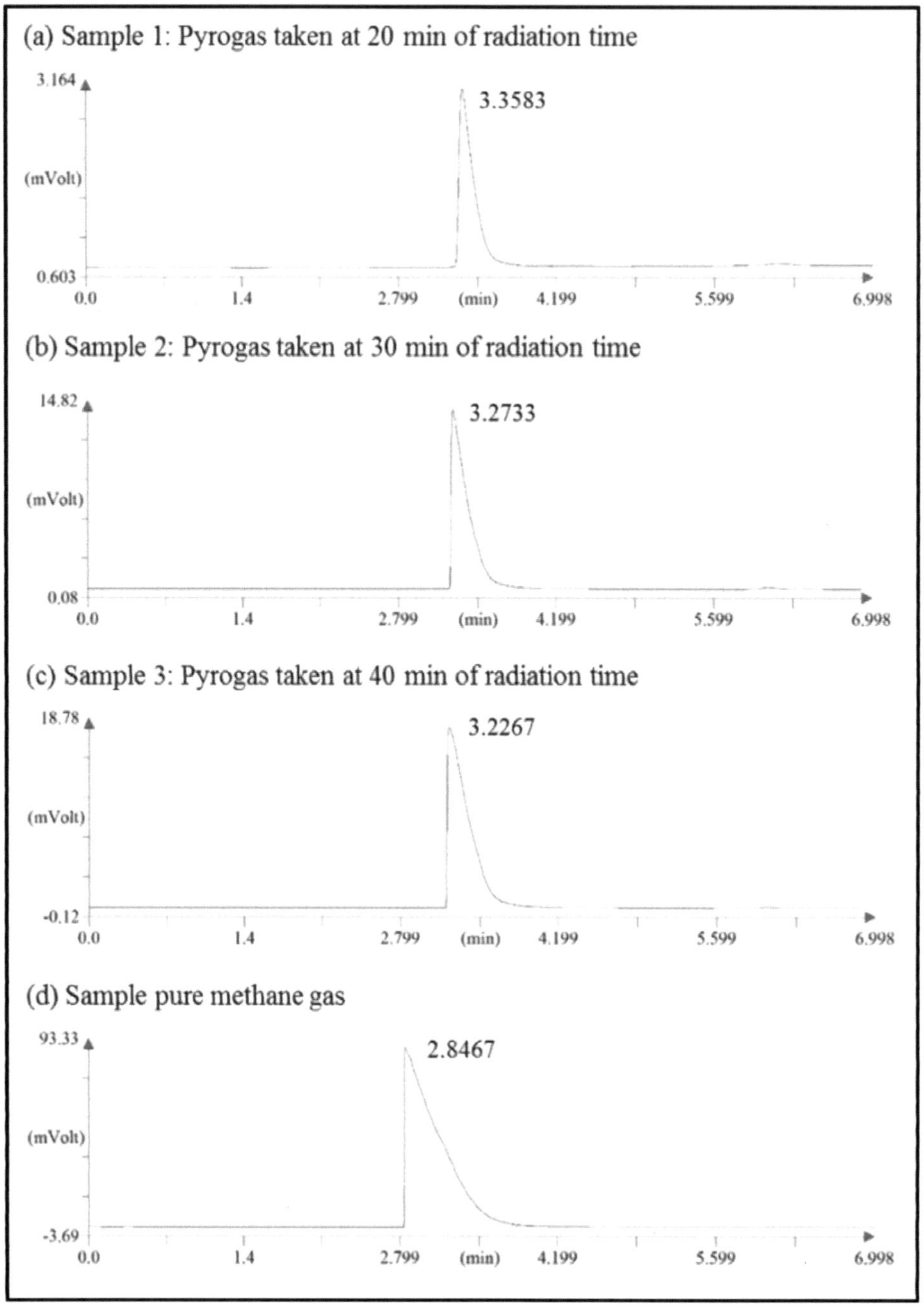

Figure 2. Comparison of the Gas Chromatography with Flame Ionization Detector Chromatogram of the Pyrogas and Pure Methane Gas.

4. Conclusion

In summary, retention time of 3.3583, 3.2733, and 3.2267 min obtained from the GC-FID of 20, 30 and 40 minutes radiation time of APS pyrolysis are proved to be methane gas. Thus, methane gas has been shown to be present in the pyrogas obtained from the pyrolysis of APS via microwave. Therefore, the research study has presented a novel technique for methane synthesis from APS via microwave assisted pyrolysis, and at the same time reduces the volume of APS in the landfills.

References

[1] Salihoglu G and Salihoglu N K 2016 *J. Environ. Manage.* vol. **169** 223–235
[2] Ruffino B and Zanetti M C 2010 in *Crete 2010*

[3] Gautam S, Bundela P, and Murumkar M 2010 *Waste Manag. Environ.* vol. **140** 57
[4] Januri Z, Abd. Rahman N, Idris S S, Matali S, Manaf A, and Fairuz S 2014 *Applied Mechanics and Materials* vol. **548** 191–195
[5] Januri Z, N. Abd. Rahman, Idris S S, Matali S, Fairuz S and Manaf A 2015 *Adv. Mater. Res.* vol. **1113** 236–241
[6] Rahman S A, Faizah S N, Abd Rahman N, Idris S S, Abu Bakar N F, Mokhtar R, Januri Z, Khalil M and Fareezuddin M 2015 *Applied Mechanics and Materials* vol. **789** 66–70
[7] Howarth R W, Santoro R and Ingraffea A "Methane and the greenhouse-gas footprint of natural gas from shale formations."
[8] Zhang J, Liang Y and Harpalani S 2016 *Appl. Energy* vol. **183** 31–42
[9] Gajaraj S, Huang Y, Zheng P and Hu Z 2017 *Biochem. Eng. J.* vol. **117** 105–112
[10] Zou L, Ma C, Liu J, Li M, Ye M and Qian G 2016 *Bioresour. Technol.* vol. **222** 82–88
[11] Córdoba V, Colavolpe M, Fernández M, Santalla E and Albertó E 2016, "Potential methane production of spent sawdust used in the cultivation of Gymnopilus pampeanus,"
[12] Marin J, Kennedy K J, and Eskicioglu C 2010 *Waste Manag.* vol. **30** 1772–177.
[13] Chan I, Srinivasan A, Liao P H, Lo K V, Mavinic D S, Atwater J and Thompson J R 2013 *Nat. Resour.* vol. **4**, 246–256
[14] Jackowiak D, Bassard D, Pauss A and Ribeiro T 2011 *Bioresour. Technol.* vol. **102** 6750–6756
[15] Jahirul M, Rasul M, Chowdhury A and Ashwath N 2012 *Energies*, vol. **5** 4952–5001
[16] Bucko S, German-Sobek M, Pavlík M, Király J and Novák M 2014 *Intensive Programme 2014 "Perspectives for the development of low-power systems using biomass,"* 250–254
[17] "ASTM D1946 - 90(2015)e1 Standard practice for analysis of reformed gas by gas chromatography," *ASTM International, West Conshohocken, PA,* 2015. .
[18] Wang C 2010 "Simultaneous analysis of greenhouse gases by gas chromatography," Shanghai
[19] Gedam V V and Regupathi I 2012 *Nat. Resour. Res.*, vol. **21** 75–82
[20] Tipler A 2010 "The determination of C2 to C5 hydrocarbons in finished gasolines using the PerkinElmer Clarus 680 GC with swafer technology,"

Parameters for Novel Production of Fruity Floral Fragrance Ester (Geranyl Butyrate) by Locally Isolated Lipase *Geobacillus thermodenitrificans* nr68 (LGT)

N. H. Nik Raikhan

Faculty of Chemical Engineering, Universiti Teknologi MARA, Shah Alam, 40450 Shah Alam, Selangor, Malaysia

raikhan7952@salam.uitm.edu.my

Abstract. Geranyl butyrate has been synthesized successfully using our locally isolated lipase *Geobacillus thermodenitrificans* nr68 (LGT) as the fragrance ester with aim to be used in a nanotechnology fragrance application. We have used and modified few parameters from the previous research and then, continued with optimization of the synthesis by looking into degree of esterification and water content in the system. Butyric acid (C4), stearic acid (C18: 0), caprylic acid (C8), linolenic acid (C18: 3), myristic acid (C14), linoleic acid (C18: 2) and oleic acid (C18: 1) were used in the substrate selection. The yield of geranyl butyrate before the optimization was 31.68±0.01%. The optimum parameters for the synthesis of geranyl butyrate were recorded as temperature of 65°C, shaking rate at 200 rpm, 5.0 ml of geraniol and 0.40 ml of butyric acid and 4.0 ml of n-butanol and 0.40 ml of oleic acid. After the optimization, geranyl butyrate synthesis was increased by 297% as to compare with the value before the parameters were optimized. We also have significantly reduced water content as a byproduct of the esterification and managed to run the system a success. The ability thermotolerant lipase from *Geobacillus thermodenitrificans* (LGT) in this synthesis is novel to Malaysian fragrance industry.

1. Introduction

Fragrance and aromatic compounds are used widely in most food industries, dairy products and detergent. The extraction of both compounds from plants has been recognized to be involving high cost as well as the source is limited to certain kinds of plants. Thus, the compounds are now widely produced by microbial sources through a process called biotransformation [1] in order to overcome the above two issues. Biotransformation activities for the production of these aromatic compounds have started last decade [2]. The process is supported very well by the use of extracellular lipases and the enzyme high affinity in organic solvent. Thermostable enzyme with high specification is now chosen at the industrial level because it is very stable very at elevated temperatures and is easy to handle when it comes to involving the organic solvent. Locally isolated species is focused as the potential enzyme producers as to create a novel production and with better control over the large scale production [3]. Lipase from *Geobacillus thermodenitrificans* (LGT) has been used to study the substrate specification for the purpose of producing potential fragrance; geranyl butyrate and butyl oleate.

2. Materials and Method

2.1 Chemicals

Butyric acid (C4), stearic acid (C18: 0), caprylic acid (C8), linolenic acid (C18: 3), myristic acid (C14), linoleic acid (C18: 2) and oleic acid (C18: 1) are products of Sigma-Aldrich. Other chemicals were terpene alcohol (Sigma), molecular sieve (Merck), skimmed-milk (Island Farms), copper (II)

Content from this work may be used under the terms of the Creative Commons Attribution 3.0 licence. Any further distribution of this work must maintain attribution to the author(s) and the title of the work, journal citation and DOI.
Published under licence by IOP Publishing Ltd

acetate (Sigma-Aldrich), pyridine acetate (Sigma-Aldrich), benzene (Sigma), heptane (Merck), tetrahydrofuran (Sigma-Aldrich), ethyl acetate (99.8%, Sigma-Aldrich), isooctane (Sigma), toluene (anhydrous, 99.8%, Sigma-Aldrich), hexane (Sigma) and heptane (Sigma).

2.2 Preparation of freeze-dried lipase enzyme

Method by [1] has been adapted. The 48h *Geobacillus thermodenitrificans* fermentation broth was centrifuged using High Speed Centrifuge Refrigerated 6500 model; Kubota. Crude lipase enzyme activity of 16.68U/ml was concentrated by removing up to 80% of the volume of water from the extract using Viva flow machine (Cole Parmer, MASTERFLEX C/S). Concentrated lipase activity was recorded as 19.82-22.20 U/ml. Then, a total of 20% fat-free skimmed-milk was added into the concentrated enzyme followed by freeze-drying process conducted in 1500-2000 millitorr vacuum; condensing at-50ºC using a Bench Top 2K (VirTis). Collected lipase powder activity was recorded as 30.50 U/g. Figure.1 shows the freeze-dried LGT.

Figure 1: The freeze-dried LGT

2.3 Esterification and degree of esterification.

10.0% (v/v) different kind of fatty acids, 10.0% (v/v) terpene alcohol (geraniol), 10.0% (w/v) LGT, 10.0% (w/v) molecular sieve and 80.0% (v/v) organic solvent were used at 200rpm and 65ºC [4].The degree of esterification was determined every 6 h for 72h and the test was conducted in triplicate. Modification of copper soap colorimetry method by Saisuburamanian *et al.* [5] was applied. Copper acetate-pyridine reagents contained 5% copper acetate [Cu(AC)] was filtered and then adjusted to a pH of 6.1 using pyridine. It was to determine the amount of fatty acids left behind after the reaction. 0.40 ml of reaction mixture was transferred to a test tube, followed by 4.60 ml of benzene and before it was vortex to dissolve the sample. Cupric solution of 1.0ml pyridine acetate was added and vortex (2 min). The optical density at 715 nm (Spectrophotometer Spectronic UNICAM 10uV Genesys) was recorded and expressed through the exchange of free fatty acids in the μmol using 0-1000 μmol standard curve containing pure oleic acid. The pure oleic acid was prepared in 4.0 ml of heptane, followed by the addition of 100 mL copper acetate-pyridine reagent. The optical density at 715 nm was plotted. The acid concentration was determined based on the standard curve plotted. The degree of esterification was determined based on the reduction of the concentration of the fatty acids converted into esters. The degree of esterification is shown in Equation 1.

$$\frac{K_A - K_B \times 100\%}{K_A} \quad (1)$$

Where;

KA = concentration of fatty acid in the initial reaction, KB = concentration of fatty acids in the reaction

2.4 Determination of water content in the system

The water produced by the esterification was determined by using Caulometer Titramate (684KF, Metrohm, Switzerland). The mixture was centrifuged at 10,000 rpm for 5 minutes and 5.0ml of the clear layer was injected to determine the water content. It was compared with the control and expressed in unit percentage. The test was carried out thrice. Control was the esterification mixture without the addition of LGT.

2.5 Substrate for terpene ester synthesis

The substrate was geraniol (~99-95% GC grade, Fluka, Chemika). Fatty acids were used to refine the formation of terpene ester. Methanol was used as organic solvents. All tests were conducted in triplicate [1]. Geranyl butyrate synthesis before optimization process has consisted of 0.50g LGT, 0.50ml geraniol, 0.50ml butyric acid, 4.00ml heptane and 0.50g molecular sieve (0.3 nm) incubated at 65°C (200rpm). The degree of esterification and water content was determined every 6 h (72 h). All tests were conducted in triplicate. We have recorded the optimum time as 42 h. For the effect of temperature and shaking rates temperature of 55°-75°C were used and shaking rates of 0, 50, 100, 150, 200 and 250 rpm [6]. The degree of esterification and water content was determined at 42 h.

The organic solvents tested were benzene, tetrahydrofuran, ethyl acetate, isooctane, toluene, hexane and heptane. This was followed by the weight of molecular sieve (0.25g, 0.50g, 0.75g, 1.00g and 1.25g). The choice of solvents and weight of molecular sieve was done according to Ibrahim *et al.* [7]. The optimum condition and synthesis of geranyl butyrate under optimum condition was 4.00ml of heptane, 0.25g of molecular sieve, 0.50g of LGT, 0.50ml of geraniol and 0.5ml of butyric acid at 200 rpm and 70°C [1].

3. Result and Discussions

The study of terpene ester synthesis was conducted using freeze-dried LGT. The skimmed milk has not interrupted the esterification because it has a fat-free response to the system. Terpene ester formation from all studies was confirmed by thin layer chromatography (TLC). Figure 2 shows geranyl butyrate obtained after the synthesis where A is geraniol, B is butyric acid, C is the product (the geranyl butyrate) and D is the reaction mixture for the synthesis of geranyl butyrate after 72h using freeze-dried LGT.

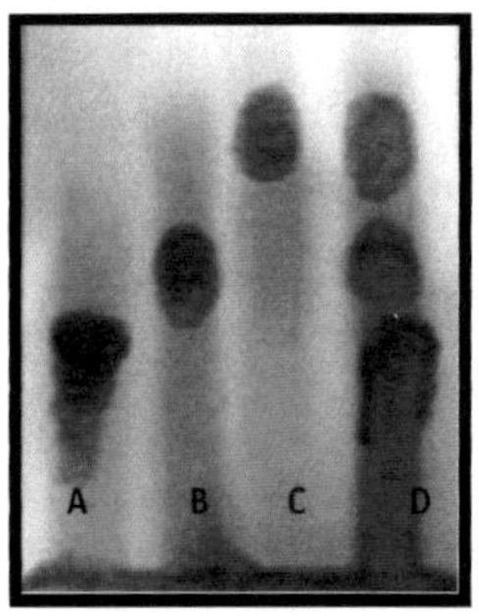

Figure 2: Thin layer chromatography (TLC) shows the separation of the components resulting from the synthesis of geranyl butyrate by LGT.

Figure 3 indicates degree of esterification of geraniol with 7 types of fatty acids, the maximum degree of esterification of 35.33 ± 0.48% was recorded with butyric acid. The esterification of geraniol with linoleic acid (C18:2), stearic acid (C18:0) and linolenic acid (C18:3) resulted lower values; 20.88±0.22%, 13.92±0.28% and 4:41±0.55, respectively. According to Maha *et al.* [8] and Abbas and Comeau [9], lipase activity increases with the increasing of fatty acid chain length and the number of linear carbon atoms to the related acid. However, it is not significant with caprylic acid (C8), myristic acid (C14) and oleic acid (C18: 1); there was no esterification happen with geraniol. This condition is believed to be associated with the presence of high quantity of water in the reaction system (0.50-0.93%). Therefore, geraniol may have shown low affinity towards the caprylic acid (C8), myristic acid (C14) and oleic acid (C18:1). Catalysis in organic solvents was significant with only 1-2% water content. Using this quantity or less, certain enzymes may switch the interaction into esterification via hydrolysis which reduces the yield. This condition may have happened to particular substrates with LGT. The study of geranyl butyrate synthesis before the optimization process is illustrated in Figure.4. Heptane was used based on the non-polar hydrophobic characteristic. Heptane has a Log P value of 4.0 and methanol with -0.76. Since a solvent with a high value of Log P seeks to reduce the water content and improve the degree of esterification (through reaction reversible obstruction of ester hydrolysis) better, the hexane was chosen. During the first 6h, the degree of esterification was recorded as 23.87±0.77%. From 12 to 36 h, the values were recorded as consistent (28.67±0.57%-

29.72±0.15%). The highest degree of esterification was recorded at 42 h with 31.68±0.09%. The esterification began to decline at the 72h with water content recorded as 0.833±0.043%. Synthesis declination can be attributed to an increment in the water content of the system. The first product in ester synthesis by most lipase is water and water activities are a standard use for detecting the enzyme hydration in different solvents; therefore, this situation is suggested to serve for the integrity of the enzyme to allow catalysis running significantly. The geranyl butyrate production has been optimized improve the degree of esterification.

The parameters such as temperature, shaking rates, types of organic solvent, amount of molecular sieve, and ratio of geraniol to butyric acid and LGT concentrations were optimized to enhance the formation of ester. Then, the geranyl butyrate production was carried out under optimum conditions and was compared with the results prior to optimization. LGT has demonstrated a significant effect of temperature on its synthesis of geranyl butyrate. Most microbial lipases used under high temperature in biotechnological applications of detergents, textile industry and the production of surfactant in the presence of organic solvents were mostly reported as a significant success. Figure.5 shows the effect of temperature on the esterification, which is projecting significant values of 38.73±0.784% to 43.89±1.76%. The highest yield of geranyl butyrate was recorded at 70°C with water content of 0.84±0.04%. This result is on surprising since LGT has been isolated from a hyperthermotolerant species [1] indicating its excellent protein stability at both high temperature as well as in most organic solvent [10]. At 65°C, the yield was recorded as 43.65±1.036% with water content of 0.79±0.01%. Basically, the yield at 55-75°C didn't reflect big difference. The water content in the system was quite high suggesting as the effect of high temperatures; therefore, we believed that the synthesis of geranyl butyrate has might be impaired. We strongly believed that without these water effects, the yield will be higher. In most non-aqueous enzymatic system, the water should be controlled to be as low as 0.01% [11], and increment of water content in even very small quantities shall give a huge change to the activity of enzyme catalysis. The enzyme is inactive in the system without water, but, catalytic degrees will increase with the degree of hydration [12]. Thus, the water accumulated in the system of LGT at those temperatures is likely to have a major effect on the yield of geranyl butyrate.

The effect of temperature on the protein enzyme will become more apparent with the presence of high water content. Denaturation effect on the enzyme was found to occur in an environment of water and high temperatures. Studies by Negishi *et al.* [13] have shown that lipase PL from *Alcaligenes* sp. has a maximum reaction at 70°C with water content at 0.5% (w/w) of the weight of the enzyme. According to this researcher, when the water content was recorded as 0.1% (w/w) over the lipase PL's weight, the enzyme's maximum temperature has rose to 85°C. This situation significantly showed that denaturation process caused by temperature is lower when the water content is low. Most commonly, various types of chemical changes, mono-molecular clotting and non-reversible aggregation take place very quickly at a high temperature and enzyme becomes inactive in lot more situation resulting in failure in synthesis. Since most of these processes involve water and therefore, is not happening in an environment without water the use of organic solvent has been a standard choice to run esterification. The protein structure and the resolution of lipases have been confirmed to be better in organic solvents rather than in aqueous solution. Thus, the unfolding of the enzyme is not easily occurring; consequently, increase the thermostability of the enzyme. The binding force between lipase and water molecules is high featuring to more susceptibility of most enzymes towards deactivation at high temperatures. The effect of shaking rates to the esterification by LGT is illustrated in Figure 6. Maximum yield of the geranyl butyrate was collected at 200 rpm (49.0±0.1%). The water content was low. System containing hydrophobic and hydrophilic substrates is homogenized by the shaking progress. While the rate and duration of the reaction by enzyme is influenced by a variety of factors related to the substrate and enzyme, the process of mixing the substrate and enzyme is important to spread the heat and result in a reaction environment. Unfortunately shaking rate at 250 rpm greatly affects the esterification with yield of only 12.03±0.1% due to potential of enzyme deactivation or side chain demolished by the activities. Nik Raikhan[1] has stated about 'excess levels of movement denature the enzyme'. This occurs due to the effect of hacks that are produced during the shaking or stirring and entrainment of air into the medium on the surface of the air and liquid [1].

The enzyme particles will form small fragments in excess of string, then, the water molecules cannot be maintained at the enzyme and thus water will accumulate in the reaction system and interfere with the esterification [14]. Conversely, at a low shaking rate (<100 rpm), aggregation of enzyme-substrate-water applied the mass transfer which makes it difficult for the esterification reaction goes well. Water had a major impact on the characteristics highlighted by lipase. The effect will change the hydration of enzyme hence indirectly alter the nature of the reaction medium and/or enzyme support system [1]. The effects of organic solvent on the degree of esterification of geranyl butyrate was studied using organic solvent having Log P as follow; heptane (Log P=4.0), hexane (Log P=3.5), isooctane (Log P=4.5), tetrahydrofuron (Log P=0.45), toluene (Log P=2.5), benzene (Log P=2.0) and ethyl acetate (Log P=3.51). The organic solvent that has a value of log P has high hydrophobic effect and can usually provide an activity or good stability and reduces the deactivation of enzyme. To identify the effects of organic solvent on this synthetic ester, the parameters of the organic solvent itself should be identified carefully. Log P logarithm coefficient is defined as the separation of a component in a dual-phase system octanol-water [15].

Figure 7 shows the effect of organic solvents on the esterification of geranyl butyrate by LGT. Based on Figure 7, the maximum yield of esterification was recorded with heptane of 48.05±0.3% and water content of 0.028±0.01%. Benzene performed the second highest yield (40.23 ± 0.21%) even theoretically solvent with low Log P value (Log P=2.0) usually doesn't work well for esterification. The water content was 0.149 ± 0.01%. Instead, hexane that has higher Log P value (Log P=3.5) gave a lower yield of geranyl butyrate (20.96±0.10%) even the water content recorded in the system was very low (0.045±0.01%). These results are believed not to be subjected to the h of study but were caused by the characteristics of organic solvents used for each system. Above all, the isooctane with the highest value of Log P (Log P=4.5) didn't produced enough yield of at least what was promoted by the toluene (Log P=2.5). This situation illustrates that Log P is not a major factor in influencing the synthesis of geranyl butyrate by LGT. Molecular sieves are molecules used to absorb water in inorganic or organic systems. Figure 8 shows the effect of molecular sieve on the yield of geranyl butyrate by LGT. The addition of more than 0.25 g molecular sieve has improved the degree of esterification of about 55%-60%, with moderate water content. However, the yield in the system without the molecular sieve was very not significant (4.29±0.367%). The maximum yield was recorded with 0.25g molecular sieve (91.56±0.1%) with water content of 0.412±0.01%. These results indicated that the molecular sieve's function is only effective when is used in suitable amount. The reaction medium without or with the use of molecular sieve will cause an increment of water content that would interfere with the esterification system (Figure 8).

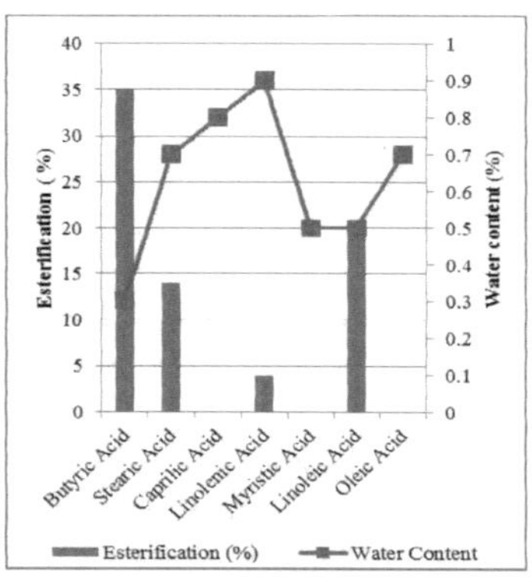

Figure 3: Esterification between geraniol with various types of fatty acids. Values are expressed as the mean of triplicate readings.

Reaction condition:

0.50ml geraniol, 0.50ml various fatty acids, 4.00ml of methanol, 0.50g molecular sieve & 0.50g LGT
200 rpm, 65°C, 72h

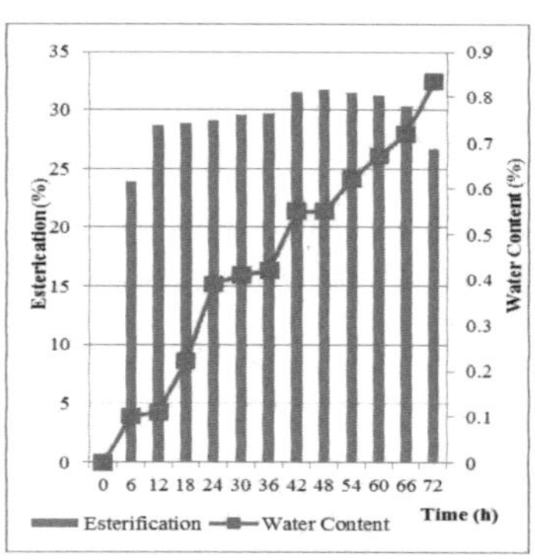

Figure 4: Profile of geranyl butyrate synthesis by LGT (before optimization).

Reaction condition:

0.50ml geraniol, 0.50ml butyric acid, 4.00ml heptane, 0.50g molecular sieve & 0.50g LGT
200 rpm, 65°C, 72h, volume of analysis: 0.40 ml

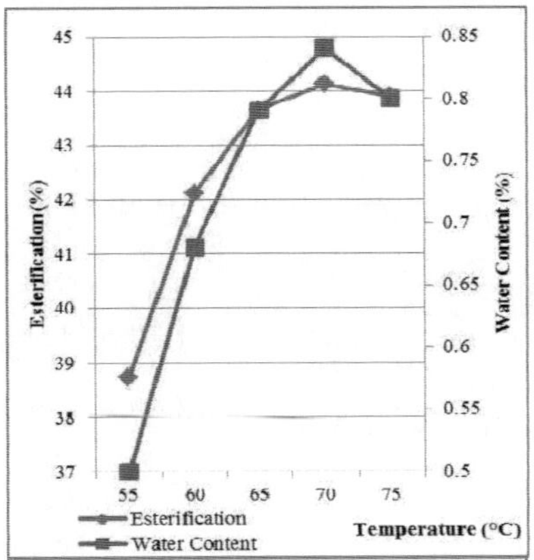

Figure 5: Effect of temperatures on the synthesis of geranyl butyrate by LGT.

Reaction condition:

0.50ml geraniol, 0.50ml butyric acid,4.00ml methanol, 0.50g molecular sieve & 0.50g LGT

200 rpm, 55-75°C, 42h, volume of analysis: 0.40 ml

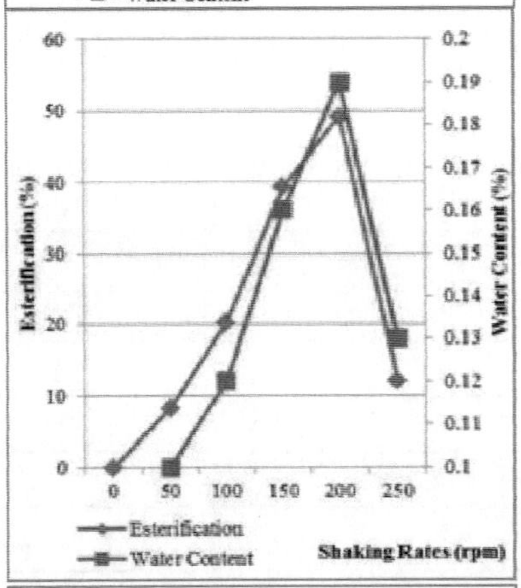

Figure 6: The effect of shaking rates to the establishment geranyl butyrate by LGT.

Reaction condition:

0.50ml geraniol, 0.50ml butyric acid, 4.00ml methanol 0.50g molecular sieve & 0.50g LGT

0, 50-200 rpm, 70°C, 42h, volume of analysis: 0.40 ml

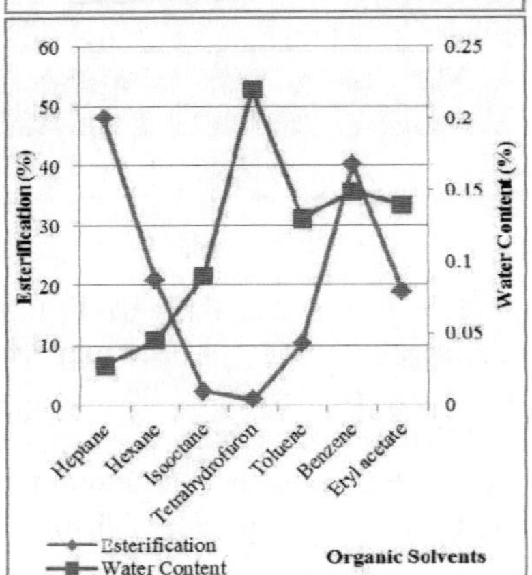

Figure 7: The effects of organic solvent on the esterification of geranyl butyrate by LGT.

Reaction condition:

0.50ml geraniol, 0.50ml butyric acid, 4.00ml heptane, hexane, isooctane, tetrahydrofuron, toluene, benzene & ethyl acetate, 0.50g molecular sieve & 0.50g LGT

200 rpm, 70°C, 42 h, volume of analysis: 0.40 ml

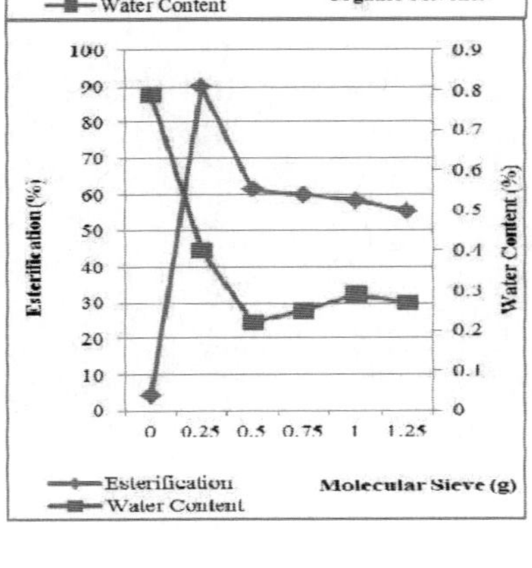

Figure 8: The effects of molecular sieve on the esterification of geranyl butyrate by LGT.

Reaction condition:

0.50ml geraniol, 0.50ml butyric acid, 4.00ml heptane, 0, 0.25, 0.50, 0.75, 1.00 and 1.25g molecular sieve & 0.50g LGT

200 rpm, 70°C, 42 h, volume of analysis: 0.40 ml

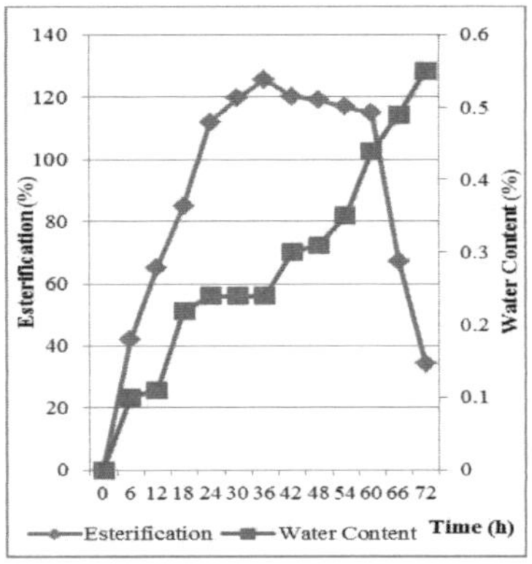

Figure 9: Profile of geranyl butyrate synthesis using optimized parameters catalyzed by freeze-dried LGT in 72 h.

Reaction condition:

0.50ml geraniol, 0.50ml butyric acid, 4.00ml heptane, 0.25g molecular sieve & 0.50g LGT

200 rpm, 70°C, 72 h, volume of analysis: 0.40 ml

Table 1 listed details of the system of LGT catalyzation before and after optimization. The yield of geranyl butyrate increased 297% after optimization, a great achievement.

Table 1: Parameters before and after optimization of geranyl butyrate synthesis catalysed by LGT

Parameters	Before-optimization	Optimized Condition
1. Solvent	Heptane	Heptane
2. Molecular sieve	0.50 g	0.25 g
3. Shaking rate	200 rpm	200 rpm
4. Temperature	65°C	70°C
5. H of optimum production	18 h starting 42 to 60 h	38 h starting 24 to 60 h
6. Yield	31.68±0.01%	125.79± 0.03% **(297% better yield)**
7. Water content	0.39 ± 0.01%	0.196± 0.01%

4. Conclusion

By using the optimized parameters of 4.00ml heptane, 0.25g molecular sieve, 0.5g LGT, 0.50ml geraniol and 0.50ml butyric acid, temperature at 70°C and the shaking rate at 200 rpm, we managed to increase the production of geranyl butyrate by 297%, a very excellent result. The ability of thermotolerant lipase from *Geobacillus thermodenitrificans* (LGT) in this synthesis is novel to Malaysian fragrance industry and we are significantly already extended this research to another field, the nanotechnology application using our locally synthesis geranyl butyrate in fragrance industry.

Acknowledgement

The authors gratefully acknowledge the Acculturation Grant Scheme RAGS/1/2014/TK05/UITM//6 from the Ministry of Education, Malaysia and the UiTM Research Management Centre (RMC) for the financial support of this research to take place. Humongous thanks go to the team who has been supporting this work.

References

[1] Nik Raikhan N H 2010 Thermotolerant lipase from *Geobacillus thermodenitrificans*, The Application in Esterification and Enantiomeric Reaction. PhD Thesis. Universiti Sains Malaysia, Pulau Pinang Malaysia.

[2] Pablo D M, Aristotelis X, Haralambos S, Jose V S 2004 *Enzyme and Microbial Technology* **35** 277-283.

[3] Nik Raikhan N H, Khairul Izwan A R 2017 International Journal of Conservation Science 8(3) 537-544.

[4] Claon P A, Akoh C C 1994 *Enzyme and Microbial Technology* **16** 835–838.

[5] Saisuburamanian N, Krithika L, Dileena K P, Sivasubramanian S, Puvanakrishnan R 2004 *Analytical Biochemistry* **330(1)** 70-73.
[6] Nik Raikhan N H 2016 *Journal of Life Sciences and Technologies* 4**(2)** 70-74.
[7] Ibrahim C. O, Nishio N, Nagai S 1987 *Agricultural and Biological Chemistry* **51** 2153-2159.
[8] Maha K C, Sylviane P, Didier T, Thomas D. 1998 *Journal of the American Chemical Society* **75(12)** 1201-1206.
[9] Abbas H, Comeau L 2003 *Enzyme Microb Technol* **32** 589–595. Doi: 10.1016/s0141-0229(03)00022- x.
[10] Nik Raikhan N H, Nur Hidayati O, Mohd Fiqri M 2017 *International Journal of Conservation Science* 8**(4)** 685-694.
[11] Yang F, Russell A J 1996 *Biotechnol Bioeng* **49(6)**709-716.
[12] Koskinen Ari, Klibanov A 1996 Enzymatic Reactions in Organic Media.
[13] Negishi S, Shirasawa S, Arai Y, Suzuki J, Mukataka S *Enzyme and Microbial Technology* **32** 66-70.
[14] Carpio C, Escobar F, Batista-Viera, F. Ruales J 2008 *Food and Bioprocess Technology* DOI: 10.1007/s11947-008-0164-9.
[15] Laane C, Boeren S, Vos S, Veeger C 1987 Biotechnology and Bioengineering **30(1)** 81-87. DOI: 10.1002/bit.260300112.

IOP Conf. Series: Materials Science and Engineering **358** (2018) 012031 doi:10.1088/1757-899X/358/1/012031

Enantioselectivity and Thermostability of a Novel Hyperhermotolerant Lipase from *Geobacillus Thermodenitrificans* nr68 (Lip.nr-68) on Secondary Racemic Alcohols Acetylation

N R Nik Him[1,a] and D Ibrahim[2,b]

[1]Bioprocess Engineering Department, Faculty of Chemical Engineering, Universiti Teknologi MARA (UiTM), 40450 Shah Alam, Selangor. MALAYSIA.
[2]Industrial Biotechnology Research Laboratory, School of Biological Sciences, Universiti Sains Malaysia, 11800 Pulau Pinang. Malaysia

[a]raikhan7952@salam.uitm.edu.my and [b]darah@usm.my

Abstract. In our previous work, a new lipase enzyme has been purified from a species identified as a Gram negative *Geobacillus thermodenitrificans* nr68, isolated from a hot spring in Malaysia with growth temperature of 48°C. This new lipase, called Lip.nr-68 has been characterized as a hyperthermotolerant protein with high stability at 65°C and has been showing excellent characteristics that are very much comparable yet better than some of those of well-known industrially-used lipases. It shows high activity against long-chain triglycerides with molecular weight of the purified enzyme estimated to be 33.5 kDa using SDS-PAGE analysis. This paper is focusing on hyperthermotolerant Lip.nr-68 performance in promoting for enantioselectivity activities towards three secondary racemic alcohols namely 1-phenylethanol, 1-cyclohexilethanol and 1-(naft-2-il) ethanol by acetylation with vinyl acetate. Lip.nr-68 has been confirmed to show high and usual enantioselectivitiy according to the Kazlauskas Rule towards all secondary racemic alcohols and has significantly approved as an enantiomer selective biocatalyst towards 1-phenylethanol and 1-cyclohexylethanol at 65°C. Lip.nr-68 has showed a reduction of *(R)* and *(S)* enantiomers as well as the production of 68-98% *ee* and almost 94% yield of 3-4 mg/ml for 1-cyclohexilethanol.

1. Introduction

The ability of most lipases to promote for the biotransformation process in the non-conventional medium like organic solvent and supercritical mediums is not to be underestimated. Lipases could greatly perform biotransformation in a macro-aqueous system with regio and enantioselective characteristics; making it a main pedestal that has become significant to relatively lots of present enzymatic study for the production of enantiomerically pure alcohols; which has been well recognized to be vital intermediates for the synthesis of pharmaceuticals compounds, agrochemicals composites, pheromones, flavours and fragrance, liquid crystals molecules and chiral auxiliaries in asymmetric synthesis [1],[2]. Therefore, enantioselectivity performance, enzymes stability and enzyme inhibition should be promoted to productively support the occurrence of pure racemic molecules. According to Itoh [3], the development of efficient methods for the synthesis of enantiomerically pure alcohols is of tremendous importance. The ability of lipases to dissolve racemic molecules has been adapted into lots of important applications such as in dairy and industrial biotechnology, oils processing, surfactants production as well as the production of pure enantiomeric molecules [4] and we are still looking into better performance of the technology. This paper is published to bring forward the excellent quality of our prime hyperthermotlerant *Geobacillus thermodenitrificans* nr68 lipase (Lip.nr-68) towards the

Content from this work may be used under the terms of the Creative Commons Attribution 3.0 licence. Any further distribution of this work must maintain attribution to the author(s) and the title of the work, journal citation and DOI.
Published under licence by IOP Publishing Ltd

enantiomeric reactions on three types of secondary racemic alcohols listed as 1-phenylethanol, 1-cyclohexilethanol and 1-(naft-2-il) ethanol. The main highlight that differ Lip.nr-68 from other most microbial lipases involved in acetylation of racemic mixtures is the stability at 65°C for 96 hours with an excellent *ee* and yield. Enantioselectivitiy of Lip.nr-68 towards the racemic mixtures was studied to accomplish and confirm the selectivity towards its enantiomer that has been applied as a model by Kazlauskas Rules [5].

2. Material and Methods

2.1 Source of enzyme, lipase activity test and source of racemic alcohols

The source of enzyme was a freeze-dried lipase from *Geobacillus thermodenitrificans* nr68 (Lip.*nr*-68) with activity of 30.50 U/g. The inorganic salts and materials for biomass production were purchased from Sigma, Aldrich or Fluka. Lipase activity was determined according to method by Nik Raikhan & Nurul [6] using *p*-nitrophenyl palmitate (*p*NPP, Sigma) as substrate. The 1-phenylethanol, 1-cyclohexilethanol and 1-(naft-2-il) ethanol were the products of Aldrich.

2.2 Nuclear magnetic resonance (NMR) analysis, Gas chromatography analysis (GC) and thin layer chromatography (TLC)

The NMR spectra were recorded in $CDCl_3$ on a Bruker DRX- 500 spectrometer (at 500MHz for 1H and 125MHz for 13C spectra). The δ scale for the analysis was recorded in ppm. The IR spectra (film) were taken on a Specord 2000 Series Spectrophotometer and the wavenumbers of the absorptions are reported in cm^{-1}.GC analyses were carried out on HP 5890 or Agilent 4890D instruments equipped with FID detector and Hydrodex-β-PM 25m×0.25 mm column (25m×0.25 mm, 0.25 μm film with permethylated-cyclodextrin; Macherey&Nagel) or 0.25 or Betadex 225 fused silica capillary column (30m×0.25mm×0.25 μm film; Supelco) using H_2 carrier gas (oven: 100°C, injector: 250°C, detector: 250°C, head pressure: 10 psi, 50:1 split ratio). Optical rotations were determined on a Perkin-Elmer 241 polarimeter [4]. TLC was carried out on Kieselgel 60 F254 (Merck) sheets cut into 6.0 cm. Movement of solvent (hexane-acetone 10:4, v/v) was allowed up to 5.5 cm [7].

2.3 Enantiomer selective reaction by Lip.nr-68 towards racemic alcohol (rac-1a-c) at room temperature

20.0 mg freeze-dried Lip.*nr*-68 was added into a mixture of 1.0 ml hexane, 0.50 ml vinyl acetate and 0.50 ml tetrahydrofuran followed by 20.0 mg of racemic substrate *rac*-**1a-c** [8]. The sealed glass vials contained the catalyst was centrifuged at 1000 rpm at room temperature (20°C) using the times listed in Table 1. The conversions were checked by TLC (hexane–acetone 10:4, v/v). At the reaction time indicated in Tables 1, Table 2 and Table 3 below, the enantiomeric compositions and the ratio of the product and residual substrate fractions were analyzed by GC on chiral stationary phase (Tables 2 and Table 3). Tables 1, Table 2 and Table 3 list results only for those reactions which exhibited more than 5.0% of conversion after 24 hours.

Table 1. Chromatography Analysis for *rac*-**1a:** *(S)*-**1a** and *(R)*-**2a**

	Alcohol (*S*)-1a	Alcohol (*S*)-1a	Acetate (*R*)-2a	Acetate (*R*)-2a
Enzyme	Lip.*nr*-68	Lip.*nr*-68	Lip.*nr*-68	Lip.*nr*-68
Appearance	Colorless oil	Colorless oil	Colorless oil	Colorless oil
T (°C)	Room T	65	Room T	65
Time (hour)	96	24	96	24
Yield	129 mg, 65%; E.e.: 39% by GC; $[\alpha]_D^{22}$=18.8 (*c*1.0, $CHCl_3$)	101 mg, 53%; E.e.: 65% by GC; $[\alpha]_D^{22}$= -57.7 (*c* 1.0, $CHCl_3$)	63.5 mg, 25%; E.e.: 97% by GC; $[\alpha]_D^{22}$= +125 (*c* 1.0, $CHCl_3$)	85 mg, 35%; E.e.: 97% by GC; $[\alpha]_D^{22}$= +123 (*c* 1.0, $CHCl_3$)
IR	3252, 3118, 2976, 2952, 1476, 1430, 1378, 1304, 1180, 1008, 900, 750, 690	3352, 3008, 2976, 2952, 1666, 1290, 1368, 1204, 1080, 1208, 850, 760, 690	1740, 1456, 1376, 1244, 1208, 1064, 984, 760, 700	1740, 1456, 1376, 1244, 1208, 1964, 944, 770, 700
Other details	^{1}H NMR: 1.513 (d, *J* = 6.9 Hz, 3H), 3.38	^{1}H NMR: 1.513 (d, *J* = 6.9 Hz, 3H), 3.38	^{1}H NMR: 1.574 (d, *J* = 7.4 Hz), 21.103	^{1}H NMR: 1.574 (d, *J* = 7.4 Hz), 21.103 (s,

	(s, 1H), 4.861 (q, J = 6.3 Hz, 1H), 7.40 (mc, 5H, Ar H); ^{13}C NMR: 25.034, 69.843, 125.071, 126.862, 127.970, 145.525	(s, 1H), 4.861 (q, J = 6.3 Hz, 1H), 7.40 (mc, 5H, Ar H); ^{13}C NMR: 25.034, 69.843, 125.071, 126.862, 127.970, 145.525	(s, 3H), 5.926 (q, J = 6.9 Hz, 1H), 7.30-7.40 (m, 5H); ^{13}C NMR: 21.416, 22.283, 72.241,125.860, 127.632,128.28, 141.422,169.96	3H), 5.926 (q, J = 6.9 Hz, 1H), 7.30-7.40 (m, 5H); ^{13}C NMR: 21.416, 22.283, 72.241,125.860, 127.632,128.28, 141.422,169.96

Table 2. Chromatography Analysis for *rac*-**1b:** (*S*)-**1b** and (*R*)-**2b**

	Alcohol (*S*)-1b	Alcohol (*S*)-1b	Acetate (*R*)-2b	Acetate (*R*)-2b
Enzyme	Lip.*nr*-68	Lip.*nr*-68	Lip.*nr*-68	Lip.*nr*-68
Appearance	Colorless oil	Colorless oil	Colorless oil	Colorless oil
T (°C)	Room T	65	Room T	65
Time (hour)	96	24	96	24
Yield	330 mg, 96%; E.e.: 50% by GC; $[\alpha]_D^{22}$= +1.6 (c 1.0, $CHCl_3$)	215 mg, 80%; E.e.: 99.5 % by GC; $[\alpha]_D^{22}$= +2.1 (c 1.0, $CHCl_3$)	278 mg, 78%; E.e.: 91% by GC; $[\alpha]_D^{22}$= +9.5 (c 1.0, $CHCl_3$)	399 mg, 93.8%; E.e.: 92% by GC; $[\alpha]_D^{22}$= +7.2 (c 1.0, $CHCl_3$)]
IR	3610, 3550-3250, 2960-2900, 1450	3610, 3550-3250, 2960-2900, 1450	2928, 1736, 1244	2928, 1736, 1244
Other details	^{1}H NMR: 0.73-1.88 (m, 12H), 1.16 (d, J = 6.0 Hz, 3H),3.57 (dq, J = 6.0 Hz, 1H); ^{13}C NMR: 20.43, 26.21, 26.30, 26.60, 28.41, 28.82, 45.27, 72.22	^{1}H NMR: 0.73-1.88 (m, 12H), 1.16 (d, J = 6.0 Hz, 3H),3.57 (dq, J = 6.0 Hz, 1H); ^{13}C NMR: 20.43, 26.21, 26.30, 26.60, 28.41, 28.82, 45.27, 72.22	^{1}H NMR: 1.09 (d, J = 6.405 Hz, 3H), 1.96 (s, 3H), 4.75 (q, J = 6.405 Hz, 1H), 1.65 (m, 11H); ^{13}C NMR: 17.05, 21.3, 26.04, 26.23, 28.5, 42.57, 74.56, 170.08.	^{1}H NMR: 1.09 (d, J = 6.405 Hz, 3H), 1.96 (s, 3H), 4.75 (q, J = 6.405 Hz, 1H), 1.65 (m, 11H); ^{13}C NMR: 17.05, 21.3, 26.04, 26.23, 28.5, 42.57, 74.56, 170.08.

Table 3. Chromatography Analysis for *rac*-**1c:** (*S*)-**1c** and (*R*)-**2c**

	Alcohol (*S*)-1c	Acetate (*R*)-2c
Enzyme	Lip.*nr*-68	Lip.*nr*-68
Appearance	White powder	Colorless oil
T (°C)	Room temperature	Room temperature
	96	96
Yield	342 mg, 71%; E.e.: 29% by GC; $[\alpha]_D^{22}$= -15 (c 1.0, $CHCl_3$)	342 mg, 16%; E.e.: 99% by GC; $[\alpha]_D^{22}$= +125.5 (c 1.0, $CHCl_3$)
IR	3610, 3550-3250, 3060, 3020, 2980, 2930, 1450, 860	1064, 1244, 1372, 1740, 2984
Other details	^{1}H NMR: 1.5 (d, J = 6.5Hz, 3H), 1.95 (br s, 1H), 4.95 (q, J = 6.5Hz, 1H), 7.4 (m, 3H), 7.72 (m, 4H); ^{13}C NMR: 24.52, 70.05, 123.7, 123.8, 125.8, 126.15, 127.7, 128.0, 128.4, 132.9, 133.36, 143	^{1}H NMR: 1.55 (d, J = 6,59 Hz, 3H), 2.02 (s, 3H), 5.98 (q, J = 6.593 Hz, 1H), 7.5 (d, 3H), 7.8 (t, 4H); ^{13}C NMR: 21.62, 22.17, 72.5, 124.09, 125.02, 126.06, 126.23, 127.66, 128.02, 128.35, 133.02, 133.18, 139.00, 170.5

2.4 Enantiomer selective reaction and thermostability of catalytic reaction by Lip.nr-68 towards 1-phenylethanol (rac-1a) and 1-cyclohexilethanol (rac-1b) at 65°C

The above method was repeated with 20.0 mg of 1-phenylethanol (*rac*-**1a**) and 1-cyclohexilethanol (*rac*-**1b**) and temperature of 65°C using the times listed in Table 2. For the thermostability test, times in the Table 3 were used. Again, the conversions were checked by TLC (hexane–acetone 10:4, v/v). The enantiomeric compositions and the ratio of the product and residual substrate fractions were analyzed by GC on chiral stationary phase. GC retention time: R_t (Hydrodex-β-PM column, 65-118°C/min, 2.5°C/min,12 psi): 17.6 min, (*R*)-**2a**; 18.5 min, (*S*)-**2a**; 18.9 min, (*R*)-**1a**; 19.5 min, (*S*)-**1a**; GC molar response factor of acetate/alcohol (*rac*-**2a**/*rac*-**1a**): 1.23 [was used to calculate c and *E* in Table 1, 2 and 3]; R_t (Hydrodex-β-PM column, 100-112°C, 1°C/min, 12 psi): 7.5 min, (*R*)-**1b**; 7.6

min, (*S*)-**2b**; 8.2 min, (*R*)-**2b**; 8.9 min, (*S*)-**1b**; GC molar response factor of acetate/alcohol (*rac*-**2b**/*rac*-**1b**): 1.25 [was used to calculate c and *E* in Table 1, 2 and 3]; R_t (Beta-Dex 225 column, 130–185°C, 2°C/min, 12 psi): 23.1 min, (*R*)-**1c**; 23.4 min, (*S*)-**1c**; 24.1 min, **2c**; GC molar response factor of acetate/alcohol (*rac*-**2c**/*rac*-**1c**): 1.21 [was used for calculations of c and *E* in Table 3]. Organic solvent was distilled out by using the rotary evaporation and residue was separated using vacuum chromatography (silica gel, hexane: acetone 10:0.5, v/v) to get the alcohol (*S*)-**1a-c** and acetate (*R*)-**2a-c**.

3. Results and Discussion

Kazlauskas *et al.* [5] has proposed and make used of a very impressive model by the active site of a hydrolase enzyme that has the capability to attach to groups of *small*R and *big*R in alcohols. Fig.1a show the site of the *small*R and *big*R on the secondary alcohols which commonly appear to be the favorable active sites for hydrolase enzyme. According to the Kazlauskas *et al.* [5], when the substitute of *small*R and *big*R differs greatly in their sizes, the resolution for enzyme activity will become absolute. Alcohols with empirical rules that shows hydroxyl group on the upper side will predict the favorable enantiomer for both lipase and esterase to be the one on the right side (Fig.1a).This means, the enzyme has to favor the (*R*)-selectivity when the priority is favoring the *big*R compares to the *small*R. Enantioselectivity of an enzyme give arise to the ability on the enzyme to recognize chiral substrate during the formation of enzyme-substrate complex. Fig.2 shows kinetic resolution for the secondary racemic alcohols *rac*-**1a-c** by the freeze-dried lipase enzyme from *Geobacillus thermodenitrificans* nr68 (Lip.*nr*-68). The (*R*)-selectivity of Lip.*nr*-68 reactions towards the 1-phenylethanol (*rac*-1a), 1-cyclohexilethanol (*rac*-**1b**) and 1-(naft-2-il) ethanol (*rac*-**1c**) will be using the model by Kazlauskas *et al.* [5]. The reaction will be decided as to follow the model or otherwise is named as Anti-Kazlauskas (*R*)-selectivity.

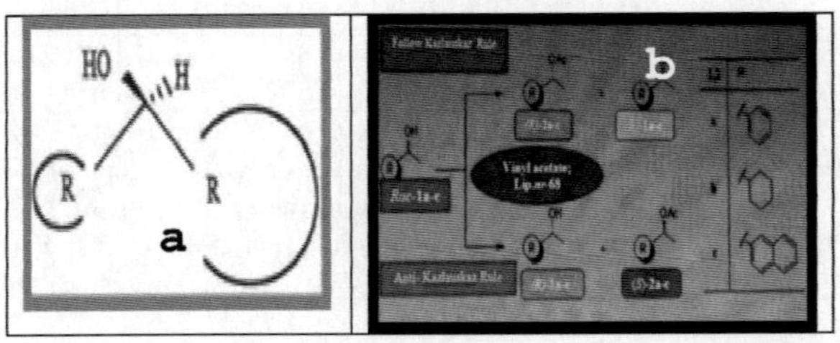

Figure 1. (a) Group of *small*R and *big*R in the alcohol base on the Kazlauskas *et al.*, (1991). [5]. (b) Kinetic resolution for racemic secondary alcohols *rac*-**1a-c** by Lip.*nr*-68 (Modified from [8]).

3.1 Lip.nr-68 reactions towards rac-1a-c at room temperature

Room temperature was recorded as 20°C. The objective of this reaction is to record the favorable substrate of the Lip.*nr*-68 among the given secondary racemic alcohols (*rac*-**1a-c**) and to determine the ability of the Lip.*nr*-68 to follow/not-to-follow the Kazlauskas Rules during the acetylation. Thin layer chromatography results of the Lip.*nr*-68 reactions towards all three *rac*-**1a-c** have confirmed acetate production. Methods by Poppe and Novàk [7] were applied to measure the acetate production from *rac*-**1a-c.** The reaction has applied the Lip.*nr*-68's potential towards the acetylation by replacing the hydrogen atom of a hydroxyl group with an acetyl group (CH_3CO). All *rac*-**1a-c** were then going through biocatalytic reaction by the Lip.*nr*-68 using method by Nagy [8] at 20°C for 96 hours. Results from gas chromatography confirmed the acetylation by Lip.*nr*-68.

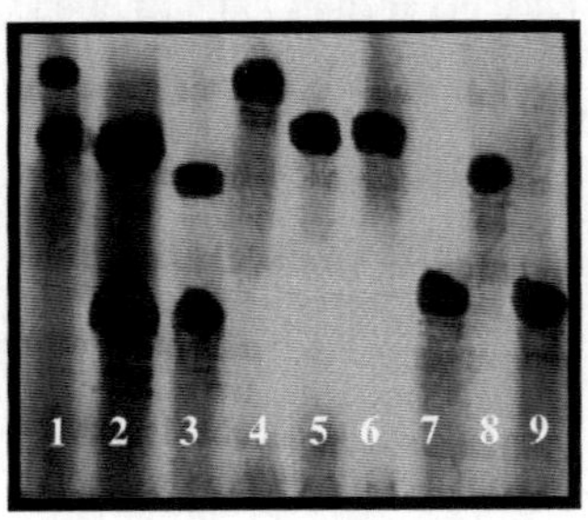

Figure 2. Thin layer chromatography (TLC) for reaction of Lip.nr-68 on the racemic substrates (rac-1a-c) after 96 hours at room temperature (20°C). Lane 1: 1-phenylethanol (rac-1a); Lane 2: 1-cyclohexilethanol (rac-1b); Lane 3: 1-(naft-2-il) ethanol (rac-1c); Lane 4: Phenylethanol acetate marker; Lane 5: 1- phenylethanol marker; Lane 6: cyclohexilethanol acetate marker; Lane 7: 1-cyclohexilethanol marker; Lane 8: 1-(naft-2-il) ethanol acetate marker; Lane 9: 1-(Naft-2-il)ethanol marker.

Table 4 shows the kinetic resolution for the racemic substrate *rac*-**1a-c** by the Lip.*nr*-68 at room temperature; 20°C. Retention time for the product analysis and details on the alcohol (*S*)-**1a-c** and acetate (*R*)-**2a-c** are listed in Table 1-3. The percentage of substrate to product conversion (c) was 27.5%, 50.7% and 18% for each of 1-phenylethanol (*rac*-**1a**), 1-cyclohexilethanol (*rac*-**1b**) and 1-(naft-2-il) ethanol (*rac*-**1c**), respectively. The highest enantioselectivity (*E*) was recorded with *rac*-**1a** (76) followed by *rac*-**1b** (65) and finally *rac*-**1c** (2). According to Heinsman *et al.* [9], the high number of ties between the reactions with chiral centers will increase the potential of both enantiomers to come together into the active site due to its high conformational of flexibility. Therefore the value of enantiomer selectivity, *E* for both compounds is expected to be low. *R* enantiomers were recorded to be higher than *S* enantiomers in all *rac*-**1a-c**. For *rac*-**1a**, the *ee* value was recorded as 57.6%, rac-**1b** as 10.4% and *rac*-**1c** as 27%. This results shows that Kazlauskas Rules [5] has been chosen by Lip.*nr*-68 during the selectivitytowards the secondary alcohols. The enzyme that acts as a chiral catalyst will usually react faster with one of the enantiomers. In terms of thermodynamics, the isomer (*R*)- and isomer (*S*)- found in the racemic mixtures are representing two kinds of reagents (A and B) which are competing for a single type of enzyme. Thus, the distinction of the Gibbs free energy (ΔΔG #), among degrees of removal for enantiomer either to act faster or slower will determine the effectiveness of the potential of the kinetic resolution. The purity of the yield commonly depends so much on the (c) value [4].

Table 4. Kinetic resolution for racemic substrates *rac*-**1a-c** by Lip.*nr*-68 at 20°C.

Substrates	Reaction Times (h)	c[a] (%)	Alcohol (*S*)-1a-c; *ee*[a] (%)	Acetate (*R*)-2a-c *ee*[a] (%)	*E*[b]
1-phenylethanol (*rac*-**1a**)	96	27.5	38.7	96.3	76
1-cyclohexilethanol (*rac*-**1b**)	96	50.7	79.8	90.2	65
1-(naft-2-il) ethanol (*rac*-**1c**)	96	18	7	34	2

Note for Table 1. *ee* is an enantiomeric excess or non-reacted substrate. [a]Values for *ee* of (*R*)-**2a-c**, (*S*)-**1a-c** and percentage of product formation (c) was determined by the Hydrodex β-PM column on GC. [b]Values of the enantiomeric selection or enantioselectivity (*E*) was calculated from c and ee_{2b} using method by Chen *et al.* [11]. It was confirmed by calculation of enantiomeric excess between substrate and product ($ee_{1a\text{-}c}$ and $ee_{2a\text{-}c}$; [12]). Freeze-dried Lip.*nr*-68 activity was recorded as 30.5 U/g.

The vinyl acetate in the study of acetylation of racemic secondary alcohols *rac*-**1a-c** by Lip.*nr*-68 was used as a chiral acyl donor. The enantioselectivity process taking place during the transesterification of amine or racemic alcohols (*R* and *S*) in the organic solvent, generally rely on chiral acyl donor such as vinyl and isopropenyl acetate (IPA), or anhydride to avoid the problem of reverse reaction. For the production of esters in the kinetic resolution through transesterification by lipase on alcohol in a particular organic solvent, the reverse reaction of the optical active compound with the non-chiral alcohol at the acyl donor reagent has been reported to be a major issue since the purity of the optical yield will decrease significantly [4]. Thus, acyl donor reagent with non-reversible quality such as enol-acetate (vinyl acetate and isopropyl acetate) is introduced and has been widely used for the acetylation of the chiral alcohols by lipases in the organic solvents. In addition to vinyl acetate and isopropyl acetate, 1-etoksivinyl ester which is a result of new donors in the process of resolution of racemic biotacalysis was superior than most of the main reagents such as vinyl ester and ester isoprophenyl [10]. Moreover, many empirical studies have shown that the kinetic resolution in biocatalysis is also affected by the type of organic solvent used in the system.

3.2 Lip.nr-68 reactions towards rac-1a and rac-1b at 65°C

Acetylation of racemic secondary alcohols *rac*-**1a** and *rac*-**1b** were performed at 65°C. Fig. 4 shows the thin layer chromatography (TLC) for reaction after 4 and 24 hours. Acetate production was observed through the clear lanes on Fig.3, by comparing with the given markers. Positive results tell the potential of Lip.*nr*-68 on acetylation reaction of H_2 where atom was substituted at the hydroxyl group with an acetyl group (CH_3OH) to produce acetate. Therefore, acetylation of *rac*-**1a** and *rac*-**1b** has been carried at 65°C based on the method by Nagy [8]. Gas chromatography results (GC) shows that both of the alcohols are suitable for the acetylation using Lip.*nr*68.

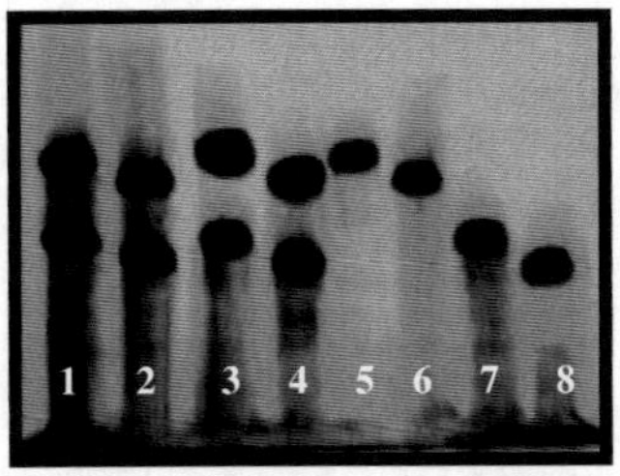

Figure 3. Thin layer chromatography (TLC) for reaction of Lip.nr-68 on rac-1a and rac-1b after 4 and 24 hours at 65°C. Lane 1: 1-phenylethanol (rac-1a, 4 hours); Lane 2: 1-cyclohexilethanol (rac-1b, 4 hours); Lane 3: 1-phenylethanol (rac-1a, 24 hours); Lane 4: 1-cyclohexilethanol (rac-1b, 24 hours); Lane 5: phenylethanol marker; Lane 6: cyclohexilethanol acetate marker; Lane 7: 1- phenylethanol marker; Lane 8: 1-cyclohexilethanol marker.

Table 5 shows the kinetic resolution of racemic substrate *rac*-**1a-b** by Lip.*nr*-68 at 65°C. The retention time for the analysis of the yield of rac-**1a-b** using gas chromatography and details on the alcohol (*S*)-**1a-b** and acetate (*R*)-**2a-b** are listed in Table 1 and Table 2. For *rac*-**1a**, the conversion of substrate-product; (c) was recorded as 21.7% and 39.7% after 4 hours and 24 hours, respectively. As for *rac*-**1b**, the percentage of substrates promoted by the Lip.*nr*-68 after 4 hours was 39.2% and was 51.1% at the 24th hour. Results after 4 hours of reactions indicated that rac-1a was going through better acetylation to compare to rac-1b with E of 65 over 55 but the enantiomeric selection or enantioselectivity (*E*) of *rac*-**1a-b** at 65°C was both very high after 24 hours with value over 100. Again, *R* enantiomer is shown to exceed over the *S* enantiomer in all the GC results. For 1-phenylethanol (*rac*-**1a**), the enantiomeric excess (*ee*) or (*S*)-**1a** was 26.8% and 65.8%, respectively at 4h and 24h, while for (*R*)-**2a** was 96.1% and 97.3%. The enantiomeric excess (*ee*) of (*S*)-**1a** at 4 and 24 hours was 67.8% and 99.5%, the rest of the data can be accessed from Table 5.

These results have clearly shown that the Lip.*nr*-68 has promoted for acetylation of the both alcohols at 65°C following the Kazlauskas Rule [5]. According to research by Bouzemi *et al.* [13], primary alcohols having hydroxymethyl group on the secondary stereogemic *S* alcohol and the *R* configuration is preferred for the production of acetic acid by acylation catalyzed by most lipases. This group has been experimenting a series of arylalkylcarbinol derivatives, where the substrates were deracemized through sequential combination of *Candida antarctica* lipase B (CAL-B) and the (*S*)-acetates were obtained in 71%–99% *ee* and 76%–89% yields. Another excellent research was the use of YS lipase from *Pseudomonas fluorescens* with isoprophenyl acetate as acylation agent in diisopropylether by Naemura *et al.* [14]. Based on the feature selection enantiomers, active site model of lipase YS has been proposed to identify the primary or secondary enantiomer which reacts more quickly to the acetylation process. Escorcia [15] reported that about 300 microorganisms were analyzed for the production of chiral secondary alcohols by enantioselective reduction of an alkyl aryl ketone and about 60 cultures showed a reduction of ketones with *R* and *S* enantiomers as well as the production of 92-99% *ee* and 95% yield of 1-4 g/L.

Table 5. Kinetic resolution for racemic substrates *rac*-**1a** and *rac*-**1b** by Lip.*nr*-68 at 65°C.

Substrates	Reaction time (h)	c[a] (%)	Alcohol (*S*)-1a-b; *ee*[a] (%)	Acetate (*R*)-2a-b, *ee*[a] (%)	*E*[b]
1-phenylethanol (*rac*-**1a**)	4	21.7	26.8	96.1	65
1-phenylethanol (*rac*-**1a**)	24	39.7	65.8	97.3	>100
1-cyclohexilethanol (*rac*-**1b**)	4	39.2	67.8	93.5	55
1-cyclohexilethanol (*rac*-**1b**)	24	51.1	99.5	92.8	>100

Note for Table 2. *ee* is an enantiomeric excess or non-reacted substrate. [a]Values for *ee* of (*R*)-**2a-c,** (*S*)-**1a-c** and percentage of product formation (c) was determined by the Hydrodex β-PM column on GC. [b]Values of the enantiomeric selection or enantioselectivity (*E*) was calculated from c and ee_{2b} using method by Chen *et al.* [11]. It was confirmed by calculation of enantiomeric excess between substrate and product ($ee_{1a\text{-}c}$ and $ee_{2a\text{-}c}$; [12]). Freeze-dried Lip.*nr*-68 activity was recorded as 30.5 U/g.

3.3 Thermostability of Lip.nr-68 reactions towards rac-1a and rac-1b at 65°C

Acetylation study on the secondary racemic alcohols *rac*-**1a-b** by Lip.*nr*-68 was accomplished with exposure of Lip.*nr*-68 at 65°C for the purpose of to investigate the effect to the acetylation. Acetylation was maintained to take place for 24 hours after the enzyme was exposed at different hours to the temperature. A total of 20.0 mg of free-substrate freeze dried Lip.*nr*-68 in 1.0 ml of hexane, 0.50 ml of vinyl acetate and 0.50 ml tetrahydrofuron was incubated at 65°C in a small glass container with a lid at 1000 rpm for 24, 48, 72, 96 and 120 hours. After each of the subsequent hours, 20.0 mg of racemic *rac*-**1a** and *rac*-**1b** was added. The analysis by gas chromatography (GC) has been evaluated and the results are listed in Table 6.

Table 6. Kinetic resolution for substrates *rac*-**1a** and *rac*-**1b** for thermostability of Lip.*nr*-68 at 65°C.

Substrate	Free-substrate enzyme exposure (h)	Reaction time (h)	c^a (%)	(*S*)-1a-b ee^a (%)	(*R*)-2a-b ee^a (%)	E^b
1-phenylethanol (*rac*-**1a**)	24	24	36.7	54.3	92.1	42
1-cyclohexilethanol (*rac*-**1b**)	24	24	44.1	76.7	88.4	34
1-phenylethanol (*rac*-**1a**)	48	24	37.7	68.7	96.3	>100
1-cyclohexilethanol (*rac*-**1b**)	48	24	49.6	64.0	96.1	>100
1-phenylethanol (*rac*-**1a**)	72	24	40.7	66.7	98.3	>100
1-cyclohexilethanol (*rac*-**1b**)	72	24	51.9	65.2	97.6	>100
1-phenylethanol (*rac*-**1a**)	96	24	49.1	67.8	93.9	>100
1-cyclohexilethanol (*rac*-**1b**)	96	24	65.5	66.9	94.1	>100
1-phenylethanol (*rac*-**1a**)	120	24	51.1	69.5	95.2	77
1-cyclohexilethanol (*rac*-**1b**)	120	24	60.2	68.4	93.2	65

Note for Table 3. *ee* is an enantiomeric excess or non-reacted substrate. [a]Values for *ee* of (*R*)-**2a-c,** (*S*)-**1a-c** and percentage of product formation was determined by the Hydrodex β-PM column on GC. [b]Values of the enantiomeric selection or enantioselectivity (*E*) was calculated from c and ee_{2b} using method by Chen *et al.* [11]. It was confirmed by calculation of enantiomeric excess between substrate and product ($ee_{1a\text{-}c}$ and $ee_{2a\text{-}c}$; [12]). Freeze-dried Lip.*nr*-68 activity was recorded as 30.5 U/g.

As expected, (*R*) enantiomer was shown to exceed over the (*S*) enantiomer in all the GC results (Table 6). However, Lip.*nr*-68 has been performing a very excellent thermostability in organic solvent. This enzyme has been exposed for 120 hours and the enantiomeric selection or enantioselectivity values (*E*) were recorded as more than 100 for both *rac*-**1a** and *rac*-**1b** acetylated by Lip.*nr*-68 exposed to 65°C at 48-96 hours (Table 6). The highest conversion rate (c) for both *rac*-**1a** and *rac*-**1b** by Lip.*nr*-68 was 49.1% and 65.5%, achieved after 24 hours by lipase exposed for 96 hours. These results indicated that the enzyme Lip.*nr*-68 has higher enantioselectivity against the substrate *rac*-**1a** over *rac*-**1b** when exposed beforehand without a substrate at the same temperature and time. The GC results also showed that the number of (*R*) enantiomer was found to exceed over (*S*). Enzymatic transformations for organic substrates are often used for the provision of optical active material [16],[17].

4. Conclusion

Research shows that Lip.*nr*-68 is an effective biocatalyst with high enantioselectivity towards all three (3) secondary racemic alcohols namely 1-phenylethanol (*rac*-1a), 1-cyclohexilethanol (*rac*-1b) and 1-(naft-2-il) ethanol (*rac*-1c) by acetylation with vinyl acetate. The thermostability was recorded maximum at 65°C for 96 hours, while the reaction has followed the Kazlauskas Rules.

Acknowledgment

The authors gratefully acknowledged the Ministry of Higher Education for the funding of RAGS (RAGS/1/2014/TK05/UITM//6) and UiTM Research Management Centre (RMC) for the funding in publishing this paper. Humongous thanks goes to the Universiti Sains Malaysia and the Budapest University of Technology and Economics, Hungary for all the facilities and guides during the research.

References

[1] Adams T B, McGowen M M, Williams M C 2007 *Food Chem. Toxicol.* **45** 171–201
[2] Pollard D J,Woodley M J 2007 *Trends Biotechno.* **25** 66–73

[3] Itoh N, Isotani K, Nakamura M, Inoue K, Isogai Y, Makino Y 2012 *Appl. Microbiol. Biotechnol.* **93** 1075–1085
[4] Nik Raikhan N H PhD Thesis. Universiti Sains Malaysia, Pulau Pinang Malaysia 2010
[5] Kazlauskas R J, Weissfloch A N E, Rappaport A T, Cuccia L A 1991 *Journal of Organic Chemistry* **5** 2656-2665
[6] Nik Raikhan N H, Nurul S A 2017 *Key Engineering Materials* **735** 136-142
[7] Poppe L, Novák L 1992 *Selective Biocatalysis*. Wiley-VCH Verlag GmbH, ISBN 10: 1560811889 ISBN 13: 9781560811886
[8] Nagy V, Toke E R, Keong L C, Szatzker G, Ibrahim D, Che Omar I, Szakacs G, Poppe L 2006 *Journal of Molecular Catalysis B: Enzymatic* **39** 141–148.
[9] Heinsman N, W J T, Orrenius S C, Marcelis C L M, de Sousa Teixeira A, Franssen M C R, van der Padt A, Jongejan J A, de Groot Æ 2008 *Biocatalysis and Biotransformation* **16** 145-162
[10] Kita Y, Takebe Y, Murata K, Naka T, Akai S 2000 *Journal of Organic Chemistry* **65** 83-88
[11] Chen C S, Fujimoto Y, Girdaukas G, Sih C J 1982 *Journal of the American Chemical Society* **104(25)** 7294-7299
[12] Rakels J L, Straathof A J, Heijnen J J 1993 *Enzyme and Microbial Technology* **15(12)** 1051-1056
[13] Bouzemi N, Grib I, Houiene Z, Aribi-Zouioueche L 2014 *Catalyst* 4(3) 215-225
[14] Naemura K, Fukuda R, Murata M, Konishi M, Hirose K, Tobe Y 1995 *Tetrahedron Asymmetry* **6(9)** 2385-2394
[15] Escorcia A M, Molina D, Daza M C, Doerr M 2013 *J. Mol. Catal. B.* **98** 21–29.
[16] Koeller K M, Wong C H 2001 *Nature* **409** 232–240.
[17] Patel R P 2000 *Stereoselective Biocatalysts* New York: Marcel Dekker Inc. 1-30

Bioethanol Production from Empty Fruit Bunch using Direct Fermentation by an Actinomycete *Streptosporangium roseum*

N R Nik Him and and T. Huda

Bioprocess Engineering Department, Faculty of Chemical Engineering,
Universiti Teknologi MARA (UiTM), 40450 Shah Alam, Selangor. MALAYSIA.

[a]raikhan7952@salam.uitm.edu.my

Abstract. Study on the production of bioethanol using palm oil empty fruit bunch (EFB) has been performed using actinomycete *Streptosporangium roseum*. Positive result of bioethanol production was recorded using Iodoform test followed by confirmation with GC-FID using a polar capillary column (PEG-type, 10m x 0.53, with autosampler) and n-propanol as internal standard. The first and second round distillation has produced azeotrope (85-15% ethanol-water) and the third round has concentrated the ethanol to 96.1%. Therefore, the process was accomplished by using molecular sieves that selectively absorbed the final excess water. Direct fermentation using *Streptosporangium roseum* has shown to be a very potential way to catalyst for the synthesis of bioethanol from EFB.

1. Introduction

The mainstream of oleochemical industry has moved away from the used organic solvents and emulsifiers and being replaced by a diversity of direct reactions on mixed waste-substrates and biomasses with the help of range of microbes. The product varies into forms of biofuels and bioethanols which sometimes refer to as agro fuels; the liquid or gaseous fuels derived from catalysis reaction takes place in fermentation. In the second generation of biofuels, production of bioethanols derived mostly from excess of starch crops like sugarcane bagasse, corn and paddy husks. Feedstock for bioethanol production may as well comes from other non-derived major food sources for example grasses and trees. Potential of bioethanol as fuels for vehicles is not a dream since long ago.

It can be used in many kinds of vehicles in its pure form even though its critical used is as additive in gasoline [1] that helps to improve vehicle emissions and increase octane formation. Direct fermentation perhaps, is one way to run fermentation in lesser time cycles compare to enzymatic-non-cell related fermentations. This process will start with plant materials follow by conversion to ethanol which involves identification of starting plant material, isolation and development of bacterial and fungal strains, and design of appropriate protocols for efficient conversion of plant material to sugar monomers [2].This paper will be discussing the production of a lignocellulosic bio-ethanol as an alternative energy production of biodiesel using Malaysian local isolate in a direct fermentation using empty fruit bunch

2. Material and Methods

2.1 Substrate preparation, culture maintenance and inoculum preparation

Empty fruit bunch (EFB) was collected from The Malaysian Palm Oil Board Malaysia (MPOB). The EFB was dried in an oven at 50°C for 48 hours to reduce moisture content

Content from this work may be used under the terms of the Creative Commons Attribution 3.0 licence. Any further distribution of this work must maintain attribution to the author(s) and the title of the work, journal citation and DOI.
Published under licence by IOP Publishing Ltd

followed by grinding to 0.5 mm size, sieving and weighing. The EFB was sterilized at 121°C for 20 min [3]. *Streptosporangium roseum* is a collection from our local isolated species with high lipase and cellulose activity and growing at 65°C with pH 7.0. It was maintained at 4°C on slant agar. The inoculum was prepared in nutrient broth (NB) with cell concentration maintained at 5.0 x 10^5 cells/ml [4].

2.2 Solid state fermentation

The solid-state fermentation (SSF) procedure was adapted from a study conducted by Lim *et al.* [10]. The 5.0 g of grinded sterilized EFB was added with 0.075 g of yeast extract which amounted to 1.5% (w/w) and 0.05 g (1% w/w) of carboxymethycellulose (CMC), followed by inoculation of 5.0 x 10^5 cells/ml of *Streptosporangium roseum* cell suspension. The amount of 8.0 ml of sterile distilled water was added to adjust the moisture concentration to 160% (v/w). Using a sterile spatula, the mixture was then mixed thoroughly and incubated at 60°C. Samples were prepared in triplicates and were withdrawn daily for 7 days. A small quantity of raw and fermented EFB was observed under the Scanning Electron Microscope (SEM). The samples was first mounted on aluminium stubs and placed on a sputter coater to coat the sample with a thin layer of gold before observing under the SEM [5].

2.3 Analysis

The sugar extraction was modified from Simple Contact Method [5]. The fermented EFB was added with 100.0 ml distilled water (dH_2O) and shake at 150 rpm (2 hours, 30°C) to allow sugar diffusion into the water. Then, the mixture content was filtered using Whatman No 1 filter paper (150 mm diameter) to separate the liquid from the solid followed by centrifuging at 5000 rpm, 4°C for 30 minutes using Sigma Refrigerated Benchtop. The supernatant from the centrifugation was analyzed for glucose content method using Dinitrosalicylic (DNS) acid adapted from Suresh [6]. The same procedure was carried out with stock sugar solution for Glucose Standard Curve.

2.4 Enzyme assay

Lipase activity was assayed by olive oil emulsion according to Nik Raikhan [7]. Substrate was prepared by homogenizing 30 mL of olive oil with 70 mL of emulsification reagent [8, 9]. One unit of lipase represented the release of 1µmol of fatty acid per min respectively under the above assay condition [8]. All assays are repeated thrice to gain the mean ± *sd* of three applications.

2.5 Bioethanol production, iodoform test and ethanol estimation

Method has been adapted and modified from Kuan *et al.* [10]. 5.0 g of fermented EFB and 1.0 ml of deionized water were mixed in a 25 mL Erlenmeyer flask, followed by the addition of n-hexane (25%, w/w of the fermented EFB) as a co-solvent. The reaction mixture was incubated in the temperature range of 66°C using a water bath with orbital shaking at 150 rpm for 48 hours. The *n*-hexane was removed using a rotary evaporator (Büchi KRvrTD 65/45, Flawil, Switzerland) with the bath temperature at 60°C. Then, 1.0 ml of distillate was added into 2.5 ml iodine along with 1.0 ml NaOH in a test tube. Positive result is recorded with cloudy formation and yellow precipitation and antiseptic smell after few minutes.For the ethanol estimation, standard curve of 1-10% ethanol in double distilled water was prepared with final volume of 25.0 ml followed by distillation. 10-15 ml of distillates was added into 25.0 ml of 3.4% chromic acid. The volume was added to 50.0 ml using double distilled water. The mixture was heated at 80°C for 15 minutes and absorbance was determined at 580 nm. Absorbance reading of samples was also determined at 580 nm and the standard curve was used to decide the ethanol percentage.

2.6 Gas chromatography (GC) and distillation

Method has been modified from Helena *et al.* [11]. The GC used was a HP 5890 Series II with FID. The analyses were performed using a polar capillary column (PEG-type, 10m x 0.53, with autosampler) and n-propanol as internal standard. The temperature of the FID was 260ºC, and the injector temperature was 70°C. The oven temperature was set up to 35°C for 5 min, followed by an increase to 175ºC at 8ºC per minute. The carrier gas was helium with a flow of 5.0 mL/min. The injection of the sample was performed using a 10 μL Hamilton syringe with a removable needle (needle gauge 22S). The volume of injection was 1.0 μL, with a split ratio of 1:50. The distillation process is illustrated in Figure 5.

3. Results and Discussion

Figure 1 depicts the growth of *Streptosporangium roseum* on solid state fermentation and concentration of glucose extracted. The actinomeycete cells feed on the inducer; CMC as the carbon source, and releasing extracellular enzymes that has the ability to breakdown and digest the polysaccharides into monosaccharide, which in this case was used to convert the empty fruit bunch (EFB) into glucose. The cellulose activity was recorded as 25.8 U/ml, quite a high amount for SSF. Grinding the EFB was done to assist the break into smaller pieces of the EFB and to increase the surface area for microorganism to act upon. In the industrial approach, the size reduction of biomass is vital to provide a large surface area for enzyme to act on during enzymatic hydrolysis [12]. Smaller size of biomass increases the accessibility of cellulosic component to be degraded by *Streptosporangium roseum.* The production of reducing sugar was monitored every day for a week. Upon centrifugation during sugar extraction, the biomass was separated and the supernatant which contained the glucose was further tested using the Dinitrosalicylic (DNS) acid test. The dinitrosalicylic acid reagent assay utilizes the dinitrosalicylic acid salt, Rochelle salt, phenol, sodium bisulphite and sodium hydroxide. The sudden heat treatment will force the reaction to a halt, as enzyme loses its activity when exposed to extreme temperature and pH. The pH of the mixture has also changes due to the addition of DNS reagent into the mixture [13]. The reducing sugars contain free carbonyl group (C=O).

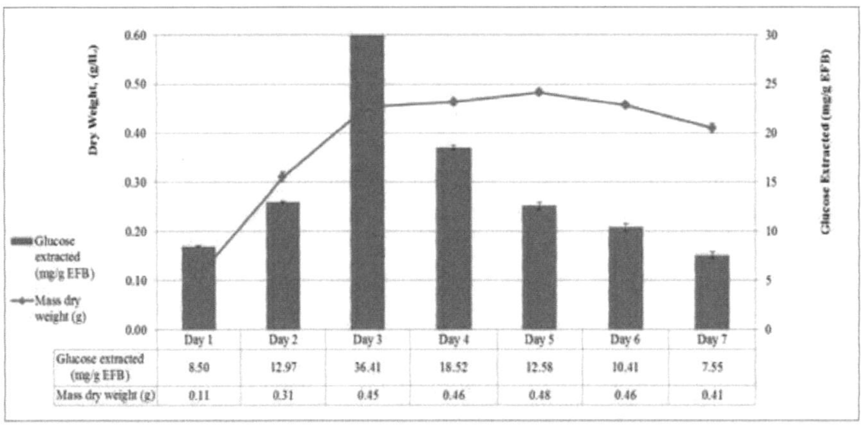

	Day 1	Day 2	Day 3	Day 4	Day 5	Day 6	Day 7
Glucose extracted (mg/g EFB)	8.50	12.97	36.41	18.52	12.58	10.41	7.55
Mass dry weight (g)	0.11	0.31	0.45	0.46	0.48	0.46	0.41

Figure 1: Growth profile of *Streptosporangium roseum* on EFB and concentration of glucose extracted for 7 days of cultivation. The solid substrate fermentation was carried out at room temperature, 60 ± 2°C, 0.5 mm EFB, 160% (v/w) moisture content, 1.5% (w/w) of yeast extract, 1.0% (w/w) carboxymethyl cellulose, inoculum size of 2.0 ml of 1.0 x10^9 spores/ml.

IOP Conf. Series: Materials Science and Engineering **358** (2018) 012032 doi:10.1088/1757-899X/358/1/012032

Figure 2(A) and 2(B) illustrates the surface area of EFB before and after fermentation, respectively. The sample was coated with metals to provide an electrically conductive layer, to restrain surface charges, and to minimize the damage caused by the radiation as well as increasing electron emission [14]. After 7 days of fermentation (Figure 2B) EFB was covered with *Streptosporangium roseum* spores with particles of produced glucose. During this stage, the *Streptosporangium roseum* has produced an enzyme that degrades the parenchyma cells that contains the cellulose and hemicellulose materials into fermentable sugars. These alterations were caused by the enzymatic hydrolysis process in which the enzyme was produced by the *Streptosporangium roseum* itself [5]. The crude lipase from *Streptosporangium roseum* has showed an excellent activity of 31.60 U/ml.

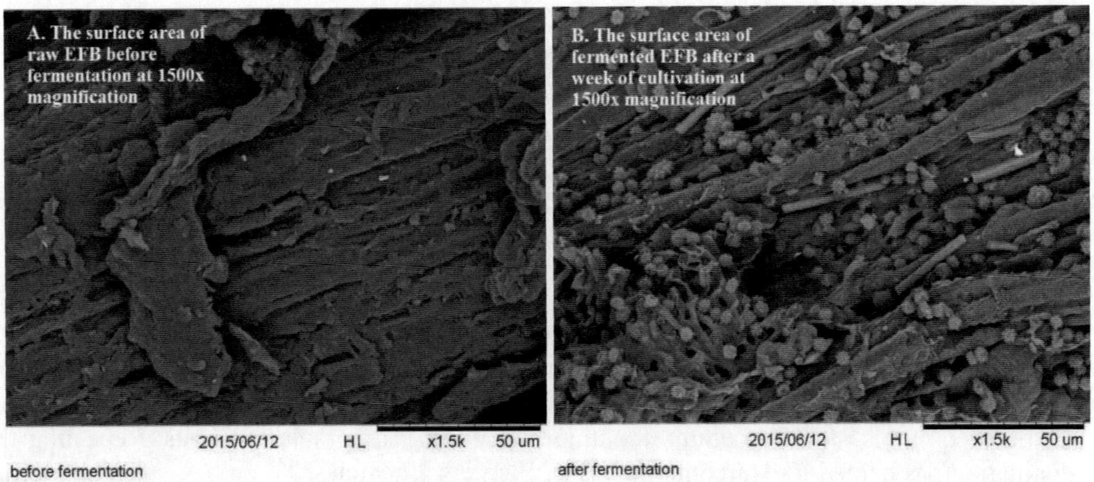

Figure 2: Scanning Electron Microscope (SEM) magnifications for (A) surface area of raw EFB before fermentation by *Streptosporangium roseum* at 1500x and (B) the surface area of the already fermented EFB with tiny particles of produced glucose or namely glucose extracted from empty fruit bunch.

3.1 Bioethanol production

Lipase from *Streptosporangium roseum* has been introduced to catalyst the journey of glucose-to-ethanol conversion through a specific fermentation with the below equation (Equation 1).

$$\underset{\text{Glucose}}{n\,C_6H_{12}O_6} \xrightarrow{\text{Lipase}} \underset{\text{Ethanol}}{2n\,C_2H_5OH} + CO_2$$

Ethanol (C_2H_5OH) produced from EFB by lipase *Streptosporangium roseum* has been detected positive with iodoform test (cloudy formation with yellow precipitation and antiseptic smell) and was confirmed by GC-FID. Chromatogram in Figure 3 shows the azeotrope of ethanol-water in the distillate sample with clear peaks. Details on the produced ethanol will be presented as quantitative data in other academic report and will not be discussed in detail here. C_2H_5OH was separated from the mash by distillation. It was a physical process where H_2O and components of C_2H_5OH were separated by differences in boiling point. The final collected ethanol in the 'ethanol barrel' was detected to be having a formation of an azeotrope, a mixture of two different liquids and in this case, was the mixture of the ethanol and water. The addition of deionized water into the reaction mixture was done because lipase requires a certain amount of water to be active [15] and to catalyze the reaction. Since the mixture has a lower boiling point than either of the components, the water cannot be totally separated by a one time-run distillation process.

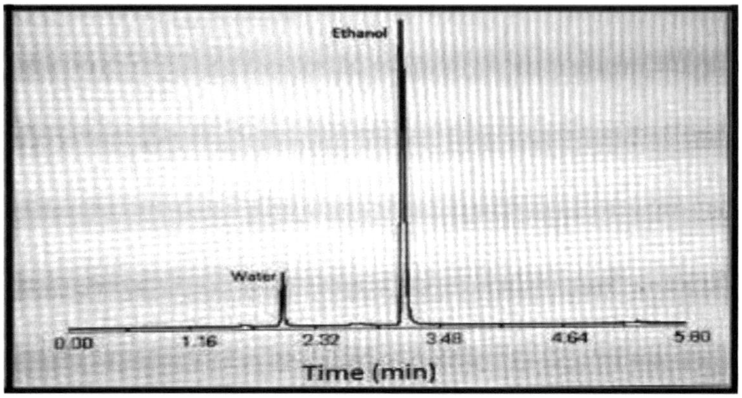

Figure 3: Representative chromatogram obtained from azeotrope of ethanol-water in the'ethanol barrel' with peak identification and retention times of the quantified compounds.

Therefore, this extraction has been repeated 3 times with the collected azeotrope (85-15% ethanol-water). In the phenomenon of boiling, pure water boils and becomes condense as water vapor at 100°C. Since alcohol boils at 78°C, most of the vapor will be alcohol at this temperature because water does not boil until 100°C. This leaves less alcohol and more water in the azeotrope so the temperature continues to climb as the solution boils. Repeating this distillation has helped to overcome the issue. Figure 4 illustrates the process. The production of pure, water-free (anhydrous) ethanol requires a dehydration step following distillation.

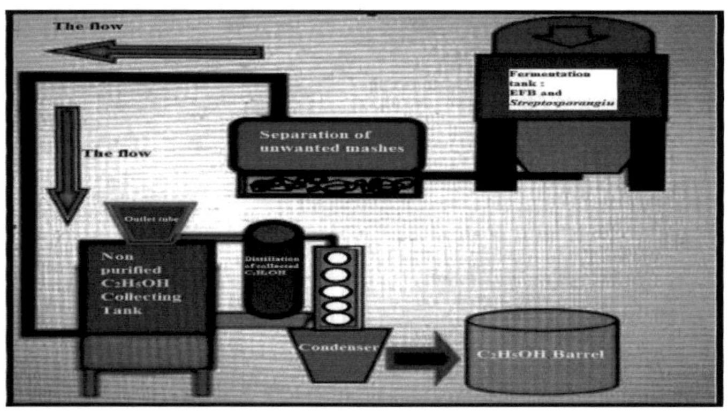

Figure 4: Distillation was used to separate a composite mixture into the base element of the produced ethanol. The process has involved the change of state from liquid to gas, and then subsequent condensation to return the now separated ethanol into liquid state again.

Dehydration, a relatively complex step in ethanol fuel production, is accomplished using molecular sieves that selectively absorb water on the basis of the difference in molecular size between water and ethanol to concentrate the ethanol from 96.1% to 100%. The percentage of ethanol and water products is subjected to minor changes depending on the lipase efficiency and then followed by the distillation process. Theoretically, the maximum conversion efficiency of glucose to ethanol is 55-57 percent on a weight basis. However, some glucose is used by the lipase enzyme for the production of metabolic products other than ethanol. In practice, between 42 to 49 percent of glucose could be converted to ethanol. By taking a 45%-fermentation conductivity of lipase *Streptosporangium roseum* in catalytic efficiency, 1,000 kilograms of fermentable sugar (glucose extracted from EFB will be producing about 570 liters of pure ethanol. This means if we are aiming for 1,000 liters of ethanol, we will need to use about 1,800 kilograms of EFB. The whole mash or fermentation products typically will contain between 50 and 100 grams of ethanol per liter which is about 5.0 to 10% weight per volume assuming that fermentation is a 100% complete.

4. Conclusion

The present research shows that EFB has the potential to be developed as an economical alternative energy source in a glucose form using lipase *Streptosporangium roseum.* This lipase has shown to be a very potential enzyme as catalyst for the synthesis of bioethanol and has been proved to catalysts the system successfully. The team is now working with the properties of the synthesized biodiesel, and planning for entrapping the enzyme for reuse method.

Acknowledgment

The authors gratefully acknowledge the UiTM for the funding in this research. Humongous thanks go to the team who has been supporting the work and this research.

References

[1] Noureddini H, Gao X, Philkana R S 2005 *Bioresour Technol* **6(7)** 769-777
[2] Mostafa S E 2010 *Journal of Advanced Research* **1(2)** 103-111
Vandenberghe L P, Soccol C R, Pandey A, Lebeault J. M. 2000 *Bioresource Technology* **74(2)** 175-178
[3] Nik Raikhan 2016 N H *Journal of Life Sciences and Technologies* 4**(2)** 70-74
[4] Lim S H, Ibrahim C O 2012 *International Journal of Biochemistry and Biotechnology* **1(1)** 7-11
[5] Suresh P V, Chandrasekaran M 1998 *World Journal of Microbiology and Biotechnology* **14(5)** 655-660
[6] Nik Raikhan N H 2003 Production of Hyperthermotolerant Lipase from Actinomycete *Streptosporangium.* M.Sc. Thesis, Universiti Sains Malaysia, Penang, Malaysia
[7] Mustranta A, Forssell P, Poutanen K 1993 *Enzyme and Microbial Technology* **15(2)** 133- 139
[8] Jesus M F C P, Branco R N, Sant Anna Jr G L, Freire D M, Silva Jr J G 1993 *Brazilian Journal of Chemical Engineering* **16(2)** 113-118
[9] Kuan I C, Lee CC, Tsai B H, Lee S H, Lee W T, Yu C 2013 *Energies* **3(6)** 2052-2064
[10] Helena P, Paula G, Susana C, Helena C, Agostinho S, Teresa M, Fernando R, Félix C, Maria L B 2009 *Journal of Chromatographic Science* **47** 272-278
[11] Mais U, Esteghlalian A R, Saddler J N 2002 Mansfield S. D., Enhancing the enzymatic hydrolysis of cellulosic materials using simultaneous ball milling. Humana Press p.815-832.
[12] Rathore S S S 2009 Decision Making in Dry-grind Ethanol Industry Using Near-infrared Spectroscopy (Doctoral Dissertation, University of Ilinois).
[13] Sawyer L, Grubb D, Meyers G F 2008 *Polymer microscopy.* 2 Springer Science & Business Media p.138
[14] Kaieda M, Samukawa T, Kondo A, Fukuda H 2001 *J. Biosci. Bioeng* **91** 12–15

Molecular Dynamics Simulation of *Mahkota Dewa (Phaleria Macrocarpa)* Extract in Subcritical Water Extraction Process

N A Hashim, S K Abdul Mudalip, N Harun, R Che Man, S Z Sulaiman, Z I M Arshad and S M Shaarani

Faculty of Chemical Engineering & Natural Resources, Universiti Malaysia Pahang, Lebuh Raya Tun Razak, 26300 Gambang, Pahang, Malaysia

kholijah@ump.edu.my

Abstract. *Mahkota Dewa (Phaleria Macrocarpa)*, a good source of saponin, flavanoid, polyphenol, alkaloid, and mangiferin has an extensive range of medicinal effects. The intermolecular interactions between solute and solvents such as hydrogen bonding considered as an important factor that affect the extraction of bioactive compounds. In this work, molecular dynamics simulation was performed to elucidate the hydrogen bonding exists between *Mahkota Dewa* extracts and water during subcritical extraction process. A bioactive compound in the *Mahkota Dewa* extract, namely mangiferin was selected as a model compound. The simulation was performed at 373 K and 4.0 MPa using COMPASS force field and Ewald summation method available in Material Studio 7.0 simulation package. The radial distribution functions (RDF) between mangiferin and water signify the presence of hydrogen bonding in the extraction process. The simulation of the binary mixture of mangiferin:water shows that strong hydrogen bonding was formed. It is suggested that, the intermolecular interaction between $O_{H2O}{\bullet\bullet}H_{MR4(OH1)}$ has been identified to be responsible for the mangiferin extraction process.

1. Introduction

Plants with therapeutic performances or known as herbs have been used as an alternative health care since ancient time. This practice has been recognized by the World Health Organization (WHO) [1,2]. *Mahkota Dewa* or scientifically known as *Phaleria macrocarpa* is one of the popular herbal plants in Malaysia due its medicinal properties [3]. *Mahkota Dewa* originated from *Thymelaceae* family and growth in the topical areas of Papua Island, Indonesia [3,4]. The bioactive ingredients in this plant contains antihistamine, antioxidant, antidiabetic and anticancer effect [5]. The example of bioactive ingredients in *Mahkota Dewa* fruits are alkaloid, saponin, flavonoid, polyphenol and mangiferin [4]. Among these compounds, mangiferin has a broad range of pharmacological effects including antidiabetic, anti-HIV, anticancer, immunomodulatory and antioxidant [4,6]. Solvent extraction, conventional methods which often used to extract the bioactive compounds from various plants. The organic solvents employed during the process are as hexane, methanol or ethanol [7]. However, this process reported as impractical due the need of further purification step as well as low extraction yield [7]. Subcritical water extraction (SWE), involve the use of water as solvent under a pressurized condition reported as a non-toxic and environmental friendly technique. As water heated under pressure to subcritical temperatures, there are significant changes to its polarity, often expressed simply as the dielectric constant. When temperature increased to sufficiently high temperatures, the dielectric constant of water decreases and observed to mimic a range of organic and hydro-organic solvents [8]. The application of SWE can be found for extraction of various herbal products such ginger, algae, olive leaves and also oregano [9-11]. However, to the best of our knowledge, the

Content from this work may be used under the terms of the Creative Commons Attribution 3.0 licence. Any further distribution of this work must maintain attribution to the author(s) and the title of the work, journal citation and DOI.
Published under licence by IOP Publishing Ltd

mechanism of mangiferin extraction using water is yet to be reported. This mechanism is quite impossible to be seen using experimental approach, unless model using a suitable software.

Molecular dynamics simulation is a powerful tool use to model the molecules and generate information at the microscopic level [12]. Recently, molecular dynamics simulation has been employed in various studies such as extraction, crystallization, and membrane development [2,13-15]. This approach is ideally suitable to examine the intra and intermolecular interaction with different materials based on radial distribution function (RDF) analysis [12,16-18]. Adam and co-workers has applied molecular dynamic simulation to simulate the patchouli oil extraction behaviour in different solvents namely acetone, ethanol and hexane. The study suggested that different solubility behaviour was due to the different of hydrogen bonding arrangement between it solute and solvent molecules [2]. This paper is aimed to model the intermolecular interaction, mainly hydrogen bonding, during the extraction of bioactive compound namely mangiferin from *Mahkota Dewa* fruits in water using molecular dynamic simulation. The simulation was performed at 373 K and 4.0 MPa [19] using COMPASS force field and Ewald summation method available in Material Studio 7.0 simulation package.

2. Methods

2.1. Simulation details

The MD simulation of mangiferin extraction were performed using Accelrys Material Studio® 7.0 software (Accelrys, Inc., San Diego, USA) which installed in HP Z420 Workstation with Windows 7 Professional operating system. The three-dimensional (3D) molecules structure of mangiferin and water downloaded from Chemspider databases and imported into the software. The partial labeling of molecules shown in Figure 1 was used in this simulation work for the purpose of molecular recognition.

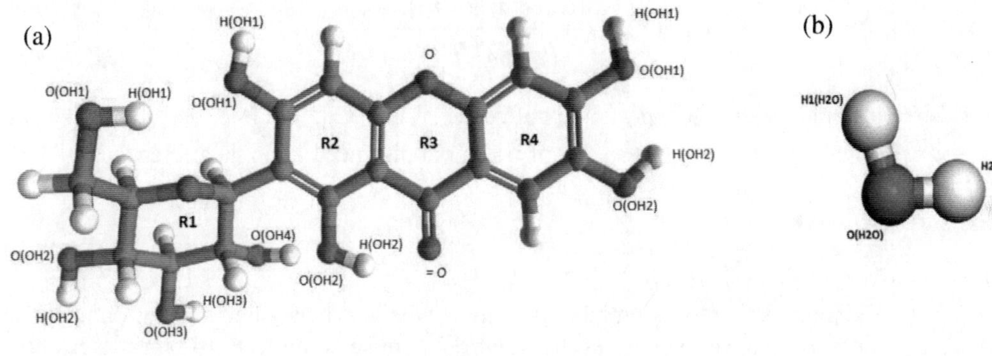

Figure 1. Chemical structure of (a) Mangiferin and (b) Water. Color representation: White - hydrogen, Red - oxygen and Grey - carbon [23].

Each molecule was optimized using geometry optimization to produce a stable molecular geometry for subsequent simulation steps. A smart algorithm was used to calculate the energy and atomic force with a fine convergence quality. The cubic periodic boundary simulation box for pure water and mixture of water and mangiferin were constructed and minimize at 373 K and 4.0 Mpa by using amorphous cell calculation module. The minimization step is crucial in minimizing the energy produce and to ensure simulation box ready for the dynamic stage. The number of molecules, densities of pure components and binary system as well as size of the simulation boxes is tabulated in Table 1. The density of binary mixture was calculated using the following equation:

$$\boldsymbol{Density, \rho = A\, B^{-\left(1-\frac{T}{C}\right)^{n}}} \tag{1}$$

where A, B, C and n is the regression coefficient of chemical compound and T represent the temperature in Kelvin (K). The simulation was initiated by equilibrating the system under constant number of molecules, volume and temperature (NVT) ensemble for 1000 ps where molecules are allowed to evolve from random starting to stable configuration with energy conservation. The simulation was continued by dynamic mode with constant number of molecules, pressure and temperature (NPT) ensemble using Nose thermostat [2] and Berendsen barostat [2] for 1000 ps. The time step was set at 1.0 fs. The cut-off distance for Lennard-Jones potential was set to 12 Å which was close to half of the cell length. A condensed-phase optimized molecular potentials for atomistic simulation studies (COMPASS) force field and Ewald summation techniques with an accuracy of 0.0001 kcal/mol were used to model the intermolecular interaction in the system [14-15,18].

Table 1. Simulation data.

System	Number of Molecule	Density [g/cm^3]	Equilibrated box size, A × B × C
Pure system:			
Water	1000	0.9529 [18]	32.56 × 32.56 × 32.56
Mangiferin	50	1.8430 [18]	31.54 × 31.54 × 31.54
Binary system:			
Mangiferin: Water	50:1000	0.9953	40.26 × 40.26 × 40.26

2.2. *Radial distribution analysis*

Atoms from the trajectory files of the simulation labeled and then each pair of labeled atoms was analyzed for its radial distribution functions (RDF). The labeled atom was based on the atom that has the ability to initiate the hydrogen bond. RDF is essential to describe the micro-structure of the material and describes the variation of specific atomic density as a function of a distance from one reference atom [13,20]. In other words, RDF counts the number of two-atom species with specific distances and can be defines as:

$$g_{xy}(r) = \frac{[N_y(r,r+dr)]}{\rho_y 4\pi r^2 dr} \quad (2)$$

where r is spherical radius distance from the reference atom [13], ρ_y is a density of a y atom, $[N_y(r,r+dr)]$ is number of y atoms in a shell of width r at distance r and x is reference atom [18].

3. Results and discussion

3.1. *Validation of MD simulation method*

The Simulation validation method is essential to determine whether the applied simulation method and COMPASS force able to produce reliable results. Figure 2 compares the RDF of pure water obtained in this work with those reported in literature [16]. The possible bonding formation between O_{H2O}••$H1_{H2O}$, O_{H2O}••O_{H2O} and $H1_{H2O}$••$H2_{H2O}$ obtained in this work shows first peaks at 1.75 Å, 2.75 Å, and 2.25 Å, respectively. It can be seen that, both simulated and literature RDF pattern show quite a similar trend. Based on these values, it can be affirmed that O_{H2O}••$H1_{H2O}$, which shows the nearest neighbor interaction, is stronger and more intense compared to O_{H2O}••O_{H2O} and $H1_{H2O}$••$H2_{H2O}$, thus suggested to represent the strength of hydrogen bonding in pure water. A slight difference in RDF intensity, g(r) for O_{H2O}••O_{H2O} was probably due to differences in number of water molecules used during simulation as Abdul Mudalip et al. [16] simulate 250 water molecules meanwhile, this study used 1000 water molecules.

Table 2 shows the comparison between simulated data and the setting parameters of water and mangiferin:water. The percentage of deviation for average density is low, which is less than 6%. Sun [21] who had performed a MD for 150 organic structures using COMPASS force field reported a maximum absolute error of 6% for the simulated density. Since the deviation obtained in this work is quite small, it can be suggested that the use of the COMPASS force field and the Ewald summation method can generate reliable simulation results.

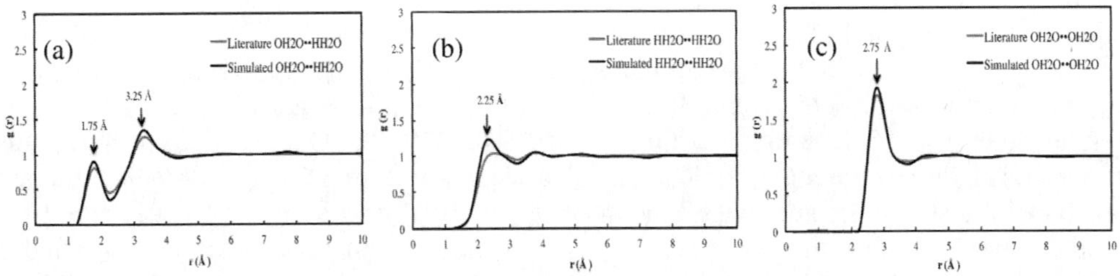

Figure 2. Comparison of RDF for (a) O_{H2O}••$H1_{H2O}$ (b) $H1_{H2O}$••$H2_{H2O}$ (c) O_{H2O}••O_{H2O} from this study and Mudalip et al. [16] .

Table 2. Simulation result of *Mahkota Dewa* extraction.

System	Water	Mangiferin/water
Average Density [g/cm^3]		
Simulated Value	0.8970	0.9960
Setting Value	0.9529	0.9953
Deviation [%]*	5.87	0.07

*Deviation = [(Simulated value - Calculated value)/ Calculated value] x 100

3.2. RDF analysis in binary mixture

The strength of the intermolecular interactions between solutes and solvents determines the solubility and amount of a solute extracted in a particular solvent. In this study hydrogen bond is one of the intermolecular interactions that present since it is the primary intermolecular forces present in water. Hydrogen bond is an attractive force exists in between hydrogen in a molecule with an electronegative atom of different molecule. The electronegative atom includes oxygen, nitrogen, and fluorine, which have a partial negative charge. Analysis of RDF can provide atomistic representation of intermolecular interaction that exists in binary mixture [14]. In binary mixture, the intermolecular interactions exist between molecules of solvent-solvent, solute-solvent, and solute-solute. During solvation, the solute must be surrounded, or solvated, by the solvent. Solutes are successfully dissolved into solvents when solute-solvent interactions are stronger than the solute-solute or solvent-solvent interactions [2].

Figure 3 (a) shows the RDF of solvent-solvent, which is O_{H2O}••$H1_{H2O}$ and O_{H2O}••O_{H2O} in binary mixture of mangiferin:water during subcritical extraction process. The RDF of O_{H2O}••$H1_{H2O}$ obtained shows strong hydrogen bonding interaction at 1.75 Å and 3.25 Å, with intensities of 2.33 and 2.02 respectively. Meanwhile, for O_{H2O}••O_{H2O} RDF observed at radial distance of 2.75 Å with an intensity of 3.91. Both O_{H2O}••$H1_{H2O}$ and O_{H2O}••O_{H2O} RDF pattern was in agreement with the RDF pattern for the simulated pure water with slight different of intensity, g(r). The same RDF pattern indicates that the presence of a mangiferin as solute in the binary system does not modify the long-range pure solvent structure and higher intensities indicates that there was strong hydrogen bonding interaction between water molecules in the binary system. The solute-solute interaction in between O_M••H_M and O_M••O_M shows in Figure 3 (b).

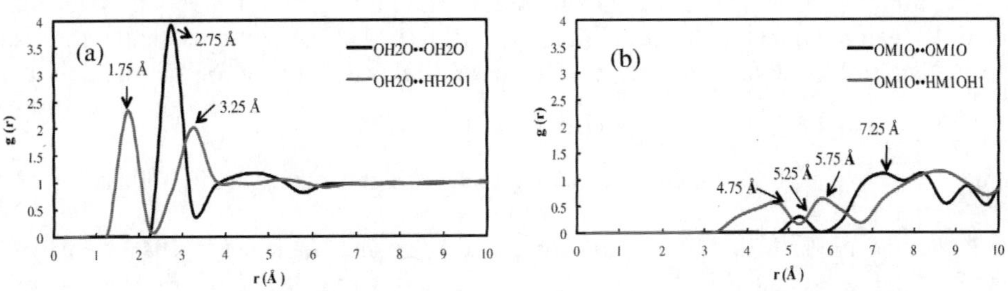

Figure 3. (a) Solvent-solvent, O_{H2O}••$H1_{H2O}$ and O_{H2O}••O_{H2O} interactions in binary system (b) Solute-solute interaction in binary system with O_M••H_M and O_M••O_M.

The first peaks observed at 4.75 Å and 5.75 Å for O_M••H_M and 5.25Å and 7. 25 Å for O_M••O_M. The radial distribution function of both interaction are less structured with no sharp peak and thus this interaction is not significant in this simulation and experimental extraction process. The RDF for the interaction of O_{H2O}••H_M and H_{H2O}••O_M illustrated in Figure 4 (a) and (b). The interactions between water atoms which is O_{H2O} with the hydrogen atom from mangiferin $H_{MR2(OH1)}$, $H_{MR2(OH2)}$, $H_{MR4(OH1)}$ and $H_{MR4(OH2)}$ recorded RDF of 1.75 Å and 3.25 Å. The RDF of O_{H2O}••$H_{MR4(OH1)}$, recorded the sharpest peak with the intensity of 2.76 at radial distance of 1.75 Å and RDF of O_{H2O}••$H_{MR2(OH1)}$ shows intensity of 2.70, which is 0.06 less from O_{H2O}••$H_{MR4(OH1)}$. The second peak of 3.25 Å recorded the intensity of 1.2. As the radial distance became longer, the intensity reach 1.0 that indicates there is no long-range orders interaction [22]. Since the O_{H2O}••$H_{MR4(OH1)}$ shows the sharpest peak, it can be suggested that it can represent the strength of solute-solvent interaction in the mangiferin extraction. Meanwhile in the Figure 4(b) interaction between $H1_{H2O}$ with $O_{MR1(OH1)}$, O_{MR1O}, $O_{MR2(OH2)}$, O_{MR3O}, and $O_{MR3=O}$ shows lower intensity, g(r) with RDF of 1.75 Å and 3.25 Å. The highest intensity shows by interaction of $H1_{H2O}$•• $O_{MR1(OH1)}$, with intensity of 1.42 at 1.75 Å radial distance and 1.43 intensity at 3.25 Å. Overall, the interaction between of $H1_{H2O}$•• O_M shows that the intensity of the interaction is low compared to interaction between O_{H2O}••H_M. Thus, this study was focusing more on interaction of O_{H2O}••H_M.

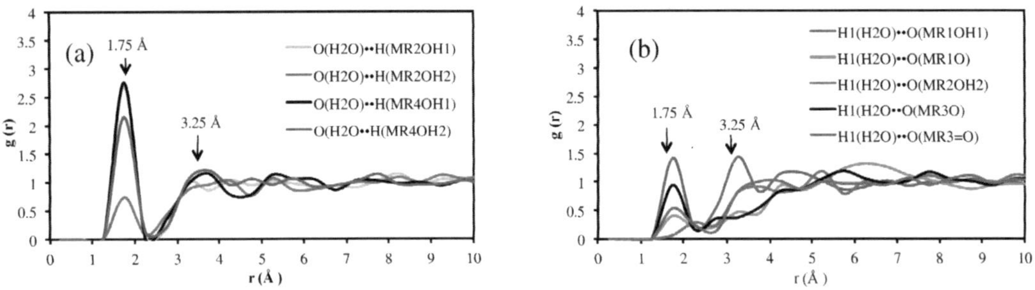

Figure 4. (a) and (b) Solvent-solvent interactions and radial distribution function (RDF) between O_{H2O} with H_M in the binary system at 373 K

4. Conclusion

The molecular point of view of mangiferin in subcritical water extraction process has been successfully revealed using molecular dynamics simulation technique. The simulated RDF pattern for pure water using the COMPASS force field and Ewald summation method show an agreement with literature with slight differences in RDF intensity. The simulation of the binary mixture of mangiferin:water shows that strong hydrogen bonding formed and the intermolecular interaction between O_{H2O}••$H_{MR4(OH1)}$ has been identified to represent the strength of solute-solvent interactions in mangiferin extraction from *Mahkota Dewa* fruits.

Acknowledgement

The authors are grateful for the financial support from the Ministry of Education Malaysia for Research Acculturation Grant Scheme (Grant No. RDU151402) and Universiti Malaysia Pahang Postgraduate Research Grants Scheme (Grant No. RDU170322). We also thank the Faculty of Chemical and Natural Resources Engineering, Universiti Malaysia Pahang for the research facilities.

References

[1] Hamid A A, Aiyelaagbe O O, Usman L A, Ameen O M and Lawal A 2010 *African J. Pure Appl. Chemostry* **4** 142–51

[2] Law K S and Soon L K 2013 *Evidence-Based Complement. Altern. Med.* **2013** 1–10

[3] Altaf R, Zaini M, Asmawi B, Dewa A, Sadikun A and Umar M I 2013 *Pharmacognosy Rev.* **7** 73–80

[4] Anggraini T and Lewandowsky P 2015 *Intl. J. on Advance Science Engineering Information Technology* **5** 59–62

[5] Tandrasasmita O M, Lee J S, Baek S H and Tjandrawinata R R 2010 *Cancer Biol. Ther.* **10** 814–24

[6] Mirza R H, Chi N and Chi Y 2013 *J. Nutr. Ther.* **2** 74–9
[7] Kuba´tova´ A, Miller D J and Hawthorne S B 2001 *J. Chromatogr. A* **923** 187–94
[8] Alghoul Z M, Ogden P B and Dorsey J G 2014 *J. Chromatogr. A.* 1486 42–9
[9] Castro Puyana M, Herrero M, Mendiola J A, and Ibanez E 2013 *Woodhead Pub. Limited* **16** 534–60
[10] Kulisic T, Radonic A, Katalinic V and Milos M 2004 *Food Chem.* **85** 633–40
[11] Frenkel D and Smit B 2002 *Understanding Molecular Simulation* (California,USA: Academic Press)
[12] Adam F, Hana A B S, Tajuddin S N, Yusoff M M 2016 *Asian J. Chem.* **28** 1253–57
[13] Dimitroulis C, Kainourgiakis E, Raptis V and Samios J 2015 *J. Mol. Liq.* **205** 46–53
[14] Mo Y, Zhang H and Xu J 2014 *J. of Chem. and Pharmaceutical Research* **6** 1534–39
[15] Schatschneider B and Chronister E L 2016 *Molecular Simulation* **34** 1159–66
[16] Mudalip S K A, Abu Bakar M R, Jamal P, Adam F and Alam Z M 2016 *Asian J. Chem.* **28** 853–58
[17] Adam F, Siti Hana A B, Yusoff M M and Tajuddin S N 2014 *J. Chem. Eng. Data* **59** 183–88
[18] Yaws C L 2008 *Thermophysical Properties of Chemicals and Hydrocarbons* (USA: William Andrew)
[19] Mark P and Nilsson L 2001 *J. Phys. Chem. A* **105** 9954–60
[20] Mudalip S K A, Bakar M R A, Adam F and Jamal P 2013 *Int. J. Chem. Eng. Appl.* **4** 124–28
[21] Sun H 1998 *J. Phys. Chem. B* **5647** 7338–64
[22] Mudalip S K A, Abu Bakar M R, Adam F, Jamal P, and Alam Z M *Jurnal Teknologi* **79** 21–7
[23] Chemical-Structure @ www.chemspider.com 2015

IOP Conf. Series: Materials Science and Engineering **358** (2018) 012034 doi:10.1088/1757-899X/358/1/012034

Bacterial Adhesion on the Titanium and Stainless-Steel Surfaces Undergone Two Different Treatment Methods: Polishing and Ultrafast Laser Treatment

N Chik[1], W S Wan Md Zain[16], A J Mohamad[2], M Z Sidek[3], W H Wan Ibrahim[1], A Reif[4], J H Rakebrandt[4], W Pfleging[4] and X Liu[5]

[1]Faculty of Chemical and Natural Resources Engineering, Universiti Malaysia Pahang, Pahang, Malaysia
[2]Faculty of Electrical and Electronics Engineering, Universiti Malaysia Pahang, Pahang, Malaysia
[3]Faculty of Mechanical Engineering, Universiti Malaysia Pahang, Pahang, Malaysia
[4]Institute for Applied Materials, Karlsruhe Institute Technology, 76344 Eggenstein-Leopoldshafen, Germany
[5]School of Engineering, University of Warwick, Coventry, CV4 7AL, UK

n.atiqahchik@gmail.com

Abstract. Bacterial adhesion has become a significant problem in many industries causing billions of dollars for its complicated removal treatment and maintenance. In this study, metal surfaces undergone treatment with ultrafast laser with varies power. The microstructure produced on its original surfaces were expected to prevent the adhesion of *Escherichia coli* (*E. coli*) ATCC 8739 and *Staphylococcus aureus* (*S. aureus*) ATCC 6838. The laser treatment was performed at 380 fs pulse duration, 515 µm central wavelength and a repetition rate of 200 kHz. Stainless steel AISI 316L was treated with an average laser power of 0.04 W (SS-0.04) and 0.11 W (SS-0.11), while Grade 5 titanium alloy was tested with high laser power 0.11 W (T-0.11). The adhesion was observed after 16 hours and the number of adhering bacteria was counted per cm^2. The result achieved shows that, increasing the average laser power is leading to an enhanced *S. aureus* adhesion while *E. coli* adhesion is reduced which is due to the hydrophobicity interaction and difference in surface texture. Meanwhile, the laser treatment showed significant reduction of the bacterial adhesion on its surface compared to the polished surfaces. Thus, ultrafast laser texturing can be suggested as a promising method to reduce the bacterial adhesion, which reduced the adhesion of >80% for *E. coli* and >20% for *S. aureus*.

1. Introduction

Bacteria are ubiquitous and easily adhere to the exposed surfaces. This natural phenomenon creates problems to many industries such as contamination in medical devices, food industry, pharmacy sector, marine science and bio-fouling or bio-corrosion of industrial equipment [1,2]. The severity of the adhesion occurs when adhered bacteria form biofilm, which is an assemblage of microbial cells that is irreversibly and often embedded in a matrix of extracellular polymeric substances (EPS) [3]. Biofilm and eradication of the pathogen can be extremely difficult to be removed, time-consuming and expensive [3]. Thus, many studies have been performed in order to investigate the factors that contributed to the bacterial adhesion and also all the possible methods for avoiding the formation of biofilm on certain surfaces [1]. The adhesion of bacteria on material surfaces is governed by many factors, including bacterial characteristic (e.g., hydrophobicity, surface charge), surface properties (e.g., roughness, wettability) and environment condition (e.g., pH, temperature) which involve the

Content from this work may be used under the terms of the Creative Commons Attribution 3.0 licence. Any further distribution of this work must maintain attribution to the author(s) and the title of the work, journal citation and DOI.
Published under licence by IOP Publishing Ltd

physicochemical and also the molecular interactions [4,5]. The net interaction forces between a cell and a flat surface in the adhesion process are well described by DVLO (Derjaguin, Verwey, Landau and Overbeek) theory [5]. Meanwhile, studies had emphasized that surface roughness and wettability affects the rate of bacterial adhesion [1,4], but, some researchers also report that the final finishing process for the metal surfaces had significant contribution and influenced the bacterial adhesion [6]. Reports on the polished surfaces with respect to bacterial adhesion are countless and established, but limited reports were found on the effect of laser treated surface especially with respect to ultrafast laser radiation. The laser treatment process was carried out in order to realize laser-induced periodical surface structure (LIPSS) on the metal surfaces by controlling the average laser power [7].

In this study, the bacterial characteristics such as surface hydrophobicity and grams type were studied where they were included as factors that affect the rate of adhesion on the metallic surfaces. Besides that, titanium and stainless steel have been tested due to its vast usage as medical implant materials and in industrial purposes [1,8]. Therefore, we investigated the adhesion rate of *E. coli* and *S. aureus* on the nano-textured surface (LIPSS) produced by ultrafast laser ablation, aiming at assessing the potential of using laser materials processing for producing surfaces with low bacterial adhesion rate which can help both, medical and industrial sectors to overcome their problems, while exploring the effect of the average laser power to the rate of bacterial adhesion on the metal surfaces.

2. Methods

2.1. Bacteria and growth condition

The bacterial strains used for this study were *E. coli* ATCC 8739 and *S. aureus* ATCC 6838, obtained from culture collection of Central Laboratory, Universiti Malaysia Pahang. For long term preservation, the bacterial cultures were maintained in Luria Bertani (LB) broth with 20% (v/v) of glycerol, in - 80 °C freezer. Prior to the experiment, the stock culture was plated on LB agar at 30 °C for 24 hours. Single loopful of bacterial cells was transferred to a 250 mL shake flask containing 50 mL of LB broth for inoculation, followed by incubation at 30 °C, 150 rpm for 18 hours. Then, the inoculums were harvested by centrifugation at 5000 rpm x g for 5 min. The absorbance of the bacterial suspension (cell pellets and phosphate buffer saline (PBS)) for both cultures were adjusted to ~1.0 M (10^6 bacterial/ml) before adhesion test was started.

2.2. Metal surfaces treatment

In this study stainless steel AISI 316L and Grade 5 titanium alloy were used. The surface of both metals were treated using laser ablation and polishing technique. The metals used in this study were cut into small pieces (1 cm x 1 cm x 3 mm). Laser treatment was performed at Karlsruhe Institute of Technology (KIT, Germany) using a micromachining workstation (PS450-TO, Optec, Belgium) equipped with an ultrafast fiber laser (Tangerine, Amplitude Systemes, France) operating with an average power of 35 W. Laser surface texturing was carried out under ambient air with a central wavelength of 515 μm, a laser pulse duration of 380 fs and a repetition rate of 200 kHz. The scanning speed of the laser beam onto the metals was maintained at 20 mm/s while the average laser power was varied. Stainless steel has been exposed to the laser beam by applying an average laser power of 0.04 W (SS-0.04) and 0.11 W (SS-0.11). While titanium only undergone processing with laser power of 0.11 W (T-0.11). The laser beam for LIPSS generation was guided through a beam expander (2-fold) and the scan head (Newson Engineering BV, Belgium) was used together with an f-theta lens with a focal length of 100 mm [7]. After the laser treatment, the achieved surface structure was characterized by scanning electron microscopy (SEM) (Carl Zeiss, SEM EVO 50). Polishing technique also has been applied to both metals. Silicon carbide (SiC) papers with grit 180 and 1500 were used to polish the stainless steel and titanium by using 6-in. grinder-polisher (Adolph and Buehler, Chicago). All metal undergone polishing technique were labelled as P-SS for stainless steel and P-T for titanium.

2.3. Bacterial characterization

Gram staining method, bacterial size analyzer and bacterial adherence to hydrocarbon (BATH) technique were carried out for the bacterial characterization. Bacterial smear from 12 hours culture was covered with crystal violet for 1 minute, gram's iodine also for 1 minute and then, followed by decolourization using 95% of ethyl alcohol. Lastly, safranin was added for 1 minute, washed with distilled water and examined under light microscope. Bacteria were viewed using SEM at 100x magnification and the sizes were measured using ImageJ software. Values were taken as average from 5 bacteria images. Prior to the test, procedures that included centrifugation of bacterial cell (15 ml) at 5000 rpm x g for 5 min, cell washing with sodium chloride (NaCl) solution twice, re-suspension the bacterial suspension in 15 ml of PBS (0.1 M, pH 7.4) solution were performed . The samples then were fixed in 2% of glutaraldehyde (1 hour, 4 °C), centrifuged at 5000 rpm x g for 5 minutes and were dehydrated in an ethanol series; 30%, 50%, 70%, 80%, 90%, 100% (each step for 10 minutes excepting 100% ethanol treatment was for 30 minutes).

Surface hydrophobicity of bacterial cells has been determined by using BATH technique. During BATH technique, several steps included centrifugation of bacterial cell and cell washing process as described in bacterial size analyzer test were performed and the absorption (A_0) was adjusted until ~1.0 (106 bacteria/ml) using UV-Vis spectrophotometer (Hitachi, U-1800 spectrophotometer). Then, 4 ml of bacterial suspension were vigorously mixed with 1 ml of n-hexane, n-hexadecane and xylene separately for 2 min. The water and organics phase were allowed to separate and after 15 min the absorbance (A) aqueous phase was measured [1,9]. The percentage of bacteria hydrophobicity was measured by the following equation (1):

$$\%\text{ of hydrophobicity} = \left(1 - \left(\frac{A}{A_0}\right)\right) \times 100 \qquad (1)$$

2.4. Bacterial adhesion test

Adhesion test was carried out in a glass container (6 cm x 9 cm x 7 cm) containing a baby cradle-like holder for holding the metal slide (2.5 cm x 7.6 cm x 0.1 cm) in the upright-vertical position. The metals slide that carry stainless steel or titanium was suspended in 70 ml of bacterial solution (absorbance adjusted to ~1.0 M) and shake at 70 rpm for 16 hours (laser structured metals) and 4 hours (polished metals). Samples were taken and dried in the incubator at 30 °C overnight and then were examined under the fluorescence microscope (Olympus FluoView Ver. 1.3). Prior to viewing, adhered bacteria were stained with Syto9 dye for 5 min, followed by flushing with generous amount of distilled water. ImageJ was used to calculate the number of bacterial adhere on metal per cm^2 based on the image captured by the fluorescence microscope.

3. Results and discussion

3.1. Bacterial characteristics

Based on the gram staining image (Figure 1a and 1b) *E. coli* ATCC 8739 was stained red, categorized as gram negative bacteria, while *S. aureus* ATCC 6838 was stained purple and is a gram positive bacteria. This staining response is based on the chemical and structural makeup of the cell wall of bacteria. Besides that, some proteins associated with the cell wall and the cell membrane of the bacterial are responsible for bacterial cell-surface hydrophobicity where hydrophobicity can influence the rate of bacterial adhesion [10] on a metal surface. The size of the bacteria were determined from the SEM images (Figure 1c and 1d) of the free cell (non-adhered bacteria), taking as average from 100 of bacteria that were selected randomly, with the help of the ImageJ software. The length of *E. coli* ranges between 0.98 μm – 2.24 μm, while *S. aureus* was between 0.41 μm – 0.73 μm. The non-adhering cell showed that the *E. coli* and *S. aureus*, are rod and spherical shape, respectively. The surface hydrophobicity for *E. coli* and *S. aureus* were determined using different types of hydrocarbon which were hexane, hexadecane and xylene. Cells are classified as highly hydrophobic when the hydrophobicity is greater than 70%, moderate at 50% - 70% and subsequently hydrophilic when it is lower than 50% [1]. Referring to Table 1, *E. coli* has hydrophilic surface as the percentage of hydrophobicity for all the hydrocarbons is varied between 4.42 % - 23.43 %. In contrast, *S. aureus* shows high affinity towards hydrocarbon because of its high hydrophobic surface (>80%) due to the presence of the protein constituents (e.g., fibrinogen, fibronectin) at the cell surface (peptidoglycan) [10].

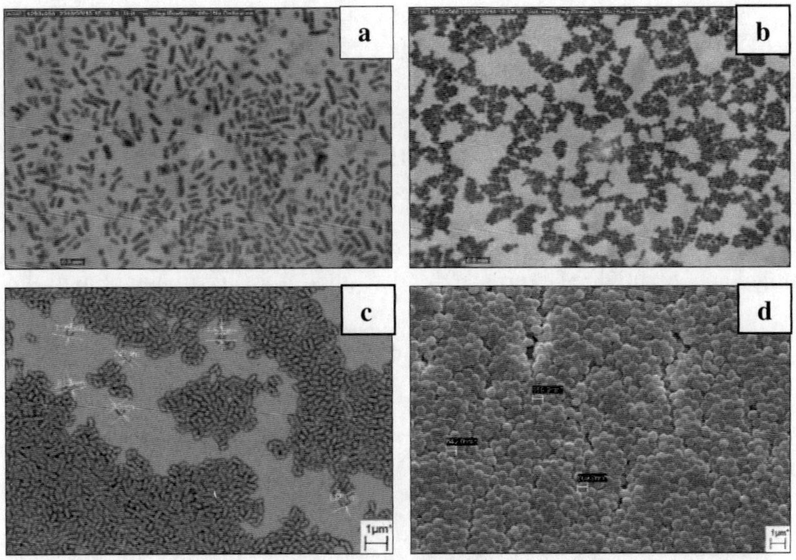

Figure 1. a) Gram stained of *E. coli* ATCC 8739 (red colour), b) gram stained of *S. aureus* (purple/violet colour) viewed under light microscope with 100x magnification, c) shape of *E. coli* under SEM at 4000x magnification, d) shape of *S. aureus* from SEM images at 4000x magnification.

Table 1. Average of bacterial cell hydrophobicity at exponential phase (12th hour) determined with hexane, hexadecane and xylene.

Hydrocarbon	**% Hydrophobicity**	
	***Escherichia coli* ATCC 8739**	***Staphylococcus aureus* ATCC 6838**
Hexane	4.42 ± 0.86	92.57 ± 1.20
Hexadecane	11.90 ± 4.75	75.18 ± 7.41
Xylene	23.43 ± 0.87	88.78 ± 1.51

3.2. Assessment of bacterial adhesion

Stainless steel AISI 316L and Grade 5 titanium alloy have been treated by ultrafast laser radiation at different average laser power (0.04 W and 0.11 W). SEM images of the surface textures after the laser treatment are presented in Figure 2. The laser treatment produced a so-called LIPSS texture. The difference between surfaces treated at different average laser power was the additional appearance of irregular nano-sized metallic particles which were produced during the ablation especially for an average laser power of 0.11 W. Various sizes of metal particles were attached on the ripple structures (Figure 2b and 2c). This creates additional submicron structures that might add to increase active surface area and roughness to the original surfaces. All metals that have been laser processed were used in the adhesion test and studied under fluorescence microscope (Figure 3) for calculating the number of bacterial adhered per cm^2. Prior to the viewing process, the samples were stained by SYTO9 dye which diffuses into the cells and fluoresces upon binding nucleic acids in green fluorescence [2].

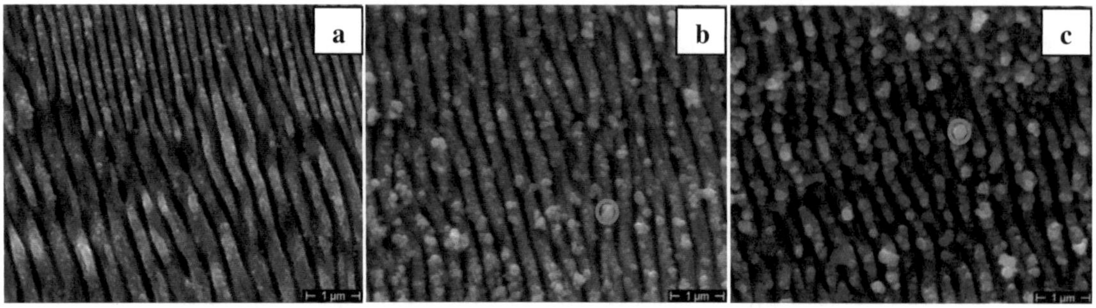

Figure 2. SEM images of the surface texture with 5000x magnification, a) SS-0.04 b) SS-0.11 c) T-0.11.The circled area highlighted nano-sized of irregular grains formed on top of the LIPSS.

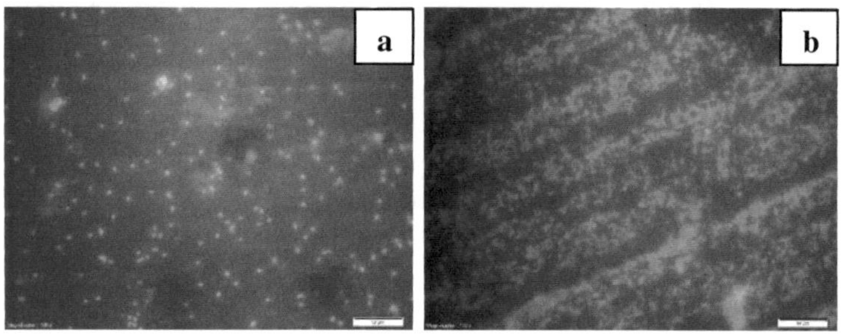

Figure 3. Fluorescence image of *E. coli* and *S. aureus* after 16 hours adhesion at 100x magnification, a) *E. coli* on T-0.11 and b) *S. aureus* on SS-0.11.

The SS-0.04 and SS-0.11 were subjected to bacterial adhesion for 16 hours. The results in Figure 4a indicate that the adhesion of *S. aureus* was doubled from 396 x $10^5/cm^2$ to 703 x $10^5/cm^2$ for surfaces which were laser structured with higher average laser power. In contrary, the adhesion of *E. coli* was decreased from 20 x $10^5/cm^2$ down to 7 x $10^5/cm^2$, contributed to 65% reduction of adhesion on surface structured with high average power. Stainless steel is a hydrophobic surface [11] which attracts *S. aureus* and repel *E. coli* due to hydrophobic and hydrophilic interaction. Hydrophobic bacteria (*S. aureus*) showed higher affinity towards hydrophobic surface, while the hydrophilic bacteria (*E. coli*) will repel from hydrophobic surface [4]. Besides that, the surface with additional nano-sized particles (Figure 2b and 2c) will increase the surface roughness [12] and presumably had contributed to increase the hydrophobicity which can attract the hydrophobic bacteria. Thus, increased the adhesion rate of *S.aureus* to stainless steel 10-100 times higher than *E.coli* (Figure 4a) which varies accordingly to the effect of laser power. Meanwhile, the different surface texture with and without nano-sized particles, reduced the adhesion of *E. coli* on stainless steel. It is assumed that the irregular metallic particles prevent the attachment of *E. coli* and also limit the contact surface between *E. coli* and the original surface. Therefore, the adhesion of *E. coli* on stainless steel can be reduced by increasing the average power during laser structuring. Referring to the Figure 4b, 43 x $10^5/cm^2$ of *E. coli* adhered on T-0.11 which is 6 times higher than on SS-0.11 (7 x $10^5/cm^2$). It is expected that the degree of titanium hydrophobicity in this study is lower than the stainless-steel hydrophobicity which is supported by Ludecke et al. [13]. Ludecke et al., stated that all titanium surfaces were moderately hydrophilic. Thus, *E. coli* which found to have hydrophilic cell surface in this study was adhered more on the titanium surface. SEM images of *E. coli* and *S. aureus* on the metallic surfaces were shown in Figure 5. Most of the bacteria were attached on top of the submicron-sized particles, and the size of bacteria were much larger compared to the gap between neighbouring nano-ripples, which seem to prevent the adhesion on surfaces without nano-particles. The adhesion of bacteria on the laser structured surfaces was further compared to the polished metals treated with SiC paper (180 and 1500 grit). Figure 6 shows that the laser structured surfaces recorded lower adhesion for both *E. coli* (>80%) and *S. aureus* (>10%) compared to polished surfaces. Herein, the result suggests that ultrafast laser material processing can be used to reduce the bacterial adhesion as the periodicity of the LIPSS structure is smaller than bacterial size (Figure 5), which can inhibit the penetration of bacteria to the original surfaces. Moreover, the presence of the additional nano-particles on the LIPSS pattern, reduced the contact area between the bacteria and the metal surface, thus can either limits or weaken the strength of adhesion [8].

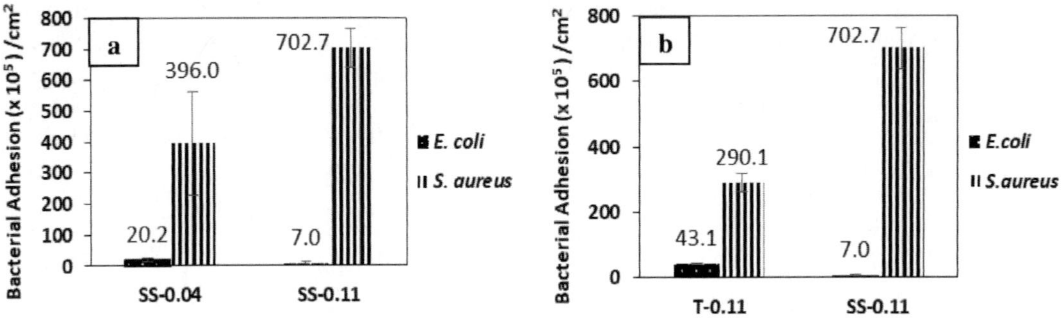

Figure 4. The total of bacterial adhesion $(x10^5)/cm^2$ on a) SS-0.04 and SS-0.11 b) T-0.11 and SS-0.11.

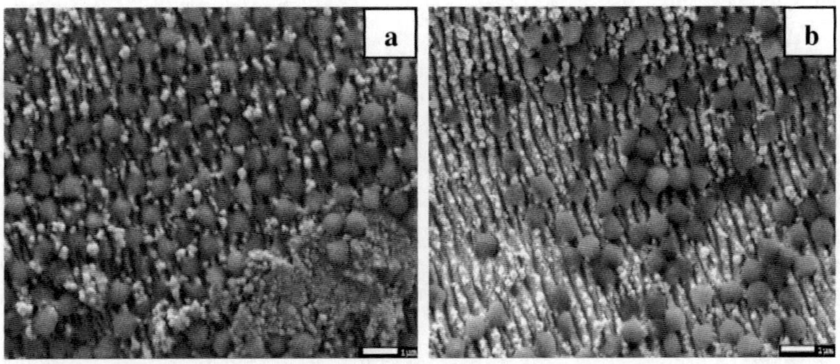

Figure 5. SEM images of bacterial on laser treated surface with 10000x magnification a) *E. coli* on T-0.11 b) *S.aureus* on SS-0.11.

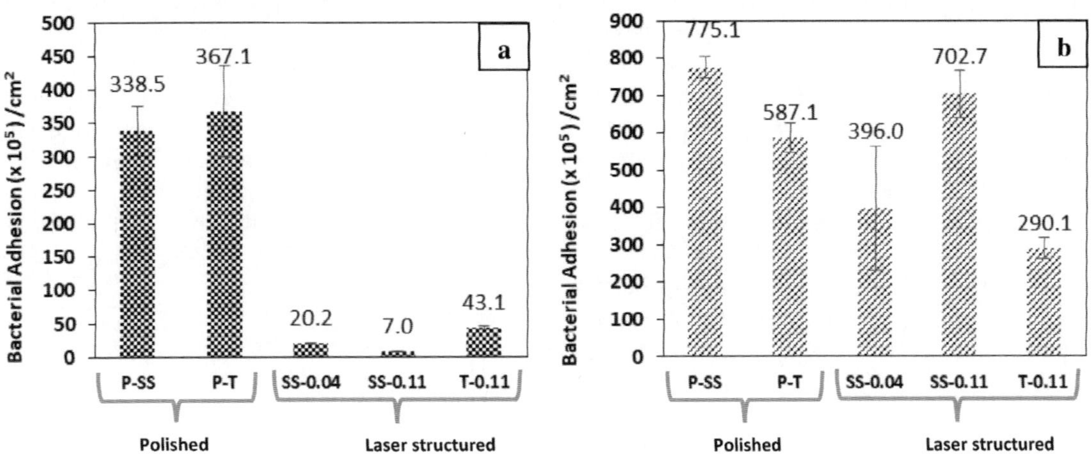

Figure 6. Bacterial adhesion on polished and laser structured surface, a) *E. coli* b) *S.aureus.*

4. Conclusion

Increasing the average laser power from 0.04 W to 0.11 W resulted in two different types of LIPSS structure. The type of LIPSS structure with and without nano-particles has a significant impact on the bacteria adhesion which varies differently for both *E. coli* and *S. aureus*. LIPSS with nano-particles led to an increased adherence of *S. aureus* on AISI 316L stainless steel by 77%, while the adhesion of *E. coli* was reduced by 65%. On the other hand, *E. coli* was observed adhered 6 times higher on Grade 5 titanium alloy compare on AISI 316L stainless steel. Presumably the different of surface hydrophobicity with the existence of laser generated nano-particles on the metal surfaces (at higher average laser power) was the governing factor that affected the rate of adhesion for both *E. coli* and *S. aureus* on these surfaces. Overall findings showed that, all the metals that undergone laser treatment had lower bacteria compared to polished metallic surfaces where *E. coli* showed >80% of reduction and *S. aureus* >10% for both stainless steel and titanium.

Acknowledgements

This project has received funding from the European Union's Horizon 2020 research and innovation program under the Marie Sklodowska-Curie grant (MSCA-RISE) agreement number 644971. In addition, the support for laser materials processing by the Karlsruhe Nano Micro Facility, a Helmholtz research infrastructure at KIT, is gratefully acknowledged.

References

[1] Bohinc K, Drazic G, Abram A, Jevsnik M, Jersek B, Nipic D, Kurincic M and Raspor P 2016 Metal surface characteristics dictate bacterial adhesion capacity *International Journal of Adhesion and Adhesives* **68** 39-46

[2] Sheng X, Ting Y P, Yen and Pehkonen S O 2007 Force measurements of bacterial adhesion on metals using a cell probe atomic force microscope *Journal of Colloid Interface Science* **310** 661-9

[3] Kokare C R, Chakraborty S, Khopade A N and Mahadik K R 2008 Biofilm: Importance and applications *Indian Journal of Biotechnology* **8** 159-68

[4] Katsikogianni M and Missirlis Y F 2004 Concise review of mechanisms of bacterial adhesion to biomaterials and of techniques used in estimating bacteria interactions *European Cells and Material* **8** 37-57

[5] M Al-Abbas F, Spear J R, Kakpovbia A, M Balhareth N, Olson D L and Mishra B 2012 Bacterial attachment to metal substrate and its effect on microbiologically-influenced corrosion in transporting hydrocarbon *Journal of Pipeline Engineering* **1** 63-72

[6] Giorgi C D, Furlan V, Demir A G, Tallarita E, Candiani G and Previtali B 2016 Laser micro-polishing of stainless steel for antibacterial surface applications *Procedia CIRP.* **49** 88-93.

[7] Zheng Y, An Z, Smyrek P, Seifert H J, Kunze T, Lang V, Lasagni A-F and Pfleging W 2016 Direct laser interference patterning and ultrafast laser-induced micro/nano structuring of current collectors for lithium-ion batteries *Proc. SPIE 9736* **97361B** doi:10.1117/1

[8] Cunha A, Elie A M, Plawinski L, Serro A P, Rego A M B D, Almeida A, Urdaci M C, Durrieu M-C and Vilar R 2016 Femtosecond laser surface texturing of titanium as a method to reduce the adhesion of *Staphylococcus aureus* and biofilm formation *Applied Surface Science* **360** 485-93

[9] Rosenberg M, Gutnick D and Rosenberg E 1980 Adherence of bacteria to hydrocarbons: A simple method for measuring cell-surface hydrophobicity *FEMS Microbiol Lett.* **9** 29-33

[10] Mamo W 1989 Physical and biochemical surface properties of gram-positive bacteria in relation to adhesion to bovine mammary cells and tissues *Rev. Sci. Tech.* **8** 163-76

[11] Faille C, Fontaine F and Benezech T 2001 Potential occurrence of adhering living *Bacillus* spores in milk product processing lines *J. Appl Microbiol* **90** 892-900.

[12] Tang M, Shim V, Pan Z Y, Choo Y S and Hong M H 2011 Laser ablation of metal substrates for super-hydrophobic effect *Journal of Laser Micro/Nanoengineering* **6** 6-9

[13] Ludecke C, Roth M, Yu W, Horn U, Bossert J and Jandt K D 2016 Nanorough titanium surfaces reduce adhesion of *Escherichia coli* and *Staphylococcus aureus* via nano adhesion points *Colloids and Surfaces B: Biointerfaces* **145** 617-25.

IOP Conf. Series: Materials Science and Engineering **358** (2018) 012035 doi:10.1088/1757-899X/358/1/012035

Characterization and Storage Stability Study of Bixin Extracted from *Bixa orellana* Using Organic Solvent

N N Husa[1], F Hamzah[2*] and H M Said[3]

[1,2]Biocatalysis and Biobased Material Research Group, CoRe of Green Technology & Sustainable Development, Faculty of Chemical Engineering, Universiti Teknologi MARA, MALAYSIA
[3]Kelantan Biotech Corporation Sdn Bhd, Lot 3174, Batu 23, Jalan Kota Bharu-Machang, Kampung Wek, 18500 Machang, Kelantan, MALAYSIA

fazlena@salam.uitm.edu.my

Abstract. Colorant is one of the additives to give a better appearance or improve the colour of the product. Synthetic colorant has been widely used in industrial due to its readiness in the market and its colour stability. However the arising issues related to the safety, nutrition and also the therapeutic effect has encouraged users to become more concern about the colouring component. Thus, the present research was conducted in order to produce natural colorant called bixin from *Bixa orellana* or also known as annatto seeds. The study was focusing on the effect of the solvent on the characteristic and concentration of the extracted bixin. While stability of the bixin during different storage condition was determined to further validate the effectiveness of the solvent. The result indicated that methanol and acetone gave darker of the bixin colour as compared to the water. However, the deterioration rate of the bixin in acetone and methanol were faster as compared to the water. The extracted bixin was analysed using Fourier-transform infrared spectroscopy (FTIR) and spectrophotometer analysis. The strong band for the bixin was observed at absorbance range of 1704 – 1740cm^{-1}. The study indicates that the concentration of the extracted bixin was the highest in methanol which gave 817.7 ppm of bixin. Meanwhile, bixin concentration in acetone and water was 602.9 ppm and 477.19 ppm respectively.

1. Introduction

Colorants or colour additive are the terms for all soluble or solubilized colouring agents (dyes or pigments) as well as insoluble pigments, employed to impart colour to a material. Colour has been widely used in pharmaceutical, food and cosmetic products because it work as sensorial characteristic that influences the expectations of the customer and also influences quality-related decisions during visual inspection[1]. Colouring agents can be classified by their origin as natural, synthetic or inorganic colorants. Synthetic organic dyes has been recognized for many years as the most reliable and economical colouring agents because they are superior to natural pigments in tinctorial power, consistence of strength, range and brilliance of shade, hue, stability, ease of application and cost effectiveness. However, the use of synthetic dyes is gradually receding due to an increased environmental awareness and to potential harmful effects of either toxic degraded products or their non-biodegradable nature [2]. Furthermore, the safety concern regarding to the synthetic colouring has been addressed due to the present of a carcinogenic component, high level toxicity and allergic reaction to the end user in some of the synthetic colorant [3]. Therefore, there is an increased interest

Content from this work may be used under the terms of the Creative Commons Attribution 3.0 licence. Any further distribution of this work must maintain attribution to the author(s) and the title of the work, journal citation and DOI.
Published under licence by IOP Publishing Ltd

in substituting the usage of synthetic natural to natural colouring agent as a consequence of perceived consumer preferences as well as legislative actions. There are many sources that can be utilize to produce the colorant or colour additives, such as herbs, flowers, roots and even beans [4]. Flavonoids, carotenoids and chlorophyll are the major contributors to the natural colours of the most plant. One of the less exploring plant and have higher colouring potential in Malaysia is *Bixa orellano* (*B. orellano*) or annatto. *B. orellano* is a native plant in Brazil and also grown in some tropical countries such as Peru, Mexico, Kenya, Asia and East Africa [5]. Bixin which is extracted from the outer layer seed of *B.orellana* (Figure 1a) is group of carotenoid pigments which give higher high range of colour-comprising of red, orange and yellow hues [6]. It is lack of toxicity and has high tinctorial value. Bixin consists of 25 carbons with the chemical formula $C_{25}H_{30}O_4$ with molecular weight of 394.51. Carboxylic acid and methyl ester are attached at end of the chain and the chemical structure is as shown in Figure 1(b). According to Santos et al. [7], several carotenoids, including bixin and norbixin, can be obtained from annatto seeds, depending on the extraction method used. Extraction of the bixin can be conducted by chemical using indirect extraction with solvents, direct extraction using aqueous alkali solutions and direct extraction using oil [8] or mechanical using ball milling or spouted bed [9] method. However, natural pigments generally are unstable and participate in different reactions. It is reported that the produced colour is strongly depend on the processing and also it storage condition. The chemical, toxicological, and antioxidant properties, degradations and extraction of bixin and norbixin have been extensively studied but for stability of the bixin in the powder form only a few studies can be found in the literatures. Thus in the presence study, storage capability of the powder bixin extracted using different solvent was studied. The study aims to determine the effect of different storage condition towards colour degradation of the bixin.

(a)

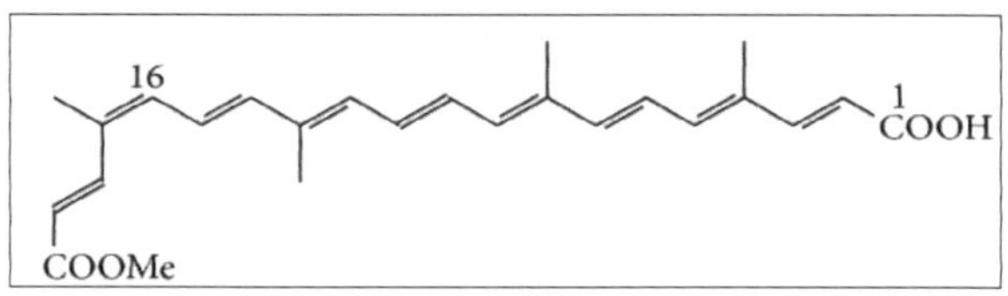

(b)

Figure 1. a) Annatto seeds in the pod b) Chemical structure of Bixin

2. Materials and Methods

2.1 *Chemicals and Reagents*

The main raw material which is *B.orellana* or annato was taken from Jeram Linang, Pasir Puteh, Kelantan with coordinate of 5.798517, 102.342852. Methanol and acetone are analytical grade reagents with > 99% purity purchased from Merck. The sodium hydroxide was purchased from R&M Chemicals, while the bixin standard was purchased from Spectrum Chemical.

2.2 *Solvent Extraction of B.orellana*

The raw material which is annatto seeds was removed from the pod. 2 g of the annatto seeds was weighed and mixed with 20 ml of solvent. The mixture of annatto seeds and the solvent was stirred using digital overhead stirrer (OS-20 Tuff) at 500 rpm until the colour was completely extracted from the seeds. The time taken for the stirring process was recorded and the same procedures were repeated for another type of solvent. The extracted colorant was filtered using Whatman paper grade 595 with the pore size ranging from 4 - 7 μm.

2.3 Characterization of Bixin

Colour analysis of the bixin was conducted using UV-Vis spectrophotometer (Perkin Lambda 750 UV/VIS Spectrophotometer) and the value was compared with the absorbance from bixin standard. Bixin standard solution was prepared by dilution of 0.05 g of standard bixin in 50 ml of solvent to produce a 1000 ppm stock solution. Then, further dilution was conducted to produce 750 ppm, 500 ppm and 250 ppm of bixin solution. The absorbance of the solution was analysed using UV-Vis spectrophotometer (Perkin Lambda 750 UV/VIS Spectrophotometer) at 540 nm to generate the standard curve. The bixin extracted from *B.orellana* was analysed using similar procedure for absorbance value. While, Fourier Transform Infrared Spectrophotometer (FTIR)- Jasco FT/IR-460 Plus in the range of 4000 to 400 cm^{-1}analysis was used to analyse the chemical functioning group presence in the extracted bixin.

2.4 Colour Stability Test

Colour stability test for bixin was conducted at two conditions which are the storage condition influencer and sunlight influencer test. 0.1 g of annatto seeds was mixed with 10 ml solvent and extracted using digital overhead stirrer at 500 rpm for 5 minutes. 5 ml of the sample was used for 4 different tests for storage condition A, B, C and sunlight influencer as tabulated in Table 1. The absorbance reading for these samples were recorded in every 3 days for 25 days. Meanwhile for the sunlight influencer condition, the sample was exposed to the sunlight for everyday and the absorbance was recorded. All the set of test was repeated triplicate.

Table 1. Colour stability test in different storage and its arrangement

Condition	Type of Vial	Storage temperature
A	**Transparent**	**Room temperature**
B	**Amber**	**Room temperature**
C	**Transparent**	**4°C**

3. Results and Discussion

3.1 Effect of the solvent on the colour and concentration of the bixin

The colour comparison between the extractions of the bixin using different types of solvents is shown in Figure 2. The results indicated that extraction of bixin using methanol as the solvent produce the brightest colorant while the distilled water produces the least bright colorant for the same mass of *B.orellana* used. Both of the methanol and acetone used producing a clear orange colorant while extraction using distilled water producing cloudy-orange colorant. This result show that methanol and acetone were extracted higher bixin amount as compared to the water. This might be due to the dissolution properties of the methanol and acetone that encouraged dilution of colouring material. According to Boeing *et al.*[10] methanol act as a better extraction solvent compared to acetone because it can simultaneously dissolving the cell membranes and the carotenoids. While Silva *et al.* [11] also claim that organic solvent such as acetone and methanol will extract higher pigment concentrations from 3.5 to 5.2% bixin. Extracted bixin was transformed into powder form using freeze and spray drying method and the output is illustrated in Figure 3. The freeze drying process producing a brighter powdered colorant compared to spray drying which produces a pale orange colour. The different temperature used in the powder formation significantly affects the colour of bixin. Fades or deteriorates of the bixin colour was occurred in spray dried process that result in pale of orange bixin powder. This might occur due to the high temperature up to 130°C used during spray dried method. According to the Wang *et al.*[12], temperatures over 80 °C negatively affect anthocyanin stability and results in degradation of the absorbance value. Similar finding was reported by Patras *et al.* [13] who demonstrated that the anthocyanins in blackberry and strawberry puree are significantly affected by thermal treatments at 70 °C for 2 min. Thus the results show that lower operating temperature able to preserve the colour from deteriorates.

Figure 2. Extract bixin using methanol, acetone and water

Figure 3. Powder form of bixin using freeze and spray dried process

The concentration of bixin in methanol, acetone and water is shown in Table 2. The results shows that the highest concentration of bixin was obtained by using methanol with a value of 817.71 ppm followed with acetone at 602.90 ppm of bixin and the least was by using distilled water with 477.19 ppm of bixin. The extraction of the bixin in the solvent is influence by solubility properties of the bixin. As reported by Pimentel and Stringheta, [14], bixin is an oil-soluble pigment and the solubility of bixin is significantly enhanced by the action of the oil present in the seeds. Thus, low concentration of the bixin in the water represents the characteristic of the oil soluble pigment. The result obtained is satisfying with the study by Boeing et al.[10], that demonstrate methanol is the better solvent for the extraction of natural colorant compared to the acetone and water. This is because methanol has the ability to dissolve the cell membrane and pigment as compared to acetone and water. Thus, the presence studies indicate that low polarity of methanol and acetone capable to extract higher concentration of the bixin as compared to water.

Table 2. Concentration prediction result of the extracted colorant

Extraction solvent	Absorbance reading	Concentration of Bixin in sample (ppm)	Polarity index
Distilled water	0.0602	477.19	10.2
Acetone	0.0167	602.90	5.1
Methanol	0.0671	817.71	5.1

3.2 Characterization of Bixin

Figure 4 shows the result for the FTIR analysis of the standard bixin which consist a strong band at 1704 cm^{-1} and a medium band on 970 cm^{-1}. Yusá-Marco *et al.*[15] state that the carotenoids compound specifically bixin band can be determined at 1721 cm^{-1}, 1603 cm^{-1}, 1149 cm^{-1} and 962 cm^{-1}. The effect of solvent on the extraction of bixin based on the functional group present in the solution is reveal in Figure 5. For the extraction with water, there are a very sharp band at 3274 cm^{-1} and 1634 cm^{-1} and the small band of bixin at 1737 cm^{-1} was detected. Besides, the presence of small bands which representing the functional group of C-O-C stretch was appeared at 1365 cm^{-1} and 1217 cm^{-1}. On the other hand, functioning group of the bixin extracted using methanol shown in Figure 6 and the presence of the bixin spectrums were detected at 1737 cm^{-1} and 1020 cm^{-1}. While, the alcohol group (O-H) representing methanol was observed at 3322 cm^{-1}. Based on the spectrum, alkyl C-H functional group was detected at small band of 2943 cm^{-1} and 2832 cm^{-1}. While, the two small bands of C-O-C were appeared at 1217 cm^{-1} and 1114 cm^{-1}. As for the bixin extracted using acetone, the presence of strong bixin band was observed at 1708 cm^{-1} and a weak band was detected at 903 cm^{-1}. FTIR spectrum of the bixin extract using acetone is illustrated in Figure 7. Other functioning group present in bixin extracted using acetone was identified at 1220 cm^{-1} that representing C-O-C group. While, a small band at 3512 cm^{-1} and 3003 cm^{-1} was frequently diagnosed as unsaturation.

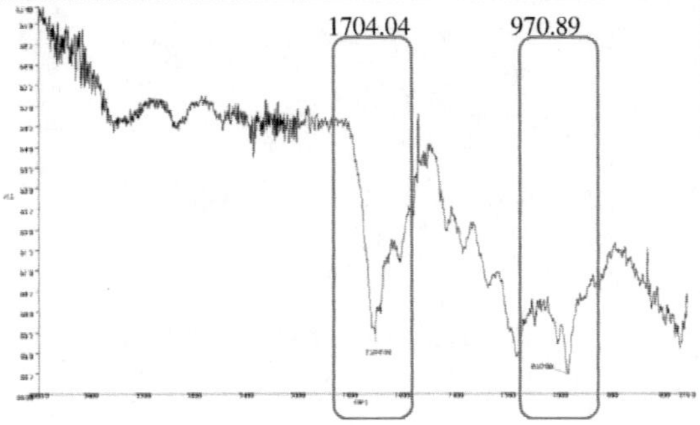

Figure 4. FTIR analysis of the standard bixin

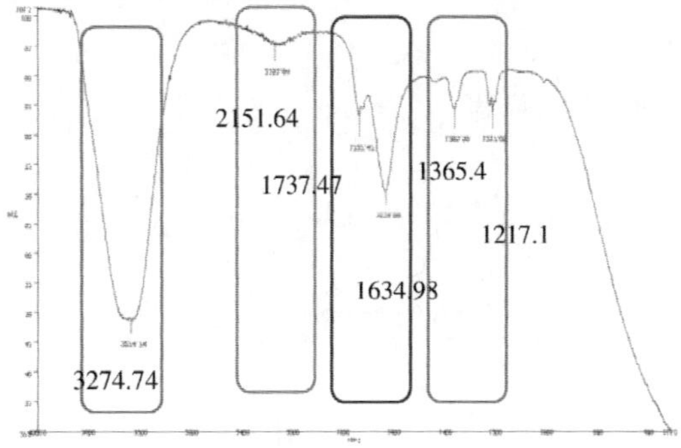

Figure 5. FTIR spectrum of annatto extract using water

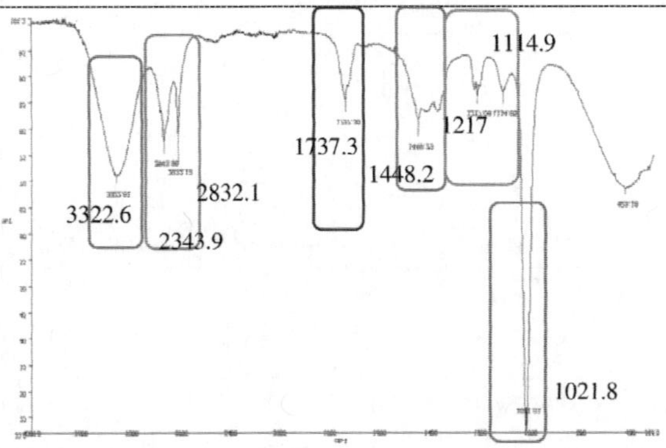

Figure 6. FTIR spectrum of annatto extract using methanol

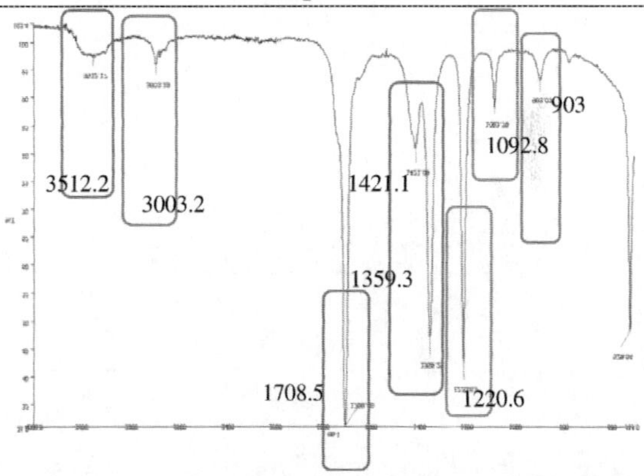

Figure 7. FTIR spectrum of annatto extract using acetone

3.3 Storage Capability of the Extract

Colour stability test for different storage condition of the bixin extract is shown in Figure 8. The deteriorate rate was calculated based on the slope obtained from the results. Figure 8 (a), shows that deteriorate of bixin at condition A. The results show that acetone able to retain the highest bixin colour after certain exposure time as compared to methanol and distilled water. However, deteriorate rate of the bixin in acetone is faster as compared to water and methanol with the value of 0.1697 abs/day. While extract in distilled water shows a more stable colour deteriorate rate based on the better linearity of graph and slope of the graph (0.0574 abs/day). While for condition B, it was observed that deterioration of the bixin show a same deteriorate pattern as storage condition A. However, deterioration rate for each of the sample was lower than condition A. This is due to the characteristic of amber vials which capable to prevent photodegradation from occurring. The deterioration rate obtained was higher for bixin in acetone followed by methanol and distilled water with the values of 0.1375, 0.1132 and 0.0667 abs/day respectively. Similar trend was recorded for bixin deterioration in condition C. However, cooling condition was able to prevent the colour deterioration process of the bixin. Deterioration slope for bixin in each solvent was lower as compared to the condition A. The value obtained was 0.1192 abs/day for bixin in acetone, 0.0713 abs/day for bixin in methanol and 0.06 abs/day for bixin in water. Then for the exposure of the bixin to sunlight studies, a similar deterioration trend was also observed. Water was the most stable solvent to prevent fast deterioration of the bixin colour. The deterioration rate for acetone, methanol and water were 0.2727, 0.1697 and 0.1645 abs/day respectively.

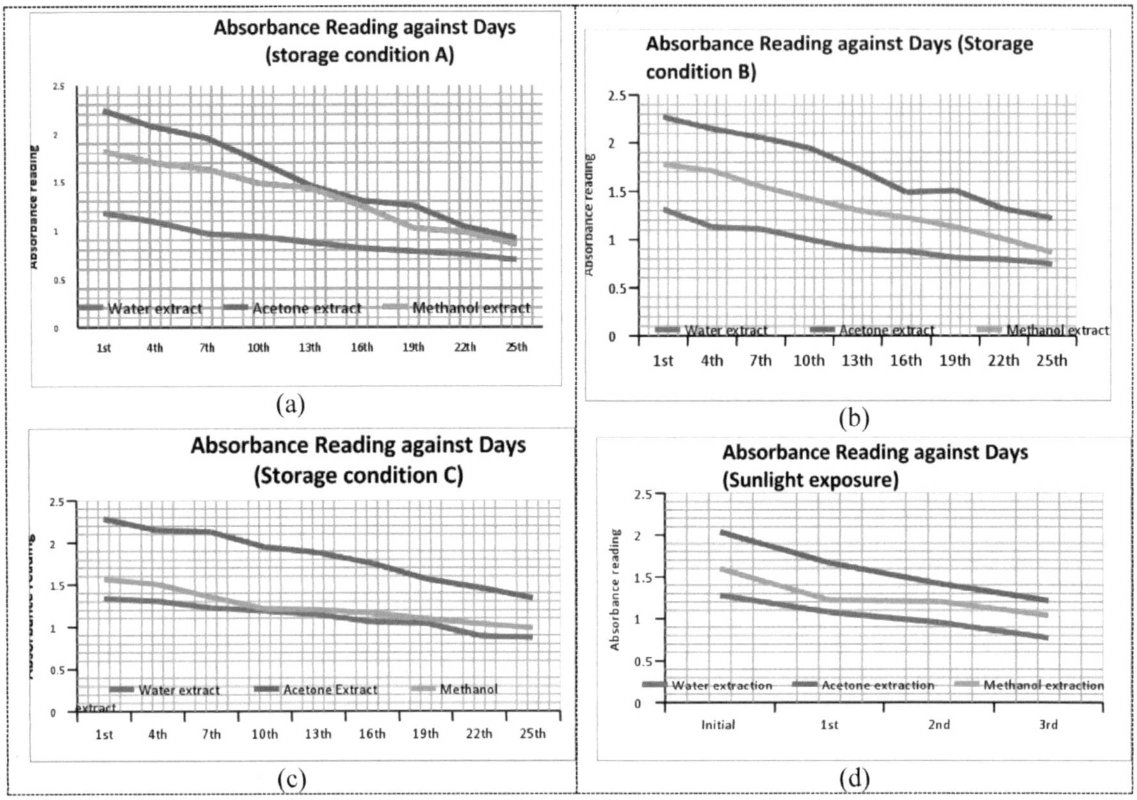

Figure 8. Colour stability in different storage condition

The deterioration rate might be influence by the characteristic of the solvent itself. Acetone have evaporation rate of >3.0 where it molecules can evaporated faster event at room temperature. This is because it molecule structure do not have pure hydrogen bonding that require less energy for it molecule to evaporate to the surrounding. This condition is explaining the highest deterioration rate of bixin. Similar characteristic is also reported for methanol. Methanol also have a boiling point near to room temperature that encourage the methanol molecule evaporate faster to the environment. While water is made up of many hydrogen bonds and higher temperature are required to break up this bond to make water evaporate. Thus at room temperature, slow evaporation will occurred and this phenomena answering the slowest degradation rate of bixin in distilled water.

4. Conclusion

Methanol, acetone and distilled water were a polar solvent that can be used to extract bixin from *B.orellana*. The present of the bixin in all of the solvent was determined at 1704 – 1740 cm^{-1}.The study indicated that methanol is the optimum solvent in bixin extraction which gave up to 817.7 ppm of bixin. On the other hand in storage stability study, bixin in the water retained its colour after 25 days of each of the storage condition. The deterioration rate shows that acetone gave the highest reduction of bixin colour as compared to the methanol and water in every storage conditions. Thus, proper storage and extraction are required in order to retain colour of the extracted bixin.

Acknowledgement

The present study was made possible through funding from Ministry of Sciences, Technology and Innovation (MOSTI) - 600-RMI/FRGS 5/3(148/2014), UiTM Internal grant 600-IRMI/DANA 5/3/REI

(0009/2016) and continuous support from Research Management Institute (RMI), of University Technology MARA is gratefully acknowledged. Authors also like to show gratitude to the facilities and constant encouragement from Faculty of Chemical Engineering, Universiti Teknologi MARA.

References

[1] Mauricio A. Rostagno and Juliana M. Prado 2013 Natural Product Extraction: Principles and Applications *Royal Society of Chemistry* RSC Publisher.

[2] Farah Maria Drumond Chequer, Gisele Augusto Rodrigues de Oliveira, Elisa Raquel Anastácio Ferraz, Juliano Carvalho Cardoso, Maria Valnice Boldrin Zanoni and Danielle Palma de Oliveira 2013 *Textile Dyes: Dyeing Process and Environmental Impact* edited by Melih Günay, ISBN 978-953-51-0892-4, InTechOpen publisher.

[3] Amchova Petra, Kotolova Hana and Ruda-Kucerova Jana 2015 *Health safety issues of synthetic food colorants* Republic Czech : Elsevier Inc p 1-9.

[4] Mixon Kerry 2014 Understanding Colorant in *The Handcrafted Soap Maker, The Journal of the Handcrafted Soap Makers Guild.* Spring Valley, California : [s.n.]

[5] Daniela de Araújo Vilar, Marina Suênia de Araujo Vilar, Túlio Flávio Accioly de Lima e Moura, Fernanda Nervo Raffin, Márcia Rosa de Oliveira, Camilo Flamarion de Oliveira Franco, Petrônio Filgueiras de Athayde-Filho, Margareth de Fátima Formiga Melo Diniz and José Maria Barbosa-Filho 2014 Traditional Uses, Chemical Constituents, and Biological Activities of Bixa orellana L.: A Review *The Scientific World Journal* **2014** p 1-11

[6] Giridhar P, Venugopalan A and Parimalan R 2014 A review on annatto dye extraction, analysis and processing- a food technology perspective *J Sci Res Reports*, **3** p 327-348

[7] Letícia F. Santos, Vanessa M. Dias, Viviane Pilla, Acacio A. Andrade, Leandro P. Alves, Egberto Munin, Viviane S. Monteiro and Sergio C. Zilio 2014Spectroscopic and photothermal characterization of annatto: Applications in functional foods *Dyes and Pigments* **110** p 72-79

[8] Kang EJ, Campbell RE, Bastian E and Drake MA 2010Invited review: annatto usage and bleaching in dairy foods *J Dairy Sci.* **93** p 3891-3901

[9] Barrozo MAS, Santos KG and Cunha FG 2013 Mechanical extraction of natural dye extract from *Bixa orellana* seeds in spouted bed In*dustrial Crops and Products* **45** p 279-282

[10] Boeing JS, Barizão EO, E Silva BC, Montanher PF, de Cinque Almeida V, and Visentainer JV 2014 Evaluation of solvent effect on the extraction of phenolic compounds and antioxidant capacities from the berries: application of principal component analysis. *Chemistry Central Journal* **48** p---

[11] Silva GF, Felix MC Gamarra, Oliveira AL and Cabral FA 2008 Extraction of bixin from annatto seeds using supercritical carbon dioxide *Brazilian Journal of Chemical Engineering* **25** (2) p 419 - 426

[12] Zhong Feng Wang, Min Zhang and Qianqian Wu 2015 Effects of temperature, pH, and sunlight exposure on the color stability of strawberry juice during processing and storage *LWT - Food Science and Technology* **60** p 1174-1178

[13] A. Patras NP, Brunton TR, Gormely F and Butler 2009 Impact of high pressure processing on antioxidant activity, ascorbic acid, anthocyanins and instrumental colour of blackberry and strawberry puree Innovative *Food Science and Emerging Technologies* **10** (3) pp. 308-313

[14] Pimentel FA and Stringheta PC, 1999 Production of powder corants of annatto through acid precipitation from extracts obtained in different extracting solutions *Brazilian Journal of Natural Dyes, Viçosa,* **3**, p. 53-57

[15] Dolores Julia Yusá-Marco, María Teresa Doménech-Carbó, Ivan Lucio Vaccarella, Antonio Fernando Batista dos Santos, Sofia Vicente-Palomino and Laura Fuster-López 2008 Characterization of colouring compunds in annato (Bixa Orellan L.) used in historic textiles by means of UV-Vis spectrophotometry and FTIR spectroscopy *Arché. Publicación Del Instituto Universitario De Restauración Del Patrimonio De La Upv* **3** p 153 – 158

IOP Conf. Series: Materials Science and Engineering **358** (2018) 012036 doi:10.1088/1757-899X/358/1/012036

Potential anti-Dengue Concoction of Carica Papaya (*C. Papaya*) Leaf and G. Mangostana (*G. Mangostana*) Pericarp and Their Bioactivity Enhancement by Fermentation: A Review

M S So'aib, J Salihon and H L Tan

Faculty of Chemical Engineering, Universiti Teknologi MARA, 40450 Shah Alam, Malaysia

sufian5129@salam.uitm.edu.my

Abstract. This review highlights the anti-dengue potency of *Carica Papaya* leaf (*CPL*) extract which was associated with platelet increase and other medicinal properties such as antiinflammatory and antioxidant. *Garcinia Mangostana*'s pericarp (GMP) extract have much commonalities with *CPL*, in addition to antiviral and immunomodulatory properties of the former. These properties may exhibit, yet unproven, analgesic, hemorrhage prevention and antiviral effects that may facilitate dengue recovery. Nevertheless, the limited bioavailability of native polyphenolic contents of both, as hinted by studies on colonic microbiome metabolism on dietary polyphenols, highlighted fermentation as viable method to enhance the functionality of the compounds. Thus, this review also highlights some relevant parameters in existing fermentation of well known fermented foods that impact their bioactivity, functionality and palatability that may applicable for the development of *CPL* and *GMP* fermentations.

1. Introduction

The dengue disease is affecting an estimated of 50-100 million people worldwide each year [1] mostly in tropical and sub-tropical regions such as Africa, America, Eastern Meditteranean, Southeast Asia and Western Pacific [2]. It is caused by dengue virus (DENV) which is transmitted by *Aedes aegypti* mosquito as its principal vector. Dengue epidemiology Malaysia is a serious problem with more than 50,000 cases resulting 122 deaths as of June 2016 alone [3]. Currently, there is no effective tetravalent antiviral drug that could inhibit antigenically distinct and constantly mutating four DENV serotypes (DENV-1, DENV-2, DENV-3 and DENV-4) [4]. The current dengue management relies on supportive care during hospitalization such as administration of analgesics, rehydration by oral or intravenous fluid intake and constant platelet and hematocrit monitoring [5] as well as preventive measures to control vector through elimination of mosquito's breeding habitat, fumigative spray and public awareness campaign [6].

Wide ranging indigenous plant species used in folk medicines had been recognized by their respective local population for anti dengue potency. Scientific studies revealed their roles in inhibiting viral mechanism, increasing platelet count and alleviating dengue symptoms such as inflammation, fever and bleeding. Sometimes, two plants are used in combination to alleviate dengue symptoms and contain viral progression [2].

Content from this work may be used under the terms of the Creative Commons Attribution 3.0 licence. Any further distribution of this work must maintain attribution to the author(s) and the title of the work, journal citation and DOI.
Published under licence by IOP Publishing Ltd

IOP Conf. Series: Materials Science and Engineering **358** (2018) 012036 doi:10.1088/1757-899X/358/1/012036

2. Anti dengue potency of *CPL* and *GMP*

2.1. CPL
Numerous animal and clinical studies culminate the efficacy of *CPL* in alleviating the dengue symptoms and facilitating the dengue patient's recovery. It was suggested that some medicinal plants, including *CPL* possess bioactive compounds such as *flavonoids*, *quercetin* and natural *chalcone* [1, 2] that inhibit DENV proteases responsible for virus-host attachment as demonstrated by strong affinity between *CPL*'s flavonoid quercetin and NS2B-NS3 proteases of DENV-2 serotype [7]. Clinical studies on serological parameters reported the increase in platelet count, white blood cell and neutrophil as a result of *CPL* juice intake [1, 4]. A thorough study on platelet producing genes, *ALOX12* and *PTAFR* genes confirmed the role of *CPL* in enhancing the expression of these genes among interventional group of 111 dengue patients as compared to control group of 117 dengue patients [5]. Further findings on the increase of platelet among clinical subjects whom administered with *CPL* extract as compared to control provided indisputable evidences of anti-dengue potency of *CPL* with respect to platelet increase to overcome thrombocytopenia [8, 9].

Other well-known medicinal properties of *CPL* such as antiinflammatory [10], antioxidant and immunomodulatory [11] that are attributed to its phytochemical compounds such as alkaloids, saponins, flavanoids, tannins, phenolics and steroids [12] may impart additional anti-dengue activity. Analogous to *Hippophae rhamnoides*, yet requires further proof [2], *CPL*'s antiinflammatory property may reduce TNF-α which responsible for inflammation of T-cell which is the suspected mechanism of vascular permeability and hemorrhagic manifestation [1], whereas its immunomodulatory activity could induce IFN-γ to mount antiviral mechanism comparative to commercial *Rivabirin* antiviral drug [10]. Studies found that normal intake of CPL extract poses no health risk as no abnormality observed on hematological and histopathological results of animal subjects which were administered with *CPL* fourteen times higher than normal dose intake [13, 14] despite the recent debate on the toxicity of *CPL* pertaining to its cyanoglycoside content.

2.2. GMP
The bioactivity specific compounds of *GMP*, most prominently its xanthones such as α-mangostin, β-mangostin and γ-mangostin are better elucidated compared to its *CPL* counterpart. The wide range of *GMP*'s medicinal values includes antiinflammatory [15], antioxidant [16], antiviral [17] and anticancer [18]. Its anti-inflammatory property may exhibit its anti-dengue activity in terms of preventing hemorrhage analogous to its *CPL* counterpart suggested earlier. Antiinflammatory activity may also create analgesic effect to alleviate muscular and joint pain since both activities involve inhibition of cyclooxygenases (COX) and its prostaglandins mediator as well nitric oxide (NO) and superoxide (•OH and $\bullet O^{2-}$) release which responsible for inflammatory and pain mechanisms in many cells including T cell [15, 19]. Additionally, antioxidant activity of *GMP* could act as superoxide scavenger to suppress the inflammation from progressing further [20]. The anti-dengue activity of GMP which may plausibly extended from its γ-mangostin antiviral activity against HIV-1 virus and its well known immunomodulatory activity [18]. The existing commercial antiviral drug such as *ribavirin* and *PegIFN* employ the same protease inhibition mechanism against HIV-1 and HCV (Hepatitis C virus). Since HCV shares strong functionality with its DENV counterpart, the potential of γ-mangostin to inhibit DENV protease similar to HIV-1 protease is intuitive [17, 21].

3. Bioavailability of *CPL* and *GMP*

Unlike its *GMP* counterpart, the bioavailability of *CPL* extract is never debated presumably due to its existing satisfactory effect. However, since polyphenols such as flavonoids and sapponin in *CPL* [11] and xanthones, gartanin, garcinone E in *GMP* [22] are major bioactive compounds of both, studies on dietary polyphenols in other plants and foods implied low bioavailability of these polyphenols compounds. Low bioavailability of around 5-10% from original intake is highlighted by the dependence of enterohepatic circulation on colonic microbiome for polyphenol breakdown [23] thus resulted lower therapeutic effects [24]. Regarding the *GMP*, low presence of total xanthones (around 2%) in serum and urine samples of volunteers who consumed xanthone-rich juice was reported while better bioavailability was achieved when xanthone was consumed with high fat western diet or ingested in the form of oil suspension gavage. Conjugated form of xanthones, as opposed to free xathone found in most serum and urine samples deliniated time consuming process of phase II metabolism by the liver to facilitate the xanthone absorption into systemic circulation [25, 26].

4. Fermentation of *CPL* and *GMP*

4.1. Enhancement of bioavailability

A few studies highlighted the use of fermentation technique to enhance bioactivity of mangosteen xanthones. Incubation of α-mangostin with fungi *Colletotrichum gloeosporioides* (EYL131) and *Neosartorya spathulata* (EYR042) resulted several phase II metabolites such as *mangostin 3-sulfate, mangostanin 6-sulfate, 17,18-dihydroxymangostanin 6-sulfate* and *isomangostanin 3-sulfate* which probably became more readily-absorbed form at the enterocytes and hepatocytes for systemic distribution [27]. The multi-fold enhancement of α and β-mangostin bioavailability by 11 to 16 times as compared to standard dry mangosteen extract was reported from the fermentation of whole mangosteen fruit using *Saccharomyces boulardii* as a starter culture [28]. Although the fermentation of *CPL* is non-existence, studies on the metabolites of various classes of polyphenol compounds, of which most *CPL* phytochemicals belong to, as a result of colonic microbiota metabolism may shed light on possible bioavailability enhancement if fermentation of *CPL* is carried out [29]. In addition, lactic acid fermentation of *Myrus communis* berries using *L. plantarum* and yeast extract was reportedly enhanced the antioxidant activity of the fruit with respect to DPPH scavenging activity, inhibition of linoleic acid peroxidation and increase of phenolic acid and flavonols [30].

4.2. Industrial fermentation of CPL and GMP

Three-stage fermentation of mangosteen pericarp was described in the production of a local nutraceutical product commencing with three weeks fermentation of rice by *Aspergillus oryzae (A. oryzae)* to produce rice brew, followed by six weeks fermentation of mangosteen pericarp soaked in rice brew and final fermentation stage that matured in nine months [31]. The use of *koji* fermentation is analogous to typical soy sauce production where it serves to provide amylase and protease enzymes for the breakdown of complex saccharide into fermentable sugars and protein respectively as precursor for subsequent lactic acid fermentation [32]. Perhaps for similar reason, yeast extract was used during *Myrus communis* berries fermentation with *L. plantarum* mentioned earlier.

Another related process employed a three-month fermentation of whole mangosteen fruits using *Saccharomyces boulardii (S. boulardi)* as starter culture at 35-55°C. The selection of *S. boulardi* as starter, which is originally the indigenous microorganism that colonises the mangosteen fruit may related to its role in hydrolysing lignin structure into simpler fermentable precursor for subsequent fermentation succession, most probably lactic acid fermentation, and also comparable to the role of *koji* fermentation. As stated earlier, the fermented mangosteen product which was co-administered with standard *GMP* extract exhibited 11 to 16 times higher bioavailability compared to standard dry mangosteen extract products [28]. We surmise the viability of CPL fermentation if appropriate selection of starter culture and saccharide source are made available.

4.3. Fermentation process parameters

Fermentation technique used for *GMP* is non-sterile process akin to other low-tech fermentation method used in many traditional foods such as *kimchi* (Korea) [33], *tempoyak* (Malaysia and Indonesia) [34], *tempe* (Malaysia and Indonesia), *tape* (Malaysia and Indonesia) *doklu* (Cote d'Ivoire) [35], *kishk* (Egypt) [36] and *saeurkraut* (international) [37]. Most of these products were originally produced by spontaneous fermentation using respective indigenous microorganisms. However, technical control parameters evolved through trial and error and iterated use of residue from successful fermentation batch as inoculum for subsequent fermentation batch known as back-slopping, yielded the best adapted strains to be used as starter culture for a better controllability of the fermentation.

The microfloral profile during fermentation, which only came to light in recent time using modern microbiological technique, culminate the prevalence of lactic acid bacteria (LAB) species, shown in table 1 as important parameters of a successful fermentation and believed to impact the palatability, nutritional value and shelf stability of many traditional fermented food products [37].

Table 1. Lactic acid bacteria (LAB) species in traditional fermented food.

Food product	Main raw material	Fermentation process	aLAB species	Ref.
Tempoyak	Durian (*Durio zabethinus*) pulp	Lactic fermentation	*Lb. plantarum, Lactobacillus sp, Weisella paramesenteroides, P. acidilactici, Lb. fructivorans, L. dextranicum, Lb. collinoides, Lb. paracasei, Lb. Plantarum, F. durionis*	[38], [39]
Tempe	Soybean	Mold fermentation	*E. faecium, L. mesenteroides ssp. Mesenteroides, Lb. delbrueckii ssp. delbrueckii.*	[40]
Tapai or tape	Cassava or glutinous rice	Alcoholic fermentation	*P. pentosaceus, Weisella sp.*	[34]
Soy sauce	Soybean	Mold fermentation (*koji*) followed by high salt (brine) concentration	*Weissella confuse, E. faecium*	[41]

aAbbreviation: Lb (Lactobacillus), P (Pediococcus), L (Leuconostoc), E (Enterococcus), F (Fructobacillus).

Raw material preparation and selection of ingredients are also critical to promote growth of desirable microorganism and protect the fermentation starter culture from spoilage microorganisms [34]. For example, production of *tempe* involved soaking, de-hulling and cooking of the raw material (soybean) followed by acidic fermentation that inhibits spoilage growth but tolerant to *Rhizopus spp.* mold. Later, *Rhizopus spp.* colonisation of the substrate outgrows other potential undesirable microorganisms and releasing inhibitive substances against spoilage microorganism. Salt addition during *kimchi* and *tempoyak* fermentation or brine fermentation during soy sauce fermentation promoted LAB growth over other spoilage microorganisms [37, 41]. Another process parameter that impact the organoleptic and functionality of fermented foods is physico-chemical succession that results from a predominant microflora that colonise the available food matrix at a particular stage of fermentation. For example, the heterofermentation of *L. fermentum* on glucose and fructose of cocoa bean's pulp during spontaneous fermentation of cocoa bean produced lactic acid and mannitol, the latter was utilised by acetic acid bacteria to produce acetic acid that is essential to modulate optimum pH for endogenous enzyme activity that gives a good sensory quality of chocolate [42].

Recent study on *tempoyak,* correlated the drop in sugar level (sucrose, fructose and glucose) and the rise of organic acids (lactic acid, acetic acid and propionic acid) with the increase in LAB population (mainly *F. durionis* and *Lb. Plantarum)* [39] which indicated substrate utilization and metabolite production respectively. In addition, yeast and bacteria are also known to produce volatile secondary metabolites such as higher alcohols, fatty acids, esters, aldehydes, ketones and thiols by utilizing the primary metabolite (organic acids and ethanol) which were pivotal for the flavor and odour of chocolate [43], soy sauce [41] and tape [36].In this section we discuss how to format the title, authors and affiliations. Please follow these instructions as carefully as possible so all articles within a conference have the same style to the title page. This paragraph follows a section title so it should not be indented.

5. Conclusion

The diverse classes of phytochemicals in *CPL* and *GMP* exhibit potential bioactivities against dengue infection and facilitate patient's recovery. Existing fermentation of GMP and other functional foods suggest that fermentation could surmount the limited bioavailability of polyphenols thus enhancing the functionality and palatability of native *CPL* and *GMP*. Thus, the development of the potential fermentation process of *CPL* and *GMP* should adapt certain artisanal elements of traditional fermented foods in terms of raw material preparation, starter culture and process condition as their process parameters to attain desired functionality and organoleptic attributes.

Acknowledgments
Authors wishing to thank University Teknologi MARA (UiTM) for research funding through 600-IRMI/DANA 5/3/REI (0002/2016) grant.

6. References

[1] Idrees S and U A Ashfaq 2012 A brief review on dengue molecular virology, diagnosis, treatment and prevalence in Pakistan *Genet. Vaccines Ther.* **10**(6) 1-10.

[2] Kadir S L A, Yaakob H and Zulkifli R M 2013 Potential anti-dengue medicinal plants: a review. *J. Nat. Med.* **67** 677–689.

[3] Dengue Situation Update in Update on the Dengue situation in the Western Pacific Region 1 2016 *World Health Organisation (WHO).*

[4] Ahmad N, Fazal H, Ayaz M, Abbasi B H, Mohammad I and Fazal L 2011 Dengue fever treatment with Carica papaya leaves extracts *Asian Pac. J. Trop. Biomed.* **1**(4) 330-333.

[5] Subenthiran S, Choon T C, Cheong K C, Thayan R, Teck M B, Muniandy P K, Afzan A, Abdullah N R and Ismail Z 2013 Carica papaya Leaves Juice Significantly Accelerates the Rate of Increase in Platelet Count among Patients with Dengue Fever and Dengue Haemorrhagic Fever *Evidence-Based Compl. Alt. Med.* **2013** 1-7.

[6] Bahagian Kawalan Pennyakit 2009 Pelan Strategik Pencegahan Dan Kawalan Denggi 2009 – 2013 *Kementerian Kesihatan Malaysia* 1-24

[7] Senthilvel P, Lavanya P, Kumar K M, Swetha R, Anitha P, Bag S, Sarveswari S, Vijayakumar V, Ramaiah S and Anbarasu A 2013 Flavonoid from Carica papaya inhibits NS2B-NS3 protease and prevents Dengue 2 viral assembly *Bioinformation* **9**(18) 889-895.

[8] Yunita F, Hanani E and Kristianto J 2012 The effect of Carica papaya L. leaves extract capsules on platelets count and hematocrit level in dengue fever *Int. J. Med. Arom. Plants* **2**(4) 573-578.

[9] Siddique O, Sundus A and Ibrahim M F 2014 Effects of papaya leaves on thrombocyte counts in dengue-a case report *J Pak Med Assoc.* **64**(3) 364-366.

[10] Patil T, Patil S, Patil and Patil S 2014 Carica Papaya Leaf Extracts – An Ethnomedicinal Boon *Int. J. Pharmacogn. Phytochem.* **6**(2) 260–265.

[11] Canini A, Alesiani D, D'Arcangelo G and Tagliatesta P 2007 Gas chromatography–mass spectrometry analysis of phenolic compounds from Carica papaya L. leaf *J. Food Compos. Anal.* **20** 584– 590.

[12] Sudhakar N and T.V. Rm 2014 Potential Medicinal Properties of Carica Papaya Linn. - A Mini Review *Int. J. Pharm. Pharm Sci.* **6**(2) 1-4.

[13] Ismail Z, Halim S Z, Abdullah N R, Afzan A, Rashid B A A 2014 Safety Evaluation of Oral Toxicity of Carica papaya Linn. Leaves: A Subchronic Toxicity Study in Sprague Dawley Rats *Evidence-Based Compl. Alt. Med.* **2014** 1-10.

[14] Afzan A, Abdullah N R, Halim S Z, Rashid B A, Raja Semail R H, Abdullah N, Jantan I, Muhammad H and Ismail Z 2012 Repeated Dose 28-Days Oral Toxicity Study of Carica papaya L. Leaf Extract in Sprague Dawley Rats Molecules **17** 4326-4342.

[15] Chen L G, Yang L L and Wang C C 2008 Anti-inflammatory activity of mangostins from Garcinia mangostana *Food Chem. Toxicol.* **46** 688–693.

[16] Yu L, Zhao M, Yang B, Zhao Q and Jiang Y 2007 Phenolics from hull of Garcinia mangostana fruit and their antioxidant activities *Food Chem.* **104** 176–181.

[17] Pedraza-Chaverri J, Cárdenas-Rodríguez N, Orozco-Ibarra M and M. Pérez-Rojas J 2008 Medicinal properties of mangosteen (Garcinia mangostana) *Food Chem. Toxicol.* **46** 3227–3239.

[18] Yu L, Zhao M, Yang B and Bai W 2009 Immunomodulatory and anticancer activities of phenolics from Garcinia mangostana fruit pericarp *Food Chem.* **116** 969–973.

[19] Coleman J W 2001 Review: Nitric oxide in immunity and inflammation *Int. Immunopharmacol.* 1397–1406.

[20] Cui J, Zhanjun Cai W H, Liu Y, Li S, Tao W and Xiang H 2010 New medicinal properties of mangostins: Analgesic activity and pharmacological characterization of active ingredients from the fruit hull of Garcinia mangostana L. *Biochem. Behav.* **95** 166–172.

[21] Lim S P, Wang Q-Y, Noble C G, Chen Y-L, Dong H, Zou B, Yokokawa F, Nilar S, Smith P, Beer D, Lescar J and Shiet P-Y 2013 Ten years of dengue drug discovery: Progress and prospects *Antiviral Res.* **100** 500–519.

[22] Obolskiy D, Pischel I, Siriwatanametanon N and Heinrich M 2009 Garcinia mangostana L.: A Phytochemical and Pharmacological Review *Phytother. Res.* **23**1047– 1065.

[23] Cardona F, Andrés-Lacueva C, Tulipania S, Tinahones F J, Queipo-Ortuño M I 2013 Benefits of polyphenols on gut microbiota and implications in human health *J. Nutr. Biochem.* **24** 1415–1422.

[24] Manach C, Scalbert A, Morand C, Rémésy C and Jime´nez L. 2004 Polyphenols: food sources and bioavailability *Am. J. Clin. Nutr.* **79** 727- 747.

[25] Chitchumroonchokchai C, Riedl K M, Suksumrarn S, Clinton S K, Kinghorn A D and Failla M 2012 Xanthones in Mangosteen Juice Are Absorbed and Partially Conjugated by Healthy Adults *The J. Nutr.* 675-680.

[26] Gutierrez-Orozco F and Failla M L 2013 Biological Activities and Bioavailability of Mangosteen Xanthones: A Critical Review of the Current Evidence *Nutrients* **5** 3163- 3183.

[27] Arunrattiyakorn P, Suksamrarn S, Suwannasai N and Kanzaki H 2011 Microbial metabolism of a-mangostin isolated from Garcinia mangostana L.*Phytochemistry* **72**: 730– 734.

[28] Mantovani M 2010 *US Patent WO2009112054 A1.*

[29] Matin L, Elisa M. Miguélez, Claudio J V and Felipe L 2015 Bioavailability of Dietary Polyphenols and Gut Microbiota Metabolism: Antimicrobial Properties *BioMed Res. Int.* **2015** 1-18.

[30] Curiel J A, Pinto D, Marzani B, Filannino P, Farris G A, Gobbetti M and Rizzello C G 2015 Lactic acid fermentation as a tool to enhance the antioxidant properties of Myrtus communis berries *Microb. Cell Fact.* **14**(67) 1-10.

[31] Tree S.B.B. *Our Processes*. 2016 [cited 2016; Available from: http://biotreebiotechnology.com/wp/?page_id=52.

[32] Caplice E and Fitzgerald G F 1999 Food fermentations: role of microorganisms in food production and preservation *Int. J. Food Microb.* **50** 131–149.

[33] Cho J, Lee D, Yang C, Jeon J, Kim J and Han H 2006 Microbial population dynamics of kimchi, a fermented cabbage product *FEMS Microbiol Lett.* **257** 262–267.

[34] Nuraida L 2015 A review: Health promoting lactic acid bacteria in traditional Indonesian fermented foods *Food Sci. Hum. Wellness* **4** 47–55.

[35] Assohoun-Djenia N M C, Djenia N T, Messaoudic S, Lhommeb E, Koussemon-Camaraa M, T. Ouassad T, Choberte J M, Onnob B and Dousset X 2016 Biodiversity, dynamics and antimicrobial activity of lactic acid bacteria involved in the fermentation of maize flour for doklu production in Côte d'Ivoire *Food Control* **62** 397–404.

[36] Holzapfel W H 2002 Appropriate starter culture technologies for small-scale fermentation in developing countries Int. *J. Food Microbiol.* **75** 197–212.

[37] Swain M R, Anandharaj M, Ray R C and Rani R P 2014 Fermented Fruits and Vegetables of Asia: A Potential Source of Probiotics *Biotech. Res. Int.* **2014** 1-19.

[38] Yuliana N 2011 Phenotypic Identification of Lactic Acid Bacteria Isolated from Tempoyak (Fermented Durian) Made in the Philippines *Int. J. Biol.* **3**(2) 145-152.

[39] Chuah L O, Shamila-Syuhada A K, Liong M T, Ahamad R, Kwai L T and Gulam R 2016 Physio-chemical, microbiological properties of tempoyak and molecular characterisation of lactic acid bacteria isolated from tempoyak *Food Microbiol.* **58** 95-104.

[40] Pisol B, Abdullah N, Khalilah Abdul Khalil K and Lilis N. 2011 Isolation and identification of lactic acid bacteria from different stages of traditional Malaysian tempeh production *Malays. J. of Microbiol.* **11**(4) 358-364.

[41] Wei C L, Chao S H, Tsai W B, Lee P S, Tsau N H, Chen J S, Lai W L, Tu J C Y and Tsai Y C, 2013 Analysis of bacterial diversity during the fermentation of inyu, a high-temperature

fermented soy sauce, using nested PCR-denaturing gradient gel electrophoresis and the plate count method *Food Microbiol.* **33** 252-261.

[42] Ho V T T, Zhao J and Fleet G 2015 The effect of lactic acid bacteria on cocoa bean fermentation *Int. J. Food Microbiol.* **205** 54–67.

[43] Ho V T T, Zhao J and Fleet G 2014 Yeasts are essential for cocoa bean fermentation *Int. J. Food Microbiol.* **174** 72–87.

IOP Conf. Series: Materials Science and Engineering **358** (2018) 012037 doi:10.1088/1757-899X/358/1/012037

Supercritical Carbon Dioxide Extraction of Selected Herbal Leaves: An Overview

I A Abd Hamid[1], N Ismail and N Abd Rahman

Faculty of Chemical Engineering Universiti Teknologi MARA, 40450 Shah Alam, Malaysia

izniatikah@yahoo.com

Abstract. Supercritical fluid extraction of carbon dioxide (SC-CO_2) is one of new alternative extraction method that has been widely used to isolate bioactive components from variety of plant materials. The method was proved to be clean and safe, compatible for the extraction of edible products such as spices, food additives, medicines and nutritional supplement products compared to traditional extraction techniques such as solvent extraction, hydro distillation and steam distillation. The SC-CO_2 extraction was known as highly influenced by its process parameter such as temperature and pressure for obtaining maximum yield. Therefore, a clear review on the optimum range of temperature and pressure for herbal leaves extraction using SC-CO_2 is necessary for future reference. The aim of this work is to analyze the effect of temperature and pressure of SC-CO_2 process without modifier on extraction yield of some selected herbal leaves i.e clubmoss, drumstick leaves, kratom leaves, mallee and myrtle leaves. The values of investigated parameters were; pressure from 8.9 to 50 MPa and temperature from 35 to 80°C. The results showed that the highest extraction yields were obtained when the pressure and temperature were above 30 MPa and 40°C. The interaction between pressure and temperature for SC-CO_2 extraction of plant leaves are crucial since the values cannot be very high or very low in order to preserve the quality of the extracts.

1. Introduction

The application of a green extraction method of supercritical carbon dioxide (SC-CO_2) in isolating bioactive compounds from plant materials has been vigorously studied. Carbon dioxide (CO_2) in its supercritical condition (i.e. above its critical pressure and temperature of 7.38 MPa and 31.1°C), is an ideal solvent for extracting functional foods, supplement and medicines since it is nontoxic, noncorrosive and chemically inert which is safer to consumer and environment. Traditionally, many herbs from plant leaves have been used as alternative treatments by extracting oil from leaves using classical extraction methods such as solvent extraction, steam distillation and hydro distillation. However, these methods possess several weaknesses such as long exposed to the hazardous chemicals in solvent extraction (i.e. methanol, ethanol, dichloromethane, hexane, ethyl acetate, acetone etc.), high operating temperature used in steam and hydro distillation process that led to the losses of many important volatile compounds, long extraction period which consumes lots of energy and cost, and impurities to the final product that can cause many health problems [1][2]. Even though these methods could extract the medicinal compounds from the plant, but the probability of the organic solvent presence in the product and the losses of volatile compounds are high [3].

Therefore, a clean extraction method of SC-CO_2 that is free from any chemical solvent and only deals with CO_2 of 99.99% purity (without modifier) has been applied for the leaves extraction process. With a mild operating temperature used in SC-CO_2 system, it will avoid the losses of important compounds during the extraction process and has a potency to recover lots of volatile components from plant leaves. In general, the content of oil in the leaves consists of a complex mixture of

Content from this work may be used under the terms of the Creative Commons Attribution 3.0 licence. Any further distribution of this work must maintain attribution to the author(s) and the title of the work, journal citation and DOI.
Published under licence by IOP Publishing Ltd

compounds including polar and non-polar compounds, whereas CO_2 is well-known to be a nonpolar molecule. However, with the special properties owned by SC-CO_2 fluid i.e. the solvating power, has made this method proficient in extracting some of polar compounds or moderately polar compounds [4]. As published by Pronyk and Mazza [5], the ability of SC-CO_2 in extraction process is highly influenced by its temperature and pressure in which a slight change of temperature and/or pressure was greatly affect the output produced. Moreover, CO_2 at its supercritical state has the ability to penetrate into the solid matrix of the sample and dissolve the desired compound on account of its dual gaseous and liquid-like properties [6]. The increase in pressure of SC-CO_2 system could increase the fluid density which will enhance the solvating power. More oil or solutes will dissolve into the supercritical fluid promoting higher extraction yield. As for the temperature, the effect on the yield may be increase or decrease since it depense on the competing effect between fluid density and solutes vapour pressure [7]. When the tempertaure is higher, the density of supercritical fluid will decrease reducing the solvent strength to dissolve compounds. However, the increment of tempereture also increase the volatility of compounds which enhancing the oil solubility into the fluid. Therefore, in this study, the effect of extraction parameters i.e. pressure and temperature towards extraction yield were reviewed on the several selected herbal leaves. The knowledge on the parameters in SC-CO_2 process is significant in order to obtain highest oil yield as well as in designing the optimum process including equipment sizing within its ideal operating conditions. This article will become a guidance to choose the best range of parameters that is suitable for the extraction of herbal leaves under SC-CO_2 condition.

2. Supercritical carbon dioxide extraction on selected herbal leaves

Clubmoss (*Lycopodium clavatum L*) is a low-growing green plant consist of high medicinal compounds and has been used as alternative therapy and homeopathic medicines by traditional practitioners. Recent studies have showed that the extract from clubmoss could treat cervical cancer disease [8]. Therefore, da Silva [9] applied supercritical carbon dioxide (SC-CO_2) method to extract oil from clubmoss plant under the following condition; pressure range from 20 to 30 MPa, temperature range from 40 to 60°C and at constant flow rate of 800 g/h (table 1). The result showed that the oil yield increases with an increase in pressure from 20 to 30 MPa and a decrease in temperature from 60 to 40°C. The influence of increasing the pressure is expected since it will increase the density of supercritical fluid as well as increasing the fluid solubility which promoted to the higher extraction yield. However, the effect of temperature on the extraction yield can be varied (increase or decrease) depending on the complex balance between the vapour pressure of solute (i.e. the volatility of the extracted compounds) and the density of SC-CO_2 fluid [10]. In this case, the effect of density dominates the effect of solute vapour pressure since the highest yield of 1.36% was obtained at the highest pressure of 30 MPa (figure 1 (a)) and at the lowest temperature of 40°C (figure 1 (b)).

Table 1. Amount of extracted oil and process parameters in the supercritical carbon dioxide extraction of the selected leaves

Leaves oil	Extraction yield (%)	Process parameter	Ref.
Clubmoss (*Lycopodium clavatum L.*)	1.36	P = 20 – 30 MPa; T = 40 - 60°C; F = 800 g/h;	[9]
Drumstick leaves (*Moringa oleifera*)	6.34	P = 30 – 50 MPa; T = 40 – 80°C; F = 2 L/min;	[11]
Kratom leaves (*Mitragyna speciose*)	1.36	P = 8.9 – 34.5 MPa; T = 40 – 80°C; F = 24 mL/min;	[12]
Mallee (*Eucalyptus loxophleba*)	4.78	P = 10 – 50 MPa; T = 40 – 80°C; F = 2 L/min;	[13]
Myrtle leaves (*Myrtus communis L.*)	4.89	P = 10 – 30 MPa; T = 35 – 50°C; F = 0.42 kg/h;	[14]

P: pressure, T: temperature, F: flowrate

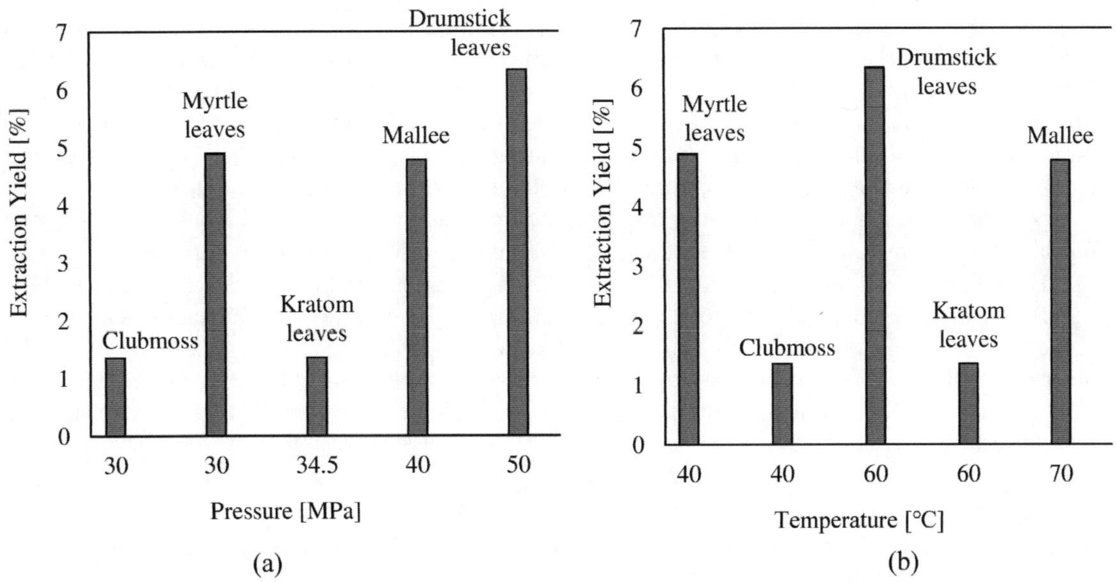

Figure 1. Pressure (a) and temperature (b) of the highest extraction yield in supercritical carbon dioxide extraction of the selected plant leaves.

Drumstick leaves (*Moringa oleifera*) contain both nutritional and medicinal values including vitamins, minerals and amino acids [15]. The plant also was consumed in food preparation such as curry. Zhao and Zhang conducted the extraction of oil from the drumstick leaves by using both traditional solvent extraction and SC-CO_2 methods [11]. The effect of different pressure (30 – 50 MPa) and, temperature (40 – 80°C) on the extraction yield were examined. As expected, when the pressure increases in isothermal condition, the extracted oil increases significantly due to the enhancement of the fluid density to solubilize more solute from the solid sample. For temperature effect, the increment from 40 to 60°C resulting in an increase in the extraction oil yield. When the temperature is high, the solute vapour pressure increased make it easier to be extracted out from solid matrix. Nevertheless, the further increment of temperature from 60 to 80°C has led to a slight decrease in the yield. This can be explained as, at that moment, the density of fluid started to reduce and thus lowered the yield. As a conclusion of this research, the maximum extraction yield can be obtained by manipulating the process parameters of temperature and pressure since the highest oil yield was achieved at 50 MPa and 60°C (figure 1). Kratom leaves (*Mitragyna speciose*) locally known as 'ketum' is a controversial plant which contains various medicinal values. The plant has been applied as traditional practice in folk medicine for the treatment of muscle pain, fever, wound, diarrhea and cough [16]. The extraction of oil from kratom leaves by using SC-CO_2 method has been conducted by Tohar et al. [12] at 8.9 to 34.5 MPa of pressure and 40 to 80°C of temperature with a constant flow rate of 24 mL/min (table 1).

In general, the extraction yield improved with the increment of pressure from 8.9 to 34.5 MPa. Conversely for the temperature effect, the yield started to decrease when the temperature was above 60°C. This phenomenon described that the further increment of temperature more than 60°C from Kratom leaves extraction will directed to a reduction in solvent density and thereby resulting to the lower yield. This is similar to the study conducted by Zhao and Zhang [11] discussed previously on the temperature effect of the extraction oil yield. In this study, the highest yield of 1.36% was obtained at the highest pressure of 34.5 MPa and at 60°C of temperature (figure 1 (a) and (b)). Zhao and Zhang [13] investigated the relationship between temperature and pressure on SC-CO_2 extraction of *Eucalyptus* leaves since the extract from the leaves contains numerous medicinal benefits such as anti-inflammatory, anti-allergenic, anti-asthmatic, antiseptic, aquaculture antiviral, anti-bacterial and anti-malarial. The laboratory experiments were carried out at a range of pressure from 10 to 50 MPa and temperature from 40 to 80°C (table 1). Based on this research, the extraction yield under a clean SC-CO_2 method also trailed the same trend for the increment in pressure. A rise in pressure from 10 to 40 MPa has risen the solvating power of the fluid and thus giving rise to more extraction yield. It is also stated that the increase in yield improved as the temperature increase from 40 to 70°C. The yield started to decrease when the pressure is further increases to 50 MPa and when the temperature is above 70°C. Such condition of yield reduction is due to the decreasing of fluid diffusion coefficient at a very high pressure. Thus, a very high pressure and temperature is not always recommended for SC-CO_2 extraction. The optimum interaction between pressure and temperature on *Eucalyptus* leaves was achieved at 40 MPa and 70°C with the highest oil yield of 4.78% and can be seen in table 1.

Zermane et al. [14] presented the result on the extraction of essential oil from Myrtle (*Myrtus communis*) leaves by using SC-CO_2 method. The extraction was carried out by varying the pressure from 10 to 30 MPa and temperature from 35 to 50°C. The optimum condition was achieved at 30 MPa and 40°C. Similar to the previous reviewed an increase in pressure leads to an increase in the oil recovery. Pressure gives a big impact to the yield since a slight change of pressure will change the solvent strength of the fluid which improving more extraction yield. As for the temperature effect, the oil yield shows a different trend. At initial stage, increasing temperature from 35 to 40°C increased the yield since raising temperature increases the solute vapour pressure, resulting in the enhancement of SC-CO_2 solubility. Then, a further increase in temperature up to 50°C decreased the yield. This is due to the reduction of the fluid density of CO_2 which reduce the solvation power. The overall effect of temperature towards extraction yield in SC-CO_2 extraction depends on whether solute vapour pressure or solvent density dominates at the particular condition.

3. Analysis of temperature and pressure on extraction yield

According to the survey of all selected plant leaves above (i.e. clubmoss, drumstick leaves, kratom leaves, mallee and myrtle leaves) that were undergone the SC-CO_2 process, the effect of the extraction pressure towards oil yield showed similar trend line. It can be explained that, as the pressure of the system increase at constant temperature, the density of supercritical fluid also increase. Density plays an important role in the extraction process since it triggers the solubility of fluid make it able to penetrate inside the solid sample to extract out the important oil at the solid surface. As can be seen in figure 1(a), most of the plant leaves achieved the maximum oil yield at pressure above 30 MPa.

Many authors agreed that the influence of temperature on the extracted yield is defined in different manners. In reviews on the extraction of the selected plant leaves above, the results of temperature effect are different. The extraction yield can be increased or decreased. Near the critical pressure, the density of fluid is very sensitive to temperature. At constant pressure, an increase in temperature will lead to a decrease in fluid density and reduces the solvent power. However, the increase in temperature will also speed up the mass transfer and enhance the yield. This is due to the increase in vapour pressure of the extractable compounds [1]. Therefore, it is hard to predict the effect of temperature in SC-CO_2 extraction.

There are other factors that can affect the extraction yield such as fluid flow rate, extraction time, sample particle size, moisture content and addition solvent as modifier. However, this paper only focus on the effect of pressure and temperature since these two factors have the biggest impact on the extraction yield based on supercritical fluid properties such as gas-like diffusivity and liquid-like density. Apart from that, most of the leaves oils consist of a complex mixture of compounds including polar and non-polar compounds. The interactive combination of pressure and temperature in SC-CO_2 system can increase the mass transfer of fluid to penetrate inside the solid sample and thus extracting more oil. The perfect combination of temperature and pressure also promotes to the increasing solute vapour pressure which is crucial for the isolation of polar substance from plant sample.

4. Conclusion

In this study, the influence of temperature and pressure on the extraction of clubmoss, drumstick leaves, kratom leaves, mallee and myrtle leaves under SC-CO_2 extraction were presented. Based on the comparison result on the effect of temperature and pressure used, the extraction yield from 1.36 to 6.34% was achieved within the range of pressure from 30 to 50 MPa and temperature from 40 to 70°C. The review of the range of parameter is necessary for the optimum design of supercritical extraction process and could be used as future reference in supercritical fluid field.

Acknowledgement
The authors are grateful to the Faculty of Chemical Engineering and Faculty of Applied Science in Universiti Teknologi MARA (UiTM), Shah Alam, Malaysia for the facilities to carry out the project and the Pure Rerama Leaf for providing the sample.

References
[1] Bimakr M, Rahman R A, Taip F S, Ganjloo A, Salleh L M, Selamat J, Hamid A and Zaidul I S M 2011 *Food Bioprod. Process.*, **89** (1) 67–72
[2] Ouzzar M L, Louaer W, Zemane A and Abdeslam H M 2015 *Chem. Eng. Trans.*, **43** 1129–1134
[3] Uquiche E, Cirano N and Millao S 2015 *Ind. Crops Prod.*, **77** 307–314
[4] Paviani L C, Saito E, Dariva C, Marcucci M C, Sanchez-Camargo A P and Cabral F A 2012 *Brazilian J. Chem. Eng.*, **29** (2) 243–251
[5] Pronyk C and Mazza G 2009 *J. Food Eng.*, **95** (2) 215–226
[6] Akanda M J H, Sarker M Z I, Ferdosh S, Manap M Y A, Norulaini N and Kadir M O A 2012 *Molecules*, **17** 1764–1794
[7] Danlami J M, Zaini M A A, Arsad A and Yunus M A C 2015 *J. Taiwan Inst. Chem. Eng.*, **53** 32–39
[8] Samadder A, Das S, Das J, Paul A, Boujedaini N and Khuda-Bukhsh A R 2013 *JAMS J. Acupunct. Meridian Stud.*, **6** (4) 180–187
[9] Da Silva G F, Gandolfi P H K, Almeida R N, Lucas A M, Cassel E and Vargas R M F 2015 *Chem. Eng. Res. Des.*, **100** 353–361
[10] Shao Q, Deng Y, Liu H, Zhang A, Huang Y, Xu G and Li M 2014 *Ind. Crops Prod.*, **60** 104–112
[11] Zhao S and Zhang D 2013 *Sep. Purif. Technol.*, **118** 497–502
[12] Tohar N, Shilpi J A, Sivasothy Y, Ahmad S and Awang K 2016 *Arab. J. Chem*
[13] Zhao S and Zhang D 2014 *Sep. Purif. Technol.*, **133** 443–451
[14] Zermane A, Larkeche O, Meniai A H, Crampon C and Badens E 2014 *J. Supercrit. Fluids*, **85** 89–94
[15] Sánchez-Machado D I, Núñez-Gastélum J A, Reyes-Moreno C, Ramírez-Wong B and López-Cervantes J 2010 *Food Anal. Methods*, **3** (3) 175–180
[16] Philipp A A, Wissenbach D K, Weber A A and Zapp J 2010 *Anal Bioanal Chem*, **396** 2379–2391

IOP Conf. Series: Materials Science and Engineering **358** (2018) 012038 doi:10.1088/1757-899X/358/1/012038

The Effects of Antifoam Agent on Dead End Filtration Process

S Mohamad Pauzi[1,3], N Ahmad[1], M F Yahya[1] and M A Arifin[2]

[1] Faculty of Chemical Engineering, Universiti Teknologi Mara
[2] Faculty of Engineering Technology, Universiti Malaysia Pahang

syazana7932@salam.uitm.edu.my[3]

Abstract. The formation of foam as a result from introducing gases during cell culture process in the bioprocess industry has indirectly affected the throughput of the product of interest. Due to that, antifoams were developed and established as one of the means to minimize the formation of foam in the cell culture. There are many types of antifoams but the silicone-type of antifoams are widely used in the bioprocess industry. Although the establishment of antifoam has aided the cell culture process, the impacts of its presence in the cell culture to the downstream process especially the dead end filtration is not widely discussed. The findings in the study emphasized on the dead end filtration performance that includes flux rate profile and the resulted filtration capacity. In this study, the concentrations of antifoam injected into the solution were varied from 0.2% v/v – 1.0% v/v and the solutions were filtered using constant flow method. The resulted maximum pressure readings and final flux rates indicated that the resistance exerted to the feed flow rate increased as the concentration of antifoam loaded in the solution increased. This later has led to the decline in the flux rates with percentage reduction between 32 – 68%. The calculated filter capacity for flux rate of 1000LMH ranged from 53 – 63L/m^2 while it is in the range of 40 – 43L/m^2 for flux rate of 2000LMH. The presence of antifoam agents in the feed load was determined to have negative effects on the dead end filtration performance and it may reduce the efficiency of the dead end filtration process.

1. Introduction

Antifoam agent or known as defoamer has been widely implemented in various industries including biological fermentation or cell culture. Since then, it has been the important compound to minimize the effects of foaming. In addition to that, there are research findings claimed that the antifoam agent has many advantages such as it can either prevent the foam generation or increase the rate of foam decay. However, the function of the antifoams is that they undermine the stability of liquid films in foam. Other than that, antifoam can boost the rate of the liquid discharge and thus increase the foam destruction [1]. Antifoam or defoamer is a chemical preservative which minimizes and avoids the creation of foams in the industrial process liquids. Method of using an antifoaming agent can be described into two; it is applied before the formation of the foam to prevent foaming or it is injected in the solution after foam formation in order to overcome it [2]. It is also widely used in various industries; bioprocess industries [3] - cell culture and microbial fermentation [4 - 5], anaerobic digesters [6] and biogas production [7]. Nonetheless, the biological effects of antifoam are poorly understood due to various range of models and inadequate data about their configuration and composition prepared by the manufacturers [8].

Since the introduction of membrane filtration over the last three decades, the usage of it has been getting more attention due to the faster evolution that it holds throughout the decades [9]. There are wide applications of membrane filtration and nowadays, much attention has been paid to the cell clarification process step with the purpose of removing the cell while obtaining the product of interest [10]. Different mode of filtration operation can be implemented during this stage of processing that are

Content from this work may be used under the terms of the Creative Commons Attribution 3.0 licence. Any further distribution of this work must maintain attribution to the author(s) and the title of the work, journal citation and DOI.
Published under licence by IOP Publishing Ltd

normal flow filtration or known as dead end filtration and crossflow filtration or known as tangential flow filtration. Both of them possess advantages and disadvantages depending on the objectives of process run and the scale of production [11].

This study covers the experimental exercises to evaluate the performance of filtration process with samples-containing antifoam agent. It focuses on the dead-end type of filters from the same membrane material with membrane area of approximately 15.2 cm^2. Apart from that, a silicone-based antifoam was used and different load of antifoam concentrations were tested. Throughout the experimental works, the data on volumetric rate and differential pressure were recorded in order to assess the changes of pressure applied on the filter membrane, especially the final maximum pressure. The volumetric flow rates were also analyzed to develop the flux rate pattern of each of the process runs.

2. Methodology

The materials used in this experiment were Lysogeny Broth (LB) broth and 10% silicone-type antifoam solution. A polyethersulfone (PES) disc filter with filter area of approximately 15.2 cm^2 from Cobetter was used for each of the process runs. The experiments were conducted using constant flow method where the initial flow rates were fixed. The initial flux rates were set for these experiments in order to normalize the constant data to filter membrane area. Two flux rates were set for the experiments, which were 1000 LMH and 2000LMH. The study was designed to have one control sample which was LB broth that did not contain antifoam agent for each of the initial flux rate and three running samples for three different antifoam concentrations loaded; 0.2%, 0.6% and 1.0%. The filter was connected to a peristaltic pump and a pressure gauge. The specific amount of antifoam were loaded in the feed containing LB broth according to the design concentration. The volume and differential pressure were observed and recorded at a specific time interval.

3. Results and Discussion

3.1 Flux rate profile

Figure 1 shows the resulted flux rate over time with initial flux rate of 1000LMH. The graph demonstrates that the flux rate decreased with time for all of the experimental runs. By comparing all of the data, the flux rates dropped as the concentration of antifoam loaded increased. The final flux rate for control sample was approximately 90LMH. Meanwhile, the recorded flux rate for antifoam concentration of 0.2%v/v, 0.6% and 1.0%v/v shows a continuous decreasing pattern where the starting flux rate of approximately 600LMH was reduced until approximately 60LMH, 50LMH and 40LMH respectively. Comparing with the final flux rate of the control sample, the filtration performance of sample containing various concentrations of antifoam has reduced by 10% each starting with 33% for 0.2% antifoam, 44% for 0.6% antifoam and 55% for 1.0% antifoam. This indicates that the filtration performance has been directly affected by increasing amount of antifoam agent loaded.

The flux rate pattern with respect to time for initial flux rate of 2000LMH is shown in Figure 2. The flux rate pattern possesses comparable behaviour to the initial flux rate of 1000LMH experimental runs that the graph verifies that the flux rate decreased with time for all of the experimental runs. However, the flux rate for all samples have decreased to similar final flux rate that ranged from 50LMH to 55LMH while the final flux rate for control sample was approximately 140LMH. This demonstrates a percentage reduction of final flux rate of around 60% for all samples containing antifoam agent that illustrates a higher declining rate as compared to the flux rates pattern for initial flux rate of 1000LMH. This indicates that higher initial flux rate can contribute to faster fouling action apart from the presence of antifoam as the fouling factor. The declining of the flux rate was supported by a study done by Liew et al. [12] as they observed a comparable flux decline profile when using PVDF microfilter to filter a yeast culture containing silicone-type antifoam agent. The reported flux reduction was approximately 42% with 1.0% concentration of antifoam added. PVDF membrane filter may pose similar attributes to PES membrane filter as both of them are hydrophobic naturally [12]. Thus, there may be some interactions between the silicone antifoam agent and the hydrophobic component of the PES membrane filter.

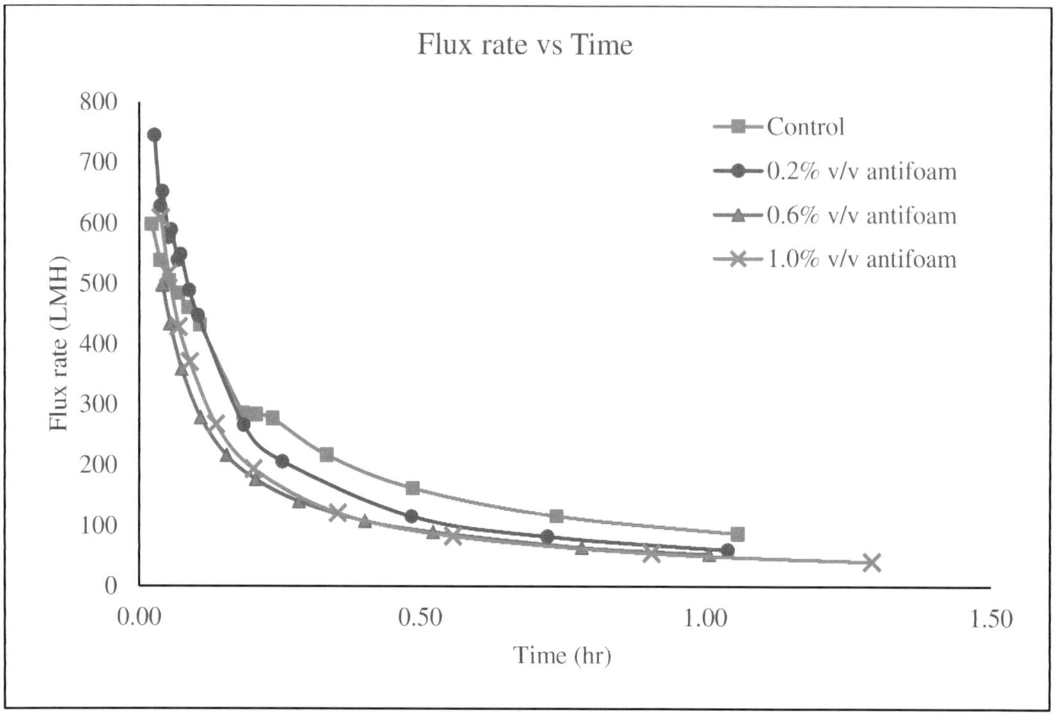

Figure 1: Flux rate versus Time (Initial Flux Rate: 1000 LMH).

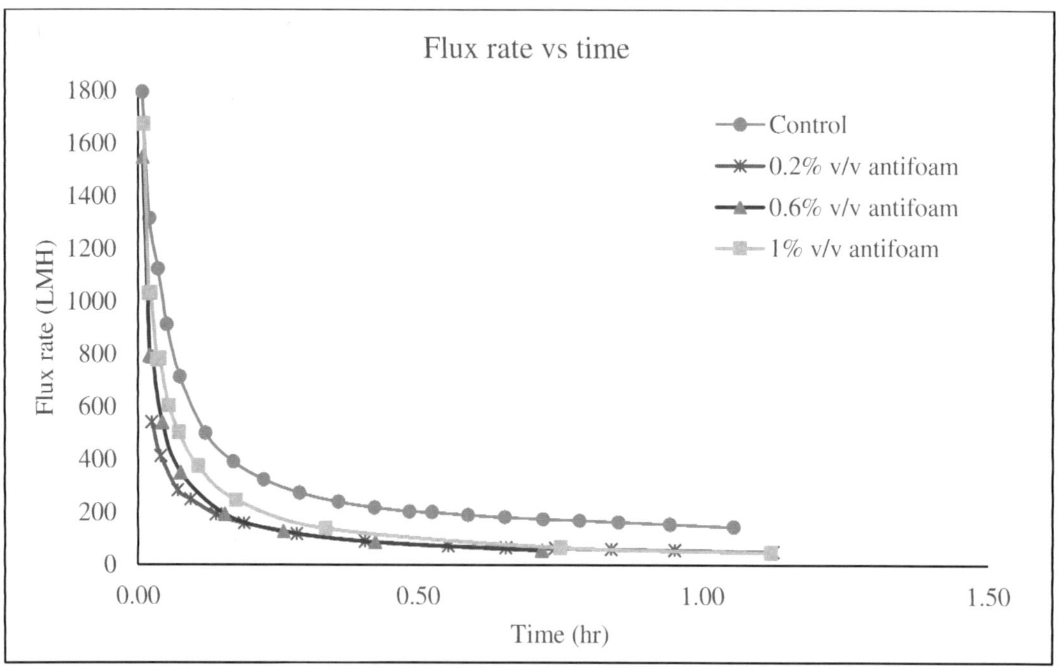

Figure 2: Flux rate versus time (Initial Flux Rate: 2000 LMH).

3.2 Filtration capacity

The relationship between resistances with the feed load of initial flux rate of 1000LMH is shown in Figure 3. The control sample demonstrated that the maximum resistance occurred during the experimental run was approximately 0.1psi/LMH with feed load of about 92L/m^2. Besides, the resistance towards the feed load increased with increasing antifoam concentration where the maximum resistance was around 0.2psi/LMH for all samples containing antifoam with loading capacity ranged from 53L/m^2 for 0.6% and 1.0% antifoam to 63L/m^2 for 0.2% antifoam. Although all of the filtration runs for samples containing antifoam agent experienced similar resistance towards the feed load, the reduction of filtration capacity slightly differs from one another where 0.2% antifoam shows a reduction around 32% while 0.6% and 1.0% antifoam shows a reduction of around 43% from the control run.

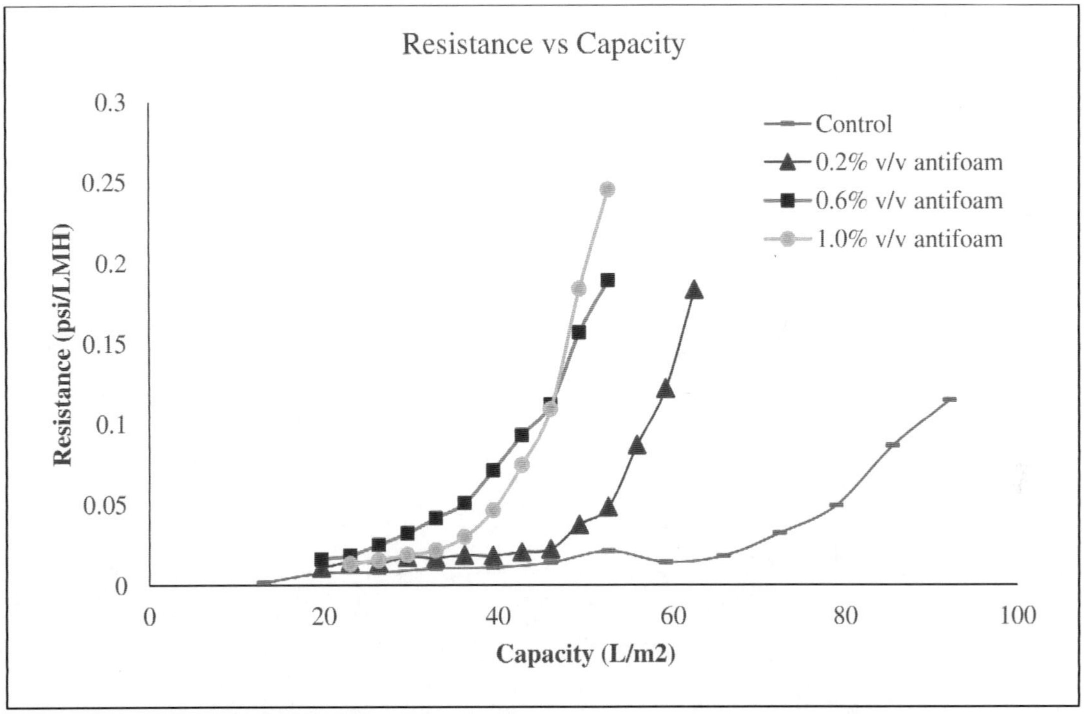

Figure 3: Resistance vs Capacity (Initial Flux Rate: 1000 LMH).

On the other hand, Figure 4 illustrates the correlation between resistances on the filter membrane with the feed load capacity of experiment runs for initial flux rate of 2000LMH. The maximum resistance on the membrane filter for control sample was 0.1psi/LMH with resulted loading capacity of 130L/m^2. The resistance towards the feed load of these experimental runs elucidated comparable behaviour with the experimental runs for 1000LMH, which they increased with increasing antifoam concentration. The maximum resistance exerted for all of samples containing antifoam agent was also 0.2psi/LMH. Due to that, the resulted maximum loading capacity for all the samples containing antifoam agent was ranged from 40L/m^2 to 43L/m^2. The filter membrane may experience similar occasion with that of 1000LMH experimental runs, but the high initial flux rate has led to faster resistance build up towards the filter membrane which then resulted in lower feed load capacity that can be processed by the filter which was proved by higher percentage reduction that is approximately 68%. The results for both initial flux rate runs have illustrated the effect of the antifoam load on the dead filtration performance. Typically, these outcomes were may be due to the cake formation on the filtration membrane over the time as the filtration mode executed was normal flow filtration where there was no filter recovery action throughout the experimental runs [13]. However, the fouling action has accelerated with the addition of the antifoam agent in the feed solution. Previous work suggested that a process solution containing silicone antifoam agent can probably be ultra-filtered at a lower temperature that is below the cloud point of the antifoam agent to reduce the fouling effect based on their study on ultrafiltration process that was run using cross flow filtration. Apart from that, they also suggested using a 100% silicone antifoam agent as impure silicone antifoam agent may contain other components that may pose hydrophobic characteristic; which is supported by the attributes of an antifoam agent that has the capability to reduce the surface tension in the liquid system [14]. Nevertheless, the resulted flux rate profile and loading capacity from this study has shown an improved technology of membrane filters as dead end filtration process does not pose the membrane recovery action throughout the process run as compared to the cross flow filtration process.

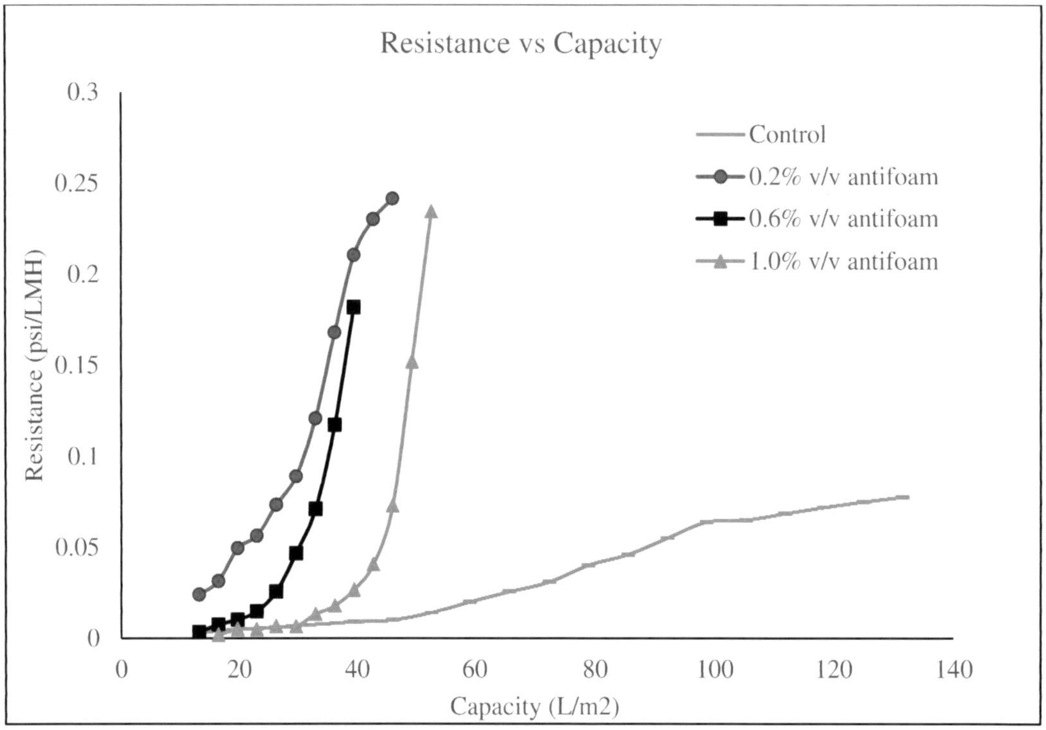

Figure 4: Resistance vs Capacity (Initial Flux Rate: 2000 LMH)

4. Conclusion

This study aimed to evaluate the dead end filtration performance based on flux rate profile and the filtration loading capacity with the presence of different concentration of antifoam agent in the feed load. The process was accomplished through the constant flow mode of filtration to filter samples containing different concentrations of antifoam agent using two initial flux rates and the resulted flow rate and pressure was observed at certain time interval. The flux rate profile shows continuous reduction of flux rate as concentration of antifoam load increased at a range of 33% to 55% for 1000LMH initial flux rate while there was a reduction of 60% of flux rate regardless of concentration of antifoam load for 2000LMH initial flux rate. Meanwhile, the calculated filtration capacity for initial flux rate of 1000LMH was in the range from 53L/m^2 to 63L/m^2 and 40L/m^2 to 43L/m^2 for initial flux rate of 2000LMH and the percentage of drop of the filtration capacity was around 32% to 43% and 68% respectively. The results of flux rate profile and filtration capacity support that the presence of silicone-type antifoam agent may contribute to the fouling action of the dead end filtration process. These values are expected to reduce more with the increase in variety of the components in the feed load passing through the filter membrane that may later increase the resistance exert towards it.

References

[1] Winterburn J and Martin P 2012 Foam mitigation and exploitation in biosurfactant production. *Biotechnology Letters* **34**(2) 187 – 195
[2] Denkov N D 2004 Mechanisms of foam destruction by oil-based antifoams. *Langmuir* **20**(22) 9463–9505 http://doi.org/10.1021/la049676o
[3] Routledge S J 2012 Beyond de-foaming: productivity the effects of antifoams on bioprocess. *Computational and Structural Biotechnology* **3**(4) 1–7

[4] Routledge S J, Hewitt C J, Bora N and Bill R M 2011 Antifoam addition to shake flask cultures of recombinant Pichia pastoris increases yield. *Microbial Cell Factories* **10**(1) 17

[5] Koch V, Rüffer H, Schügerl K, Innertsberger E, Menzel H and Weis J 1995 Effect of antifoam agents on the medium and microbial cell properties and process performance in small and large reactors. *Process Biochemistry* **30** 435–446

[6] Subramanian B, Miot A, Jones B, Klibert C and Pagilla K R 2015 A full-scale study of mixing and foaming in egg-shaped anaerobic digesters. *Bioresource Technology* **192** 461- 470

[7] Kougias P, Boe K, Tsapekos P and Angelidaki I 2014 Foam suppression in overloaded manure-based biogas reactors using antifoaming agents. *Bioresource Technology* **153** 198-205

[8] Barber W 2005 Anaerobic digester foaming: causes and solutions. *Water* **21**: 45–49

[9] Reis R and Zydney A 2007 Bioprocess membrane technology. *Journal of Membrane Science* **297**: 16 – 50

[10] Yavorsky D, Blanck R, Lambalot C and Brunkow R 2003 The clarification of bioreactor cell cultures for biopharmaceuticals. *Pharmaceutical Technology* (March): 62 – 76

[11] Mirliss M J 2002 A comparison between DE and crossflow filtration. *Water Conditioning & Purification* (July)

[12] Liew M K H, Fane A G and Rogers P L 1997 Fouling effects of yeast culture with antifoam agents on microfilters. *Biotechnology and Bioengineering* **53**: 10 – 16

[13] Walter J K, Jin Z, Jornitz M W and Gottschalk U 2011 Membrane separations. In Janson J.C. (ed.) *Protein Purification: Principles, High Resolution Methods and Application* 3rd ed (New York: John Wiley) pp 279 – 317

[14] Mc Gregor W C, Weaver J F and Tansey S P 1988 Antifoam effects on ultrafiltration. A *Biotechnology and Bioengineering* **31**: 385 – 389

Effect of Substrate Concentration and Reaction Time of *Aquilaria subintegra* Leaves Extract on Inhibition of Pancreatic Lipase

S Zainal[1,a], M Musa[1], J Idris[2] and K H Ku Hamid[1]

[1]Faculty of Chemical Engineering, Universiti Teknologi MARA, 40450 Shah Alam, Selangor Malaysia
[2]Faculty of Chemical Engineering, Universiti Teknologi MARA (UiTM) Sarawak, 94300, Kota Samarahan, Sarawak, Malaysia

drsafari@salam.uitm.edu.my[a]

Abstract. The purpose of this research are to recovery of pancreatic lipase inhibitor and to study the effect of using different concentration of substrate and reaction time on pancreatic lipase inhibitor. In this research, *Aquilaria subintegra* mature and fresh leaves was used as a sample. The research was conducted by using hydro-distillation with different concentrations, which are 100 μM, 200 μM and 300 μM and reaction times from 20, 40 and 60 minutes were studied. Based on the results obtained for the samples of phenol, flavonoid, gallic acid and quercetin were 49.30 μg/ml, 314.33 μg/ml, 12.94 μg/ml, and 5.15 μg/ml, respectively.

1. Introduction

Aquilaria also commonly known as agarwood, gaharu, aloeswood, eaglewood or Oudh is belonging to Thymelaeceae family. Aquilaria species mainly produces in Southest Asia, with Indonesia, Malaysia, Vietnam, Cambodia, Thailand, Laos and Papua New Guinea [1]. Aquilaria is fragrant resinous wood from tropical green forest and estimated of 20 species that located mainly in Asia [2]. Agarwood has been used as a traditional medicine used to treat aches, muscular pains, inflammation and anaphylaxis for over thousands of years. It's also very widely used for centuries as incense for religion purpose. Recently, agarwood has been widely used for research purposes in medical area that can treat many type of diseases. It reached with chemical constituents such as phenolic compounds, flavonoids, terpenoids, alkaloids, tannins and many more [3-5]. Agarwood can been formed through inducement process such as injury, cutting, pest or insect, disturbance, microorganism, fire, chemical or colonization. Recently, obesity problem has growing rapidly for bigger health crisis in the world every year. Obesity is when someone is overweight that can risk their health. Obesity occurs from over-eating habit especially an unhealthy diet and lack of exercise. The study found that more than 60% of men and 50% of women in the world were either overweight or obese. Obesity can lead to many health problems including heart disease, stroke, diabetes, high blood pressure, unhealthy cholesterol, cancers and many more.

There has an effective way that can prevent and solve the obesity problem which is inhibit fat absorption in human digestive system. The key for inhibit fat absorption is pancreatic lipase that inhibit the pancreatic lipase. Pancreatic lipase is an enzyme secreted from pancreas that hydrolyses and break down dietary fat molecules with converting triglyceride substrates found in ingested oils to mono-glycerides and free fatty acid. The free fatty acid can cause obesity problem with the formation of free fatty acid will contributing to insulin resistance that is major lead to obesity. There are many researchers that study about this anti-obesity by recovery of pancreatic lipase inhibitor from many

Content from this work may be used under the terms of the Creative Commons Attribution 3.0 licence. Any further distribution of this work must maintain attribution to the author(s) and the title of the work, journal citation and DOI.
Published under licence by IOP Publishing Ltd

kinds of natural resources. Many plants have been reported to inhibit lipase activity which the presence of secondary metabolites such as polyphenols, benzopyrines whose members include flavonoids, saponins, and caumarins [6]. Thus, in this research is to recover the pancreatic lipase inhibitor. *Aquilaria subintegra* (*A.subintegra*) has rich in phytochemicals content as potential of pancreatic lipase inhibitor. The bark, stem and leaves of *A.subintegra* have compound of pancreatic lipase inhibitor. Hydro distillation technique can help to extract the *Aquilaria sp* in order to get the compound of pancreatic lipase inhibitor. The different of substrate concentration, reaction time and temperature will affect the recovery of pancreatic lipase inhibitor from extraction of *Aquilaria sp*. Hence, this project is recovery of the pancreatic lipase inhibitory from extraction of *A.subintegra* leaves and to determine the effect of different substrate concentration and reaction time use on pancreatic lipase inhibitory recovery. There are some compounds that can act as pancreatic lipase inhibitor that can solved people that have obesity problem. Hence, this research project for recovery of pancreatic lipase inhibitor compound from the extraction of *A.subintegra* leaves.

In this study, the compounds and its concentration that can act as pancreatic lipase inhibitor is determined. However, there are some parameters that can affect the recovery of pancreatic lipase inhibitors of *A.subintegra* leaves extract that are substrate concentration and reaction time for the extraction of *A.subintegra* leaves. Thus, for this research purpose is to determine what are the effects of substrate concentration and reaction time from *A.subintegra* leave extraction on inhibition of pancreatic lipase recovery.

2. Experimental

2.1. Materials
The material used for this research were mature and fresh *A.subintegra* leaves.

2.2. Preparation of Leaves Samples
The drying process of the leaves was done by using Memmert oven at 60°C. This step was important in order to remove any moisture present in the leaves. Next, the fresh sample of *A.subintegra* leaves is preferred in form of grinded form in order to provide larger surface area that increasing the rate of reaction between the samples molecules and the solvent molecules in the extraction process. The grinding process was done by using the Mastar (MAS-160BL(A)-I) blender for around 1 minute until the required size of leaves were obtained. The grinded leaves were then sifted through a 250 μm sieve in order to obtain the uniform size for further procedures.

2.3 Soaking
1. The dried fiber and also the powdered *A.subintegra* samples were soak in the distilled water to extract the grinded leaves samples [7]. It was done for 24 hours at room temperature. The ratio of the grinded leaves to distilled water was 1:100 g/mL.

2.4 Ultrasonication
All dried and soaked fiber and powdered *A.subintegra* samples will further enhance using ultrasonication. The samples were placed in a NEXXsonics NS-A-18H ultrasonicator (37 KHz) at 30 °C, and ultrasonication time was 30 minutes for each sample.

2.5. Hydro-Distillation
The dried and grinded leaves samples continue in hydro-distillation process. The heating process were done by using the heating mantle of TOPS MS-06 at the atmospheric pressure. The heating process was continuously until all essential oil was obtained at the receiving flask usually the sufficient amount of hydrodistillate were between 350 mL to 400 mL. This is to ensure all phenolic compounds has been distilled [3]. After that, the leaves extract was evaporated in order to ensure there is no more concentrated crude leaf extract which is more suitable for further testing by using a Heidolph rotary evaporator (Laborota 4000 efficient). After the extraction process end, the apparatus was let to cooled to room temperature before the sample was continued for extraction [8]. The extracted samples were transferred into the tight sample bottle kept in the room temperature.

2.6. Lipase Assay

The lipase from porcine pancreas (Merck KGaA) was prepared and suspended in tris-HCl buffer (pH 7.4) with sodium hydroxide, NaOH (2.5 mmol) to result in a 200 unit/mL lipase solution. Assay buffer was p-Nitrophenyl Phosphate (p-NPP) was dissolved in isopropanol to a concentration of 100 μmol and used as substrate and different substrate concentration was used that are 200 and 300 μmol. Next, 3 mL of each extracted leaves sample was mix with 1 mL of the enzyme suspension and 1 mL of substrate solution. 6 mL of tris-HCl buffer was added into the mixture. The mixture was mixed with hydrolytic reaction followed by incubating at 37°C for three different times, which are 20 minutes, 40 minutes and 60 minutes. The acetone-ethanol mixture with 1:1 ratio was added to stop the reaction before measure the absorbance value using Hach DR 2800 spectrophotometer at wavelength 405 nm. The leaf extract absorbance value was compared with control sample or blank sample. Hence, the percentage of inhibition was determined based on absorbance value obtain with compared with control sample.

2.7. Analytical Method

The samples are analyzed by using high performance liquid chromatography (HPLC) on a Waters Millennium 32®system. To detect the present of gallic acid and quercetin in the samples. The phenolic and flavonoid is analyzed by difference method that are using Total Phenol Content and Total Flavonoid Content method. The samples are using spectrophotometer to determine the value of absorbance by using 415 nm and 760 nm of wavelength [9], and the absorbance values were determined by averaging 3 readings for each sample.

2.7.1 Limits of detection and the standard curve

The limits of detection (LOD) were calculated from the relationship between the standard deviation (SD) and the slope, using the appropriate multiplier. The calibration curve of the flavonoid standard (5.0 to 100.0 μg/mL), phenol (5.0 to 80.0 μg/mL), gallic acid (0.5 to 3.0 μg/mL), and Standard quercetin (2.5 to 25.0 μg/mL) were determined from four to five concentration points over the range of concentrations, following the Lambert-Beer law. The slope and other statistics of the calibration curves were calculated by linear regression by using Microsoft Excel.

3. Results and Discussions

3.1. Volume of Sample after Pre-treatment

The best efficiency for volume for leaves extract after hydro-distillation should be between 350 ml to 400 ml. From Table 1, the volume of the sample (After Hydro-distillation) for mature and fresh leaves are not standardize is due to the variation of extraction parameters. The volume of leaves extract after evaporation should be in between 16 to 20 ml for mature and fresh leaves.

Table 1. Volume extract after pre-treatment for mature leaves and fresh leaves

Mature Leaves			Fresh Leaves		
Samples	**After Hydro-distillation (mL)**	**After Evaporation (mL)**	**Samples**	**After Hydro-distillation (mL)**	**After Evaporation (mL)**
A1	330	20	**B1**	342	20
A2	379	17	**B2**	347	18
A3	332	18	**B3**	343	20
A4	361	20	**B4**	345	18
A5	366	20	**B5**	334	16
A6	349	20	**B6**	345	20
A7	344	19	**B7**	361	20
A8	353	20	**B8**	362	16
A9	363	20	**B9**	340	16

3.2. Total Flavonoid Content

From Table 2, the highest concentration of flavonoid content calculated by using the standard curve of flavonoid is A8 which is 354.33μg/ml while the lowest is A3 which is 191μg/ml, this finding also support by Bahrani et al. 2014 [10] and Naef 2010 [11].

Table 2. Average absorbance value of total flavonoid content for mature leaves and fresh leaves

Mature leaves		**Fresh Leaves**	
Samples	**Absorbance**	**Samples**	**Absorbance**
A1	0.274	**B1**	0.242
A2	0.204	**B2**	0.272
A3	0.173	**B3**	0.212
A4	0.285	**B4**	0.257
A5	0.211	**B5**	0.259
A6	0.273	**B6**	0.236
A7	0.189	**B7**	0.224
A8	0.320	**B8**	0.195
A9	0.289	**B9**	0.284

While, the highest concentration of flavonoid content calculated by using the standard curve of flavonoid is B9 which is 314.33 μg/ml while the lowest is B8 which is 215.44 μg/ml Flavonoid content dry has higher than fresh leaves.

3.3. Total Phenol Content

From Table 3, the highest concentration of phenol content calculated by using the standard curve of phenol is A3 which is 58.22μg/ml while the lowest is A6 which is 21.70μg/ml. The highest concentration of phenol content calculated by using the standard curve of phenol is B9 which is 49.30 μg/ml while the lowest is B3 which is 21.26μg/ml, this finding also support by Bahrani et al. 2014 [10] and Naef 2010 [11]. Phenol content of dry leaves has higher compare to fresh leaves.

Table 3. Average absorbance value of total phenolic content for mature leaves and fresh leaves

Mature Leaves		**Fresh Leaves**	
Samples	**Absorbance**	**Samples**	**Absorbance**
A1	0.232	**B1**	0.236
A2	0.258	**B2**	0.242
A3	0.333	**B3**	0.163
A4	0.169	**B4**	0.250
A5	0.189	**B5**	0.202
A6	0.165	**B6**	0.277
A7	0.206	**B7**	0.213
A8	0.253	**B8**	0.213
A9	0.180	**B9**	0.292

3.4. The Effect of Substrate Concentration and Reaction Time

From Tables 4 and 5, the best inhibition is in lower substrate concentration and short reaction time for both dry and fresh samples. This may due to long reaction will lower the efficiency of the inhibitors.

Table 4. Average absorbance value of mature leave samples for 100, 200 and 300 μM substrate concentrations and percentage inhibition

Time	**Concentration (μM Absorbance)**					
(min)	**100**	**(%)**	**200**	**%**	**300**	**%**
20	0.215	31.5	0.221	29.6	0.238	24.2
40	0.276	12.1	0.247	21.3	0.246	21.7
60	0.265	15.6	0.269	14.3	0.255	18.8

Table 5. Average absorbance value of fresh leave samples for 100, 200 and 300 µM substrate concentrations and percentage inhibition

Time	**Concentration (µM Absorbance)**					
(min)	**100**	**(%)**	**200**	**%**	**300**	**%**
20	0.215	31.5	0.237	24.5	0.261	16.9
40	0.218	30.6	0.231	26.4	0.217	30.9
60	0.282	10.2	0.261	16.9	0.246	21.7

3.5. HPLC (High Performance Liquid Chromatography)

HPLC is one of the important widely used purification techniques for the isolation of natural products such as flavonoids, gallic acid, phenolic and quercetin compounds [12].

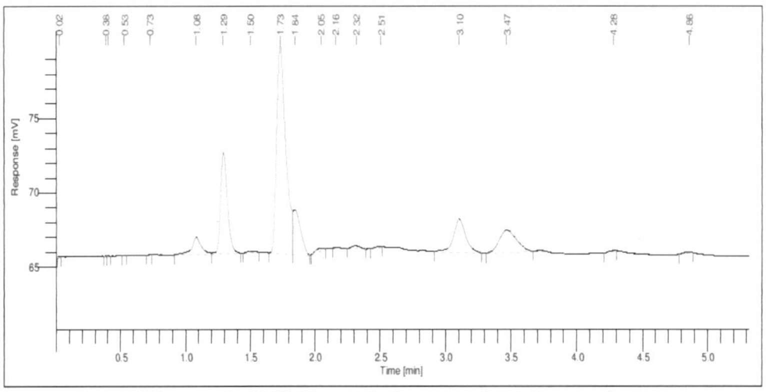

Figure 1. HPLC result for sample A5 sample

From the HPLC result of mature and fresh leaves samples from Figures 1 and 2, it shows the both samples mature leave (A5) and fresh leave (B5) contains garlic acid and quercetin compounds. Based on standard curve equation for garlic acid and quercetin compound the highest concentrations were obtained are 12.94 µg/ml and 5.15 µg/ml, respectively.

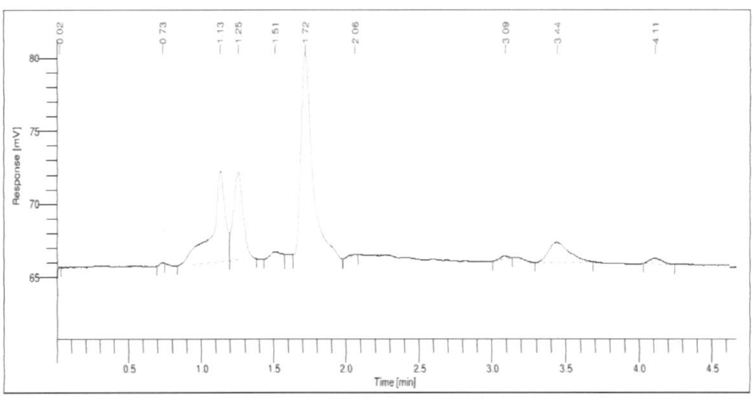

Figure 2. HPLC result for sample B5 sample

4. Conclusion

In this research, we have recovered the pancreatic lipase inhibitors from *A.subintegra* mature and fresh leaves extract. There contains phenol (49.30 µg/ml), flavonoid (314.33 µg/ml), gallic acid (12.94 µg/ml) and quercetin (5.15 µg/ml) from this research. Future work on this research will focus more about the potential of this leaves extract with more feasibility of different materials and technique to improving the efficiency of resource utilization.

IOP Conf. Series: Materials Science and Engineering **358** (2018) 012039 doi:10.1088/1757-899X/358/1/012039

Acknowledgements
The authors would like to acknowledge Faculty of Chemical Engineering, UiTM for providing analytical instruments, materials and support by researchers, academician, practitioners and staffs contributed towards understanding and thoughts and guidance.

References
[1] Pearsoon G A 2007 *Agarwoood: The Life of a Wounded Tree* (Leiden: Leiden University Newsletter) p 2
[2] Muhammad Hanif Azhari N A M 2014 *Morphological evaluation on leaf characteristics of common Aquilaria species in Malaysia* (Bangi : University Kebangsaan Malaysia)
[3] Ibrahim A 2017 Comparative *Analysis of In Vitro Bioactivities and Phenolic Content of Leaf Extracts from Six Species of Aquilaria* (Penang: Universiti Sains Malaysia) p 4
[4] Ray G, Jiratchariyakul W, Sithisarn P, and Leelamanit W 2017 *Antioxidant Activity of Aquilaria subintegra Ethanolic Leaf Extract* (Bangkok : Mahidol University)
[5] Ishihara M, Tsuneya T, and Uneyama K 1993 *J. Ess. Oil Res* **5** 283-289
[6] Danis A O 2015 *Inhibition of Pancreatic Lipase by Culinary Plant Extracts*. Sci Med central
[7] Khairudin N F, and Ku Hamid K H 2012 *J. of Tropical Forest Sci.* 24(**4**) 557–564
[8] Pornpunyapat J, Chetpattananondh P, and Tongurai C 2011 *Bangladesh J. Pharmacol* **6** 18-24
[9] Bueno F G, Machareth M A D, Panizzon G P, Lopes G C, Mello J C P, and Leite-Mello E V S 2012 *Quím. Nova* Vol.35 **4** 822-826
[10] Bahrani1 H, Mohamad J, Paydar M, and Rothan H A 2014 *Current Alzheimer Res.* **11** 1-9
[11] Naef R 2010 *Flavour and Fragrance J.* **26** 73-87
[12] Gothai S, Ganesan P, Park S, Fakurazi S, Choi D, and Arulselvan P 2016 Nutrients **8** 461

IOP Conf. Series: Materials Science and Engineering **358** (2018) 012040 doi:10.1088/1757-899X/358/1/012040

Effect of Plasticizers on Physicochemical and Mechanical Properties of Chitosan-Gelatin Films

N Mohammed Manshor, M I Rezali, J Jai and A Yahya

Faculty of Chemical Engineering, Universiti Teknologi MARA, 40450 Shah Alam, Selangor, Malaysia

norasmah@salam.uitm.edu.my

Abstract. Composite chitosan-gelatin films were produced to investigate the effect of plasticizer and composition of chitosan and gelatin on physicochemical and mechanical properties of the films. The films were prepared according to ratio of chitosan: gelatin of 1:1, 1:2 and 2:1. For each film, glycerol, sorbitol and sucrose were added as plasticizer. The film forming solution was poured on a glass plate and dried for 12 hours in an oven at 60°C. The highest tensile strength was 4.04 MPa for films of ratio 2:1 plasticized with glycerol compared to sorbitol and sucrose which were 3.94 MPa and 3.84 MPa, respectively. However, films plasticized with sorbitol at ratio of 1:2 had the highest percent elongation which was 68.20%, followed by glycerol and sucrose which were 26.51% and 24.08%, respectively.

1. Introduction

Synthetic plastics are non-biodegradable which can cause increasing of waste and lead to pollution for centuries [8]. As an alternative, edible film or coating from biodegradable materials that can be degraded easily or might be consumed directly without producing waste [2]. It is produced from polysaccharides, proteins, lipids or combination of these materials. Besides of edibility and biodegradability, these materials are abundantly available and low cost. The functionality and performance of edible film as barrier to moisture and gases depend on film composition [9]. Chitosan is an amino polysaccharide derived from the N-deacetylation of chitin which exists naturally in exoskeleton of crustacean such as crab and shrimp [15]. It is whitish in color, odorless and tasteless. Chitosan is naturally abundant biopolymers that have excellent properties such as biodegradable, non-toxic, and bio-compatible [7]. Great attention of chitosan mostly due to its ability as film forming material in food industry as food packaging and preservative because it shows satisfactory gas barrier and mechanical properties [22].

Gelatin can be obtained by the controlled hydrolysis of the fibrous insoluble collagen present in the bones and skin generated as waste during animal slaughtering and processing [15]. Like chitosan, gelatin can form films and coating with good optical properties, adequate mechanical properties and excellent gas barrier properties [16]. Gelatin also has been found to contribute as major constituent in bio-plastic industry as food coating and preservative. Plasticizing agent are generally essential to overcome the weakness of film, mainly brittleness. By reducing intermolecular forces of polymer, plasticizer reduces the rigidity of film structure and increasing the mobility of biopolymeric chains, thus improving film's mechanical properties [18]. Glycerol has been proven to enhance film flexibility, reduce film puncture strength along with elasticity and also good water vapor barrier [20]. Besides, glycerol is mainly compatible for gelatin based film in improving mechanical properties and flexibility of the films [1]. Bergo et al. [4] found that the structure of gelatin film plasticized with glycerol showed no tendency of re-crystallization due to high stability and increasing of moisture in

Content from this work may be used under the terms of the Creative Commons Attribution 3.0 licence. Any further distribution of this work must maintain attribution to the author(s) and the title of the work, journal citation and DOI.
Published under licence by IOP Publishing Ltd

the film. Fakhoury et al. [10] showed that manioc-gelatin film plasticized with sorbitol has stronger resistant through higher value of tensile strength and percent of elongation. Furthermore, it was found that amaranth flour films plasticized with glycerol give higher percent of elongation compared to films that used sorbitol [19]. The study by Isotton et al. [13] concluded that any plasticizers from the group of polyols were found to perfectly blend with starch based film. The structure was changed from malleable to brittle when using sucrose as plasticizer for starch based film [17]. The usage of sucrose as plasticizer to ß-lactoglobulin film was found to increase the transition temperature T_g of the film which means that the suitability and bonding between film and plasticizer was not recommendable [6]. Many types of plasticizer have been used successfully in formulation of edible film for the purpose to improve their characteristics. Although numerous studies have been done on composite chitosan-gelatin film, this research was focused to investigate the effect of different type of plasticizer on physicochemical and mechanical properties of the film.

2. Materials and Methods

2.1. Materials

Chitosan and gelatin as raw materials were obtained from Merck. Glycerol (Merck Chemicals), sorbitol (Sigma Aldrich) and sucrose (Systerm) acts as plasticizer.

2.2. Preparations of chitosan and gelatin films

The method was adopted from Hosseini et al. [12] with some modification. Chitosan (2% w/v) was prepared by dissolving chitosan in acetic acid (1% w/v) and stirred for four hours at temperature of 60°C. Gelatin (4% w/v) was prepared by dissolving gelatin in distilled water and stirred for half an hour at temperature of 40°C. Then, chitosan and gelatin film forming solution was mixed at ratio of 1:1, 1:2 and 2:1 and continuously stirred for another half an hour. Glycerol, sorbitol and sucrose were added to each film forming solution prior to casting on glass plate. Then it was dried in an oven at 55°C for 24 hours.

2.3. Mechanical Properties

Universal Tensile Machine (Tinius Olsen, H50KT) was used to measure tensile strength and percent elongation of the films. The test was carried out according to ASTM D-682 with cross head speed of 10 mm/min. Tensile strength was calculated by dividing the maximum load for the film to break cross-sectional area, and percent elongation by dividing film elongation at rupture to initial gauge length. Percent elongation is the ratio of extension to the length of the sample.

2.4. Physicochemical Properties

Physicochemical properties tested included moisture content, film thickness and color. Moisture content was carried out by drying each of the film in an oven at 105°C. The initial and final weights were recorded. Meanwhile film thickness was measured by using micrometer (Mitutoyo, Series 103-177) at five different points then the values were averaged. Color measurement was carried out by using chromameter where the measurement results were displayed in Hunter values of lightness, L*, redness or greenness, a* and yellowness or blueness, b*.

3. Results and Discussions

3.1. Tensile strength

Figure 1 shows the effect of chitosan: gelatin ratio on tensile strength for different type of plasticizer. It is observed that the highest tensile strength was the film plasticized with glycerol compared to sorbitol and sucrose. Similar finding was obtained by Srinivasa et al. [18], where the chitosan/fatty acid films plasticized with glycerol had greater tensile strength than sorbitol and sucrose. Glycerol was miscible easily with chitosan and became more flexible compared to other plasticizers. This is also believed that due to molecular weight of glycerol is smaller than sorbitol and sucrose; it influences the mechanical properties of the film, thus giving more flexibility than other plasticizers. Al-Hassan and Norziah [1] found that the tensile strength for sago starch/fish gelatin films plasticized with sorbitol had the highest tensile strength compared to glycerol. However, glycerol was found to be more effective compared to sorbitol when dealing with polysaccharides, protein and polysaccharide-protein based films because of higher mobility of polymer chains. Hence, the polysaccharides based film become more flexible and stretchable when using glycerol as plasticizer compared to sorbitol.

For films at ratio of chitosan: gelatin of 1:2, the tensile strength was the highest when plasticized with sorbitol compared to glycerol and sucrose. Chen et al. [6] reported that sorbitol plasticized films had slightly higher tensile strength than glycerol and polyethylene glycol (PEG) but increasing of plasticizer concentration, glycerol plasticized films showed higher tensile strength than sorbitol and PEG. From Figure 1, film plasticized with sucrose have the lowest tensile strength compared to other plasticizers. The usage of sucrose as plasticizer changed the film from malleable to brittle film [17]. Similar result was obtained by Al-Hassan and Norziah [1]. According to ratio of chitosan: gelatin, the highest tensile strength was recorded at 4.04 MPa for ratio 2:1. This shows that the higher the amount of chitosan, the higher the tensile strength of the film. The result was supported by Xu et al. [21] that found the tensile strength of composite film of chitosan and starch was increased by the increasing amount of chitosan leading to stronger films produced.

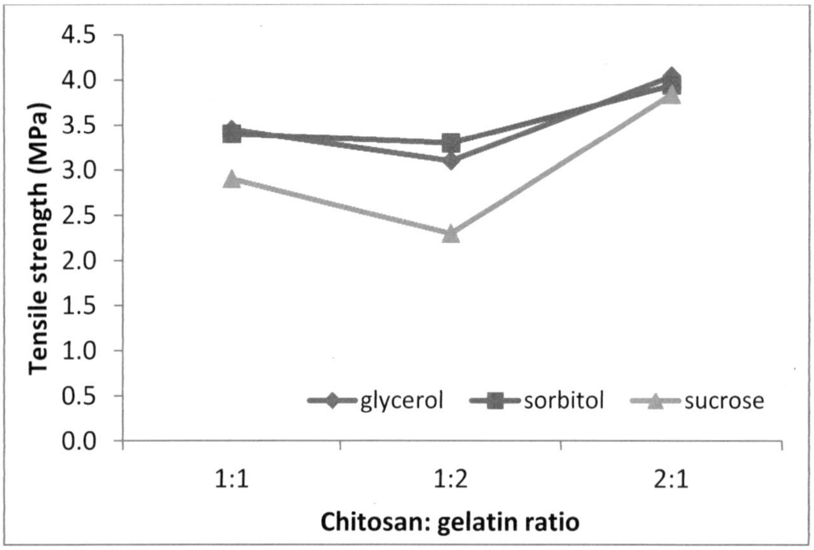

Figure 1. Effect of plasticizer on tensile strength for different chitosan:gelatin ratio.

3.2. Percent elongation

The effect of plasticizer on percent elongation for different chitosan:gelatin ratio is illustrated in Figure 2. It is shown that films plasticized with sorbitol had the highest percent elongation compared to glycerol and sucrose. Similar finding also found by Chen et.al [6], Arvanitoyannis et al. [3] and Srinivasa et al. [18] when comparing the percent elongation of films plasticized by few plasticizers such as glycerol, PEG and palmitic acid. Fakhoury et al. [10] reported that the sorbitol-plasticized films had about 20% higher percent elongation compared to glycerol-plasticized films. According to them, during the drying of films, water was evaporated and allowed the formation of starch network that bring the chains very near to each other, facilitating the formation of a denser matrix. However, several researchers had found that glycerol was the effective plasticizer which produced films with better percent elongation compared to sorbitol [5, 11]. The effectiveness of glycerol is most likely due to its small molecular weight, highest number of hydroxyl groups at the same concentration and hydrophilic nature which attracts water into the polymer matrix thus the effective level of plasticizer increases [11].

As shown in Figure 2, the highest percent elongation was obtained by glycerol plasticized film with ratio 2:1. While film plasticized with sucrose have the lowest percent elongation compared to sorbitol and glycerol. According to Fadini et al. [9], this is might be the conversion of amorphous sucrose to crystalline sucrose, resulting in brittle films that weakened and cracked. Although sucrose plasticized film exhibit lower percent elongation than others for chitosan:gelatin ratio of 1:1 and 1:2, the use of sucrose can be a great alternative plasticizer since it exhibit higher tensile and percent elongation properties compared to films without any plasticizer [14].

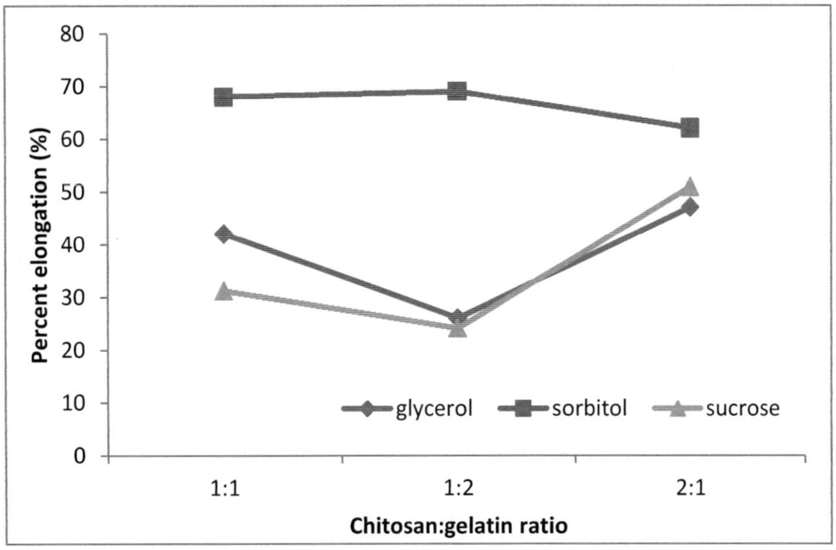

Figure 2. Effect of plasticizer on percent elongation for different chitosan:gelatin ratio.

3.3. Thickness

Table 1 shows the thickness, moisture content and color parameter for each of the film at different chitosan:gelatin ratio and plasticizer. Film plasticized with glycerol have the highest thickness which was 0.20 mm for ratio 1:1 followed by sorbitol and sucrose as shown in Table 1. This is because of hygroscopic nature of glycerol, the ability to retain moisture between its molecules resulted the film to be thicker than others. Fakhoury et al. [10] found that the thickness of chitosan-starch film was increased with the used of glycerol compared to sorbitol. Glycerol plasticized film produced thicker film because of its hydrophilic properties and higher protein content than other films [1]. Furthermore, films with ratio 1:2 have greater thickness than films with ratio 2:1. According to Hosseini et al. [12], biodegradable film produced with chitosan was found to decrease in water vapor permeability which led to thinner film produced. The results obtained were contradicted might be due to the addition of chitosan enhanced the cross-linking of gelatin and increase the free volume of the polymeric matrix.

3.4. Moisture content

Films with high chitosan content possess higher moisture content compared to films with high gelatin content as shown in Table 1. According to Al-Hassan & Norziah [1], the films produced from protein or polysaccharides material with chitosan is highly sensitive to moisture and exhibit poor water vapor resistance. Thus, the results obtained showed that the films with highest chitosan amount (ratio of 2:1) contain the highest moisture (27.87% for glycerol, 20.54% for sorbitol and 19.78% for sucrose). On the other hand, the films produced with chitosan-gelatin ratio of 1:2 shows the lowest amount of moisture content which are 19.78%, 17.24% and 17.65% for glycerol, sorbitol and sucrose, respectively. This shows that the lower the amount of chitosan, the lower the moisture content in the films. Moreover, glycerol plasticized films contains the most moisture content compared to sorbitol and sucrose. Bergo et al. [4] found that the used of glycerol among other plasticizers showed the greatest moisture content and increase significantly with the increasing amount of glycerol. Glycerol was identified having hygroscopic nature due to high likely to absorb moisture. Thus, it tends to retain water in the films than other plasticizer.

Table 1. Thickness, moisture content and color parameters for each films.

Plasticizer	Chitosan-Gelatin Ratio	Thickness (mm)	Moisture Content (%)	Color		
				L*	a*	b*
Glycerol	1:1	0.20	24.00	43.93	-0.46	1.42
	1:2	0.20	19.78	40.91	-0.44	1.42
	2:1	0.16	27.87	43.10	-0.90	3.35
Sorbitol	1:1	0.18	19.42	43.44	-0.43	1.48
	1:2	0.20	17.24	43.35	-0.41	1.69
	2:1	0.18	20.54	40.25	-0.82	2.46
Sucrose	1:1	0.16	19.23	42.79	-0.36	1.85
	1:2	0.19	17.65	40.65	-0.45	1.85
	2:1	0.14	19.78	43.07	-0.63	2.14

3.5. Color

Color of the films are expressed in term of L* (lightness or brightness), a* (redness or greenness) and b* (yellowness or blueness). From observation, overall films produced were slightly transparent. From Table 1, the yellowness of the films increases with the increasing amount of chitosan. The films produced with doubled amount of chitosan (ratio 2:1) shows higher lightness value than films with doubled amount of gelatin (ratio 1:2) except for films plasticized with sucrose. The values of a* and b* increase with the increasing amount of chitosan except for sucrose plasticized film. It indicates that, the higher the amount of chitosan in the film, the higher the tendency of films to be yellowish and greenish.

4. Conclusion

In summary, it can be concluded that the glycerol plasticized films have the highest tensile strength compared to sorbitol and sucrose at any ratio. The highest percent elongation was obtained by sorbitol plasticized films followed by glycerol and sucrose. Sucrose was found to make films from malleable to brittle despite of any ratio used. Due to hygroscopic nature, glycerol tends to promote higher moisture content for films containing double amount of chitosan and decrease the thickness. From observation, there has been no significant difference in colour for all films produced. Based on the value of L*, a* and b*, the higher amount of chitosan resulted the higher lightness, yellowness and greenness value.

Acknowledgment

The authors wish to extend their gratitude to Universiti Teknologi MARA (UiTM), Shah Alam for the financial support through 600-RMI/DANA 5/3/LESTARI (30/2015). The authors also would like to thank Faculty of Chemical Engineering, Universiti Teknologi MARA (UiTM), Shah Alam for the support, services and facilities provided.

References

[1] Al-Hassan A A and Norziah M H 2012 *Food Hydrocolloids* **26** 108-17
[2] Aruna D, Sasikala P, Rajamani L, Kavitha V, Yazhini G and Shakila Banu M 2012 *Food Sci. Qual. Manage.* **3** 1-10
[3] Arvanitoyannis I S, Nakayama A and Aiba S 1998 *Carbohydr. Polym.* **37** 371-82
[4] Bergo P, Moraes I C and Sobral P J A 2013 *Food Hydrocolloids* **32** 412-15
[5] Boesel L F 2014 *Carbohydr. Polym* **115** 356-63
[6] Chen H R, Chen H T, Liu S, Dun G.Q and Zhang Y 2014 *J. Northeast Agric. Univ. (English Edition)* **21** 67-72
[7] Dutta P K, Dutta J and Tripathi V S 2004 *J. Sci. Ind. Res.* **63** 20-31
[8] Espitia P J P, Du W, Avena-Bustillos R J, Soares N F F and McHugh T H 2014 *Food Hydrocolloids* **35** 287–96
[9] Fadini A L, Rocha F S, Alvim I D, Sadahira M S, Queiroz M B, Alves R M V and Silva L B 2013 *Food Hydrocolloids* **30**(2) 625-31

[10] Fakhoury F M, Martelli S M, Bertan L C, Yamashita F, Mei L H I and Queiroz F P C 2012 *LWT-Food Sci. and Technol.* **49** 149-54
[11] Haq M A, Hasnain A and Azam M 2014 *LWT-Food Sci. and Technol.* **55**(1) 163-69
[12] Hosseini S F, Rezaei M, Zandi M and Ghavi F F 2013 *Food Chem.* **136**(3) 1490-95
[13] Isotton F S, Bernardo G L, Baldasso C, Rosa L M and Zeni M. 2015 *Ind. Crops and Prod.* **76** 717-24
[14] Kurek M, Galus S and Debeaufort F 2014 *Food Packaging and Shelf Life* **1**(1) 56-67
[15] Pereda M, Ponce A G, Marcovich N E, Ruseckaite R A and Martucci J F 2011 *Food Hydrocolloids* **25** 1372-81
[16] Sahoo S and Nayak L 2013 *J. Pharm. Biomed. Anal.* **30** 892-901
[17] Santacruz S, Rivadeneira C and Castro M. 2015 *Food Hydrocolloids* **49** 89-94
[18] Srinivasa P C, Ramesh M N and Tharanathan R N 2006 *Food Hydrocolloids* **21** 1113-122
[19] Tapia-Blácido D R, Sobral P J A and Menegalli F C 2013 *LWT-Food Sci. and Technol.* **50** 392-400
[20] Vanin F M, Sobral P J A, Menegalli F C, Carvalho R A, Habitante A M Q B 2005 *Food Hydrocolloids* **19** 899–907
[21] Xu Y X, Kim K M, Hanna M A and Nag D 2015 *Ind.Crops and Prod.* **21** 85–92
[22] Zhong Q P and Xia W S 2008 **46**(3) 262-69

IOP Conf. Series: Materials Science and Engineering **358** (2018) 012041 doi:10.1088/1757-899X/358/1/012041

Optimization of Bread Enriched with *Garcinia mangostana* Pericarp Powder

U K Ibrahim, R Mohd Salleh, S N S Maqsood-ul-Hague, S Abd Hashib and S F Abd Karim

Faculty of Chemical Engineering, Universiti Teknologi Mara, 40450 Shah Alam, Selangor, Malaysia

ummi985@salam.uitm.edu.my

Abstract. The aim of present work is to optimize the formulation of bread enhanced with *Garcinia mangostana* pericarp powder with the combination of baking process conditions. The independent variables used were baking time (15 - 30 minutes), baking temperature (180 - 220°C) and pericarp powder concentration (0.5 - 2.0%). The physical and chemical properties of bread sample such as antioxidant activity, phenolic content, moisture analysis and colour parameters were studied. Bread dough without fortification of pericarp powder was used as control. Data obtained were analyzed by multiple regressions and the significant model such as linear and quadratic with variables interactions were used. As a conclusion, the optimum baking conditions were found at 213°C baking temperature with 23 minutes baking time and addition of 0.87% for *Garcinia mangostana* pericarp powder to the bread formulation.

1. Introduction

Recent revolutions in the world reflect extreme transformations in many aspects of human life especially in daily food consumption. Nowadays, people are looking food for the health protection that is easily accessible and nutritious. Unfortunately, present foods are mostly having minimum nutrients essential for the maintenance of human health [1]. Thus, it is necessary to develop enriched food with functional application. Interests in incorporating active ingredients such as dietary fiber and phenolic antioxidants into popular foods such as bread have grown rapidly, due to the increased consumer health awareness [2]. The main characteristic of an antioxidant is its ability to trap free radicals. Highly reactive free radicals and oxygen species are present in biological systems from a wide variety of sources. These free radicals may oxidize nucleic acids, proteins, lipids or DNA and can initiate degenerative disease. Antioxidant compounds like phenolic acids, polyphenols and flavonoids scavenge free radicals such as peroxide, hydroperoxide or lipid peroxyl and thus inhibit the oxidative mechanisms that lead to degenerative diseases.

Food byproduct, due to its phenomenal biodiversity, is a treasure house of many novel healthy food ingredients and biologically active compounds such as those found in *Garcinia mangostana* pericarp. Despite having so many health benefits, *Garcinia mangostana* pericarp has been underexploited for food purposes. Bakery products are widely used throughout the world and are the potential source of incorporating with functional ingredient. Bread was found to have advantages as the nutritional products for health through additional of functional ingredients. One of the latest enrichment has been the addition of seaweed to enhance the phytochemical content in breadsticks [3]. The authors revealed that fiber-rich seaweed bakery products are acceptable to consumers and have potential of increasing seaweed consumption among non-seaweed consumers. Thus, expanding the use of breads as medium for nutritious product could potentially be done in the near future.

Content from this work may be used under the terms of the Creative Commons Attribution 3.0 licence. Any further distribution of this work must maintain attribution to the author(s) and the title of the work, journal citation and DOI.
Published under licence by IOP Publishing Ltd

Bread making is one of the most usual ways of processing cereals. The antioxidant activity and residual free phenolic acid content of flour were found decreased by mixing, fermentation and baking [4]. Due to that, addition of functional ingredient that contains high antioxidant in dough formulation hope will enhance the antioxidant activity in bread and may have potential health advantages for consumers. Therefore, the amount of phenolic acids and antioxidant activity in bread after baking process should be investigate and optimization of the process should be done. Several studies have been carried out to find potential sources of natural antioxidant in bread [5-12]. With respect to that, this study would like to take the opportunity to find out the potential of *Garcinia mangostana* pericarp powder as a functional ingredient in bread formulation.

2. Material and methods

2.1. Dough formulations
Commercial wheat flour, sugar, salt, yeast and bread improver was obtained from local market. *Garcinia mangostana* pericarp powder was purchased from Malaysia Herbal Shop. A standard white bread recipe was prepared. All dry ingredients were weighed and mixed in a laboratory mixer for 10 minutes at low speed. The produced dough was covered with food wrapper to prevent excessive moisture loss and left to rise at room temperature for 60 minutes. Then, the dough was punched and kneads again to release the air inside the dough. Each of dough was weighed in equal size and placed in four rectangular baking containers. The dough was allowed to rest for another 60 minutes before baking. The design of experiment is presented in Table 1.

Table 1. Design experiment of bread baking

Run	Temp (°C)	Time (min)	Pericarp powder (%)	Run	Temp (°C)	Time (min)	Pericarp powder (%)
1	220.00	30.00	2.00	11	200.00	22.50	1.25
2	200.00	22.50	1.25	12	200.00	22.50	0.50
3	200.00	22.50	1.25	13	180.00	15.00	0.50
4	200.00	22.50	1.25	14	200.00	22.50	2.00
5	180.00	15.00	2.00	15	180.00	30.00	0.50
6	220.00	15.00	2.00	16	200.00	22.50	1.25
7	220.00	15.00	0.50	17	180.00	30.00	2.00
8	180.00	20.00	1.25	18	200.00	30.00	1.25
9	220.00	20.00	1.25	19	220.00	30.00	0.50
10	200.00	22.50	1.25	20	200.00	15.00	1.25

2.2. Baking conditions
Bread dough was baked at different temperatures (180 to 220°C) and different times (15 to 30 minutes). Each of the dough with different baking condition was placed in the conventional baking oven to provide uniform heat distribution over the dough during baking process. The control bread sample was baked without addition of *Garcinia mangostana* pericarp powder with the same baking conditions.

2.3. Bread sample extraction
The bread sample extraction, phenolic acid measurement and antioxidant activity were determined using modified procedure established by Das et al. [13]. Bread samples were sliced (3 cm width and 1cm thickness) and dried in an oven at 40°C for 24 h. The dried material was ground in a grinder to obtain powdered bread. 10 g of powdered bread was extracted with 100 mL of 80% aqueous methanol for 2 h at 40°C. Samples were then centrifuged at 12,000×g for 15 min. The supernatant collected was used in the assay.

2.4. Determination of phenolic acid
1 mL of bread extract was added with 0.5 mL of the Folin-Ciocalteau reagent. After 5 minutes, 1 mL of 7.5% sodium carbonate was added and the contents mixed thoroughly. The final volume should be adjusted to 10 mL with distilled water and left to stand for 30 minutes at room temperature. A colour

developed and the absorbance was measured at 765 nm in spectrophotometer using gallic acid as a standard. Calibration curve for gallic acid equivalent was determined using modified procedure established by de Amorim et al. [14].

2.5. Determination of antioxidant activity

The DPPH radical solution was prepared by dissolving 10 mg of DPPH in 100 mL of 80% methanol. 1 mL of the bread extracts were then mixed with 1 mL of DPPH solution and 4 mL of 80% methanol. The mixtures were shaken vigorously and allowed to stand at room temperature for 30 minutes. The control consisted of 1 mL DPPH + 4 mL 80% methanol. The mixture was kept in dark for 30 minutes at room temperature. The decrease in absorbance of the resulting solution was measured spectrophotometrically at 517 nm. The percentage of inhibition or the percentage of discoloration was calculated as Equation 1:

$$\left[\frac{Absorbance(\text{control})_{517\text{nm}} - Absorbance(\text{sample})_{517\text{nm}}}{Absorbance(\text{control})_{517\text{nm}}}\right]\times 100\% \tag{1}$$

2.6. Colour measurement

The crust and the crumb color parameters of the bread samples were measured following the method described by Das *et al.* [14] with some modifications. The colour parameters was obtained from chromameter, CR-400, Konica Minolta, illumination of D_{65}, by standardizing with a white plate (L* = 89.8; a* = 0.3173; b* = 0.3343). For each treatment, 3 replications were taken. The results were expressed according to the CIELAB system of colour measurement in terms of L* (lightness/darkness) ranging from 100 (white) to 0 (dark), a* (redness/greenness) ranging from positive values (red) to negative values (green), and b* (yellowness/blueness) ranging from positive values (yellow) to negative values (blue).

2.7. Moisture content measurement

Bread crumb and crust was ground using laboratory grinder/blender. Moisture content was determined by taking about 2 g of the grinded bread crumb and crust, placed in a moisture analyzer plate and dried for 30 minutes at 130°C. Percentage of moisture in the bread crust and crumb were read from the moisture analyzer.

3. Results and Discussion

3.1. Analysis of variance (ANOVA)

The acceptability of the model was determined by the lack of fit, coefficient of determination (r^2) and the Fisher's test value (F-value) obtained from the analysis of variance (ANOVA) generated by the Design Expert software. ANOVA were performed to calculate the significance of the linear, quadratic and interaction effects of the factor variables on response variables. The model equations and coefficient of determination (r^2) for each response variable were given in Table 2. In order to get the best fit model, the minimum value for r^2 should be at least 80% [15]. The significance of the coefficients was determined by using p-values ($p \leq 0.05$). From the results showed, most of response variables can be considered as good fit model as their coefficient of determination were more than 80% except for moisture content in the bread crust and crumb.

Table 2. Regression equations for physical and chemical properties of bread with different dough formulations and baking conditions

Chemical properties	Equation	*p*-values ($p \leq 0.05$)	Lack of fit	r^2
Phenolic content (bread crust)	$Y = 1.59 - 0.06\,X_1 + 0.06\,X_2 - 0.44\,X_3{*} - 0.11\,X_1^2 + 0.56\,X_2^{2}{*} + 0.14\,X_3^2 + 0.0065\,X_1X_2 + 0.056\,X_1X_3 - 0.024\,X_2X_3$	0.014	0.4215	0.8016
Phenolic content (bread	$Y = 1.57 - 0.036\,X_1 - 0.025\,X_2 - 0.090\,X_3 + 0.91\,X_1^{2}{*} + 0.13\,X_2^2 - 0.24\,X_3^2 - 0.13\,X_1X_2 - 0.010\,X_1X_3 - 0.12\,X_2X_3$	0.0065	0.2234	0.8336

crumb)				
Antioxidant (bread crust)	$Y = 86.09 - 11.47 X_1* - 0.51 X_2 + 0.15 X_3 - 8.94 X_1^{2}* - 3.08 X_2^{2}* + 0.84 X_3^2 - 0.34 X_1X_2 + 0.046 X_1X_3 + 1.07 X_2X_3$	<0.0001	0.4814	0.9801
Antioxidant (bread crumb)	$Y = 83.32 - 6.80X_1* + 0.24 X_2 - 1.12 X_3 - 0.20 X_1^2 + 8.59 X_2^{2}* - 9.97 X_3^{2}* + 0.81 X_1X_2 - 0.067 X_1X_3 - 0.32 X_2X_3$	0.0054	0.1120	0.8402
Physical properties				
Colour (bread crust)				
L*	$Y = 49.33 + 1.31 X_1 - 4.84 X_2* - 4.42 X_3* - 0.45 X_1X_2 - 6.36 X_1X_3* - 0.71 X_2X_3$	0.0005	0.4495	0.8103
a*	$Y = 12.49 + 0.70 X_1* + 2.51 X_2 - 0.77 X_3 - 0.84 X_1^2 - 3.49 X_2^{2}* - 1.37 X_3^2 + 0.29 X_1X_2 - 0.029 X_1X_3* - 1.50 X_2X_3$	0.0001	0.1495	0.9792
b*	$Y = 25.95 - 1.76 X_1 - 2.07 X_2* - 1.22 X_3 - 2.71 X_1^2 - 2.92 X_2^2 - 1.26 X_3^2 - 0.40 X_1X_2 + 4.90 X_1X_3* - 0.50 X_2X_3$	0.0017	0.1533	0.8770
Colour (bread crumb)				
L*	$Y = 56.61 - 0.85 X_1 - 0.41 X_2 - 2.39 X_3* + 0.33 X_1^2 - 0.97 X_2^2 - 2.40 X_3^2 + 0.37 X_1X_2 + 1.59 X_1X_3* + 0.62 X_2X_3$	0.0143	0.5693	0.8006
a*	$Y = 1.49 - 0.16 X_1 + 0.66 X_2 + 2.20 X_3* - 0.49 X_1^2 + 0.58 X_2^2 + 1.13 X_3^2 - 0.79X_1X_2 - 0.17X_1X_3 + 0.79 X_2X_3*$	0.0018	0.1250	0.8747
b*	$Y = 15.93 + 0.19 X_1 + 0.43 X_2 + 1.56 X_3* - 1.33 X_1^2 + 1.02 X_2^2 - 0.83 X_3^2 - 0.23 X_1X_2 - 0.30 X_1X_3 + 0.27 X_2X_3$	0.0129	0.0795	0.8054
Moisture content (bread crust)	$Y = 26.96 + 0.99 X_1 + 0.16 X_2 + 2.47 X_3* - 2.75 X_1^2 + 2.48 X_2^2 - 3.18 X_3^2 + 0.63 X_1X_2 + 0.17 X_1X_3 - 0.62 X_2X_3$	0.0320	0.1694	0.7588
Moisture content (bread crumb)	$Y = 45.36 - 0.037 X_1 - 1.79 X_2* + 0.90 X_3$	0.0092	0.8275	0.5037

X_1 is baking temperature, X_2 is baking time and X_3 is concentration of *Garcinia mangostana* pericarp
*The term is significant at $p \leq 0.05$, lack of fit ≥ 0.05 and $r^2 \geq 0.8$

3.2. Optimization process

The criteria of optimization were selected on the basis of accepted bread quality. Therefore, the aim of this study was to maximize the responses of phenolic content and antioxidant activity as well as to get the desired colour and moisture content. As a result, the optimum levels of X_1 is 213°C, X_2 is 23 minutes and X_3 is 0.87%. The predicted responses at the optimum levels can be listed as 1.75 and 1.92 mg/g phenolic content and 74.68 and 77.02% of antioxidant activity in bread crust and crumb, respectively. For colour parameters, the value of lightness (L*) is 54.04 for bread crust and 56.22 for bread crumb. While, the redness value (a*) for both crust and crumb were 12.92 and 1.35 and finally for yellowness parameter (b*) indicates 22.10 and 14.61 for bread crust and crumb, respectively. All the colour parameters showed the acceptability for bread colour quality. Additionally, the predicted values for moisture content are 24.45 and 42.38% for both crust and crumb.

Table 3 shows the values of physical and chemical properties of bread sample and control. Bread enhanced with *Garcinia mangostana* pericarp powder was found to have significant effect especially on phenolic content and antioxidant activity. From the table, it demonstrated that enhanced bread contains higher antioxidant and phenolic content than normal bread. Other properties present almost comparable in quality by showing slightly difference between enhanced and normal bread. However, bread crumb lightness (L*) value was discovered much different between the sample and control due to the dark colour of pericarp powder but in the acceptable range. In conclusion, enhancement of *Garcinia mangostana* pericarp powder into bread formulation will produce functional bread with high phenolic content and antioxidant activity without affecting the quality of the bread such as colour and moisture content. All breads containing pericarp powder showed significantly higher phenolic compounds and antioxidant activities when compared with the control.

Table 3. Comparison between Predicted, Experimental and Control Values of Bread Sample Properties under Optimal Conditions

BREAD	Phenolic Content (mg/g)		Antioxidant Activity (%)		Colour		Moisture Content (%)	
	Crust	Crumb	Crust	Crumb	Crust	Crumb	Crust	Crumb
Predicted Values	1.75	1.92	74.68	77.02	L* = 54.04 a* = 12.92 b* = 22.10	L* = 56.22 a* = 1.35 b* = 14.61	24.45	42.38
Actual Values	1.81	1.97	74.83	77.10	L* = 54.96 a* = 12.22 b* = 23.8	L* = 55.42 a* = 1.38 b* = 13.69	25.14	42.55
Values Without Pericarp Powder (Control)	0.03	0.14	54.80	49.76	L* = 56.99 a* = 10.40 b* = 29.44	L* = 73.08 a* = 1.81 b* = 11.88	33.17	44.91
% Error Between Actual and Predicted Values	3.3%	2.5%	0.2%	0.1%	L* (1.7%) a* (5.7%) b* (7.1%)	L* (1.4%) a* (2.2%) b* (6.7%)	2.7%	0.4%

Optimal Conditions: Baking temperature, X_1 = 213°C, baking time, X_2 = 23 minutes and concentration of pericarp powder, X_3 = 0.87 wt%
Percentage error acceptable range ≈ ±10%

4. Conclusion

The objective of this study is achieved by performed the optimization of bread enhancement with *Garcinia mangostana* pericarp powder. The results of this study gave useful information antioxidant activity, phenolic content, bread surface colour and moisture content since magnitude of baking conditions and dough enhancement with *Garcinia mangostana* pericarp powder was adjusted according to the optimum conditions given. It was found that baking temperature, baking time and pericarp powder content give significant effect on phenolic content, antioxidant activity, bread crust and crumb surface colour and moisture content. The optimum levels were found 213°C for baking temperature, 23 minutes for baking time and 0.87% for *Garcinia mangostana* pericarp powder percentage added to the bread formulation. Since the study produce good experimental results, it should be recommended to produce high quality functional bread using *Garcinia mangostana* pericarp powder as functional ingredient in bread dough formulation. Further research will be carried out to investigate the feasibility of a functional bread model containing *Garcinia mangostana* pericarp powder, the health beneficial effects of the product, and possible side effects to encourage their application in improving human health.

Acknowledgement

Thanks to Institute of Research Management & Innovation (IRMI), UiTM for providing internal grant [File No: 600-IRMI/MYRA 5/3/LESTARI (0145/2016)], Ministry of Education (MOE) Malaysia for research funding and all of the contributions in completing the research study.

References

[1] Gohara A K, Souza A H P, Rodrigue Â C, Stroher G L, Gomes, Souza, S T M, Visentainer N E J V, and Matsushita M 2013 Chemometric methods applied to the mineral content increase in chocolate cakes containing chia and azuki *Journal of the Brazilian Chemical Society*. 24(5) pp 771-776

[2] Sivam A S, Sun-Waterhouse D, Siew Y Q and Perera C O 2010 Properties of bread dough with added fiber polysaccharides and phenolic antioxidants: a review *Journal of Food Science* **75**(8) pp 163-174

[3] Cox S and Abu-Ghannam N 2013 Incorporation of *Himanthalia Elongata* seaweed to enhance the phytochemical content of breadsticks using response surface methodology (RSM)

International Food Research Journal **20**(4) pp 1537-1545

[4] Conforti F D and Davis S F 2006 The effect of soya flour and flaxseed as a partial replacement for bread flour in yeast bread *International Journal of Food Science and Technology* **41**(2) pp 95–101

[5] Hardoko L, Hendarto and Siregar T M 2010 Pemanfaatan ubi jalar ungu (*Ipomoea Batatas l. poir*) sebagai pengganti sebagian tepung terigu dan sumber antioksidan pada roti tawar *Jurnal Teknologi dan Industri Pangan* **XXI**(1) pp 25-32

[6] Wang R, Zhou W and Isabelle M 2007 Comparison study of the effect of green tea extract (GTE) on the quality of bread by instrumental analysis and sensory evaluation *Food Research International* **40**(4) pp 470–479

[7] Das L, Raychaudhuri U and Chakraborty R 2012 Supplementation of common white bread by coriander leaf powder *Food Science and Biotechnology* **21**(2) pp 425-433

[8] Lim H S, Park S H, Ghafoor K, Hwang S Y and Park J 2011 Quality and antioxidant properties of bread containing turmeric (*Curcuma Longa* L.) cultivated in South Korea *Food Chemistry* **124**(4) pp 1577–1582

[9] Balestra F, Cocci E, Pinnavaia G and Romani S 2011 Evaluation of antioxidant, rheological and sensorial properties of wheat flour dough and bread containing ginger powder *LWT - Food Science and Technology* **44**(3) pp 700-705

[10] Ho L H, Noor Aziah Abdul Aziz and Baharin Azahari 2013 Physico-chemical characteristics and sensory evaluation of wheat bread partially substituted with banana (*Musa acuminata* X *balbisiana* cv. *Awak*) pseudo-stem flour *Food Chemistry* **139**(1-4) pp 532–539

[11] Abdellatif Mohamed, Xu J and Singh M 2010 Yeast leavened banana-bread: formulation, processing, colour and texture analysis *Food Chemistry* **118**(3) pp 620–626

[12] Peng X, Maa J, Cheng K W, Jiang Y, Chen F and Wang M 2010 The effects of grape seed extract fortification on the antioxidant activity and quality attributes of bread *Food Chemistry* **119**(1) pp 49–53

[13] Das L, Raychaudhuri U and Chakraborty R 2013 Herbal fortification of bread with fennel seeds *Food Technology and Biotechnology* **51**(3) pp 434–440

[14] de Amorim E L C, de Almeida de Castro V T N, de Melo J G, Corrêa A J C and da Silva Peixoto Sobrinho T J 2012 Standard operating procedures (SOP) for the spectrophotometric determination of phenolic compounds contained in plant samples *Latest Research into Quality Control* pp 47-66

[15] Şakiyan Ö 2014 Optimization of formulation of soy-cakes baked in infrared microwave combination oven by response surface methodology *Journal of Food Science Technology* **52**(5) pp 2910–2917

IOP Conf. Series: Materials Science and Engineering **358** (2018) 012042 doi:10.1088/1757-899X/358/1/012042

Optimization of Progressive Freeze Concentration on Apple Juice via Response Surface Methodology

S Samsuri[1], N A Amran[1] and M Jusoh[2*]

[1]Chemical Engineering Department, Universiti Teknologi PETRONAS, 32610 Seri Iskandar, Perak, Malaysia
[2]Faculty of Chemical and Energy Engineering, Universiti Teknologi Malaysia, 81310 UTM Skudai, Johor, Malaysia

mazura@cheme.utm.my[*]

Abstract. In this work, a progressive freeze concentration (PFC) system was developed to concentrate apple juice and was optimized by response surface methodology (RSM). The effects of various operating conditions such as coolant temperature, circulation flowrate, circulation time and shaking speed to effective partition constant (K) were investigated. Five different level of central composite design (CCD) was employed to search for optimal concentration of concentrated apple juice. A full quadratic model for K was established by using method of least squares. A coefficient of determination (R^2) of this model was found to be 0.7792. The optimum conditions were found to be coolant temperature = -10.59 °C, circulation flowrate = 3030.23 mL/min, circulation time = 67.35 minutes and shaking speed = 30.96 ohm. A validation experiment was performed to evaluate the accuracy of the optimization procedure and the best K value of 0.17 was achieved under the optimized conditions.

1. Introduction

The selection of the technology used in fruit juices concentration process plays an important role in order to maintain flavour, colour and functional properties of concentrated fruit juices. The use of freeze concentration (FC) system is an excellent idea in the manufacturing of high quality fruit juices [1]. During FC, water is separated from a liquid solution by cooling or freezing it until ice crystals are formed [2]. A very pure ice crystal which contains only water is left behind a concentrated liquid product. Several FC methods have been conceived to improve the effectiveness of the concentration process such as suspension freeze concentration (SFC) and progressive freeze concentration (PFC). SFC system consists of crystallization, growth, and separation of ice crystals where it comprises a scraped surface heat exchanger (SSHE) and a recrystallizer [3]. SSHE is the most expensive processing unit in an FC plant where 30% of the total investment costs come from SSHE [4]. Although a high separation efficiency is achieved but the initial and operational costs are relatively high. Thus, the industrial future of FC is associated more with PFC because of the simpler separation step and low operational costs.

PFC is one of the other methods of FC where only a single ice crystal is formed on a cooling wall with a moving ice front [5]. In this study, a PFC system was proposed which is a spiral finned freeze concentrator [6]. This system has spiral fins as the main part, providing highest contact surface area and optimum flow characteristic. The productivity can be increased with the increase of contact surface area between the coolant and the solution through the cooling surface. In the present paper,

Content from this work may be used under the terms of the Creative Commons Attribution 3.0 licence. Any further distribution of this work must maintain attribution to the author(s) and the title of the work, journal citation and DOI.
Published under licence by IOP Publishing Ltd

PFC has been applied to apple juice where the application of this PFC system on the real solution can be known. Optimum conditions for the apple juice concentration were determined using Response Surface Methodology (RSM).

2. Materials and Methods

2.1. Materials

Fuji apples were purchased from AEON Taman Universiti and washed with water to remove any adhering substances and sliced. Juice was extracted from the apples by homogenizing in a juicer. Then, the concentration of the apple juice was tested using a refractometer. The chosen concentration of apple juice was 0.1 g/mL according to its standard concentration.

2.2. Construction of Progressive Freeze Concentrator

The research activities start with the designing new crystallizer for PFC system, as depicted in Figure 1. After the crystallizer was successfully designed and fabricated, the new crystallizer was then tested in order to check whether it can produce a layer of ice crystal. The details of designing process has been explained in Samsuri et al. [6] where it also proves that the crystallizer is capable to produce the ice crystal from glucose solution. Then, the optimization process was performed to investigate the effect of four operating conditions on the response and optimizing these operating conditions of the apple juice concentration process. STATISTICA software Version 8.0 StatSoft Inc., USA was used to allow the optimization process to be effectively conducted. Experiments were conducted according to the experimental arrangements from the Design of Experiment (DOE) tool by the software.

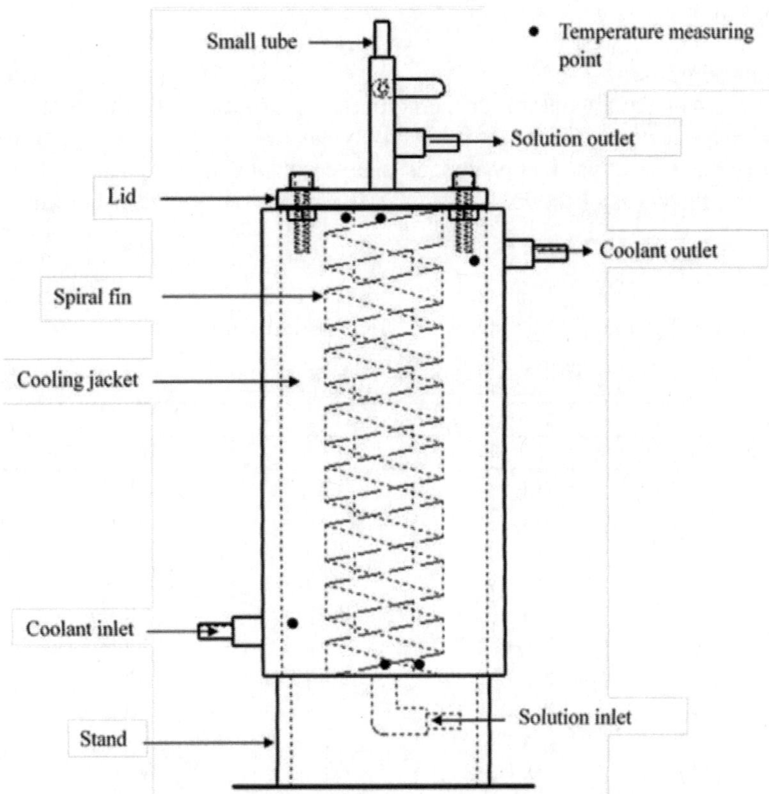

Figure 1 Schematic diagram of spiral finned crystallizer

2.3. Evaluation of system efficiency

The effective partition constant, K, is the most important parameter which portrays the effectiveness of the system. The ice purity increases when K decreases. Thus, the efficiency of the FC is high. The K between ice and liquid phase can be defined by Eq. (1):

$$K = 1 - \frac{\log\frac{C_O}{C_L}}{\log\frac{V_L}{V_O}} \quad (1)$$

where V_L and V_O are volume (mL) of concentrated and initial solution and C_L and C_O are concentration of apple juice (g/mL) in the concentrated and initial solution [7].

2.4. Experimental design

Based on the result of single factor at a time basis experiments, four variables including coolant temperature, circulation flowrate, circulation time and shaking speed were designed. The variables and their coded and uncoded values are presented in Table 1.

Table 1. Experimental range and coded value of variables.

Factors	Range and Levels				
	-α	-1	0	+1	+α
Circulation Flowrate, X_1 (mL/min)	2600	2800	3000	3200	3400
Circulation Time, X_2 (minutes)	20	40	60	80	100
Coolant Temperature, X_3 (°C)	-16	-14	-12	-10	-8
Shaking Speed, X_4 (ohm)	20	30	40	50	60

3. Results and discussion

3.1. Apple juice concentration

The designing process of new apparatus to carry out PFC process with the goal of improving the concentration performance as well as better productivity has resulted in a new crystallizer as described before. During freezing, ice crystal appeared as a layer on the inner surface of the spiral finned crystallizer wall. The thickness of the ice crystal varies for all experiments according to the operating conditions.

3.2. Model fitting

All 26 of the designed experiments and results were depicted in Table 2.

Table 2. Response (K) for each run.

Run	Manipulated Variables				K
	X_1	X_2	X_3	X_4	
1	-14(-1)	40(-1)	2800(-1)	30(-1)	0.67
2	-14(-1)	40(-1)	2800(-1)	50(+1)	0.70
3	-14(-1)	40(-1)	3200(+1)	30(-1)	0.65
4	-14(-1)	40(-1)	3200(+1)	50(+1)	0.71
5	-14(-1)	80(+1)	2800(-1)	30(-1)	0.54
6	-14(-1)	80(+1)	2800(-1)	50(+1)	0.55
7	-14(-1)	80(+1)	3200(+1)	30(-1)	0.66
8	-14(-1)	80(+1)	3200(+1)	50(+1)	0.69
9	-10(+1)	40(-1)	2800(-1)	30(-1)	0.36
10	-10(+1)	40(-1)	2800(-1)	50(+1)	0.36
11	-10(+1)	40(-1)	3200(+1)	30(-1)	0.61
12	-10(+1)	40(-1)	3200(+1)	50(+1)	0.21
13	-10(+1)	80(+1)	2800(-1)	30(-1)	0.36
14	-10(+1)	80(+1)	2800(-1)	50(+1)	0.42
15	-10(+1)	80(+1)	3200(+1)	30(-1)	0.20
16	-10(+1)	80(+1)	3200(+1)	50(+1)	0.33
17	-16(-α)	60(0)	3000(0)	40(0)	0.73

18	-8(+α)	60(0)	3000(0)	40(0)	0.32
19	-12(0)	20(-α)	3000(0)	40(0)	0.51
20	-12(0)	100(+α)	3000(0)	40(0)	0.48
21	-12(0)	60(0)	2600(-α)	40(0)	0.36
22	-12(0)	60(0)	3400(+α)	40(0)	0.26
23	-12(0)	60(0)	3000(0)	20(-α)	0.23
24	-12(0)	60(0)	3000(0)	60(+α)	0.11
25 (C)	-12(0)	60(0)	3000(0)	40(0)	0.29
26 (C)	-12 (0)	60(0)	3000(0)	40(0)	0.29

The empirical mathematical model of the predicted K (Y) as a function of coolant temperature (X_1), circulation time (X_2), circulation flowrate (X_3) and shaking speed (X_4) and their interaction using linear and quadratic regression coefficient of main factors and linear-by-linear regression coefficients of interaction is represented in Eq. (2):

$$Y = 10.68565 + 0.63149\ X_1 - 0.02411\ X_2 - 0.00424\ X_3 + 0.00922\ X_4 + 0.02002\ X_1^2 + 0.00018\ X_2^2 - 0.0001\ X_4^2 + 0.00009\ X_1X_2 - 0.00006\ X_1X_3 - 0.0011\ X_1X_4 + 0.00017\ X_2X_4 - 0.00001\ X_3X_4 \quad (2)$$

The fit of the model was evaluated by coefficient of determination (R^2) and analysis of variance (ANOVA).

3.3. ANOVA

The reliable way to evaluate the quality of the fitted model is by the application of ANOVA [8, 9]. The ANOVA in Table 3 indicates that the model for K gives good predictions. The F-value for K is 2.77, higher than the tabulated F-value ($F_{0.05,\ 14,11} = 2.74$) at 0.05 significant level.

Table 3. ANOVA for quadratic model.

Source	Sum of Squares of Error (SS)	Degree of Freedom (DF)	Mean Squares (SS)	F-value
Regression (SSR)	0.690	14	0.049	2.77
Residual	0.196	11	0.018	
Total (SST)	0.886	25		
R^2	0.779			

After the validity of the regression model has been evaluated, it is also important to identify the variables that can influence the process significantly. The sorted multiple regression results are presented in Table 4, which would be used to evaluate the significance of regression coefficients of the model for response of K, where standard error estimates the common within-group standard deviation.

From Table 4, it can be seen that the mathematical model is statistically significant ($P < 0.001$) and this contributes largely towards the response variable. The quadratic term shows a substantial significant effect at 97% confidence level (p-value < 0.03). In general, the significant order of the independent variables on K from high to low based on the F- and p-values is: coolant temperature, circulation time, circulation flowrate and shaking speed. The R^2 value was found to be 0.7792 which is acceptable because it is larger than 0.75. R^2 should be at least 0.75 for the good fit of the model [10].

Table 4. Significance of regression coefficients of the model for response of K.

Factor	Coefficient Estimation	Standard Error	F	p
X_1	0.631	0.326	23.109	0.001
X_1^2	0.020	0.008	6.297	0.029
X_2^2	0.000	0.000	5.029	0.047
X_2X_4	0.000	0.000	0.993	0.340

X_2	-0.024	0.029	0.788	0.394
X_3^2	0.000	0.000	0.675	0.429
X_1X_3	-0.000	0.000	0.505	0.492
X_1X_4	-0.001	0.002	0.434	0.524
X_3X_4	-0.000	0.000	0.260	0.620
X_4	0.009	0.061	0.2109	0.655
X_4^2	-0.000	0.000	0.102	0.756
X_2X_3	-0.000	0.000	0.028	0.870
X_3	-0.004	0.005	0.027	0.872
X_1X_2	0.000	0.001	0.011	0.920

3.4. Effect of independent variables

The three dimensional (3D) response surfaces could also demonstrate the effect of four variables on the K. It could also visualize the significant ($p < 0.05$) interaction effects of factors on the PFC process. For the effects of coolant temperature and circulation time on the K, as depicted in Figure 2(a), the solution is saturated with solutes after 80 minutes, thus causing some inclusions of the solute into the ice formed. Longer circulation time than 80 minutes is not suitable for this PFC process. Anywhere outside of these range, the K-value is high as affected by decreasing coolant temperature and circulation time. Elliptical contour obtained from the interaction between coolant temperature and circulation flowrate in Figure 2(b) shows that there is a perfect interaction between the variables [11]. It could also be seen that K-value is low when the coolant temperature is intermediate while the circulation flowrate is high. The K-value decreases with the increasing circulation flowrate. The solutes were not easily trapped and were rapidly brought away from the surface of the stagnant ice layer, causing higher amount of solutes remaining in the concentrated solution and lower solute entrapment in the ice crystal. However, a decrease in the coolant temperature beyond the optimum region resulted in an increase in the K-value, as can be seen in Figure 2(c). During this time, ice crystal had grown faster which had overtaken the movement of solute particles in the direction of the solution flow. Thus, there is a high solute inclusion in the ice formed.

From Figure 2(d), both intermediate values of circulation time and circulation flowrate yielded the best value of K in these interactions. At circulation time of more than 70 minutes, the formed ice crystal has nearly filled the entire volume of the crystallizer. Thus, the diameter of the path for solution flow has been reduced. The solutes also became saturated in the concentrated solution and this led to the contamination of solutes in the ice layer. It can be seen from the interaction between circulation time and shaking speed in Figure 2(e) that intermediate circulation time is giving the best value of K although the shaking speed is changing. The result shows that circulation time has a greater effect on K than the shaking speed, especially at intermediate circulation time. Initially, ice growth rate is fast and decreases over time due to increased heat transfer resistance and the concentration of solute interface, which then lowers the freezing point, making it more difficult to freeze. The decrease of the ice growth rate making less inclusion of solute into the ice crystal. It can be seen in Figure 2(f) that the value of K decreases as the circulation flowrate and shaking speed increase from 3000 to 3400 mL/min and from 60 to 65 ohm, respectively. Besides, the value of K increases with further increase of shaking speed. Increase of shaking will erode not only the interface containing rejected solute, but also the pure ice crystal layer formed on the inner wall, thus making the solution less concentrated where much water returned back to the solution.

IOP Conf. Series: Materials Science and Engineering **358** (2018) 012042 doi:10.1088/1757-899X/358/1/012042

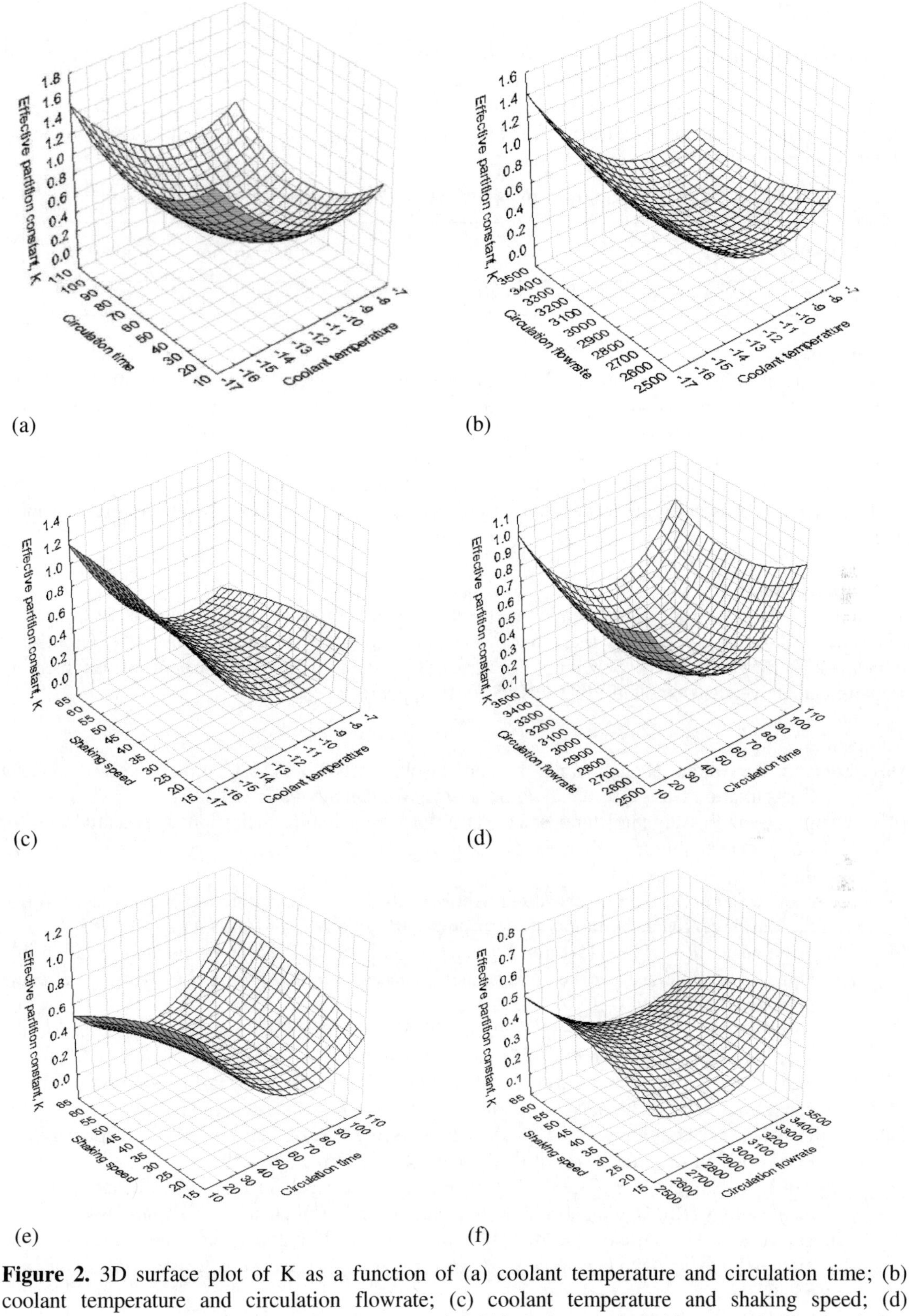

Figure 2. 3D surface plot of K as a function of (a) coolant temperature and circulation time; (b) coolant temperature and circulation flowrate; (c) coolant temperature and shaking speed; (d) circulation time and circulation flowrate; (e) circulation time and shaking speed; (f) circulation flowrate and shaking speed.

3.5. Optimization of variables and model verification

Additional experiment was carried out to validate the optimization result obtained by the RSM. Hence, Table 5 lists the optimum value for each parameter involved accordingly and the validation value is shown. K-value for validation experiment is better than prediction value and the values of prediction and validation are close enough to each other.

Table 5. Optimum conditions found by the model and verification of the model.

Response	X_1 (°C)	X_2 (min)	X_3 (mL/min)	X_4 (ohm)	Prediction	Validation
K	-10.59	67.35	3030.23	30.96	0.25	0.17

4. Conclusion

RSM was successfully used to optimize K for the PFC process on the apple juice concentration. All R^2 and F-value for K signify good accuracy of fitted model generated by the software. The experimental value is in agreement with the predicted value. Under the optimum conditions (-10.59 °C of coolant temperature, 67.35 minutes of circulation time, 3030.23 mL/min of circulation flowrate, 30.96 ohm of shaking speed), the K-value reaches 0.17 which represents the high efficiency of the PFC process. This study may provide useful tools to develop an efficient process for industrial concentration of apple juice.

Acknowledgments

The authors would like send our gratitude to Universiti Teknologi Malaysia and Ministry of Education Malaysia (MOE) for providing the financial support through Research University Grant and Fundamental Research Grant Scheme (Vot no. 04H46 and 4F224) and Chemical Engineering Department, Universiti Teknologi PETRONAS for its support.

References

[1] Samsuri S, Amran N A, Yahya N and Jusoh M 2016 Review on Progressive Freeze Concentration Designs *Chemical Engineering Communications* **203** 345-63

[2] Otero L, Sanz P, Guignon B and Sanz P D 2012 Pressure-shift nucleation: A potential tool for freeze concentration of fluid foods *Innovative Food Science & Emerging Technologies* **13** 86-99

[3] Lemmer S, Klomp R, Ruemekorf R and Scholz R 2001 Preconcentration of Wastewater through the Niro Freeze Concentration Process *Chemical Engineering & Technology* **24** 485-8

[4] Habib B and Farid M 2006 Heat transfer and operating conditions for freeze concentration in a liquid–solid fluidized bed heat exchanger *Chemical Engineering and Processing: Process Intensification* **45** 698-710

[5] Miyawaki O, Liu L and Nakamura K 1998 Effective Partition Constant of Solute between Ice and Liquid Phases in Progressive Freeze-Concentration *Journal of Food Science* **63** 756-8

[6] Samsuri S, Amran N A and Jusoh M 2015 Spiral finned crystallizer for progressive freeze concentration process *Chemical Engineering Research and Design* **104** 280-6

[7] Liu L, Miyawaki O and Nakamura K 1997 Progressive Freeze-Concentration of Model Liquid Food *Food Science and Technology International, Tokyo* **3** 348 - 52

[8] Bezerra M A, Santelli R E, Oliveira E P, Villar L S and Escaleira L A 2008 Response surface methodology (RSM) as a tool for optimization in analytical chemistry *Talanta* **76** 965-77

[9] Samsuri S and Mohd Bakri M M 2018 Optimization of fractional crystallization on crude biodiesel purification via response surface methodology *Separation Science and Technology* **53** 567-72

[10] Haaland P D 1989 *Experimental Design in Biotechnology*: Taylor & Francis)

[11] Muralidhar R V, Chirumamila R R, Marchant R and Nigam P 2001 A response surface approach for the comparison of lipase production by Candida cylindracea using two different carbon sources *Biochemical Engineering Journal* **9** 17-23

IOP Conf. Series: Materials Science and Engineering **358** (2018) 012043 doi:10.1088/1757-899X/358/1/012043

Meal Disturbance Effect on Control of Blood Glucose Level for Critically-ill Patients using In-silico Works

N F M Yusof[1], A M Som[1,2], S A Ali1 and N H Azman[1]

[1]Chemical Engineering Department, Universiti Teknologi MARA, 40450 Shah Alam, Selangor, Malaysia
[2]Accounting Research Institute, Universiti Teknologi MARA, 40450 Shah Alam, Selangor, Malaysia

ayub522@salam.uitm.edu.my

Abstract. This study was conducted to determine the effect of meal disturbance on blood glucose level of the critically ill patients and to simulate the control algorithm previously developed using in-silico works. The study is significant so as to reduce the mortality rate of critically ill patients who usually encounter hyperglycaemia or/and hypoglycaemia while in treatment. The meal intake is believed to affect the blood glucose regulation and causes the hyperglycaemia to occur. Critically ill patients receive their meal through parenteral and enteral nutrition. Furthermore, by using in-silico works, time consumed and resources needed for clinical evaluation of the patients can be reduced. Hovorka model was employed in which the simulation study was carried out using MATLAB on the virtual patient and it was being compared with actual patient in which the data were provided by Institut Jantung Negara (IJN). Based on the simulation, the disturbance on enteral glucose supplied had affected the blood glucose level of the patient; however, it remained unchanged for the parental glucose. To reduce the occurrence of hypoglycaemia and hyperglycaemia, the patient was injected with 30 g/hr and 10 g/hr of enteral glucose, respectively. In conclusion, the disturbance of meal received can be controlled through in-silico works.

1. Introduction

A critically-ill patient is defined as the patient who has high risk for actual or potential fatal health problems. They are usually highly unstable, weak and complicated; hence intense and attentive nursing cares are mostly needed by those patients. The critically-ill patient is normally diagnosed with hyperglycaemia, although the patient has no diabetes history in the past [1]. Hyperglycaemia is a phenomenon in which the body contains high blood glucose (more than 6.6 mmol/L) due to the lack of insulin in the body or the insulin present is not effectively used by the body. Its occurrence in critically-ill patients is due to the release of counter-regulatory stress hormones which are known as corticosteroids and cathecholamines [2]. Another phenomenon which is the contradictory to hyperglycaemia is known as hypoglycaemia in which patient's body contains a low blood glucose level (below 4.0 mmol/L). Hypoglycaemia also contributes to the mortality and morbidity rates of critically-ill patients when they go through an intensive glucose control [3].Unlike other patients, critically-ill patients eat their meal through enteral or parenteral nutrition or both [4-10]. Enteral nutrition is one of the ways to deliver food to the stomach or small intestine using tubes whereas parenteral nutrition is a method of delivering the nutrition into the body through veins, which is also known as intravenous feeding. In this research, the intake of the nutrition is crucial as it influences the blood glucose level of the patients and thus, affecting the glucose insulin control of the patients.

Content from this work may be used under the terms of the Creative Commons Attribution 3.0 licence. Any further distribution of this work must maintain attribution to the author(s) and the title of the work, journal citation and DOI.
Published under licence by IOP Publishing Ltd

The meal intake of the critically-ill patients disturbs the blood glucose level in the body. Whenever the patients receive nutrition by either enteral or parenteral nutrition, the blood glucose level in their body will increase and it consequently results in hyperglycaemia [11]. This problem might lead to a situation where it is difficult to regulate the blood glucose level of the patients in the range of 4.0 to 6.6 mmol/L. Hence, there is a need to study the effect of meal disturbance on the blood glucose control of the patients.Furthermore, most studies on intensive insulin therapy and glucose control protocols of critically-ill patients have been conducted clinically. However, the main concern to these studies is central to enormous times and resources required for the clinical assessment. Those clinical experiments might not be convenient to subjects of trial as they would affect patients physically and mentally. The need of time and resources during clinical evaluation indirectly represents financial burden to the researchers [12-14]. By using in-silico works or simulation method, the cost spent on the clinical evaluation can be reduced and the inconvenience that the patients experienced during the clinical assessment can be eliminated. In this study, mathematical equations which represent the blood glucose level of the patients are adopted from Hovorka et al. [15]. The study is conducted to determine the effect of meal disturbance on blood glucose level of the critically ill patients and to simulate the control algorithm developed through in-silico works using MATLAB.

2. Methodology

2.1 Data collection and extraction

Data collection and information of the actual patients required for the study were obtained from the Intensive Care Unit, Institut Jantung Negara (IJN), Kuala Lumpur. The data provided include patients' biodata namely; name, date of birth, gender, race, type of meal intake, glucose level, etc. Approval data from IJN ethic committee was received prior to carrying out this study. The extracted data were then filtered out to sort for the needed information related to the study. By using data filter mechanism available in MATLAB, the data were sized down to smaller numbers so as to comply with the scope of the research. The study focuses on critically-ill patients with the age ranges from 52 ±10 years and the weight of 75 ± 23 kg. The patients received meal through parenteral and enteral nutrition. The patient data were termed as clinical data throughout this study. Parameters and variables are taken from patient 5 of Hovorka model [15], fitted into equations, used as simulated data and then the simulation data were compared with IJN clinical data. Table 1 shows parameter values of critically-ill patient obtained from Hovorka[15] which were used in the simulation study using MATLAB.

Table 1. Parameter values of critically-ill patient (Patient 5)

Quantity	Symbol	Value
Renal glucose threshold	G_R(mmol/L)	10.3
Fraction to which endogenous insulin secretion is suppressed by the delivery of exogenous insulin	$F_{IE,basal}$ (unitless)	0.20
Fractional renal glucose clearance	k_r (min^{-1})	0.009
Plasma insulin concentration at which the fractional clearance of insulin is halved	$K_{M,I}$ (mU L^{-1})	564
Glucose distribution volume	V_G(L kg^{-1})	0.16
Insulin distribution volume	V_I(L kg^{-1})	0.12
Blood glucose concentration at which the non-insulin dependent glucose flux attains half of its maximum value	$K_{M,N}$ (mmol/L)	1
Fractional elimination rate extrapolated to zero insulin concentration	k_e (min^{-1})	0.15
EGP extrapolated to zero insulin concentration	EGP_0 (mmol kg^{-1} min^{-1})	0.0226
Enteral glucose bioavailability (unitless)	F_{GE} (unitless)	0.29
Plasma insulin concentration at which the fractional transfer rate k_{21} achieves half of its maximum value	$K_{M,T}$ (mU L^{-1})	15.8
Beta-cell responsiveness	M_I (mU kg$_{-1}$ min$_{-1}$ per mmol L^{-1})	0.323

Half-time suppression of endogenous insulin secretion	$t_{1/2,IE}$ (min)	88
Time-to-maximum of enteral glucose absorption	$t_{max,G}$ (min)	14
Pre-study insulin secretion	$U_{IE,basal}$ (mU kg^{-1} min^{-1})	0.39
Non-insulin-dependent glucose flux	F_{01}(mmol kg^{-1} min^{-1})	0.0104
Fractional transfer rate between the non-accessible and accessible glucose compartments	k_{12} (min^{-1})	0.063
Deactivation rate of remote insulin effect on glucose distribution/transport	k_{a1}(min^{-1})	0.0032
Deactivation rate of remote insulin effect on glucose disposal	k_{a2}(min^{-1})	0.0558
Deactivation rate of remote insulin effect on endogenous glucose production	k_{a3}(min^{-1})	0.0363
Insulin sensitivity of disposal	S_{ID} (min^{-1} per mU L^{-1})	0.0003
Insulin sensitivity of EGP	S_{IE} (per mU L^{-1})	0.072
Insulin sensitivity of distribution/transport	S_{IT} (min^{-1})	0.109

2.2 Glucose-Insulin Regulatory System Model Selection

Model of the glucose-insulin system or the algorithm to control the blood glucose level of critically-ill patients was selected from Hovorka et al. [15] which consisted of five sub-models as shown in Figure 1.

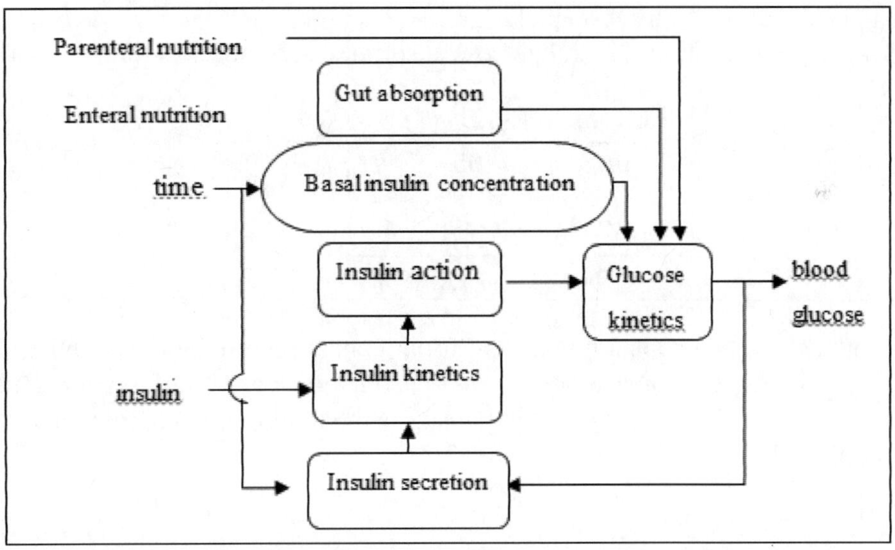

Figure 1. The glucoregulation model that combining five sub-models

a) Endogenous insulin secretion sub-model

Equations 1 and 2 represent the endogeneous insulin secretion sub-model as follows:

$$U_{\mathrm{IE}}^{u}(t) = \frac{60}{1000} W \left(M_I(G(t) - 5.5) + U_{\mathrm{IE,basal}} \left(F_{\mathrm{IE,basal}} + \left(1 - F_{\mathrm{IE,basal}}\right) e^{-t\frac{\ln(2)}{t_{1/2,\mathrm{IE}}}} \right) \right) \quad (1)$$

$$U_{\mathrm{IE}}(t) = \begin{cases} U_{\mathrm{IE}}^{u}(t), & \text{if } U_{\mathrm{IE}}^{u}(t) > 0 \\ 0, & \end{cases} \quad (2)$$

Otherwise

Where $U_{\mathrm{IE}}^{u}(t)$ and $U_{\mathrm{IE}}(t)$ are the unconstrained and actual endogenous insulin secretions at time t, respectively with unit of U h^{-1}, W is the subject's weight (kg), M_I is the beta-cell responsiveness expressing the ability of the beta cell to step up insulin secretion when glucose deviates from the entering glucose concentration of 5.5 mmol/L (mU kg^{-1} min^{-1} per mmol L^{-1}), $U_{\mathrm{IE,basal}}$ represents the basal insulin secretion (mU kg^{-1} min^{-1}), $F_{\mathrm{IE,basal}}$ is the fraction where endogenous insulin secretion is suppressed by the delivery of exogenous insulin (unitless) and $t_{1/2,\mathrm{IE}}$ is the half-time of the suppression of endogenous insulin secretion by exogenous insulin (min).

b) Insulin kinetics sub-model
Equation 3 represents the insulin kinetics sub-model as follows:

$$\frac{\mathrm{d}I(t)}{\mathrm{d}t} = \frac{1000}{60W}\frac{U_{\mathrm{IX}}(t) + U_{\mathrm{IE}}(t)}{V_I} - k_e\frac{K_{M,I}}{I(t) + K_{M,I}}I(t) \tag{3}$$

Where $I(t)$ is the plasma insulin concentration (mU L^{-1}), $U_{\mathrm{IX}}(t)$ is the infusion of the exogenous insulin (Uh^{-1}), k_e is the fractional elimination rate extrapolated to zero insulin concentration (min^{-1}), $K_{M,I}$ is the plasma insulin concentration at which the fractional clearance of insulin is divided by two (mUL^{-1}) and V_I is the insulin distribution volume (Lkg^{-1}).

c) Enteral glucose absorption sub-model
Equations 4 and 5 represent the enteral glucose absorption sub-model as follows:

$$\frac{\mathrm{d}A_1(t)}{\mathrm{d}t} = \frac{F_{\mathrm{GE}}U_{\mathrm{GE}}(t)}{60} - \frac{A_1(t)}{t_{\max,G}} \tag{4}$$

$$\frac{\mathrm{d}A_2(t)}{\mathrm{d}t} = \frac{A_1(t)}{t_{\max,G}} - \frac{A_2(t)}{t_{\max,G}} \tag{5}$$

Where A_1 and A_2 are two compartments that form a chain representing absorption of enteral glucose (g), F_{GE} is the bioavailability of enteral glucose (unitless), $U_{\mathrm{GE}}(t)$ is the administration of enteral glucose (g h^{-1}) and $t_{\max,G}$ is the time-to-maximum of enteral glucose absorption (min).

d) Insulin action sub-model
Equations 6 to 8 represent the insulin action sub-model as follows:

$$\frac{\mathrm{d}x_1}{\mathrm{d}t} = -k_{a1}[x_1(t) - I(t)] \tag{6}$$

$$\frac{\mathrm{d}x_2}{\mathrm{d}t} = -k_{a2}[x_2(t) - I(t)] \tag{7}$$

$$\frac{\mathrm{d}x_3}{\mathrm{d}t} = -k_{a3}[x_3(t) - I(t)] \tag{8}$$

And x_1, x_2 and x_3 are the (remote) effects of insulin on glucose transport, glucose disposal and endogenous glucose production (mU L^{-1}) while k_{ai}, $i = 1,\ldots, 3$, represent deactivation rate constants (min^{-1}).

e) Glucose kinetics sub-model

Equations 9 to 11 represent the glucose kinetics sub-model as follows:

$$\frac{dQ_1(t)}{dt} = -F_{01}^{c} - k_{21}(t)Q_1(t) + k_{12}Q_2(t) - U_R(t) + \frac{5.551}{W}\left[\frac{U_{GP}(t)}{60} + \frac{A_2(t)}{t_{max,G}}\right] + EGP(t) \tag{9}$$

$$\frac{dQ_2(t)}{dt} = k_{21}(t)Q_1(t) - \left[k_{12} + S_{I,MOD}(t)S_{ID}x_2(t)\right]Q_2(t) \tag{10}$$

$$G(t) = \frac{Q_1(t)}{V_G} \tag{11}$$

Q_1 and Q_2 (mmol/kg) indicate the masses of glucose in the accessible compartment and non-accessible compartment respectively, k_{12} (min^{-1}) is the transfer rate constant from the non-accessible to the accessible compartment, V_G (L kg^{-1}) is the distribution volume of the accessible compartment, $U_{GP}(t)$ is the parenteral glucose infusion (g h^{-1}), G (mmol/L) is the blood glucose concentration, S_{ID} is the insulin sensitivity of glucose disposal (min^{-1} per mU L^{-1}), $S_{I,MOD}$ is the insulin sensitivity modifier, and EGP$_0$ is the endogenous glucose production (EGP) extrapolated to the zero insulin concentration. F_{01}^{c} is the total non-insulin dependent glucose flux corrected for the ambient glucose concentration while $U_R(t)$ (mmol kg^{-1} min^{-1}) is the renal glucose clearance above the glucose threshold of G_R (mmol/L).

3. Results and discussion

3.1 Blood glucose level (BGL) of patients from clinical study and simulation

Figure 2 shows the comparison between the blood glucose level (BGL) of a critically ill patient from Hovorka work in 2008 (Patient 5) [15] with the actual patient from the clinical study (Patient X) in which the data was provided by Institut Jantung Negara.

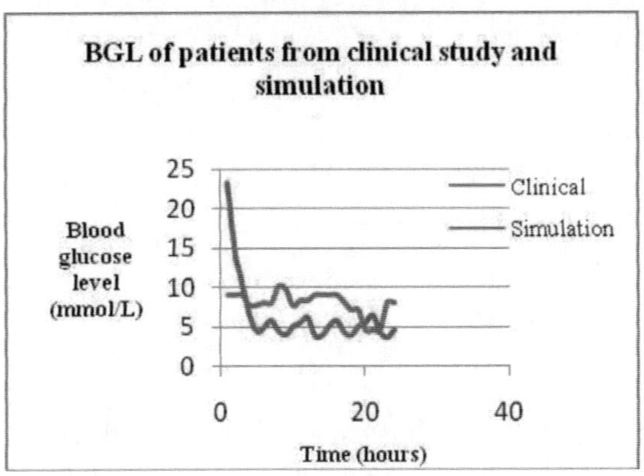

Figure 2. Blood glucose level of patients from clinical study and simulation

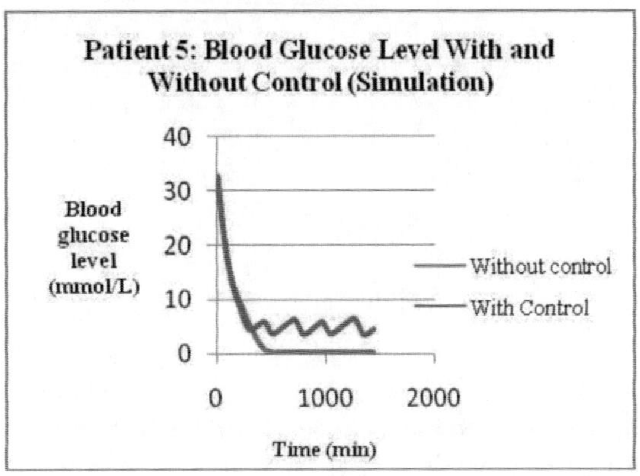

Figure 3. Blood glucose level of Patient 5 with and without control (simulation)

It can be seen that the BGL pattern for both patients are quite similar. Although at the beginning hours the BGL of patient from simulation work was quite high i.e. 23.3 mmol/L and only 9.1 mmol/L for patient from the clinical study, the difference in blood glucose concentration as well as the BGL patterns for both patients was not too large beginning at about the fifth hour. Hence, both patients i.e. Patient X and Patient 5 have been chosen to be discussed in this study. Hovorka et al. [15] did not specifically disclose the information of all six patients in his paper. However it was known that those patients consist of three males and three females, age 52 ±10 years and weight of 75 ± 23 kg.

3.2 The difference of blood glucose level for Patient 5 with and without control

Figure 3 illustrates the simulation of two types of blood glucose level of a critically ill patient in two situations; without control and with control. Both conditions with control and without control showed the same pattern at the beginning of the simulation. The blood glucose level increased slightly from 28.1 mmol/L at 0 minute to 32.7 mmol/L at 10.8 minutes. For the simulation which included control, it decreased significantly to 4.5 mmol/L at 300 minutes. After that the blood glucose level was regulated in the safe range which was between 4.4 and 6.6 mmol/L. The lowest level was about 3.5 mmol/L at time 1358 minutes while the highest level was 6.6 mmol/L from 1254 to 1268 minutes. Without any control, the blood glucose level kept decreasing to 0.2 mmol/L at 519 minutes and remained constant until 1440 minutes. This condition resulted in hypoglycaemia at which the blood glucose concentration of a person fell below 4.0 mmol/L and could cause death [3]. The control scheme mentioned above can be explained by controlling the amount of parenteral and enteral glucose which was received by the patients as their meal or nutrition. A more detailed discussion on enteral and parenteral glucose disturbance is provided in the next section.

3.3. Blood glucose level of patient 5 for different amount of enteral glucose

Figure 4 shows three different patterns of blood glucose level for three different amounts of enteral glucose (Uge); namely 30, 60 and 130 g/hr. All of them showed the same pattern at the beginning where at 0 minute the BGL was 28.1 mmol/L and it increased slightly to 32.7 mmol/L at time 10.8 until 14.5 minutes.

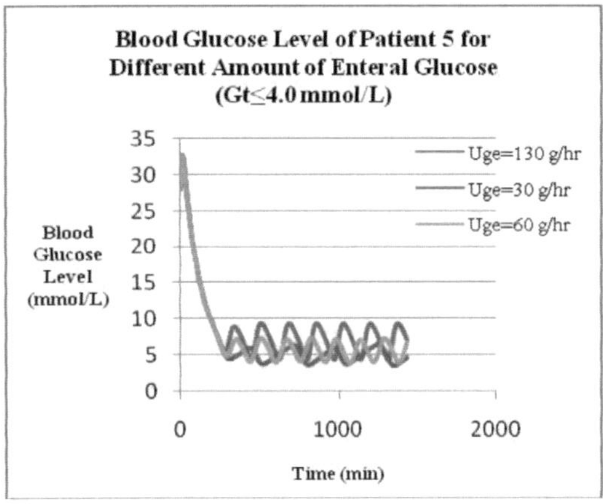

Figure 4. Blood glucose level of Patient 5 for different amount of enteral glucose (Gt≤4.0 mmol/L)

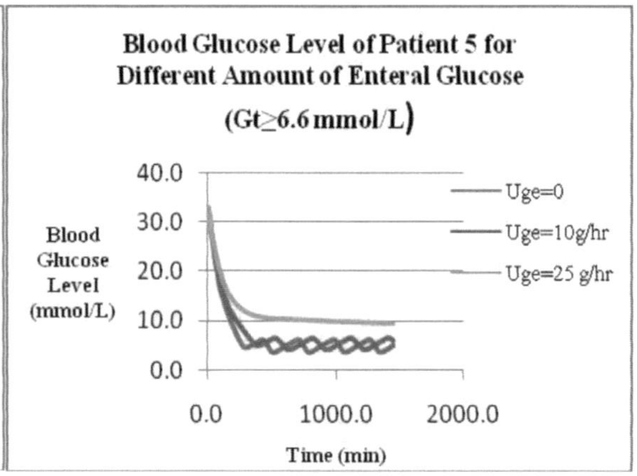

Figure 5. Blood glucose level of Patient 5 for different amount of enteral glucose (Gt≥6.6 mmol/L)

After that, the BGL decreased significantly to 5 mmol/L and began to fluctuate. It can be seen that for the highest amount of Uge, which was 130 g/hr, the BGL was higher than the other two. The BGL fluctuated within 4.3 to 9.3 mmol/L. The hyperglycaemia case could be observed in this situation. When Uge supplied was 60 g/hr, the BGL fluctuated within 3.9 to

7.5 mmol/L. Meanwhile for Uge equals to 30 g/hr, the BGL fluctuates within 3.3 to 6.6 mmol/L. Although there was occurrence of BGL fell below 4.0 mmol/L, it only occurred for a few minutes .This clearly indicated that 30 g/hr of enteral glucose was preferred to be the amount of enteral glucose that should be injected to Patient 5 to control his/her blood glucose in a safe range. Figure 5 depicts how different amount of enteral glucose affects the blood glucose control when the BGL is more than or equal to 6.6 mmol/L.

When patient was not injected with any Uge, like previously the BGL increased sharply from 28.1 to 32.7 mmol/L before it slowly decreased to 4.5 mmol/L. The BGL then fluctuated with the lowest value of BGL recorded was 3.6 mmol/L and the highest value of BGL recorded was 6.2 mmol/L. As the patient was injected with 10 g/hr of enteral glucose, the BGL pattern seemed better as the lowest value of BGL recorded was 4.1 mmol/L and the highest value of BGL recorded was 6.5 mmol/L. Meanwhile, with 25 g/hr of Uge, the patient's BGL seemed to regulate within 10 mmol/L and showed little change from time to time. It indicated that a case of hyperglycaemia might occur if 25 g/hr of Uge was supplied to the patient when his/her BGL was more than or equal to 6.6 mmol/L. Hence, for patient 5 to regulate his/her BGL within the safe range, 10 g/hr of enteral glucose was the best option

3.4. Blood Glucose Level of Patient 5 for Different Amount of Parenteral Glucose

Figures 6 and 7 show that for three different amount of parenteral glucoses (Ugp), the pattern of BGL is the same. There was no difference shown in the regulation of BGL. In Figure 6, the BGL increased shortly from 28.1 mmol/L to 32.1 mmol/L before it slowly decreased to 4.0 mmol/L at 427 minutes. The BGL increased and then decreased with the lowest value of BGL recorded 3.8 mmol/L and the highest value of BGL recorded was 6.4 mmol/L. Just like the previous figure, the BGL of Patient 5 in Figure 7 was initially 28.1 mmol/L.

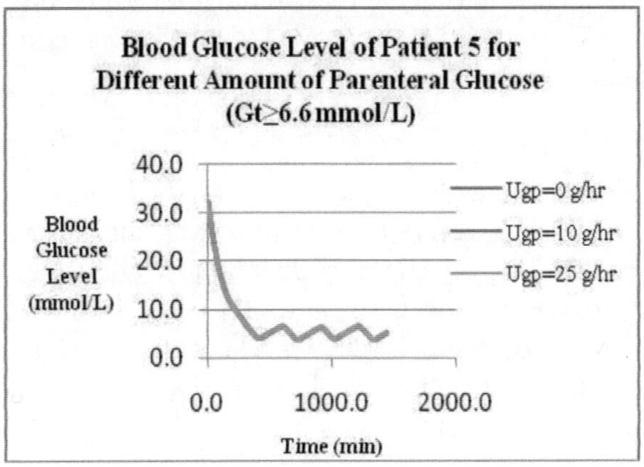

Figure 6. Blood glucose level of Patient 5 for different amount of parenteral glucose (Gt>6.6mmol/L)

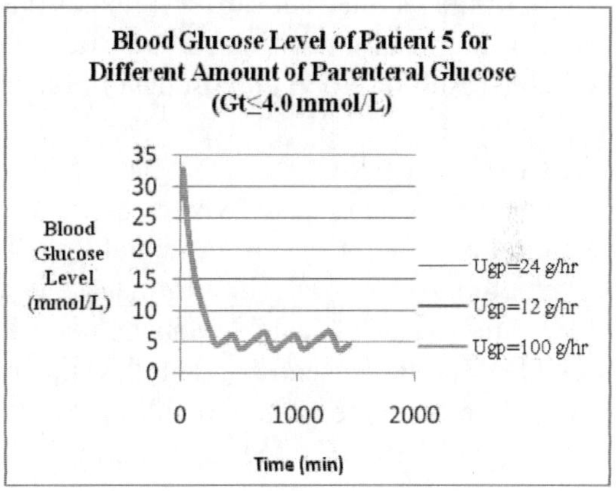

Figure 7. Blood glucose level of Patient 5 for different amount of parenteral glucose(Gt≤4mmol/L)

The BGL increased slightly to 32.7 mmol/L before it decreased dramatically to 4.5 mmol/L at 300 minutes. Starting at this point, the BGL fluctuated within the range of 3.6 to 6.6 mmol/L. Both figures showed that there was a little chance for hypoglycaemia to occur as the BGL fell below 4.0 mmol/L but none of hyperglycaemia case could be observed. However, both figures also tell us that with different amount of parenteral glucose supplied to patient 5, similar pattern of BGL could be observed. This indicates that by disturbing the parenteral glucose supplied to the patient, the BGL patterns are not affected. This occurrence opposes the normal situation where both enteral and parenteral nutritions contribute to the change in blood glucose level of the patient. It is suggested that modification on equations which involve the Uge should be carried out in the future study because these equations are very important to ensure that a well simulation work can be conducted so that the relationship between parenteral nutrition and blood glucose level of the patient can be examined.

The control strategy in MATLAB simulation might be lacking in terms of little correlation between meal disturbance and insulin infusion rate. The insulin infusion is one of the most active parameters in glucose-insulin control in Hovorka model [16, 17]. Insulin is needed to help the patient to regulate his blood glucose. However in this simulation, although insulin infusion had been considered as one of the input variables aside from enteral and parenteral glucoses, the value cannot be varied for different times. In previous work from Hovorka and others, the glucose-insulin control was generated where the reaction between insulin infusion rate and blood glucose level can be interpreted [15, 18, 19, 20, 21]. Therefore, it is recommended that in the future study, the equations involving insulin infusion in the simulation work could be improved to generate a more reliable result of blood glucose level of the critically-ill patient.

4. Conclusion

The objectives of the study on the effect of meal disturbance on blood glucose level as well as simulation work on control algorithm to control the blood glucose level using in-silico works for critically-ill patients were achieved. Based on the simulation, the disturbance on enteral glucose supplied affected the blood glucose level of the patient. To reduce the occurrence of hypoglycaemia and hyperglycaemia, the patient was injected with 30 g/hr and 10 g/hr of enteral glucose, respectively. However, the disturbance on parenteral glucose did not affect the blood glucose level of the patient. Therefore, a study should be conducted to modify the equations which involve parenteral glucose as a variable input of the model to improve the result of the simulation. Besides, it is also recommended that in the future study, the equations concerning insulin infusion rate in the simulation work could be improved to generate a more reliable result of blood glucose level of the patient.

Acknowledgement

The authors wish to acknowledge and extend their gratitude to the Intensive Care Unit, Institut Jantung Negara (IJN) for all the data provided and assistance rendered throughout this study. The authors also would like to thank Malaysian Ministry of Higher Education (MOHE) and Innovation and Research Management Institute (IRMI) of Universiti Teknologi MARA (UiTM) for the research grant through 600-RMI/ERGS 5/3(18/2011). The authors would lastly like to express their thankfulness to the Faculty of Chemical Engineering, Universiti Teknologi MARA (UiTM) for the financial support and assistance given.

References

[1] Van den Berghe G, Wouters P, Weekers F, Verwaest C, Bruyninckx F, Schetz M, Vlasselaers D, Ferdinande P, Lauwers P and Bouillon R 2001 *N. Engl. J. Med.* **345** 1359-67
[2] Kitabchi A E, Umpierrez G E, Miles J M and Fisher J N 2009 *Diabetes Care* **32** 1335-43
[3] Investigators T N-S S 2011 *N. Engl. J. Med.* **367** 1108-18
[4] Barrett M, Demehri F R and Teitelbaum D H 2015 *Curr. Opin. Clin. Nutr. Metab. Care* **18** 496-500
[5] Braunschweig C L, Levy P, Sheean P M and Wang X 2001 *Am. J. Clin. Nutr.* **74** 534-42
[6] Cahill N E, Dhaliwal R, Day G, Jiang X and Heyland D K 2010 *Crit. Care Med.* **38** 395-401

[7] Elke G, Van Zanten A R, Lemieux M, Mc Call M, Jeejeebhoy K N, Kott M, Jiang X, Day A G and Heyland D K 2016 *Crit. Care* **20** 1-14
[8] Gungabissoon U, Hacquoil K, Bains C, Irizarry M, Dukes G, Williamson R, Deane A M and Heyland D K 2015 *J. Parenter. Enteral Nutr.* **39** 441-48
[9] McClave S A and Heyland D K 2009 *Nutr. Clin. Pract.* **24** 305-15
[10] McClave S A, Martindale R G, Rice T W and Heyland D K 2014 *Crit. Care Med.* **42** 2600-10
[11] Gosmanov A R and Umpierrez G E 2013 *Curr. Diab. Rep.* **13** 155-62
[12] Chase J G, Shaw G M, Wong X W, Lotz T, Lin J and Hann C E 2006 *Biomed. Signal Proces.* **1** 3-21
[13] Wilinska M E, Budiman E S, Taub M C, Ellen D, Allen J M, Acerini C L, Dunger D B and Hovorka R 2009 *J. Diabetes Sci. Technol. (online)* **3** 1109-20
[14] Wilinska M E, Chassin L J and Hovorka R 2008 *J. Diabetes Sci. Technol. (online)* **2** 417-23
[15] Hovorka R, Chassin L J, Ellmerer M, Plank J and Wilinska M E 2008 *Physiol. Meas.* **29** 959-78
[16] Yusof N F M, Som A M, Ali S A and Anuar ALAH 2015 *Adv. Mat. Res.* **1113** 739-44
[17] Yusof N F M, Som A M, Ibrehem A S and Ali S A 2014 *Adv. Mat. Res.* **938** 299-304
[18] Hovorka R, Kremen J, Blaha J, Matias M, Anderlova K, Bosanska L, Roubicek T, Wilinska M E, Chassin L J, Svacina S and Haluzik M 2007 *J. Clin. Endocrinol. Metab.* **92** 2960-64
[19] Yusof N F M, Som A M, Ibrehem A S and Ali S A 2014 *J. Applied Sciences* **14** 1465-68
[20] Yusof N F M, Som A M, Ibrehem A S and Ali S A 2012 *IEEE-EMBS Proc. Int. Conf. on Biomed. Eng. and Sci.* 273- 78
[21] Kirubakaran V, Radkakrishnan T K and Sivakumaran N 2013 *Proc. of 12th IFAC Symposium on Computer Applications in Biotechnology* **46** 291-96

A Selective Organic-Based Corrosion Inhibitors Containing Iodide Ion as Enhancer for Protection of Carbon Steel: A Review

I M Ibrahim[1], E S Mohd Kassim[1], H Husin[1], J Jai[1], M Daud[2] and M A Hashim[1]

[1]Faculty of Chemical Engineering, Universiti Teknologi MARA (UiTM), 40450 Shah Alam, Selangor, MALAYSIA.
[2]Industrial Technology Division, Malaysian Nuclear Agency (Nuclear Malaysia) Bangi, 43000 Kajang Selangor, Malaysia

iznimariah@yahoo.com

Abstract. This paper contains a review on the effect of halide ion with a selected inhibitor which is imidazole derivatives on the efficiency of corrosion inhibition. The paper first describes the mechanism of synergistic inhibition effect among halide ions enhancer with inhibitor on the steel surface. Then the paper describes the measured inhibition efficiency and summarizes the synergistic inhibition condition of imidazoline derivatives inhibitor with iodide ions. The characteristic of synergistic inhibition effect and the relationship between the amount of iodide ion consumption and the amount of organic inhibitor consumption are also discussed. It has been shown that, the synergistic effect between imidazole derivative and iodide ion is an effective method to improve the inhibitive performance in different aqueous media.

1. Introduction

Carbon steel is used in wide of industries because of its physical characteristics, such as stiffness and high strength, however they are highly susceptible to corrosion in aggressive solution [1]. Problem attributed to the deterioration of metal by chemical reaction with the environment is called corrosion. Corrosion costs billions of dollars every year and in the middle of 2013 this cost reached more than $2 trillion or 3-5% of Gross Domestic Product (GDP) in the United States [2]. Studies have found an occurrence of synergistic corrosion inhibition effect when secondary species is added into an organic inhibitor [3]–[7]. Synergistic inhibition effect is an improved performance of organic inhibitors as compared to merely organic inhibitors on the rate of corrosion reaction [8]. It has proven to be an effective way to decrease the amount of organic inhibitor consumption and to expand the application of organic inhibitor in media. Examples of secondary species that added to organic inhibitor are halide ions (Cl^-, Br^-, and I^-), metal ions and organic inhibitor mixtures.[3], [9]–[13]. Since the synergistic inhibition effect of halide ions widely reported in the previous researchers and it has a significant impact on the modern corrosion inhibitor formulation, thus this paper will discuss the effect of iodide ion with a selected inhibitor which is imidazoline derivatives.

1.1. Synergistic mechanism of halide ions

The addition of halide ions has been found to improve the adsorption of organic inhibitor cations via ion-pair interaction. When halide anions are pre-adsorbed on metal surfaces, they created excess of negative charges on the metal surfaces. Previous studies reported that the pre-adsorption of halide ions on the metal surface has facilitated the adsorption of organic inhibitor cations by forming intermediate bridges between the positively charged metal surface and the positive end of the organic inhibitor. This

Content from this work may be used under the terms of the Creative Commons Attribution 3.0 licence. Any further distribution of this work must maintain attribution to the author(s) and the title of the work, journal citation and DOI.
Published under licence by IOP Publishing Ltd

promoted a better adsorption mechanism of the organic inhibitor cation on the metal surfaces. The interaction enhanced the inhibition efficiency to a considerable extent [14].

Among halide ions, the synergistic inhibition effect increases in the order of Cl < Br < I. This indicates that the radius of halide ions has an important role in determining the adsorption mechanism on metal surface. The largest ionic radius of iodide ion has a higher hydrophobicity property as compared to the other two halide ions. The pre-adsorption of iodide ion onto metal surface will decrease the hydrophilicity of the metal surface. Consequently, adsorption of organic molecules onto the pre-adsorbed metal surface is favoured, instead of water molecules [15]. In the presence of halide ion (X^-), the mechanism for the anodic dissolution is as given in the following equations [16]:

$$Fe + H_2O + X^- \leftrightharpoons [FeXOH]^-_{ads} + H^+ + e, \quad (1)$$

$$[FeXOH]^-_{ads} \xrightarrow{r.d.s} FeXOH + e, \quad (2)$$

$$FeXOH + H^+ \leftrightharpoons Fe^{2+} + X^- + H_2O \quad (3)$$

Generally, halide ion will participate in the formation of intermediate. The adsorbed intermediate, $[FeXOH]^-_{ads}$ on the metal surface involved in the rate determining step (r.d.s). Organic inhibitor molecules will then combine with the adsorbed intermediate to form a metal-inhibitor complex. The resulting metal-inhibitor complex will either catalyzed or inhibit further metal dissolution depending on its solubility [16]. Two possible models for inhibitor adsorption with the aid of halide ion have been described by Okafor, Liu, & Zheng [9]. The first model is the adsorption of halide ion which then further attracts inhibitor molecule to metal surface (co-operative mechanism). In this co-operative mechanism, the strong chemisorption of halide ion on the metal surface is responsible for the synergistic effect of halide ion in combination with organic molecules. Then, the organic molecules are adsorbed by coulombic attraction at the metal surface, where the halide ion is already chemisorbed. The stabilization of the adsorbed halide ion by means of electrostatic interaction with organic molecules leads to greater surface coverage and thereby greater inhibition efficiency. The second model is the exchange of adsorption site by halide ion and organic ion (competitive mechanism). In competitive adsorption, the anion and cation are adsorbed at different sites on the metal surface. The adsorption of an anion may enhance the adsorption of a cation and vice versa. A combination of both models is also possible [18]. Since the synergistic inhibition effect has a significant impact on the modern corrosion inhibitor formulation, thus this paper will discuss further the effect of iodide ion with three selected inhibitors of imidazoline derivative namely amido-imidazoline (IM), 2-undecyl-1-ethylamino imidazoline (2UEI) and 2-undecyl-1-ethylamino-1-methylbenzyl quaternary imidazoline (2UMQI).

2. Effect of iodide ion on inhibitive action of Imidazoline derivative

A study investigated the inhibition effect of amido-imidazoline (IM) on carbon steel in a 3 wt% sodium chloride and its synergistic effect in the presence of iodide ion (from KI solution) by using an electrochemical impedance spectroscopy technique and a potentiodynamic polarization measurement [19]. In the presence of 2000 mg/L I^- ion (Table 1), the inhibition efficiency, % η showed an increase of 14.8% at 50 mg/L IM when measured via the impedance technique. The increment however was only 1.8% and 2.7% at 75 mg/L IM and 100 mg/L IM, respectively. Similar trend was revealed via the polarization technique. A significant increase of 9.7% at 50 mg/L IM; but only 1.2% and 1.7% at 75 mg/L IM and 100 mg/L IM, respectively. They also found that both the adsorption of IM and IM/KI on carbon steel surface followed the Langmuir adsorption isotherm. In this case, they reported that the co-operative adsorption mechanism dominated over the competitive adsorption mechanism in an IM/KI inhibition system. The co-operative adsorption could stabilize the adsorption of IM on the surface, therefore the inhibition ability improved in the presence of KI. The co-operative adsorption mechanism was also suggested by other researchers in another imidazoline /I^- study [20].

Another study studied the inhibition effect of 80 mg/L 2-undecyl-1-ethylamino imidazoline (2UEI) on carbon steel in CO_2-saturated 3 wt% NaCl and its synergistic effect in the presence of I^- ion [17]. According to Table 2, they found that the presence of highly concentrated I^- ions did not show the synergistic inhibition effect with the 2UEI inhibitor. The increase was only 1% and 2.38% at 830 mg/L and 2000 mg/L I^- ion respectively. Without the presence of the I^- ions, the inhibition efficiency of the 2UEI inhibitor was very close to 100%. Similar investigation was conducted by using 50 mg/L 2-undecyl-1-ethylamino-1-methylbenzyl quaternary imidazoline (2UMQI) as the inhibitor on carbon steel in 1.0 M H_2SO_4 [18]. Unlike Okafor, Liu, & Zheng [9], they found that the inhibition efficiency increased by about 50% of its initial reading irrespective to I^- ion concentrations as obtained in Table 3. Furthermore, when they explored the steel surface through AFM images as shown in Figure 1, they found that the steel surfaces produced the smoothest surfaces under the condition of 50 mg/L 2UMQI+2.5 mM KI. Hence they proved that the 2UMQI/KI system is more effective inhibitor than 50 mg/L 2UMQI alone, without addition of KI.

Table 1. Inhibition efficiency (% η) for carbon steel in 3 wt% NaCl solution at 25°C in the absence and presence of IM and KI from impedance spectroscopy and potentiodynamic polarization [19].

IM	Impedance method (% η)		Polarization method (% η)	
(mg/L)	Without KI	With 2000 mg/L KI	Without KI	With 2000 mg/L KI
0	-	-	-	-
50	76.7	90.1	85.1	94.2
75	91.2	92.9	95.8	97.0
100	91.8	94.3	96.2	97.9

Table 2. Inhibition efficiency for carbon steel in in 3 wt% NaCl saturated with CO_2 at 25°C in the absence and presence of 2UEI and KI using potentiodynamic polarization [17].

2UEI (mg/L)	Polarization method (% η)		
	0 mg/L KI	830 mg/L KI	2000 mg/L KI
0	-	-21.8	-51.0
80	95.0	96.0	97.4

Table 3. Inhibition efficiency for carbon steel in 1 M H_2SO_4 at 25°C in the absence and presence of 2UMQI and KI using impedance spectroscopy and potentiodynamic polarization [18].

2UMQI	Impedance method (% η)				Polarization method (% η)			
(mg/L)	0 mg/L KI	16.6 mg/L KI	166 mg/L KI	415 mg/L KI	0 mg/L KI	16.6 mg/L KI	166 mg/L KI	415 mg/L KI
0	-	93.6	96.5	98.1	-	88.2	93.2	96.6
50	61.6	98.6	99.5	99.6	63.4	97.5	98.2	98.9

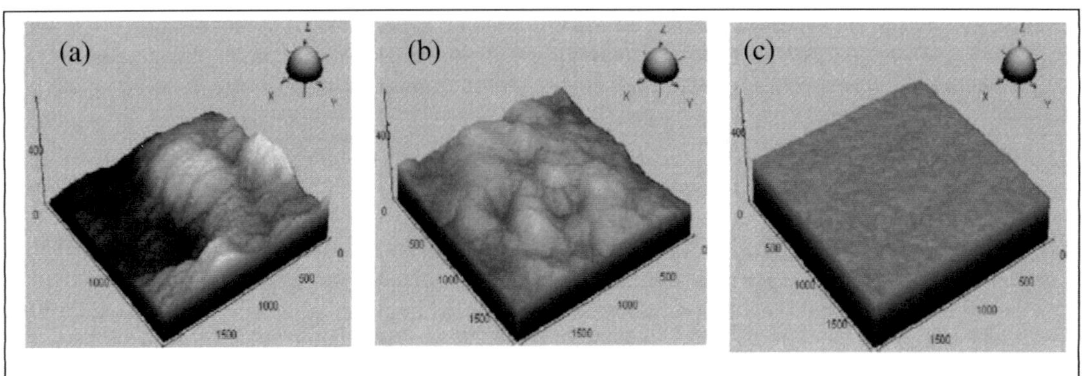

Figure 1. AFM images of carbon steel immersed in 1.0 M H_2SO_4 (a) without inhibitor (b) containing 50 mg/L 2UMQI and (c) containing 50 mg/L 2UMQI + 2.5 mM KI after 1 hour of immersion at 25 °C [18].

Figure 2 is plotted in order to observe the effect of I^- ion at different concentrations of inhibitor systems. Heydari & Javidi [19] studied at a fixed amount of I^- ion, Okafor, Liu, & Zheng [17] and Okafor & Zheng [18] studied at varies amount of I^- ion. It is observed that a more evident increment of the inhibition efficiency between inhibitor-only system and inhibitor/I^- system, particularly when the inhibitor is presence at low concentration. According to Figure 2, without the presence of I^- ion, the inhibition efficiency is 63.4% and 85.1% for 50 mg/L 2UMQI inhibitor and 50 mg/L IM inhibitor, respectively. In the presence of 50 mg/L 2UMQI + 16.6 mg/L I^- ion, the inhibition efficiency is 97.5%. While in the presence of 50 mg/L IM + 2000 mg/L I^- ion, the inhibition efficiency is 94.2%. The inhibition efficiency remains at a value close to 100% at increasing I^- ion.

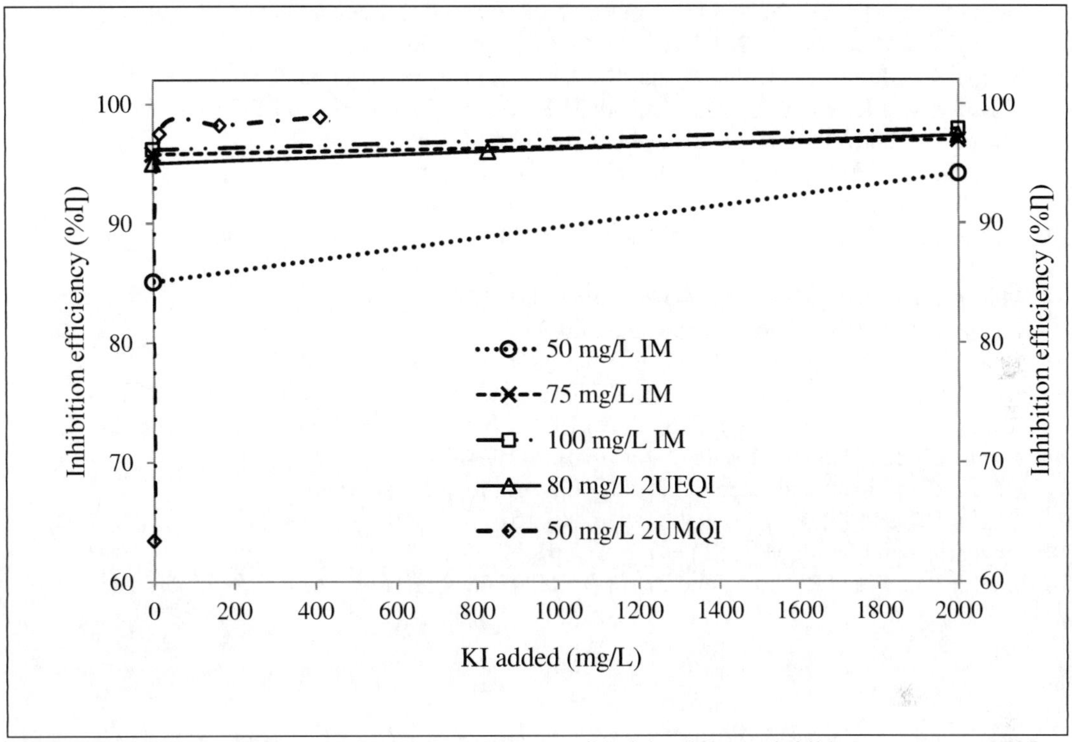

Figure 2. Effects of [I^-] ion on inhibition efficiency (% η) in different inhibitor systems at 25°C [17]–[19].

This means that there is a synergistic inhibition effect of the I^- ion in the imidazoline derivative inhibitors. However, only a slight increase of inhibition efficiency has occurred when inhibitors are presence at high concentrations. This is shown from the potentiodynamic polarization measurements of 75 mg/L and 100 mg/L IM systems and 80 mg/L 2UEQI systems. The inhibition efficiencies are already close to 100% despite no I^- ion is present. In this case, it is concluded that the synergistic inhibition effect of the I^- ion is predominant in a low concentration inhibitor system. This will become a benefit to manufacturing industries since they will be able to reduce corrosion related problem issues at lower operation costs. This is possible when they use enhancer such as iodide ion together with a small amount of corrosion inhibitor.

3. Conclusions

It has been shown that synergistic effect between organic inhibitors and iodide ion is an effective method to improve the inhibitive performance for metals in different aqueous media. The results from a series of investigations revealed that the adsorption of organic inhibitors was stabilized by the presence of iodide ion (I^-) in the solutions and enhanced the inhibition efficiency significantly due to synergistic effect. The synergistic effect increases with increases in KI concentration in the mixture. It has proven to be an effective way to decrease the amount of organic inhibitor consumption.

Acknowledgment
The authors would like to thank the Ministry of Education Malaysia for financial support via Fundamental Research Grant Scheme (600-RMI/FRGS 5/3 (10/2014)). The authors also gratefully acknowledged use of services and research facilities available at the Faculty of Chemical Engineering, Universiti Teknologi MARA, Malaysia and MTeG, Malaysian Nuclear Agency (Nuclear Malaysia).

References
[1] Al Juhaiman L A 2016, *Int. J. Electrochem. Sci.* **11** 2247–2262
[2] Koch G, Varney J, Thompson N, Moghissi O, Gould M, and Payer J 2016
[3] Wang L, Qu M, Yang Y, Peng L, and Ma S 2016 *Int. J. Electrochem. Sci.* **11** 9307–9325
[4] Ali A I and Mahrous Y S 2017 *Rsc Adv.*, 7 **38** 23687–23698
[5] Solmaz R, Altunbaş E, Döner A, and Kardaş G 2011 *Corros. Sci.***53** 3231–3240
[6] Hosseini M, Mertens S F L, and Arshadi M R 2003 *J. Corros. Sci.*, **45** 1473–1489
[7] Morad M S 2007 *J. Appl. Electrochem.***37** 661–668
[8] Li X, Deng S, Fu H, and Xie X 2014 *Corros. Sci.* **78** 29–42
[9] Wang W and Free M L 2004 *Corros. Sci.***46** 2601–2611
[10] Yang L, Xu Y, Zhu Y, Liu L, Wang Y, and Huang Y 2016 *Int. J. Electrochem. Sci.* **11** 6943–6958
[11] Manssouri M 2015 *J. Mater. Environ. Sci.* **6** 631–646
[12] Laamari M R 2016, *Arab. J. Chem.* **9** S245–S251
[13] Khan M Z H, Aziz M A 2016 *Anti-Corrosion Methods Mater.* **63** 308–315
[14] Khamis A, Saleh M M, Awad M I, and El-Anadouli B E 2013 *Corros. Sci.* **74** 83–91
[15] Ridhwan A M, Rahim A A, and Shah A M 2012 *Int. J. Electrochem. Sci.***7**, 8091–8104
[16] Farag A A and Hegazy M A 2013 *J. Corros. Sci.* **74** 168–177
[17] Okafor P C, Liu X, and Zheng Y G 2009 *Corros. Sci.* **51** 761–768
[18] Okafor P C and Zheng Y 2009 *J. Corros. Sci.*, **51** 850–859
[19] Heydari M and Javidi M 2012 *Corros. Sci.* **61** 148–155
[20] Okafor P C, Liu C B, Liu X, and Zheng Y G 2009 *J. Appl. Electrochem.***39** 2535–2543

Effect of Organic Oxygen Scavenger on Performance of Pyrrole as Corrosion Inhibitor

E S Mohd Kassim, I M Ibrahim, J Jai, M S So'aib, N Ahmad Zamanhuri, H Husin, and M A Hashim

Faculty of Chemical Engineering, Universiti Teknologi MARA,
40450 Shah Alam, Selangor, Malaysia

ezansuhailah@yahoo.com

Abstract. The inhibitory effect of pyrrole in the presence of methyl ethyl ketoxime (MEKO) and erythorbic acid (EA) on the corrosion of carbon steel in static of condition 3.5 wt% NaCl solution were studied using Linear Polarization Resistance (LPR) method. Experimental results found that the inhibition effect of pyrrole increased with the increase of oxygen scavenger concentration.The inhibition efficiency was observed to be about 67% after addition of erythorbic acid (EA) into saline solution containing 100 ppm of pyrrole compared by adding MEKO which recorded about 59%. The addition of oxygen scavenger could reducing the corrosion rate of carbon steel by reacting with dissolved oxygen in the solution and thus further to protect metal surface.

1. Introduction

Corrosion is worth investigating in oilfield application because corrosion problem represent a large portion of the total cost for oil and gas producing companies every year worldwide[1]. Corrosion problem are usually connected with operating problems and equipment maintenance and resulting in severe economic losses [2]. According Garcia et al; the economic costs linked to the corrosion of natural gas sweetening and oil refining plants range between 10 % and 30 % of the maintenance budget [3]. Corrosion inhibitor is a chemical compound which when added to an aqueous system in a small quantity, serves to retard the corrosion rate of a metal exposed to that aqueous media [4]. Compounds containing nitrogen, oxygen, sulphur and phosphorus in system have particularly been reported as efficient corrosion inhibitor [5]. These compounds prevent corrosion by blocking the active corrosion sites either by getting adsorbed, or by forming a protective layer or an insoluble complex on the metal surface [6]. However, in order to improve the inhibitive force and reduce the cost, synergistic effect is an effective method [7]. The previous study had been proven that pyrrole as effective corrosion inhibitors for carbon steel when immersed in saline media [8]. In order to better perform in term of efficiency, the synergistic effect of pyrrole and the addition at various concentrations of methylethy ketoxime (MEKO) and erythorbic acid (EA) were carried out in the investigative experiment. The objective of research is on the inhibition effect of pyrrole with different addition of concentration of oxygen scavengers into 3.5 wt% NaCl solution at room temperature (25°C) and static condition. This study was conducted using linear polarization resistance (LPR) method.

Content from this work may be used under the terms of the Creative Commons Attribution 3.0 licence. Any further distribution of this work must maintain attribution to the author(s) and the title of the work, journal citation and DOI.
Published under licence by IOP Publishing Ltd

IOP Conf. Series: Materials Science and Engineering **358** (2018) 012045 doi:10.1088/1757-899X/358/1/012045

2. Experimental

2.1 Material and material preparation

The chemical composition of specimen was analyzed using Inductively Couple Plasma (ICP) and CHNS analyzer. The percentage composition of carbon steel used in this research is as follow: C, 0.06%; Al, 0.092%; Mn, 0.550%; Ni, 0.056%; Cu, 0.350%; Zn, 0.025%; Pb, 0.070%; Cr, 0.064% and iron as remainder. The commercial grade carbon steel was cut into dimension 1 cm x 1 cm x 0.5 cm and the embedded into cold mounting polymer. The nichrome wire was used to solder to the specimen as an electrical connection. The exposed surface of specimen (area 1 cm^2) was ground and polished as previous reported. A 3.5 wt% sodium chloride solution was used as the aqueous solution. A 100 ppm pyrrole was used as test corrosion inhibitor and different concentration of MEKO and EA were then added to evaluate the synergistic corrosion inhibition effect on addition to the test inhibitor.

3. Linear polarization resistance (LPR) measurement and surface morphology analysis

In the Linear polarization resistance (LPR) experiment, the working electrode was immersed in beaker containing typically about 900 ml of NaCl solution. All the experiments were performed at room temperature (25 °C) and static condition. The set-up of experiment was equipped with platinum auxiliary electrode (AE) and a reference saturated calomel electrode (SCE) which was connected to working electrode (WE). The LPR measurements were performed by using a VoltaLab, Model PGP 201 and controlled by corrosion analysis software model Voltamaster 4. The working electrode was immersed for 1 hour in experimental solution to obtain stable open circuit potential (OCP).The polarization resistances value, R_p was determined after 100 ppm pyrrole injected into 3.5 wt% NaCl with an arbitrarily chosen concentration (50 ppm and 200 ppm) of MEKO and EA. The inhibition efficiency was evaluated from the measured value of R_p according to Equation 1;

$$IE\% = \frac{R_P - R_{P0}}{R_P} \times 100\% \quad (1)$$

Where R_p and R_{po} are the polarization resistance value with and without addition of MEKO. The surface morphology of carbon steel was examined using Field Emission Scanning Electron Microscope, FESEM (DSM 982 Gemini Supra 40 VP) and Atomic Force Microscope, AFM (XE-100, Park System).

4. Result and Discussion

Comparison of R_p values with time when immersed in 3.5 wt % NaCl solution containing 100 ppm pyrrole over 1 hour period with selected concentration (50 ppm and 200 ppm) of MEKO and EA is shown in Figure 1. The immersion of 1 hour period indicates that inhibitor can protect the corrosion in short period [9]. It is very clear from the Figure that the value of R_p was increased in the presence of pyrrole. The value of R_p is smaller in the uninhibited solution compared to the inhibited, suggested that pyrrole suppressed the dissolution of carbon steel specimen in the aggressive environment [10].Addition of MEKO to pyrrole is found to increase R_p by almost twice times that of the pyrrole alone. For instance, the addition of 50 ppm and 200 ppm MEKO to 100 ppm pyrrole increased R_p from 1.510 kohm.cm^2 to 2.910 kohm.cm^2.

However, after addition concentration of 200 ppm EA, the R_p values was increased to 3.489 kohm.cm^2.The result also shows that combination of pyrrole and MEKO or EA can give the higher corrosion resistance compared to uninhibited solution. It is also reported by I.M Ibrahim et al (2013); that the combination or synergistic effect of fatty amide and hydrazine increased the corrosion resistance and inhibited the corrosion process excellently compared uninhibited solution [11]. In this experiment, Methyl ethyl ketoxime (MEKO) and Erythorbic acid (EA) were selected as an oxygen scavenger which had ability to eliminate the dissolved oxygen molecules in the solution[12]. Oxygen scavenger was added to deplete the maximum possible amount of dissolved oxygen in solution in order to enhance the inhibition efficiency[13]. Methyl ethyl ketoxime (MEKO) is an organic scavenger that reacted to oxygen to produce methyl ethyl ketone (MEK). The reaction with dissolve oxygen is known to occur as shown below.

IOP Conf. Series: Materials Science and Engineering **358** (2018) 012045 doi:10.1088/1757-899X/358/1/012045

$$C_4H_9NOH + \frac{1}{2} O_2 \longrightarrow C_4H_9O + N_2O + H_2O \quad (2)$$

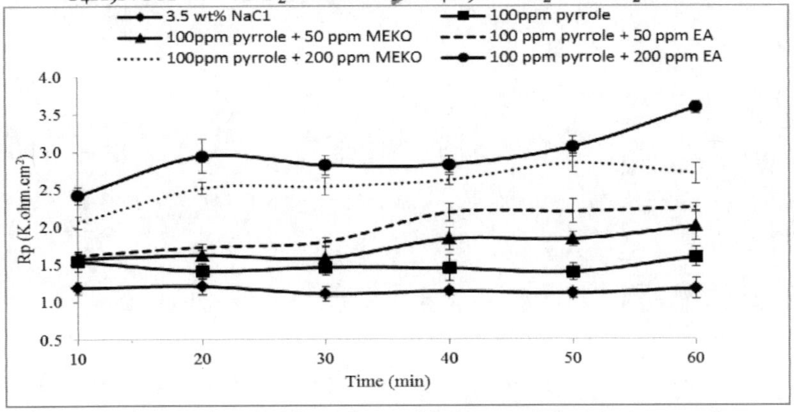

Figure 1. Comparison of R_p values with immersion time in pyrrole-inhibited 3.5 wt% NaCl solution between MEKO and EA

The advantages of methyl ethyl ketoxime (MEKO) include a higher thermal stability than other scavengers. MEKO can break down to smaller organic, carbon dioxide and nitrogen compound [14]. Erythorbic acid (EA) is an isomer of vitamin and reported to the react quickly with dissolve oxygen is to form organic acids and water. The reaction with dissolved oxygen as shown below [15].

$$C_6H_8O_6 + 1/_2O_2 \longrightarrow C_6H_6O_6 + H_2O \quad (3)$$

By acting as oxygen scavenger, amount of dissolved oxygen molecules available become depleted. With increasing the concentration of MEKO or EA into solution, more oxygen molecules available would react with MEKO or EA thus further protected to the metal surface. Since oxygen molecules that involved in cathodic reaction were decreased, the anodic reaction also will be decreased correspondingly. Table 1 shows the effect of addition oxygen scavengers on the inhibition efficiency that contained 100 ppm of pyrrole. It is observed that after addition of 50 ppm and 200 ppm of MEKO in the inhibited solution, the inhibition efficiency increased. At the higher concentration of 200 ppm EA, the inhibition efficiency reached about 67% compared to addition MEKO into solution which recorded about 59%. It is can be concluded that the addition of MEKO or EA into the inhibited system is intended to enhance their inhibition efficiency due to interaction between pyrrole and both oxygen scavengers [16].

Table 1. Comparison inhibition efficiency of pyrrole addition MEKO and EA in 3.5 wt % NaCl solution

	100 ppm pyrrole	**100 ppm pyrrole + 50 ppm MEKO**	**100 ppm pyrrole + 50 ppm EA**	**100 ppm pyrrole + 200 ppm MEKO**	**100 ppm pyrrole + 200 ppm EA**
Inhibition efficiency (%)	30.72	45.26	39.98	59.55	67.21

4.1 FESEM

The FESEM images of carbon steel specimen after 60 minutes immersion in 3.5 wt% of NaCl solution in the absence and presence of oxygen scavengers are shown in Figure 2 at magnification of 500. In Figure 2(a), the sample was immersed in 3.5 wt % NaCl was severely damaged with a lot of surface corroded. Figure 2 (b) shows the sample was containing 100 ppm pyrrole in addition of MEKO. In the presence of MEKO, the surface of steel is less corroded and looks better compared to Figure 2 (a). After the combination of pyrrole with EA, the surface is also looks smoother as shown in Figure 2 (c). It is indicated the pyrrole and both of oxygen scavenger could be work together to cover the metal surface thus giving it a high degree of protection against corrosion.

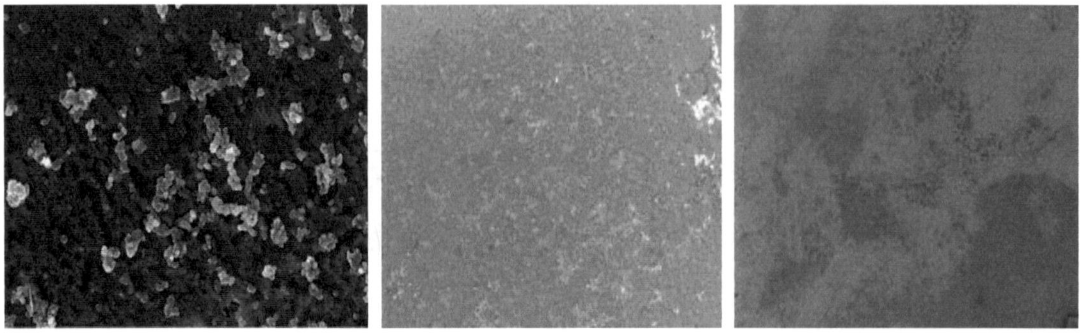

(a) (b) (c)

Figure 2.FESEM images of carbon steel after immersion in (a) 3.5 wt % NaCl (b)100 ppm pyrrole +200 ppm MEKO (c) 100 ppm pyrrole +200 ppm EA

Table 2..EDX analysis on samples.

	Element (%)				
Sample	Fe	O	Na	Cl	C
Carbon steel immersed in 3.5 wt% NaCl solution	79.65	17.23	1.37	0.64	1.11
Carbon steel in 100 ppm pyrrole + 200 ppm MEKO	91.75	5.99	0.69	0.38	1.99
Carbon steel in 100 ppm pyrrole + 200 ppm EA	95.19	1.99	0.69	0.38	1.75

The EDX analyses are shown in Table 2 show the elemental composition that contained in sample. In Table 2, the data shows the percentages of Fe, O, Na, C1 and C represent the chemical composition on the metal surface. The main points that wanted to highlight here is the percentage of oxygen in sample. The percentage oxygen in sample (b) and (c) are lower compared to (a) was proven that ability of MEKO and EA act as oxygen scavenger to eliminate the dissolved oxygen available in solution thus to reduce corrosion rate.

4.2 AFM

To support the result, the selected of specimen was analyzed by AFM. The three dimensional AFM images of carbon steel specimen after 60 minutes immersion in inhibited solution in the absence and presence of MEKO at static condition and room temperature are as shown in Figure 3.

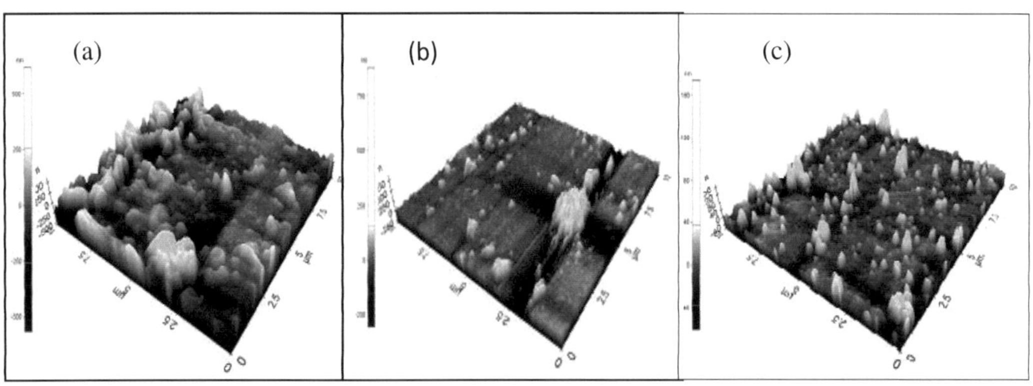

Figure 3. AFM image of carbon steel after 60 minutes immersion in (a) 3.5 wt% NaCl (b) 100 ppm pyrrole + 200 ppm MEKO (c) 100 ppm pyrrole + 200 ppm EA

Figure 3(a) show surface morphology of the specimen immersed in 3.5 wt% NaCl solution. It shows a lot of corrosion products deposit on metal surface. The average roughness of surface (Ra) was 97 nm. Figure 3(b) is specimen immersed in 100 ppm pyrrole into 3.5 wt % NaCl and give value of Ra is 51 nm. However, after combination of 200 ppm MEKO and EA to the inhibited solution as shown in Figure 3(b) and 3(c), the surface become smoother compared to previous figure .The Ra value is 26 nm and 12.305 nm respectively.

5. Conclusion

The inhibition efficiency of pyrrole can be further improved by introducing of an oxygen scanvenger that is Metylethyl ketoxime (MEKO) and erythorbic acid (EA). By added the different concentrations of oxygen scavenger into the solution, the amount oxygen molecules available in solution become depleted hence can reduce the corrosion rate. The combination of pyrrole with oxygen scavengers has been found to enhance the efficiency of inhibitive effects due to multiple effects of effective corrosion inhibition [17].

Acknowledgement

The authors express they're thanks to Ministry of Education Malaysia for the financial support through Fundamental Research Grant Scheme (600-RMI/FRGS 5/3 (10/2014). The author also gratefully acknowledged using services and other research facilities available at the Faculty of Chemical Engineering, Universiti Teknologi MARA, Malaysia

References

[1] Ibrahim Salama 2011 *Corrosion Inhibitors in the Oilfield.*
[2] Finšgar M and Jackson J 2014 *Corros. Sci* **86** 17–41.
[3] Garcia-Arriaga V, Alvarez-Ramirez J, Amaya M, and Sosa E 2010 *Corros. Sci* **52** 2268–2279
[4] Mobin M, Parveen M, and Rafiquee M Z A 2013 *Arab. J. Chem* **10** S1364-S1372
[5] Musa A Y, Mohamad A B, Kadhum A A H, Takriff M S, and Tien L T 2011 *Corros. Sci.* **53** 11 3672–3677
[6] Ibrahim T H and Zour M A 2011 *Corros. Sci* **6** 6442–6455
[7] Krim O,Bouachrine M, Hammouti, Elidrissi A, and Hamidi M 2008 *Port. Electrochim. Acta* **26** 3 283–289
[8] Ibrahim I M, Yunus S, and Hashim M A 2013 *Int. J. Sci. Eng. Res* **4** 2 1–12
[9] Manickam M, Sivakumar D, Thirumalairaj B, and Jaganathan M 2016 *Hindawi Publ. Corp* **2016**.
[10] Solomon M M and Umoren S A 2015 *J. Environ. Chem. Eng* **3** 3 1812–1826
[11] Ibrahim I M, Daud M, Jai J, and Hashim A 2015 *Int. J. Eng. Technol* **7** 3 844–851
[12] Paper C, Takahashi J, and Water K 2016 *A New Patented Oxygen Scavenger to Replace Hydrazine in Boiler Water Treatment Applications*
[13] Tretheway K R and Chamberlain J 1995 *Corrosion for Science and Engineering*, Longman Si
[14] John D. Zupanovich Chem Treat 2002 *Oxidation and degradation product of common oxygen scavengers*
[15] Boiler Systems 2004 *Association of Water Technologies Technical Manual* 58–95
[16] Arukalam I O 2014 *Carbohydr. Polym.* **112** 291–299
[17] Ridhwan M A, Rahim A A, and Shah A M 2012 *Corros. Sci* **7** 8091–8104

IOP Conf. Series: Materials Science and Engineering **358** (2018) 012046 doi:10.1088/1757-899X/358/1/012046

Synthesis of reduced Graphene Oxide (rGO) using different treatments of Graphene Oxide (GO)

M F Zainuddin[1], N H Nik Raikhan[2], N H Othman[3] and W F H Abdullah[4]

[1, 2] Bioprocess Engineering Department, Faculty of Chemical Engineering, Universiti Teknology MARA (UiTM), 40450 Shah Alam, Selangor. Malaysia
[3] Oil and Gas Engineering Department. Faculty of Chemical Engineering, Universiti Teknologi MARA (UiTM), 40450 Shah Alam, Selangor. Malaysia
[4] Electrical Engineering Department, Faculty of Electrical Engineering,Universiti Teknologi MARA (UiTM), 40450 Shah Alam, Selangor. Malaysia

nik.raikhan@gmail.com

Abstract. In this work, a combined chemical and mechanical method was used for the production of graphene instead of chemical method only. The use of mechanical sonication was to assists exfoliation graphene oxide (GO) besides the used of chemical reagents. Then, the reduction of GO into graphene was carried out using L-ascorbic acid. The effects sonication cause synthesis of GO undergoes peeling graphitic layer and at the same time expose the layer with oxidizing agent. The properties of GO and reduced-graphene oxide (rGO) using various routes were investigated using XRD and FTIR. The main characteristics peak was observed at 7° and 9° for GO prepared using sonication and without sonication, respectively. The decreased of 2-theta degree of GO prepared using sonication indicates that the d-value becomes bigger. Estimation of average diameter rGO (with sonication) was 24.49 nm while rGO (without sonication) was 126.2 nm. The reduction of both GO was then carried out using an environmentally reducing agent, ascorbic acid. It was found that the L-ascorbic acid was effective in removing oxygenated functional groups. The conductivity values obtained for rGO-s was 7640 S/m while rGO-ws was is 678 S/m.

1. Introduction

Graphene is one of new material discovered by reseacher from Manchester University in 2004 and won Noble Prize in Physics for their pioneering work in year 2010 [1]. Over the past few years, graphene based flexible thin-film transistors and supercapacitor electrode has attracted great research interest due to their unique properties such as high capacity to store energy. Graphene can be synthesised from graphite and several methods such as flake fragmentation [2] and growing [3] methods. Most of the methods proposed aims to improve the quality, properties and quantity of graphene production. Method has been divided into two categories with relatively different ways that is bottom up and top down graphene [4]. In fact of that, top-down process have three method that is mechanical exfoliation, chemical exfoliation and chemical synthesis whereby single graphene sheet fragmented by separation or exfoliation with grapene oxide as intermediate phase. While, bottom-up graphene method more on some quality properties, average size and thickness produced by pyrolysis, epitaxial growth, chemical vapor deposition (CVD) and other [5].

Graphene oxide (GO) is utilized as precursor for a large-scale production of graphene-based especially in chemical exfoliation and CVD methods [6]. GO thin layer is insulator materials due to owing many oxygenated functional group and defect attribute to tunability and feasibility

Content from this work may be used under the terms of the Creative Commons Attribution 3.0 licence. Any further distribution of this work must maintain attribution to the author(s) and the title of the work, journal citation and DOI.
Published under licence by IOP Publishing Ltd

incorporation [7]. The hydrophilic structure can easily disperse in water to give stable solution for single layer. Reviewed by Huang et al. [7] on graphene composite which summarize properties and combination of various chemical reported before and attraction on combination functional groups on polymer with GO. The covalent functionalization of graphene derivative is mainly based on reaction of functional groups on GO and rGO surface. This behaviour attractive on some research on GO and rGO for combination with polymer for backbone of graphene, combination with nano-metal oxide composite [8-10], combination conducting polymer for electrical enhancement [11-14], and bio-tribological systems [15]. Application in electrical field, rGO are commonly synthesized and used in nano-scale because of it is high in surface area contact and porosity are excellent in coating process for sensor.

Quality of rGO in almost cases is come through from the degree of reduction attainable and restoration of relevant properties (eg: electrical conductivity) [16]. While, GO quality from degree of reduction parallel with existing reactive time and temperature [6] to assists expanding graphite layer for exfoliation. Other than that, the oxidative-treatment GO do attachment of oxygen-containing functional groups during oxidation process increase interlayer between graphitic layer, thus, weakening van der Waals forces. The development method reagent used in reduction graphene oxide is very highly by using hydrazine hydrate [16-19], sodium borohydrate [16, 19, 20], glucose [21], hydroquinone [22] and many more. Due to lack of quality interm of getting high conductivity and high surface area (small size particle) for demand graphene, new innovation method are combine with chemical and mechanical methods with drop of sodium nitrate reagent because it is highly fuming carcinogenic vapour. The innovation is reduction reagent by using eco-friendly reagent uses in food and pharmaceutical like ascorbic acid. L-ascorbic acid has potential as reducing agent thus serves double function in reduction of hydrophilic GO to hydrophobic rGO and conversion residual Mn (VII) ions to soluble Mn (II) ions.

Addition mechanical treatment in several part in the way to get graphene believed that highly potential in assists to breakdown van der Waals forces between graphitic layers. Ultrasonic treatment of graphite during oxidizing agent can peel apart the oxidized layers and expose the inner layers to the chemical oxidants. Typically, performance of sonication exert on GO suspension are proportional to time and also can be temperature of suspension. Temperature variable important in expand the graphitic layer and also remove oxygenated groups on basal plane. Repeated sonication process are proposed before is unsucessful by using bath sonication but nonetheless resulted in nanoscale graphene. Another addition treatment by using ball mill effect on graphite are not much research on graphene but in CNTs are various interm to study understanding dispersion CNTs and effect on CNTs structure which reported length of CNTs shortened significantly with increasing milling time [23]. Combination mechanochemical synthesis reduced graphene for effectiveness and size has been investigated.

2. Materials and methods

2.1 Chemicals and materials

Graphite with a purity of 99% was purchased from COMAK. Sulphuric acid (H_2SO_4, 98% purity) and hydrogen peroxide (H_2O_2, with 30% purity) were supplied by R&M Chemicals. Potassium permanganate ($KMnO_4$) and L-ascorbic acid with purity 99% was obtained from Sigma Aldrich. Ethylene glycol (Merck) was purchased from Merck and used as dispersion solvent.

2.2 Preparation of GO using sonication method

Graphene oxide was prepared by using improved Hummer's method as it is one of the eco-friendly method due to the removal of carcinogenic chemical, sodium nitrate ($NaNO_3$) [24]. Typically, 5 g of graphite were ball milled for several hours to weaken Van der Waals interaction between graphitic layers. Ball milled graphite powder were mixed with 200 ml H_2SO_4 in a beaker under ice bath condition and stirred for 1 hour. Then, 30 g of $KMnO_4$ was added slowly and the stirring was continued for another 24 hours for oxidation reaction to complete. The mixture was stirred for 2 hours to homogenize the solution and appear dark brown solution. 100 ml distilled water was then added slowly and the solution was sonicated for more than 5 hours at 30°C. After the sonication procedure, the mixture was heated at 90°C for 1 hour and sufficient distilled water was added and stirs. Next, 30 ml hydrogen peroxide was added to stop the reaction. Suspension was divided into two equal parts; one washed to obtain GO and the other was further processed for preparation of rGO. After washed by HCl and distilled water, adjust pH almost to neutral. Finally, centrifuge suspension with 10,000 rpm for 20 min and GO precipitate were dried at oven.

2.3 Preparation of reduce graphene oxide (rGO)

The deoxygenation of graphene oxide to reduced graphene oxide (rGO) was carried out using L-ascorbic acid (vitamin C) as reducing agent. 10 g L-ascorbic acid was first dissolved in 100 ml of distilled water. The L-ascorbic acid solution was mix together with GO-s and GO-ws. Then, the mixture was heated at temperature 95°C for 1 hour. After heating, immediately sonicate for 15 minutes. The resultant black precipitate were simply filtered with vacuum pump and further were washed with 1M hydrochloric acid solution and distilled water to neutral pH. Finally, filter and dry in oven overnight.

2.4 Characterization and measurement

2.4.1 X-ray Diffraction (XRD) analysis

X-ray diffraction (XRD) patterns were obtained using D/Max 2200V/PC, Rigaku, Japan where CuKα as the radiation source with wavelength (λ= 0.15 nm). The XRD scans were carried out in a 2θ range from 5° to 80° at a scanning rate of 2°/ min using a voltage of 40 kV and current of 40 mA.

2.4.2 Fourier transformed infrared (FTIR) study

Fourier transform infrared (FTIR) spectroscopy were recorded using Perkin Elmer Spectrum One in order to study the chemical functional groups of graphite, GO and rGO after the synthesis process.

2.4.3 Dynamic light scattering (DLS)

Particle size and zeta potential analysis of the rGO were carried out using a Nano-ZS Zetasizer (Malvern, UK) in electronic light scattering mode. 0.5 mg/ml rGO suspension in ethylene glycol was prepared and sonicated for 10 minutes to form homogenous solution.

3. Result and Discussion

X-ray diffraction (XRD) analysis was first carried out to observe any crystal structure changes after graphite has been oxidised to GO, followed by the reduction of GO to rGO. Based on figure 1, it can be seen that graphite shows a high intensity and sharp peak at 2θ= 26.7°. This indicates that crystallinity morphology and large crystallite size (with interlayer space is about 3.35 Å) of graphite. When graphite expose to oxidizing agent ($KMnO_4$) and intercalant agent (H_2SO_4), the graphitic interlayer d-spacing will increase drastically to weaken bond between layers. These situation called graphite intercalated compound (GIC). After the oxidation treatment of graphite to GO, the peak dissapeared and new peak appears at lower 2 theta degree which indicates that the graphite was completely convert to GO [15]. A new and sharp peak appears at 2θ= 9° for GO-s and 2θ= 11.3° GO-ws. Both are in the range of common peak between 2θ= 9°-12 ° [25,26]. Higher 2 theta degree and intensity when no ultrasonication was used may due to the merging of individual graphene oxide flake size more than 100 nm [27] attribute from agglomeration.

In addition, GO-ws usually forms polycrystalline material which contain wide width before going to lower of the graph. Consist of many single crystal oriented are not completely dissociate during method without sonication cause X-rays to be scattered. By introducing sonication during the preparation of GO, it reduces agglomeration of particles and weakening the Van der Waals bonding between each graphitic layer and at the same time force oxygenated group entering to the deep area. Value of d-spacing for GO (with sonication) and GO (without sonication) is 9.551 Å and 7.789 Å respectively. Evidently, existing sonication would give impact to graphitic layer in expanding to weaken the bonding and exfoliate layer by layer. Normally, rGO and graphite can be seen almost the same about their peak ranging in 2θ= 25-27° due deoxygenated functional group with differs in peak sharpness and particle size. Degree of reduction of rGO (L-AA) and rGO (L-AA/S) is relatively at different peak width and location of peaks. The main rGO (L-AA) peaks characterized by intense and broad at 2θ= 26.0° and adjacent weak peak located at 2θ= 20.1° and 21.0 ° and 2θ= 44.0°. rGO (L-AA/S) shows less intense and broad XRD peaks at 2θ= 26.0° with appear small peak adjacent at theta 2θ= 26.0°. Amorphous crystallite particles can be estimated this rGO based on wider peak appears but in theory of adjacent two peaks are come from orientation different physical shape can be ascribe difference in number of atoms in plane. Stated crystalite size is 48 nm and 47 nm for rGO-ws and rGO-s respectively. The reduced sheet decreases, both around 3.34 Å is very close to graphite to mention this deduction intersheet hydrophobic were removed water molecule as expander during oxidation. Effectiveness in removal oxygenated functional group will reduce *d*-spacing sheet and hydrophobic properties appears.

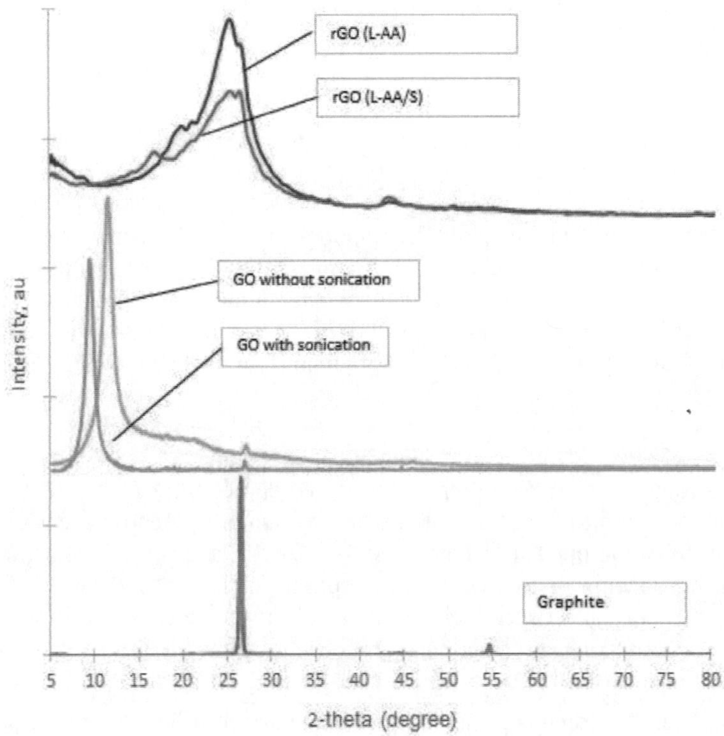

Figure 1. XRD pattern of graphite, GO-s, GO-ws, rGO-s and rGO-ws

Figure 2- compares the functional group of graphite, GO and rGO using FTIR spectrometer. It can be seen that the graphite used does not any oxygenated functional group. Both GO contains O-H bond at 3000-3600 cm^{-1}. However, a broad and intense band was observed for GO-s while GO-ws contains less intense band at 3000-3600 cm^{-1} (O-H stretching vibration) with intermolecular single H-bonds. The high frequency of the spectrum results of vibration or rotation of atom that may raise electron to higher energy levels. Besides the O-H bond, others bond such as at 1710 cm^{-1} was observed indicates occurance of oxidation ; C=O (carbonyl group). The presence of the oxide linkage with carbon atom reflected some interaction between graphite basal plane and $KMnO_4$, H_2SO_4 and water molecules. At wavelength 1600 cm^{-1} shows that contain C=C (aromatic) molecule were still present after treatment because they stable at this condition. 1120 cm^{-1} (O-H bending vibration from hydroxyl groups). Sulfur functional group found at 1030 cm^{-1} and alkenes groups with bending vibration at 890 cm^{-1}. At 550cm^{-1} (C-Cl stretching vibration from alkyl halid group) may be contribution from washing. For all the reduced GO (Figure 2b), the intesities of the band associated to oxygen funtional groups strongly decrease. rGO (L-AA) contain small amount oxygenated functional group with broad (O-H stretching vibration) and (C-O stretching vibration) at 3300-3700 cm^{-1} and 1300 cm^{-1} respectively. Aromatic

(C=C stretching vibration), 1600 cm-1 were found in rGO (L-AA) indicates stretch band shows the remaining sp^2 character [28]. Sulfur and chloride functional group are still appears but in low intensity in fingerprint area.

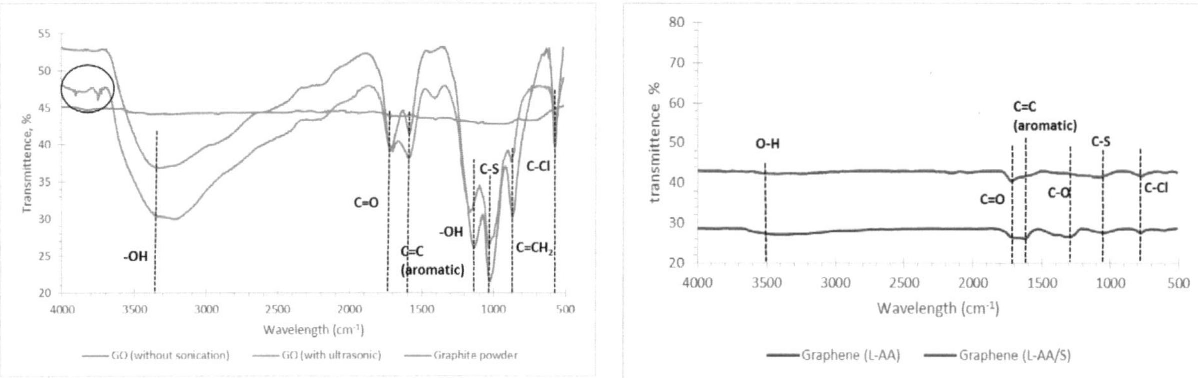

a b

Figure 2. (a) FT-IR specta of graphite and GO-s and GO-ws (b) comparison FT-IR spectra between rGO-s and rGO-ws

Dynamic light scattering zetasizer plays an important role to determine the surface conductivity and particle size. Prior to testing, a solution 0.5 mg/ml rGO in ethylene glycol solvent was prepared [6]. Ethylene glycol is the best solvent to disperse rGO that remove hydrophobicity characteristic. The average diameter particle sizes was 24.49 nm and 126.2 nm for rGO-s and rGO-ws respectively. While, conductivity shows by rGO (L-AA/S) is more than (L-AA) which is 7640 S/m over 678 S/m.

4. Conclusion

In this work, it was found that the rGO prepared using combined ultrasonic and chemical method has better properties in compared to those untreated with ultrasonic. Evidently, GO-s give positive results as common review formation peak at 9° compared to GO-ws. Phenomena occurring GO-ws may due to lacking of assist in opening or expanding the graphitic interlayer spacing that just only focus on oxidation reagent. It seems like depending on chemical are not effectively in expand the graphitic interlayer. Relative amount oxygen functional group was estimated from intensities of the oxygen related band in FTIR measurements. GO-s give high d-spacing interlayer than GO-ws with value 9.551 Å and 7.789 Å. The results strengthen that the graphitic surface influences the structural, optical and electrical properties of the GO. GO synthesis in term of exfoliation control would be useful in obtaining good efficiency to reduce. As a result, rGO-s is containing high conductivity than rGO-ws.

Acknowledgement

The authors would like to thank Universiti Teknologi MARA, UiTM for all the research facilities. The work is partially supported by Ministry of Education Malaysia under the Research Acculturation Grant Scheme (Project code: RAGS/1/2014/TK05/UITM//6) and Niche Research Grant scheme (Project code: 600-RMI/NRGS 5/3 (6/2013).

Reference

[1] Dresselhaus M S and Araujo P T 2010 *J. Nano. Chem.* **4**, 6297
[2] Chen G, Weng W, Wu D, Wu C, Lu J, Wang P and Chen X 2004 *J. Carbon* **42**, 753
[3] Pei S and Cheng H M 2012 *J. Carbon* N. Y. **50**, 3210
[4] Ryu S, Han M Y, Maultzsch J, Heinz T F, Kim P, Steigerwald M L and Brus L E 2008 *J.Nano Lett.* **8**, 4597
[5] Alam S, Nizam B and Maksudul U 2016 *J.Int. Nano Lett*

[6] Xu C, Yuan R and Wang X 2014 *J. Carbon N. Y.* **71**, 345
[7] Huang X, Qi X, Boey F and Zhang H 2012 *J.Chem. Soc. Rev.* **41**, 666
[8] Yun Y 2016 *J. Electchem. Sci* **11**, 2778
[9] Wang H, Liu Y, Li M, Huang H, Xu H M, Hong R J and Shen H 2010 *J. Optoelectron. Adv. Mater. Rapid Commun.* **4**, 1166
[10] Lee M S, Lee K, Kim S Y, Lee H, Park J, Choi K H, Kim H K, Kim D G, Lee D Y, Nam S and Park J U 2013 *J. Nano Lett.* **13**, 2814
[11] Dinh P and Van Hieu N 2011 *J. Appl. Surf. Sci.* **257**, 9817
[12] Ammar A, Al-Enizi A M, Al Maadeed M A and Karim A 2015 *J. Chem.* **9**, 274
[13] Ratautaite V, Topkaya S N, Mikoliunaite L, Ozsoz M, Oztekin Y, Ramanaviciene A and Ramanavicius A 2013 *J.Electroanalysis* **25**, 1169
[14] Wessling B and Wessling B 2016 *J. Polymer* **2**, 786-798
[15] Li J, Zeng X, Ren T and van der Heide E 2014 *J. Lubricants* **2**, 137
[16] Fernández-Merino M J, Guardia L, Paredes J I, Villar-Rodil S, Solís-Fernández P, Martínez-Alonso A and Tascón J M D 2010 *J. Phys. Chem.C* **114**, 6426
[17] Preetha K C and Remadevi T L 2015 *J. Mater. Sci. Semicond. Process.* **39**, 178
[18] Moon I K, Lee J, Ruoff R S and Lee H 2010 *J.Nat. Commun.* **1**, 73
[19] Shin H J, Kim K K, Benayad A, Yoon S M, Park H K, Jung I S, Jin M H, Jeong H K, Kim J M, Choi J Y and Lee Y H 2009 *J. Adv. Funct. Mater.* **19**, 1987
[20] Gao W, Alemany L B, Ci L and Ajayan P M 2009 *J. Nat. Chem.* **1**, 403
[21] Velasco-Soto M A, Pérez-García S A, Alvarez-Quintana J, Cao Y, Nyborg L and Licea-Jiménez L 2015 *J. Carbon N. Y.* **93**, 967
[22] Wang G, Shen X, Wang B, Yao J and Park J 2009 *J. Carbon N. Y.* **47**, 1359.
[23] Bastwros M, Kim G, Zhu C, Zhang K, Wang S, Tang X and Wang X 2014 *J.Compos* **60**, 111
[24] Chen J, Yao B, Li C and Shi G 2013 *J. Carbon N. Y.* **64**, 225
[25] Zhu C, Guo S, Fang Y and Dong S 2010 *J. Nano Mole* **4**, 2429
[26] Abdolhosseinzadeh S, Asgharzadeh H and Seop Kim H 2015 *J. Sci Rep* **5**, 10160
[27] Rohilla S, Kumar S, Aghamkar P, Sunder S and Agarwal A 2011 *J. Magn. Magn. Mater.* **323** 897
[28] Ahmad A F, Humaizah F, Moin A, Munawar H, Mohd K, Rahman I A, Mohamed F, Hua C C, Ramli S and Radiman S 2013 *J. Analyt Sci* **17**, 475

IOP Conf. Series: Materials Science and Engineering **358** (2018) 012047 doi:10.1088/1757-899X/358/1/012047

Effect of Graphene Oxide (GO) on the Surface Morphology & Hydrophilicity of Polyethersulfone (PES)

N F D Junaidi, N A Khalil, A F Jahari, N Z K Shaari, M Z Shahruddin, N H Alias and N H Othman

Department of Oil and Gas, Faculty of Chemical Engineering, Universiti Teknologi MARA 40450 Shah Alam Selangor Darul Ehsan, Malaysia

nurhidayati0955@salam.uitm.edu.my

Abstract. Membrane has been widely used in water and wastewater treatment. One of the major issues related membrane separation is concentration polarization or fouling, which can lead to a decline of flux and premature failure of membrane. However, fouling can be controlled by modification of membrane properties such as morphology and hydrophilicity. In this work, a modification of polymeric membrane, polyethersulfone (PES) was carried out using graphene oxide in order to attain high antifouling characteristics. Graphene oxide (GO) was added at different compositions ranging from (0.1 wt%-1.0 wt%). GO was synthesized using modified Hummers' method and characterized using XRD and FTIR prior to using it as additive for the PES membrane. The prepared PES-GO composite membranes were characterized using FTIR and SEM, Contact angle measurement and pure water flux test were then conducted to investigate the hydrophilicity of the PES-GO membranes. It was found that the additions of GO has significantly improved the hydrophilicity of the membranes.

1. Introduction

An ideal membrane is a membrane that can produces a maximum permeate flux with the maximum solute rejection, reducing both capital and operating costs [1]. Membranes can be classified into two categories based on the materials, which are organic (polymeric) and inorganic (ceramic or metallic) [2]. According to Ahmad *et al.* [1], various polymeric membrane can be used in water and waste water treatments, such as cellulose acetate (CA), polysulfone (PS), polyethersulfone (PES), polyacrylonitrile (PAN), polyvinylidene fluoride (PVDF), polyvinyl alcohol (PVA) and polyimide (PI). Among that, PES has been considered as the most significant polymeric membrane used for water application due to its transparent and amorphous properties, high transition glass temperature (Tg=225 ºC), high mechanical, thermal, and chemical resistances thus, ideal for asymmetric membranes with different pore sizes and surfaces [1]. However, there are some limitation of PES membranes owing to their hydrophobic surfaces, which boost the attachment of hydrophobic organic substances on the membrane surfaces or inside the membrane pores [1–4].

A few studies were performed in order to enhance the permeability and rejection rates and simultaneously reducing the fouling problems. Among the techniques proposed to improve the hydrophilicity and flux of the PES membranes were blending, grafting, chemical treatment, and coating. The blending/additives additions technique was considered as the best practice for surface improvement such as hydrophilicity, surface roughness, surface charge, and the pore size due to its convenient operation and mild conditions [1,5]. Graphene oxide (GO) is a 2D nanomaterials that has shown numerous advantages properties such as, high hydrophilic nature, promote negative surface charges, better miscibility with polymers, superior electron transport and high mechanical stability [4]. Ouyang *et al.*[6], specified that GO contains extensive amount of covalently attached oxygen-containing groups, such as hydroxyl, epoxy groups, carbonyl, and carboxyl groups, which create

Content from this work may be used under the terms of the Creative Commons Attribution 3.0 licence. Any further distribution of this work must maintain attribution to the author(s) and the title of the work, journal citation and DOI.
Published under licence by IOP Publishing Ltd

highly hydrophilic material. Lee et al. [7] prepared polysulfone/GO membrane for treatment of wastewater in membrane bioreactor. The membrane experienced higher resistance to both antifouling and anti-biofouling properties. The existence of GO in membrane material would encourage hydrophilicity which would certify high water permeation and obstruct biofouling due to the low interfacial energy between a surface and water. GO synthesized using Modified Hummers' method is better in compared to Brodie and Staudenmaier methods as it consists of fewer steps and the quality can be controlled easily [8].

2. Methodology

2. 1 Material

All chemicals used in the experiments were reagent grade. Graphite powders (Mw=12.01 g/mol, COMAK) were used as received. PES (Mw= 22,000 g/mol) and N, N Dimethylacetamide (DMAc) were purchased from Merck USA and used as polymer and solvent in preparation of cast solution. H_2SO_4 (98%, R&M Chemicals), $NaNO_3$ (84.99 g/mol, SYSTERM), $KMnO_4$ (158.05 g/mol, R&M Chemicals) and H_2O_2 (30%, Merck) were used as received. HCL and distilled water were used for washing method. Deionized water (DI) was used in sample preparation and for pure water flux (PWF) measurement.

2. 2 Graphene Oxide (GO) Synthesis

GO was prepared by using modified Hummer's method [9]. Preparation of GO consist of 6 stages, start with ice bath stage, stirring, stop reaction, washing, centrifuging and drying. In ice bath stage, 5 g of graphite powder, 2.5 g of sodium hydroxide, 200mL of concentrated sulphuric acid and 30 g of potassium permanganate were added into the beaker for 5 hours at temperature below 15 ºC. Then, ice bath was removed and the mixture was stirred 20 hours at room temperature. Then, the mixture slowly heated to 70 ºC for 2 hours and 90 ºC for another 1 hour with addition of 100mL Distilled Water. Hydrogen peroxide was added to stop the reaction. The resulting mixture was continually washed by HCl dilution for several times. Followed by centrifuging and drying for 24 hours at 60-70 ºC to form GO sheet.

2. 3 Fabrication of Composite Polyethersulfone – Graphene Oxide (PES-GO)Membranes

The PES-GO membranes were fabricated via phase inversion (immersion precipitation). The casting solution was prepared by using polyethersulfone (PES) and N,N-Dimethylacetamide (DMAc) as polymer and solvent, respectively. GO was used as a hydrophilic modifier/additive to improve the flux rate of membrane [10]. The addition of additives onto the PES membranes was carried out using blending technique forming composite PES-GO membranes. Different amounts of GO were added according to the Table 1 below and bare PES was prepared as a control experiment.

Table 1. Composition of additive and solvent to fabricate membrane for PES-GO Membrane.

Membrane	PES (wt%)	DMAc (wt%)	GO (wt%)
PES-0.1GO	17.0	82.9	0.1
PES-0.5GO	17.0	82.5	0.5
PES-1.0GO	17.0	82.0	1.0
Bare PES	17.0	83.0	0.0

GO was first dispersed in DMAc solvent and sonicated for 1 hour to obtain a homogenous casting solution. Then, PES was added slowly to GO/DMAc solution and stirred for 24 hours. PES was dried at 70 ºC for 24 hours before used. Next, the prepared casting solution was cast onto the glass plate using glass rod, immediately immersed in distilled water coagulation bath for 24 hours to complete phase inversion process and dried under room conditions for another 24 hours.

2. 4 Characterization of GO and PES-GO membranes

X-Ray diffraction, XRD (Rigaku) analysis was first carried out to confirm the formation of graphene oxide from graphite. The measurement was performed in scanning range of 4 0 Kv, 40 mA, 5-80 °, (2

°/min). The distribution of GO in membrane and its functional group was observed using Fourier Transform Infrared (FTIR, Perkin Elmer). The spectra were recorded in the wave length range of 500-4000 cm^{-1}. The morphology of the membrane were then observed by using Scanning Electron Microscope, SEM (Hitachi, S-3400N).The hydrophilic behaviour of membranes was tested using water contact angle goniometer (AST Product INC, VCA-3000s). Pure water flux (PWF) test was then conducted to determine the permeation of water. The test was carried for at room temperature and 6 bar and finally, the flux was calculated by using equation (1) [10].

$$J = \frac{V}{A\Delta t} \tag{1}$$

Where J is pure water flux (L/m^2h), V is volume of permeate (L), A is effective area of membrane (m^2) and Δt is the sampling time (h).

3. Result &Discussion

3. 1 GO

Figure 1 compares the XRD spectra for graphite and GO in order to confirm the formation of GO from graphite. It was observed that the graphite has a very intense and sharp peak at 2θ = 27.5°. After the oxidation and exfoliation process using modified Hummer's method, the sharp reflection of graphite was disappeared and a new peak was observed at 2θ = 10.18°. The disappearance of graphite peak of represents that graphite was altered after oxidation to completely formed graphene oxide. This is consistent with works by Tissera *et al*, [10] that the occurrence of oxygenated functional groups and interpolated water molecules in the graphite caused high interlayer space in the synthesized graphite oxide. It can be indicated that the graphite was fully oxidized due to no diffraction peak corresponding to graphite at XRD spectrum of synthesized graphene oxide.

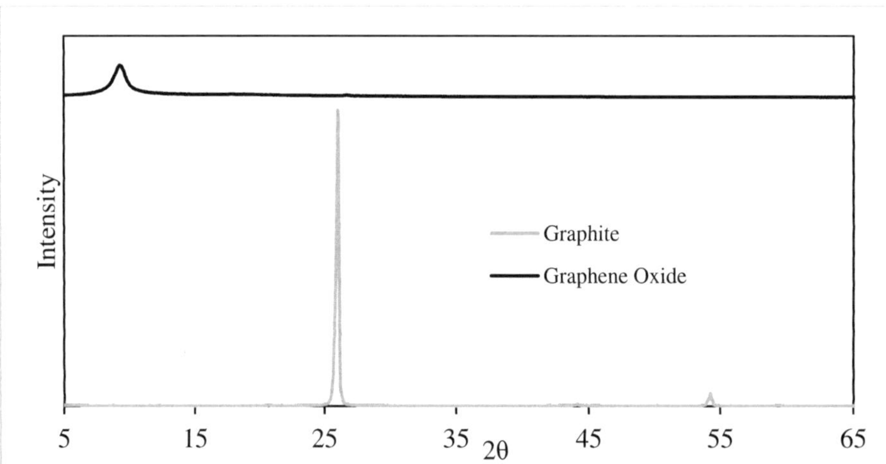

Figure 1. XRD pattern for Graphite and Graphene Oxide.

3.2 *Properties of Composite PES-GO Membranes*

Figure 2 shows the FTIR spectra for bare GO, PES membrane and composites PES-GO membranes. This analysis was conducted to observe the compositions and to identify the functional groups of membranes upon the introduction of GO. From Figure 2, it can be confirmed that GO is a hydrophilic material with a number of functional groups such as hydroxyl (O-H = 3500cm^{-1}), carbonyl (C-C) similar with worked done by Yu *et al* [11]. After the introduction of GO, this spectrum indicates the GO was entrapped and formed as one part of PES structure which responsible for enhancing the hydrophilic characteristics of the membrane. There are large amounts of oxygen-containing functional groups, including carboxyl, hydroxyl, and carbonyl, on GO is due to oxidation which responsible for hydrophilic characteristics of GO.

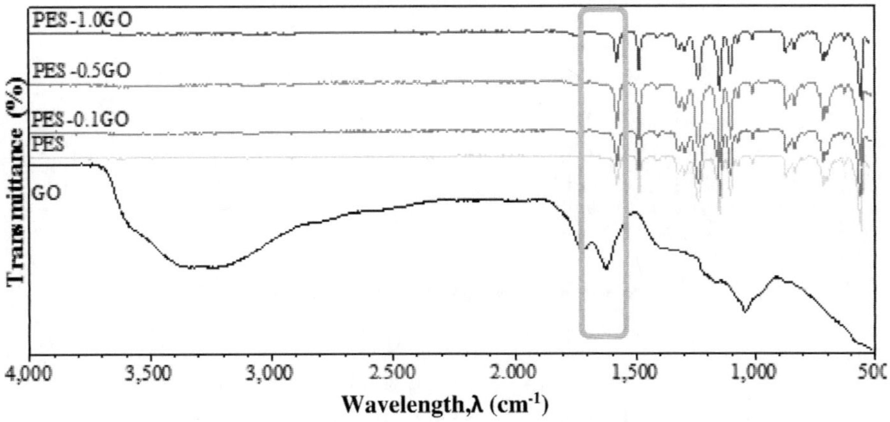

Figure 2. FT-IR of GO, PES, PES-0.1GO, PES-0.5GO and PES-1.0GO.

3.3 Hydrophilicity properties of the composite PES-GO membranes

Figure 3 displays the contact angle and pure water flux values for the composite PES-GO membrane. Lower water contact angle indicates that the membranes are more hydrophilic. From the contact angle plot, it can be seen that the contact angle decreased from 59° to 46° when 1.0 wt% of GO was added into the PES membrane. This value is significantly low in compared to Jin *et al* [12], which the contact angle was 78°to 55°. The pure water flux was also observed to increase with the amount of GO achieving 123 L/m^{2}h. This value was significantly higher in compared to other results in Safarpour *et al.* [5], where Titanium Oxide was used. The high increment of PWF might be due to the highly hydrophilic properties of GO.

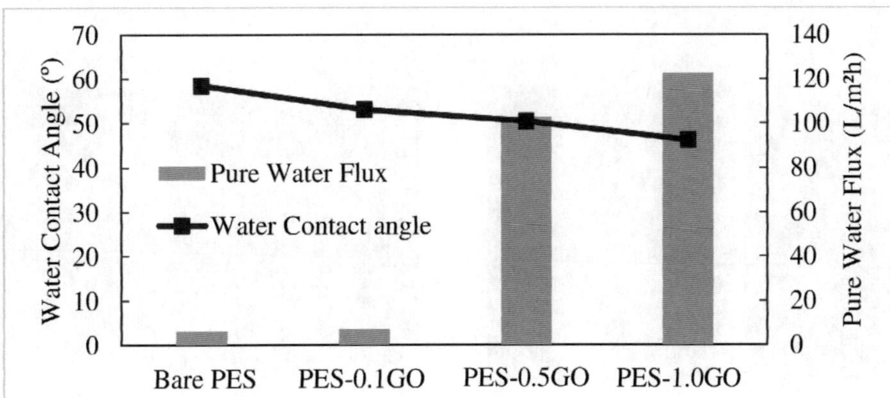

Figure 3. Pure water flux test water contact angle of composite PES-GO membranes.

Figure 4 shows the stability of pure water flux for 1 hour. The result shows that the pure water flux was in a stable mode for all the membranes. This indicates that the membranes are stable with time. A stable membrane is required due to highly hydrophilic which has lower contact angle value and produced high water permeate [12].

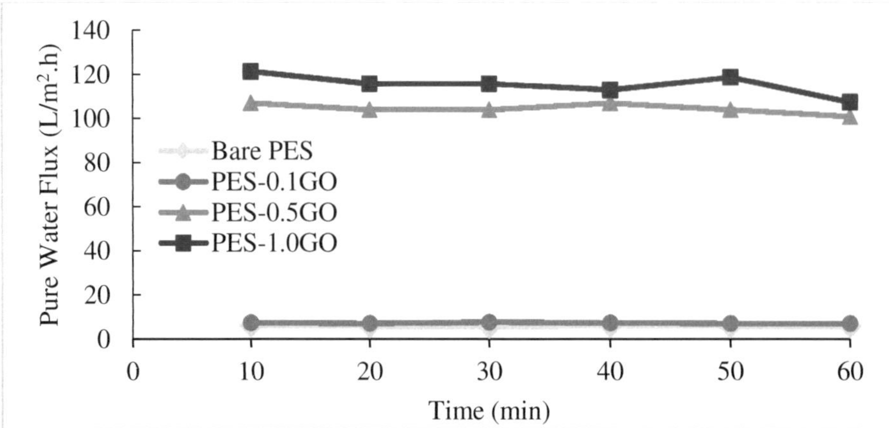

Figure 4. Pure water flux stability.

3.4 Morphology of Composite PES Membranes

Figures below represents the cross-sectional image of Bare PES membrane, PES-0.1GO, PES-0.5GO and PES-1.0GO respectively. Hydrophilic additives can be used to improve the pure water flux of the membrane and at the same time minimizing the fouling issues [4]. From the PWF and contact angle studies, it was observed that the addition of GO can significantly affected the hydrophilicity of the PES membrane due to present of hydroxyl and carbonyl functional group in GO. The morphology of the bare PES membranes shows the interactions of polymer and solvent and hence produced more porous membrane. As a result, the size of fingerlike structure amplified. It is seen that with the addition of GO, the surface roughness on the membrane surface increases progressively causing in the formation of numerous pores [12].

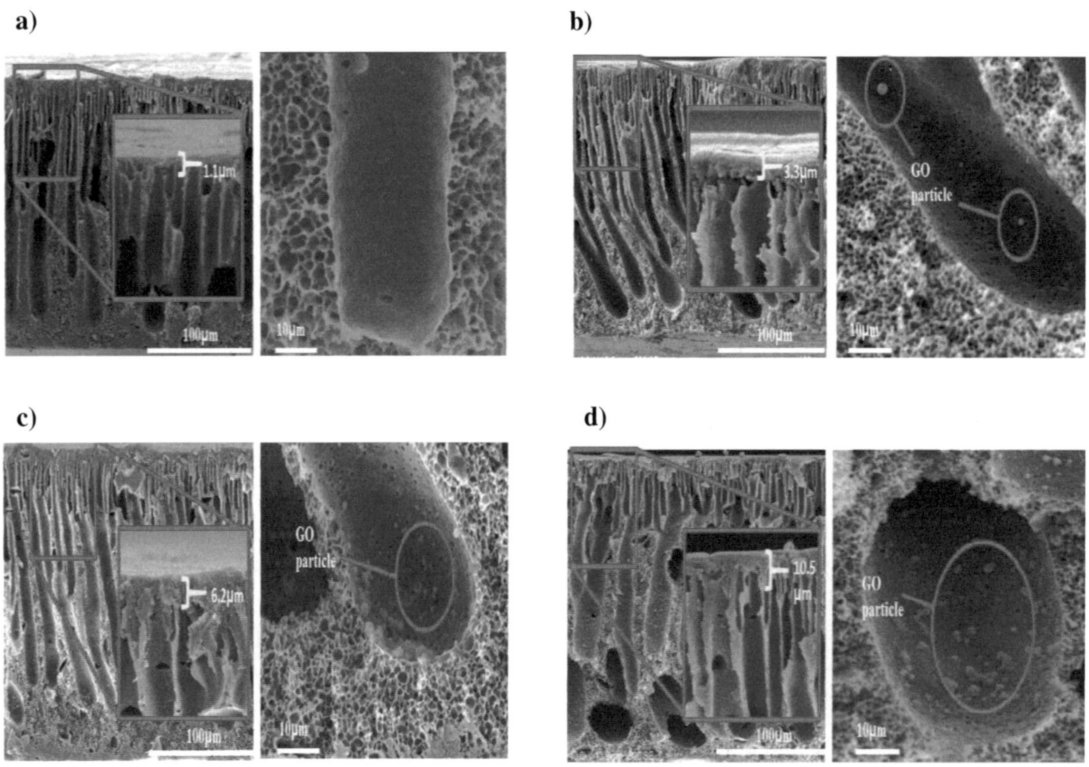

Figure 5. SEM images of a) bare PES membrane, b) PES-0.1GO, c) PES-0.5GO, and d) PES-1.0GO

Figure 5 indicated that as more GO added, the finger-like structure bend slightly and the bottom part become bigger. This is due to high hydrophilicity of GO that attract more water during phase inversion. The thickness of membrane skin layer also increase with addition of GO. The presence of more GO particle can be spotted as more GO added. Razaee *et al.* [4] mentioned that the faster the exchange rate of solvent and non-solvent in the coagulation process, the larger pores, more finger like pores and more channels.

4. Conclusion

Graphene oxide was successfully synthesized using modified Hummer's method and used as additives for the modification of PES membranes. Based on the contact angle and pure water flux test, it was found that the composite PES-1.0GO membrane exhibited the most hydrophilic characteristic due to the high amount of GO. From the SEM image, it can be seen that GO was homogeneously dispersed in the PES membrane. Based on the result obtained, it is strongly believed that GO can serve as a highly hydrophilic modifier for PES membrane fabrication, thus minimizing the fouling problems.

Acknowledgments

The authors would like to acknowledge Ministry of Higher Education (MOHE) for the FRGS research funding (FRGS/1/2015/TK05/UITM/03/2), which makes this important research viable and effective.

References

[1] Ahmad A L, Abdulkarim A A, Ooi B S and Ismail S 2013 *Chem. Eng. J.* **223** 246–267
[2] Jutta T 2014 *VTT Technology* **188** 54
[3] Igbinigun E, Fennell Y, Malaisamy R, Jones K L and Morris V 2016 *Memb. Sci. J* **514** 518–526
[4] Rezaee R, Nasseri S, Mahvi A H, Nabizadeh R, Mousavi S A, Rashidi A, Jafari A and Nazmara S 2015 *Environ. Heal. Sci. Eng. J.* **13:16** 1–11
[5] Safarpour M, Vatanpour V and Khataee A 2015 *Desalination* **393** 65-78
[6] Ouyang G, Hussain A, Li J and Li D 2015 *RSC Adv* **5**, 70448-70460
[7] Lee J, Chae H R, Won Y J, Lee K, Lee C H, Lee H H, Kim I C and Lee J min 2013 *Memb. Sci. J.***448** 223–230
[8] Mahmoud K A, Mansoor B, Mansour A and Khraisheh M 2015 *Desalination* **356** 208–25
[9] Aditya K S, Lukka T Y, Arthanareeswaran G, Matsuura T and Ismail A F 2016 *Chem. Eng. J.* **286** 528–537
[10] Tissera N D, Wijesena R N, Perera J R, Silva K M N De and Amaratunge G A J 2015 *Appl. Surf. Sci.* **324** 455–463
[11] Yu L, Zhang Y, Zhang B, Liu J, Zhang H and Song C 2013 *Memb. Sci J.* **447** 452–462
[12] Jin F, L W, Zhang C, Li Z, Su R, Qi W, Yang Q and He Z 2013 *RSC Adv* **3** 21394–21397

Formation of Sol Gel Dried Droplets of Carbon Doped Titanium Dioxide (TiO_2) at Low Temperature via Electrospraying

S U Halimi [1], S Abd Hashib [1], N F Abu Bakar [1*], S N Ismail [1], M Nazli Naim [2], N Abd Rahman[1] and J Krishnan[3]

[1]Faculty of Chemical Engineering, Universiti Teknologi MARA, MALAYSIA
[2]Food and Process Department, Faculty of Engineering, Universiti Putra Malaysia (UPM), MALAYSIA
[3]Department of Chemical Engineering, SSN College of Engineering, Tamilnadu, INDIA

syafiza0358@salam.uitm.edu.my/ fitrah@salam.uitm.edu.my

Abstract. The high band gap energy of TiO_2 and inconsistency in particles size has imposed a significant drawback on TiO_2 applications. Dried droplets of carbon-doped TiO_2 fine particles were produced by using electrospraying technique. The C-doped TiO_2 particles were prepared by hydrolysis of titanium isopropoxide with the addition of carbon precursor followed by electrospraying the suspension in stable Taylor cone-jet mode. Coulomb fission of charged droplets from the electrospraying technique successfully transformed dispersed liquid C-doped TiO_2 particles into solid. The deposited C-doped TiO_2 droplets were collected on aluminium substrates placed at working distances of 10 to 20 cm from the tip of the electrospray needle. The collected C-doped TiO_2 droplets were characterized by using FESEM, UV-Vis, FTIR and XRD. By increasing the working distance, the average droplets size of the deposited C-doped TiO_2 was reduced from ±163.2 nm to ±147.56 nm. UV-Vis analysis showed a strong absorption in the visible-light region and about 93 nm red shift of the onset spectrum for C-doped TiO_2. The red shift indicates an increase in photocatalytic efficiency by reducing the TiO_2 band gap energy from 3.0 eV to 2.46 eV and shifting its activity to the visible-light region. FTIR analysis indicated the presence of Ti-C and C-O chemical bonding in the C-doped TiO_2.

1. Introduction

Titanium dioxide (TiO_2) is a promising material that can be used in the degradation of pollutants due to its highly photosensitive, chemically stable, biologically inert and non-toxic in nature. Unfortunately, TiO_2 has a drawback which is high band gap energy (3.2eV) that is activated under ultraviolet light (λ<387nm). The high band gap energy is due to oxygen vacancies and localization of Ti 3d states [1,2]. Other than that, TiO_2 is also thermally unstable which leads to easy phase transformation [3], depending on its synthesis method. To improve TiO_2 properties, methods need to be developed for controlling the droplet size, surface are and morphology, as well as doping with other elements to enhance the photochemical response of TiO_2 towards the visible-light region.

The properties of TiO_2 can be altered by doping, where an impurity is intentionally introduced into TiO_2. There are three opinion regarding the modification mechanism of the doped TiO_2. First is narrowing the TiO_2 bandgap by substituting oxygen in TiO_2 lattice [4], second, introducing an impurity energy level above the valence band [5], and lastly the formation of oxygen vacancy sites as blockers for reoxidation [6]. All three mechanisms lead to improvement of the TiO_2 properties. In this

Content from this work may be used under the terms of the Creative Commons Attribution 3.0 licence. Any further distribution of this work must maintain attribution to the author(s) and the title of the work, journal citation and DOI.
Published under licence by IOP Publishing Ltd

IOP Conf. Series: Materials Science and Engineering **358** (2018) 012048 doi:10.1088/1757-899X/358/1/012048

study, carbon element [7] is chosen as the doping element to produce C-doped TiO_2. Carbon is a promising doping element due to its significant overlapping of electron state between the oxygen near the valence band of TiO_2 [8]. Carbon also has high adsorption capacity, it can be used in both air pollutant degradation and waste water treatment as well as five times more active than nitrogen and other doping element under visible-light irradiation [9]. Methods of producing TiO_2 also have been studied in order to produce C-doped TiO_2 with the desired particle size, morphology and structural characteristic.

Various methods have been used to produce C-doped TiO_2 particles, such as chemical precipitation [10,11,12], sol-gel [13,14] and electrospray drying. In the chemical precipitation method, C-doped TiO_2 suspension needs to be heated rapidly for the hydrolysis process to occur. Precipitates need to be washed and then heated again to produce particles. As a result, the C-doped TiO_2 particles produced from this method are highly aggregated and lead to a wide distribution of particle sizes. In the sol gel method, control of the hydrolysis process is essential in order to obtain a homogenous oxide network. This requires long aging process to form the gel, high calcination temperature to remove organic molecules and grinding process to produce C-doped TiO_2 particles. As a result, the desired anatase phase of C-doped TiO_2 particles was difficult to obtain because of repetition of the heating process and the high calcination temperature. Besides, the produced C-doped TiO_2 also has inconsistent particle sizes due to the grinding process. Therefore, electrospraying technique is introduced in this study as a method to produce fine and uniform C-doped TiO_2 particles without heating at high temperature, aging or grinding process as required in other methods.

Electrospraying or electrohydrodynamic spraying is a method of liquid atomization by electrical forces [15]. It has the ability to produce monodisperse particles, reduce molecular aggregation and operated at ambient condition [16]. The size of the electrospray droplets ranges from hundreds of micrometers down to tens of nanometers. The droplets are highly charged up to a fraction of the Rayleigh limit, which is the maximum charge carried by a droplet. The charge will overcome the liquid surface tension and leads to fission of the droplets. Charged droplets are self-dispersing within the working distance due to the fission process, resulting in the absence of droplets agglomeration [17,18]. Therefore, the electrospraying technique is proposed in this study to produce visible light active C-doped TiO_2 particles.

2. Materials and Methods

2.1 Preparation of C-doped TiO_2 suspension

C-doped TiO_2 suspension was prepared by adding titanium isopropoxide (TTIP) (98%, Sigma-Aldrich Chemie, GmBH) drop-wise as a precursor into 50 mL of deionized water. The mixture was stirred at 500 rpm under room temperature until it turned to cloudy yellow. The cloudy suspension was then dispersed into a mixture of 50 mL ethanol and 10 mL acetic acid (99%, Merck KGaA, Darmstad Germany). Different concentrations (0.25 M, 0.5 M, 0.75 M and 1.0 M) of acetylacetone (98%, Merck KGaA, Darmstad Germany) as a dopant precursor were added to the suspension. The suspension was stirred for 1 hour and sonicated (QSonica Sonicators, Q700) at amplitude of 50 kHz for 30 min. The same procedure was used to prepare the un-doped TiO_2 suspension by using TTIP.

2.2 Electrospray deposition of C-doped TiO_2 suspension

Electrospray equipment was set up as shown in Figure 1. The prepared C-doped TiO_2 suspension was loaded into a 25 mL syringe with a needle size of 0.15 mm inside diameter on a syringe pump (SP-1000, Next Advance Inc). The needle tip and counter electrode were connected to an electrostatic generator (G60 EMCO High Voltage). An AC-DC converter (M10-QS305, MCP Lab Electronics) was used to convert alternate current to direct current before connecting it to the electrostatic generator. Positive and negative charge voltage was applied at the needle tip and counter electrode respectively. The flow rate of the C-doped TiO_2 suspension was set at 2.0 mL/h. Voltage was applied in a range of 1.23 to 1.28 kV to maintain the formation of a Taylor cone meniscus at the needle tip. Dispersed C-doped TiO_2 droplets were collected on a flat surface grounded substrate at various working distances (WD) of 10 cm, 15 cm and 20 cm from the counter electrode. The substrate containing deposited droplets were dried in an oven at 60-100 °C for 1 hour.

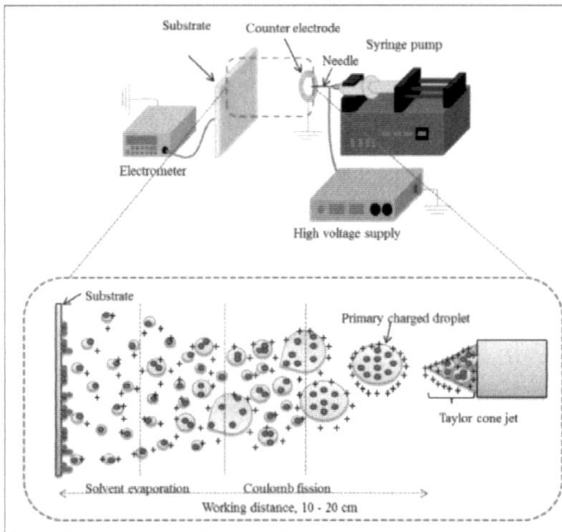

Figure 1. Electrospray setup and mechanism

2.3 Estimation of primary charged droplet size

The primary charged droplet size of C-doped TiO_2 can be estimated by using a scaling law to investigate the droplet fission behaviour of electrospraying, proposed by Hartman et al. [19] Equation (1), Fernandez de la Mora [20] Equation (2) and Ganan-Calvo [21] Equation (3) respectively, as follows:

$$d_d = (\frac{\rho \varepsilon_o Q^3}{\gamma K})^{1/6} \quad (1)$$

$$d_d = 1.66\varepsilon_r{}^{-1/6}(\frac{Q\varepsilon_r\varepsilon_o}{K})^{1/3} \quad (2)$$

$$d_d = 1.2164(\frac{Q\varepsilon_r\varepsilon_o}{K})^{1/3} \quad (3)$$

Where d_d, ρ, ε_o, Q, γ, K and ε_r are the primary droplet size (m), liquid density (kg/m^3), electrical permittivity of vacuum (C^2/Nm), liquid flow rate (m^3/s), surface tension (N/m), conductivity of liquid (S/m) and relative permittivity respectively. Assuming evaporation of droplets occurs without substantial loss of mass by any other mechanism during electrospaying, the theoretical dried droplet size (d_p) can be calculated according to Equation (4).

$$d_p = d_d\theta^{1/3} \quad (4)$$

Where d_p is the theoretical dried droplet size and θ is the volume fraction of material in the solution. The number of fission was determined by dividing the volume of primary charged droplet size over the volume of dried droplets from the diameter of the droplets size of FESEM images. The volume of the droplets is assumed to be spherical. The electrosprayed particles on the substrate were scraped out and characterized using Field Emission Scanning Electron Microscopy (FESEM), UV-Visible Spectroscopy (UV-Vis), Fourier Transform Infrared Spectroscopy (FTIR) and using X-Ray Diffraction (XRD). UV-Visible Spectroscopy (UV-Vis) was used to study the changes in absorption wavelength for both C-doped and un-doped TiO_2 samples. Absorbance data for all samples were recorded at wavelength in a range from 200 nm to 800 nm and presented in a graphical form of absorbance versus wavelength. Fourier Transform Infrared Spectroscopy (FTIR) was used to study changes in functional group for the electrosprayed C-doped TiO_2 samples. The scan spectra were recorded in the range from 500 cm^{-1} to 4000 cm^{-1}. The crystal phase of the electrosprayed C-doped TiO_2 samples was determined using X-Ray Diffraction (XRD) analysis. The analysis was performed at 40 kV and 150 mA at an angle of 2θ from 20° to 80°. The scan speed was 1° /min.

3. Results and Discussion

Figure 2 shows the relationship between the zeta potential and particles size in suspensions with various pH. High zeta values were recorded at pH 2 and 13, at 27.98 mV and -45.65 mV respectively. The zeta potential is a term referring to the electrical potential at the boundary of a double layer of particles surface in suspension. Particle suspensions with zeta potential values approximately or closer to ±30 mV have high degrees of stability [22]. In terms of the particle size in suspensions, at pH 2 and 13, small C-doped TiO_2 particles were obtained with an average size of 249.12 nm and 278.25 nm respectively. Prior to electrospraying, the C-doped TiO_2 suspension was stabilized at pH 2 because this is close to the suspension's initial pH of 3.2. In the acidic condition, the C-doped TiO_2 particles in suspension carried a positive charge as the H^+ ions surrounded its surface. This condition is similar to the initial condition of the suspension before acid was added to reduce the pH. Thus, only a little acid is required to add into the suspension.

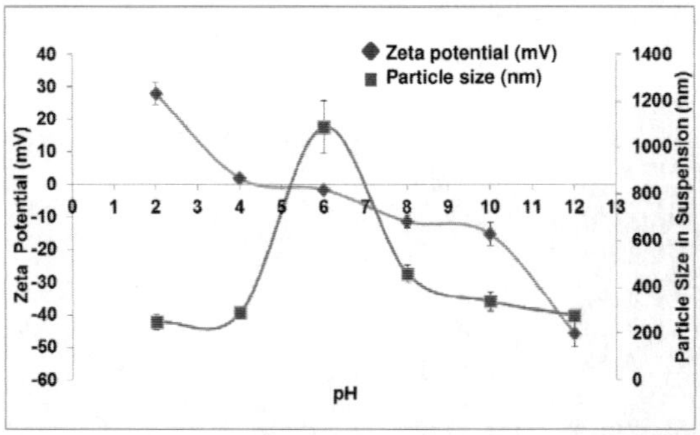

Figure 2. Zeta potential and particles size in suspension versus pH

In this study, the isoelectric point obtained for the C-doped TiO_2 suspension was at pH 4.3. Since no work was found for characterizing the zeta potential of C-doped TiO_2 suspension, the isoelectric point for the C-doped TiO_2 is referred to TiO_2 suspension which has a pH range from 4.0 to 7 [23]. At the isoelectric point, H^+ and OH^- ions are neutralized and there is no charge surrounding the C-doped TiO_2 particles surface. Therefore, C-doped TiO_2 particles tend to flocculate due to the attractive Van der Walls force and later sedimentation occurred. As shown in Figure 2, at pH values near the isoelectric point, the biggest average particle size was recorded at 1088 nm. This is undesirable for C-doped TiO_2 suspension to undergo electrospraying, because it is in an unstable condition. Hence, the C-doped TiO_2 suspension was stabilized at pH 2 prior to electrospraying.

From Figure 3, morphological changes of the C-doped TiO_2 droplets were observed as the working distance (WD) was varied from 10 cm to 20 cm. A wide distribution of dried droplets ranging from 20 to 640 nm with average droplets size of 163.2 nm was observed at 10 cm WD. The droplets size distribution decreased to 20-500 nm and 20-360 nm for WDs of 15 and 20 cm with average droplets size of 162.9 and 147.6 nm respectively. Large C-doped TiO_2 droplets were collected at WDs of 10 cm and 15 cm, which indicated that less fission of primary charged droplets had occurred. The sizes of primary charged droplet, the deposited droplet size and numbers of fission events were estimated and shown in Table 1. Table 2 shows the numbers of fission that occurred during electrospraying increases when the WD increases from 10 cm to 20 cm for each of the scaling law models. From the experimental number of fissions, the number of fission is high for large size of primary charged droplet in agreement with the Fernandez de la Mora model [24], since the conductivity value of C-doped TiO_2 suspension was sufficiently high (i.e. 0.8 S/m) which near to high conductivity condition as in the proposed model. The number of fission from the experiment is in agreement with theoretical number of fission calculated using Fernandez de la Mora model, equation (2). Where the number of fission value at the furthest distance is the closest to the theoretical results derived for the conductivity solution by Fernandez de la Mora [25].

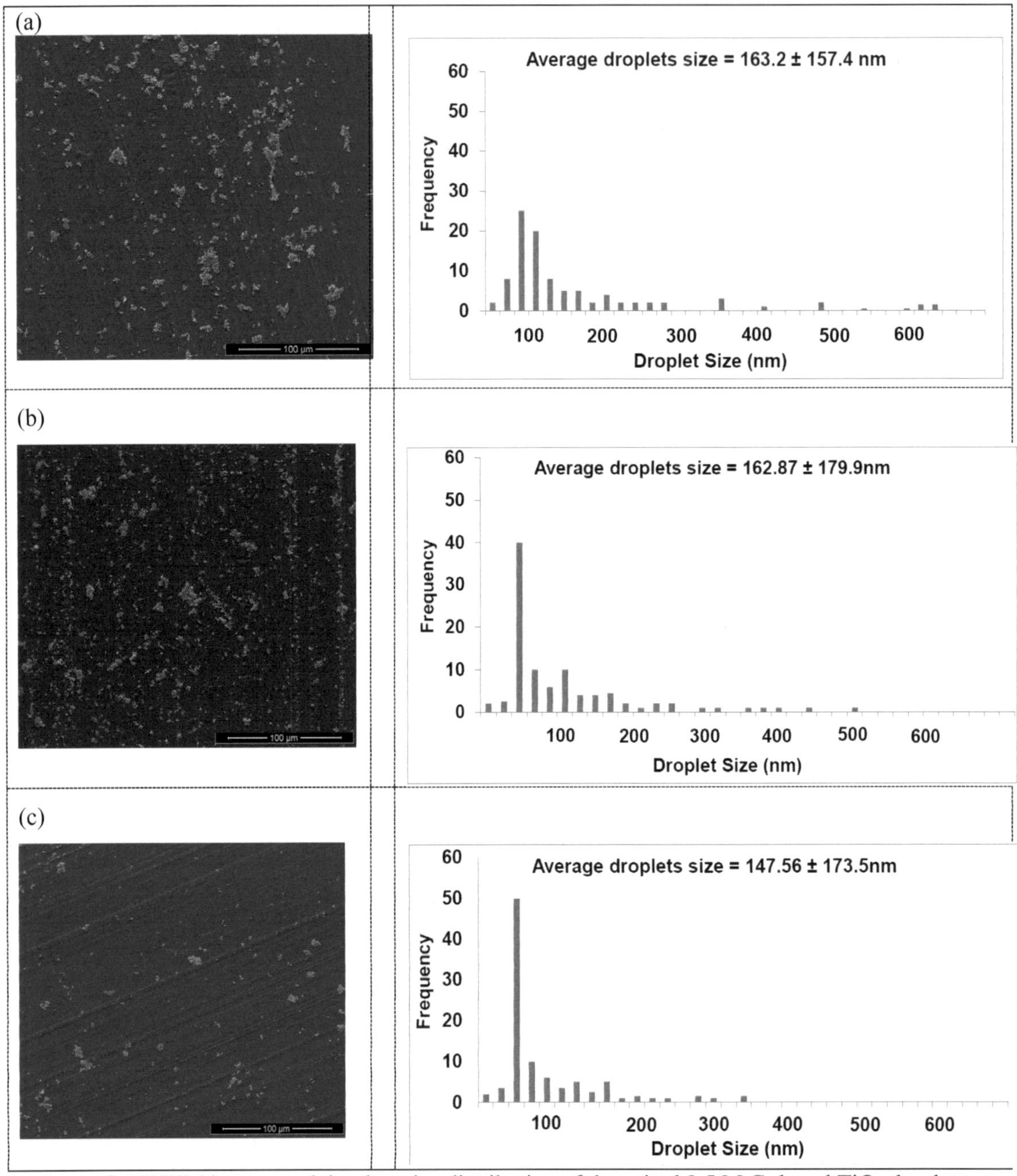

Figure 3. FESEM images and droplets size distribution of deposited 0.5 M C-doped TiO_2 droplets at various WD. (a) 10 cm, (b) 15 cm (c) 20 cm

With increase in WD, the droplets have longer flight and solvent evaporation times, and thus have more time for fission to occur. Hence, this leads to the production of smaller average droplets size. Coulomb fission or explosion of droplets occurs when charged droplets reach the Rayleigh limit. The Rayleigh limit is the maximum amount of charge that a droplet can carry on its surface [26]. When the droplet reaches the Rayleigh limit, it becomes unstable because of the high density of charge on its surface and it breaks up into smaller droplets. The loss of mass during fission is too small to affect the droplet size [27]. The mechanism of Coulomb fission during electrospraying ensure the primary charged droplet underwent fission along the WD before deposited on the substrate [25, 38]. Hence, electrospraying technique could produce a narrow distribution of C-doped TiO_2 droplets with an average droplets size of 147 nm.

Table 1. Estimation of primary and deposited droplets sizes.

Scaling law model	Primary droplet size, d_d (nm)	Theoretical deposited droplets size, d_p from Equation (4) (nm)	Theoretical number of fission
Hartman	2755	642.83	79.02
Fernandez de la Mora	641.46	149.67	79.79
Ganan-Calvo	991.32	231.31	79.03

Table 2. Estimation of experimental number of fission.

Scaling law model	Experimental number of fission, WD = 10 cm	Experimental number of fission, WD = 15 cm	Experimental number of fission, WD = 20 cm
Hartman	4811	4837	6503
Fernandez de la Mora	61	61	82
Ganan-Calvo	224	225	303
Average particle size, $*d_p$ (nm)	163.2	162.9	147.6

$*d_p$ is average deposited droplets size from FESEM image of the respective working distance (WD)

UV-Vis analysis of C-doped samples was performed, and the absorbance spectrum was obtained as shown in Figure 4. The cut-off wavelength for the un-doped TiO_2 sample was ~413 nm with calculated band gap energy of 3.0 eV. The absorbance spectrum for un-doped TiO_2 corresponded well with the anatase phase TiO_2 absorbance spectrum from previous work, which has a band gap energy of ~3.2 eV [28,29]. For the C-doped TiO_2 samples, the absorption spectrum was red-shifted and the cut-off wavelength was found at 420, 506, 493 and 440 nm for different dopant concentrations of 0.25, 0.5, 0.75 and 1.0 M respectively. An increase in the cut-off wavelength in this study, clearly indicated a decrease in the band gap energy for C-doped TiO_2 samples [30]. The band gap energy can be calculated from the UV-Vis absorbance spectrum and the relation between the band gap energy and cut-off wavelength in the absorption spectra is shown in Equation (5) [31]:

$$Band\ gap\ energy\ (E_{bg}) = \frac{hC}{\lambda} \quad (5)$$

where h is Plancks constant (6.626 x 10^{-34} Js), C is the speed of light (3.0 x 10^8 m/s), λ is the cut-off wavelength of UV-Vis spectra (m) and the conversion factor is 1 eV which is equal to 1.6 x 10^{-19} J. The calculated band gap energy for doped TiO_2 is shown in Table 3.

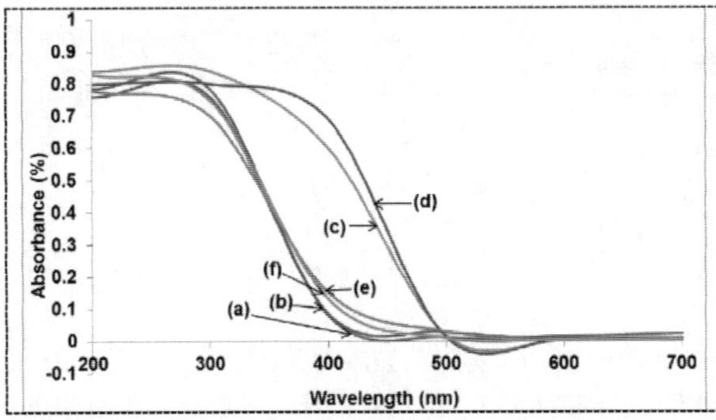

Figure 4. UV-Vis absorption spectra for un-doped and C-doped TiO_2 at different dopant concentration. (a) undoped electrosprayed TiO_2, (b) 0.25M electrosprayed C-doped TiO_2, (c) 0.5M electrosprayed C-doped TiO_2, (d) 0.75M electrosprayed C-doped TiO_2, (e) 1.0M electrosprayed C-doped TiO_2, (f) 0.5M unelectrosprayed C-doped TiO_2.

From Table 3, it is observed that the band gap energy of C-doped TiO2 is lower than the un-doped TiO2. The presence of carbon in C-doped TiO2 samples creates interface states which result in low band gap energy of TiO2 [32]. The 0.5 M electrosprayed C-doped TiO2 has the lowest band gap energy compared to other electrosprayed C-doped TiO2 samples, which is 2.46 eV. The lowest band gap energy confirms that 0.5 M electrosprayed C-doped TiO2 has the best properties to be used as photo-catalyst and able to degrade pollutant faster under visible light. The observed variation of the cut-off wavelength and the pattern of band gap energy depend on the carbon content in the C-doped TiO2 with varying dopant concentrations. This is because accommodation for carbon substitution in the TiO2 was limited when the dopant concentration increased. At high dopant concentration which is more than 0.5 M, excess of carbon presents in the samples. The excess carbon in doped TiO2 might hinder the excitation of electron from valence band to conduction band when the C-doped TiO2 irradiated under visible light.

Table 3. Band gap energy for un-doped and C-doped TiO_2 sample at different concentration

Sample	Cut off wavelength (m)	E_{bg} (J)	E_{bg} (eV)
Un-doped electrosprayed TiO_2	4.13×10^{-7}	4.81×10^{-19}	3.00
0.25M electrosprayed C-doped TiO_2	4.2×10^{-7}	4.73×10^{-19}	2.96
0.5M electrosprayed C-doped TiO_2	5.06×10^{-7}	3.93×10^{-19}	2.46
0.75M electrosprayed C-doped TiO_2	4.93×10^{-7}	4.03×10^{-19}	2.52
1.0M electrosprayed C-doped TiO_2	4.4×10^{-7}	4.52×10^{-19}	2.83
0.5 M un-electrosprayed C-doped TiO_2	4.3×10^{-7}	4.62×10^{-19}	2.89

Hence, the excess carbon in the 0.75 M and 1.0 M electrosprayed C-doped caused high band gap energy. In summary, carbon in TiO_2 acts as an impurity to form interface states that effectively lower the band gap. As for the 0.5 M unelectrosprayed C-doped TiO_2 sample, the band gap energy obtained is 2.89 eV and the band gap is higher than for the 0.5 M electrosprayed C-doped TiO_2. As discussed earlier, electrosprayed C-doped TiO_2 has a narrow distribution of fined-sized droplets and its high surface area provided more active sites. Hence, the efficiency of electrosprayed C-doped TiO_2 under visible-light irradiation is increased.

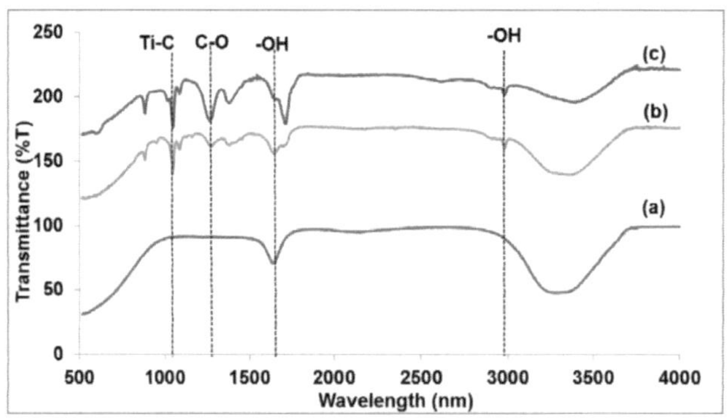

Figure 5. FTIR results for C-doped TiO_2 samples at different dopant concentration. (a) undoped TiO_2, (b) 0.5M unelectrosprayed C-doped TiO_2, (c) 0.5 M electrosprayed C-doped TiO_2

The bonding characteristics of the functional groups in the C-doped TiO_2 samples were identified using FTIR. Figure 5 shows the absorption peaks for un-doped and doped TiO_2 samples. New strong peaks were formed for all the doped TiO_2 samples at around 1044.5, 1263.8, 1650 and 3387 cm^{-1} [33]. Two peaks at 1044.5 cm^{-1} and 1263.8 cm^{-1} represent Ti-C and C-O bonds respectively. The peaks around 1650 cm^{-1} and 3387 cm^{-1} represent stretching vibration of OH bonds from surface water molecules and hydroxyl groups in the samples. In the photocatalytic mechanism, hydroxyl groups were trapped in holes that were generated when TiO_2 was irradiated under light. The hydroxyl radicals formed can suppress the recombination of electron-holes [34]. Therefore, the presence of carbon bonding and hydroxyl groups in electrosprayed C-doped TiO_2 shows the improvement in TiO_2 properties.

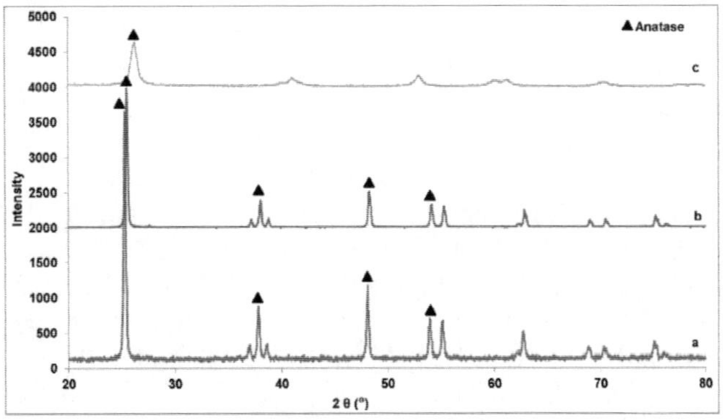

Figure 6: XRD pattern for C-doped TiO_2 samples. (a) undoped electrosprayed TiO_2, (b) 0.5 M electrosprayed C-doped TiO_2 (c) 0.5 M unelectrosprayed C-doped TiO_2

Figure 6 shows the crystalline phase of electrosprayed C-doped TiO_2. Only anatase phase was present in the C-doped TiO_2 at a diffraction angle of 23°, 37.6°, 48° and 54° [35]. No signals of rutile and brookite phase were detected. The presence of carbon as a doping element in TiO_2 does not induce new crystalline forms other than anatase phase [36]. Acetylacetone used as a source of carbon doping is an organic compound to chelate the titanium atoms and is also responsible for the formation of anatase phase. The intensity of the anatase phase for both 0.5 M electrosprayed and unelectrosprayed C-doped TiO_2 was lower than the un-doped TiO_2. This could be due to the low crystalline nature of the carbon element compared to titanium [37]. The reduced intensity of the anatase phase in 0.5 M C-doped TiO_2 is a sign of the presence of carbon in the TiO_2 lattice. However, the peak and intensity of anatase for the un-electrosprayed 0.5 M C-doped TiO_2 was weak compared to the electrosprayed sample. In the electrospraying technique, the produced C-doped TiO_2 particles were dried properly, since the mechanism of droplet fission from electrospraying has been proven to facilitate solvent evaporation, so that only dried 0.5 M C-doped TiO_2 particles were collected. For the un-electrosprayed 0.5 M C-doped TiO_2, the presence of impurities may cause the weak anatase peak formation in the XRD results. Hence, a high drying temperature or calcination process was needed for the un-electrosprayed 0.5 M C-doped TiO_2 to enhance the formation of the anatase phase.

4. Conclusions

This work demonstrated that the TiO_2 suspension is most stable at pH 2, at which it has a high zeta potential and small particles size in suspension, which is 27.98 mV and 249.12 nm. A stable condition was crucial prior to electrospraying in order to avoid sedimentation of particles during the process. The direct current (DC) voltage applied in the electrospraying method can produce small size droplets with a narrow size distribution and an average droplets size of ±147.56 nm. Increasing the distance from the needle tip to the substrate from 10 cm to 20 cm significantly reduces the deposited droplets size because of Coulomb fission. The fission mechanism was utilized for drying and produces small size droplets with a narrow size distribution. Heating at low temperature was applied for doping the TiO_2 with carbon helps to improve the TiO_2 properties. It lowers the TiO_2 band gap to 2.46 eV and it can be activated under visible-light illumination.

Acknowledgement

The authors would like to express their gratitude to Universiti Teknologi MARA (UiTM) Shah Alam, Malaysia for funding this work. This study was supported by LESTARI Grant 600-IRMI/MYRA 5/3/LESTARI (0143/2016) and Research Entity Initiative (REI) Tier 5, 600-RMI/DANA 5/3 REI (2/2013).

References

[1] Shen M, Wu Z, Huang H, Yukou D, Zhigang Z, and Ping Y 2006 Carbon-doped anatase TiO_2 obtained from TiC for photocatalysis under visible light irradiation *Journal of Materials Letters* **60** 693-697. doi: 10.1016/j.matlet.2005.09.068

[2] Cong Y, Zhang J, Chen F, and Anpo M 2007 Synthesis and Characterization of Nitrogen-Doped TiO_2 Nanophotocatalyst with High Visible Light Activity *Journal of Physical Chemistry C,* **111**(19) 6976-6982. doi: 10.1021/jp0685030

[3] Zhang W, Zou L, and Lianzhou W 2009 Photocatalytic TiO_2/adsorbent nanocomposites prepared via wet chemical impregnation for wastewater treatment A review *Journal of Applied Catalysis A: General* **371** 1-9. doi: 10.1016/j.apcata.2009.09.038

[4] Asahi R, Morikawa T, Ohwaki T, Aoki K, and Taga Y 2011 Visible-Light Photocatalysis in Nitrogen-Doped Titanium Oxides (Vol. 293). www.sciencemag.org: Toyota Central R&D Laboratories, Nagakute, Aichi 480-1192, Japan

[5] Irie H, Watanabe Y, and Hashimoto K 2003 Nitrogen-Concentration Dependence on Photocatalytic Activity of TiO_2-xNx Powders *Journal of Physical Chemistry B* **107**(23) 5483-5486. doi: 10.1021/jp030133h

[6] Aziz A, Cheng C K, Ibrahim S, Matheswaran M, and Saravanan P 2012 Visible light improved, photocatalytic activity of magnetically separable titania nanocomposite *Chemical Engineering Journal* **183**, 349-356.

[7] Park J H, Kim S, and Allen J B 2006 Novel Carbon-Doped TiO_2 Nanotube Arrays with High Aspect Ratios for Efficient Solar Water Splitting *Nano Letters* **6**(1) 24-28. doi: 10.1021/nl051807y

[8] Sun Y, Murphy C J, Karla R. Reyes-Gil, Enrique A. Reyes-Garcia, Justin P. Lilliy, and Daniel Raftery 2008 Carbon-doped In_2O_3 films for photochemical hydrogen production *International Journal of Hydrogen Energy* **33** 5967-5974. doi: 10.1016/j.ijhydene.2008.07.100

[9] Huang Y, W Ho, Shuncheng L, Lizhi Zhang, Guisheng Li, and Jimmy C Y 2008 Effect of Carbon Doping on the Mesoporous Structure of Nanoctrystalline Titanium Dioxide and Its Solar-Light-Driven Photocatalytic Degradation of NOx. *Langmuir* **24** 3510-3516. doi: 10.1021/la703333z

[10] Chi B, Zhao L, and Tetsuro Jin 2007 One-Step Template-Free Route for Synthesis of Mesoporous N-Doped Titania Spheres *Journal of Physical Chemistry C* **111** 6189-6193. doi: 10.1021/jp067490n

[11] You X, Chen F, Jinlong Zhang, and Masakazu Anpo 2005 A novel deposition precipitation method for preparation of AG-loaded titanium dioxide *Catalysis Letters* **102**(3-4) 247-250. doi: 10.1007/s10562-005-5863

[12] Li Y, Hwang D S, Nam Hee L, and Sun-Jae K 2005 Synthesis and characterization of carbon-doped titania as an artificial solar light sensitive photocatalyst *Chemical Physics Letters* **404** 25-29. doi: 10.1016/j.cplett.2005.01.062

[13] Su C, Hong B Y, and Tseng C M 2004 Sol-gel preparation and photocatalysis of titanium dioxide *Catalysis Today* **96** 119-126. doi: 10.1016/j.cattod.2004.06.132

[14] Zhu Y, Zhang L, Chong Gao, and Lili Cao 2000 The synthesis of nanosized TiO_2 powder using a sol-gel method with $TiCL_4$ as a precursor *Journal of Material Science* **35** 4049-4054.

[15] Jaworek A 2007 Electrospray droplet sources for thin film deposition *Journal of Material Science* **42** 266-297. doi: 10.1007/s10853-006-0842-9

[16] Jaworek A and Sobczyk A T 2008 Electrospraying route to nanotechnology: An overview *Journal Of Electrostatics* **66** 197-219. doi: 10.1016/j.elstat.2007.10.001

[17] Hwang D, Lee H, Sung-Yeon J, Seong Mu J, Dongho K, Yongsok S and Dong Young K 2011 Electrospray Preparation of Hierarchically-structured Mesoporous TiO_2 Spheres for Use in Highly Efficient Dye-Sensitized Solar Cells *Applied Materials & Interfaces* **3** 2719-2725. doi: 10.1021/am200517v

[18] Jayasinghe S N 2008 Electrospray Self-assembly: An emerging jet-based route for directly forming nanoscaled structures *Journal of Physica E* **40** 2911-2915. doi: 10.1016/j.physe.2008.02.005

[19] Yuteri C U, Hartman R P A and Marijnissen J C M 2010 Producing Pharmaceutical Particles via Electrospraying with an Emphasis on Nano and Nano Structured *KONA Powder and Particle Journal* **28** 91-115.

[20] Kulkarni P, Baron P A and Klaus Willeke. 2011 Scaling Laws, Required Liquid Physical Properties, and Operating Domain *Aerosol Measurement: Principle, Technique and Applications* (Vol. Third Edition): John Wiley & Sons.

[21] Ganan-Calvo A M 1998 The Surface Charge In Electrospraying: Its Nature and Its Universal Scaling Laws *Journal Of Aerosol Science* **30**(7) 863-872. doi: 0021-8502/99/$

[22] Greenwood R 2003 Review of the measurement of zeta potentials in concentrated aqueous suspension using electroacaoustics *Advances in Colloid and Interface Science* **106** 55-81. doi: 10.1016/S0001-8686(03)00105-2

[23] Schneider C, Hanisch M, Bastian Wedel, Arben Jusufi and Matthias Ballauff 2011 Experimental study of electrostatically stabilized colloidal particles: Colloidal stability and charge reversal *Journal of Colloid and Interface Science* **358** 62-67. doi: 10.1016/j.jcis.2011.02.039

[24] Rohner T C, Lion N and Hubert H. Girault 2004 Electrochemical and theoretical aspects of electrospray ionisation *Physical Chemistry* **6** 3056-3068.

[25] Zolkepali N K, Noor Fitrah Abu Bakar, M. Nazli Naim, Nornizar Anuar, Nurul Fadhilah Kamalul Aripin, Mohd Rushdi Abu Bakar, I. Wuled Lenggoro, Hidehiro Kamiya 2016 Formation of Fine and Encapsulated Mefenamic Acid Form I Particles for Dissolution Improvement via Electrospray Method *Particulates and Science Technology.* doi: 10.1080/02726351.2016.1246496

[26] Gu W, Heil P E, Hyungsoo C and Kyekyoon K 2007 Comprehensive model for fine Coulomb fission of liquid droplets charged to Rayleigh limit *Applied Physics Letters* **91**(064104). doi: 10.1063/1.2767774

[27] Saalah S, M. Nazli Naim, Mohd. Noriznan Mokhtar, Noor Fitrah Abu Bakar, Masao Gen, and I. Wuled Lenggoro 2014 Transformation of cyclodextrin glucanotransferase (CGTase) from aqueous suspension to fine solid particles via electrospraying *Journal of Enzyme and Microbial Technology* **64-65** 52-59. doi: 10.1016/j.enzmitec.2014.06.002

[28] Scanlon D O, Dunnil C W, John Buckeridge, Stephen A. Shevlin, Andrew J. Longsdail, Scott M. Woodley, C. Richard A. Catlow, Michael J. Powell, Robert G. Palgrave, Ivan P. Parkin, Graeme W. Watson, Thomas W. Keal, Paul Sherwood, Aron Walsh, and Alexey A. Sokol 2013 Band alignment of rutile and anatase TiO_2 *Nature Materials* **12** 798-801. doi: 10.1038/NMAT3697

[29] Reddy K M, Manorama S V and A Ramachandra Reddy 2002 Bandgap studies on anatase titanium dioxide nanoparticles *Materials Chemistry and Physics* **78**, 239-245.

[30] Wang G, Xu L, Jun Zhang, Tingting Yin and Deyan Han 2011 *Enhance Photocatalytic Activity of TiO_2 Powders (P25) via Calcination Treatment.* Hubei Normal University, Huangshi, Hubei, China. International Journal of Photoenergy

[31] PerkinElmer 2009 UV/Vis/NIR Spectrometer *Simple Method pf Measuring the Band Gap Energy Value of TiO_2 in the Powder Form using a UV/Vis/NIR Spectrometer*. Shelton, CT USA: PerkinElmer.

[32] Lin Y T, Weng C H, Yu-Hao Lin, Ching-Chang Shiesh and Fang-Ying Chen 2012 Effect of C content and calcination temperature on the photocatalytic activity of C-doped TiO_2 catalyst *Journal of Separation and Purification Technology* **116** 114-123. doi: 10.1016/j.seppur.2013.05.018

[33] Khalil K M S, Baird T, Mohamed I. Zaki, Ahmed A. El-Samahy and Aida M. Awad 1998 Synthesis and characterization of catalytic titanias via hydrolysis of titanium (IV) isopropoxide *Journal of Colloids and Surfaces A: Physicichemical and Engineering Aspects* **132** 31-44.

[34] Dong F, Wang H and Zhongbiao Wu 2009 One-Step "Green" Synthetic Approach for Mesoporous C-Doped Titanium Dioxide with Efficient Visible Light Photocatalytic Activity *Journal of Physical Chemistry C* **113** 16717-16723. doi: 10.1021/jp9049654

[35] Wang H, Wu Z and Yue Liu 2009 A Simple Two-Step Template Approach for Preparing Carbon-Doped Mesoporous TiO_2 Hollow Microspheres *Journal of Physical Chemistry C* **113** 13317-13324. doi: 10.1021/jp9147693

[36] Anson-Casaos A, Tacchini I, Andrea Unzue and M. Teresa Martinez 2013 Combined modification of a TiO_2 photocatalyst with two different carbon forms *Applied Surface Science* **270** 675-684.
[37] Yue L, Hai-qiang W and Wu Zhong-biao 2007 Characterization of metal doped-titanium dioxide and behaviours on photocatalytic oxidation of nitrogen oxides *Journal of Environmental Sciences* **19** 1505-1509.
[38] Naim M N, Abu Bakar N F, Iijima M, Kamiya H, and Lenggoro I W, Electrostatic Deposition of Aerosol Particles Generated from an Aqueous Nanopowder Suspension on a Chemically Treated Substrate *Japanese Journal of Applied Physics* **49** (2010) 06GH17 1-6

IOP Conf. Series: Materials Science and Engineering **358** (2018) 012049 doi:10.1088/1757-899X/358/1/012049

Performance of Kaolin Clay on the Concrete Pavement

M E Abdullah[1], R P Jaya[2,7], M N A Shahafuddin[3], H Yaacob[4], M H Wan Ibrahim[5], F M Nazri[6], N I Ramli[7] and A A Mohammed[8]

[1,5] Faculty of Civil and Environmental Engineering, Universiti Tun Hussein Onn Malaysia, 86400 Batu Pahat, Johor Bahru, Malaysia
[2,3,4] Faculty of Civil Engineering, Department of Geotechnics and Transportation, Universiti Teknologi Malaysia, 81310 Johor Bahru, Malaysia
[6] School of Civil Engineering, Universiti Sains Malaysia, Engineering Campus, 14300 Nibong Tebal, Penang, Malaysia
[7] Faculty of Civil Engineering and Earth Resources, Universiti Malaysia Pahang, 26300 Gambang, Pahang, Malaysia
[8] Faculty of Engineering, University of Nottingham, Malaysia Campus, Jalan Broga, 43500 Semenyih, Selangor, Malaysia

ramadhansyah@utm.my

Abstract. This paper investigates the performance of concrete pavement containing kaolin clay with their engineering properties and to determine the optimum kaolin clay content. The concrete used throughout the study was designed as grade 30 MPa strength with constant water to cement ratio of 0.49. The compressive strength, flexural strength and water absorption test was conducted in this research. The concrete mix designed with kaolin clay as cement replacement comprises at 0%, 5%, 10% and 15% by the total weight of cement. The results indicate that the strength of pavement concrete decreases as the percentage of kaolin clay increases. It also shows that the water absorption increases with the percentage of cement replacement. However, 5% kaolin clay is found to be the optimum level to replace cement in a pavement concrete.

1. Introduction
Concrete pavement has been utilized for many areas such as highways, airports, parking lots, industrial facilities, and various types of infrastructure [1]. Concrete pavements if designed properly and built from durable materials according to the specifications, can provide long duration of service with little or almost no maintenance at all [2]. Nowadays, the introduction of new materials as concrete ingredient is getting popular due to strict environmental legislations [3]. Research on different materials as substitution and replacement for cement such as kaolin clay, palm oil ash (POFA) and rice husk ash have been used in construction [4,5,6]. In Europe, a code of practice has been enforced for the use of waste materials in cement and dolomitic lime manufacture [5]. This is to solve waste problem and to improve the sustainability of operation by reducing the use of natural raw materials. With continuous production worldwide, the current utilization of waste by-product is believed to rise with the realization of environmental benefits that contributes to the protection and sustainability of future construction [6]. One possible source is kaolin clay, presented as fine particles, is beneficial to some properties of PCC such as compressive strength and segregation.

Content from this work may be used under the terms of the Creative Commons Attribution 3.0 licence. Any further distribution of this work must maintain attribution to the author(s) and the title of the work, journal citation and DOI.
Published under licence by IOP Publishing Ltd

2. Material and methods

2.1. Cement
Ordinary type I Portland cement (Blue Lion Cement) (OPC) was used as the major binder in the production of moderately strong concrete. The OPC was chemically analyzed using an X-ray fluorescence (XRF) apparatus, in accordance with the procedures in BS EN 197-1 [9]. According to Dewi et al. [10] the chemical composition of OPC was 70% calcium oxide, 17.8% silicon dioxide, 3.9% aluminium oxide, 3.2% iron oxide, 1.5% magnesium oxide and 3.6% sulfur trioxide.

2.2. Kaolin
The only pozzolanic material that used in this research is the kaolin clay. The pozzolanic shall conform to the prescribed chemical composition requirements and physical requirements as stated in the ASTM C618 [11]. Kaolin is neither the by-product of an industrial process nor is it entirely natural.

2.3. Aggregate
Local natural sand from granite was used as fine aggregate in concrete mixtures. In addition, crushed granite with a single size of 20 mm was used as coarse aggregate. The coarse and fine aggregates both had a specific gravity of 2.66 and water absorption of 0.48 and 0.86%, respectively.

2.4. Compressive strength test
Compressive strength is commonly considered the most important property of concrete. In this study, the compressive strength test of all the concrete mixes was performed on 100×100×100 mm cubes. The specimens were compressed by a compression machine with a maximum capacity of 3000kN and a loading rate of 150kN/min. The reported compressive strength was the average of the three samples. The test was conducted according to the British standard test method BS EN 12390-3 [12].

2.5. Flexural strength test
The flexural strength test of specimens was carried out using the prims flexural test. Three pavement concrete prims specimens of 100×100×500 mm were prepared. These specimens were tested at 7, 14, and 28 days and then reported. The mean readings of maximum stress applied to the concrete prims by 3-point load test using flexural-testing machine were recorded, and the data were used to calculate the flexural strength of the pavement concrete prims containing the prepared blended binders. The flexural strength test was carried out in accordance with BS EN 12390-5 [13].

3. Results and Discussion

3.1. Compressive strength
Figure 1 demonstrates the compressive strength of Portland cement pastes when affected by the replacement percentage. The concrete exhibits an upward projection in compressive strength as the concretes ages. The ordinary Portland cement concrete continued to gain strength over time with value for all percentage of cement replacement. With an increase in age, the degree of hydration increase as will the compressive strength. For instance, the control samples recorded is 37.51 N/mm^2 at age of 7 days. At the age of 28 days, the kaolin replaced concrete continues to gain strength over time. The highest strength recorded is 46.05 N/mm^2 for control mix at 28 days which is still higher than any of the kaolin replaced concrete mix. With all kaolin replacement, there is a reduction in cement amount for the concrete paste. This is in turn reducing the released amount of C-H gel from the hydration process. The C-H gel does react with the admixtures and rapidly removes calcium hydroxide from the system and accelerate the ordinary Portland cement hydration [14]. It can be said that all different blended concrete pavement decrease in compressive strength when they are compared to the control sample's compressive strength. Therefore, an optimum replacement recorded is for 5% replacement.

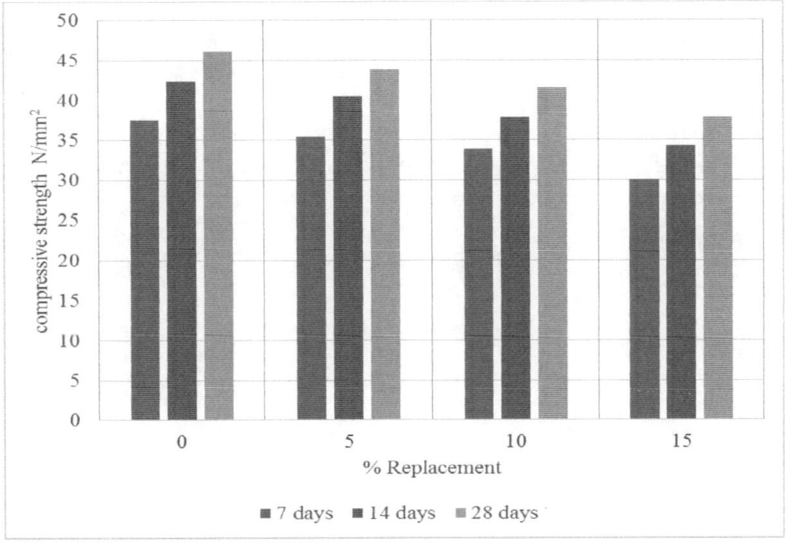

Figure 1. Strength of pavement concrete at different kaolin replacement

3.2. Flexural Strength

Figure 2 shows the flexural strength with different percentage of replacement. The concrete exhibits an upward trend in flexural strength with an increase in age and curing. At 28 days, the control samples recorded with flexural strength of 3.82 MPa. While for 5% of cement replacement, the concrete recorded flexural strength of 3.80 MPa. Figure 2 shows that with all cement replacement, it exhibit lower flexural strength than the control mixes. The flexural strength decreased slightly to 3.75 MPa with 10% replacement. However, at 15% replacement, the flexural strength drops up to 3.73 MPa. At 5% replacement, the flexural strength of concrete was achieving 3.80 MPa. The flexural strength is found to decrease by the addition of admixtures due to the decrease in bond strength as they form a rigid structure comparatively. This phenomenon is decrease in flexural strength of concrete.

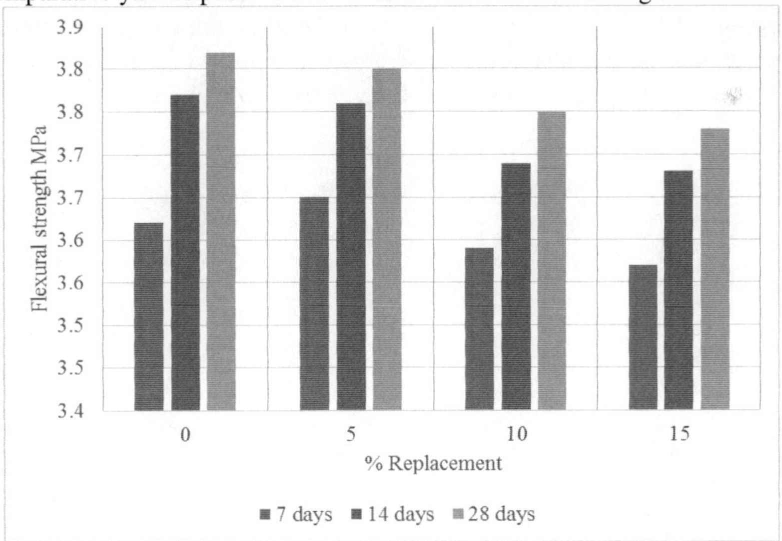

Figure 2. Flexural strength of pavement concrete at different kaolin replacement

3.3. Water Absorption

Water absorption is measured by measuring the increase in mass as a percentage of dry mass. Water absorption has significant effects to the durability of pavement concrete. Table 1 shows the water absorption at 28 days with different percentage of replacement. It can be observe that the water absorption is directly proportional to percentage of cement replacement. The water absorption of the control sample is 1.14% and continues to increase for each replacement with the exception for the 15% cement replacement. However, at 15% replacement of kaolin clay, the water absorption was about

1.51%. The increase in water absorption indicates higher amount of void in the concrete. Kaolin clay suspends in the paste. This will increase the spacing between fine aggregate. When high amount of kaolin, any more than 3%, air void began to level out.

Table 1. Water absorption with different % kaolin replacement

% replacement	Water absorption		
	WW	DW	% different
0	2.301	2.275	1.14
5	2.279	2.248	1.37
10	2.320	2.281	1.71
15	2.287	2.253	1.51

3.4. Relationship between Strength and Water Absorption

Figure 3 shows the relationship between compressive strength and water absorption. It can be seen that the higher the water absorption, the lower the strength of the concrete. The results also show that the two lines directly related. There is an observable decreases in compressive strength from 46.05 N/mm^2 to 37.83 N/mm^2 as the water absorption increases from 1.14% to 1.51%. The drop in compressive strength can be due to the amount of kaolin clay in the concrete mix. As the amount of kaolin clay increase, the more water is absorbed by the kaolin. It increase the water absorption but due to the lack of cement to bind the aggregate together. Therefore, compressive strength decreases because of the increase in water absorption. The structure of kaolin is affected by the adsorption of heavy metals. The displacement of H+ ions and the adsorption of the cations of Pb, Zn or Cd could create swelling, internal stress in the molecule, flocculation, a decrease in shear strength and an increase in hydraulic conductivity and compressibility [15]. All those changes listed above indicate the creation of empty spaces in the clay structure.

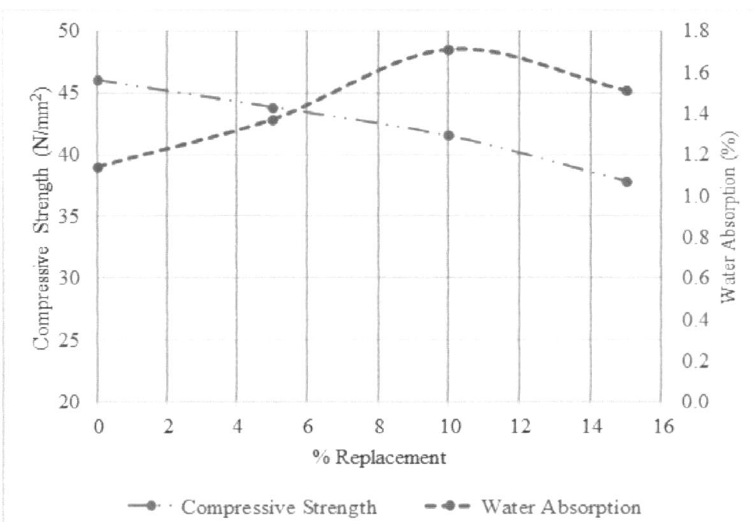

Figure 3. Relationship between strength vs. water absorption

4. Conclusions
Admixture of kaolin can decrease the compressive strength of concrete. However, among kaolin clay replacement, 5% has the highest compressive strength compared to the other two values of replacements. Furthermore, kaolin clay also decreases the flexural strength of concrete. But, 5% of the binder replacement has the highest flexural strength when compared to the other two values of replacements. In addition, water absorption has significant effects to the durability of pavement concrete. It can be seen that the water absorption continuously increase with increment kaolin clay, except at 15% cement replacement.

Acknowledgement
The support provided by Malaysian Ministry of Higher Education (MOHE), Universiti Teknologi Malaysia (research grant number Q.J130000.2522.18H05 and Q.J130000.2522.14H30), Universiti Tun Hussein Onn Malaysia and Fundamental Research Grant Scheme (research grant number 1458) for this study is highly appreciated.

References
[1] Asi M, Qasrawi H Y and Shalabi F I 2007 *Can. J. Civil Eng.* **34** 902–911.
[2] Shafiq N, Nuruddin M F, Khan S U and Ayub T 2015 *Const. Build. Mater.* **81** 313-323.
[3] Ilić B, Radonjanin V, Malešev M, Zdujić M and Mitrović A 2016 *Appl. Clay Sci.* **123** 173-181.
[4] Lotfy A, Karahan O, Ozbay E, Hossain K M A and Lachemi M 2015 *Const. Build. Mater.* **83** 102-107.
[5] Khalid N H A, Hussin M W, Mirza J, Ariffin N F, Ismail M A, Lee H.-S, Mohamed A and Jaya R P 2016 *Const. Build. Mater.* **102** 950–960.
[6] Jayanti D S, Mirza J, Jaya R P, Abu Bakar B H, Hassan N A and Hainin M R 2016 *P. I. Civil. Eng-Mar. Eng.* **169** 76-85.
[7] Liu F, Meng L-y, Ning G-F and Li-Juan Li 2015 *Const. Build. Mater.* **95** 207-217.
[8] Anastasiou E K, Liapis A and Papayianni I 2015 *Resour. Conserv. Recy.* **101** 1-8.
[9] BS EN 197-1 2011 *Cement. Composition, specifications and conformity criteria for common cements.* BSI, London, UK.
[10] Dewi S J, Ramadhansyah P J, Norhidayah A H, Md. Maniruzzaman A A, Hainin M R and Mohd Haziman W I 2014 *Appl. Mech. Mater.* **554** 203-207.
[11] ASTM C618 2015 *Standard Specification for Coal Fly Ash and Raw or Calcined Natural Pozzolan for Use in Concrete*. ASTM International, West Conshohocken, PA.
[12] BS EN 12390-3 2009 *Testing hardened concrete. Compressive strength of test specimens.* BSI, London, UK.
[13] BS EN 12390-5 2009 *Testing hardened concrete. Flexural strength of test specimens.* BSI, London, UK.
[14] Abu Bakar B H, Ramadhansyah P J and Megat Azmi M J 2011 *Mag. Concrete Res.* **63** 313-320.
[15] Chen C, Habert G, Bouzidi Y, Jullien A and Ventura A 2010 *Resour. Conserv. Recycl.* **54** 1231–1240.

Biodegradation Behaviour of Thermoplastic Starch Films Derived from *Tacca leontopetaloides* Starch under Controlled Composting Condition

A M Mohd Amin, S Mohd Sauid, K H Ku Hamid and M Musa

Faculty of Chemical Engineering, Universiti Teknologi MARA, Shah Alam, Malaysia

suhaila_sauid@salam.uitm.edu.my

Abstract. The biodegradation study of thermoplastic starch (TPS) films derived from *Tacca leontopetaloide*s starch; namely TPS/GLY, TPS/ACE and TPS/BCHR were investigated under controlled composting conditions. A manual set-up test rig in laboratory scale was built according to ISO 14855-1: 2012. The biodegradation percentage was determined by measuring the amount of CO_2 evolved using titration method and validated by automatic system (Arduino UNO System) that detected the CO_2 evolved. After 45 days under controlled composting condition, results indicated that TPS/GLY degraded the fastest, followed by TPS/BCHR and the TPS/ACE had the slowest degradation. The biodegradation process of TPS/GLY, TPS/ACE and TPS/BCHR also exhibited two stages with different degradation speeds. From these results, it indicated that chemical modification of the TPS films by adding acetic acid and rice husk bio-char to the thermoplastic starch can have a major impact on the biodegradation rate and final biodegradation percentage.

1. Introduction

The large use of synthetic non-degradable polymer and plastic materials produced from petrochemicals has led to serious environmental pollution. The solution to solve this problem is to use the biodegradable polymers (biopolymer); that is a renewable and can be the alternative material for replacement of many petroleum-based products. One type of the biodegradable biopolymer is thermoplastic starch that is produced from processed starch in the presence of plasticizers and under the influence of heat and shear [1]. Biodegradable polymer normally degrades into natural and harmless product results from the naturally occurring action of microorganism such as bacteria and fungi [2] such as carbon dioxide and water. In this study, the thermoplastic starch (TPS) was produced from *Tacca Leontopetaloides* starch. *T. leontopetaloides*, an annual herb of yam family contained starch in its tuber is a non-staple food readily available in Malaysia. The *T. leontopetaloides* starch was plasticised with glycerol in the present of heat to produce the TPS films with required properties. However, significant disadvantage of thermoplastic starch (TPS) is poor water resistance, making its application become limited. Thus, chemical modification was conducted to the TPS films by adding acetic acid and rice husk biochar in the formulation to improve its properties. However, the biodegradable properties of these TPS films after its end used or at ultimate disposal had not been investigated in detail. Therefore it is necessary to study the biodegradation behaviour of the TPS films in the natural environments.

Content from this work may be used under the terms of the Creative Commons Attribution 3.0 licence. Any further distribution of this work must maintain attribution to the author(s) and the title of the work, journal citation and DOI.
Published under licence by IOP Publishing Ltd

2. Materials and Method

2.1. Materials
T. leontopetaloides starch used was collected from Mersing, Johor. Glycerol, acetic acid and sulphur were purchased from Merck (M) Sdn Bhd. Biochar of rice husk was supplied by Sendi Enterprise in Sekinchan, Selangor.

2.2. Preparation of TPS Films
Thermoplastic starch (TPS) films were prepared by dissolving *T. leontopetaloides* starch in distilled water to produce starch solution and heated under constant stirring for 30 min. Then, 10% (v/v) of glycerol was periodically added until solution completely gelatinized at 85°C to 90°C. The plasticized starch solution was then casted onto the polyacrylic plate and dried in ventilated oven at 45°C until constant weight was obtained. Then, the plasticized dried films were peeled off from the polyacrylic plate and milled using two-roll mill machine. 0.2% (w/v) sulphur was added during the milling process until even thickness was obtained. The curing temperature was set at 55°C to melt-blend the components and the speed of the roll-mill was adjusted at 5 rad/s. The sheet film obtained was denoted as TPS/GLY. Similar procedure was repeated to produce TPS/ACE where 5% (v/w) diluted acetic acid (10% v/v) was added to the plasticized starch solution. For TPS/BCHR film, the plasticized TPS/ACE film was supplemented with 5% rice husk biochar during melt-blend using two-roll machine until even thickness of film was obtained.

2.3. Compost Characterization
Mature compost was obtained from a plant nursery. The compost was screened using 10 mm sieve and large solids such as stone and wood were manually removed. The compost properties were determined according to the methods summarized in table 1.

Table 1. Compost characterization method

Properties	Test method
pH	pH was measured by mixing compost with water at weight ratio (inoculum: water = 1:5) according to ISO 10390:2005 [8]
Moisture content	Water (moisture) content by mass of compost was determined by ASTM D2216-98. [9]
Total solids	The compost was dried at about 105°C until constant mass was obtained.
Volatile solids	Compost was preheated, weighed and ignited at 550°C and reweight.
C/N Ratio	Standard test methods for instrumental determination of carbon, hydrogen, and nitrogen in laboratory samples according to ASTM D5373:2002.

2.4. Biodegradation Testing Rig
In this experiment, the test rig composed of three parts: I) a pressurized air control system which was to control aeration rate, to produce air saturated vapor and to remove carbon dioxide (CO_2) from the compressed air; II) a composting system consisted of 2L special vessel containing test materials and inoculum compost and III) CO_2 trap system to evaluate the carbon dioxide evolved from the biodegradation of the test materials in the composting system. The schematic diagram of the set-up is illustrated in Figure 1.

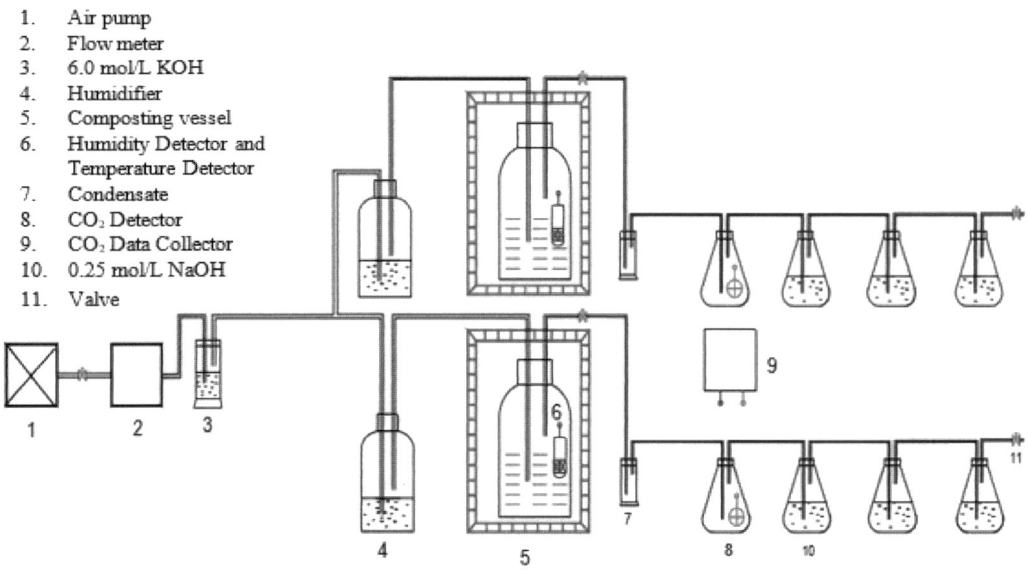

Figure 1. The test system of biodegradation for TPS films

2.5. Procedure

Compost with moisture content of 50 to 55% were mixed with test material (measurement 10mm x 10mm x 1mm) in 2L special vessel according to the ratio of 6:1 (w:w dry mass). Two special vessels were used; one containing compost with test materials and another one as background control containing compost only. Both vessels were aerated at 30 to 40 mL/min throughout the experiment to ensure adequate oxygen for biodegradation process and placed in the water bath for test environment of 58°C±2°C in dark condition. Humidity and temperature in the vessels were monitored using the humidity and temperature detector. The CO_2 evolved from the biodegradation vessels was measured twice daily to estimate the cumulative CO_2 production.

2.6. Biodegradation percentage determination

Titration method (Method A): The CO_2 gas produced from biodegradation of test material or inoculum compost was trapped in 0.25 mol/L NaOH. To evaluate the amount of CO_2 absorbed in the NaOH solution, 20 mL aliquots were removed from 500 mL of NaOH solutions. Then, acid-base titration was used to measure the amount of CO_2 evolved using 0.25M hydrochloric acid (HCl) with double indicator, Phenolphtalein and Methyl-Orange. The total amount of CO_2 evolved was calculated according to Du et al. [3] method by implementing equation (1).

$$CO_2(g) = V_2 \times C_{HCl} \times 44 \times \frac{500\ mL}{20\ mL} \quad (1)$$

Automatic method (*Method B*): The CO_2 gas evolved was detected by the CO_2 detector supported by microcontroller of Arduino UNO System. The board of microcontroller provide a real time clock module that capable to receive the data logging from CO_2 detector for each second. The raw data logging of CO_2 evolved was then collected using formatted Secure Digital (SD) Card that attached on the circuit of microcontroller board. Then, the raw data logging of CO_2 produced at 6 hours, 12 hours and 18 hours were averaged and used to calculate the quantity of CO_2 produced in gram from the biodegradation test using Eq. 2.

$$CO_2\ (g) = \frac{ppm\ CO_2}{10{,}000} \times liter\ of\ free\ air \times \frac{1\ m^3}{1000\ L} \times \frac{1.987\ kg}{m^3} \times \frac{1000g}{1kg} \times \frac{12}{44} \quad (2)$$

The theoretical amount of CO_2 ($ThCO_2$), g/flask was calculated using Eq. 3 and the biodegradation percentage (D_t) was calculated using Eq. 4 respectively.

$$ThCO_2 = M_{TOT} \times C_{TOT} \times \frac{44}{12} \quad (3)$$

$$D_T = 100\% \times \frac{(CO_2)_T - (CO_2)_B}{ThCO_2} \quad (4)$$

M_{TOT} is the total dry solids in test sample added into composting flask; C_{TOT} is the proportion of total organic carbon in the total dry solids in test sample; 44 and 12 are the CO_2 molecular mass and atomic mass of carbon, ${CO_2}_T$ is the cumulative amount of CO_2 evolved in each composting flask containing test materials; ${CO_2}_B$ is the mean cumulative amount of CO_2 evolved in blank flask.

3. Results and Discussion

3.1. Properties of compost.

Table 2 shows the properties of compost used in the controlled composting condition. As shown, the pH is slightly alkaline with the moisture content of 50.35%. The volatile solids indirectly indicating the organic matter and the % shown is on the total dry solids. The C:N ratio showing that the compost used is mature compost as it is ranged between 10 to 20 : 1 [4].

Table 2 The properties of compost

Chemical properties	Value
pH	7.94
Moisture (%)	50.35
Total dry solid (%)	49.38
Volatile dry solid (%)	24.22
C:N ratio	13.58

3.2. Biodegradation of TPS films.

The biodegradation curves of TPS/GLY, TPS/ACE and TPS/BCHR are shown in Figure 2 according to method A and Method B respectively. Two phases were observed in TPS/GLY AND TPS/BCHR biodegradation curves which can be defined as slow degrading phase which occurred during the first 20 days. The slopes were gradual at these days showing that the degradation rates were lower. The second phase, fast degradation phase occurred after 20 days as steeper slopes were observed. However, TPS/ACE showed linear curves throughout the 45 days of composting.

After 45 days of incubating the test materials with the compost mixture in the controlled composting condition, the degradation percentage TPS/GLY, TPS/ACE and TPS/BCHR were 70.1%, 23.89% and 54.04% respectively when determined using Method A. The percentages of degradation were validated as these values were similar when determined using Method B (69.8%, 24.1% and 50.1% respectively). Comparing the rate of degradation and degradation percentage for all types of test materials, TPS/ACE had the slowest values. These results indicating that chemical modification by adding acetic acid in the formulation affected the action of microorganisms in the compost to degrade the TPS/ACE film. The acetic acid was added as cross linking agent to improve the TPS/ACE film performance. It reacted with the hydroxyl group of starch creating new chemical bonds between molecular starch chains thus making its hydrophilic character reduced [5-7]. During the composting, the moisture content of the compost was kept constant at 50 to 55%. When moisture diffused into the test materials, it will cause the film to swell and stimulate the microbial growths thus enhancing biodegradation. However, the incorporation of the acetic acid reduced the moisture absorption and therefore resulting to slower the rate of degradation and degradation percentage. Similar observations also encountered by [5]. Among the three types of films, TPS/GLY had the highest degradation rate. This was predictable since moisture absorption expected to be higher which promoting higher microbial degradation compared to the other two films.

For TPS/BCHR, the degradation rate was in between TPS/GLY and TPS/ACE. Table 3 shows the physical and chemical properties of the rice husk biochar used. From Table 3, the ash content was almost 50% of the content making the biochar relatively inert. Addition of acetic acid also made the film had lower water absorption, and yet the degradation rate was higher compare to TPS/ACE. Possible explanation could be due to the pH of the biochar. As it was slightly alkaline, it altered the pH of films. Furthermore, the biochar also has source of carbon and nitrogen that may had gave the microorganism source of nutrients.

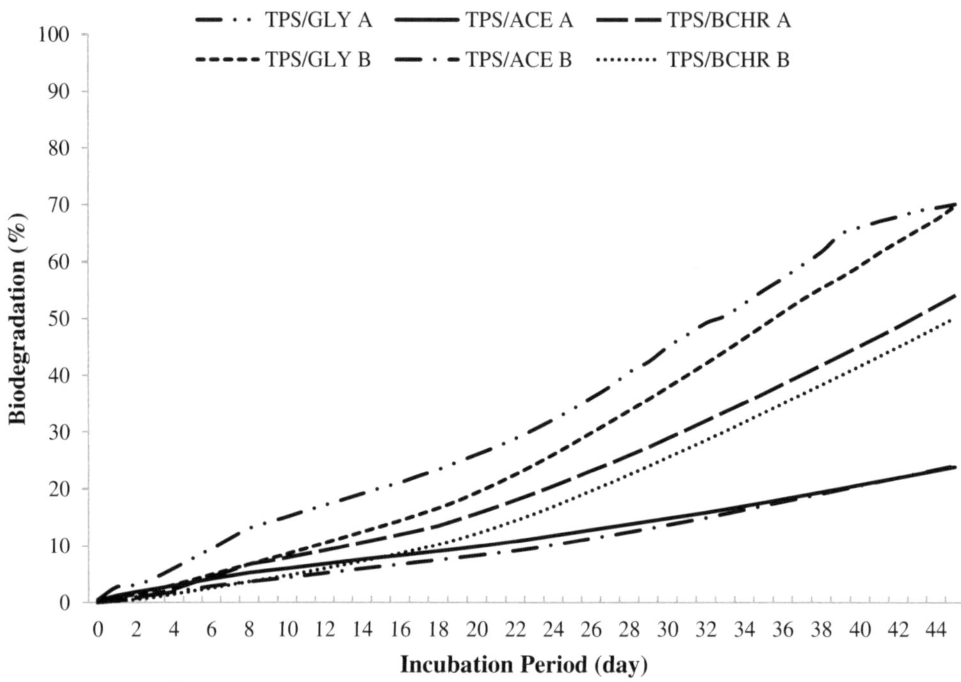

Figure 2 Biodegradation curve of TPS/GLY, TPS/ACE and TPS/BCHR under controlled composting condition (A – Method A, B – Method B).

Table 3. Physical and Chemical Properties of Rice Husk Biochar

Moisture % (w/w)	pH	Volatile Matter (%)	Ash (%)	C (%)	H (%)	N (%)	BET Surface Area (m^2/g)
8.15	8.20	13.64	48.20	26.23	6.88	0.28	186.99

4. Conclusion

The biodegradability of three types of TPS samples namely TPS/GLY, TPS/ACE and TPS/BCHR were investigated. It was observed that, TPS/GLY had the highest percentage of degradation followed by TPS/BCHR and TPS/ACE had the slower degradation rate. Chemical modification by the addition of acetic acid and biochar significantly affected the degradation percentage. The microorganisms that were responsible for the degradation of these films could be isolated and identified in future work.

Acknowledgment

The authors are thankful and grateful to Ministry of High Education (MOHE) and Universiti Teknologi MARA, (UiTM) for supporting and providing funds for this paper through the research Grant no. RMI/RAGS5/3 (84/2013).

References

[1] Qiang Z, Jian T, Richard C M Y, Albert C K M, Robert K Y L and Cunjiang S 2008

Polym. Degrad. Stab. **93** 1571-76
[2] Leon J and Leszeck M 2009 *Thermoplastic Starch:A Green Material For Various Industries* (Germany:Wiley-VCH)
[3] Du Y L, Cao Y, Lu F, Li F, Cao Yi, Wang X L and Wang Y Z 2008 *Polym. Test.* **27** 924 - 30
[4] Bernal M P, Alburquerque J A and Moral R 2009 *Bioresour. Technol.* **100** 5444–53
[5] Seligra P G, Jaramillo C M, Famá L and Goyanes s 2016 *Carbohydr. Polym.* **138** 66-74
[6] Ghanbarzadeh B, Almasi H, and Entezami A A 2011 *Ind. Crops Prod.* **33** 229 - 35
[7] Olssona E, Hedenqvist M S, Johanssona C and Järnströma L 2013 *Carbohydr. Polym.* **94** 765-72
[8] NF ISO 10390, 2005. *Soil Quality, Determination of pH*. AFNOR, 7
[9] American Society for Testing and Materials, 1998. *Standard Test Method For Laboratory Determination of Water (Moisture) Content of Soil and Rock*, ASTM D2216-98. Philadelphia, USA

IOP Conf. Series: Materials Science and Engineering **358** (2018) 012051 doi:10.1088/1757-899X/358/1/012051

Effects of Synthesis Method on Electrical Properties of Graphene

M F I Ahmad Fuad[1], H H Jarni, W N Shariffudin, N H Othman and 'A N Che Abdul Rahim

Faculty of Chemical Engineering, Universiti Teknologi MARA, 40450 Shah Alam, Selangor

fazrilirfan@salam.uitm.edu.my[1]

Abstract. The aim of this study is to achieve the highest reduction capability and complete reductions of oxygen from graphene oxide (GO) by using different type of chemical methods. The modification of Hummer's method has been proposed to produce GO, and hydrazine hydrate has been utilized in the GO's reduction process into graphene. There are two types of chemical method are used to synthesize graphene; 1) Sina's method and 2) Sasha's method. Both GO and graphene were then characterized using X-Ray Powder Diffraction (XRD) and Fourier Transform Infrared Spectrometry (FT-IR). The graph patterns obtained from XRD showed that the values of graphene and GO are within their reliable ranges, FT-IR identified the comparison functional group between GO and graphene. Graphene was verified to experience the reduction process due to absent of functional group consist of oxygen has detected. Electrochemical impedance spectrometry (EIS) was then conducted to test the ability of conducting electricity of two batches (each weighted 1.6g) of graphene synthesized using different methods (Sina's method and Sasha's method). Sasha's method was proven to have lower conductivity value compare to Sina's method, with value of 6.2E+02 S/m and 8.1E+02 S/m respectively. These values show that both methods produced good graphene; however, by using Sina's method, the graphene produced has better electrical properties.

1. Introduction

Graphene oxide (GO) consists of oxygenated graphene sheets with oxygen-containing functional groups on their basal planes and edges. Thus, GO shows excellent hydrophilicity, strong chemical activity, and uniformity of dispersion in water, making it more easily to be handled and deposited on membrane. A successful use of GO ceramic hollow fiber composite membrane has been recently reported for dehydration of dimethyl carbonate and high water flux is obtained [6]. It is also reported that membranes made with pure GO exhibited a lot of special properties for separation due to their 2D channels, hydrophilic surface as well as the defects on GO sheets [7]. In this work, the graphene oxide membrane was synthesized and deposited onto the alumina hollow fibre substrate using a vacuum assisted technique. The performance of the membrane to treat oily wastewater was then investigated. Graphene, a one-atom thick planar sheet of sp^2 bonded carbon atoms packed in a honeycomb lattice, is considered to be the mother of all graphitic materials like fullerenes, carbon nanotubes, and graphite [1]. Graphene also has been known for its ability to conduct an electricity and good mechanical

Content from this work may be used under the terms of the Creative Commons Attribution 3.0 licence. Any further distribution of this work must maintain attribution to the author(s) and the title of the work, journal citation and DOI.
Published under licence by IOP Publishing Ltd

properties[2-4]. It also known for its zero band gap properties that lead to good conductivity properties for graphene.

This study is conducted to synthesis graphene oxide by using modified Hummer's method [3][4]. This modified method is used as it is simpler and less harm to environment. After that it will be followed by an experiment that to carry out reduction of graphene oxide to synthesis graphene with different type of chemical method based on Sina's method [3] and Sasha's method [4]. Both methods using hydrazine hydrate as reducing agent which is the most widely used reducing agent as a way of control to reduce GO [5]. Preparation of graphene via chemical is found approximately around 70 years before, where this method is being widely used at the moment. Transformation of graphene oxide to graphene can be observed clearly through colour change of the mixture from brown (graphene oxide) to black (graphene) [3][6]. Next, characterization analysis was conducted to investigate which method could synthesis a good graphene.

2. Experimental

2.1 Preparation of Graphene Oxide (GO)

Graphene Oxide was prepared from graphite flake (Sigma Aldrich) using modified Hummers method [3][4][7]. 5 g of graphite flake (Sigma Aldrich) was added with 2.5 g sodium nitrate ($NaNo_3$) into 1000ml beaker. 200ml of sulfuric acid (H_2SO_4) was then added into solution. Solution was stirred for 1 hour. 30g of potassium permanganate ($KMnO_4$) was added slowly within 2 hours at temperature was maintained below 15° C. The solution was keep on stirred at room temperature for 12 hours after that. Solution was heated till 70° C for 2 hours before added 100ml of distilled water slowly. Temperature was increased up to 90° C then and stirred for 1 hour. 100ml of water was added again slowly before 30ml of hydrogen peroxide (H_2O_2) was added. GO solution then washed with acidic hydrochloric acid (36%, Merck). GO solution centrifuged by 1000 rpm in 30 minutes and dried in oven at 70° C to gain GO in (solid) thin film form.

2.2 Reduction of Graphene Oxide using Sina's Method

GO solution was prepared by dissolving 10 mL of hydrazine hydrate (80%, Merck) in 100 mL of distilled water. The solution then was added slowly into GO solution at room temperature. Reduction process then conducted at 95° C for 1 hour. Resultant of black precipitate was filtered by cellulose filter paper. 1M of hydrochloric acid (36%, Merck) used to wash filtered precipitate for neutralized the pH value. Those precipitate was dried in the oven for about 3 hours with temperature at 70° C to obtain graphene powder.

2.3 Reduction of Graphene Oxide using Sasha's Method

100 mg GO powder was added into 250 mL round bottom flask. 100 mL of distilled water then was added to dissolve GO powder into GO solution. GO solution then was sonicated using ultrasonic bath cleaner for 45 minutes (till the dispersion is clear). 1 mL of 32.01 mmol hydrazine hydrate (80%, Merck) was prepared and added into the solution. Solution was left in 100° C of oil bath for 24 hours. Precipitate from the solution was filtered to isolate it. 100 mL of distilled water and 100 mL was used to wash precipitate for three times. Precipitate was dried in oven for 3 hours with 70° C to obtain graphene powder.

2.4 Characterization Test

This characterization test is conducted to identify properties for each GO and graphene synthesis. XRD test is conducted using (XRD Sigma 3-18K), with range angle from 5° till 60°, at 2 minute for each angle. Functional group data for each material was gained from Fourier transform infrared (FT-IR) study on the GO and graphene powder by (FTIR Spectrum One). FT-IR was performed over the wave number range of 4000-500 cm $^{-1}$ for each sample. Electrochemical impedance spectrometry (EIS, HIOKI 1) was conducted at room temperature then to measure conductivity for each GO and graphene. The test was implemented V = 1.00 V with auto range and normal speed as it perimeter.

Figure 1. a) Visual of solution from GO , b) graphene (Sina's method) and c) graphene (Sasha's method)

3. Results and Discussions

3.1 XRD Analysis

Graph pattern for each GO, graphene (Sina) and graphene (Sasha) is analyze to identify each (d-spacing) and peak value for 2° (degree). XRD pattern of GO and graphene by Sina's and Sasha's method are presented in Figure 2. It clearly showed that typical broad peak position for GO was 2° = 9.76° (d-spacing ~ .0905 nm). Compared with GO, both graphene shifted to higher 2° angles; ii - 26.58 (d-spacing ~ 0.335nm) and iii – 24.88 (d-spacing 0.358nm). It was convinced that there was a decrease in the average interlayer spacing of both graphene based on d-spacing values [5]. Furthermore, from the broaden peak of graphene and lower d-spacing compare to GO, it showed (sp^2 carbon) reestablished during reduction due to removal of oxygen-containing functional groups and the stacking of graphene layers. This proved formation of few layer graphene [8].

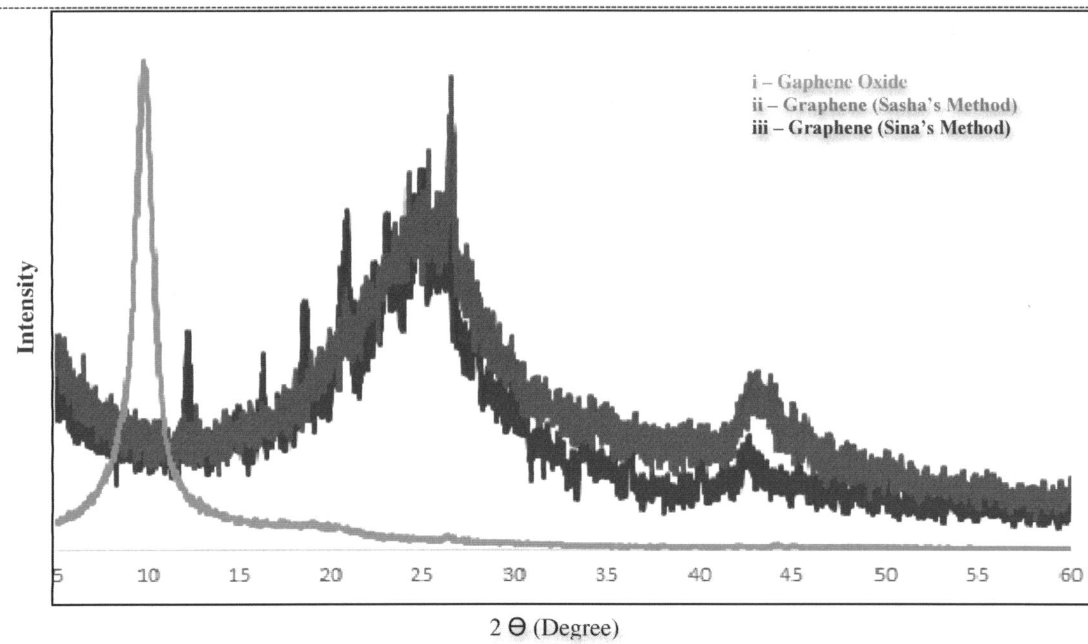

Figure 2. Graphene oxide and Graphene comparison of XRD data

3.2 FTIR Analysis

IOP Conf. Series: Materials Science and Engineering **358** (2018) 012051 doi:10.1088/1757-899X/358/1/012051

Figure 3 represent FT – IR spectroscopy of the GO and graphene samples. From the FT – IR spectra of GO it showed that the absorption bands corresponding to C – H deformation vibration at 2886.42 and 2864.58 cm^{-1} (carbogen). The C=O (carboxyl and carbonyl) form at vibration of 1719 cm^{-1} and followed by remaining sp^2 skeletal vibration C=C at 1626 cm^{-1}. After GO undergo reduction process, characteristic absorption has been decreased. It showed =C – H character due to a new absorption band at 1625 cm^{-1} and also C – H at absorption band 1540 cm^{-1}. Another stretching vibration happens at 1440 cm^{-1} that showed character of C=C which is related to graphite structure of graphene [8].

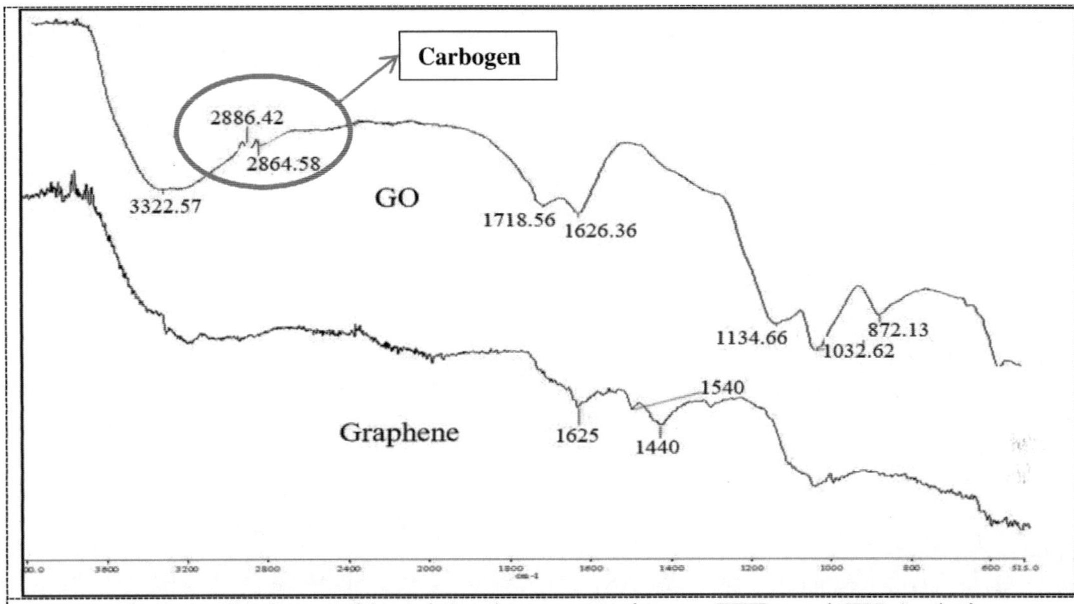

Figure 3. Graphene oxide and Graphene comparison on FTIR graph EIS Analysis

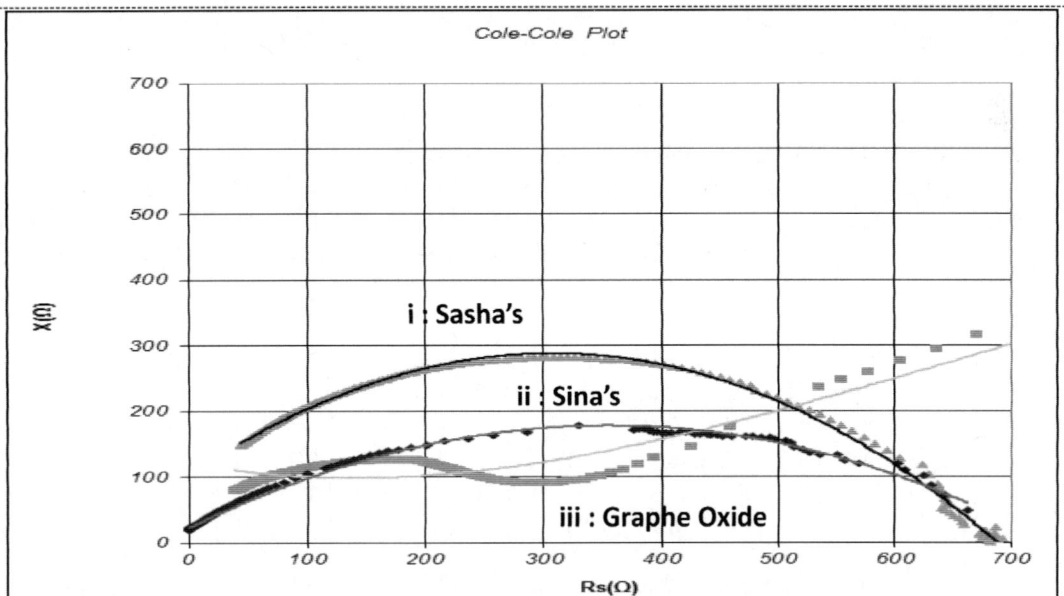

Figure 4. EIS data comparison for GO and Graphene

From Figure 4 above, it showed three pattern of graph produced from EIS test. From that graph it has labelled as (i), (ii) and (iii), where it showed graph pattern of graphene (Sasha's method), graphene (Sina's method) and GO respectively. From the calculation based on the result acquired, it has indicated that conductivity value for (i) is approximately 6.2E+02 S/cm then followed by (ii) graph that indicate value of 5.7E+02 S/cm. From these data, it showed graphene that synthesis by Sasha method (i) is better than Sina method (ii) as it produces higher conductivity value. While for (iii) which is GO there is approximately 2.9E+02 S/cm for its conductivity value, where it can be concluded as low a conductivity. However, by using a constant weight of 0.16 g for each graphene, it showed result of 8.1E+02 S/cm for graphene synthesis by Sina's method. Data was tabulate in Table 1 below to simplify understanding for conductivity value for each material.

Table 1. Conductivity and Resistivity Data for Graphene .

Material	Conductivity [S/cm]	Resistivity [Ω]
Graphene Oxide	2.9E+02	3.4E-03
Graphene (Sina's Method at 0.09g)	5.7E+02	1.8E-03
Graphene (Sina's Method at 0.07g)	2.5E+02	4.1E-03
Total Graphene (Sina's Method at 1.6g)	8.1E+02	1.2E-03
Graphene (Sasha's Method 1.6g)	6.2E+02	1.6E-03

4. Conclusions

As conclusion chemical method has been proved able to achieve highest reduction capability even not a complete reduction of oxygen from GO. Chemical method using hydrazine hydrate as reducing agent for graphene synthesis is split into (a) Sina Abdolhosseinzadeh and (b) S. Stankovich. All the data of comparison has been proved as the characteristic test taken as for GO is reliable compare to [1-10].

Acknowledgements

This research is fully supported by LESTARI grant, 600-IRMI/DANA 5/3/LESTARI (0079/2015). The authors fully acknowledged Universiti Teknologi MARA for the approved fund which makes this important research viable and effective.

References

[1] Sur U K 2012 *Graphene: A Rising Star on the Horizon of Materials Science* (Int. J. Electrochem., vol. 2012), p 1–12.
[2] Alwarappan S and Kumar A 2014 *Graphene-Based Materials* (CRC Press vol. 320)
[3] Abdolhosseinzadeh S, Asgharzadeh,H and Kim H S 2015 *Fast and fully-scalable synthesis of reduced graphene oxide* (Sci. Rep., vol. 5, p. 10160)
[4] Stankovich,S, Dikin D A, Piner R D, Kohlhaas K A, Kleinhammes A, Jia Y, Wu Y, Nguyen S T, and Ruoff 2007 R S *Synthesis of graphene-based nanosheets via chemical reduction of exfoliated graphite oxide* (Carbon N. Y., vol. 45, no. 7, p. 1558–1565)
[5] Moon I K, Lee J, Ruoff R S, and Lee H 2010 *Reduced graphene oxide by chemical graphitization* (Nat. Commun., vol. 1, no. 6, p. 1–6)
[6] Chua C K and Pumera M 2014 *Chemical reduction of graphene oxide: a synthetic chemistry viewpoint* (Chem. Soc. Rev., vol. 43, no. 1, p. 291–312)
[7] Rani A, Nam S, Oh K A, and Park M 2010 *Electrical Conductivity of Chemically Reduced Graphene Powders under Compression* (Carbon Letter vol. 11, no. 2, pp. 90–95)
[8] Zhang X, Li K, Li H, Lu J, Fu Q, and Chu Y 2014 *Graphene nanosheets synthesis via chemical reduction of graphene oxide using sodium acetate trihydrate solution* (Synth. Met., vol. 193, pp. 132–138)
[9] Chen W, Yan L, and Bangal P R 2010 *Preparation of graphene by the rapid and mild thermal reduction of graphene oxide induced by microwaves* (Carbon N. Y., vol. 48, no. 4, pp. 1146–1152)

[10] Weng W and Wu C 2003 *Exfoliation of graphite flakes and its nanocomposites* (Carbon N. Y., vol. 41, pp. 619–621)
[11] Xu C, Yuan R, and Wang X 2014 *Selective reduction of graphene oxide* (Carbon N. Y., vol. 71, no. 1, p. 345)
[12] Wan W, Zhao Z, Hu H, Gogotsi Y, and Qiu J 2013 *Highly controllable and green reduction of graphene oxide to flexible graphene film with high strength* (Mater. Res. Bull., vol. 48, no. 11, pp. 4797–4803)

Characterization and Activation of Indonesian Natural Zeolite from Southwest Aceh District-Aceh Province

Y Yulianis[1], S Muhammad[1,2], K Pontas[1,2], M Mariana[1,2,3] and M Mahidin[1,2,3]

[1] Postgraduate Program of Chemical Engineering, Syiah Kuala University, Banda Aceh 23111, Indonesia
[2] Department of Chemical Engineering, Syiah Kuala University, Banda Aceh 23111, Indonesia
[3] Environmental and Natural Resource Research Center, Syiah Kuala University, Banda Aceh 23111, Indonesia

mahidin@unsyiah.ac.id

Abstract. This study aims to identify the effect of activation processes of Indonesian zeolite from Southwest Aceh District, Aceh Province on the physical characteristics and chemical contents changes. The work was conducted by downsizing of natural zeolite into nano particle size, treating it physically (heated up to 105°C) and chemically (soaked with 0.5 M HCl for 1 hour), and finally calcining it at the temperature of 350° C for 2 hours. The natural and activated nano zeolites were then characterized by using SEM, BET, XRD, XRF and FTIR in order to examine their characters and chemical contents. The characterization results showed that the activated nano zeolite has better appearances than the natural one. The XRD analysis showed that the main minerals of zeolite are quartz and calcite clinochlore. Further, the XRF analysis showed that there are elements of magnesium, calcium and potassium which can be as a cation exchange with other metal elements. Based on the identified properties, this zeolite showed a good performance to be used as an adsorbent in waste water treatment process, especially after activated.

1. Introduction

Zeolite is a natural material that can be utilized in various applications including as adsorption, catalyst, building industry, agriculture, soil remediation and energy (Wang and Peng, 2010). Zeolite was first discovered in 1795 by A F Cronstedt, a mineralogist from Sweden. Based on his research, zeolite can release water rapidly when heated so it looks like a stone boiling. Zeolite name comes from the Greek "zeo" which means to boil and "lithos" means stone (Izci and Izci, 2007).

Zeolite contained in tufa rocks are formed from the sedimentation of volcanic ash that has been experiencing a process of alteration. Alteration process took place on a new depositional environment causing materials change from volcanic glass into mineral zeolite. The geology zeolite formed in a different environment, precipitated zeolite derived from volcanic sediment on the lake environment is alkaline and diagenetic alteration of volcanic ash, hydrothermal deep-sea sediments (Ibrahimi and Sayyadi, 2015). In this study the zeolite obtained from the Province of Aceh, Southwest Aceh District. The data from the Department of Mines and Energy of Southwest Aceh District informs that in this area has encountered outcrops zeolite, but the reserve and quality of that mineral zeolite is still unknown. Based on the Tapaktuan geological map sheet from the Geological Research and Development Centre, Bandung 1982, there are several volcano formations; among them is Tapaktuan Volcano Formation. From geological maps and field observations, in several locations of Southwest

Content from this work may be used under the terms of the Creative Commons Attribution 3.0 licence. Any further distribution of this work must maintain attribution to the author(s) and the title of the work, journal citation and DOI.
Published under licence by IOP Publishing Ltd

Aceh District there are outcrops of the mineral zeolite. Different environmental conditions affect the color of zeolites, which are green, greenish-white, red and white meat, brown bluish gray.

Zeolites are hydrated silicates of alumina compounds with alkali and alkaline earth cations. The structure of natural zeolite formed by alumina silica atoms tetrahedral form three-dimensional structures that are connected with one another through the sharing of oxygen atoms so as to form cavities and channels that contain water molecules and cations alkali and or alkaline (Mirzaei et al., 2016). Zeolite structure, divided into three components, namely the framework of alumina silicates, empty space interconnected containing metal cations and water molecules. The water contained within the zeolite pores can be removed by heating to push the water in the zeolite pores (Lestari, 2010). Natural zeolite is a zeolite obtained directly from nature, has a cheaper price than the zeolite synthesis. Natural zeolites have advantages including that of low cost because of the availability of natural zeolite with an abundance of cation exchange capacity, acts as a molecular sieve, have a high surface area due to the structure of the porous, has structural stability was good although in acidic conditions, capable of neutralizing the acid solution through H^+ ion exchange of cations in the solution with the zeolite.

Based on the fact that there are a number of advantages of natural zeolites attract the interest of researchers in exploiting natural zeolite as an adsorbent to absorb heavy metals. Erdem et al. (2004) used natural zeolite to adsorb the Co^{2+}, Cu^{2+}, Zn^{2+} and Mn^{2+}ions in industrial waste. They found that the natural zeolite types clinoptilolite has a good ability to adsorb these ions; an optimum ions adsorption sequence was $Co^{2+}>Cu^{2+}>Zn^{2+}>Mn^{2+}$. The adsorption of heavy metal ions in the acid mine water by using natural zeolite adsorbent was conducted by Motsi et al. (2009). In that study, natural zeolite applied as an adsorbent without any chemical modification to adsorbs Fe^{2+}, Cu^{2+}, Zn^{2+} and Mn^{2+} in the acid mine drainage. They observed the effect of zeolite mass, solution pH, competing ions and heat treatment. They concluded that the adsorption of heavy metal ions from solution follows the sequence $Fe^{3+}>Zn^{2+}>Cu^{2+}>Mn^{2+}$. Behind its leading, natural zeolite has also the disadvantage since its content of impurities (alkali and alkaline earth metals) and the water content that can reduce the absorption capacity.

Potential resources of zeolites in Southwest Aceh District, Aceh Province characterization needs to be done to reveal the physical and chemical properties of natural zeolite particles in order to optimize the utilization of zeolite and to improve the performance with the physical and chemical activation to remove impurities contained in the zeolite pores. Improving performance is one of the natural zeolite mineral resource conservation efforts that enhance the value-added minerals. The objectives of this paper were to investigate the influence of activation treatments (both physically and chemically) on the physical characters and chemical contents changes of Indonesian zeolite from Southwest Aceh District.

2. Methodology

The experimental procedure consists of materials and methods, zeolite activation and characterization.

2.1. *Materials and methods*

Equipment used in this study are jaw crusher, ball mill, sieve shaker, oven dryer (Memmeth DIN 12880-KI), X-Ray Difraction (XRD, Shimadzhu 6000), X-Ray Florence (XRF, Spectroxepos), FTIR (Shimaddzu Prestige 21), stirrer plate (Barnstead Thermolyne Cimarec), furnace, drop pipette, analytical scale (Shimadzu AW-220), measuring pipette 25 mL, measuring cylinder 100 mL, erlenmeyer 250 mL and distilled water bottle. Meanwhile, materials used in this study were natural zeolite from Southwest Aceh District, Hydrochloric acid (HCl) and distilled water.

2.2. *Experimental and analytical*

2.2.1. Activation of zeolite. Zeolite activation procedure is shown in Figure 1. At first, natural zeolite was cleaned and dried, and then pulverized by using jaw crusher and ball mill. Product of ball mill further screened using sieve shaker until the residue of zeolite particles were obtained in nano size. To remove the impurities, nano zeolite was washed with distilled water for 1 hour at temperature of 70°C, and then precipitated. Next, the sediment formed was dried at temperature of 105°C. After that, it was activated with 0.05 M HCl at the ratio of zeolite to acid by 1:3; the mixture was stirred for 1 hour at

IOP Conf. Series: Materials Science and Engineering **358** (2018) 012052 doi:10.1088/1757-899X/358/1/012052

temperature of 70°C, and re-precipitated. The sediment formed at this step was dried again at 105°C, then washed with distilled water for 1 hour at temperature of 70°C and precipitated. Finally, the sediment formed here then dried at temperature of 105°C and calcined at 350°C during 2 hours.

Calcination is done in order to evaporate the trapped water in the pores of the zeolite and to burn the organic compounds contained in the zeolite so that the space between the pore becomes relatively clean. This is an effort to make the surface area of zeolite increases that consequently increase the adsorption capacity of the zeolite as well (Utubira et al., 2006).

2.2.2. Characterization of zeolite. The physical characteristics and chemicals contents for both natural and activated nano zeolites were examined by Scanning Electron Microscopy (SEM), Brunauer Emmett Teller (BET), X-Ray Diffraction (XRD), X-Ray Florence (XRF), and Fourier Transform Infrared (FTIR).

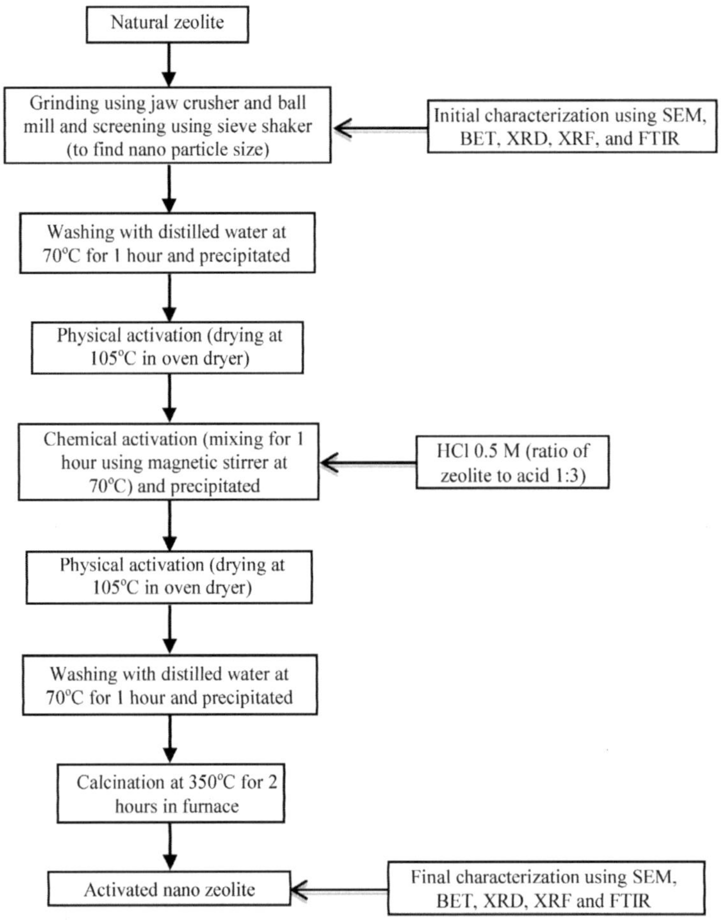

Figure 1. Schematic flow of the process

A scanning electron microscope (SEM) is used to obtain information about the surface topography and composition. The main SEM components include: source of electrons, column down which electrons travel with electromagnetic lenses, electron detector, sample chamber, computer and display to view the images. The zeolite sample is mounted on a stage in the chamber area and both the column and the chamber are evacuated by a combination of pumps. The position of the electron beam on the sample is controlled by scan coils situated above the objective lens. These coils allow the beam to be scanned over the surface of the sample. This beam scanning, as the name of the microscope suggests, enables information about a defined area on the sample to be collected. As a result of the electron-sample interaction, a number of signals are produced. These signals are then detected by appropriate detectors. Further, the specific surface area was measured using SHIBATA APP.SA-1100 at the nitrogen flow rate of 150 ml/min. The APP.SA method is based on the amount of nitrogen adsorbed on

IOP Conf. Series: Materials Science and Engineering **358** (2018) 012052 doi:10.1088/1757-899X/358/1/012052

the surface of solid particle. Measurement performed at 77.362 K with sample mass 0.2772 g. Thereafter, the sample of nano zeolite powder put in XRD instrument which connected with the computer to view the data. The XRD analysis is based on constructive interference of monochromatic X-rays and a crystalline sample: The X-rays are generated by a cathode ray tube, filtered to produce monochromatic radiation, collimated to concentrate, and directed toward the sample. The interaction of the incident rays with the sample produces constructive interference (and a diffracted ray) when conditions satisfy Bragg's Law ($n\lambda = 2d \sin \theta$). This law relates the wavelength of electromagnetic radiation to the diffraction angle and the lattice spacing in a crystalline sample. The XRD data from the computer analyzed with MATCH software to find the crystal's structure and the composition of mineral.

XRF (X-ray fluorescence) is a non-destructive analytical technique used to determine the elemental composition of materials. XRF analyzers determine the chemistry of a sample by measuring the fluorescent (or secondary) X-ray emitted from a sample when it is excited by a primary X-ray source. The sample of nano zeolite powder put in XRF instrument. The sample is irradiated with high energy X-rays from a controlled X-ray tube. When an atom in the sample is struck with an X-ray of sufficient energy (greater than the atom's K or L shell binding energy), an electron from one of the atom's inner orbital shells is dislodged. The atom regains stability, filling the vacancy left in the inner orbital shell with an electron from one of the atom's higher energy orbital shells. The electron drops to the lower energy state by releasing a fluorescent X-ray. The energy of this X-ray is equal to the specific difference in energy between two quantum states of the electron. The measurement of this energy is the basis of XRF analysis.

Finally, in order to investigate the properties changes achieved due to activation treatments, the functional groups analysis was attempted by a Fourier Transform Infrared FT-720 spectrometer (FTIR). Samples preparation in the FTIR test was done by the KBr pellets method. The functional groups absorbance of the activated nano zeolites were then compared to natural one and previous data published elsewhere.

3. Results and discussion

The observation results and discussion are out lined in this section. Discussion will be focused on the morphology, surface area and chemical contents for both natural and activated nano zeolites.

3.1. Morphology of zeolite

Characterization by SEM used to determine the morphology of natural and activated nano zeolites. Figure 2a is the SEM images with 10,000X enlargement of the natural nano zeolite.

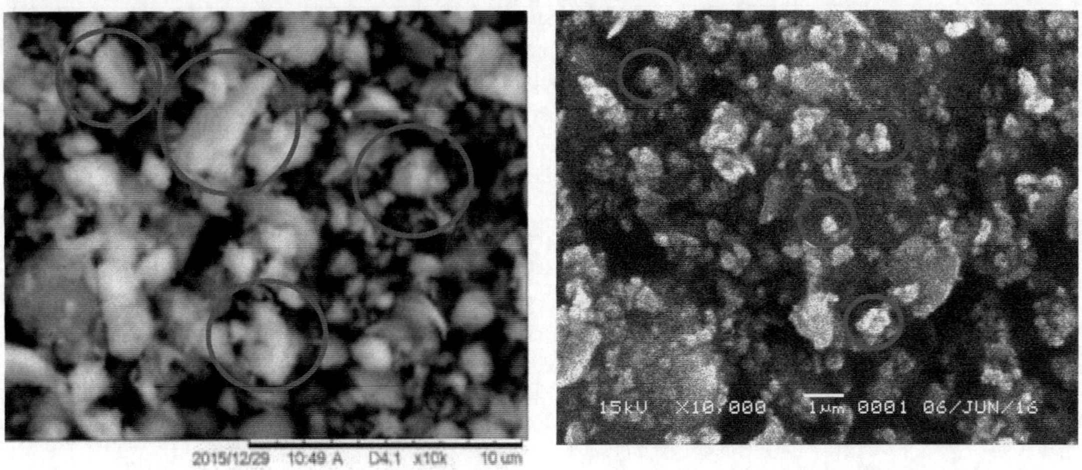

(a) SEM image of natural nano zeolite (b) SEM image of activated nano zeolite

Figure 2. SEM images of zeolite before and after activated.

Morphology of natural zeolite showed a flaky heterogeneous form, as indicated by the circle. Moreover, Figure 2b is the result of SEM image of activated nano zeolite at the same scale enlargement. In opposite, the morphology of nano zeolites after activation presented a grains-like form, as well as has been investigated by Noroozifar et al. (2014) and Mirzaei et al. (2016). They stated that the appearance of the nano zeolite SEM image has a drusy texture, i.e. it looks like a flake crystal with high micro porosity (as highlight by the circle).

3.2. Surface area

Further characterization of zeolite was done by BET to determine the surface area. The results show that the BET surface area of nano-pore of the activated nano zeolite surface area is 12.238 m^2/g, pore volume is 0.054 cm^3/g and pore size is 177.261 Å or 17.726 nm (Table 1). From the view point of surface area, pore volume and pore size the activated nano zeolite is not quite significant from the natural one. This phenomenon might prove the fact that both natural and activated zeolite was characterized in the nano particle sizes. Those BET data also showed that the nano-pore size of zeolites before and after activation can be classified as a meso-porous size. According to IUPAC, meso-porous diameter lies in the range of 2-50 nm. Nano particle sizes have a higher of ratio of surface area to pore volume compare to similar particles which has larger particle size. This makes the nano-particles more reactive in the adsorption behavior.

Table 1. Data of surface area

	Surface Area (m^2/g)	**Pore Volume** (cm^3/g)	**Pore Size** (Å)
Natural nano zeolite	10.469	0.052	200.300
Activated nano zeolite	12.238	0.054	177.261

3.3. Chemical contents

The result of XRD (Figure 3a) shows that the natural nano zeolite has a composition of mineral quartz (SiO_2), clinochlore [$(Mg, Fe)_6$ (Si, Al) $4O_{10}$ $(OH)_8$], calcite ($CaCO_3$), as shown by sharp intensity in the region 2θ = 26.65° (d = 3.34 Å), which is a characteristic of mineral quartz (SiO_2) and reflection with the greatest intensity 2θ = 29.43° (d = 3.03 Å) is a characteristic of mineral calcite ($CaCO_3$), and mineral clinochlore [$(Mg, Fe)_6$ (Si, Al) $4O_{10}$ $(OH)_8$] provides reflection with the greatest intensity 2θ = 18.76° (d = 4.7 Å). The XRD results also show the crystalline form is monoclinic. The result of XRD for activated nano zeolite (Figure 3b). It showed that the mineral content is dominated by mineral quartz (SiO_2), calcite ($CaCO_3$), and reduced mineral composition clinochlore [$(Mg, Fe)_6$ (Si, Al) $4O_{10}$ $(OH)_8$]. This is because the zeolite has been through the process of activation and heating at high temperatures (Ates and Hardacre, 2012).

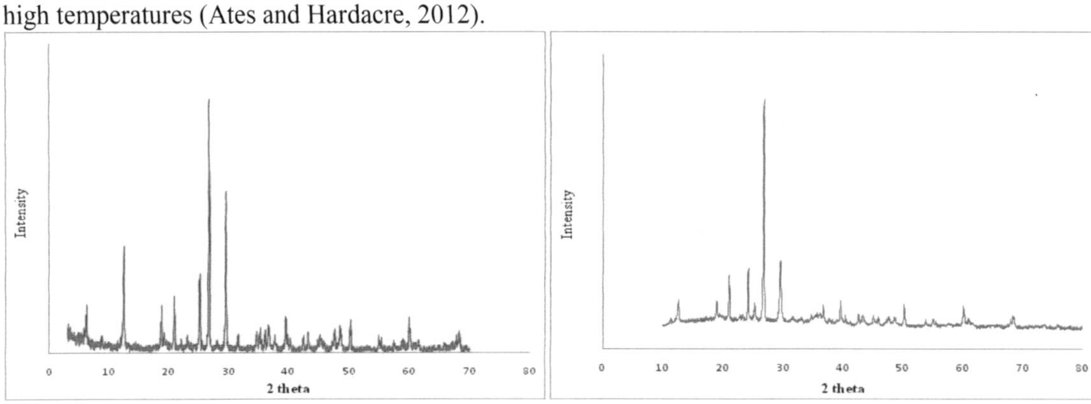

(a) XRD pattern of natural zeolite (b) XRD pattern of activated nano zeolite

Figure 3. XRD patters of zeolite before and after activated.

The characterization results using XRF (Table 2) shows the content of silica and alumina is the main composition of the zeolite. Other elements contain in the zeolite are magnesium, calcium and potassium. These elements can be exchanged with other ions, both with metal or non-metal ions. Whereas the rest elements are an impurities that occupied the zeolite pores. The impurities can be reduced by heating and chemical activation of zeolites such as by acid, base or salt treatment (Margeta et al., 2013). Ratio of Si/Al in the natural and activated nano zeolites are 3.99 and 3.87, respectively. It is clear that the ratio of Si/Al in the zeolite before and after activation did not change so much; it shows that the natural zeolite has been successfully activated without damaging the structure of the zeolite.

Table 2. Data of XRF characterization

Natural nano zeolite		Activated nano zeolite	
Compound	% Weight	Compound	% Weight
SiO_2	47.32	SiO_2	39.01
Al_2O_3	10.46	Al_2O_3	8.9
Fe_2O_3	8.87	Fe_2O_3	8.21
MgO	8.84	MgO	6.96
SO_3	0.16	SO_3	0.23
K_2O	0.45	K_2O	0.36
CaO	11.54	CaO	13.9
Na_2O	0.07	Na_2O	0.11
P_2O_5	0.161	P_2O_5	0.161
TiO_2	0.377	TiO_2	0.362
Mn_2O_3	0.163	Mn_2O_3	0.146
Cr_2O_3	0.078	Cr_2O_3	0.135

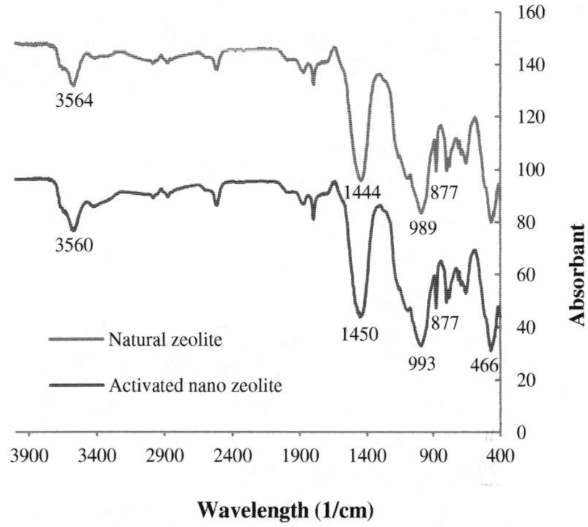

Figure 4. FTIR patterns of natural and activated nano zeolites.

The characterization results using FTIR (Figure 4) showed that the natural zeolite has an early peak at 3564 cm-1 and nano zeolites after activation decreased to 3560 cm^{-1}, this represents the absorption peak of OH group is in the range 3700-1600 (Elaipoulus et al ., 2010). Decline in the peak of O-H group in the nano zeolites after activation is due to loss of water molecules inside the zeolite. This proves that the nano zeolite after activation has lower impurities than the natural zeolite. In addition, there is the peak of 877-993 cm^{-1} which is the asymmetric stretching vibration of Si-O and Al-O; interpreted the bond absorption at the main structure unit of zeolite (Barczyk et al., 2014). Therefore, the carbonate ion in the natural zeolite appears at the peak of 1444 cm^{-1} and for the activated zeolite at 1450 cm^{-1}. Wave number carbonate lies in the range of 1450-1410 cm^{-1} (Coates, 2000). The identified wave numbers recorded in this investigation compared to previous studies shown in Table 3. The results of FTIR measurements show that the functional groups contained in the sample exhibits the character of zeolite framework structure. Both zeolites, before and after activation, provide almost the same wave numbers. This may imply that the activation of zeolites (physically and chemically) did not change the basic framework of the zeolite structure group.

Table 3. FTIR spectra readings on zeolites

Functional group	Bonded	Wavelength number (cm^{-1})		
		Natural zeolite	Activated nano zeolite	Reference
Un-bonded O-H	O-H	3564	3560	3800-1600***
Carbonate ion	Ca-O	1444	1450	1450-1410**
Asymmetric stretching vibration of Si-O and Al-O	Si-O-Si and Si-O-Al	877 and 989	877 and 993	1300-800*
Bending vibration of Si-O or Al-O	Si-O or Al-O	-	466	466-439***

Source: *Elaipoulus et al. (2010); **Coates (2000); ***Barczyk et al. (2014)

4. Conclusions

Based on the results and discussions, some conclusions can be written as following:

1. Morphology of the activated nano zeolite significantly changed due to an activation process.
2. Surface area, chemical content and the basic framework of the zeolite were almost un-changed after treated by the proposed process.
3. Impurities contained in the zeolite can be reduced through activation process (physically and/or chemically) and calcination.

References

[1] Ates A and Hardacre C 2012 *J. Coll. Interface Sci.* **372** 130
[2] Barczyk K, Mozgawa, W and Król M 2014 *Spect. Acta Part A: Molecular and Biomolecular Spect.* **133** 876
[3] Coates J 2000 Encyclopedia of Analytical Chemistry ed R A Meyers (New York: John Wiley & Sons) pp 10815–10837
[4] Elaiopoulos K, Perraki T and Grigoropoulou E 2010 Microporous and Mesoporous Materials **134** 29
[5] Erdem E, Karapinar N and Donat R 2004 J. Coll. Interface Sci. **280** 309
[6] Ibrahimi M M and Sayyadi A S 2015 Int. J of Geology Agric. Env. Sci. **3** 1
[7] Izci E and Izci A 2007 Turkish J. Chemistry **31** 523
[8] Lestari D Y 2010 Prosiding Seminar Nasional Kimia dan Pend. Kimia 2010 (Yogyakarta) p 1 (in Indonesia)
[9] Margeta K, Logar N Z, Šiljeg M and Farkas A 2013 *Natural zeolites in water treatment –how effective is their use (Water Treatment)* ed W Elshorbagy and R K Chowdhury (Rijeka: InTech) chapter 5 pp 81-112
[10] Mirzaei N, Hadi M, Gholami M, Fard R F and Aminabad M S 2016 *J. the Taiwan Inst. Chem. Eng.* **59** 186
[11] Motsi T, Rowson N A and Simmons M J H 2009 *Int. J. Mineral Process.* **92** 42
[12] Noroozifar M, Khorasani-Motlagh M and Naderpour H 2014 *Microporous and Mesoporous Materials* **197** 101
[13] Utubira Y, Wijaya K, Triyono and Sugiharto E 2006 *Indo. J. Chem.* **6** 231
[14] Wang S and Peng Y 2010 *Chem. Eng. J.* **156** 11

IOP Conf. Series: Materials Science and Engineering **358** (2018) 012053 doi:10.1088/1757-899X/358/1/012053

Characterization of Vinyl Ester Composites Filled with Carbonized *Jatropha* seed shell: effect of accelerated weathering

N A Sri Aprilia[1]*, H P S Abdul Khalil[2], Amri Amin[3], Cut Meurah Rosnelly[1], Ummi Fathanah[1] and Mariana[1]

[1]Chemical Engineering Department, Engineering Faculty, University of Syiah Kuala, Banda Aceh, Indonesia
[2]School of Industrial Technology, Universiti Sains Malaysia, Penang, Malaysia
[3]Engineering Faculty, University of Abulyatama, Aceh Besar, Indonesia

sriaprilia@unsyiah.ac.id

Abstract. The effect of accelerated weathering test of carbonized *jatropha* seed shell filled vinyl ester biocomposites was investigated. In this study, four loading of carbonized *jatropha* seed shell and one without loading of vinyl ester biocomposites were used. The samples exposure at several circles time in QUV chamber. The durability of vinyl ester biocomposites filled carbonized *jatropha* seed shell changes in mechanical properties and weight loss during exposure in UV and condensation. The tensile test and flexural indicated decrease with increasing of carbonized *jatropha* seed shell loading. The SEM fracture surface of biocomposites looks rough and some carbonized out of the matrix.

1. Introduction

Although inorganic fillers currently dominated the polymer industry, biomass filler have become more accepted in recent years and have made significant in roads into specialty market. This material applications continues to increase as in the construction, transportation, industrial and consumer applications. In other words, this material can be improved strength and resistance to environmental [1]. Loading of biomass in material composites will effect on the strengthening and flexibility of the materials [2]. Durability of environment is an important aspect of overall performance of material biocomposites used in both interior and exterior application where sunlight and condensation could have detrimental effects. The fact that interaction between biomass filler and matrix is not only important for the the properties of the composites but also for conserving the integrity of the biocomposites by maximizing filler-matrix adhesion [3]. Accelerated weathering for composites are the way of exposure testing that attempt to rapidly occur the changes material composites on outdoor and indoor exposure equipment [4]. Accelerated weathering for biocomposites are necessary to ensure the materials and products meet their expected functionality and lifetime. The strength and flexibility of the composite materials useful for applications outdoors. Because of they will be influence on the weather and the material composites to be easily damaged [5]. The effect of lighting from an artificial light source and this may be attended by water spraying and circular experimental conditions, because of changes in temperature, relative humidity, and irradiation energy. The degradation processes of materials composites by exposing to artificially intensive weathering condition, allowing a fast

Content from this work may be used under the terms of the Creative Commons Attribution 3.0 licence. Any further distribution of this work must maintain attribution to the author(s) and the title of the work, journal citation and DOI.
Published under licence by IOP Publishing Ltd

evaluation of the service durability of material composites in a rationally short period of time. In this study, accelerated weathering for carbonized *jatropha* seed shell filled vinyl ester composites has been investigated durability of biocomposites with the effect of UV exposure and condensation. Accelerated weathering for carbonized *jatropha* seed shell filled vinyl ester has been studied for carbonized particle loading. The analysis for this study includes change in physical and mechanical properties and morphology of vinyl ester composites.

2. Experimental

Biomass filler was prepared from *jatropha* seed shell obtained from Aceh Province, Indonesia. The size of biomass filler after carbonization is > 300 mesh. Vinyl ester resin with 42% styrene monomer content, methyl ethyl ketone peroxide (MEKP), and cobalt napthenate were purchase from Zarm Scientific & Supplies Sdn. Bhd. Malaysia.

2.1. Accelerated weathering processes

The weathering accelerated chamber model MEGASOL was used in this study (James H. Heal & Co. Ltd, Halifax England) with burner life 1500 hours, megasol light & weathering fastness tester. Vinyl ester composites was prepared followed the previous method [2]. The composites were in size 160 x 20 mm samples for flexural test and 120 x 15 mm for tensile test were exposed in UV cabinet for 96, 240, 360 and 720 h using a cycle of 2-h UV exposure followed by 30 minutes condensation at 29°C and humidity 32% RH, which is cycle recommended in ASTM D 4329-1999. After exposure, the panels were allowed to equilibrate back to ambient condition and reweighed. Analysis for flexural, tensile and morphology follow the mechanical and morphology analysis for vinyl ester composites below.

2.2. Mechanical Properties

Mechanical properties include tensile strength and flexural properties of biocomposites were performed using the INSTRON 5582 Universal Testing Machine (USA). The size of biocomposites sample by dimensions 120 mm x 20 mm x 7 mm (ASTM D638). Tensile properties including tensile strength, tensile modulus were acquired from the recorded data. Five specimens were tested and the average value was tabulated. ASTM D790 standard method were followed for flexural properties including flexural strength and tensile modulus and crosshead speed was 2 mm/min [1].

2.3. Scanning electron microscopy

Scanning electron microscopy (SEM) (model EVO MA10, Carl-ZEISS SMT, Germany) was used for tensile fracture surface morphology of biocomposites. The SEM accelerating voltage is 20 KV. To avoid electrostatic charge, the specimens of fractured surface were mounted on aluminum stubs and sputter coated with thin layer of palladium and gold.

3. Results and discussion

3.1 Weight change before and after weathering

Figure 1 shows the weight changes of carbonized *jatropha* seed shell before and after accelerated weathering. All of the composites would have small changes in their weight after exposure for 96 hours. The higher weight loss after exposure for 360 hours in weathering chamber were composites with carbonized particle loading 40%, the change is about 0.25%. After 360 hours exposure to UV and condensation condition, all carbonized vinyl ester composites had lost weight below 1%. The temperature of the surface of the specimens exposed, increased step by step from normal levels to be used when performing the test temperature by increasing weathering (room temperature), this cause weight loss from the composites, although UV exposure followed condensation for the composite [6].

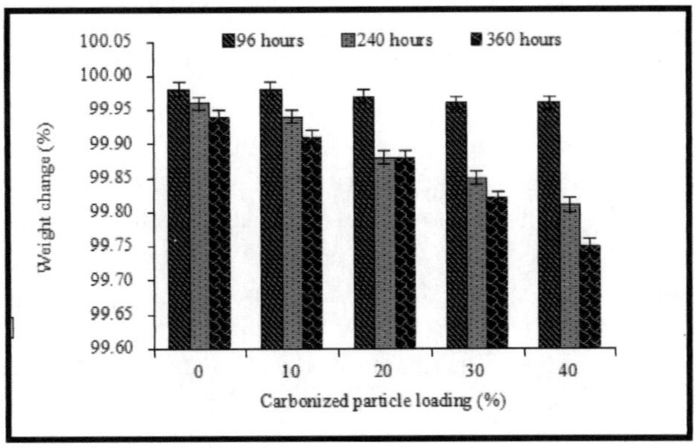

Figure 1. Weight change of carbonized *jatropha* seed shell filled vinyl ester composite

3.2 Tensile properties for weathering test

The change of tensile strength for carbonized *jatropha* seed shell filled in vinyl ester biocomposites compared to neat vinyl ester were illustrated in Figure 2a. Neat vinyl ester increase on tensile strength after weathering for 96 hours but slightly change when exposed for 240 hours and 360 hours. But, for carbonized *jatropha* seed shell filled vinyl ester biocomposites, the tensile strength slightly decreases after weathering is accelerated.

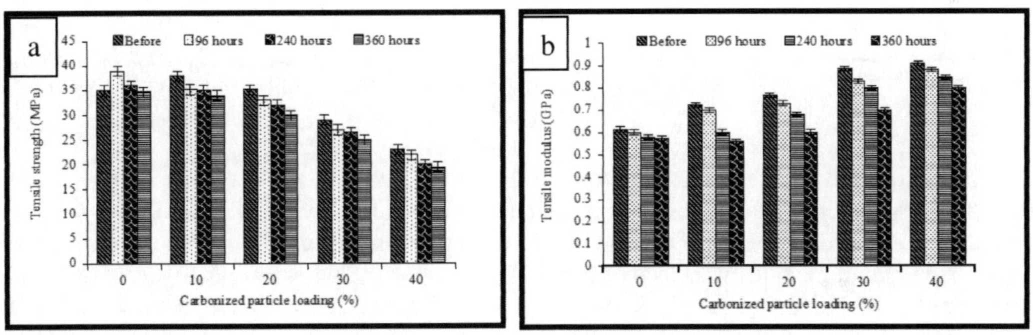

Figure 2. Change of tensile strength (a) and Tensile modulus (b) before and after accelerated weathering

The change of tensile strength of the carbonized *jatropha* seed shell filled vinyl ester composites are about 0.01 to 0.15% at carbonized particle loading from 0 to 40% respectively to vinyl ester matrix. The higher carbonized particle loading, the larger the losses in tensile strength during exposure to weathering. With increasing weathering time, the tensile strength of carbonized *jatropha* seed shell filled vinyl ester composites continuously lost their strength. It is decreases because of embrittlement of the vinyl ester matrix and could be because of the degradation of biofiller-matrix interfacial bonds [7]. Figure 2b show the tensile modulus of neat vinyl ester composite and carbonized *jatropha* seed shell filled vinyl ester composites. Tensile modulus also decreases after UV and condentation condition for all vinyl ester composites. Losses in tensile modulus 0.06% for neat vinyl ester and 0.12 to 0.22% for carbonized particle loading in the vinyl ester matrix composites with increasing weathering time. This suggests that the decrease in molecular weight vinyl ester matrix-chain reaction is reduced due to the scission induced by photo-oxidation from UV radiation [8].

Karbhari et al. (2003) [9] implied that degradation on the surface components polymeric which have been shown to affect mechanical properties are disproportionate; as deficiency resulting from ultraviolet radiation can be compounded by the action of temperature, freeze-thaw, wind-borne abrasives, moisture, and other environmental components. However, with UV exposure, carbonized *jatropha* seed shell filled in vinyl ester matrix can damage, e.g. tensile and flexural strength. According to Maxwell et al. (2005) [10], UV radiation is known to degrade the polymer materials, although only the outer layer is likely to be affected. Signs of photo-degradation include embrittlement (cracking of the surface), changes in color, and the loss of transparency. These factors caused the decrease in mechanical properties of materials.

3.3 Flexural properties for weathering test

Figure 3 show the change of flexural properties of carbonized *jatropha* seed shell filled vinyl ester composites. Flexural strength (Figure 3a) of vinyl ester composites, decrease after exposure to UV radiation and condensation with water spray with addition of weathering time. The losses of flexural strength is 0.02% for neat vinyl ester and 0.07 to 0.27% for carbonized particle loading in vinyl ester composites. The loss in strength was likely due to moisture penetration into the matrix, which degraded the filler–polymer interface. This decreases the stress transfer efficiency from matrix [11]. At high temperature, the rate of moisture absorption and accelerated degradation properties of materials were increase [10] and [4].

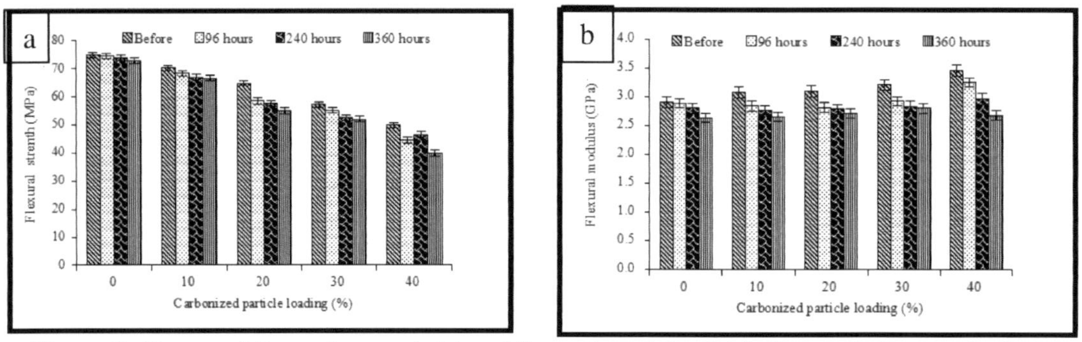

Figure 3. Change of flexural strength (a) and flexural modulus (b) before and after weathering accelerated

After exposure to UV radiation for 2h followed by condensation for 30 minute, the flexural modulus (Figure 3b) of vinyl ester composite will decrease. The degradation was most hard for composites exposed to condensation and cyclic UV radiation. After 360 hours of cyclic exposure the transverse flexural modulus was decreased by 0.09% for neat vinyl ester composite, 0.12 to 0.29% for carbonized particle loading in vinyl ester composites. The influence of temperature on the radiation aging effect of materials investigated for several polymer materials for technical applications and for being used as reference material in the accelerated weathering tests. The higher carbonized particle loading after exposure to weathering test makes composites ductility more than small carbonized particle loading. The mechanical properties decrease is due to the embrittlement of the polymeric matrix and to degradation of fiber-matrix interfacial bonds [7].

3.4. Scanning electron microscopy

Figure 4 and 5 show the SEM image of neat vinyl ester and carbonized *jatropha* seed shell filled vinyl ester composites before and after exposure to accelerated weathering chamber. From the Figure 4(a) and 4(b), the SEM image of vinyl ester composite exhibits a regular and smooth fracture surface with no significant defects. After exposure for 360 hours, the surface looks rough and smooth less (Figure 4b and Figure 5b). Degradation of composite fracture surface was less distinguishable after UV radiation and condensation to 360 hours and the carbonized filler already defects on vinyl ester matrix (Figure 5b). This could mean that it is still in the early stages of degradation by photo-oxidation of the accelerated. Accelerated weathering causes the formation of surface defects on matrix of vinyl ester, and some white particles can be observed due to the formation of carbonized particle out of the matrix [12], which allows further degradation (including biodegradation) in the environment [7]. According to Al-Shabanat (2011) [13], it is obvious that there are cracks and fissures and micro can be seen on the surface of a composite of degraded greatly increased after exposure and some of the particles can be observed which might be caused by a large number of fragments of the fracture surface [14]. This entire surface damage from UV exposure during spontaneous as a result the surface layer of composites, which leads to shrinkage [5]. Although cracks on surface are clear in all composites after being accelerated weathered for 360 hours, the severity of cracks varies from one another where the damage was done by weathering on matrix vinyl ester contained carbonized particle filler of *jatropha* seed shell.

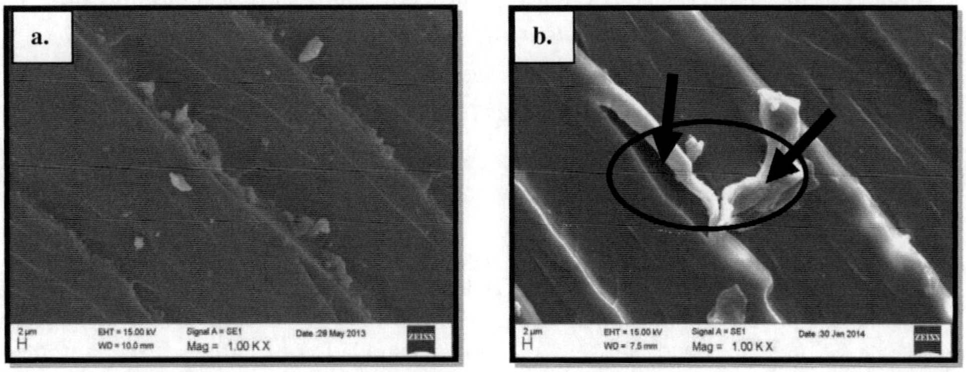

Figure 4. SEM image of neat vinyl ester composite:(a) before and (b) after exposure 360 hours

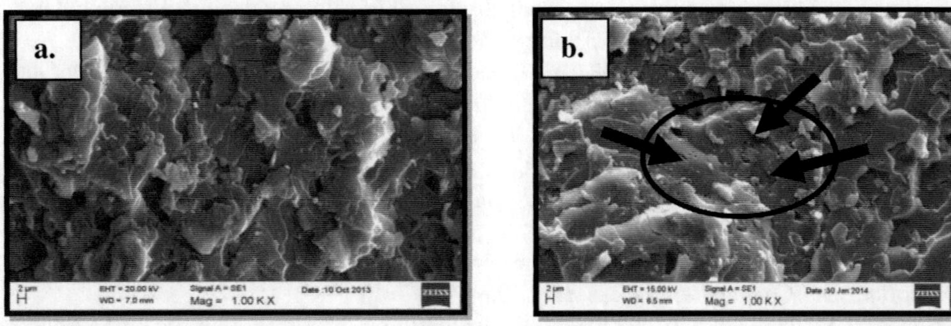

Figure 5. SEM image carbonized *jatropha* seed shell filled vinyl ester composite: (a) before and (b) after exposure 360 hours

4. Conclusions

The weight change of carbonized *jatropha* seed shell occurred after exposure in weathering chamber. Losing weight is happening this is because surface temperature specimens affected increases step by step than the normal rate used when doing test weathering by increasing temperature. The mechanical properties of carbonized *jatropha* seed shell filled vinyl ester biocomposites decrease when subjected in UV light and condensation by exposure time. When ultraviolet radiation, the biocomposites can compound by the action of temperature, moisture, wind-borne abrasives, freeze-thaw, and other environmental components. The higher carbonized particles loading after exposure in weathering test make composites ductility. The mechanical properties decrease is due to embrittlement of the polymeric matrix and to degradation of fiber-matrix interfacial bonds. The degradation of the composite fracture surface was less distinguishable after UV radiation and condensation. Accelerated weathering causes the formation of surface defects on matrix of vinyl ester, some white particles being observed due to formation of carbonized particle out of the matrix, which allows further degradation (including biodegradation) in the environment.

References

[1] Aprilia N A S, Hossain Md S, Abdullah C K., Abdul Khalil H P S, Rosamah E, Dungani R, Davoudpour Y, and Zaidul S M 2015 Environmental Durability of Vinyl Ester Composites Filled with Carbonized Jatropha Seed Shell *Bioresources* **10** (2) p 2350-2359

[2] Aprilia N A S, Abdul Khalil H P S, Bhat A H, Dungani R and Hossain Md S 2014 Exploring Material Properties of Vinyl Ester Biocomposites Filled Carbonized Jatropha Seed Shell *Bioresources* **9** (3) p 4888-4898

[3] Peng Y, Liu R, and Cao J 2015 Characterization of surface chemistry and crystallization behavior ofpolypropylene composites reinforced with wood flour, cellulose, and lignin during accelerated weathering *Applied Surface Science* **332 p** 253–259

[4] Shimizu K., Tokuta Y, Oishi A, Kuriyama T, and Kunioka M 2016 Weatherability of Polypropylene by Accelerated Weathering Tests and Outdoor Exposure Tests in Japan, Hindawi Publishing Corporation *Journal of Polymers* p 1-14.

[5] Yadong Lv, Huang Y, Yang J, Kong M, Yang H, Zhao J, and Li G 2015 Outdoor and accelerated laboratory weathering of polypropylene: A comparison and correlation study, Polymer Degradation and Stability **112** p 145-159.

[6] Mehta G, Mohanty A.K, Drzal L T, Kamdem D P, and Misra M 2006 Effect of Accelerated Weathering on Biocomposites Processed by SMC and Compression Molding *J Polym Environ* **14** p 359–368

[7] Darie R N, Ruxandra B, and Carmen A T 2013 Influence of accelerated weathering on the properties of polypropylene/polylactic acid-eucalyptus wood composites *International Journal of Polymer Anal. Charact.* **18** p 315–327

[8] Kumar B G, Singh R P, and Nakamura T 2002 Degradation of carbon fiber-reinforced epoxy composites by ultraviolet radiation and condensation *Journal of Composite Material* **36** (24)

[9] Karbhari V M, Chin J W, Hunston, Benmokrane, Juska T, Morgan R, Lesko J, Sorathia, and Reynaud D 2003 Durability gap analysis for fiber-reinforced polymer composites in civil infrastructure, *Journal of Composites for Construction* **7** p 238-247.

[10] Maxwell A S, Broughton W R, Dean G, and Sims G D 2005 Review of accelerated ageing methods and lifetime prediction techniques for polymeric materials, *NPL Report DEPC MPR 016*, National Physical Laboratory, United Kingdom.

[11] Stark N M, and Matuana L M 2007 Characterization of weathered wood plastic composite surfaces using FTIR spectroscopy, contactangle, and XPS *Polymer Degradation and Stability* **92** p 1883-1890

[12] Turku I, and Kärki T 2016 Accelerated weathering of fire-retarded wood–polypropylene composites *Composites: Part A* **81** p 305–312

[13] Al-Shabanat M 2011 Study of the Effect of Weathering in Natural Environment on Polypropylene and Its Composites: Morphological and Mechanical Properties *International Journal of Chemistry* **3** (1).

[14] Yang T-C, Noguchi T, Isshiki M, and Jyh-Horng W 2014 Effect of titanium dioxide on chemical and molecular changes in PVC sidings during QUV accelerated weathering *Polymer Degradation and Stability* **104** p 33-39.

IOP Conf. Series: Materials Science and Engineering **358** (2018) 012054 doi:10.1088/1757-899X/358/1/012054

Correlation Study of PVDF Membrane Morphology with Protein Adsorption: Quantitative Analysis by FTIR/ATR Technique

N Ideris[1], A L Ahmad[2], B S Ooi[2] and S C Low[2]

[1]Faculty of Chemical Engineering, Universiti Teknologi MARA, 40450, Shah Alam, Selangor, Malaysia
[2]School of Chemical Engineering, Universiti Sains Malaysia, 14300, Nibong Tebal, Pulau Pinang, Malaysia

norhidayah7229@salam.uitm.edu.my

Abstract. Microporous PVDF membranes were used as protein capture matrices in immunoassays. Because the most common labels in immunoassays were detected based on the colour change, an understanding of how protein concentration varies on different PVDF surfaces was needed. Herein, the correlation between the membrane pore size and protein adsorption was systematically investigated. Five different PVDF membrane morphologies were prepared and FTIR/ATR was employed to accurately quantify the surface protein concentration on membranes with small pore sizes. SigmaPlot® was used to find a suitable curve fit for protein adsorption and membrane pore size, with a high correlation coefficient, R^2, of 0.9971.

1. Introduction

Adsorption of protein molecules on a sorbent surface is an important phenomenon in many fields of technology, particularly in biomedical and pharmaceutical fields. Many studies have shown how this process is greatly influenced by the characteristics of the specific protein, sorbent surface, and physicochemical environment [1]. Several reviews on protein adsorption on artificial membrane surfaces have been published, by focussing either on protein-repellent [2, 3] or protein-adsorptive [4, 5] membranes for applications such as ultrafiltration and anion-exchange chromatographic separation processes. One such application is the biochemical immunoassay, where a higher amount of protein adsorption with strong binding is necessary for achieving accurate results. These assays are generally classified as homogeneous or heterogeneous. In the latter, one protein constituent is immobilized on a solid sorbent surface, while the other components are delivered via the solution phase. The output signal of this type of assay usually consists of an observable colour change such as that observed in the widely used enzyme-linked immunosorbent assay (ELISA) [6].

A sorbent surface used for protein adsorption in immunodetection is polymeric membrane, commonly made of nitrocellulose or polyvinylidene fluoride (PVDF) membrane [7]. In the present study, microporous flat PVDF sheets were synthesized with micropores in the range 0.2–1.12 μm. An accurate analysis of the dependence of protein concentration adsorbed onto these membrane surfaces on the size of the micropores was necessary. Previous studies to quantify protein adsorption have been mainly based on the concept of a linear relationship between protein concentration and colour intensity [8]. However, such methods do not provide an actual quantity of protein, they only allow for a comparative analysis [9].

Content from this work may be used under the terms of the Creative Commons Attribution 3.0 licence. Any further distribution of this work must maintain attribution to the author(s) and the title of the work, journal citation and DOI.
Published under licence by IOP Publishing Ltd

Other approaches have also been employed for measuring protein adsorption, including evanescent wave techniques such as Fourier transform infrared/attenuated total reflectance (FTIR/ATR) [9-11], surface plasmon resonance (SPR) [12, 13], and total internal reflection fluorescence (TIRF) [14, 15]. Among these techniques, TIRF provides a higher surface sensitivity and can measure competitive protein adsorption. However, the proteins need to be labelled with a fluorophore such as fluorescein isothiocyanate (FITC), which increases the number of steps and cost to the process. In addition, the difficulty of waste disposal is often complicated and expensive. FTIR/ATR and SPR techniques have the capability to measure protein surface concentration. Furthermore, FTIR/ATR provides additional information about the structure and conformation of protein molecules [1].

In the present study, FTIR/ATR was employed for its advantages in the analysis of thin films and homogenous solutions. This technique was first proposed by Harrick and de Pru in 1966 [16]. It is particularly sensitive to light-absorbent species near the internal reflection element (IRE), making it a potential method for quantitative measurement of adsorbed species. The evanescent wave, which picks up information from a surface in ATR spectrometry, penetrates the surface to a depth of a few micrometers; this depth is known as the depth of penetration, d_p. The ATR measurement has a constant d_p as long as the wavelength, incident angle, and refractive index of the IRE and sample remain the same [17]. Because the thickness of the adsorbed protein layer is considered to be much smaller than d_p, it is expected that it will experience an almost identical evanescent wave intensity, regardless of the pore size on the membrane surface [18, 19]. Multi-bounce FTIR/ATR spectroscopy has better sensitivity compared to the single-bounce spectroscopy; however, for highly absorbent materials such as polymers, the single-bounce spectroscopy is the most suitable setup for the ATR measurement.

Because the output signal of the immunoassay analysis is identified by an observable colour change of the sorbent surface, this study focussed on achieving a greater understanding of the correlation between the PVDF membrane (sorbent surface) characteristics and protein adsorption. The knowledge gained from this study will contribute to the development of higher efficiency PVDF membranes for biomedical applications.

2. Materials and methods

2.1 Materials

Bovine serum albumin (BSA) and deuterium oxide, D_2O, with an isotopic purity of 99.9 atom %D was obtained from Sigma-Aldrich (St Louis, MO, USA). Potassium phosphate (monobasic and dibasic anhydrous) were obtained from Merck (Germany) and were deuterated in D_2O for the preparation of a phosphate buffer solution. The buffer (0.05 M) was maintained at pH 7 in order to replicate the physiological conditions for immunoassays. All chemicals were used without further purification.

2.2 Protein Adsorption on PVDF Membrane

A symmetrical, microporous PVDF membrane was prepared via non-solvent-induced phase inversion in accordance with a previous published work [20].The different PVDF membranes were cut into 1cm × 1cm sections, and the upper surface was marked with a pencil. The samples were then soaked in ethanol for 15 s at room temperature, and placed on pre-wetted filter paper. A 3 μL aliquot of protein (BSA, bulk concentration, C_b = 3 mg/mL) in the phosphate buffer (pD 7 in D_2O) was applied to the membrane surface and exposed to air for 30 minutes to allow the adsorption process to take place. Then, the samples were washed with the phosphate buffer to remove any unbound protein molecules within the PVDF membrane matrix. Finally, the membrane was left for air-dried for another 30 min before carried out the adsorption measurement using ATR-FTIR [21]. At least three samples from each membrane were used to determine the protein adsorption in order to confirm the reproducibility of the experiment.

2.3 Membrane–Protein Adsorption Correlation

The obtained ATR spectra were quantitatively analysed using the theory of Harrick and de Pru [16, 22]. An FTIR spectrometer (OMNIC iS10 Nicolet, Thermo Scientific, USA) was equipped with a Smart iTR diamond ATR accessory, which was used with an incident angle of 42° and single-bounce optics.The final protein adsorption (Γ) is expressed as followed, in the unit of mol/cm^2,

$$\Gamma = \frac{\frac{A}{N} - \varepsilon C_b d_e}{1000\,\varepsilon\left(2\frac{d_e}{d_p}\right)} \tag{1}$$

where, Γ is the protein adsorption, A is the area of absorbance, N is the total number of internal reflections, ε is the absorption coefficient,C_b is the bulk concentration, d_e is the effective thickness and d_p is the depth of penetration.The absorption coefficient (ε) of the protein solution was found to be 106.342×10^3 cm/g. The ε value was subsequently used in the analysis of the ATR spectra, as described above.

In this study, PVDF membranes with mean pore sizes; 0.28 μm, 0.37 μm, 0.55 μm, 0.72 μm and 1.12 μm were used to determine membrane–protein adsorption correlation. All the membrane samples possessed a high porous structure with porosity value of more than ~60%. Based on the morphology characterisation, five highly porous membranes with different pore structure were selected and the protein measurement adsorption was carried out as described previously. Then, the correlation between protein adsorption and membrane pore structure was further explained and evaluated using the SigmaPlot® software, version 12.3.

3. Results and discussion

The membrane morphology must be able to capture and retain various proteins for its application in the biomedical field, such as in protein assay tests. Generally, the quantity of protein adsorbed onto the membrane surface is directly related to the surface pore structure and porosity of the membrane. In the present study, five different membrane morphologies comprised of different surface pores, D_p were selected and examined to develop a correlation between membrane morphologies and adsorbed protein.

The quantitation of the protein adsorbed on PVDF membranes with different morphology was analysed by the FTIR/ATR approach. In this method, the amount of the protein adsorbed was calculated based on the area of the absorbance region of interest. The spectral range selected was that corresponding to the amide I′ region (1705–1595 cm^{-1}) because of the structure of the protein backbone [22]. Thus, the protein was prepared in deuterated water to avoid the strong peaks at 3360 cm^{-1}(H–O stretching band) and 1640 cm^{-1}(H–O–H bending vibration), which would interfere with the peak of interest. Figure 1 (left) shows the ATR spectra of all the membrane samples before the adsorption of the protein. As can be seen, the spectra exhibit almost identical trends across the entire wavelength range, with no peaks evident at from 1500 to 1700 cm^{-1}. After protein adsorption measurement, it was proved that all the membranes were able to retain certain amount of protein on membrane surfaces as the existence of peak at ~1653 cm^{-1}. It can be seen that the highest peak intensity was observed for the membrane with the smallest pore size with average peak intensity of 0.0140.

On increasing the pore size, the peak intensity reduced to 0.0136 and 0.0120 at 0.37 μm and 0.55 μm respectively, and becoming almost the same at 0.75 μm and 1.123 μm with peak intensity of 0.0116 and 0.0114 respectively. The area of absorbance or area under the curve within the range 1705–1595 cm^{-1} (amide region) was then used in Equation 1 for protein surface concentration measurement using the C_b value of 3.0 mg/mL. The ε value was obtained from the absorbance measurements using the Beer's law. As the peak of interest was at approximately 1653 cm^{-1}, the value of d_p was calculated at λ = 1/1653 cm^{-1}, θ= 42°, n_1= 2.4 (IRE surface), and n_2= 1.42, with the assumption that protein adsorption had a negligible effect on the refractive index of the PVDF membrane (n_2). The refractive index of the IRE surface (diamond) and PVDF membrane polymer were obtained from Thermo Scientific and Solvay Solexis, respectively. A value of 1.284 μm was obtained for d_p, which was then used to calculate d_e.

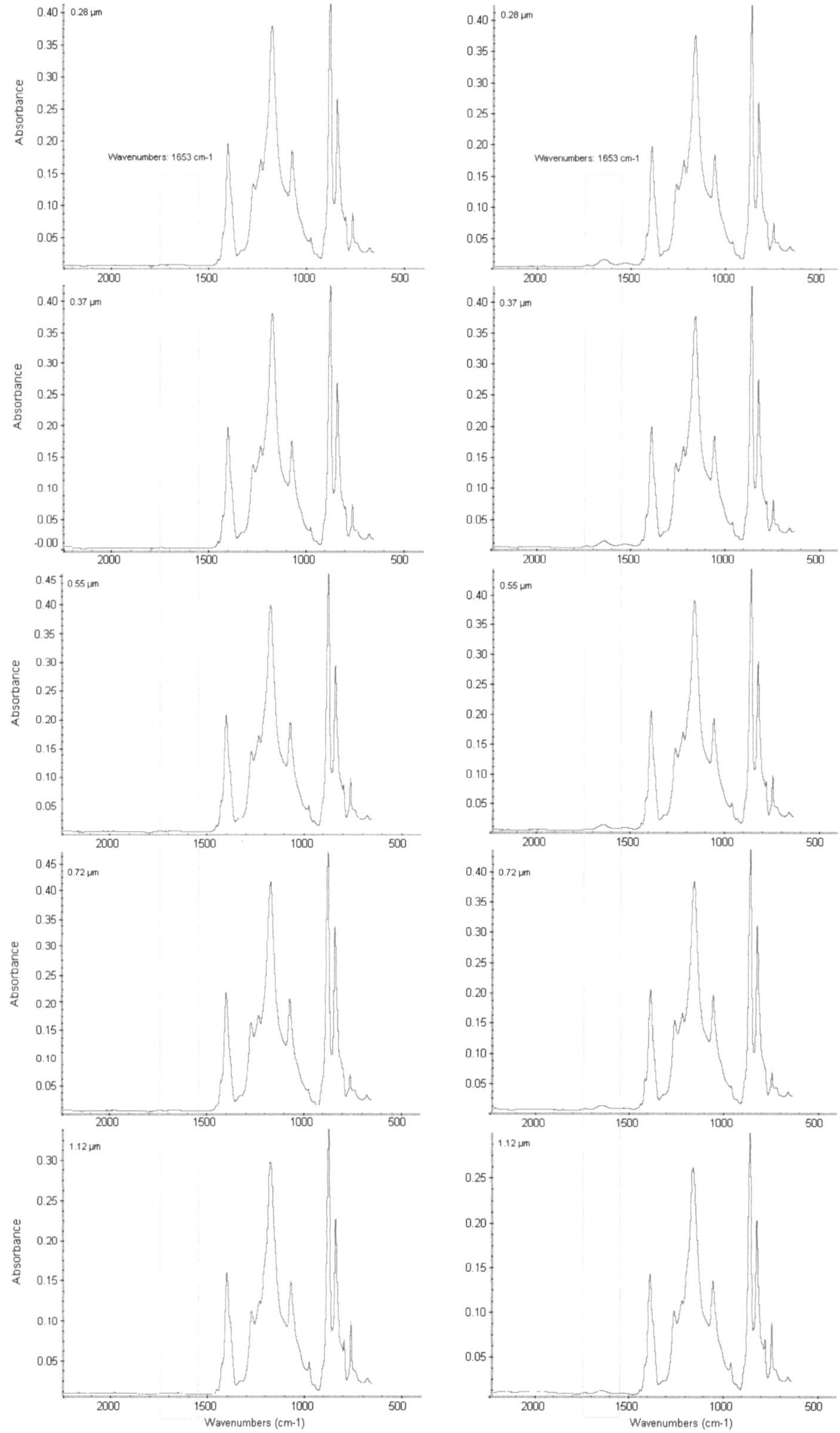

Figure 1. ATR peak intensity for PVDF membrane samples before (left) and after (right) protein adsorption

The correlation between the proteins adsorbed on different membrane surfaces with various pore sizes was next evaluated. As shown in Figure 2, the membranes with lower pore sizes retained more protein at an initial applied protein concentration of 3.0 mg/ml and the amount of protein was gradually decreased as the pore size is increased. It is also shows that the correlation between pore size and protein adsorption was non-linear. Variation of protein adsorption was observed to be less significant for smaller values of pore sized which is from 0.28 μm to 0.37 μm. However, as the pore size increased to 0.55 μm, the protein adsorption decreased until it reached almost a plateau at 1.123 μm. Thus, in order to develop a correlation between these two factors, various expressions were examined in order to find a suitable curve fit for the values of protein adsorption (Γ) and pore size diameter (D_p). SigmaPlot® version 12.3 was used to develop a curve fit for the function of pore size diameter (D_p) and protein adsorption (Γ). Equation 2 showed an acceptable correlation coefficient with a high value of R^2 (0.9971)

$$\Gamma = 2.5995x10^{-9} + (6.452x10^{-10})/[1+10^{\wedge}(0.4622 - D_p)(-6.1141)] \quad (2)$$

Because of the same protein adsorption method was used, these findings were highly dependent on the membrane pore size. As the amount of the protein adsorbed was found to be related to surface area availability, an appropriate combination of small pores and high porosity is required for obtaining microporous membranes with a high degree of interconnectivity. Generally, the membranes with smaller pore size at equal porosity would maximise the internal surface area for interactions between the membrane surface and the protein. On smaller Dp, the immobilised protein was able to bind and diffuse firmly and horizontally within the smaller pore matrix because more interconnecting polymer structures were formed within the membrane. In addition, smaller Dp allowed a longer time for the membrane–protein interaction to took place as the diffusion resistance of the protein solution is much higher in smaller pore size [8].

As previously mentioned before, the protein adsorption was decreased notably at 0.55 μm and becomes lowest at Dp of 1.123 μm. This is mainly due to the incapability of the protein to tightly immobilize onto the membrane. Bigger pore size with lower interconnecting polymer matrix will reduce affinity forces of membrane-protein interaction and cause the protein to detach during the washing process. Besides, lower interconnecting matrix will cause the deposited protein solution to diffuse vertically due to the gravity effect [23]. For this reason, the resulting amount of protein adsorption was low as compared to the membrane with smaller pore size and not much difference in amount of protein adsorption, Γ can be observed even when membrane pore size, Dp was further increased.

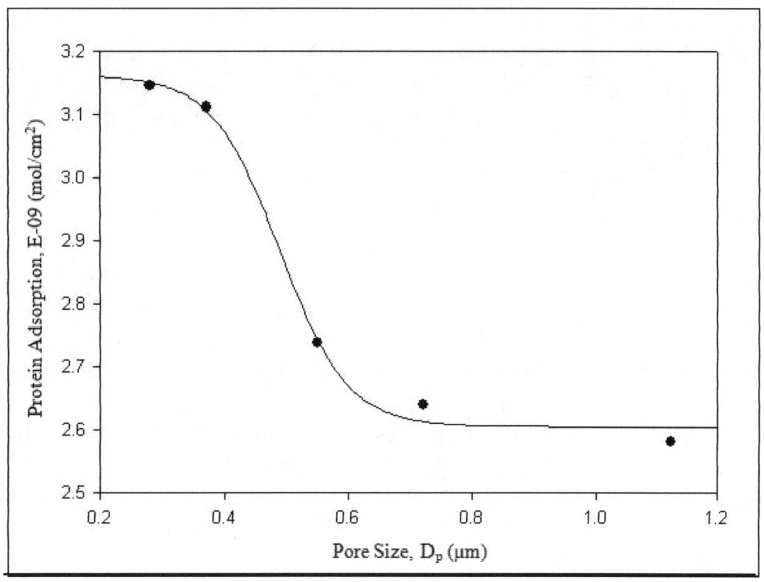

Figure 2. Curve fit of protein adsorption, Γ (mol/cm^2) with respect to various pore size, D_p (μm). Protein conditions: BSA, pD 7, 3 mg/mL

4. Conclusion

In summary, this study systematically investigated the relationship between the protein adsorption and the membrane pore size. Measurement of protein adsorption by FTIR/ATR demonstrated that it was possible to obtain accurate data within a small range of membrane pore size. The results indicated that protein adsorption on a symmetrical, highly microporous membrane was significantly governed by pore size. A good correlation between the adsorption and the pore size was observed, with a high R^2 value obtained for the fitted data. It is expected that the estimation of protein adsorption on different PVDF membranes pore size will be useful for a wide range of applications that involve proteins, particularly quantitative immunoassays.

Acknowledgements

The authors acknowledge UniversitiSains Malaysia, USM (Vice-Chancellor Award 2010, USM-PGRS Grant (8034059), USM Membrane Cluster (8610012), ERGS Grant (6730004) and UiTM LESTARI Grant (600-IRMI/MyRA/5/3/LESTARI 0106/2016) for financial support.

References

[1] Kingshott, P. and Hocker, H. 2006 *Adsoprtion of proteins: Measurement methods, in: A.T. Hubbart (Ed.) Encyclopedia of Surface and Colloid Science*, (Florida: Taylor & Francis), p 669.

[2] Li Y and Chung T-S 2008, *Journal of Membrane Science* **309** (1-2) 45

[3] Zhou J, Meng S, Guo Z, Du Q and Zhong W 2007 *Journal of Membrane Science* **305** (1-2) 279

[4] Ghosh R 2003) *Biochemical Engineering Journal* **14** (2) 109

[5] He, D. & Ulbricht, M. (2008) Preparation and characterization of porous anion-exchange membrane adsorbers with high protein-binding capacity, *Journal of Membrane Science*. 315 (1-2):155-163.

[6] Debnath, M., Prasad, G.B.K.S. & Bisen, P.S., Immunoassay, in: Molecular Diagnostics: Promises and Possibilities, Springer Science+Business Media, 2010, pp. 171-180.

[7] Yonan, C.R., Duong, P.T. & Chang, F.N. (2005) High-efficiency staining of proteins on different blot membranes, *Analytical Biochemistry*. 338:159-161.

[8] Ahmad, A.L., Ideris, N., Ooi, B.S., Low, S.C. & Ismail, A. (2016) Impact of membrane pore structure on protein detection sensitivity of affinity-based immunoassay, *Polish Journal of Chemical Technology*. 18 (2):97-103.

[9] Tantipolphan, R., Rades, T., McQuillan, A.J. & Medlicott, N.J. (2007) Adsorption of bovine serum albumin (BSA) onto lecithin studied by attenuated total reflectance Fourier transform infrared (ATR-FTIR) spectroscopy, *International Journal of Pharmaceutics*. 337 (1-2):40-47.

[10] Xie, J., Riley, C., Kumar, M. & Chittur, K. (2002) FTIR/ATR study of protein adsorption and brushite transformation to hydroxyapatite, *Biomaterials*. 23 (17):3609-3616.

[11] Yang, Q., Zhang, Y., Liu, M., Ye, M., Zhang, Y. & Yao, S. (2007) Study of fibrinogen adsorption on hydroxyapatite and TiO2 surfaces by electrochemical piezoelectric quartz crystal impedance and FTIR/ATR spectroscopy, *Analytica Chimica Acta*. 597 (1):58-66.

[12] Akasaka, H., Konishi, M., Gawazawa, N., Ohshio, S., Toda, I. & Saitoh, H. (2012) Detailed evaluation of protein adsorption on alumina surface through surface plasmon resonance phenomenon, *Applied Surface Science*. 258 (24):9526-9530.

[13] Gaspar, A. & Gomez, F.A. (2013) Application of surface plasmon resonance spectroscopy for adsorption studies of different types of components on poly(dimethylsiloxane), *Analytica Chimica Acta*. 777 (0):72-77.

[14] Wang, L., Xu, G., Shi, Z., Jiang, W. & Jin, W. (2007) Quantification of protein based on single-molecule counting by total internal reflection fluorescence microscopy with adsorption equilibrium, *Analytica Chimica Acta*. 590 (1):104-109.

[15] Xu, Z. & Marchant, R.E. (2000) Adsorption of plasma proteins on polyethylene oxide-modified lipid bilayers studied by total internal reflection fluorescence, *Biomaterials*. 21 (10):1075-1083.

[16] Harrick, N.J. & du Pré, F.K. (1966) Effective thickness of bulk materials and of thin films for internal reflection spectroscopy, *Applied Optics*. 5 (11):1739-1743.

[17] Ohta, K. & Iwamoto, R. (1985) Lower limit of the thickness of the measurable surface layer by Fourier transform infrared attenuated total reflection spectrometry, *Analytical Chemistry*. 57 (13):2491-2499.

[18] Ohta, K. & Iwamoto, R. (1985) Experimental proof of the relation between thickness of the probed surface layer and absorbance in FTIR/ATR spectroscopy, *Applied Spectrosopy*. 39 (3):418-425

[19] Tompkins, H.G. (1974) The physical basis for analysis of the depth of absorbing species using internal reflection spectroscopy, *Applied Spectroscopy*. 28 (4): 335-341.

[20] Ahmad, A.L., Ideris, N., Ooi, B.S., Low, S.C. & Ismail, A. (2012) Synthesis of polyvinylidene fluoride (PVDF) membranes for protein binding: Effect of casting thickness, *Journal of Applied Polymer Science*. 128 (5):3438-3445.

[21] Andersson, P.O., Lundquist, M., Tegler, L., Borjegren, S., Baltzer, L. & Osterlund, L. (2007) A novel ATR-FTIR approach for characterisation and identification of ex situ immobilised species, *A European Journal of Chemical Physics and Physical Chemistry*. 8 (5):712-722.

[22] Yokoyama, Y., Ishiguro, R., Maeda, H., Mukaiyama, M., Kameyama, K. & Hiramatsu, K. (2003) Quantitative analysis of protein adsorption on a planar surface by Fourier transform infrared spectroscopy: lysozyme adsorbed on hydrophobic silicon-containing polymer, *Journal of Colloid and Interface Science*. 268 (1):23-32.

[23] Low, S.C., Shaimi, R., Thandaithabany, Y., Lim, J.K., Ahmad, A.L. & Ismail, A. (2013) Electrophoretic interactions between nitrocellulose membranes and proteins: Biointerface analysis and protein adhesion properties, *Colloids and Surfaces B: Biointerfaces*. 110:248-253.

IOP Conf. Series: Materials Science and Engineering **358** (2018) 012055 doi:10.1088/1757-899X/358/1/012055

A Study on Anti – Fouling Behaviour and Mechanical Properties of PVA/Chitosan/TEOS Hybrid membrane in The Treatment of Copper Solution

N A Sulaiman, N Z Kassim Shaari[1] and N Abdul Rahman

Department of Chemical and Process Engineering, Faculty of Chemical Engineering, Universiti Teknologi MARA (UiTM), Malaysia

norinzamiah@salam.uitm.edu.my[1]

Abstract. In a wastewater treatment by using membrane filtration, fouling has been one of the major problems. In this study, the anti-fouling behaviour of the fabricated thin-film composite membrane were studied during the treatment of water containing copper ion. The membranes were prepared from a polymer blend of 2wt.% chitosan with 10 wt.% poly(vinyl alcohol) (PVA) and then it was cross – linked with tetraethylorthosilicate (TEOS) through sol-gel method. The membrane had been evaluated for its resistance against organic fouling where humic acid had been chosen as organic foulant model which represent the natural organic matter (NOM) in water or wastewater. The dead-end filtration experiments were carried out by using 50 ppm of copper solution with and without the presence of humic acid as feed solution, which was passed through two types of thin film composite membranes. The possible reversible fouling was evaluated by using relative flux decay (RFD) and relative flux recovery (RFR) calculations. The percentage of copper ion removal was evaluated by using Atomic Absorption Spectroscopy (AAS). Based on the results, with the presence of humic acid, the membrane incorporated with silica precursor (TEOS) showed lower flux decay (3%) and higher flux recovery (76%), which show that the formulated hybrid membrane possesses the anti fouling property. The same trend was observed in the mechanical properties of hybrid membrane, where the presence of TEOS has improved the tensile strength and flexibility of the membrane. Therefore, the fabricated thin film composite with the anti-fouling properties and good physical flexibility has potential to be used in the treatment of wastewater containing heavy metal as it could result in good saving in term of operational cost.

1. Introduction

Heavy metals are regularly assumed to be hazardous and highly toxic to human being. They have tendency to accumulate on human body and easily effect on human being indirectly [1][2]. Excess amount of heavy metal accumulate in living things, it can cause serious health syndromes such as kidney damage, liver damage and many else. Furthermore, heavy metals can easily enter a water supply by industrial and consumer waste, or even from acidic rain breaking down soils and releasing heavy metals into streams, lakes, rivers, and groundwater [2]. Therefore, many treatment technologies hass been applied to overcome this problem such as chemical precipitation, ion – exchange and membrane separation. However, there is still some limitations of these processes such as production of high sludge and high operating cost. Fenglian Fu et. al (2011) [3][4] reported that membrane filtration are the most frequently studied for the treatment of heavy metal from industrial wastewater because of its higher efficiency of heavy metal removal. Membrane filtration is becoming a common method for heavy metal removal due to its simplicity, no solvent is consumed and the quality of removed particles

Content from this work may be used under the terms of the Creative Commons Attribution 3.0 licence. Any further distribution of this work must maintain attribution to the author(s) and the title of the work, journal citation and DOI.
Published under licence by IOP Publishing Ltd

are maintained throughout the process [3][4]. However, there are limitations which affect the membrane performance that is fouling on membrane surface. Fouling can affect the separation performance of membrane and decreases the permeate flux [5][6]. Yan Hao et. al (2011) reported that higher membrane fouling phenomena occurs in the treatment of industrial wastewater because it contains of multiple substances especially metal ions. Thus, many researcher has developed new membranes that have high hydrophilicity to overcome the fouling problem [7]. In this study, a new thin-film composite membrane has been developed. Generally, the thin-film composite membrane has its own class compared to a normal membrane as it consists of more than one layers of membrane (top selective layer and bottom porous substrate). Each of the layers can be independently controlled and optimized to achieve desired selectivity and permeability while offering excellent mechanical strength and compression resistance [8]. One of the methods to enhance hydrophilicity of membrane is by incorporating hydrophilic polymer in membrane formulation [7].

Basically, fouling on membrane filtration largely depend on their interaction among membrane surface and types of foulant in water[7]. Natural organic matter (NOM) is a complex matrix of organic material that is frequently found in surface water with three major substances - hydrophobic (humic substances), hydrophilic and transphilic fraction [9]. The major fraction of natural organic matter (NOM) is humic acid (HA) which contributes to problems like odour, taste, colour and acidity to the water. Humic acid (HA) has high ability of metal complexation capacity and interactive properties with organic pollutants which has caused an increase in transport and distribution of inorganic and organic micro-pollutants [9]. Through an extensive research on organic fouling, it has been reported that membrane fouling is caused by the attachment of colloidal and soluble organic matter to the membrane [10]. Therefore, fouling should become worse when humid acid is combined with metal ions due to an increase in the ionic strength between the humic acid substance [6]. Therefore, in this study, thin-film composite membranes which were formulated from chitosan, PVA and TEOS as the thin layer and coated on polysulfone membrane were tested on the anti fouling behaviour by using humic acid as the foulant model. The anti-fouling behaviour was determined by using relative flux decay (RFD) and relative flux recovery (RFR) calculations.

2.Materials and Method

2.1 Chemicals.

Poly(vinyl alcohol) (PVA) with a hydrolysis degree of 87-89% (MW : 85,000 – 124,000), polysulfone (PSF) resin pellet (MW: 44,000 – 53,000), tetraethylorthosilicate (TEOS) with 99% purity and hydrochloric acid with 37% purity as catalyst and commercial chitosan (deacetylation degree 84.8 ± 1.2 %), humic acid (CAS Number 1415-93-6) and copper nitrate [$Cu(No_3)_2.5H_2O$] were used in this study. These chemicals were obtained from Sigma Aldrich, Malaysia. 1-methyl-2-pyrrolidone (NMP) with purity of 99% was obtained from Merck, Malaysia. Deionized water was also used as a solvent.

2.2 Formulation of TFC Membrane

The method to prepare thin film composite membrane for heavy metal removal was according to Norin et. al (2012) with some modification [11][12][13]. The formulation consists of two important steps, preparation of polysulfone polymer solution for porous support membrane and formulation of hybrid membrane by using sol –gel method. The hybrid layer was coated on the polysulfone membrane. There were three types of membranes that were produced and the formulation was shown in Table 1. In this study, the membrane had an effective surface area of 17.3494 cm^2 in a flat sheet configuration and the mode of filtration is according to dead-end filtration [13].

2.3 Fouling Performance of TFC Membrane

In this study, the anti-fouling performance testing is performed using the dead-end filtration rig at 8 bar. The anti-fouling performance of the fabricated thin film composite membrane was evaluated by using copper solution as the feed solution with and without the presence of humic acid. 13wt% polysulfone membrane and hybrid membrane with different concentration of TEOS were used in the experiment. Firstly, 1000 ppm of humic acid stock solution was prepared by dissolving 10 gram of humic acid powder in 10000 ppm of NaOH. The prepared stock of humid acid solution was stored in a volumetric flask at room temperature in a dark place [14]. Next, 0.5 grams of copper nitrate was diluted with deionized water to prepare 1L of 500 ppm copper solution in a volumetric flask. For the membrane fouling performance, the membrane was evaluated through the filtration of copper stock solution with and without the presence of humic acid. A mixture of 50 mL of 50 ppm copper nitrate solution with 50 mL of each humic acid solution was prepared and used immediately. Each filtration process of the membrane was conducted in 3 stages. Firstly, pure water flux is recorded for 0.5 hour and the stabilized flux was denoted as J_0. Then, the membrane was tested with a copper-humic acid solution for 2 hours and the collection of permeate was made at every 30 minute interval and the volume collected was recorded. The stabilized flux is denoted as *Jp*. Then, the membrane was cleaned by back-wash method. The back-washing method is done by immersing the membrane in 50 mL of deionized water and being stirred in a shaker at 100 rpm for 30 minutes. Lastly, pure water flux was recorded again for 0.5 hour and the final flux was denoted as J_1. Relative flux Decay (RFD) and Relative Flux Recovery (RFR) were calculated by using equations (2) and (3) respectively [7]. Samples from permeate solution was analyzed for copper concentration by using Atomic Absorption Spectroscopy (AAS). The permeate flux, mL/cm^2.min^{-1} was determined analytically from the total volume of permeate collected against time by using Eq. (1).

$$flux, f : \frac{\Delta v}{A.\Delta t} \tag{1}$$

$$RFD : \left[\frac{(J_o - J_p)}{J_o}\right] \times 100 \tag{2}$$

$$RFR : \left(\frac{J_1}{J_p}\right) \times 100 \tag{3}$$

Where Δv (mL) is the volume of the permeate sample, A is an effective membrane area in cm^2, and Δt is time permeate. The rate of percentage removal (R %) of copper ions was calculated from the concentration of copper ions in the feed and permeates by using Eq. (4).

$$R\% : \left[\frac{(C_f - C_p)}{C_f}\right] \times 100 \tag{4}$$

Where C_f is the concentration feed solution, C_p is the concentration of permeate copper ions solution

Table 1. Formulation of membranes.

Sample	Membrane name	PSF (wt%)	NMP	PVA	CS (wt%)	TEOS (wt%)
13wt% polysulfone (porous support)	M1	13	87	0	0	0
PVA/Chitosan/0wt%TEOS coated on PSF support membrane	M2	13	87	10	2	0
PVA/Chitosan/3wt%TEOS coated on PSF support membrane	M3	13	87	10	3	3

PSF: Polysulfone, NMP: N-methyl-2-pyrrolidone, PVA: Poly (vinyl alcohol), CS: chitosan, TEOS: tetraethylorthosilicate, wt%: weight percent

2.4 Mechanical analysis

The samples were prepared in the dry states and the analysis was carried out using Instron Universal testing model. The hybrid membranes were cut into a rectangular shape of 100 mm length and 40 mm width. The speed of the machine was 5mm/min and the gauge length was set at 25 mm according to the standard method ASTM D 882 for evaluation on tensile of thin film [17].

3. Results and Discussions

3.1. Interaction of humid acid with Cu^{2+} ions during performance testing

In this study, heavy metal solution was mixed with the organic foulant (humic acid) to examine the membrane fouling performance. Humic acid was chosen as organic pollutant in the heavy metal solution to simulate the actual environment of the feed solution [15][7][16][4]. Figure 1 (a) (b) show the permeate fluxes, *(J)* of the PSF membrane and TFC membrane with and without addition of nano-silica, tetraethylorthosilicate (TEOS) during the filtration of humid acid solution and without humid acid solution towards copper ion solution. Correspondingly, the data of the prepared membrane from the experiment are summarized in Table 2. As shown in Figure 1 and Table 2, for feed solution without humic acid, membrane M3 that incorporated with silica precursor (3wt% TEOS) had the lowest initial permeate flux due to an increase of selective interaction of between water molecules and hydroxyl groups of the hybrid membranes [13]. However, it has higher flux recovery as compared to M1 and M2 where 33% of flux drop after the 2 hours filtration and a 95% flux recovery after the membrane cleaning. This result indicates that the membrane containing TEOS exhibited almost fully recoverable to its initial flux. For the feed solution with humic acid, interaction of humid acid (HA) with copper solution has caused all membranes have lower flux recovery as compared to that without the presence of humic acid. These results show natural organic matter (NOM) represented by humic acid has really caused the organic fouling towards the membranes.

Table 2. Relative flux decay (RFD) and relative flux recovery (RFR) of the TFC membranes towards filtration of copper ion solution with and without humic acid.

HA (ppm)	Type of membrane	Membrane name	J_0 (mL/min.m^2)	J_p (mL/min.m^2)	J_1 (mL/min.m^2)	*RFD*	*RFR*	% copper ions removal
0	13wt% PSF	M1	0.04054	0.00845	0.02344	56.5166	30.8057	93.67
	0wt% TEOS	M2	0.00596	0.00653	0.00538	29.0323	90.3226	98.56
	3wt% TEOS	M3	0.00404	0.00192	0.003843	33.9286	95.2381	99.7
1000	13wt% PSF	M1	0.02651	0.01153	0.013833	38.7681	52.1739	93.26
	0wt% TEOS	M2	0.00605	0.003458	0.004227	26.2698	69.8444	99.6
	3wt% TEOS	M3	0.00480	0.003843	0.003650	3.00000	76.0000	92.72

From Table 2, with the presence of humic acid, the membrane incorporated with TEOS (M3) had much lower flux decays (3%) as well as higher flux recovery (76%) as compared to M1 and M2. According to Jio Na et. al (2012), at first stage of filtration, concentration polarization phenomenon is significant as it plays a major role and then gradually it is replaced by the membrane pore blockage resistance and gel resistance. After that, the fouling resistance of membrane will continue to grow higher. Humic acid has properties that can adsorb strongly onto polysulfone (PSF) ultrafiltration membrane surface. This is because humic acid has a high active surface and the hydrophilicity of -COOH and -OH functional groups. Besides that another characteristics of humid acid is the hydrophobicity of central aromatic and fat molecules groups [4][5][15][16]. In a solution, humic acid exists in the shape of loose micellar aggregates which show the characteristics of colloidal particles. Since humic acid can be easily absorbed onto membrane surface and into membrane pores, the hydrophobicity of polysulfone membrane made the interaction of membrane and humic acid increases, resulting in a heavy absorption and flux decline significantly [4]. This will cause the membrane fouling resistances continued to grow higher by turns. Thus, it is proven from the results based on

Figure 1 (a) and (b) that the membrane with absence of humid acid as foulant model had low flux decay towards copper ion solution as well as higher flux recovery rate. However, their recovery rate decreases with interaction of humid acid on membranes but still higher as compared with flux decay. Besides that, the composite membrane (M3) exhibits better anti fouling performance than other membranes even with the presence of humic acid in feed solution. These results can be attributed to the hydrophilicity and long term stability of the membrane. These indicates that the improvement of TFC membrane can effectively counter back the fouling caused by NOM, especially in the case towards copper ion solution. According to Yan Hao et. al (2011), the flux will reduce drastically in the presence of metal ions due to intermolecular bonds production between organic compound and carboxylic acid group from humic acid. Therefore, it is the possible reason for the increase in fouling with the combination of copper ion and humic acid. Although the fouling has been enhanced in this experiment, the rejection copper ion through the membranes is high (> 90%). It was found that the copper ion removal efficiency of the MI, M2 and M3 were not significantly different where more than 90% removal was achieved.

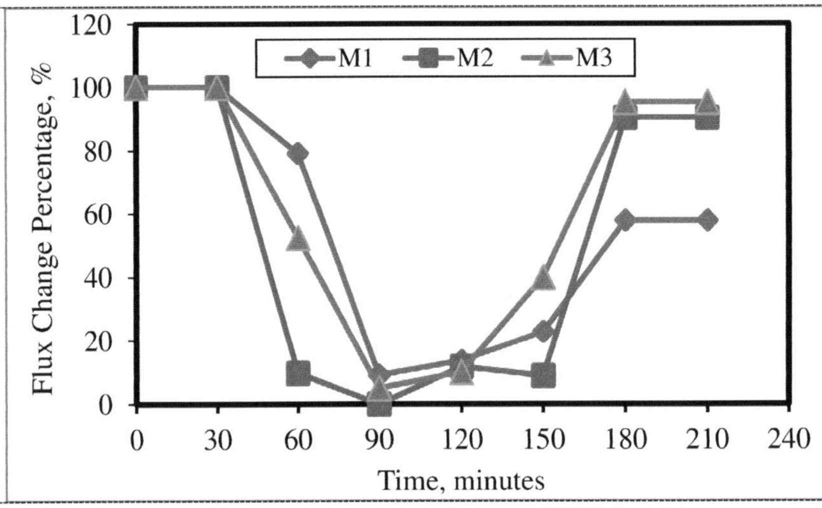

(a)

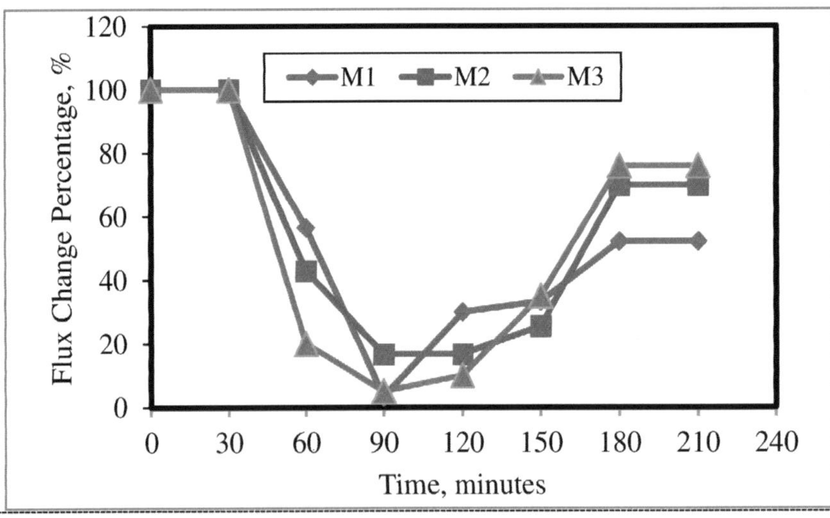

(b)

Figure 1. Flux change percentage vs time from filtration of copper ion solution (a) without humic acid solution, and (b) with humic acid solution by using different membranes

3.2. Mechanical Strength of TFC Membrane

The composite membrane should be tough enough to withstand pressure during filtration. Therefore, mechanical properties is an important parameter to identify their stability of the pressure driven membrane [7].

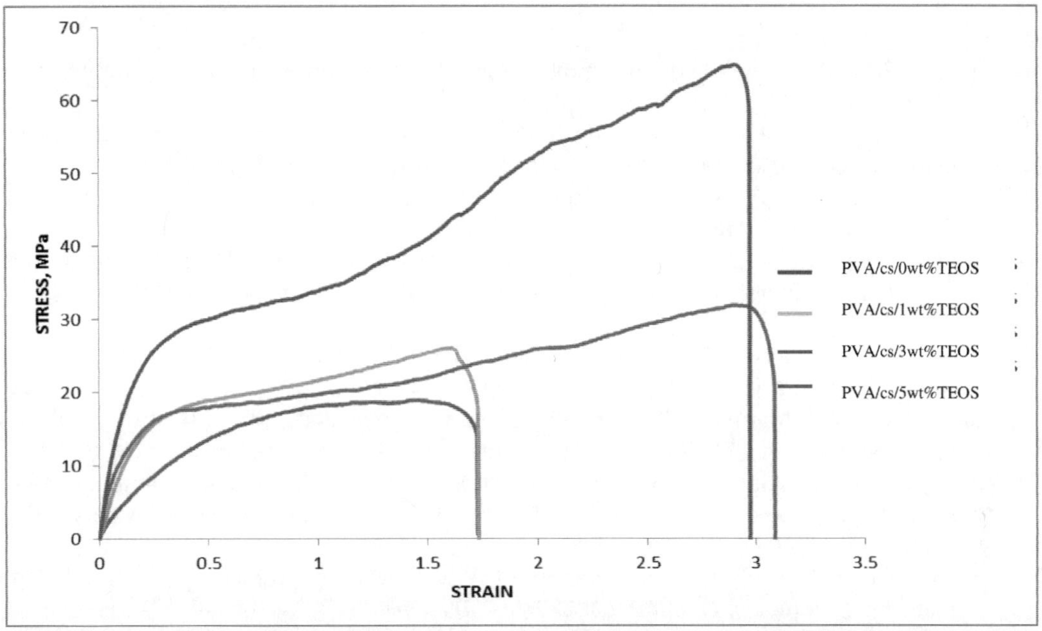

Figure 2. Stress vs Strain of hybrid membrane with different amount of TEOS Concentration

As shown in Figure 2, the tensile strength and its elastic modulus increased with the increase in TEOS concentration. In general, TEOS content will improve the mechanical properties of hybrid membrane [18], where more TEOS content in membrane made the membrane becomes more rigid. Thus, it needs more strength to fracture. From the tensile strength result, it also showed that silica nano precursor such as TEOS can disperse uniformly throughout the polymer matrix and it interacts with polymer chain through hydrogen bonding. Thus, the polymer matrix becomes trapped between silica precipitates [19]. The increasing of tensile strength with increasing TEOS content is resulted from the interaction of the hydrogen bond or other bond formed between PVA/CS and TEOS after the sol –gel process. As reported by Xiojuan Lu et. al (2008), the tensile strength of the membrane increases with increasing of TEOS content but excess amount of TEOS content will cause hybrid homogeneity decreases, thus the flexibility or tensile strain will decrease. Based on Figure 2, the tensile strain increases at higher TEOS content from 1wt.% TEOS to 3wt.% TEOS but at 5wt.% TEOS, the strain starts to decrease. This is due to hybrid membrane that has too much TEOS content can cause the film become more rigid and show a decrease in flexibility [17]. Generally, TEOS network restricts the movement of the polymer molecules, which weaken the flexibility of the membrane. As the conclusions, from this study, TFC membrane with hybrid membrane consists of 3wt.% TEOS is sufficient to deal with high pressure during filtration and has good stability. In addition, membrane incorporated with 3wt.% TEOS also strong enough to withstand the transmembrane pressure during filtration and it is no easily broken during the cleaning or backwashing of the membrane.

4. Conclusion

The fabricated TFC membrane consists of chitosan/PVA/TEOS as the hybrid layer and polysulfone as the support membrane has the anti-fouling properties even when humic acid is used as a foulant model. It is exhibited through high RFR value which reaches up to 76% while RFD value is as low as 3%, with high percentage removal of copper ion (> 90%). Furthermore, the thin film composites possess good physical flexibility, which makes it potentially being used in the treatment of wastewater containing heavy metal as it could result in good saving in term of operational cost.

Acknowledgments
The author would like to thank to Ministry of Higher Education (MOHE) and Universiti Teknologi MARA (UiTM) for awarding research grant, RMI/FRGS 5/3 (92/2014) for the financial support to conduct this research.

References
[1] Barakat M A 2011 New trends in removing heavy metals from industrial wastewater *Arab. J. Chem.* **4** 361–377
[2] Renge V C, S. Khedkar V and Pande S V 2012 Removal of heavy metals from wastewater using low cost adsorbents : a review *Sci. Revs. Chem. Commun.* **2** 580–584.
[3] Fu F and Wang Q 2011 Removal of heavy metal ions from wastewaters: A review *J. Environ. Manage.* **92** 3 407–418
[4] Na J and Yonggang Z 2012 Singapore The Effect of Humic Acid on Ultrafiltration Membrane Fouling *International Conference on Computer Technology and Science (ICCTS2012)* **47** 2–8
[5] Katsoufidou K, Yiantsios S G and Karabelas A J 2005 A study of ultrafiltration membrane fouling by humic acids and flux recovery by backwashing **266** 40–50
[6] Hao Y, Moriya A, Maruyama T, Ohmukai Y and H. Matsuyama, 2011 Effect of metal ions on humic acid fouling of hollow fiber ultrafiltration membrane *J. Memb. Sci.* **376** 247–253
[7] Zhu X, Loo H, and Bai R 2013 A novel membrane showing both hydrophilic and oleophobic surface properties and its non-fouling performances for potential water treatment applications *J. Memb. Sci.***436** 47–56
[8] Lau W J, Ismail A F, Misdan N and Kassim M A 2012 A recent progress in thin film composite membrane : A review *Desalination* **287**190–199
[9] Zularisam A W, Ismail A F and Salim R 2006 Behaviours of natural organic matter in membrane filtration for surface water treatment a review *Desalination* **194** 211–231
[10] Xiao K, Sun J, Mo Y, Fang Z, Liang P, Huang X, Ma J, and Ma B 2014 Effect of membrane pore morphology on microfiltration organic fouling: PTFE/PVDF blend membranes compared with PVDF membranes *Desalination* **343** 217–225
[11] Kassim Shaari N Z and Abdul Rahman N 2012 Thin film composite membrane with hybrid membrane as the barrier layer : Preparation and characterization *IEEE Colloq. Humanit. Sci. Eng. Res. (CHUSER 2012)* 615–620
[12] Sulaiman N A, Kassim Shaari N Z and Abdul Rahman N 2016 Removal of Cu (II) and Fe (II) ions through thin film composite (TFC) with hybrid membrane *J. Eng. Sci. Technol* SOMCHE 2015 36–49
[13] Karpudewan M and Keong C C 2013 *Jurnal Teknologi* **2** 1–6
[14] Nyström M, Ruohomäki K and Kaipia L 1996 Humic acid as a fouling agent in filtration *Desalination* **106** 79–87
[15] Baran G and Makovníková J 2003 The influence of humic acid quality on the sorption and mobility of heavy metals *Plant Soil Environ.* **49** 65–571.
[16] Pas A 2013 Influence of copper (II) ions on stability of dissolved humic acids – coagulation studies *Acta Agrophsica* **20** 253–267.
[17] Bahrami S B, Kordestani S S, Mirzadeh H and Mansoori P 2002 Poly (vinyl alcohol) - Chitosan Blends: Preparation, Mechanical and Physical Properties *Iran. Polym. J.* **1**12. 139–146.
[18] Shi F, Ma Y, Ma J , Wang P and Sun W 2012 Preparation and characterization of PVDF / TiO 2 hybrid membranes with different dosage of nano-TiO 2 *J. Memb. Sci.*, **389** 522–531.
[19] Wu Y, Wu C, Li Y, Xu T and Fu Y 2010 PVA-silica anion-exchange hybrid membranes prepared through a copolymer crosslinking agent *J. Memb. Sci.*, **350** 322–332.

IOP Conf. Series: Materials Science and Engineering **358** (2018) 012056 doi:10.1088/1757-899X/358/1/012056

High Concentration Protein Ultrafiltration: a Comparative Fouling Assessment

Y P Lim[1] and A W Mohammad[2]

[1] Faculty of Chemical Engineering, Universiti Teknologi MARA, Malaysia
[2] Department of Chemical and Process Engineering, Faculty of Engineering and Built Environment, Universiti Kebangsaan Malaysia, Malaysia

yingpei@salam.uitm.edu.my, wahabm@vlsi.eng.ukm.my

Abstract. In this paper, the predominant fouling mechanism via pH manipulation in gelatin ultrafiltration (UF) at constant operating pressure was studied. Two 30 kDa molecular weight cut off (MWCO) UF membranes with different hydrophilic/hydrophobic properties were tested at solution pH near gelatin isoelectric point (IEP), pH below and above gelatin's IEP. The resistance-in-series model was used to determine quantitatively the contribution of each filtration resistance occurred during gelatin UF. The governing fouling mechanisms were investigated using classical blocking laws. The results demonstrated that concentration polarization remain as dominant fouling resistance in gelatin UF, but exceptional case was observed at pH away from gelatin's IEP, showing that combined reversible and irreversible fouling resistances contributed around 57% and 37%, respectively to the overall fouling resistances. Under all experimental condition tested, permeate flux decline was accurately predicted by all the models studied. Fouling profile was fitted well with "Standard Blocking", "Intermediate Blocking" and "Cake Filtration" model for regenerated cellulose acetate (RCA) membrane and "Cake Filtration" model for polyethersulphone (PES) membrane.

1. Introduction

With the continuous improvement of membrane and the establishment of membrane technology as an efficient separation unit in a wide range of applications in food industry, gelatin is one of the animal products that has benefited greatly by UF process. The application of UF system to perform initial concentration in gelatin production has greatly improved the product quality, reduced emissions and labour costs as well as water usage. In view of the remarkable advantages offered by UF, the application of UF system as an alternative to the classical methods in gelatin concentration has been studied by several authors [1-3]. However, one of the major drawbacks in the application of membrane technology for protein concentration and purification is that the flux declines over time due to membrane fouling. While the instantaneous impact of fouling is observed through decreasing in permeate flux rate and selectivity, the ultimate effect may cause permanent loss in membrane's permeability due to irreversible fouling and the shorter membrane lifetime. Nevertheless, many researchers have studied protein fouling in UF but there is still major discrepancy in regards to the underlying mechanisms governing protein fouling and the fouling rate. Different flux decline mechanisms are suggested in the literatures, e.g., osmotic pressure governed flux decline [4], gel or cake layer controlled flux decline [5] as well as membrane pores blocking, such as, complete blocking, standard blocking and intermediate blocking. [6]. Similar types of fouling mechanisms are researched for other types of solutions, e.g. MF/UF of beer [7], cheese whey [8] and natural organic matters [9]. Hence, by studying fouling behaviour and its associate mechanism, protein fouling could be mitigated by minimizing the attractive forces between membrane-protein and protein-protein. This, in turn, would reveal higher performance characteristics (i.e. retention and flux) and yield higher operational

Content from this work may be used under the terms of the Creative Commons Attribution 3.0 licence. Any further distribution of this work must maintain attribution to the author(s) and the title of the work, journal citation and DOI.
Published under licence by IOP Publishing Ltd

life during protein UF. Thus, the scope of the present study is to obtain quantitative data on the effect of solution pH on filtrate flux during gelatin filtration through RCA and PES membrane and to develop a more fundamental understanding of the fouling mechanism using these results.

2. Experimental

2.1 Materials

Ultrafiltration experiments were conducted using Ultracel® composite RCA membranes and Biomax® PES membranes, both provided by Millipore Corp. (Bedford, MA). Pharmaceutical grade gelatin (4 % w/v) obtained from Halagel (M) Sdn. Bhd. was used as model protein without further purification. The characteristics of gelatin used were presented elsewhere [10]. The pH of the feed stream was adjusted with dropwise addition of 3 M HCl or NaOH and measured with a Metler Toledo pH meter (model FG2 FiveGo™). All reagents used in the experiments were of analytical grade unless otherwise stated. Pure water was purified with Heal Force® UPW System NW UF Series by Nison Instrument (Shanghai) Limited with resistivity of 18 MΩ-cm.

2.2 Ultrafiltration Experiment

The filtration experiments were carried out by using a dead-end stirred cell filtration system (Amicon model 8200) from Millipore Co., USA at pressure 2 bars and temperature 40°C. Fresh membrane hydraulic resistance (R_m) was obtained by pure water flux measurement (J_o) using Equation (1)

$$J_o = \frac{\Delta P}{\mu R_m} \tag{1}$$

After that, the stirred cell were emptied and rapidly refilled with gelatin solution for UF process and total filtration resistance ($R_t = R_m + R_{cp} + R_{rf} + R_{if}$) was determined using Equation (2).

$$J_v = \frac{\Delta P}{\mu(R_m + R_{cp} + R_{rf} + R_{if})} \tag{2}$$

After 60 minutes, the filtration was stopped and was replaced by ultrapure water (UPW). Pure water flux after release of applied pressure (J_1) was measured using Equation (1) and concentration polarization resistance (R_{cp}) was calculated using Equation (3)

$$R_{cp} = R_t - \frac{\Delta P}{\mu J_1} \tag{3}$$

Next, the cell and the solution reservoir were fully emptied and gently rinsed to remove any labile gelatin. The cell was then filled with UPW and re-pressurized, and pure water filtration was again performed (J_2). The reversible fouling resistance (R_{rf}) was calculated using Equation (4)

$$R_{rf} = \frac{\Delta P}{\mu J_3} - \frac{\Delta P}{\mu J_2} \tag{4}$$

The membrane was then rinsed in a 0.1 M NaOH solution so that weakly adsorbed gelatin on the membrane surface could be desorbed and the pure water flux was again measured (J_3). The irreversible fouling was calculated using Equation (5).

$$R_{if} = \frac{\Delta P}{\mu J_3} - \frac{\Delta P}{\mu J_0} \tag{5}$$

The equation for classical blocking laws as shown in table 1, adapted from [11] was used to identify the predominant fouling mechanism occurred during gelatin UF at different pH. The permeate volume versus time data were fitted into the model using Matlab R2007a and the best fit parameter was obtained by minimizing the sum of square residuals (SSR).

Table 1. Classical Blocking Laws.

Classical Blocking Laws	Equation	Parameters
Cake filtration	$V = \frac{1}{K_c J_o}\left(\sqrt{1+2K_c J_o^2 t} - 1\right)$	K_c (s/m^2)
Intermediate blocking	$V = \frac{1}{K_i}\ln(1 + K_i J_o t)$	K_i (m^{-1})
Standard blocking	$V = \left(\frac{1}{J_o t} + \frac{K_s}{2}\right)^{-1}$	K_s (m^{-1})
Complete blocking	$V = \frac{J_o}{K_b}(1 - \exp(K_b t))$	K_b (s^{-1})

3. Results and discussion

3.1 Resistance-in-series model

The total filtration resistance (R_t) was maximum at pH 5.3, minimum at pH 4 and intermediate at pH 6.8 as shown in table 2. Particularly, irreversible resistance (R_{if}) was fifth fold higher at gelatin's IEP. At IEP, gelatin molecule has net neutral charge, therefore more easily precipitated or adsorbed and the deposited layer is relatively compact. Kilduff et al. [12] detected that the specific cake resistance was highest at neutral pH due to negligible electrostatic repulsion among foulants, which was in good agreement with our study.

Table 2. Quantification of various fouling resistance at different pH for RCA and PES membrane.

pH	Membrane Type	R_{cp}/R_m	R_{rf}/R_m	R_{if}/R_m	R_t/R_m
pH < gelatin's IEP (pH 4)	RCA	20.3	0.50	0.12	20.9
	PES	3.02	3.48	1.84	8.34
pH near to gelatn's IEP (pH 5.3)	RCA	40.7	0.51	0.55	41.8
	PES	45.30	4.19	3.45	52.9
pH > gelatin's IEP (pH 6.8)	RCA	17.1	0.21	0.11	17.4
	PES	5.14	2.10	1.74	8.98

Generally, the fouling resistance trend is observed at $R_{cp} > R_m > R_{rf} > R_{if}$. R_{cp} contributed to a greater extent (> 91%) to the total filtration resistance during gelatin UF except for PES membrane at pH 4. High R_{cp}/R_m ratio will favor gel forming tendency or increase osmotic pressure at membrane-solute interface and cause decreased in driving force [13]. Thus, it is suggested that improved mass transfer or hydrodynamic conditions at the membrane surface would be beneficial to achieve higher flux. For PES membrane, combination of R_{rf} and R_{if} show contribution of 57% at pH 4 and 38.5% at pH 6.8 to the final global resistance, higher than 7% at pH 5.3. At pH near gelatin's IEP (pH 5.3), R_r/R_m and R_f/R_m for PES membrane were significantly higher compared to RCA membrane. The main driving force was hydrophobic attraction between PES membrane and gelatin. At pH 4 and 6.8, R_r/R_m and R_f/R_m for RCA were comparatively higher for PES membrane. Similar observations were reported by Luey et al. [14] during bLG adsorption on hydrophilic and hydrophobic silicon surfaces. They found that electrostatic forces play a major role in the protein adsorption on hydrophilic surfaces while non-electrostatic interactions dominate on hydrophobic surfaces. At pH 4, the surface charge of the gelatin was positive and the membrane surface was negatively charged, so electrostatic attraction result in a rigid cake on the membrane surface. Whereas at pH 6.8, electrostatic repulsion induced by the like-charged membrane and gelatin (negative). In addition, the stronger electrostatic repulsion between already adsorbed gelatin fouling layer and freshly transported gelatin from the bulk – further reduces protein deposition on the surface and inside the pores. This explain high R_{cp}/R_m and R_{rf}/R_m values at basic pH and indicates that gelatin fouling relies more on electrostatic forces rather than hydrophobic forces at pH away from gelatin's IEP. Therefore, it can be inferred that in electrostatic interactions

IOP Conf. Series: Materials Science and Engineering **358** (2018) 012056 doi:10.1088/1757-899X/358/1/012056

controlling scenario, enhancement in membrane hydrophilicity does not beneficial in minimizing fouling phenomena in gelatin UF.

3.2 Classical Blocking Models

An example of fitting experimental data and model prediction is shown in figure 1 and figure 2 for RCA and PES membrane at pH 6.8. Table 3 shows the SSR values and the best fitted parameters for both membranes at varying pH. The bold sign in the table shows the dominant fouling mechanism. In general, membrane hydrophilicity affected the nature of fouling mechanism occurred during gelatin UF. For RCA membrane, it was found that the fouling mechanism is affected by the change in solution pH and the sign of gelatin charge. For the range of pH studied here, gelatin charge changed from positive (at pH 4) to negative (at pH 6.8) passing through neutral at IEP (pH 5.04). Whereas the zeta potential for RCA membrane was -5.4 mV. It is postulated that fouling mechanism changes from external fouling to in-pore fouling in response to protein net charge, and thus it was mainly controlled by electrostatic interactions forces for hydrophilic type UF membrane. As for PES membrane, interaction between gelatin molecules and the membrane surface and inner pores may occur immediately due to the amphoteric nature of gelatin molecules and the hydrophobic nature of PES membrane. Therefore, PES membrane was prone to fouling through solute attachment on the porous surface, resulting in partial pore blockage or cake layer formation. This result was in agreement with higher water flux after external cleaning (showing lower Rf/Rm value) for PES membrane as compared to RCA membrane, indicating that the removal of the major part of fouling (especially surface fouling through cake formation and solute attachment).

Table 3. The model fit error (SSR) and fitted parameters using classical blocking laws.

Membrane Type	pH	Model fit error, SSR (x 10^{-4})				Fitted Parameters
		Standard	Complete	Intermediate	Cake	
RCA	4	2.79	7.65	**1.05**	1.82	$K_i = 28.2\ m^{-1}$
	5.3	**0.34**	0.54	0.60	1.74	$K_s = 25.1\ m^{-1}$
	6.8	3.41	8.42	1.08	**0.20**	$K_c = 1.82 \times 10^6\ s/m^2$
PES	4	55.0	116	22.0	**2.34**	$K_c = 3.37 \times 10^5\ s/m^2$
	5.3	1.52	2.31	1.01	**0.46**	$K_c = 3.20 \times 10^7\ s/m^2$
	6.8	15.0	49.0	**1.85**	4.91	$K_i = 17.2\ m^{-1}$

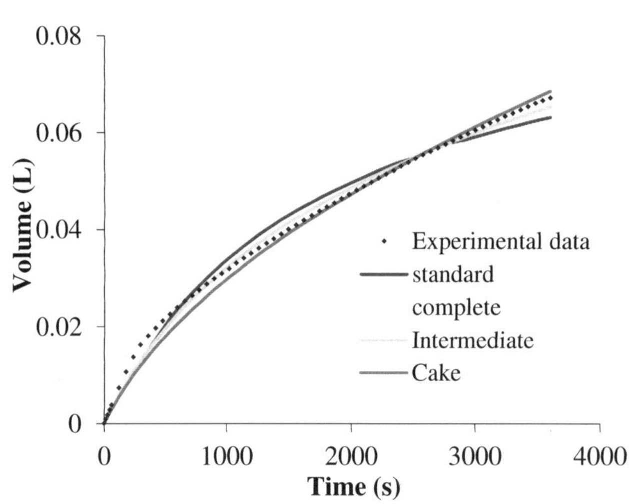

Figure 1. Fitting of experimental data to classical blocking laws for RCA membrane at pH 6.8.

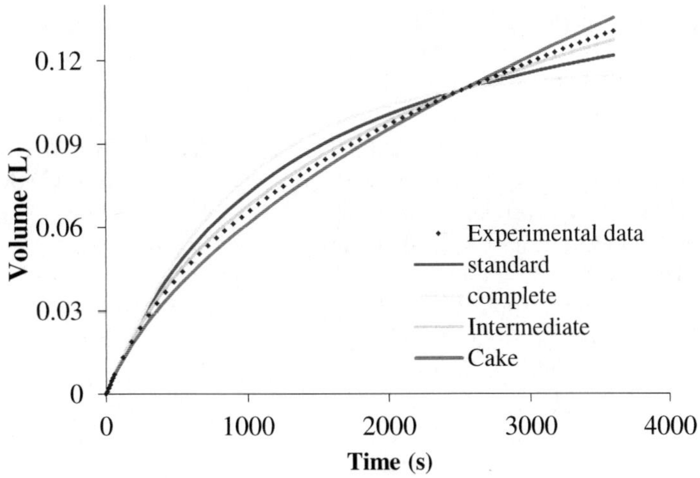

Figure 2. Fitting of experimental data to classical blocking laws for PES membrane at pH 6.8.

From table 3, it was obvious that cake formation, intermediate blocking and standard blocking mechanism were fitted well to all the experimental runs, but complete blocking mechanism displayed the highest deviations. Since gelatin is a type of denatured proteinaceous solution exists with various molecular weights which ranging from 10 kDa to 400 kDa [15], the sizes of the gelatin solutions therefore maybe smaller, same size or even larger than the membrane pore sizes. Therefore, all the blocking mechanisms could happen, but cake formation shall be the dominant fouling mechanism occurred during the filtration process.

4. Conclusion

Quantification of membrane fouling and determination of dominant fouling mechanism during gelatin UF for both RCA and PES membrane were successfully carried out. Generally, effects of membrane hydrophobicity have an impact on the contribution of different fouling resistance. RCA membrane demonstrated that concentration polarization resistance was the main cause for membrane fouling (>90%) at all pHs range. For PES membrane, R_{cp} remain as dominant fouling resistance in gelatin UF at pH 5.3, but exceptional case was observed at pH 4 and pH 6.8, showing that combined R_{rf} and R_{if} contributed around 57% and 37%, respectively to the overall fouling resistances. The best-fitted model for gelatin UF experimental data using RCA membrane was standard blocking, intermediate blocking and cake filtration, depending on pH. Fouling mechanism changes from in-pore blocking to external fouling in response to protein net charge, and thus it was mainly controlled by electrostatic interactions forces. On the other hand, PES membrane was prone to fouling through solute attachment on the porous surface, resulting in cake layer formation. This was confirmed by best fitted cake filtration model to flux decline data in all pH range tested.

Acknowledgement

The authors would like to acknowledge the financial grant funded by Ministry of Science, Technology and Innovation (MOSTI) Malaysia via grant Technofund TF0206A084 and Universiti Kebangsaan Malaysia via grant UKM-GUP-KPB-08-32-129.

References

[1] Dutré B and Trägårdh G 1995 *Journal of Food Engineering* **25(2)** 233-244.
[2] Simon A, Vandanjon L, Levesque G and Bourseau P 2002 *Desalination* **144** 313-318.
[3] Wang Y and Shi B 2011 *Food and Bioproducts Processing* **89(3)** 163-169.
[4] Peeva L, Gibbins E, Luthra SS, White LS, Stateva RP and Livingston AG 2004 *J. Membr. Sci.* **236** 121-136.

[5] Karode SK 2001 *J. Membr. Sci.* **188 (1)** 9-20.
[6] Kosinstev S, Holdich RG, Cumming IW, Starov VM 2002 *J. Membr. Sci.* **208 (1-2)** 181-192.
[7] Blanpain P and Lalande M 1997 *Filtration & Separation* **34(10)** 1065-1069.
[8] Aimar P, Taddei C, Lafaille JP and Sanchez V 1988 *J. Membr. Sci* **38(3)** 203-221.
[9] Taniguchi M, Kilduff JE and Belfort G 2003 *Env. Sci. & Tech* **37(8)** 1676-1683.
[10] Lim YP and Mohammad AW 2011 *Food and Bioprocess Technology* **4(2)** 304-311.
[11] Bolton G, LaCasse D, Kuriyel R 2006 *J. Membr. Sci.* **277(1-2)** 75-84.
[12] Kilduff JE, Mattaraj S and Belfort G 2004 *J. Membr. Sci.* **239(1)** 39-53.
[13] Guo W, Ngo HH and Li J 2012 *Bioresource Technology* **122** 27-34.
[14] Luey JK, McGuire J and Sproull RD 1991 *J. Col. Inter. Sci.* **143(2)** 489-500.
[15] Cole CGB. 2000 *Gelatin. In Encyclopedia of Food Science and Technology*, ed Frederick JF (New York: John Wiley and Sons) p 1183-1188

Preparation of Mixed Ionic Electronic Conducting (MIEC) Membrane Supported on Al_2O_3 Substrate: Effects of Substrate Morphology

'A S Sihar, M F Mohd Rodzi, A N Che Abd Rahim, M Z Shahruddin, N H Alias and N H Othman

Department of Oil and Gas Engineering, Faculty of Chemical Engineering, Universiti Teknologi MARA, 40450 Shah Alam, Selangor, Malaysia.

nurhidayati0955@salam.uitm.edu.my

Abstract. Two types of solvents [N-methyl-2-pyrrolidone (NMP) and dimethyl sulfoxide (DMSO)] were used in the preparation of alumina (Al_2O_3) dope suspensions whereby the dope was allowed to experience phase inversion in the coagulation bath. The green body of alumina discs were sintered at 1450°C and the morphologies of prepared alumina-NMP and alumina-DMSO were studied. To improve the catalytic activity of the oxygen transport membrane for oxygen separation purposes, surface modifications of the membranes were carried out using different mixed ionic electronic conducting (MIEC) oxides catalysts (LCF 91 and LSF 55). The oxide powders and viscous sol were synthesised using sol-gel method and calcined at 950°C to form pure perovskite structure of MIEC materials and studied using XRD while the specific surface area of the calcined powder was measured using BET (Brunauer-Emmett-Tellet) technique. The honey-like MIEC sols were used for surface modification and conducted via dip-coating method and the morphology of the surface-modified alumina membranes were analysed using SEM with EDS mapping. From the study, it was found that the amount of MIEC oxide deposited onto alumina-DMSO is higher than alumina-NMP.

1. Introduction

Since the start of the industrial revolution in the 18th century, human activities have massively upraised the greenhouse gases concentration in the atmosphere. This unpleasant circumstance has led to the invention of new technologies that can capture and sequestrate CO_2, which is known as carbon capture and storage (CCS) [1,2]. Oxy-fuel combustion using oxygen instead of air is one of the available CCS technologies. By eliminate nitrogen from the combustion process, a pure stream of CO_2 and H_2O can be formed. The uses of MIEC-based ceramic membrane that allow only oxygen to pass through across its membrane wall can be considered as the potential replacement to conventional method like cryogenic distillation and pressure swing adsorption as it can reduce cost, energy and more adapting to green technology and sustainable approach [3]. Surface modification of MIEC materials onto ceramic substrates has been extensively studied recently. In our previous work, $La_{0.6}Sr_{0.4}Co_{0.2}Fe_{0.8}O_{3-\delta}$ was used as surface-modified materials onto the same material hollow fibre substrates and the deposition was carried out via vacuum asissted technique [4]. In this study, alumina, Al_2O_3, was used as the substrates material for the deposition of lanthanum-based MIEC oxides. Alumina membrane shows little oxygen permeation activites. Therefore, to improve the oxygen permeability, mixed ionic electronic conducting (MIEC) materials such as lanthanum cobalt ferrite oxide, $LaCo_{0.9}Fe_{0.1}O_{3-\delta}$ (LCF 91) and lanthanum strontium ferrite oxide, $La_{0.5}Sr_{0.5}FeO_{3-\delta}$ (LSF 55) were selected in this study as catalyst for surface modification on alumina disc substrates. Based on study by Jin *et al.* [5], it has been learnt that the viscosity of the catalytic sol plays an important role in deposition properties since

Content from this work may be used under the terms of the Creative Commons Attribution 3.0 licence. Any further distribution of this work must maintain attribution to the author(s) and the title of the work, journal citation and DOI.
Published under licence by IOP Publishing Ltd

less viscous sol could infiltrate better through the membrane and higher viscosity sol will leads to non-homogenous deposition and micro-cracks on the membrane surface [5]. Thus, the viscosity of the catalytic sol must be properly controlled in order to form a homogenous deposition.

The oxygen permeation rate can also be improved by using porous substrates for the deposition of MIEC oxides [6]. By fabricating ceramic substrates which having huge number of finger-like and sponge-like structures, they could provide a good support for the catalysts deposition. Thus, the objective of this research was to prepared LCF 91 and LSF 55 perovskite MIEC oxides supported on different types of alumina disc substrates fabricated from two different solvent (DMSO or NMP) and the morphological of both MIEC membranes were studied.

2. Experimental

2.1. Preparation of MIEC Oxides

The MIEC oxides were prepared using sol-gel method [7]. Stoichiometric amount of metal nitrates such as $La(NO_3)_3.6H_2O$, $Sr(NO_3)_2$, $Co(NO_3)_2.6H_2O$ and $Fe(NO_3)_3.9H_2O$ were used according to the required MIEC oxides formula and dissolved in distilled water. Chelating agent, citric acid (anhydrous, Merck), and polymerization agent, ethylene glycol (> 99.9%, Merck) were then added to the metal nitrates solution with a molar ratio of ethylene glycol to citric acid to metal nitrates of 9:3:1. The solution was stirred and heated at 80°C for 5 h till brownish sol was formed. Then, heated MIEC sol was further evaporated in oven at 80°C for over 40 h to allow polyesterification process and form honey-like viscous of MIEC sol. The calcination of the sol was carried out at 750, 850 and 950°C for 4 h to investigate the development of perovskite structure.

2.2. Preparation of Al_2O_3 Substrates

Alumina disc substrates were prepared using phase inversion-sintering technique [8]. 0.36 wt% of dispersant, Arlacel (Uniqema, USA) was first dissolved in 42.19 wt% solvent prior to the addition of 52.23 wt% alumina powder (Al_2O_3) (< 1 µm alumina, Alfa Aesar). Dimethyl sulfoxide (DMSO) or N-methyl-2-pyrrolidone (NMP) (≥ 99.9%, Merck) were chosen as the solvent to be used in this work. The suspension was roll-milled for 48 hours using tri-roller mill to obtain homogeneous suspension. Then, 5.22 wt% of polymer binder, polyethersulfone (PESf) (Veradel A-301, Solvay Advanced Polymer) was added to the suspension. The added polymer-suspension mixture is further roll-milled for 48 hours. In order to fabricate alumina disc to be used as the substrate for the preparation of MIEC-based membrane, the suspension was poured into the steel plate disc mold. The mold was immediately immersed into the coagulation bath contained tap water to allow the phase inversion process to take place. The prepared alumina discs were then dried at room temperature least for 24 hours prior to sintering process at 1450°C. Two steps sintering process where first the temperature of furnace was increased from room temperature to 600°C at rate of 2°C/min and held for 2 h followed by an increment to 1450°C at rate of 5°C/min and held for 4 h was used in this work.

2.3. Preparation of MIEC Membrane Supported on Al_2O_3 Substrates

In order to deposit the MIEC oxide membrane onto the porous alumina disc substrates, a dip coating technique was used. The alumina discs were dip-coated into the prepared MIEC sol before being calcined in a furnace at 950°C for 4 hours. The amount of MIEC oxide deposited onto the membrane was then calculated using the equation below;

$$\text{Percentage of weight gain, \%} = \frac{W_2 - W_1}{W_1} \times 100 \qquad (1)$$

Whereby, W_1 is the initial weight of disc substrate and W_2 is the surface-modified alumina disc substrate.

2.4. Characterizations

The crystallite structure of MIEC oxides were first analysed using X-ray diffractometer (XRD) (D/Max 2200/PC, Rigaku, Japan) with scanning rate of 5°/min from 2θ range from 10° to 80°. Then the rotary viscometer (LVDV-II, Brookfield, USA) was used to measure the viscosity of the MIEC sol, conducted at room temperature (20°C) with rotary speed of 70 rpm. After the calcination of the MIEC oxides, the surface area of the powder was measured using (3Flex Surface Characterization, Micromeritics, USA) using nitrogen gas with degas temperature of 120°C for 2 h. In order to observe the morphology of the alumina disc substrates prepared using different solvents and the homogeneity of the MIEC deposition onto the disc, SEM (S-3400N, Hitachi, Japan) with EDS analysis (XFlash 6|10, Bruker, USA) was used.

3. Results and Discussion

3.1. Mixed ionic electronic conducting (MIEC) oxides powders

Figure 1 shows the XRD patterns for LCF91 and LSF55 oxides calcined at different temperatures; 750°C, 850°C and 950°C. At 750°C and 850°C, the perovskite structure did not fully develop and the secondary phases can still be observed. However, as the calcination temperature increases, the secondary phases of strontium carbonate in the MIEC were disappeared as they were decomposed during the calcination process. Besides that, higher calcination temperature was found to produce higher and intense peaks, which indicates the increases of particles size, as the grain particles fused to form larger size of particles or oxides. In addition, the peaks are slightly shifted to the right (in the oval-highlighted areas) for both of MIEC oxides, which indicates that the crystal lattices were decreases as the temperature goes up [5].

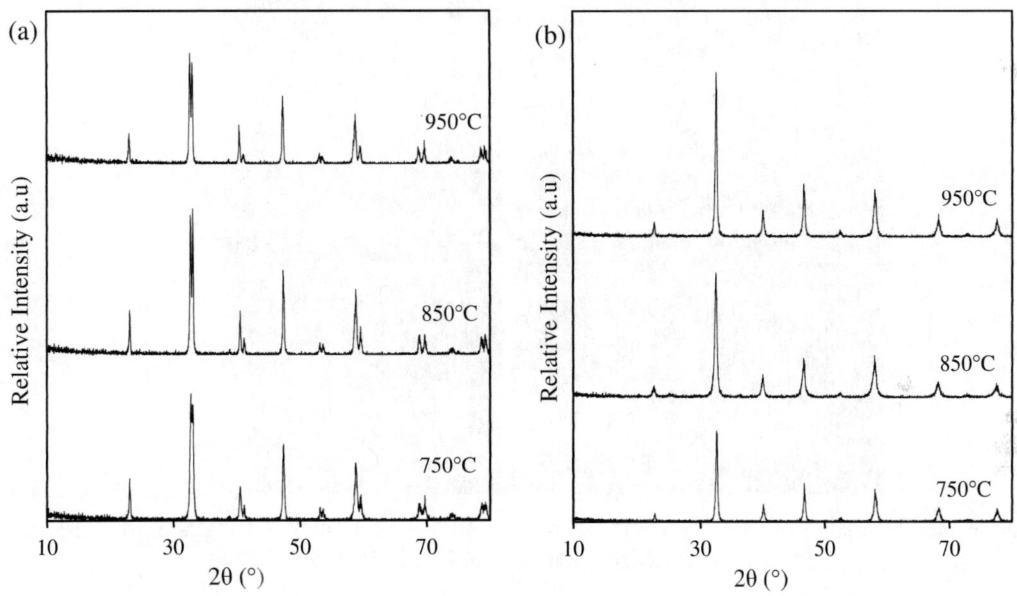

Figure 1. XRD patterns of (a) LCF 91 and (b) LSF 55 calcined at different temperature.

For the preparation of MIEC membrane supported on alumina disc substrates, 950°C was chosen as the calcination temperature. The BET surface area for LCF and LSF are found to be around 1.57 and 2.86 m^2/g respectively. The BET surface area for both oxides are considered low and it is consistent with the work done by Roseno et al. [9]. Dense structure with low surface area is normally possess by MIEC perovskite-type oxides. The adsorption and desorption process of this materials are normally controlled by the surface oxygen exchange rates and this will allow the oxygen ions and electrons transported into or out of those dense structure aggregated grains [10].Therefore, there is a need to deposit the MIEC oxides onto the porous substrates to increase the active surface area and simultaneously the oxygen flux for membrane.

3.2. MIEC solutions

The viscosity of LCF91 sol was observed to be higher than LSF 55, in which 78.09 cp in compared to 63.34. This value is consistent in a range of suitable viscosity for deposition as suggested by Jin et al. [5].This is because if the MIEC sol viscosity is below than 5 cp, it could lead to formation of holes and non-homogenous coating due to discontinuities in the membrane. Meanwhile, if the catalytic sol is too viscous (higher than 80 cp), formation of cracks due to non-homogeneity of MIEC deposition onto the support [5]. Thus, MIEC sols with honey-like texture were chosen to be deposited onto alumina disc substrate. The dip-coating technique allow the deposition of 3.4 wt% LCF & 2.94 wt% LSF onto the disc substrates (refer Figure 3).

3.3. Effect of solvents on the morphology of alumina substrates

Figure 2 shows the SEM images of LCF membrane supported on alumina substrates, which was prepared using (a) DMSO and (b) NMP solvents. It was found that the alumina disc prepared using DMSO (1.46 mm) was thicker in compared to NMP (1.14 mm). Interestingly, the morphology of DMSO was better in compared to NMP, in which an asymmetric structure consists of straight finger-like structure was observed with longest finger-like pore of 1.34 mm (approx. length). Meanwhile, asymmetric structure with sponge-like layer across the entire cross-section was observed for NMP substrates. As the compositions of suspensions used to prepare the discs were fixed in this study, it was believed that the solvent power plays a very important role in determining the final morphology of the substrates.

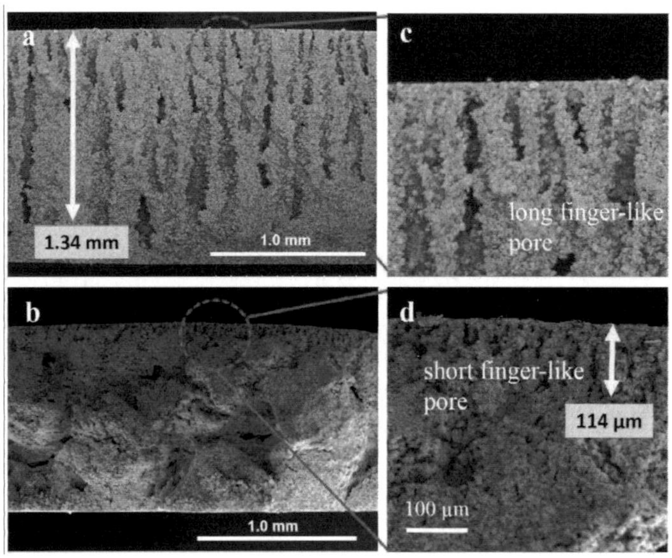

Figure 2. SEM images of LCF membrane supported on different alumina substrates prepared using solvent DMSO (a and c) and NMP (b and d).

The dope prepared using NMP was in compared to DMSO and therefore, during the phase inversion process (the convection of solvent and coagulant), the dope experienced faster water influx in compared to the solvent outflow due to the high concentration of coagulant at the interface, resulting the formation of short pore channels at the top (approx. length of 114 μm) and dense sponge-like structure at the bottom of the substrates [11]. Figure 3 compares the amount of MIEC (LCF 91 and LSF 55) deposited onto the disc substrates. Generally, both type of discs can act as good substrates for the deposition of MIEC. It was found that the amount of MIEC oxides deposited on alumina-DMSO substrates was higher than alumina-NMP. This might be due to the existence of finger-like structures that can entrapped more catalytic sol. Based on the study conducted, alumina-DMSO disc were chosen as the substrates for the deposition of MIEC oxides due to its feature which consisting opened finger-like structure.

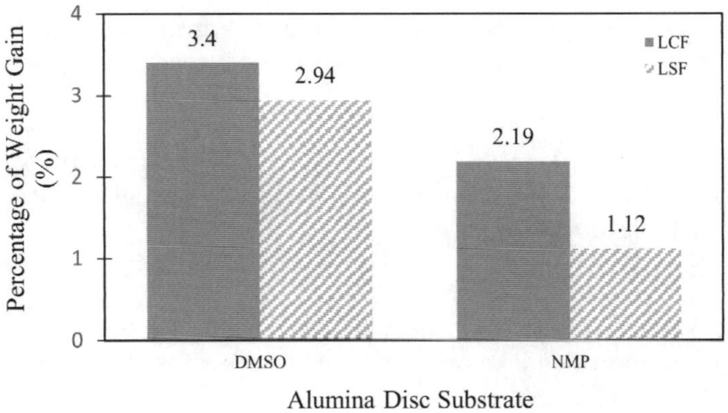

Figure 3. Percentage of MIEC oxides weight gain after calcination.

3.4. Surface modification of alumina substrates using LCF and LSF oxides

Figure 4 compares the dispersion of LCF 91 and LSF 55 oxides on alumina-DMSO disc substrate.

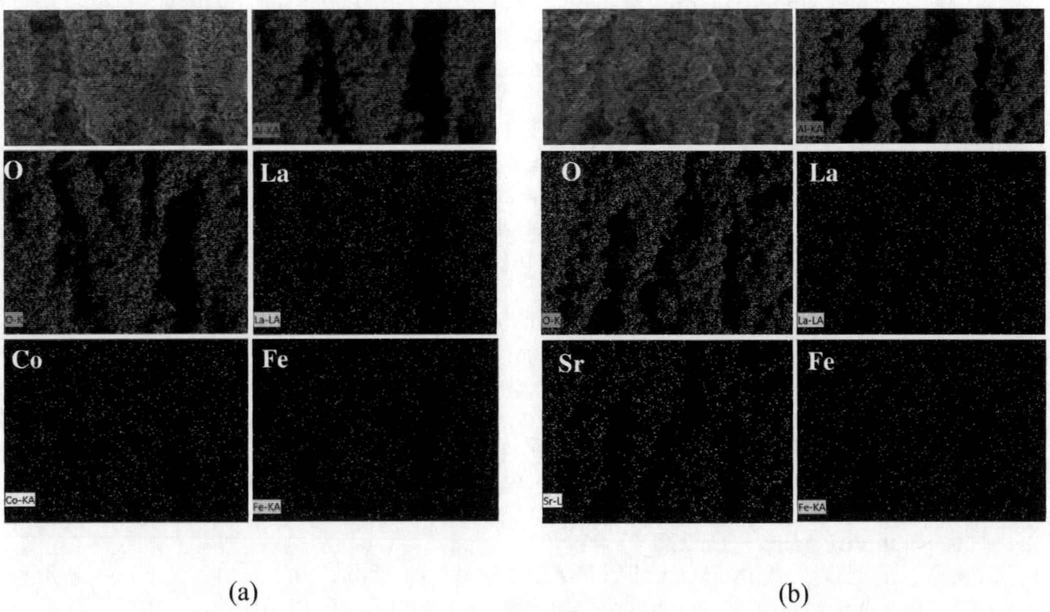

Figure 4. SEM images and EDS mappings of (a) LCF and (b) LSF oxide supported on alumina-DMSO.

The SEM images show that the MIEC oxides were well-deposited across alumina cross-section. In general, both MIEC were homogenously deposited into the finger- and sponge-like structure of the substrates, covering the Al_2O_3 particles. As the alumina was used as the substrates, the EDS mapping shows a high amount of Al and O element. In compared to the work done by Abdullah et al. [12], the deposition of MIEC membrane in this work was more uniformed.

4. Conclusion

Two types of MIEC oxides (LCF 91 and LSF 55), were successfully prepared using sol-gel method and calcined at 950°C to obtain pure perovskite structure. However, as the MIEC oxides possess low surface area due to high calcination temperature, the MIEC oxides need to be deposited onto the porous alumina disc membrane to obtain higher oxygen permeation. The amount of MIEC oxides deposited was in a range of 1.1 – 3.4 %, in which the use of alumina substrates allows higher amount of MIEC oxides to be deposited. This might be due to the long and uniform finger-like. From the EDS mapping, all the respective elements for both MIEC oxides shows a uniform and well dispersed across the alumina-DMSO disc substrates thickness. Therefore, this membrane can be tested for oxygen permeation in the future study to evaluate its performance as oxygen transport membrane.

Acknowledgments
The authors would like to acknowledge Ministry of Higher Education (MOHE) for the RAGS research funding (RAGS/1/2015/TK0/UITM/02/11) and Universiti Teknologi MARA which make this important research viable and effective.

5. References

[1] Wennersten R, Sun Q and Li H 2015 The future potential for Carbon Capture and Storage in climate change mitigation - An overview from perspectives of technology, economy and risk *J. Clean. Prod.* **103** 724–36
[2] Zhu J, Liu Z, Guo S and Jin W 2015 Influence of permeation modes on oxygen permeability of the multichannel mixed-conducting hollow fibre membrane *Chem. Eng. Sci.* **122** 614–21
[3] Xing Y 2012 *Development of Thin Film Oxygen Transport Membranes on Metallic Supports*
[4] Othman N H, Shahruddin M Z, Sihar A S, Wu Z and Li K 2016 In-Situ Catalytic Surface Modification of Micro-Structured $La_{0.6}Sr_{0.4}Co_{0.2}Fe_{0.8}O_{3-\delta}$ (LSCF) Oxygen Permeable Membrane Using Vacuum-Assisted technique *MATEC Web Conf. 69* **05002** 1–6
[5] Jin W, Li S, Huang P, Xu N and Shi J 2000 Fabrication of $La_{0.2}Sr_{0.8}Co_{0.8}Fe_{0.2}O_{(3-\delta)}$ mesoporous membranes on porous supports from polymeric precursors *J. Memb. Sci.* **170** 9–17
[6] Han N, Zhang S, Meng X, Yang N, Meng B, Tan X and Liu S 2016 Effect of enhanced oxygen reduction activity on oxygen permeation of $La_{0.6}Sr_{0.4}Co_{0.2}Fe_{0.8}O_{3-\delta}$ membrane decorated by K_2NiF_4-type oxide *J. Alloys Compd.* **654** 280–9
[7] Wang Z, Yang N and Meng B 2008 Preparation and oxygen permeation properties of highly asymmetric $La_{0.6}Sr_{0.4}Co_{0.2}Fe_{0.8}O_{3-\alpha}$ perovskite hollow-fiber membranes *Ind. Eng. Chem. Res.* 510–6
[8] Lee M, Wang B, Wu Z and Li K 2015 Formation of micro-channels in ceramic membranes - Spatial structure, simulation, and potential use in water treatment *J. Memb. Sci.* **483** 1–14
[9] Roseno K T C, Brackmann R, da Silva M A and Schmal M 2016 Investigation of LaCoO3, $LaFeO_3$ and $LaCo_{0.5}Fe_{0.5}O_3$ perovskites as catalyst precursors for syngas production by partial oxidation of methane *Int. J. Hydrogen Energy* **41** 18178–92
[10] Sunarso J, Baumann S, Serra J M, Meulenberg W A, Liu S, Lin Y S and Diniz da Costa J C 2008 Mixed ionic-electronic conducting (MIEC) ceramic-based membranes for oxygen separation *J. Memb. Sci.* **320** 13–41
[11] Shao X, Dong D, Parkinson G and Li C-Z 2014 Microstructure control of oxygen permeation membranes with templated microchannels *J. Mater. Chem. A* **2** 410
[12] Abdullah N, Rahman M A, Othman M H D, Ismail A F and Jaafar J 2016 Preparation and Characterization of Dual Layer Thin Layer Lanthanum Strontium Cobalt Ferrite / Alumina Hollow Fiber Membrane using Dip-coating and Brush-coating Techniques *Sains Malaysiana* **45** 1715–21

Breakthrough Adsorption Study of Crude Oil Removal Using Buffing Dust

L W N Setyaningsih[1], A T Yuliansyah[2], A Prasetyo[2], S K Arimanintan[2] and D R Putri[2]

[1]Department of Chemical Engineering, Islamic University of Indonesia, Yogyakarta, INDONESIA
[2]Department of Chemical Engineering, Gadjah Mada University, Yogyakarta, INDONESIA

lucky.nuzulia@uii.ac.id

Abstract. The utilization of leather industry solid waste as adsorbent to separate oil from water emulsions of surfactant flooding process is a solution that is relatively inexpensive. This study was conducted aiming to obtain a mathematical model that is appropriate for the adsorption process of crude oil by buffing dust in emulsion phase with a continuous adsorption method. Variations in the column adsorption experiments were carried out, such as: flow rate of feed of water-crude oil-surfactant, the concentration of crude oil in the feed, and mass of adsorbent used. Data were evaluated using three models: Adams Bohart, Thomas and Yan. Best results are obtained on the following conditions, the feed flow rate of 60 mL/minute, the crude oil concentration in feed is 1.5% volume and the mass of adsorbent used was 10 g. The values of kinetic constant and adsorption capacity obtained from Yan Model was 21.7774 mL/mg/minute and 220.9581 mg/g with the relative error obtained is 5.4424%.

1. Introduction

Buffing dust is another issue faced by every leather industry, therefore it need an effort to deal with the solid waste pollution. This solid waste contains chromium as 27 g/kg of buffing dust, synthetic fats, oils, tanning agents, chemical dyes. Buffing dust is a microfined solid particulate impregnated with trivalent chromium (Cr^{3+}) and potentially converted into hexavalent chromium (Cr^{6+}) during the incineration at temperature 800°C[1]. This waste causes clinical problems like respiratory disease, allergic dermatitis, kidney malfunction and lung cancer if ingested or inhaled by human [2]. Disposal of buffing dust should not be done carelessly, because there is a standard of maximum chromium concentration that permitted, which is in soil were 714 mg / kg of soil and in groundwater were 0.25 mg / L[1].

The density of buffing dust is relatively small, 0.05 g/cm^3, which indicates that the buffing dust has a large pore volume. Buffing dust generated from the leather industry has a porous surface and a good candidate to be used as an adsorbent [3]. Buffing dust can be used as an adsorbent to separate oil from water emulsions of surfactant flooding process. The utilization of leather industry solid waste as adsorbent is a solution that is relatively inexpensive because the adsorbent used is waste that has been unused and can obtained at a low price or even free, and the process is simple.

Similar studies done by Gammoun [3] was analyzed factors that affect the sorption of crude oil using buffing dust in batch systems. The study resulted several conclusions as follows, buffing dust has high selectivity to crude oil sorption than water sorption, adsorption of crude oil into buffing dust occurs spontaneously and was not influenced by time, the sorption capacity of crude oil by buffing dust about

Content from this work may be used under the terms of the Creative Commons Attribution 3.0 licence. Any further distribution of this work must maintain attribution to the author(s) and the title of the work, journal citation and DOI.
Published under licence by IOP Publishing Ltd

IOP Conf. Series: Materials Science and Engineering **358** (2018) 012058 doi:10.1088/1757-899X/358/1/012058

0.9844 g / g adsorbent. Other results showed that the salinity in sea water has no effect on the adsorption process.

This study was conducted aiming to obtain a mathematical model that is appropriate for the adsorption process of crude oil by buffing dust in emulsion phase with a continuous adsorption method. Variations in the column adsorption experiments were carried out, such as: flow rate of feed of water-crude oil-surfactant, the concentration of crude oil in the feed, surfactant concentration, and mass of adsorbent used. The kinetic and dynamic behaviours of the adsorption process were then evaluated to study the adsorption mechanism of crude oil molecules onto buffing dust.

2. Material and Method

The material used in this research were buffing dust that were supplied by PT. Nike Indonesia, this solid waste was not purified prior to use, crude oil used in this research was obtained from wells P-14 Rantau, Sumatera, Indonesia, Sodium Lauryl Sulfate were supplied by CV. Brataco Chemical, Yogyakarta, Indonesia.

The research method is as follows, the feed with certain concentration of crude oil (1%, 1,5% and 2% volume) and surfactants (2% volume) as much as 4 L was loaded into the container and then flowed with the desired flow rate of (60 mL/min, 100 mL/min, 125 mL/min) into adsorption pond with a diameter of 4 cm containing buffing dust and marbles. Every few minutes 10 mL sample was taken to measure the crude oil concentration by using spectrophotometer UV-Vis. The flow was stopped when the column has been saturated. All the experiments were carried out at room temperature. Data were evaluated using three models: Adams Bohart, Thomas and Yan. Model Adams-Bohart written in linear form as follows[4]:

$$ln\frac{C}{C_0} = k_{AB}\,C_0\,t - k_{AB}\,q_{AB}\,\frac{z}{U} \tag{1}$$

Where C is the concentration of crude oil in the effluent after passing through the adsorption column, C_0 is the concentration of crude oil in the effluent before adsorption, k_{ab} is the kinetic constant on the model of Adams-Bohart, t is time of draining waste, q_{ab} is the adsorption capacity, z is the position of the axial direction in columns and U is the flow rate of the waste through the column. Thomas Model is written as follows[5]:

$$ln\left(\frac{C_0}{C} - 1\right) = \frac{k_{Th}}{Q} q_{Th} W - k_{Th} C_0 t \tag{2}$$

Where k_{th} is the kinetic constant of Thomas model, Q is the discharge of liquid through the column, q_{th} is the adsorption capacity and W is mass of buffing dust. Yan Model in pokhrel and viraraghavan[6] proposed an empirical equation that can overcome Thomas model weaknesses, especially thomas models in predicting the effluent concentration at time of zero. Yan Model is written as follows:

$$ln\left(\frac{C}{C_0 - C}\right) = a \ln V - a \ln f \tag{3}$$

$$f = \frac{k_Y\,q_Y\,W}{Q} \tag{4}$$

$$a = \frac{k_Y\,C_0}{Q} \tag{5}$$

Where V is the volume of waste, k_Y is the kinetic constant on Yan model, and q_Y is the adsorption capacity.

IOP Conf. Series: Materials Science and Engineering **358** (2018) 012058 doi:10.1088/1757-899X/358/1/012058

3. Result and Discussion

The result of oil sorption on buffing dust using continuous system presented in form of curves between ratio of the outlet crude oil concentration to inlet crude oil concentration as a function of time (C/C_0 vs t).

3.1. Effect of initial crude oil concentration

The effect of crude oil concentration in the feed is shown in figure 1. It shown that the lower concentration of crude oil provides breakthrough curve which increasingly ramps, it indicates that the volume of solution that can be passed through the column until the occurrence of the saturation was higher as the decreasing concentrations of crude oil. This is due to a decrease in the concentration of crude oil in the feed led to more displacement than the adsorbate molecules from body fluids to the surface of the adsorbent so that the mass transfer coefficient and diffusion will decrease. At the highest crude oil concentration which is 2% by volume, the adsorbent reached saturation in the shortest period of time which is less than 20 minutes. The higher the concentration of crude oil in the feed will give acceleration to reach saturated conditions by buffing dust. This is due to the increasing number of adsorbent surface which is covered by adsorbate molecules. Adsorption capacity will rise with increasing concentration of crude oil in the feed, due to large differences in concentrations that produce a driving force of the adsorption process. At the beginning of the curve it shows that C/C_0 incline sharply, suggesting that in these experiments the axial diffusion is quite affecting.

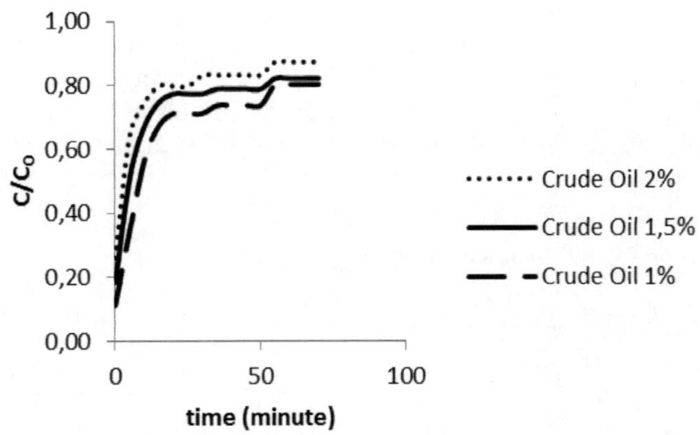

Figure1. Breakthrough curves for crude oil sorption on buffing dust at different initial crude oil concentration

3.2. Effect of emulsion flowrate

The effect of solution flow rate shown in figure 2. The increasing flow rate of the feed stream generating capacity in reducing crude oil was higher and the time to reach saturation was faster. Figure 2 show that the greater the velocity of solution will produce a ramps breakthrough curve. The amount of solution that can be streamed in the column until it reaches saturation will increase along with rising mass used buffing dust. This is because the molecules adsorbed crude oil increased. At low flow rates adsorption capacity is also low due to the retention time in the column is not sufficient for the diffusion of crude oil into the pores of bufing dust. It is because the molecules of crude oil left column before equilibrium conditions occur.

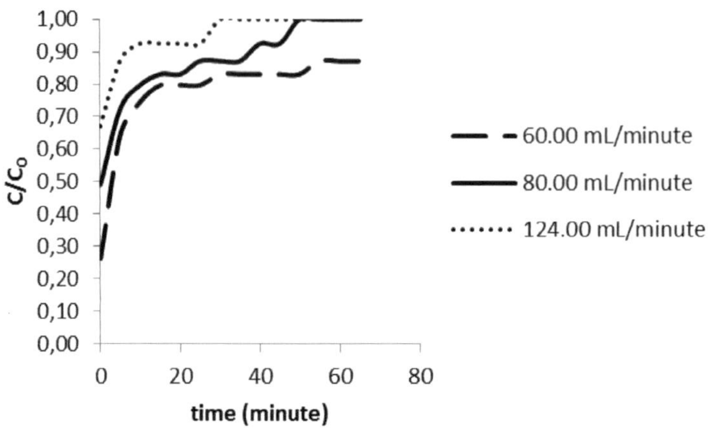

Figure 2. Breakthrough curves for crude oil sorption on buffing dust at different emulsion flowrate

3.3. Effect of mass of adsorbent

The effect of mass of adsorbent shown in figure 3. Figure 3 shows that the greater the mass of buffing dust used, the longer time required to reach the saturation and the greater the volume of solution that may be passed through would be. This is due to the greater mass of adsorbent, the surface area available for adsorption of crude oil on the surface of the adsorbent more. The mass of adsorbent used will affect the high bed, the lower the mass of adsorbent used the height of the bed will be decreased so that the residence time of molecules of crude oil in the column is shorter and therefore the time available was not enough for the molecules of crude oil to diffuse into the pores of the adsorbent.

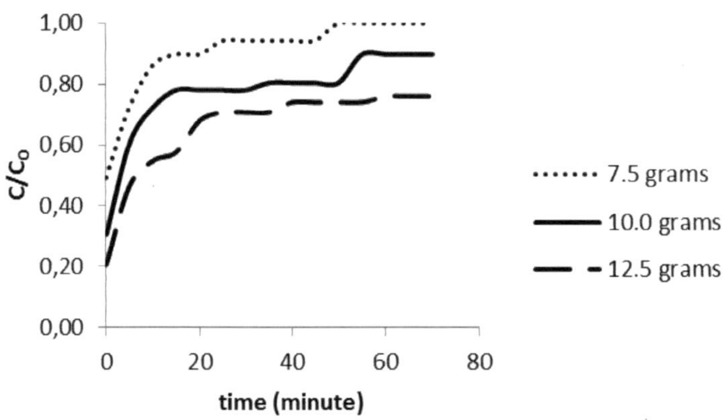

Figure 3. Breakthrough curves for crude oil sorption on buffing dust atdifferent mass of adsorbent

3.4. Breakthrough curve study

The data were fitted to three kinetics models and verified with the sum of error squares (SSE %). Comparison between the experimental data with the three models shows that Yan Model fit to a whole range of values of C/C_0. Thomas Model is just fit for the area of final breakthrough curves, whereas the Adams Bohart model is not good enough to describe the region early breakthrough curve. The calculation result of k_{ab} and q_{ab} from equation 1, k_{th} and q_{th} from equations 2 and k_y and q_y from equation 3 are summarized in Table 1. Table 1 show that the concentration of crude oil increased the value of adsorption capacity of the column using Yan and Thomas Models (q_{th} and q_y) decreased, while the Adams-Bohart Model has tendency to increase. In another study showed the different trend, where adsorption capacity increase with increasing solute concentration. It is because that in high difference concentrations provide greater driving force to the adsorption process in order to achieve higher capacity. Similarly, the kinetic constant for Yan and Thomas Models (k_{th} and k_y) decreased when the crude oil concentration increased, while the kinetic constant of Adams-Bohart Model increased. This may occur because of the emulsion solution characteristic. The emulsion contains surfactant that make emulsion become more stable, which decreases the adsorption capacity of adsorbent. Also emulsion becomes less hydrophobic and therefore the contact between adsorbent and solute decreases[7].

Table 1. Parameter of Yan, Thomas and Adams-Bohart Model for Crude oil Concentration Variation

Crude Oil Concentration (%vol)	Yan			Thomas			Adams-Bohart		
	k_Y	q_Y	SSE %	k_{Th}	q_{Th}	SSE %	k_{AB}	q_{AB}	SSE %
2%	15.25	15.04	8.05	0.01	-206.36	271.37	0.003	171.69	15.07
1.5%	21.78	20.96	5.44	0.02	-42.29	233.40	0.004	145.37	10.90
1%	36.53	40.11	7.65	0.03	83.73	168.50	-0.0001	-3260	32.01
	SSE		7.05	SSE		224.43	SSE		19.33

The SSE of Yan Models is 7.048%. smaller than SSE resulted in Thomas and Adams-Bohart Models. Table 2 shows that the kinetic constant of Adams-Bohart. Thomas and Yan Model increases with the increasing of flow rate of the solution stream. This indicates that the kinetic constant of Adams-Bohart, Thomas and Yan showed that mass transfer resistance in the liquid is not negligible. On the other hand, with flowrate increasing the adsorption capacity using Yan and Adams-Bohart increase while the Thomas model result different trend where lower flow rate would increase the adsorption of crude oil on the buffing dust.

Table 2. Parameter of Yan. Thomas and Adams-Bohart Model for Flowrate Variation

Flowrate (mL/min)	Yan			Thomas			Adams-Bohart		
	k_Y	q_Y	SSE %	k_{Th}	q_{Th}	SSE %	k_{AB}	q_{AB}	SSE%
60	15.2866	15.0731	6.7355	0.0145	-169.021	266.8574	0.0033	159.8112	11.5702
80	34.6530	34.1430	11.1886	0.0227	-201.646	92.1001	0.0033	220.8782	14.1670
120	63.5292	57.8637	8.8749	0.0303	-534.428	59.9419	0.0017	315.6504	7.8136
	SSE		8.9321	SSE		139.6331	SSE		11.1836

As the mass of adsorbent increased. the value of adsorption capacity increased for both of Thomas and Adams-Bohart Model, while the value of adsorption capacity for Yan Model decreased. This was due to the greater mass of adsorbent the closer the distance between the grains buffing dust would be. Thus there were overlaps resulting adsorbent surface in contact with the liquid is getting smaller. The kinetic constant for Yan and Thomas Models decreased with increasing of mass of adsorbent, but increased for Adams-Bohart Models.

The assumptions used in the Adams-Bohart Model that axial dispersion is negligible cannot be applied in this experiment. The experimental results show that the macroscopic properties of the buffing dust affect the kinetic constant so that the phase in the column cannot be considered pseudo homogeneous. Likewise, the assumptions used in Thomas model, in which the axial dispersion is negligible and Thomas model derivation are based on a second order reaction kinetics cannot be applied in this experiment because the mass transfer of prisoners greatly influence the determination of kinetic constants. Meanwhile the model formulated to improve Thomas model in describing the area of early breakthrough curve.

Best results are obtained on the following conditions, the feed flow rate of 60 ml/min, the crude oil concentration 1.5% volume and the mass of adsorbent used was 10 g. The values of kinetic constant and adsorption capacity obtained from Yan Model were 21.7774 mL/mg/minute and 220.9581 mg/g with the relative error obtained is 5.4424%.

Table 3. Parameter of Yan, Thomas and Adams-Bohart Model for Mass of Adsorbent Variation

Mass of buffing dust (grams)	Yan			Thomas			Adams-Bohart		
	k_Y	q_Y	SSE %	k_{Th}	q_{Th}	SSE. %	k_{AB}	q_{AB}	SSE. %
7.5	30.177	34.9355	9.1653	0.0289	-233.183	301.4891	0.0025	167.384	8.2739
10	21.0065	19.7754	8.8695	0.0161	-65.258	258.2961	0.0036	164.7672	11.7994
12.5	13.7276	15.4143	9.7012	0.0119	93.0787	179.5293	0.0038	192.7947	16.4755
	SSE		9.2453	SSE		246.4382	SSE		12.1829

4. Conclusion

Comparison between the experimental data with the three models shows that Yan Model fit to a whole range of values of C/C_0. Thomas Model is just fit for the area of final breakthrough curves. Whereas the Adams Bohart model is not good enough to describe the region early breakthrough curves.

References

[1] Sekaran G, Shanmugasundaram K A, Mariappan M 1998 Characterization and Utilisation of Buffing Dust Generated by the Leather Industry *J. Hazard Mater* B **63** 53-68

[2] Swarnalatha S, Srinisvasulu S, Srimurali M and Sekaran G 2008 Safe Disposal of Toxic Chrome Buffing Dust Generated from Leather Industries *J. Hazard Mater* **150** 290-299

[3] Gammoun G, Soufiane T, Abderrahman A, Mohammed A, and Miguel D L G 2007 Decontamination of Water Polluted with Oil Through The Use of Tanned Solid Wastes *J. Environ Eng Sci* **6** 553-559

[4] Ghomshe S M, Mousavi S M, Soltanieh M and Kordi AK. 2011 Batch and column study of haloacetic acids adsorption onto granular activated carbon *Sci Res Essays* **6** 3553-3560

[5] Sekhula M M, Okonkwo J O, Zvinowanda C M, Agyei N N, and Chaudhary A J 2012 Fixed bed column asodrption of Cu (II) onto maize tassel-PVA beads *Journal of Chemical Engineering and Process Technology*, **3** 1-5

[6] Pokhrel D and Viraraghavan T 2008 Arsenic removal in an iron oxide-coated fungal biomass column: analysis of breakthrough curves *Bioresource Technol* **99** 2067-2071

[7] Wang D, Elisabeth M, Robert P, and Lin Y S 2012 Adsorption of Oils from Pure Liquid and Oil-Water Emulsion on Hydrophobic Silica Aerogels *Sep Purif Technol* **99** 28-35

IOP Conf. Series: Materials Science and Engineering **358** (2018) 012059 doi:10.1088/1757-899X/358/1/012059

CaO Nanocatalyst for Transesterification Reaction of Palm Oil to Biodiesel: Effect of Precursor's Concentration on the Catalyst Behavior

N Hassan*, K N Ismail, K H Ku Hamid and Abdul Hadi

Faculty of Chemical Engineering, Universiti Teknologi MARA, 40450 Shah Alam, Malaysia

*noraakinah@yahoo.com, knoor@salam.uitm.edu.my, Kuhalim@salam.uitm.edu.my, hadi9598@salam.uitm.edu.my

Abstract. Depletion of fossil fuel sources in a few decades due to industrialization and motorization has led to a keen interest in the production of alternative fuels like biodiesel. Research on the development and improvement of more efficient transesterification process for biodiesel production has attain great attention in the last decade. The using of low cost catalyst is one of the main focuses on the biodiesel production. As a basic heterogeneous catalyst, CaO has been examined in the transesterification of vegetable oils for biodiesel production. In this research, calcium oxide (CaO-*X*) catalysts were prepared by sol-gel method at different Ca^{2+} precursor concentration (X = 1.0, 1.5, 2.0 M). The crystalline structure and morphology of the synthesized catalysts were characterized by means of x-ray diffraction (XRD) and N_2 adsorption-desorption analysis. All the synthesized catalysts were then applied to transesterification reaction of palm oil to produce biodiesel. The characterization by x-ray diffraction demonstrate CaO-1.0 was partially hydrated due to the incomplete reaction during synthesis. As a matter of fact, formation of H_2O on the surface of CaO causes lower basic strength of the catalysts, thus responsible in lowering the catalytic activity. It is demonstrated that CaO-2.0 exhibits mesoporous structure with least chemisorb amount of H_2O on the catalysts surface has a very active catalytic activity. It was found that 2.0M of calcium precursor has high catalytic activity and 81% FAME yield was obtained within 3h reaction.

1. Introduction

Biodiesel, a clean renewable fuel, has as of late been considered as the best possibility for a diesel fuel substitution on the grounds that it can be utilized as a part of any pressure ignition motor without the requirement for adjustment. There are four primary ways to produce biodiesel which are direct use and blending of raw oils, micro-emulsions, thermal cracking and transesterification [1]. Transesterification is a process in which fat or oil reacts with an alcohol to form esters and glycerol. It can be catalyzed by both homogeneous catalysts and heterogeneous catalysts [2], [3]. Biodiesel catalyzed by heterogeneous catalysts has advantages over homogeneous catalysts from both economic and environmental standpoints. Using heterogeneous catalysts can reduce the cost of biodiesel production because they more readily separate from the liquid and have a long lifetime

Content from this work may be used under the terms of the Creative Commons Attribution 3.0 licence. Any further distribution of this work must maintain attribution to the author(s) and the title of the work, journal citation and DOI.
Published under licence by IOP Publishing Ltd

[4], [5]. Solid base catalysts, calcium oxide has been extensively investigated due to its high basicity, low solubility in the organic solvents, and low cost [6].

Commonly calcium oxide, CaO are produced via thermal treatment (decomposition) or could be synthesized by the sol – gel method without using any surfactants and complicated tools. The sol-gel method is based on inorganic polymerization reactions and has a cutting age over the common solid-state routes because it allows molecular mixing of constituents leading to chemical homogeneity, and that shorter diffusion distances for reactants lead to lower sintered temperature for crystallization [7], [8]. In a sol – gel process, the precursor solution is converted into an inorganic solid by dispersion of colloidal particles in a liquid (sol) and conversion of sol into rigid phase (gel) by hydrolysis and condensation reactions. The method is well applicable for the synthesis of nanoparticles of oxides of different metals like Sn, Ti, V, Zr, Ta, Nb, Hf, In, Fe, Cr, Ni, Mn, Sm, W, Li, Al, in aqueous, non – aqueous (organic), mediums with or without surfactants. The method allows better control on the properties of nanoparticles such as volatilization, contamination, phase separation, etc. The advantage of being able to control these properties lead to the formation of highly pure and homogenous multi component compounds [9].

Despite its low cost and easy preparation, there is a problem that related to CaO catalyst employed in the reaction of producing biodiesel which is the low catalytic activity of CaO in transesterification of biodiesel [10]–[12]. This problem could affected the yield of biodiesel. Therefore, this research addresses the method that can be used to increase the catalytic activity in biodiesel production. In this research, the study was focused on the synthesized heterogeneous catalysts of calcium oxide, CaO via sol – gel method with different concentration which are 0.5M, 1.0M, 1.5M and 2.0M. Besides that, this study was also focused on the study of the physical properties of CaO, the characterization of CaO and also the production of biodiesel from palm oil and methanol catalyzed by prepared CaO in laboratory scale. The crystallinity, phase structure and crystallography of CaO were characterized using X-ray diffraction analysis (XRD). Whereas, the surface area, pore volume, pore sizes and isotherm profile of the catalyst were characterized by N_2 adsorption-desorption analysis.

2. Experimental

2.1 Materials and Catalyst Preparation

The sol-gel sample of calcium oxides (CaO) were prepared using a modified Pechini procedure reported elsewhere [10], [11], [13]. Briefly, equimolar of an aqueous solution (1.0M) containing calcium nitrate tetrahydrate and citric acid was prepared. For this purpose, approximate weight of $Ca(NO_3).4H_2O$ (Merck, Malaysia) was completely dissolved in approximately 80mL of deionized water and added slowly to thus prepared citric acid/deionized water solution. The mixture was heated at 80±2°C under vigorous stirring until a viscous gel (translucent pale yellow) was formed. The gel was dried overnight at 130±5°C. The sample was then crushed in porcelain mortar before calcined in a muffle furnace at 850°C for 5h with a step of 5°C/min heating rate. Other samples were prepared accordingly to the same procedure with different molarity of 1.5M and 2.0M

2.2 *Catalysts Characterization*

The crystallinity behavior of the catalysts was recorded using Rigaku Ultima IV diffractometer with Cu Kα (0.15418nm) radiation source over 2θ range from 20° to 70° with a 1.2°/min operating at 40kV and 30mA. The textural properties were obtained using N_2 adsorption-desorption (Autosorb-1 Quanta Chrome Instrument TUSA).

2.3 Transesterification of palm oil with methanol over CaO

The catalytic activity of transesterification of palm oil was carried out under reaction condition of 60°C reaction temperature, 3h reaction time, 9:1 methanol to oil molar ratio, and 4 wt% catalyst under 600 rpm stirring. First mixture was prepared by dissolving methanol and catalysts for 30min. Then, 25g of oil was heated up to the reaction temperature in 50mL conical flask using hot plate equipped with temperature probe sensor. The mixture of methanol/catalyst was added into heated oil and heated up to the reaction temperature in a specified reaction time. The catalyst was separated from the product by centrifuging at 10 000 rpm for 5 min and finally separated by separation funnel. The biodiesel yields were calculated from the equation below:

$$Yield\ (\%) = \frac{weight\ of\ biodiesel\ produced}{weight\ of\ palm\ oil}\ x\ 100 \qquad (1)$$

3. Result and Discussion

3.1 Crystallinity behavior of CaO-X Catalysts

The phase and purity of the synthesized catalysts were characterized by powder x-ray diffraction (XRD) analysis. Figure. 1 shows the x-ray diffraction patterns of the samples CaO-1.0, CaO-1.5 and CaO-2.0 respectively.

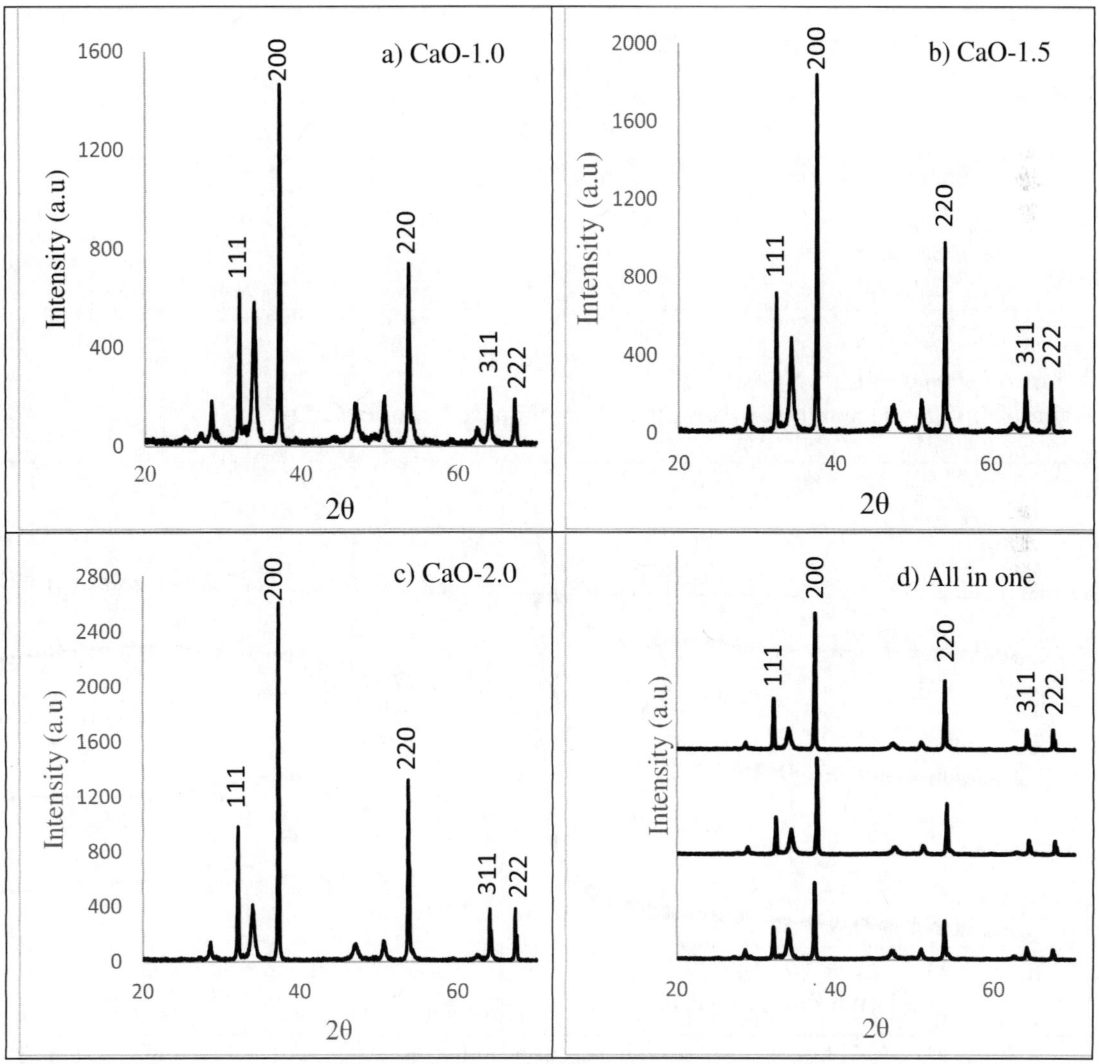

Figure 1. XRD patterns for CaO catalysts with different precursor concentrations: a) CaO-1.0, b) CaO-1.5, c) CaO-2.0, and d) All in one.

It can be seen that they are nearly identical patterns with traces of impurities are observed. All the diffraction peaks of the catalysts can be indexed to a cubic structure of CaO at 2θ = 32.41°, 37.60°, 54.23°, 64.18°, and 67.86° (PDF 00-001-1160, *a* = 4.7970Å and *c* = 4.7970Å) with high crystallinity. For all the catalysts, traces of the impurities are also identified at peaks corresponding to portlandite, $Ca(OH)_2$ at 2θ = 28.78°, 34.20°, 47.31°, 50.98°, and 62.73° (PDF 00-002-0968, *a* = 3.5790Å and *c* = 4.9060Å) are present at relatively high intensity (>30%). It is clear that CaO sample are not pure. It seems that chemisorption of H_2O on CaO surface is very significant due to the reaction with moisture in the exposed environment during cooling process after calcination, storage and handling of the catalysts. This explains that sample CaO-1.0 is possibly damaged since the diffractogram of CaO-1.0 exhibits strong peaks corresponding to $Ca(OH)_2$ (I = 69.1%) compared to sample CaO-2.0 which almost pure CaO was obtained. Diffractogram of all catalysts exhibits a strong orientation along *c*-axis (200). The intensity of plane (200) increases with the increase of the precursor concentration. The crystallite size is calculated based on Scherrer's formula corresponding to full width of half maximum (FWHM) of the most intense peak (200). Table 1 shows the crystallite size of the catalysts at different concentration. It is clear that the average crystallite size of the catalysts increases in the order of CaO-1.0 < CaO-1.5 < CaO-2.0 according to the intensity of the most intense peak (200).

Table 1. The crystalline size for CaO catalysts.

Sample Code	[Precursor] (M)	Crystallite size (nm)
CaO-1.0	1.0	46
CaO-1.5	1.5	51
CaO-2.0	2.0	61

3.2 Morphologies of CaO-X catalysts

Figure. 2 (a) shows the nitrogen adsorption-desorption isotherms of CaO-*X*.

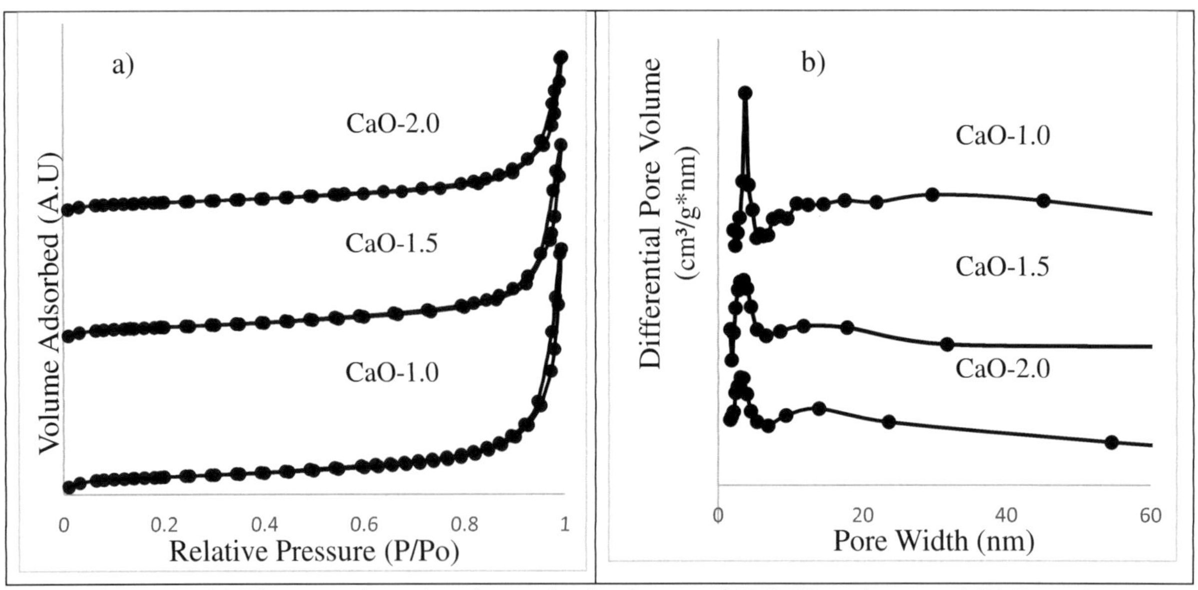

Figure 2. (a) Nitrogen adsorption-desorption isotherms of CaO-*X* catalysts, and (b) Pore size distribution curve of CaO-*X*.

According to the International Union of Pure and Applied Chemistry classification, all the CaO-*X* catalysts exhibit the type-IV isotherm with type-H4 hysteresis loop, which means that the prepared catalysts possess mesoporous structure [14]. The pore-size distribution shown in Figure. 2(b) of all catalysts is calculated using BJH model. The pore size distribution of all catalysts display nearly identical pattern and exhibit average pore size in the range of 21.78-28.48 nm. Textural properties of CaO-*X* catalysts are summarized in Table 2. The surface area and the average pore volume of CaO-*X* catalysts decreases with increasing precursor concentration. This result is well consistent with XRD result, where at higher concentration of precursor solution, larger crystallite size is obtained with reduced in surface area.

Table 2. Textural properties of CaO-*X* catalysts

Sample	Surface area (m^2/g)	Pore Volume (cm^3/g)	Dpore (nm)
CaO-1.0	11.19	0.058	28.48
CaO-1.5	9.19	0.049	23.56
CaO-2.0	7.74	0.037	21.78

3.3. Transesterification of Palm Oil over CaO

Figure. 3 shows the FAME yield obtained in the transesterification reaction of palm oil over CaO-*X* samples. It is observed that the yield of biodiesel increases with the increase of the concentration of the precursor. The yield increases in the order of 64% < 75% < 81% with respect of the concentration of the precursor.

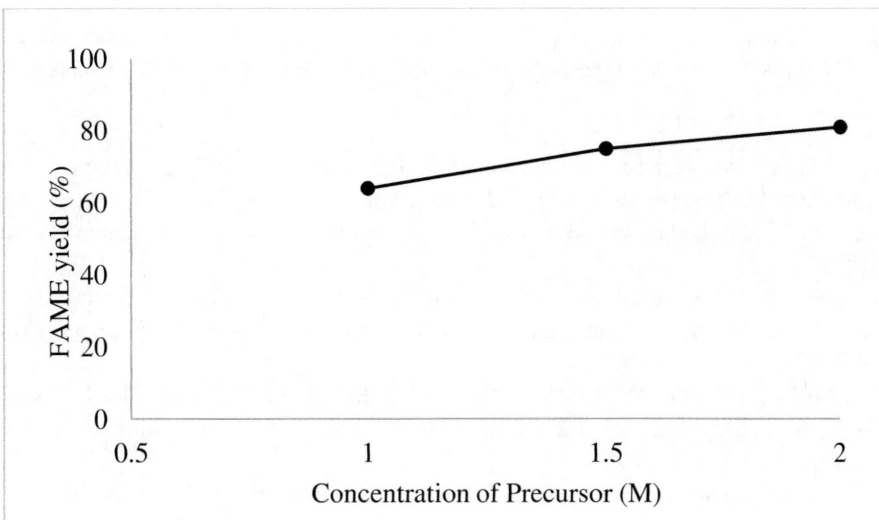

Figure 3. FAME yield of CaO-*X* catalyzed palm oil.

The trend can be explained by the relation with the XRD and N_2 adsorption-desorption results. From previous characterization, XRD results indicate that the fresh catalysts CaO-*X* is partially hydrated and sample CaO-1.0 has high possibility of damaged. This observation elucidate CaO is rapidly undergoes hydration processes during cooling event from calcination process, storage and handling. There was expectation that catalysts with higher surface area, smaller particle size could promote the catalytic activity in the conversion of product. Of note, all catalysts exhibit mesopores which are suitable for the accessibility or large molecules of free fatty acid. However, in this case of study, CaO-1.0 which exhibit good physical properties unable to promote the rate of reaction due to the poisoned active surface sites by hydration of H_2O on CaO core surfaces [15]. Besides, Alonso et al. also pointed out that low basic strength of $Ca(OH)_2$ is not strong enough to promote the transesterification reaction of fatty acids, thus suggesting that CaO-1.0 is practically has lower catalytic activity (lowest FAME yield of 64%) [16]. By increasing the precursor concentration, lower hydroxyl was formed on the catalysts surface. Catalysts CaO-2.0 emerge as a very active catalysts with least traces of hydroxyl, in 3h of reaction time a yield of 81% is obtained.

3. Conclusion

Metal oxides of calcium, CaO-*X* catalysts were prepared by sol-gel method at different calcium precursor concentrations (X = 1.0, 1.5, 2.0 M) to modify the morphology and physical properties of CaO-*X* catalysts. They were then applied in the transesterification of palm oil to biodiesel. Among the CaO-*X* (X = 1.0, 1.5, and 2.0 M) catalysts, CaO-1.0 catalyst showed the smallest crystallite size with larger surface area. However, the catalytic activity of CaO-1.0 was deteriorated by the poisoned active surface sites by hydroxyl due to the incomplete reaction during synthesis of the catalyst. In the transesterification of palm oil, yield of the biodiesel increased with respect to the concentration of the reactant. With respect to the sample of high concentration of reactant (CaO-2.0), a very intense peaks from CaO are visible indicates small amount of hydroxyl was formed on the CaO surface. It was found that 2.0M of calcium precursor has high catalytic activity as a heterogeneous catalyst and it was obtained 81% FAME yield within 3h reaction. This results demonstrate strong basic sites present at the surface of CaO causes significant activation of the catalytic activity in transesterification reaction. Thus, it is very reasonable to assume that hydroxylation of CaO is detrimental for the catalytic activity.

Acknowledgement

The authors acknowledge the financial support from Ministry of Higher Education Malaysia for Fundamental Research Grant Scheme (FRGS, Project Number : 600-RMI/FRGS 5/3 (91/2013)). We would also thank to Universiti Teknologi Mara (UiTM) for supporting the research.

References

[1] Leung D Y C, Wu X and Leung M K H 2010 A review on biodiesel production using catalyzed transesterification *Appl. Energy*, vol. **87** no. 4 1083–1095

[2] Gryglewicz S 2015 Rapeseed oil methyl esters preparation using heterogeneous catalysts vol. 8524

[3] Daud N M, Sheikh Abdullah S R, Abu Hasan H and Yaakob Z 2014 Production of biodiesel and its wastewater treatment technologies: A review *Process Saf. Environ. Prot.*, vol. **94** 487–508

[4] Aransiola E F, Ojumu T V, Oyekola O O, Madzimbamuto T F and Ikhu-Omoregbe D I O 2014 A review of current technology for biodiesel production: State of the art *Biomass and Bioenergy* vol. **61** 276–297

[5] Islam A, Taufiq-Yap Y H, Chu C M, Chan E S and Ravindra P 2012 Studies on design of heterogeneous catalysts for biodiesel production *Process Saf. Environ. Prot.* vol. **91** no. 1–2 131–144

[6] Taufiq-Yap Y H, Lee H V, Hussein M Z and Yunus R 2011 Calcium-based mixed oxide catalysts for methanolysis of Jatropha curcas oil to biodiesel *Biomass and Bioenergy* vol. **35** no. 2 827–834

[7] Epifani M, Melissano E, Pace G and Schioppa M 2007 Precursors for the combustion synthesis of metal oxides from the sol-gel processing of metal complexes *J. Eur. Ceram. Soc.* vol. **27** no. 1 115–123

[8] Wang B, Yan R, Lee D H, Zheng Y, Zhao H and Zheng C 2011 Characterization and evaluation of Fe2O3/Al 2O3 oxygen carrier prepared by sol-gel combustion synthesis *J. Anal. Appl. Pyrolysis* vol. **91** no. 1 105–113

[9] Bail A 2013 Investigation of a molybdenum-containing silica catalyst synthesized by the sol-gel process in heterogeneous catalytic esterification reactions using methanol and ethanol

Appl. Catal. B Environ. vol. **130–131** 314–324

[10] Zhang X, Li Z, Peng Y, Su W, Sun X and Li J 2014 Investigation on a novel CaO-Y2O3 sorbent for efficient CO2 mitigation *Chem. Eng. J.* vol. **243** 297–304

[11] Bakan F, Laçin O and Sarac H 2013 A novel low temperature sol-gel synthesis process for thermally stable nano crystalline hydroxyapatite *Powder Technol.* vol. **233** 295–302

[12] LuoC, Zheng Y, Zheng C, Yin J, Qin and Feng B 2013 Manufacture of calcium-based sorbents for high temperature cyclic CO2 capture via a sol-gel process *Int. J. Greenh. Gas Control* vol. **12** 193–199

[13] Wang B, Song X, Wang Z and Zheng C 2014 Preparation and Application of the Sol–Gel Combustion Synthesis-Made CaO/CaZrO3 Sorbent for Cyclic CO2 Capture Through the Severe Calcination Condition *Chinese J. Chem. Eng.* vol. **22** no. 9 991–999

[14] Neimark A V and Sing K S W 2008 Characterization of Solid Catalysts *Book Chapter*

[15] Zabeti M, Wan Daud W M A and Aroua M K 2009 Activity of solid catalysts for biodiesel production: A review *Fuel Process. Technol.* vol. **90** no. 6 770–777

[16] Alonso D M, Mariscal R, Granados M L and Maireles-Torres P 2009 Biodiesel preparation using Li/CaO catalysts: Activation process and homogeneous contribution *Catal. Today* vol. **143** no. 1–2 167–171.

IOP Conf. Series: Materials Science and Engineering **358** (2018) 012060 doi:10.1088/1757-899X/358/1/012060

Evaluation on the Presence of Nano Silver Particle in Improving a Conventional Water-based Drilling Fluid

H Husin[1,5], N Ahmad[2], N Jamil[2], O H Chyuan[3] and A Roslan[4]

[1]Department of Petroleum Engineering, Universiti Teknologi PETRONAS, 32610 Seri Iskandar, Perak, Malaysia

[2]Faculty of Chemical Engineering, Universiti Teknologi MARA, 40450 Shah Alam, Selangor, Malaysia

[3]Department of Mechanical Engineering, Faculty of Engineering, University of Malaya, 50603 Kuala Lumpur, Malaysia

[4]Faculty of Chemical Engineering, Universiti Teknologi MARA Cawangan Johor, Kampus Pasir Gudang, 81750 Masai, Johor, Malaysia

hazlina.husin@utp.edu.my

Abstract. Worldwide demand in oil and gas energy consumption has been driving many of oil and gas companies to explore new oil and gas resource field in an ultra-deep water environment. As deeper well is drilled, more problems and challenges are expected. The successful of drilling operation is highly dependent on properties of drilling fluids. As a way to operate drilling in challenging and extreme surroundings, nanotechnology with their unique properties is employed. Due to unique physicochemical, electrical, thermal, hydrodynamic properties and exceptional interaction potential of nanomaterials, nanoparticles are considered to be the most promising material of choice for smart fluid design for oil and gas field application. Throughout this paper, the effect of nano silver particle in improving a conventional water based drilling fluid was evaluated. Results showed that nano silver gave a significant improvement to the conventional water based drilling fluid in terms of its rheological properties and filtration test performance.

1. Introduction

The success in drilling operation in oil and gas production is highly depending on the performance of drilling fluid used. Drilling fluid is design to ensure the rotary drilling of subterranean formations is conceivable and economical with minimal negative impact to environment [1]. Drilling activities will meet great technical challenges especially in deep water operation. Under high pressure and high temperature (HPHT) conditions, performance of drilling fluid used in drilling activities could be diminished from its design parameters, hence led to technical problems such as pipe sticking, loss circulation and invasion to formation. Up to date, nanotechnology is considered as an advance technology in energy industries. The main difference between drilling fluids containing nanoparticles and the conventional base fluid is due to the very small sizes of the particles dispersed in them [2]. In this paper, nano silver was used as an additive in water based drilling fluid formulation towards enhancing the performance of the drilling fluid. The performance of the water based drilling fluid such as the changes on density, rheological properties and filtration test was evaluated.

Content from this work may be used under the terms of the Creative Commons Attribution 3.0 licence. Any further distribution of this work must maintain attribution to the author(s) and the title of the work, journal citation and DOI.
Published under licence by IOP Publishing Ltd

IOP Conf. Series: Materials Science and Engineering **358** (2018) 012060 doi:10.1088/1757-899X/358/1/012060

2. Materials and Methodology

2.1. Materials

Nano silver particle was supplied by Terra Techno Engineering Company. It is in a form of powder and grey in color. Its average particle size is less than 100 nm and has a surface area of 3.7 m^2g^{-1}.

2.2. Preparation of water based drilling fluid

Table 1 shows the composition that was used to prepare a water based drilling fluid [3]. The selected materials follow the ANSI/API 13A/ISO 13500 Specification for drilling fluid materials standard. Four samples of water based drilling fluid were prepared using a mixer, brand Hamilton. Every time an addition of component was made, a continuous stirring for about five minutes was conducted to make sure the added components are thoroughly mixed. Step wisely, fresh water and prehydrated gel were added, followed by soda ash, potassium chloride and caustic soda, Hydrozan and Hydro star NF® , barite and lastly nano silver particle (0.1 – 1.0 g). After the last component was added, the final mixture was stirred for one hour to confirm homogeneity, through visual observation. All tests were conducted according to ANSI/API 13B-1/ISO 10414-1 standard.

Table 1. Elements in water based drilling fluid formulation.

Material	Function	Quantity (gram)
Fresh Water	Base fluid	287.22
Prehydrated Gel	Viscosifier	22.56
Soda Ash	Hardness control	0.20
Potassium chloride	Water activity and cation exchange	39.13
Caustic Soda	pH controller	0.70
Hydrozan	Viscosifier	1.50
Hydro star NF®	Fluid loss	141.49
Barite	Weighing agent	3.50

2.3. Drilling fluid density measurement

Density of water based drilling fluid (or mud weight) was measured using a drilling fluid balance of accuracy to within 0.01 gcm^{-3}. The lid of the balance was removed from the balance cup and drilling fluid sample was placed inside the cup. The cup was covered with the lid and some drilling fluid was ensured to overflow from a hole of the lid to make sure the cup was fully filled with drilling fluid. Then, the arm balance was placed on the base, with the knife-edge resting in the fulcrum. The rider was moved along graduated arm until the vial was centered. Density measurement was recorded at the left hand edge of the rider.

2.4. Drilling fluid rheology measurement

Rheology properties of water based drilling fluid were determined by using a viscometer, brand Grace Instrument, US, model M3600 [4]. Drilling fluid sample was placed in the viscometer cup. The knurled knob was turned between the rear support posts to raise or lower the rotor sleeve until it was immersed in the sample to the scribed line. A dial reading at speed 300 RPM and 600 RPM was recorded. Plastic viscosity (PV) measurement was calculated by taking a dial reading at 600 RPM minus a dial reading at 300 RPM. A yield point (YP) measurement was calculated by subtracting PV value from the 300 RPM dial reading.

2.5 Filtration test

Filtration test was conducted at standard room temperature by using a high pressure, high temperature filter press equipment (HPHT filter press). Volume of filtrate of the sample was measured at pressure of 100 and 500 psi, after the sample was left for 30 minutes in a standard cell.

3. Results and discussion

Figure 1 shows that the presence of nano silver gave an insignificant effect on mud weight of water based drilling fluid. The mud weight of water based drilling fluid without the presence of nano silver is

9.0 ppg (i.e: pounds per gallon) as stated in ANSI/API 13A/ISO 13500 standard. Consequently, in the presence of nano silver at 0.1 g, 0.5 g and 1.0 g, the density is 9.1 ppg, 9.1 ppg and 9.3 ppg, respectively. In drilling operation, mud weight is designed slightly higher than formation pressure to ensure maximum penetration rate [5]. Too high mud weight will cause loss circulation of formation, stuck of drill pipe, decrease on penetration rate and formation damage. In general, the presence of nano silver gave ambiguous effect on mud weight due to negligible dimension of its particle size. Hence, the presence of nano silver would not affect the common drilling operation.

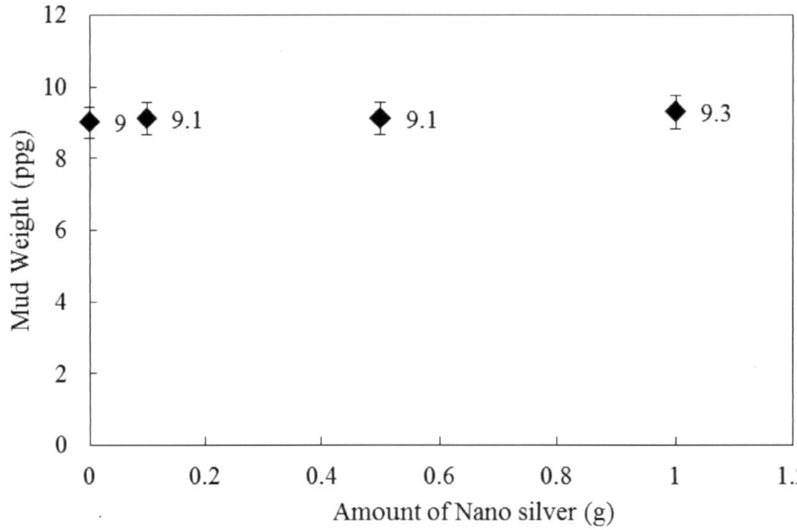

Figure 1. Density measurement of water based drilling fluid with and without the presence of nano silver. Data shown are with 5% error bar.

Plastic viscosity (PV) is defined as a resistance of fluid to flow [6]. PV represents the viscosity of drilling fluid. Figure 2(a) shows the viscosity of water based drilling fluid with and without the presence of nano silver.

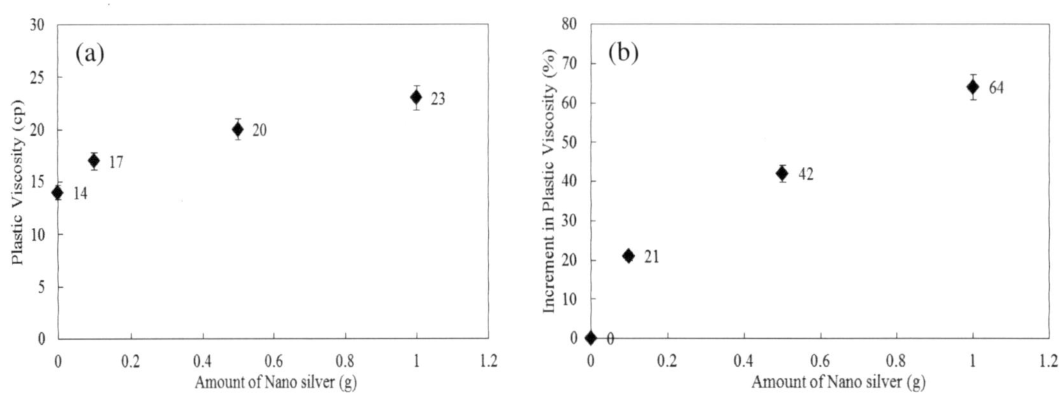

Figure 2(a)-(b). Plastic viscosity measurement of water based drilling fluid with and without the presence of nano silver. Data shown are with 5% error bar.

Significant increment due to the presence of nano silver as shown in figure 2(a) is reflected in figure 2(b), in terms of percentage. Figure 2(b) shows that, with only 0.1 g nano silver, the increment of PV is up to 21%. At 0.5 g, the increment is almost doubled (i.e. 42%) while at 1.0 g, the percentage of increment is 64%. The difference between 21% and 64% increment is deemed caused by the unique physical properties of nano particles. Nano silver is naturally insoluble in water and does not combine readily with other components in the water based drilling fluid formulation to form compounds. Therefore, due to these physical properties of nano silver, specifically (i) high solid content and (ii) high stability of waterborne particles, PV measurement was observed to be increased without the formation of additional attractive force. The Dejaguin–Landau–Verwey–Overbeek (DLVO) theory is a basis for theoretical calculations explaining nanoparticles behaviour in solution. According to DLVO theory, the stability of particles is determined by the net electrostatic surface interactions of the particles and their Van der Waals forces.

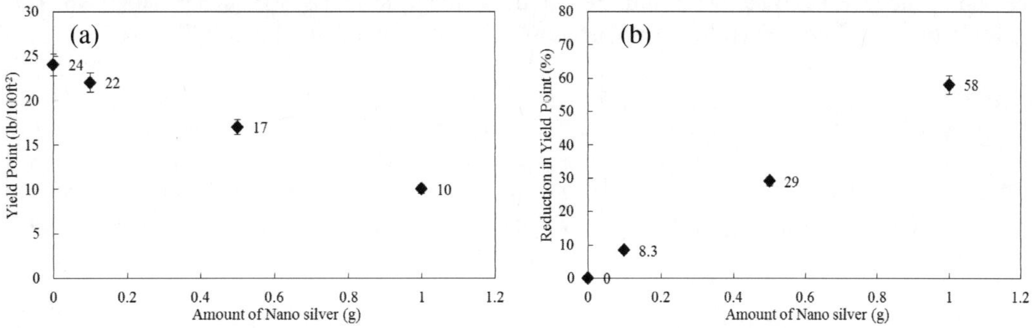

Figure 3(a)-(b). Yield point measurement of water based drilling fluid with and without the presence of nano silver. Data shown are with 5% error bar.

The net electrostatic surface interaction between the nano silver particles is further investigated by measuring the YP values. YP is the initial resistance to flow caused by the electrochemical forces between the particles [7]. Technically, it indicates how much pump pressure that must be applied during drilling operation in order to bring rock cuttings up to surface level. Meanwhile, in terms of surface interaction, high YP means attractive surface force is exhibited and low YP means steric surface force is exhibited. When YP was determined as a function of nano silver weight (n_{Ag}), a reduction of YP in the form of linear relationship $YP = -13.71\ n_{Ag}+23.734$ was reported (figure 3(a)). The percentage of reduction is shown in figure 3(b). In the presence of 0.1 g nano silver, the reduction of YP is 8.3%. At 0.5 g, the reduction is more than tripled (i.e. 29%). At 1.0 g, the percentage of reduction is in the magnitude of 5th (i.e. 58%).

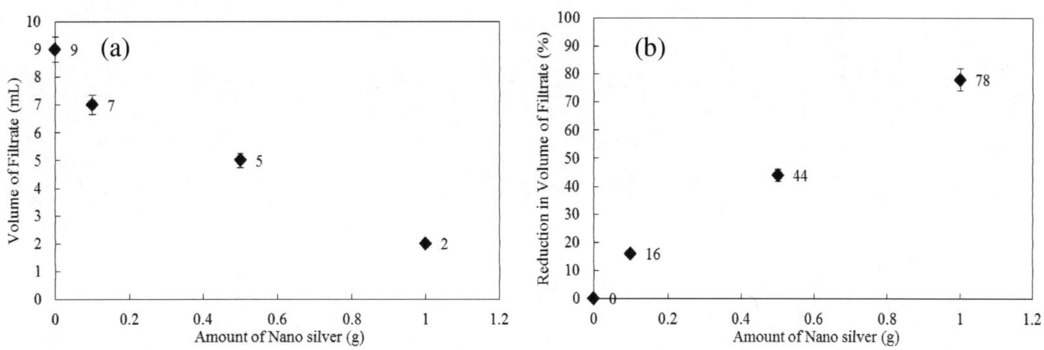

Figure 4(a)-(b). Volume of filtrate measurement of water based drilling fluid with and without the presence of nano silver. Data shown are with 5% error bar.

The effect of reduced YP revealed the existence of steric force formed under fixed ionic strength condition, between the nano silver particles in a colloidal 'water based drilling fluid' system. In general, water based drilling fluid with a presence of nano silver has a lower YP as compared to a conventional type. This means that only low force or stress is needed in order for the drilling fluid to flow. It is indeed highly desirable to have a low YP to allow sand and shale cuttings to settle out and entrained gas to escape, minimize swabbing effect during pulling the string out of hole and permit the circulation to be started at low pump pressure (i.e. economically feasible) [7].

Figure 4(a) shows the effect of nano silver on the volume of filtrate (i.e. fluid loss) of water based drilling fluid whilst the percentage reduction is shown in figure 4(b). Without the presence of nano silver, the fluid loss is 9 mL. At 0.1 g, 0.5 g and 1.0 g nano silver, the fluid loss is 7 mL, 5 mL and 2 mL, respectively. Drilling fluid with low volume of filtrate is highly desirable when designing drilling fluid. In drilling operation, high fluid loss would cause invasion of drilling fluids and borehole instability hence could reduce reservoir potential and reduce the amount of productivity. Here, it is observed that nano silver can acts as a fluid loss additive since fluid loss is reduced in the presence of nano silver. Common fluid loss additives are bentonite, starch, and organophillic lignite.

4. Conclusion

Overall, nano silver shown an influential effect on the performance of water based drilling fluid. According to density, rheology and filtration studies, nano silver proves that it can be used as an additive in the formulation of water based drilling fluid. Specially in improving the performance of conventional water based drilling fluid, nano silver increased the plastic viscosity up to 64%, reduced both yield point and fluid loss by 58% and 78%, respectively.

Acknowledgement

We wish to acknowledge Universiti Teknologi MARA for the financial support (Grant No. 600-RMI/DANA 5/3/REI (7/2014)). We wish to thank the referees for their contribution in making this a better paper.

References

[1] Fattah K A and Lashin A 2016 *J. African Earth Sci.* **117** 345–57
[2] Hassani S S, Amrollahi A, Rashidi A, Soleymani M and Rayatdoost S 2016 *J. Pet. Sci. Eng.* **146** 183–90
[3] Akpabio J U, Inyang P N and Iheaka C I 2015 *Int. J. Eng. Technol.* 29-35
[4] Ariffin T S T, Yahya E and Husin H 2016 *Procedia Eng.* **148** 1149-55
[5] Awele N 2014 *Aalborg Uni. Esbjerg (Master Thesis, MSc. Oil and Gas Tech.)*
[6] Dhiman A S 2012 *Dalhousie Uni., Halifax, Nova Scotia (Master Thesis, Master of Eng.)*
[7] Baumert M, Allouche E and Moore I 2005 *Int. J. Geomech.* **5** 339-49

IOP Conf. Series: Materials Science and Engineering **358** (2018) 012061 doi:10.1088/1757-899X/358/1/012061

Effect of Anions on Nanofiber Formation of *β*-sheet Propensity Amphiphile Peptide

H Shamsudeen[1], H L Tan[1] and Z Eshak[2]

[1, 2*] Faculty of Chemical Engineering, Universiti Teknologi Mara, 40450 Shah Alam, Selangor, MALAYSIA

[2] Faculty of Pharmacy, UiTM Puncak Alam Campus, 42300 Selangor, Malaysia

hueyling@salam.uitm.edu.my

Abstract. Peptide self-assembly forms different nanostructures under simple alteration in the solution environment. Understanding the mechanism of the assembly will help us to control and tailor functional nanomaterials. This study aims to investigate the influence of anions on the self-assembly morphology and shape using a synthetic peptide of FFFFKK. Circular Dichoism (CD) and Environmental Scanning Electron Microscope (ESEM) were used to determine the secondary structure and self-assembly morphology, while Image J imaging software was used to measure diameter size. In the absence of anion, FFFFKK formed anti-parallel *β*-sheet that adopted sizeable fibrillar structure with a minimal increment over the first 7 hours of assembly. Irregular structure was observed in the presence of Iodide ion (I^-) with a less stable secondary structure such as *β*-turn and *β*-loop. In the presence of perchlorate ion (ClO_4^-), needle-like structure was observed with predominantly *β*-sheet structure. Our study showed that peptide morphology can be controlled by using different anions with careful selection of amino acid residues in peptide sequence.

1. Introduction

Molecular self-assembling is the association of molecular building blocks that spontaneously organized into ordered macroscopic structure by non-covalent bond without external control [1, 2]. Over the years, the preference to study the behavioral growth of self-assembled peptide nanostructures (SPNs) has expanded [3, 4] due to its biocompatibility and chemical diversity. Many short amphiphilic peptides are commonly used as scaffolds to produce inorganic materials that are in one-, two- or three-dimensions at nano-scale [5, 6]. These peptides have also been successfully used to fabricate materials with well-defined size and shape that have applications in various industry such as drug delivery, wound healing and tissue engineering [7-9]. The assembly mechanism occurs within the existence of several factors including peptide-peptide interaction [2]. However, interruption of external stimuli such as anions and changes in pH or temperature will cost the interaction of either forming a functional or non-functional structure [2]. These self-assembling structures are relatively discrete and unstable. Therefore, the ability to control the rational design peptides into targeted structures at the early stages of self-assembly should be further developed. This study provided insights into the role of anions on the structure formation over the first seven hours of the assembly. We designed a synthetic peptide, Ac-FFFFKK-$CONH_2$ that consist of four hydrophobic amino acids, phenylalanine, (F) and two positively charge hydrophilic amino acids, lysine, (K) at pH 7. The study was carried out at room temperature in solutions with (I^- or ClO_4^-) or without anions. The assembly growth monitoring and characterisation were performed at the intervals of three to four hours. This study provided insights to enhance our mechanistic understanding of the peptide self-assembly and offer the strategy of controlling the assembly growth.

Content from this work may be used under the terms of the Creative Commons Attribution 3.0 licence. Any further distribution of this work must maintain attribution to the author(s) and the title of the work, journal citation and DOI.
Published under licence by IOP Publishing Ltd

2. Materials and methods

2.1 Materials
Sodium iodide and sodium perchlorate were purchased from Sigma-Aldrich. FFFFKK was synthesized based on the standard Fmoc solid-phase synthesis strategy [1] and purified by reverse-phase high-performance liquid chromatography indicating the high purity of at least 98%. F is phenylalanine represent a neutral charged hydrophobic amino acid and K is lysine, a positively charged hydrophilic amino acid.

2.2 Sample Preparation
Purified peptide was dissolved in solutions containing either distilled water or anions (130 mM of I^- or $ClO4^-$) to create solutions at a final concentration of 1.0 mM, that is above the critical self-assembling concentration (CSAC) of both the peptides. Concentration of salt solutions were 130 mM, similar to the standard salt concentration in human blood that is between 135 – 145 mM [10]. Solutions were stirred continuously resulting homogenous solutions before further characterization.

2.2.1 Environmental Scanning Electron Microscope (E-SEM)
Samples for E-SEM were recorded on a JEOL JSM-6701F electron microscope. The magnifications used to observe the growth were varied according to the practicalities on observing the samples from as low as 130x to 40000x. A small volume (10 μL) of fresh or aged peptide solution was dropped onto an aluminum tape glued on top of the SEM sample stub. Once absorb the sample were copper coated by a sputter coater. Then, the samples were placed on the stage of the E-SEM and inserted into the vacuum chamber to be viewed.

2.2.2 Circular Dichroism (CD)
The CD measurements were performed on a JASCO's J-1500 spectrometer using a 1 mm path length quartz cuvette. The CD spectra were recorded at room temperature with wavelengths ranging from 180 to 260 nm and a scan speed of 3s interval on average. The bandwidth was set to 2 nm [11] and an integrated Hg lamp was used as the light source. Before the reading of the sample was taken, the solvent background was subtracted and the spectra could be smoothed [11]. The resultant CD signals were expressed as mean residue ellipticity, [θ] which is in deg. cm^2. $dmol^{-1}$ unit versus wavelength.

2.2.3 Digital Image Analysis
Diameter of the structure was measured with the aid of Image J 1.43 (NIH) imaging software using eSEM images. The analysis was carried out by setting the measure scale to suit with the 2D eSEM image. Other detail and required measurements were set up in this step. Next, the boundaries of each structure on the image were detected. The boundary can easily be detected from binary images or other types, which were obtained from the threshold SEM images. Measurements were performed on regular structures only.

3. Results and discussion
Critical self-assembly concentration (CSAC) of FFFFKK was determined prior to the morphology study and secondary structure analysis. An analogy to CSAC has been drawn to the critical micelle concentration (CMC) of amphiphilic molecules such as surfactants. An amphiphilic peptide may display similar concentration-dependent self-assembly behavior, leading to the formation of peptide nano-structures [12]. We performed conductivity measurement to determine the CSAC values of FFFFKK as a function of peptide concentration. A series of peptide solutions with concentrations ranging from 0.02 to 2.0 mM was prepared by diluting the stock solution with distilled water. The conductivity of the peptide solution at different concentration was recorded by a conductivity meter. Below the CSAC, the addition of peptide molecules to the solution causes an increase in the number of charge carriers and consequently, an increase in the conductivity. Above the CSAC, further addition of peptide molecules increases the micelle concentration while the monomer concentration remains approximately constant. Since micelle is much larger than a peptide monomer, it diffuses more slowly

through the solution and is a less efficient charge carrier. A break on the plot of conductivity against peptide concentration is the CSAC. The plotted results suggested that FFFFKK has CSAC value of 0.25 mM (data not shown). In such case, peptides were able to self-associate since all the work was performed at a concentration of 1.0 mM.

Table 1. Average diameter of the structure formed in peptide solutions containing distilled water, iodide ion or perchlorate ion.

	FFFFKK (Diameter)/ *μm*		
Time/hour	**Water**	**Iodide ion**	**Perchlorate ion**
3	2.8	Irregular structure	1.1
7	3.2	Irregular structure	1.4

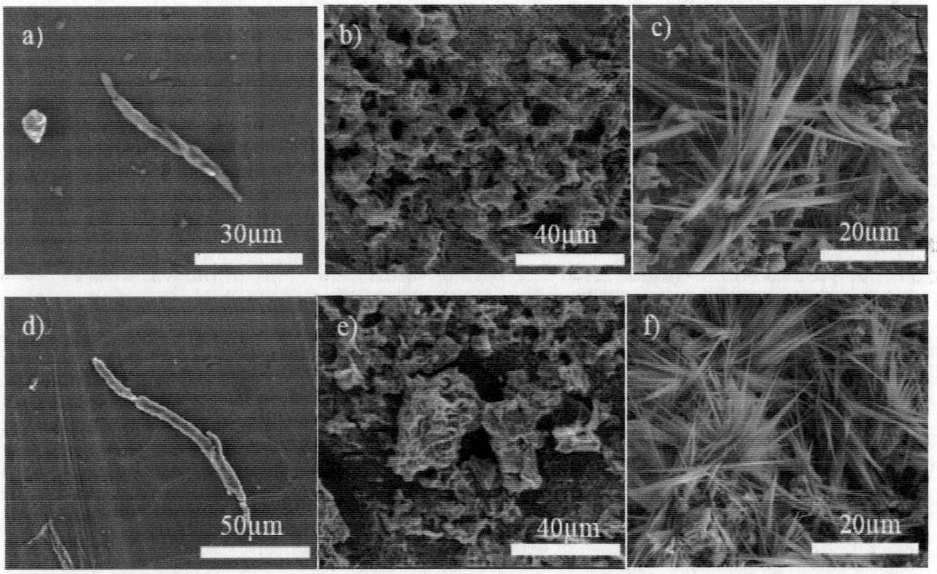

Figure 1. The ESEM images of FFFFKK after 3 hours of growth in: (a) Water, (b) iodide ion, (c) perchlorate ion; after 7 hours of growth in: (d) water, (e) iodide ion (f) perchlorate ion.

SEM images were measured to examine the self-assembly structures visually in the samples at different time intervals. Significant changes in the morphology and size were observed when FFFFKK dissolved in different aqueous milieu, either in the form of fibril, needle-like or amorphous structure as shown in Figure 1. In Table 1 is shown the average diameter of the structure formed when FFFFKK is dissolved in different solutions. FFFFKK formed fibrillar structure in water with diameter of 2.8 μm in the first three hours. It grew to 3.2 μm after 7 hours in the same solution with very low density. FFFFKK formed needle-like structure in solution containing ClO_4^- with lower diameter between 1.1 to 1.4 μm. Screening effect from anion to overcome the electrostatic is prominent in this case in which higher density of needle-like structure was observed. The formation of needle-like structure is most likely due to the higher tendency of ClO_4^- to induce micelisation [16] than I^-. Irregular and porous structure was found in the presence of iodide ion indicating the structure was not stable. Diameter measurement was not able to perform due to the limitation of the Image J imaging software.

Phenylalanine (F) is a strong hydrophobic amino acid that has a high tendency in forming β-sheet structure due to its hydrophobic effect [15]. CD results depicted in Figure 2 suggest that different packing geometries of β-sheets, β-loop and β-turn were present during the peptide assembly. In the distilled water, assembly structure remains to be anti-parallel β-sheet for 7 hours with proximate minimum at 195 and 220 nm respectively. In solution containing iodide ion, β-loop structure was observed within the first three hours with an aproximate negative minimum at 200 and 220 nm. The structure was later evolved to β-turn with weak signal at 195 and 210 nm. In the presence of perchlorate ion, FFFFKK formed needle-like structure with weak β-sheet signal at 195 and 215 nm initially. A strong β-sheet signal with negative minimum at 215 nm was observed with higher density of needle-like structure at the later hours.

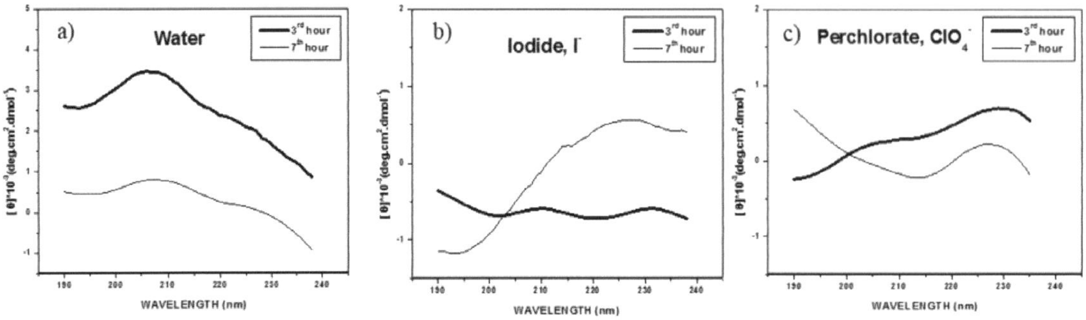

Figure 2. The CD spectra of FFFFKK during the first 7 hours in a) distilled water, b) sodium iodide and c) sodium perchlorate

Phenylalanine has the ability to form a stable hydrogel despite its aromatic residue [16]. Subsequently, the observation of different supramolecular architecture with the presence of anions indicated that the protonated lysine residues were neutralized. Drastic changes in morphology and shape when FFFFKK was in solutions containing I^- or ClO_4^- was most likely attributed to the complex interplay between neutralized electrostatic repulsion of charged lysine side chains and attraction energy between the structures [18]. The anions (I^- and ClO_4^-) were effective in screening the charge-charge interaction between the peptide monomers through anion binding [19, 20]. Anions were able to penetrate closer to the hydrophobic region of the peptide in which the water/hydrophobic interface of the peptide enhances the adsorption of large and polarizable ions such as I^- and ClO_4^- [21]. Observation of less stable structures such as β-loop and β-turn in the presence of I^- indicating that the self-assembly structures were prone to structural changes due to weak hydrogen bond and unstable NH-O angle [21]. The presence of needle-like structure in solutions containing ClO_4^- is most likely due to higher micelle affinity of ClO_4^-, which favors the formation of β-sheet structure that eventually forms needle-like structure [14].

4. Conclusion

In this study, we have demonstrated the influence of anions on FFFFKK have the potential in producing different higher order self-assembled architectures in solution containing perchlorate ion. These structural changes have been well-correlated with the interplay of the non-covalent forces including hydrophobic interaction and electrostatic repulsion. Formation of β-sheet structure drives the formation and growth of fibrillar structure. It is noticeable that the presence of anions (I^- and ClO_4^-) has a remarkable effect on the morphology. FFFFKK formed irregular structures in solution containing I^-. The peptide formed needle-like structures in large amount with the presence of perchlorate ion. Relatively less stable structures such as β-turn and β-loop was found in the early hours of assembly. With different preferences in promoting secondary structures, systematic changes in peptide sequences and charge distribution enable us to study the linkage between molecular architecture and self-assembled structure. Further study will be carried out to monitor the morphological changes of the peptide, including others in the series (FFKKFF and KFFFFK) at longer hours (up to 48 hours). This study provides insights to the future design of self-assembled nanomaterials in a rational way, such as scaffolds for tissue engineering and bone implant.

Acknowledgments

The authors would like to thank the Faculty of Chemical Engineering of Universiti Teknologi MARA for the financial support.

References

[1] Zhang S, Marini DM, Hwang W, Santoso S 2002 *Curr Opin Chem Biol* **6(6)** 865-871

[2] Mendes AC, Baran ET, Reis RL, Azevedo HS 2013 *Nanomed Nanobiotechnol* **5(6)** 582–612
[3] Xu H, Wang J, Han S, Wang J, Yu D, Zhang H 2008 *Langmuir* **25(7)** 4115–4123
[4] Huang R, Su R, Qi W, Zhao J, He Z 2011 *Nanotechnology* **22(24)** 245609
[5] Han S, Kim D, Han S, Kim NH, Kim SH, Lim Y 2012 *J Am Chem* **134(38)** 16047–16053
[6] Caplan MR, Schwartzfarb EM, Zhang S, Kamm RD, Lauffenburger DA 2002 *Biomaterials* **23(1)** 219–227
[7] Huang R, Su R, Qi W, Zhao J, He Z 2011 *Nanotechnology* **22(24)** 245609
[8] Shamsudeen H, Tan HL. 2015 *Adv Mater Res* **1113** 586–593
[9] Tan HL 2012 *The Chem Eng (IChemE)* March 2012 48–51
[10] Wang T, Zhong X, Wang S, Lv F, Zhao X 2012 *Int J Mol Sci* **13(11)** 15279–15290
[11] Tadros TF 2005 Physical Chemistry of Surfactant Solutions 19-51
[12] Savin T, Doyle PS 2007 Soft Matter **3(9)** 1194
[13] Hong Y, Fung S, Bezaire J, Keyes C 2003 *Biophys J* **85(1)** 537–548
[14] Zou D, Tie Z, Lu C, Qin M, Lu X, Wang M 2010 *Biopolymers* **93(4)** 318–329
[15] Li Y, Wang Y, Huang G, Ma X, Zhou K, Gao J 2014 *Angew Chem Int Ed Engl* **53(31)** 8074–8078
[16] Zandomeneghi G, Krebs MRH, Mccammon MG, Fändrich M 2004 *Protein Sci.* **13(12)** 3314-3321
[17] Cai, S., Singh BR 1999 *Biophys Chem* **80** 7-20
[18] Characteristic_IR_Band_Positions: http://www2.lbl.gov/mmartin/bl1.4/IRbands.html
[19] Zhao Y, Wang J, Deng L, Zhou P, Wang S, Wang Y 2013 *Langmuir* **29(44)** 13457–13464
[20] Zhou P, Deng L, Wang Y, Lu JR, Xu H 2016 *J Colloid Interface Sci.* **464** 219–228
[21] Galant NJ, Song HC, Jli I, Viskolcz B, Csizmadia IG, Knak Jensen SJ 2014 *Chem Phys Lett* **593** 48–54

IOP Conf. Series: Materials Science and Engineering **358** (2018) 012062 doi:10.1088/1757-899X/358/1/012062

The Effect of Nano Loading and Ultrasonic Compounding of EVA/LDPE/Nano-magnesium Hydroxide on Mechanical Properties and Distribution of Nano Particles

I A Azman[1], R M Salleh[1], S M Alauddin[1] and M I Shueb[2]

[1]Faculty of Chemical Engineering Universiti Teknologi MARA,(UiTM),Shah Alam, Selangor, Malaysia
[2]Agensi Nuklear Malaysia, Bangi, Selangor, Malaysia.

ruzitah@salam.uitm.edu.my

Abstract. Blends of Ethylene Vinyl Acetate (EVA) and Low-Density Polyethylene (LDPE) are promising composite which have good mechanical properties to environmental stress cracking. However, they lack fire resistant properties, which limits it usage in wire and cable industry. In order to improve flame retardancy ability, a range of nano-magnesium hydroxide (nano-MH) loading which is from 0 phr to maximum of 20 phr with ultrasonic extrusion 0-100 kHz frequencies have been introduced. Ultrasonic extrusion was used to improve the distribution of nano-MH. It was found that, 10 phr of nano loading with 100 kHz ultrasonic assisted has greater tensile strength compared to the nanocomposite without ultrasonication. Further increase of nano MH loading, will decrease the tensile properties. Better elongation at break was observed at10 phr nano-MH with the frequency of 50 kHz. The sample of 20 phr of nano-MH assisted with 50 kHz ultrasonic exhibits good flexural properties while 10 phr of nano-MH without the ultrasonic assisted demonstrates good in izod impact properties. From the evaluation of mechanical properties studied, it was found that 10 phr of nano-MH has shown the best performance among all the samples tested for EVA/LDPE/nano-MH composites. Transmission Electron Microscopy (TEM) has been conducted on 10 phr sample with different frequencies in order to observe the distribution of nano-MH particles. The sample with 100 kHz frequency shows more uniform dispersion of nano-MH in EVA/LDPE composites. This investigation indicates that the ultrasonic technology can enhance the mechanical properties studied as well as the dispersion of nano particles in the composite.

1. Introduction

The Polyolefins are the major group of commercial thermoplastics which have a wide range of applications due to their better combination of properties and low cost. Blends of Low-Density Polyethylene (LDPE) and Ethylene Vinyl Acetate (EVA) are widely used in consumer products such as cable and wire insulation [1][2]. It has good mechanical properties such as flexibility and toughness, good resistance to chemicals, and good process ability [3]. However, it has some drawbacks, such as low melting temperature, low thermal stability and high flammability, same goes with the ethylene vinyl acetate (EVA) applications that have been limited by its existing flammability. Therefore, flame retardants are introduced in order to improve the copolymer blends. Currently, the wire and cable industry are going through dramatic changes that require improvement in the design of wire and cable to compete with others in the industry. An alternative way to inhibit ignition and lower heat release rate during combustion is by adding additives such as nano magnesium hydroxide (nano-MH) [4]. The addition of these nanoparticle additives will affect the mechanical properties of the composite [5]. Apart from that, the success of nanoparticle applications depends on the ability of the nanoparticle to disperse evenly in the matrix. Since the powders are dry, the nanoparticle cannot disperse well in the copolymer

Content from this work may be used under the terms of the Creative Commons Attribution 3.0 licence. Any further distribution of this work must maintain attribution to the author(s) and the title of the work, journal citation and DOI.
Published under licence by IOP Publishing Ltd

matrix which leads to agglomeration [6]. Hence, the ultrasonic has been introduced into the extrusion process to enhance the dispersion of the nano particle [7].

Therefore, the aims of this research are to study the mechanical properties of LDPE/EVA at different loading of nano-magnesium hydroxide and ultrasonic frequency. Subsequently, the effect of nanoparticle distribution will be studied by applying ultrasonic extrusion with varying frequencies on the optimum loading of nano-MH in EVA/LDPE blend.

2. Experimental Procedure

2.1. Material

LDPE was purchased from Asia Polymer Corporation, Taiwan with density 0.9185 g/cm3 and melt flow index 8.3 g/10 min, EVA (Exxon's Escorene UL02528) with melt flow index of 2.0g/10 min. Nano-MH was supplied by Sigma Aldrich, Switzerland, have an average thickness of 20 nm and length of 90 nm.

2.2. Composite preparation

In mixing process, the LDPE/EVA/nano-MH composites were prepared by using extruder with ultrasonic assisted. It had three controlled temperature zones which were set 170°C with die zone at 160°C and the extruder screw speed is 20 rpm. Ultrasonic was set between 0-100 kHz. The extruding ratio of the sample was based on the 100 phr of LDPE with 1:1 ratio of EVA and the nano MH was varied with 5 different ratios ranging from 0, 5, 10, 15 and 20 phr. The extruded samples were moulded in a compression mould at 170oC under pressure of 10 MPa for 6 minutes after 6 minutes. Then the mould was transferred to a cold press and then cooled at the same pressure. By using manual cutter machine, the rectangular sheet was being cut into required ASTM samples. 10 replicates for each formulation was being cut. The collected sample will undergo mechanical and dispersion testing.

2.3. Testing

2.3.1. Mechanical testing. The tensile strength elongation at break and young modulus, were measured with a universal testing machine from Instron Co. (Canton, MA) in accordance with ASTM D 638M (at a speed of 50 mm/min). The same machine with flexural tester was being used to obtain the flexural strength and flexural modulus with respect to the ASTM D790 (80 span mm, speed 50 mm/min). Izod impact (ASTM D256) strength was being measured by using Conventional Izod Impact Tester with an 4J impact load used. 2.54 mm notching depth have been notched at the sample. The sample was saoked in liquid nitrogen for 1 minute before being attach and tested at the machine. The purpose of soaking the sample in liquid nitrogen is to harden and toughen the sample in order to have an impact response.

2.3.2. Transmission electron microscope (TEM). TEM test was conducted to the LDPE/EVA/nano-MH samples to observe the nano magnesium hydroxide dispersion and intercalation in polymer matrix. 100kV of acceleration voltage was set at the TEM machine to observe the morphologies of the nano-MH particles in LDPE/EVA matrix. The sample was slice into ultrathin film form with a degradation temperature (Tg) of 55°C.

3. Results and discussion

3.1. Mechanical Properties

3.1.1. Tensile strength. Figure 1 shows the tensile strength for EVA/LDPE composites with different nano loading from 0, 5, 10, 15 and 20 phr at different frequency of ultrasonic. From the graph, it can be clearly seen that at 10 phr of nano loading with 100 kHz ultrasonic improved tensile strength compared to the nanocomposite without ultrasonication and 50 kHz of ultrasonic frequency. However, with further increased of nano MH, the tensile property decreased and reduced to 7MPa at 20phr. This is due to the decrease of interfacial interaction between nanoparticle and matrix [8]. Thus, deteriorate the tensile strength of the composite.

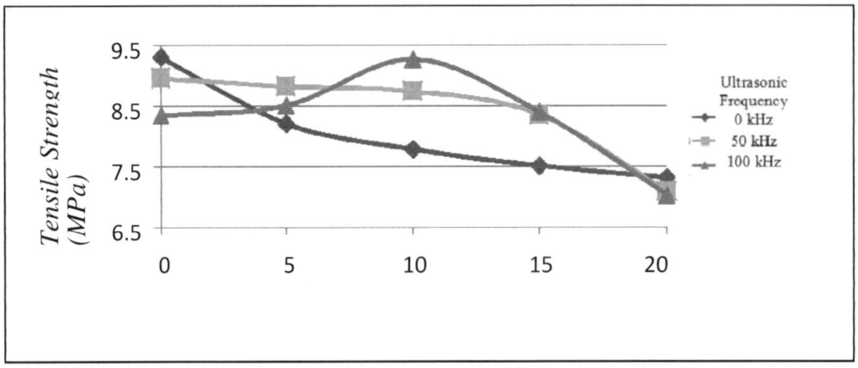

Figure 1. Tensile strength for varying ultrasonic frequency

3.1.2. Elongation at break. Figure 2 shows the flexural strength for EVA/LDPE composites with different nano loading from 0, 5, 10, 15 and 20 phr at different frequency of ultrasonic. From the graph, without ultrasonic, the elongation at break was slightly decreased when increased the nano loading from 5 phr to 20 phr. While for 50 kHz ultrasonic frequency, the elongation at break is higher at 10 phr of nano loading and decreased after the futher increased of nano-MH. Similar trend was observed at higher ultrasonic frequency (100kHz), but the elongation at break increment at 10phr of nano-MH loading is small. It was observed that, increasing the nano-MH loading from 10 to 20phr decrease the elongation ability of EVA-LDPE composite. This is due to the intercalation of EVA-LDPE matrix into the layer of nano-MH particles which will limit the mobility of the polymer chain to move freely Hence, lowering the elongation at break [9].

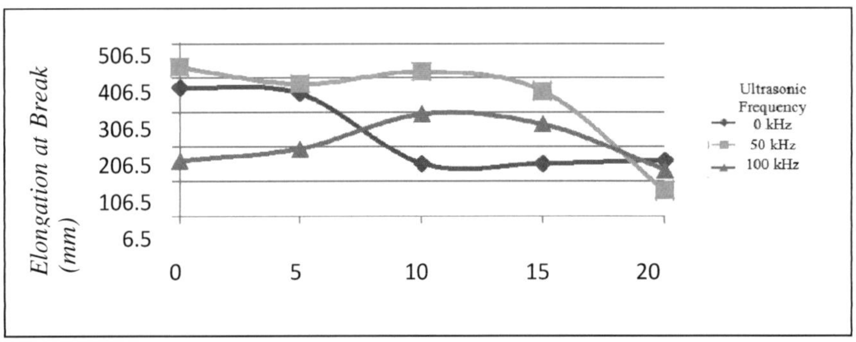

Figure 2. Elongation at break varying ultrasonic frequency

3.1.3. Youngs modulus. Figure 3 shows the Young Modulus for EVA/LDPE composites with different nano loading from 0, 5, 10, 15 and 20 phr with different frequency of ultrasonic.

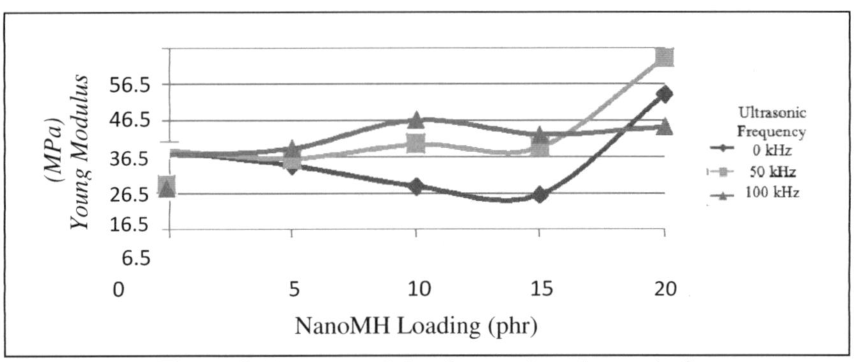

Figure 3. Young modulus for varying ultrasonic frequency

From the graph, it can be clearly seen that at 15 phr of nano loading with different frequency, the graph increased when reached at higher nano loading, 20 phr. However, with 50 kHz ultrasonic frequency, it shows the highest Young Modulus at 20 phr of nano loading. This indicates that, addition of nano-MH particle into EVA/LDPE matrix enhance the stiffness Young Modulus [9]. The increase in Youngs modulus with nano-MH loading suggests an efficient stress transfer between the nanoparticle and the matrix after treatment the fiber. Another reason is good dispersion or mixing between the nano-MH and the EVA/LDPE matrix [10].

3.1.4. Flexural strength. Figure 4 shows the flexural strength for EVA/LDPE composites with different nano loading from 0, 5, 10, 15 and 20 phr with different frequency of ultrasonic. From the graph, at 10phr with 100 kHz ultrasonic frequency, shows higher flexural strength (8.1 MPa). A similar trend is observed where, with increasing nano loading, the flexural strength increases. The flexural strength of the nanocomposite increases when 20phr nano-MH added due to compatibility factor which is affected from the better dispersion of nano particles.

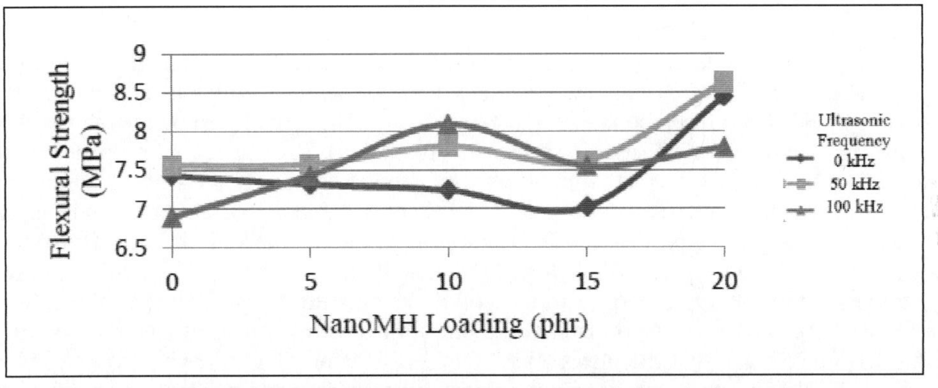

Figure 4. Flexural strength for varying ultrasonic frequency

3.1.5. Izod impact. Figure 5 shows the izod impact strength for EVA/LDPE composites with different nano loading from 0, 5, 10, 15 and 20 phr with different frequency of ultrasonic. From the graph, it can be clearly seen that at 10 phr of nano loading with 100 kHz ultrasonic assisted was the lowest in izod impact strength compared to the nanocomposite without ultrasonication and 50 kHz of ultrasonic frequency. Overall, at 5 phr of nano loading it can be clearly seen that with 50 kHz of ultrasonic frequency shows higher izod impact strength compared to others.

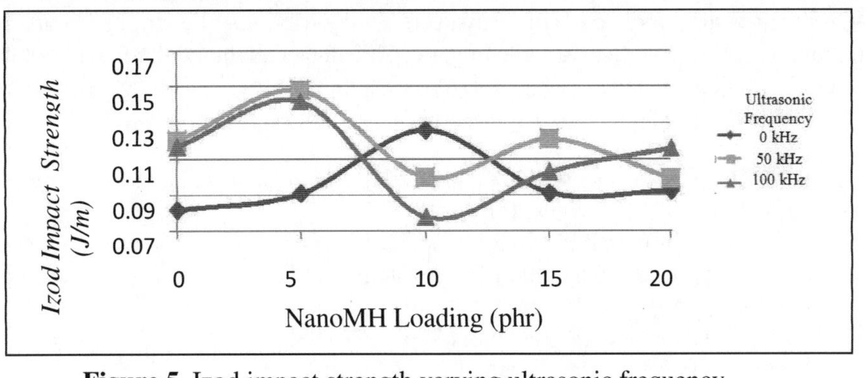

Figure 5. Izod impact strength varying ultrasonic frequency

3.2. Dispersion of nanoparticles

The distribution of the nanosized particles plays an important role in nanocomposite properties as nanocomposite properties are strongly dependant on the final morphology of the material. However, dispersion of the nanosized particles remains a challenge for nanocomposites due to the strong interaction of the nanoparticles.

IOP Conf. Series: Materials Science and Engineering **358** (2018) 012062 doi:10.1088/1757-899X/358/1/012062

Figure 6. Dispersion of 10phr Nano-Magnesium Hydroxide with varying Ultrasonic Frequency (a)100 kHz, (b)50 kHz and (c) without ultrasonic with 100 kV acceleration of electron and 35000 X magnification.

The nano-MH is a good additive for promoting the flame retardancy ability of the nanocomposite. However, the distribution of the nano-MH is an important parameter that affects the final properties of the composite. A good distribution can be achieved by various methods, with the simplest method which is ultrasonication [11]. Therefore, ultrasonic assisted extrusion was conducted on the optimum blending which is 10phr of nano loading tested as shown in Figure 6(a), (b) and (c) with different ultrasonic frequencies. It can be clearly seen that, with 10 phr of nano loading at 100 kHz of ultrasonic frequency is sufficient to show a good distribution of nano-MH that contributes to excellent tensile strength and Youngs modulus properties of the composite which is verified by Transmission Electron Microscopy (TEM) images as seen in Figure 6(a). While for 10 phr nano loading with 50 kHz of ultrasonic frequency, the nanoparticles were still agglomerates in the copolymer matrix as can be seen in Figure 6(b). As shown in Figure 6(c), for 10 phr nano loading without ultrasonic, there was a bundle of agglomeration occurs which can be proven that the ultrasonic assisted is a better way to have a good distribution in nanoparticles [12]. When increased the nano loading, aggregation will start to occurs which leads to weakening of the property of the composite. The optimum properties will be obtained when the dispersion is uniform [13].

4. Conclusion

A rapid intercalation of nano composite inside a matrix was achieved through ultrasonic process. This can be seen in TEM analysis showing a better dispersion of nanoparticles. Enhanced dispersion will theoretically improve the efficiency of nano-MH as flame retardant. Nano-MH loading at 10phr has the highest tensile strength compared to other loadings. The elongation at break decreases as the increase of nano-MH added to the nanocomposite. However, Young Modulus, Flexural Strength and Izod Strength showing reverse result from tensile and elongation. It can be concluded that, 10 phr of nano-MH loading is sufficient in finding a balance of the various mechanical properties studied as well as cost efficiency.

References

[1] Sabet M., Hassad A., and M. U. Wahit, 2007. *Regional Annual Fundamental Science Seminar (RAFSS 2007)*, 28-29 May 2007.
[2] Coudreuse A., Noireaux P., Noblat R., Basfar A., 2010, *Journal of Fire Science*, **28**, 497-507.
[3] Ahamad A., Patil C. B., Mahulikar P. P., Hundiwale D. G. and Gite V. V. *Journal of Elastomers and Plastics*, 2012, **44**, 3, 251-261.
[4] Gui H., 2007, *Comp Sci and Techno.*, **67**, pp. 974–980,
[5] Mahmood B., Ghazanfar M., Farhood Z., Mohammad A. B., 2007, *Nukleonika*, **52**(2):77-81.
[6] Chen Y., Zhou T., Fang H., Li S., Yao Y., and He Y., 2015, *Procedia Eng.*, **102**, 388–394.
[7] Pegel S., Ha L., Kretzschmar B., Villmow T., and Po P. , 2008, *Polymer*, **49**, 3500–3509.
[8] Liu S. P., Ying J. R., Zhou X. P., Xie X. L., and Mai Y. W. , 2009, *Compos. Sci. Technol.*, **69**, 11–12, 873–1879.
[9] Bee S., Hassan A., Ratnam C. T., Tee T., Tin L., and Hui D, 2014, *Compos. Part B*, **61**, 41–48A.
[10] Huda M. S., Drzal L. T. and Misra M., 2005, *Ind Eng Chem Res*, **44**, 5593-5601.

[11] Xie X., Mai Y., and Zhou X., 2006, *Materials Science and Engineering: R: Reports,* **49**, 89–112.

[12] Tarawneh M. A., Ahmad S. H., Rasid R., Yahya S. Y., Yong S., and Noum E., 2011, *Journal of Reinforced Plastic and Composites*, **30**, 524-532 .

[13] Zhang F., Zhang H., and Su Z. , 2007, *Appl. Surface Sci.* **253**, 7393–7397.

IOP Conf. Series: Materials Science and Engineering **358** (2018) 012063 doi:10.1088/1757-899X/358/1/012063

Green Synthesis of Ag, Cu and AgCu Nanoparticles using Palm Leaves Extract as the Reducing and Stabilizing Agents

N A N Mohamad, N A Arham, J Junaidah, A Hadi and S A Idris

Fakulti Kejuruteraan Kimia, Universiti Teknologi MARA, Shah Alam 40450, Selangor, Malaysia

junejai@salam.uitm.edu.my

Abstract. This paper reports the green synthesis of Ag, Cu and AgCu nanoparticles at room temperature using palm leaves extract. The purpose of this study is to eliminate the use of chemicals in the synthesis of nanoparticles and evaluate the efficiency of the palm leaves extract as the reducing and stabilizing agents. The palm leaves extract was added to metal salt solution and continuously stirred until reaction completed. The produced nanoparticles were analyzed using atomic absorption spectroscopy (AAS), Fourier transform infrared spectroscopy (FTIR), energy dispersive X-ray spectroscopy (EDS), scanning electron microscopy (SEM) and transmission electron microscopy (TEM). The analyses revealed that palm leaves extract has efficiently reduced the silver ions, but not the copper ions. During synthesis of AgCu nanoparticles, simultaneous reduction was occurred leading to formation of alloyed nanoparticles. Biomolecules from the palm leaves extract adsorbed on the surface of nanoparticles forming a capping layer thus stabilized the nanoparticles. The produced Ag and Cu nanoparticles were predominantly spherical with the particle size of Cu nanoparticles were larger than Ag nanoparticles. The AgCu nanoparticles closely resembled the Ag nanoparticles due to high Ag content with average size of 13nm. Therefore, palm leaves extract has a potential to be a good reducing and stabilizing agents.

1. Introduction

Ag and Cu nanoparticles were proven to have an efficient antibacterial activity [1,2]. Both of them also have good electrical conductivity, but Cu nanoparticles are easily oxidized, while Ag nanoparticles are expensive. Combination of Ag and Cu nanoparticles is one of the alternatives to overcome those limitations. Shin et al. [3] observed that AgCu nanoparticles exhibit enhanced structural stability compared to Ag or Cu nanoparticles. Valodkar et al. [4] found that AgCu nanoparticles have effective antibacterial activity against *Escherichia coli* and *Staphylococcus aureus*, while Li et al. [5] reported that AgCu nanoparticles successfully improved the electrical migration of ink in printing conductive lines.

Green synthesis is a favourable method to produce nanoparticles at room temperature and pressure without use of hazardous chemicals. Through green synthesis, plant extract act as reducing agent thus reduce the metal ions into nanoparticles. Various plant parts have been previously used in the synthesis of Ag and Cu nanoparticles such as mango peel [6], pineapple leaf [7], latex [8] and *Magnolia kobus* leaf [2]. Although synthesis of AgCu nanoparticles has been extensively explored, but to the best of our knowledge none of the approached involved used of plant extract. Research team from Universiti Putra Malaysia [9] found that palm leaves extract contained high polyphenols content. The high polyphenols content in plant extract may act as both reducing and stabilizing agents for nanoparticles [10]. Work by Valodkar et al. [4] confirmed the role of hydroxyl groups from phenolic compounds as reducer and stabilizer in synthesis of AgCu nanoparticles. Herein, we demonstrate formation of AgCu

Content from this work may be used under the terms of the Creative Commons Attribution 3.0 licence. Any further distribution of this work must maintain attribution to the author(s) and the title of the work, journal citation and DOI.
Published under licence by IOP Publishing Ltd

nanoparticles through simultaneous reduction of Ag and Cu ions using palm leaves extract as the reducing and stabilizing agents.

2. Materials and methods

2.1. Materials

$AgNO_3$ and $Cu(NO_3)_2$ were of analytical grade purchased from Merck Malaysia. Ag and Cu standards for AAS were purchased from Sigma-Aldrich Malaysia. Palm leaves from *Elaeis guineesis* species were collected from oil palm plantation located at Pasir Panjang, Sekinchan, Selangor, Malaysia. The extract was prepared following the previous work by Arham et al. [11].

2.2. Synthesis of Ag, Cu and AgCu nanoparticles

Palm leaves extract (10mL) was added to 1mM of metal salt solution (90mL); $AgNO_3$ was used for synthesis of Ag nanoparticles and $Cu(NO_3)_2$ was used for synthesis of Cu nanoparticles. For synthesis of AgCu nanoparticles, palm leaves extract (10mL) was added to equally mixed 1mM $AgNO_3$ and 1mM $Cu(NO_3)_2$ (90mL). The reaction took placed at room temperature under constant stirring.

2.3. Characterizations

Atomic absorption spectroscopy (AAS) model Hitachi Z-2000 was used to monitor the conversion of metal ions into nanoparticles. The solutions were centrifuge at 14000rpm for 20mins and the supernatant was analyzed for unreacted ions. After the reaction reached its saturation, the nanoparticles solutions were centrifuged at 10000rpm for 15mins and redispersed in distilled water. This centrifugation step was repeated thrice and the collected particles were dried overnight at 60°C. Palm leaves extract and the dried nanoparticles were subjected to Fourier transform infrared (FTIR) spectroscopy using model Perkin Elmer Spectrum One FTIR spectrometer. Surface images of the nanoparticles were captured using scanning electron microscope (SEM) model Quanta 450 FEG ESEM. Compositional study was carried out using energy-dispersive X-ray spectroscopy (EDX) attached to the SEM. High resolution images of the nanoparticles were captured using transmission electron microscope (TEM) model Tecnai G^2 20. EDX attached to TEM was used to obtained line profile of a single particle.

3. Results and discussions

The formations of Ag and Cu nanoparticles were monitored every 30mins using AAS. Figure 1 shows the conversion of Ag and Cu ions with respect to time. There is significant increase in the conversion of Ag ions upon 30mins addition of palm leaves extract.

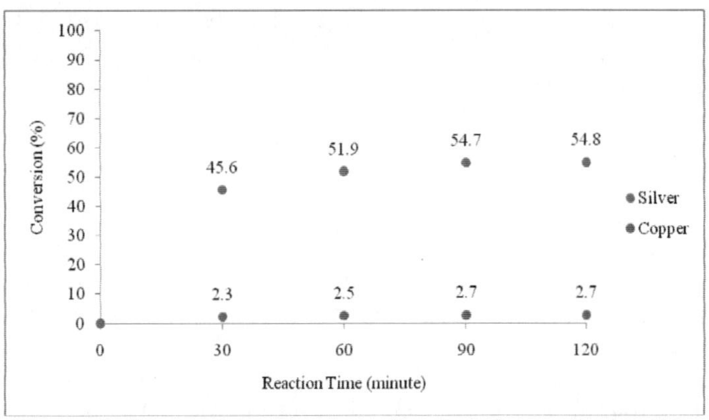

Figure 1. Conversion versus reaction time for synthesis of Ag and Cu nanoparticles.

Previous work on synthesis of Ag nanoparticles using Tulsi leaves extract recorded a significant conversion only after 60mins of reaction [12]. Therefore, the palm leaves extract has efficiently reduced the Ag ions and fairly rapid formation of Ag nanoparticles was achieved. The conversion of

Ag ions begins to achieve constant from 90-120mins suggesting saturation in the reaction. However, the palm leaves extract poorly reduced the Cu ions. This might be due to difference in the reduction potential of Ag and Cu ions. The reduction potential of Ag is larger than Cu (Ag: 0.8V, Cu: 0.34V) [13], signifying that Ag has greater tendency to be reduced. After 2hrs, about 54.8% of Ag ions and 2.7% of Cu ions have been reduced by the same amount of palm leaves extract. Figure 2 shows the conversion of Ag and Cu ions with respect to time for the synthesis of AgCu nanoparticles. Detailed analysis within the first 30mins of reaction suggests a rapid conversion in which significant increasein the conversion of ions occurred upon 5mins addition of the palm leaves extract.

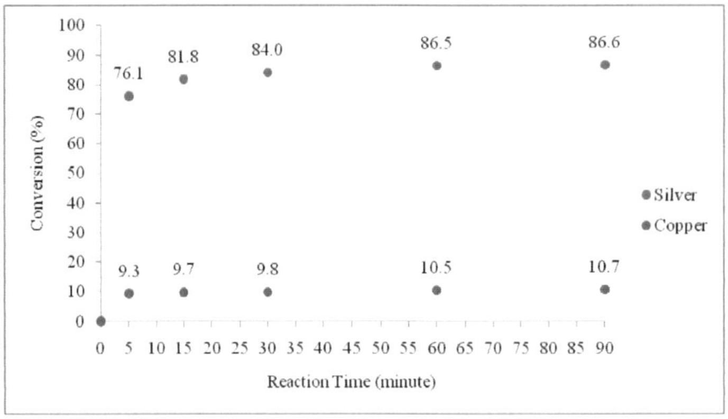

Figure 2. Conversion versus reaction time for synthesis of AgCu nanoparticles.

This result indicates that reduction of Ag and Cu ions by the palm leaves extract occurred simultaneously. However, due to difference in reduction potential, the conversion of Ag ions was higher compared to Cu ions. The simultaneous reduction has accelerated the reduction process. The conversion of both ions during synthesis of AgCu nanoparticles reached saturation after 60mins compared to 90mins during synthesis of Ag and Cu nanoparticles. Previous work by Tamuly et al. [14] also discovered that synthesis of AuAg nanoparticles completed faster than synthesis of Au and Ag nanoparticles. In addition, the final conversion has increased compared to the recorded conversion during synthesis of Ag and Cu nanoparticles. After 90mins, about 86.6% of Ag ions and 10.7% of Cu ions were reduced. However, detail study is required to understand the kinetics behind the reaction. Finally, based on the total reduction of Ag and Cu ions, the produced AgCu nanoparticles were estimated to contain 89.2% of Ag and 10.8% of Cu. This composition is almost similar to standard composition of coin silver which contained 90% of Ag and the remaining 10% is Cu [15].

From figure 3, the spectrum for palm leaves extract indicates presence of hydroxyl (3307cm^{-1}), carbonyl (1738cm^{-1}), amide I (1635cm^{-1}), aromatic amine (1366cm^{-1}) and aliphatic amine (1229 and 1217cm^{-1}). Ag nanoparticles show peaks at 3274cm^{-1}, 1739cm^{-1}, 1510cm^{-1}, 1366cm^{-1}, 1236cm^{-1}, 1217cm^{-1}, 2878cm^{-1} and 1014cm^{-1}, while Cu nanoparticles show peaks at 3242cm^{-1}, 1606cm^{-1}, 1352cm^{-1}, 1209cm^{-1}, 1152cm^{-1}, 2956cm^{-1} and 1048cm^{-1}. AgCu nanoparticles show peaks at 3261cm^{-1}, 1606cm^{-1}, 1366cm^{-1}, 1217cm^{-1}, 2917cm^{-1} and 1047cm^{-1}. Based on the spectra of all three types of nanoparticles, the significant reduction in the intensity of hydroxyl and carbonyl groups suggested that these functional groups might involve in the reduction of the metal ions. Presence of other functional groups suggested that the nanoparticles were capped with biomolecules from the palm leaves extract, hence stabilized nanoparticles were produced.

EDX spectra in figure 4 confirms presence of Ag and Cu elements in the produced Ag, Cu and AgCu nanoparticles. Presence of C and Si might be originated from the palm leaves extract that adsorbed on the surface of nanoparticles. Previous work on green synthesis of Au nanoparticles also discovered the presence of C and Si elements [16]. The finding from EDX is in accordance with FTIR analysis in which the biomolecules from the palm leaves extract was responsible as the stabilizing agent. It was also observed that the signal for Ag was stronger than Cu indicating greater amount of Ag compared to Cu, thus verifying the findings obtained through AAS analysis. Line profile of a single AgCu nanoparticle depicted in figure 5 suggests that the produced nanoparticle composed of two alloyed metal elements. The intensity of Ag was greater than Cu, thus further supported the finding from AAS that the produced nanoparticles contained high amount of Ag.

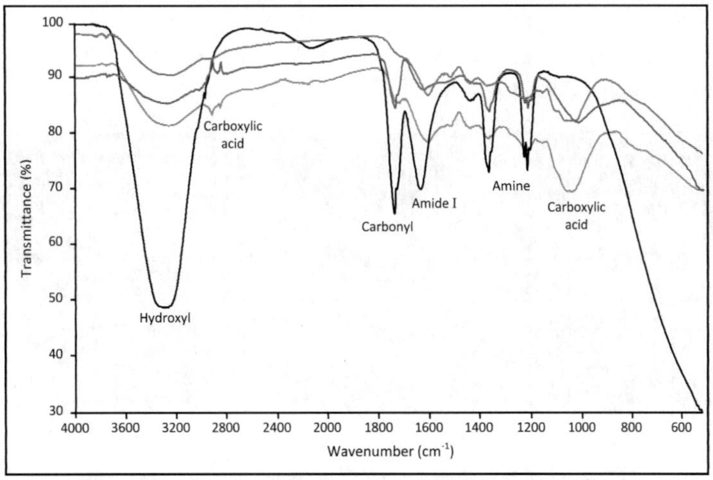

Figure 3. Comparison between FTIR spectra of palm leaves extract and the nanoparticles.

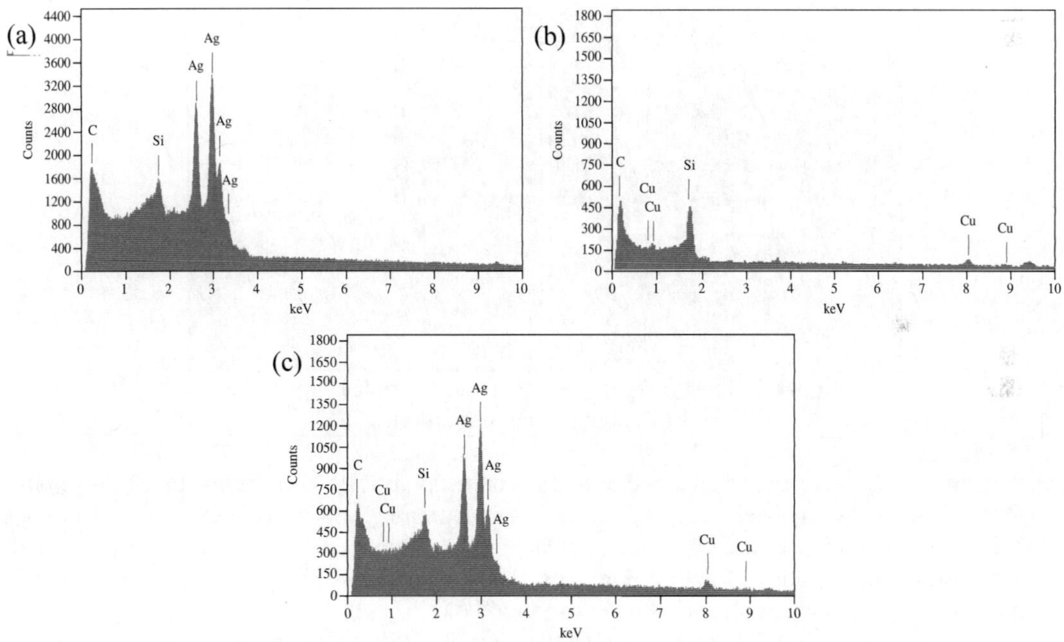

Figure 4. EDX spectra of (a) Ag, (b) Cu and (c) AgCu nanoparticles.

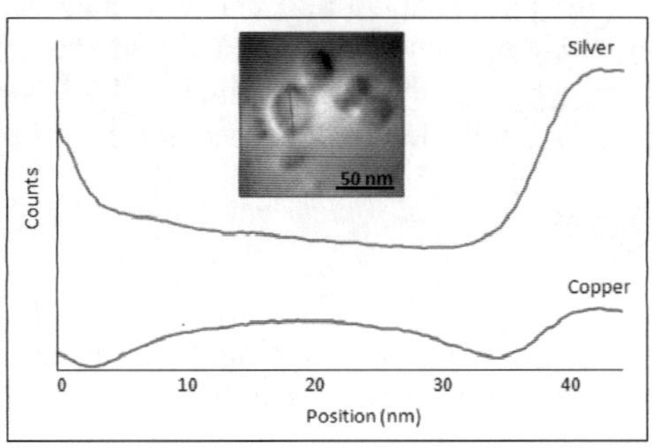

Figure 5. Line profile of AgCu nanoparticles.

Figure 6 shows the SEM images of Ag, Cu and AgCu nanoparticles captured at 100000x magnification. The Ag nanoparticles were mostly spherical but some agglomeration was observed for Cu nanoparticles. The size of Cu nanoparticles was larger compared to Ag nanoparticles. Similar finding was reported through chemical synthesis of Ag and Cu nanoparticles [17]. The AgCu nanoparticles on the other hand closely resemble to the Ag nanoparticles. This might be due to the large Ag content in the produced AgCu nanoparticles.

Figure 6. SEM images for (a) Ag, (b) Cu and (c) AgCu nanoparticles.

TEM image of AgCu nanoparticles in figure 7 revealed that the nanoparticles were predominantly spherical and monodispersed. This finding was comparable to AgCu nanoparticles synthesized through chemical method [18].

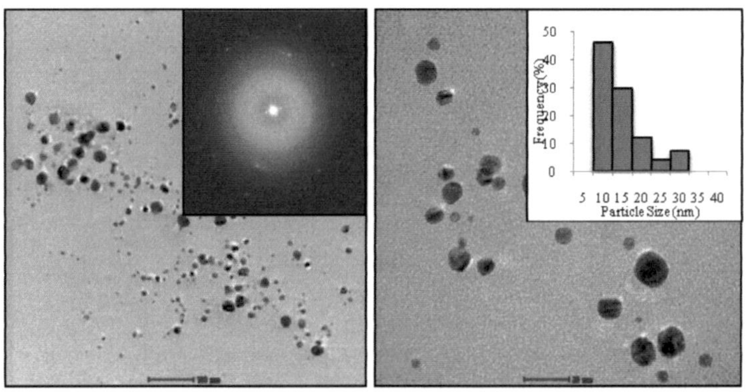

Figure 7. TEM images for AgCu nanoparticles captured at low and high magnifications.

A faint thin layer was observed on the surface of nanoparticles confirming the adsorption of stabilizer from the palm leaves extract. The presence of capping layer was also reported by Tamuly et al. [14]. The uniform contrast for each nanoparticle suggests a homogenous electron density within the particle, thus resemble to alloy structure. Previous study on green synthesis of AuAg nanoparticles through simultaneous reduction also resulted in formation of alloy structure [19,20]. The inset selected area electron diffraction (SAED) image showed a ring-like diffraction pattern indicating that the produced AgCu nanoparticles were crystalline [21]. The measured interplanar distance is 0.21nm which could be indexed to (111) plane of AgCu alloy. The average size of at least 100 particles is 13nm. Based on the inset histogram, the AgCu nanoparticles lie in the range of 5-30nm, almost similar to the range of polymer-stabilized AgCu nanoparticles which is 10-30 nm [18]. This finding indicates that palm leaves extract is not only an efficient reducing agent, but also has a potential to be a good stabilizing agent.

4. Conclusion

Ag, Cu and AgCu nanoparticles were successfully produced through reduction with palm leaves extract. The rapid reduction process and the resulting monodispersed nanoparticles proved the efficiency of the palm leaves extract as reducing and stabilizing agents. It is the novelty of this study to discover formation of silver rich alloy through simultaneous reduction method. The average size obtained was 13nm, comparable to the polymer-stabilized AgCu nanoparticles.

Acknowledgments
This research was funded by ERGS grant no. ERGS/1/2013/TK05/UITM/02/04. All assistances and facilities from Faculty of Chemical Engineering, UiTM are gratefully acknowledged.

References

[1] Nazeruddin G M, Prasad N R, Waghmare S R, Garadkar K M and Mulla I S 2014 *J. Alloys Compd.* **583** 272-277
[2] Lee H J, Song J Y and Kim B S 2013 *J.Chem. Technol. Biotechnol.* **88** 1971-77
[3] Shin K, Kim D H, Yeo S C and Lee H M 2012 *Catal. Today* **185** 94-98
[4] Valodkar M, Modi S, Pal A and Thakore S 2011 *Mater. Res. Bull.* **46** 384-389
[5] Li Y S, Lu Y C, Chou K S and Liu F J 2010 *Mater. Res. Bull.* **45** 1837-43
[6] Yang N and Li W H 2013 *Ind. Crops Prod.* **48** 81-88
[7] Emeka E E, Ojiefoh O C, Aleruchi C, Hassan L A, Christiana O M, Rebecca M, Dare E O and Temitope A E 2014 *Micron* **57** 1-5
[8] Harne S, Sharma A, Dhaygude M, Joglekar S, Kodam K and Hudlikar M 2012 *Colloids Surf.* B **95** 284-288
[9] Ismail M K 2007 http://www.ibd.utm.my/rcentre/download_view.php?download=4387
[10] Kharissova O V, Dias H R, Kharisov B I, Pérez B O and Pérez V M J 2013 *Trends Biotechnol.* **31** 240-248
[11] Arham N A, Mohamad N A N, Jai J, Krishnan J and Yusof N M 2013 *Int. J. Sci. Eng.* **5** 95
[12] Ramteke C, Chakrabarti T, Sarangi B K and Pandey R A 2012 *J. Chem.*
[13] Petrucci R H, Herring F G, Madura J D and Bissonnette C 2010 *General Chemistry Principles and Modern Applications* (Pearson College Division)
[14] Tamuly C, Hazarika M, Borah S C, Das M R and Boruah M P 2013 *Colloids Surf.* B **102** 627
[15] Davis J R 1990 *Properties and Selection: Nonferrous Alloys and Special-Purpose Materials* (ASM International)
[16] Wang Y, He X, Wang K, Zhang X and Tan W 2009 *Colloids Surf.* B **73** 75-79
[17] Tiwari A D, Mishra A K, Mishra S B, Kuvarega A T and Mamba B B 2013 *Carbohydr. Polym.* **92** 1402-07
[18] Zhang D and Liu X 2013 *J. Mol. Struct.* **1035** 471-475
[19] Mondal S, Roy N, Laskar R A, Sk I, Basu S, Mandal D and Begum N A 2011 *Colloids Surf.* B **82** 497-504
[20] Jacob J, Mukherjee T and Kapoor S 2012 *Mater. Sci. Eng.* C **32** 1827-34
[21] Mani U, Dhanasingh S, Arunachalam R, Paul E, Shanmugam P, Rose C and Mandal A B 2013 *Progress in Nanotechnology and Nanomaterials* **2** 21-25

The Effect of Particles Shape and Size on Feedstock Flowibility and Chemical content of As-sintered NiTi Alloys

R A Abdul Kadir[1*], R Razali[2], N H Mohamad Nor[1], I Subuki[2] and M H Ismail[1+]

[1]Centre for Advanced Materials Research (CAMAR),
Faculty of Mechanical Engineering, University Teknologi Mara, 40450 Shah Alam, Selangor , Malaysia.
[2]Faculty of Chemical Engineering, University Teknologi Mara, 40450 Shah Alam, Selangor , Malaysia.

*ros_atikah90@yahoo.com, [+]hussain305@salam.uitm.edu.my

Abstract. This paper presents a comparative study of two different titanium powders in fabrication of NiTi alloys by metal injection moulding (MIM) route. Two batches of powder mixture consisted of Ni-Ti and Ni-TiH_2 with atomic ratio (at%) of 50-50 were prepared. TiH_2 powder was used as a substitution for pure Ti powder owing to its relatively cheaper cost and has been claimed favourable in producing less impurity uptake in sintering process. The binder system used for both mixtures comprised of composite binder of palm stearin (PS) and polyethylene (PE) at weigth ratio (wt%) of 60-40. The flow behaviour of the mixtures was analysed using a capillary rheometer at different shear rates and temperatures. The results showed that owing to irregular shape of TiH_2 compared to Ti powder, the viscosity of the feedstock was significantly higher, thus required greater temperature in order to improve the mouldability of the feedstock. Nevertheless, both feedstocks exhibited pseudoplastic, a shear thinning behavior with shear rate and temperature, desirable properties for injection moulding process. Samples prepared with Ni-Ti feedstock were sintered in a high vacuum furnace, while Ni-TiH_2 feedstock was sintered in a tube furnace under a flowing of Argon gas. The results showed that the impurity contents (Carbon and Oxygen) for both feedstocks were almost comparable, suggesting NiTi alloy samples prepared with TiH_2 powder is an attractive route for manufacturing of NiTi alloys.

1. Introduction
NiTi alloy is considered as one of the promising candidates used in biomedical applications, especially for bone implants due to its unique shape memory effect (SME) and pseudo-elasticity (PE) which is not found in many conventional metallic materials such as stainless steel and titanium alloys. In the other hand, mechanical behaviour of the alloy exhibits very close properties to bone compared to others [1]. Processing of NiTi alloy by powder metallurgy (PM) route, specifically by metal injection moulding (MIM) has been extensively studied by Ismail [1]. One of the important attributes in the MIM is the flow behaviour of the prepared feedstock via rheological test. The success of feedstock flow through die orifice during rheological test is strongly influenced by powder characteristics in terms of size and shape, relative fraction between powder and binder and processing parameters such as temperature and pressure. These attributes are significantly important to ensure the part is successfully moulded with uniform density which leading to further densification via debinding and sintering processes.

Content from this work may be used under the terms of the Creative Commons Attribution 3.0 licence. Any further distribution of this work must maintain attribution to the author(s) and the title of the work, journal citation and DOI.
Published under licence by IOP Publishing Ltd

In processing of NiTi alloy by PM route, the use of TiH_2 as substitution to pure Ti powder has been investigated by several researchers [2-3] owing to several advantages such as (i) cheaper than pure Ti powder, (ii) capable to be sintered in argon flow atmosphere despite expensive route of vacuum sintering (iii) better phase homogeneity via sintering process and (iv) low impurity uptake particularly oxygen and carbon. Most of the TiH_2 powder used by researchers is irregular in shape, considered as sufficient when conventional compaction and sintering method is employed as irregular shape powder provides better powder inter-locking mechanism. However, in MIM route, owing to the flow of powder-binder mixture via small flow path in mould cavity, starting powders with spherical shape is highly required as it greatly reduces friction during moulding, thus improve mouldability. Irregular powder will increase inter-particle and reduce the packing density compared to spherical shape. This circumstance will probably lead to powder-binder separation during injection moulding and further sintering process cannot be preceded. Generally, the powder size of less than 20 μm is required and the irregular shape powder is undesirable for injection moulding process[1]. The irregular size of TiH_2 will increase the inter-particle friction and reduce the packing density compared to spherical shape. It is believed that processing improvement can be done by tailoring the powders blend with different size and shape where different particles size produce better moulding capacity when mixing the finer spherical shape with irregular coarse powder [4-5].

In this proposed research, the effect of TiH_2 substitution in processing of NiTi by MIM route is investigated, taken into account the feedstock rheology, injection moulding, sintering and impurity contents after the sintering process. The low impurity uptake, particularly O_2 and C after sintering process is significantly important as it leads to better mechanical and corrosion resistance of the NiTi alloy. The result is compared with previously done formulation using elemental Ni and Ti powder mixtures [6].

2. Materials and Method

Figure 1 shows the SEM images of the elemental Ti, TiH_2 (after ball-milled) and Ni powders used in the present work. Clearly, different morphology can be observed; Ti powder shows spherical shape with very fine surface finish while the TiH_2 powder is irregular and the surface texture is rough. The Ni powder is also spherical in shape but the surface is rougher than the Ti powder. The average particle size measured by a Malvern® particle size analyser were Ti =10.86 μm, TiH_2 = 26.66 μm and Ni = 7.87 μm. Due to greater size of the as-received TiH_2, the powder was ball milled at a speed of 300 rpm in a ball mill to reduce the particle size, resulting the mean particle size of 9.77 μm.

Figure 1. SEM of starting elemental powders: (a)Ti (b) TiH_2 and (c) Ni

Feedstock preparation was done by mixing the powder mixtures and binder system using an internal mixture, Haake Rheomix at mixing temperature of 150°C for 2 hours. The flow behaviour of the feedstock was then characterized using a Rosand RH2000 capillary rheometer. The tungsten carbide die having a diameter and length of 1 mm and 16 mm, respectively was used resulting in an aspect ratio (L/d) of 16. The pelletized feedstock was loaded into the preheated chamber at a constant heating rate of 5°C/min and at the temperature range of 110-150°C for feedstock A, while 150 - 170°C for feedstock B at a constant shear rate within the range 5 – 8000 s^{-1}. Tensile shape sample in accordance with ASTM 638-type V standard [7] were then injection moulded using a manually operated injection moulding at a temperature range between 130°C and 170°C. All moulded samples were then subjected to solvent extraction using n-heptane solution at 50°C to remove the primary binder of palm stearin. Samples prepared from feedstock A were sintered in a high vacuum furnace at a sintering temperature of 1150°C as reported in previous work [6], while samples from feedstock B were sintered in an argon gas flow tube furnace at 1100°C as investigated by Chen [3]. The chemical contents of the sintered parts were determined using a Leco Melt Extraction Machine (LECO TCH 600) to determine the impurity levels (Carbon and Oxygen) for each sample.

3. Results and Discussion

3.1 Viscosity vs Shear Rate

Figure 2 shows the result of flow behavior for (a) Ni-Ti feedstock and (b) Ni-TiH_2 feedstock, clearly both feedstock formulations demonstrate almost similar pattern of flow; viscosity decreasing with shear rate. This common behaviour for injection moulding materials is also known as shear thinning or pseudoplastic in which increasing of shear allows small particles to fill within the gap between large particles and feedstock homogeneity and fluidity are improved [8,9]. It is also noticed that the value of viscosity for Ni-TiH_2 feedstock is significantly higher, approximately 30 times greater than the Ni-Ti feedstock, indicating greater friction of the feedstock attributed from the irregular shape and rough surface of TiH_2 powder. Ismail [1] has highlighted in his work that the recommended viscosity for feedstock is the range of 10 and 1000 Pa.s at the injection moulding operating temperature. It was also observed that the required temperature for Ni-Ti feedstock to exhibit good flow was relatively lower, in the range of 130 and 150°C, while for Ni-TiH_2 feedstock was in the range of 150 and 170°C. This shows that higher temperature is needed during actual injection moulding in order to reduce the inter-particle friction as well as bonding between particle and binder. However, too high moulding temperature will cause other problems, especially powder-binder separation and flashing, thus feedstock may not be proceeded to injection moulding process.

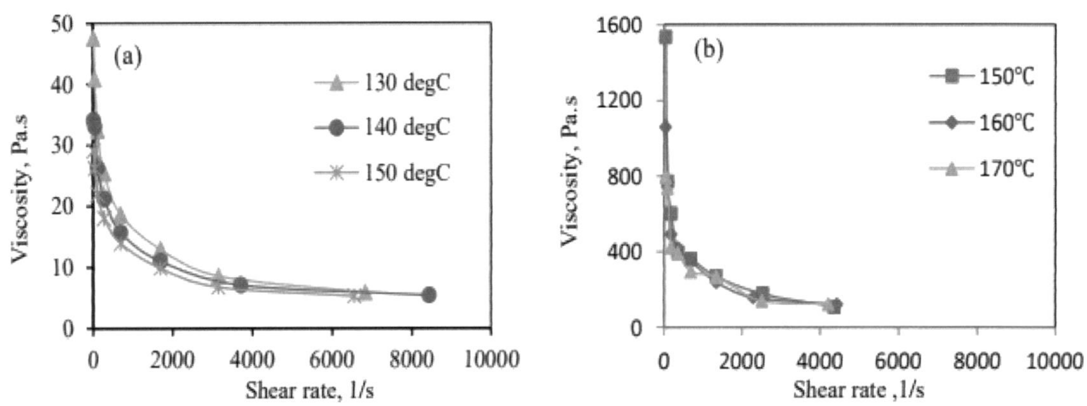

Figure 2. Variation in viscosity and shear rate for a) Ni-Ti feedstock, b) Ni-TiH_2 feedstock

3.2 Shear sensitivity index (n)

One of the information that can be extracted from the rheological analysis is the level of shear sensitivity or generally referred as *n* index. It is determined from the slope of logarithmic plots of shear stress versus shear rate. The *n* index values below than 1 is generally reported by many researchers [1, 6, 8-11], indicating pseudoplastic flow which is the required flow in injection moulding process. In the rheological finding by Ismail [1] using Ni and Ti powder mixtures, the optimal range of *n* was in the range of 0.5 and 0.7. Feedstock with greater than 0.7 exhibited very viscous flow resulted in short-shot defects, while too low *n* values resulted in jetting flow leading to binder separation. Figure 3 compares the *n* values for both feedstocks, clearly Ni-Ti feedstock exhibited better flowbility in comparison with Ni-TiH_2 feedstock owing to spherical shape Ti powder. The results also show that increasing the barrel temperature resulted in *n* values increases, corresponding to better flowability as suggested by Yimin et al. [10]. They claimed that greater value of *n* will result in less sensitivity of viscosity changes with shear rate.

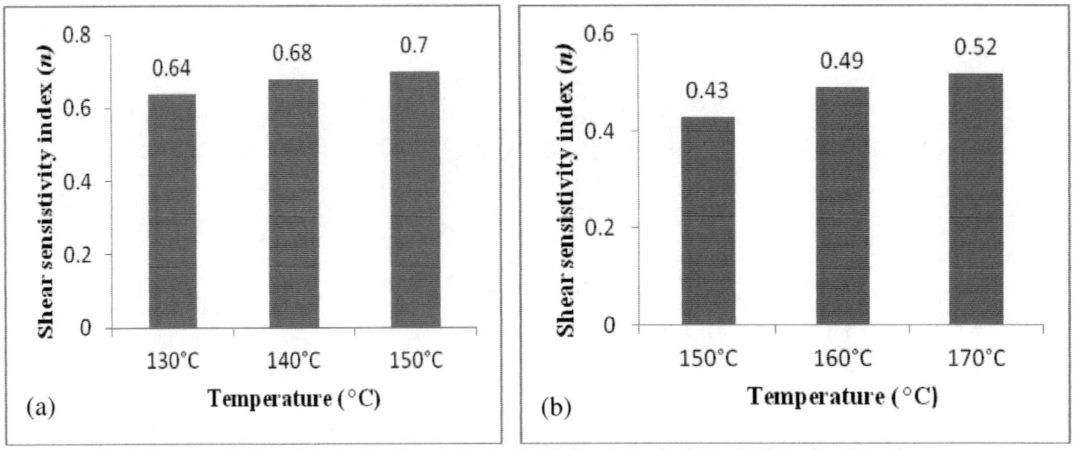

Figure 3. Sensitivity Index *(n)* a) Ni-Ti feedstock, b) Ni-TiH_2 feedstock

3.3 Activation Energy (E)

Another aspect that commonly investigated is the flow activation energy of the feedstock. Table 1 compares the *E* values for two batches of feedstock. Low *E* value is preferable which indicates viscosity is not strongly influenced by temperature and desirable for injection moulding. Low temperature sensitivity on viscosity is desirable for a good quality of MIM feedstock. Agote el al. [11] have stated that the value of *E* determines the sensitivity of the viscosity to temperature which generally should be as small as possible in order to avoid any sudden viscosity changes [10] . The flow activation energy of evaluated feedstock can be attained from the slope of graph ln against (1/T). The result clearly shows that Ni-TiH_2 feedstock exhibited higher *E* value, indicating higher sensitivity during moulding attributed to higher viscosity of the feedstock.

Table 1. *Flow activation energy for both feedstock*

Feedstock	Flow activation energy (kJ/mol)
A (Ni-Ti)	16.1
B (Ni-TiH_2)	55.72

3.4 Injection Moulding and Chemical Contents

Both feedstocks were successfully injection moulded to tensile shape as shown in Figure 4 in the range of 140 and 170°C. It was observed that owing to irregular shape of TiH_2 powder and higher viscosity, a greater temperature was used for feedstock B to ensure good flowbility. After sintering process the samples were tested using LECO TCH 600 to determine impurities containing oxygen and carbon of the as-sintered sample. The sintering environment is different where feedstock A (Ni-Ti) was sintered using high vacuum environment and feedstock B (Ni-TiH_2) using low cost tube argon furnace. Table 2 shows the impurity contents for the as-sintered parts in comparison with other related published research using MIM of elemental Ni and Ti powders. It shows that both feedstocks have low chemical contents that suitable for biomedical implant. It can be seen that the impurity of oxygen in the present work is almost double and the carbon is comparable with the previous work. The lower impurity contents were because of employing a powder with lower initial impurity contents [1, 12]. However, all previous studies using MIM route show an increase of impurity from the initial powder, particularly oxygen. Thus, maintaining the low oxygen content comparable to that contained in the initial powder is almost difficult. Furthermore, the use of binder system which mainly comprised of hydro-carbon and several stages involved in MIM give the greater tendency for the impurities to be introduced prior to sintering process. However, the carbon concentration can be reduced to a minimum value if suitable binder can be identified and optimized process parameters are used.

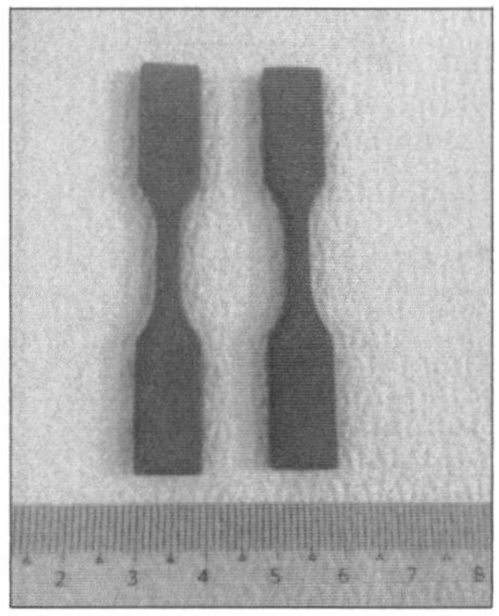

Figure 4. Green parts after been removed from mould cavity

Table 2. Comparison of impurity contents between present and previous works

Reference	Conditions			Impurity contents (wt%)	
	Composition	Sintering Temperature (°C)	Atmosphere	O_2	C
Feedstock A (Ni-Ti)	50.0at.% Ni - 50.0at.%Ti	1150	Vacuum	0.498	0.117
Feedstock B ($Ni-TiH_2$)	50.0at.% Ni - 50.0at.%TiH_2	1100	Argon	0.426	0.119
[1]	50.9at.% Ni-49.1at.%Ti	1150	Vacuum	0.200	0.062
[12]	50.9at.% Ni-49.1at.%Ti	1100	Vacuum	0.3524	0.112

From the present work, results show that sintering of NiTi alloy starting with elemental powders of Ni and TiH_2 in an argon tube furnace following metal injection moulding process seems to be attractive for several reasons; (i) low cost of TiH_2 and argon tube sintering and (ii) comparable impurity uptake with vacuum sintered condition (Feedstock A). Further improvement is in progress in order to minimize the impurity uptake, thus improving the mechanical behaviour and biocompatibility of the sintered NiTi alloy.

4. Conclusions

It can be concluded that spherical shape powder has better rheological behaviour compared to irregular powders. Both feedstock exhibited shear thinning which is the desired flow behaviour for injection moulding. Feedstock A containing Ni-Ti exhibited better rheological properties in terms of viscosity range, low shear sensitivity index and activation energy in comparison with Feedstock B containing $Ni-TiH_2$. This resulted in low injection temperature during injection moulding. From the sintering point of view, samples prepared with TiH_2 powder and sintering in an argon atmosphere exhibited a comparable impurity uptake with samples sintered in vacuum condition, thus providing better insights for future works on NiTi alloy processing by MIM route.

Acknowledgement

The authors would like to thank to young lecturer scheme scholarship, Ministry of Education (MOE) and Universiti Teknologi Mara for awarding research grant 600-RMI/FRGS 5/3 (72/2015) for financial support throughout the research.

IOP Conf. Series: Materials Science and Engineering **358** (2018) 012064 doi:10.1088/1757-899X/358/1/012064

References

[1] Ismail M H (2012). *Porous Niti Alloy by Metal Injection Moulding (MIM) Using Partly Water Soluble Binder System* (Doctoral dissertation, University of Sheffield).

[2] Bertheville B, Neudenberger M and Bidaux J E 2004 *Materials Science and Engineering: A*, **384** 143-50

[3] Chen G, Cao P and Edmonds N 2013 *Materials Science and Engineering: A*, **582**, 117-125.

[4] Contreras J M, Jiménez-Morales A and Torralba J M 2007 *Materials Science Forum* 365–8.

[5] Chen G, Cao P, Wen G, Edmonds N and Li Y 2013 *Intermetallics* **37** 92–99.

[6] Razali R, Abdullah Z, Subuki I, Ismail M H and Muhamad N 2014 *App. Mech. Mat.* **575** 78-82

[7] Azuddin M, Choudhury I A and Taha Z 2014 *J Brazillian Society of Mech. Sci. Eng.* **37** 79-86.

[8] Abolhasani H and Muhamad N 2009 *Int. J.Mech. and Mater. Eng.* **4** 294–9

[9] Thavanayagam G, Pickering K L, Swan J E, and Cao P 2015 *Pow. Tech.* **269** 227-32

[10] Yimin, L, Liujun L, and Khalil K 2007 *J. Mat. Proc. Tech.* **183** 432-9

[11] Agote I, Odriozola A, M Gutierrez, Santamaria A, Quintanilla J, Coupelle P and Soares, J. 2001 *J. Euro. Cer. Society* **21** 2843–53

[12] Schöller E, Krone L, Bram M, Buchkremer H P and Ståaver D 2005 *J. Mat. Sci.* 40 4231-3

IOP Conf. Series: Materials Science and Engineering **358** (2018) 012065 doi:10.1088/1757-899X/358/1/012065

Cocrystal Screening of Ibuprofen with Oxalic Acid and Citric Acid via Grinding Method

M F Othman[1], N Anuar[1], S Ad Rahman[2] and N A Ahmad Taifuddin[1]

[1] Faculty of Chemical Engineering, Universiti Teknologi MARA (UiTM), MALAYSIA
[2] Faculty of Chemical and Natural Resources Engineering, Universiti Malaysia Pahang (UMP), MALAYSIA

fitri@salam.uitm.edu.my

Abstract. Ibuprofen is a Class II Biological Safety Class (BSC) drugs used for relief of arthritis, as an analgesic and possesses the effect of antiplatelet. The major problem involves in ibuprofen is it has a low solubility and high permeability thus causes an unsatisfactory therapeutic effect to humans. Thus, in this work, alteration of ibuprofen's physicochemical properties is conducted by means of cocrystallization technique. Co-crystallizations of ibuprofen were prepared with selected coformers using dry grinding and liquid assisted grinding (LAG) techniques in different molar ratios while ethanol and propanol were used as a solvent. The new crystalline forms were identified and characterized by differential scanning calorimetry (DSC), powder X-ray diffraction (PXRD) and fourier transform infrared spectroscopy (FTIR). Analysis for Ibuprofen-Citric acid (IBP-CA) system, co-crystal was successfully formed in 1:2, 1:3, 2:1 and 3:1 molar ratios for neat grinding method although the co-crystal produced is unstable. Meanwhile, for Ibuprofen-Oxalic acid (IBP-OA) system, the co-crystal formation was identified only in 1:1, 1:2 and 1:3 molar ratios for the neat grinding method. LAG method shows that co-crystal formation was unsuccessful in both solvents for IBP-CA, while IBP-OA co-crystal was formed in the molar ratio 1:1, 2:1 and 3:1 in ethanol, and 2:1 and 3:1 in propanol.

1. Introduction

Co-crystal is structurally homogenous crystalline materials contain two or more neutral building blocks that are present in definite stoichiometric amounts [4]. While pharmaceutical co-crystal can be defined as co-crystal with one of the co-crystal components as an API and the other components are called as co-formers or excipient [5]. In other words, it is a crystalline complex of two or more neutral molecular components tied together in the crystal lattice through interactions of the non-covalent bond, essentially hydrogen bonding [6]. The resulting multi-component crystalline phase retains the intrinsic activity of the parent API.

Ibuprofen is an active pharmaceutical ingredient (API) known as non-steroidal anti-inflammatory drug (NSAID with low solubility and high permeability thus causes an unsatisfactory therapeutic effect to humans. Alteration of the physicochemical properties of crystalline active pharmaceutical ingredients (APIs) by introducing a pharmaceutically suitable compound is a present strategy in drug formation and dosage optimization [1]. The use of co-crystallization technology in active pharmaceutical ingredient (API) crystallizations allows the technology to bring improve pharmaceutical products to the market place as well as for the improved drug delivery [2]. It has garnered much attention in the pharmaceutical industry since the formation of co-crystals can improve the properties of an active pharmaceutical ingredient (API) such as solubility, bioavailability, storage,

Content from this work may be used under the terms of the Creative Commons Attribution 3.0 licence. Any further distribution of this work must maintain attribution to the author(s) and the title of the work, journal citation and DOI.
Published under licence by IOP Publishing Ltd

stability, and manufacturability [3]. In this paper, Ibuprofen is selected as the API and the coformer used are Oxalic acid and Citric acid. Cocrystal formation between IBP and coformers are conducted using neat grinding and liquid assisted grinding method. The main objective of this paper is to explore the cocrystal formation between Ibuprofen and selected coformers as a method to improve the physicochemical properties of the model drug. Analysis of thermal and crystallinity properties of cocrystal are also examined.

2. Material and Method

2.1. Material

Ibuprofen (IBP, MW = 206.29 g/mol) was purchased from Shasun Pharmaceutical Limited, India (Batch No.: IB 11070248) (SN Grade). Citric acid (CA, MW = 192.124 g/mol), Oxalic acid (OA, MW = 90.0349 g/mol), Ethanol (MW = 46.07 g/mol) and Propanol (1-propanol, MW = 60.1 g/mol) were obtained from Sigma-Aldrich (Selangor, Malaysia). All compounds were used as received without further purification.

2.2. Method

2.2.1. Sample Preparations

The samples were prepared using neat grinding and liquid assisted grinding method. Additional three drops of solvent, Ethanol was added for the liquid assisted grinding method. Co-crystal blends of Ibuprofen-Citric acid and Ibuprofen-Oxalic acid were prepared by using oscillatory ball milling (Mixer Mill MM400, Reysch GmbH, Germany). The ball milling samples were produced by placing Ibuprofen-Citric acid and Ibuprofen-Oxalic acid at the molar ratios of 1:1, 1:2, 1:3, 2:1 and 3:1 into 25 ml milling jars with four 6 mm in diameter of stainless steel balls. Milling is carried out at 30 Hz for 30 min before it was characterized using various analytical techniques including Differential Scanning Calorimetry (DSC), X-ray Powder Diffraction (XPRD) and Fourier Transform Infrared (FTIR) Spectroscopy.

2.2.2. Differential Scanning Calorimetry (DSC)

DSC thermograms were obtained using DSC 820 (Mettler Toledo Inc., Columbus, USA) under nitrogen gas flow of 10 mL/min. Sample powders (approx. 8.5 mg) were analysed in 30 μL non-hermetic sealed aluminium pans. All samples were analysed between 25 °C and 250 °C at a heating rate of 10 °C/min and the melting point was recorded.

2.2.3. X-ray Powder Diffraction (XPRD)

XPRD was performed using x-ray diffraction analyser (Rigaku D/Max-2000, Woodlands, USA) using Cu Kα radiation (λ = 1.542 Å) in continuous scanning mode. An acceleration voltage and current of 40 kV and 30 mA were used. Data were collected over an angular range of 5o and 50o 2Θ with a scan speed of 0.5° over 2°/min.

2.2.4. Fourier Transform Infrared (FTIR) Spectroscopy

FTIR analytical method used FTIR spectrophotometer (Perkin Elmer Spectrum One, USA) that conducted the optical detection using $LiTaO_3$ detector with CsI beamsplitter that has a wavelength range of 7,800-225 cm^{-1}. For each sample, the scan is conducted in the range of 500 to 3500 cm^{-1}.

3. Result and Discussion

3.1. Thermal Analysis of Ibuprofen-Citric Acid and Ibuprofen-Oxalic Acid Co-Crystal Screening

The screening of the Ibuprofen-Citric acid (IBP-CA) and Ibuprofen-Oxalic acid (IBP-OA) formation was carried using the molar ratio 1:1, 1:2, 1:3, 2:1 and 3:1. The ground products from the experiment were analyzed using DSC. With DSC, the heat flow into and out of the sample and all the transitions or reactions due to energy changes involved can be measured [7]. Figure 1(a) shows the thermogram analysis of Ibuprofen-Citric acid (IBP-CA) sample produced using the neat grinding method. Based on figure 1, pure Ibuprofen shows the melting peak at 75.82°C and Citric acid shows the melting peak at 151.59°C while the melting temperature of Oxalic acid is at 187-195°C. The melting points of pure substances were in good agreement with the value published in the literature [3,8,9]. There was another peak appeared from the thermogram of pure Oxalic acid at 101.14°C. This is the point where Oxalic acid shows dihydrate characteristic. This is due to an adsorption of water molecules on Oxalic acid particle [10].Figure 1 also shows that there is a possibility for the formation of co-crystal for the molar ratio of 1:2, 1:3, 2:1 and 3:1 from the presence of a peak at the beginning of the thermogram. However, peaks at the beginning of the thermogram show the unstable co-crystal formation. The peaks for the pure components are also observed where there is a slight shift in the peaks to a lower temperature, occurred as a result of interaction induced by thermal energy between the drug and the coformer during the DSC scan of the sample [11]. Meanwhile, only peaks for pure components are observed for molar ratio 1:1 indicating that the formation of co-crystal is unsuccessful. For Ibuprofen-Oxalic acid (IBP-OA) system, the possibility of co-crystal formation was identified only in 1:1, 1:2 and 1:3 molar ratios whereas only peaks for pure components are observed for molar ration 2:1 and 3:1.

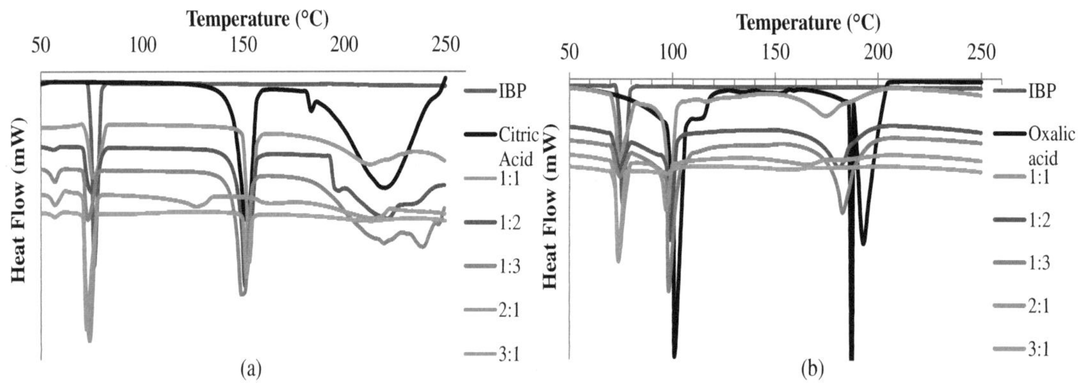

Figure 1. Differential scanning calorimetry (DSC) analysis of (a) IBP-Citric acid and (b) IBP-Oxalic acid in different molar ratio, using neat grinding method showing the possibility of the formation of co-crystal.

Figure 2(a) and (b) shows the DSC thermogram for IBP-CA and IBP-OA respectively at different ratio prepared using liquid assisted grinding method with ethanol used as the solvent. From the thermogram, there is no new peaks are observed for IBP-CA while there are new peaks observed for IBP-OA at ratio 1:1. 2:1 and 3:1. However, the peaks for pure components are also observed giving an early indication that mixtures of pure components and co-crystal exists. Grinding of IBP-CA assisted with propanol also shows no formation of cocrystal occur as shown in figure 3(a) as no new peaks are observed. Liquid assisted grinding of IBP-OA using propanol shown in figure 3(b) shows a probability for mixtures of pure components and co-crystal to exist at ratio 2:1 and 3:1 as small peaks are observed between IBP and Oxalic acid. The melting temperature of IBP-CA is slightly increase as the ratio increase as shown in figure 3(a). This indicates a significantly increase on intermolecular bonding of pure component with the present of ethanol and propanol solvent. This may potentially be the reason of unsuccessful attempt to form IBP-CA cocrystal via LAG method [12]. The IBP and the co-formers used in this work have more than one pKa values, reflecting the ionization constants of acidic and basic moieties of the molecules. According to the 'rules of three', the possibility of salt formation between drug molecule with wide array coformer candidates, their pKa differences should be more than 3 unit ($\Delta pKa > 3$). The pKa value of Ibuprofen is 4.9. Citric acid has three differences of pKa values which are 3.08, 4.74 and 5.40 while Oxalic acid is 1.23 and 4.19 [13]. Thus, there will be a various difference of pKa, (ΔpKa) between Ibuprofen-Citric Acid and Ibuprofen-Oxalic Acid. Table 1 below shows the ΔpKa values according to the type of mixtures. It shows that the differences in values of Ibuprofen-Citric Acid are less than 3. Therefore, Ibuprofen and Citric acid have a greater probability of forming cocrystal instead of salt formation. While for Ibuprofen-Oxalic Acid, one of the value shows the greater value to form salt because of it more than 3. Thus, the mixture of Ibuprofen-Oxalic Acid can be in cocrystal form or maybe in salt form [14].

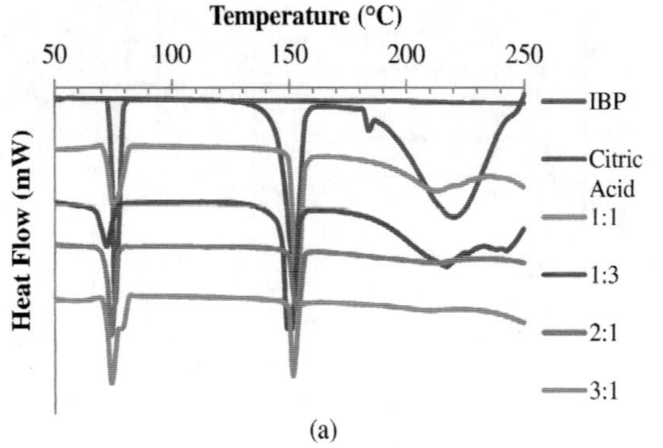

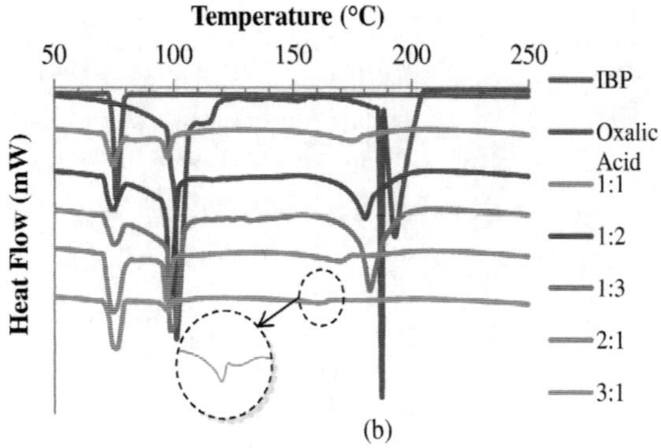

Figure 2. Differential scanning calorimetry (DSC) analysis of (a) IBP-Citric acid and (b) IBP-Oxalic acid in different molar ratio, using liquid assisted grinding method with ethanol solvent.

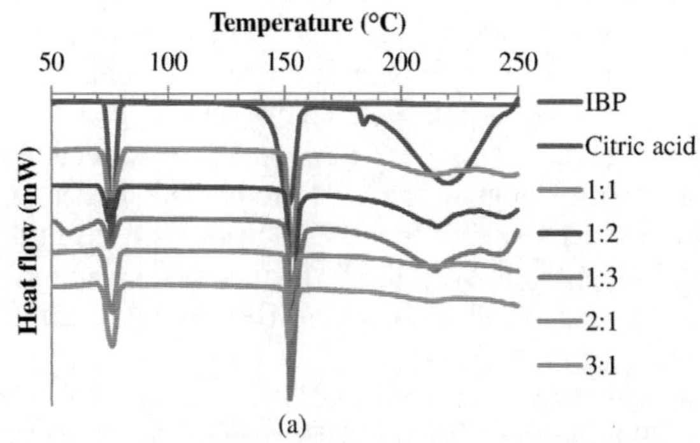

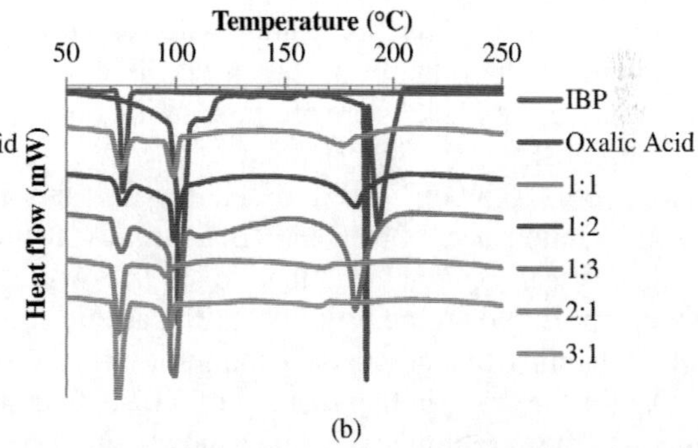

Figure 3. Differential scanning calorimetry (DSC) analysis of (a) IBP-Citric acid and (b) IBP-Oxalic acid in different molar ratio, using liquid assisted grinding method with propanol solvent.

Table 1. Difference in pKa, (ΔpKa) values of Ibuprofen-Citric Acid and Ibuprofen-Oxalic Acid

Type of Mixtures	(ΔpKa)
Ibuprofen-Citric Acid	1) 4.91 – 3.08 = 1.83
	2) 4.91 – 4.74 = 0.17
	3) 4.91 – 5.40 = -0.49
Ibuprofen-Oxalic Acid	1) 4.91 – 1.23 = 3.68
	2) 4.91 – 4.19 = 0.72

3.2. Crytalline Phase Transformation and Proton Transfer Determination

Further investigation was conducted by characterizing the product using XPRD and FTIR to confirm the formation of co-crystal. Figure 4(a) and (b) shows the XPRD and FTIR analysis for Ibuprofen-Citric acid using the dry grinding method. The ratios between IBP and CA were varied stoichiometrically i.e. 1:1, 1:2, 1:3, 2:1 and 3:1. In figure 4(a), the XRPD data show that new peaks were detected in the ratio of 1:1, 1:3, 2:1 and 3:1. New peaks show the probability of co-crystal formation. New peaks are still detected for ratio 1:1 even if the DSC thermogram shows no new peak for melting point. No new peaks for 1:2 are observed in the XRPD profile even if there is a peak detected using DSC. The inconclusive result is obtained based on the analysis for 1:1 and 1:2 ratio.

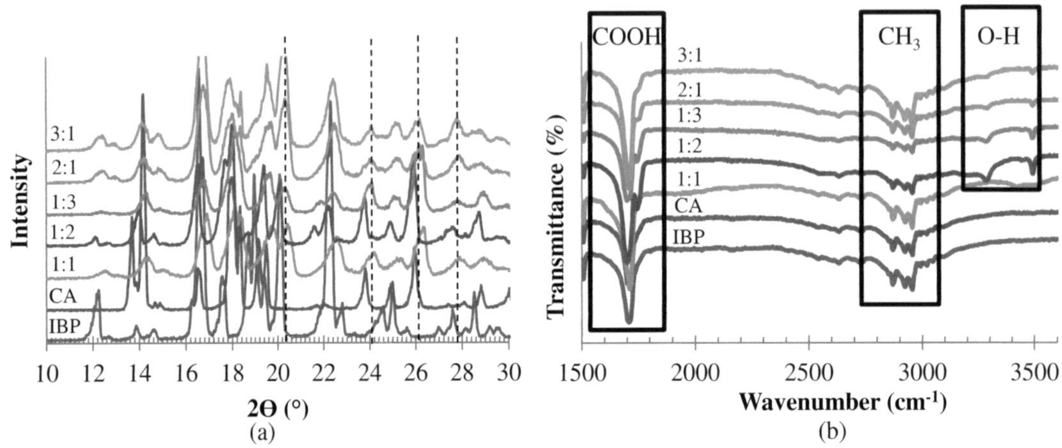

Figure 4. (a) XPRD profile and (b) FTIR spectrum for IBP-CA at different mole ratio using the dry grinding method.

Meanwhile, figure 4(b) shows the trend of FTIR spectrum for IBP-CA at varies ratio for the dry grinding method. The spectrums with band lower than 1500cm^{-1} are not shown because it is considered as the fingerprint region that is rarely diagnostically useful and rarely used to identify the functional group [15]. There are possibilities of proton transfer at both pure component (IBP and CA) as both consists of –COOH bond in the molecular structure. There are new peaks at wavenumber between 3200-3500cm^{-1} at ratio 1:2, 1:3, 2:1 and 3:1 that belongs to the O-H group. This indicates a new interaction occurs between IBP and CA at the respective ratio. Since no solvent is used in dry grinding, it can be concluded that no solvates were observed, however, the formation of salt cannot be concluded. Based on the 'rules of three', according to the ΔpKa values between IBP and CA, there is a high probability that co-crystal occurs for ratio 1:2, 1:3, 2:1 and 3:1.

As for the dry grinding method of IBP-OA, new peaks are observed for ratio 1:1, 1:2 and 1:3. This supports the preliminary result shown in DSC thermogram figure 1(b) where new melting temperatures were observed. Figure 5(a) shows the XRPD profile having at least 0.2° difference between OA and pure components for IBP-OA at ratio 1:1, 1:2 and 1:3 using the dry grinding method. There are no new peaks are observed for the ratio of 2:1 and 3:1 for IBP-OA using dry grinding. New peaks are observed for every ratio used for IBP-OA prepared using liquid assisted grinding with both ethanol and propanol solvents as depicted in figure 5(a) and (b) even though DSC only shows new peaks for ratio 1:1, 2:1 and 3:1 for ethanol and 2:1 and 3:1 for propanol. This is probably due to the phase transformation of the molecules. Meanwhile, the FTIR spectrum of 1:1, 1:2 and 1:3 IBP-OA with ethanol and ratio 2:1 and 3:1 IBP-OA with propanol follows the same trends as depicted in figure 4(b) where the functional group of COOH and CH_3 are present in the compound, with no clear indication of OH group present. Thus, the formation of salt cannot be clearly determined.

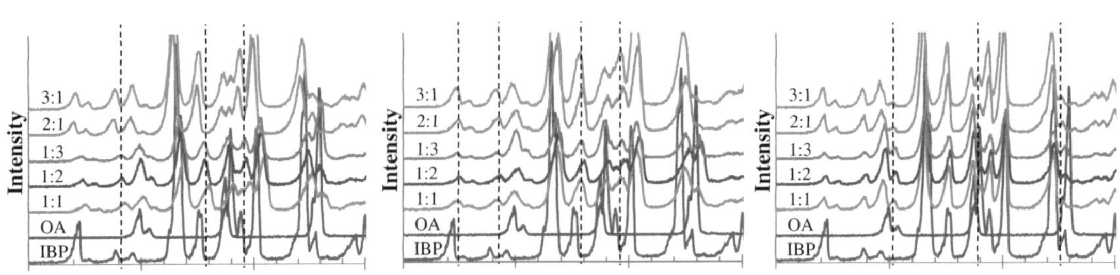

Figure 5. XPRD profile for IBP-OA using (a) dry grinding, (b) liquid assisted grinding using ethanol and (c) liquid assisted grinding using propanol at different mole ratio.

IOP Conf. Series: Materials Science and Engineering **358** (2018) 012065 doi:10.1088/1757-899X/358/1/012065

4. Conclusion

Formation of Ibuprofen-Citric acid (IBP-CA) and Ibuprofen-Oxalic acid (IBP-OXA) co-crystal was detected in the molar ratio 1:2, 1:3, 2:1 and 3:1 and 1:1, 1:2 and 1:3 respectively using the neat grinding method. Meanwhile, no IBP-CA co-crystal was formed using LAG method for both solvents. IBP-OXA co-crystal was formed in the molar ratio 1:1, 2:1 and 3:1 in ethanol, and 2:1 and 3:1 in propanol. Confirmation of co-crystal was made using XRPD analysis and FTIR spectrum determination where there is a high possiblity that co-crystal was formed for IBP-CA at ratio of 1:2, 1:3, 2:1 and 3:1, while salt formation was inconclusive for IBP-OA at ratio of 1:1, 1:2 and1:3 using neat grinding method. New peaks were determined using XRPD for IBP-OA in ethanol and propanol indicating phase transformation occurs, however, the salt formation cannot be clearly determined. For future work, solution crystallization with controlled parameters can be employed in order to assess a further details on cocrystal formation.

Acknowledgement

The authors would like to express their gratitude to Universiti Teknologi MARA for funding of this work. This work is part of grants (600-RMI/RAGS 5/3 (211/2014))

References

[1] Sanphui P, Sudalai Kumar S and Nangia A. 2012. *Cryst. Growth Des.* **12** 4588 – 4599
[2] Jasud S, Warad A, Rahul A, Jagdale G, and Zinjad S 2013. *WJPPS.* **2** 4682 – 4697
[3] Mukaida M, Watanabe Y, Sugano K and Terada K 2015. *Eur. J. Pharm. Sci.* **79** 61 – 67
[4] Aakeröy C B, Fasulo M E and Desper J 2007. *Mol. Pharm.* **4** 317 – 322
[5] Qiao N, Li M Schlindwein W, Malek N, Davies A and Trappitt G 2011. *Int. J. Pharm.*, **419** 1 – 11
[6] Gangadhar Z A, Bhaskar D A and Bhaskar G S 2014. *WJPPS.* **4** 1392
[7] Giron D 1986. *J Pharm Biomed Anal*, **4** 755 – 825
[8] Uchida H, Yoshida M, Kojima Y, Yamazoe Y and Matsuoka M 2005 *J. Chem. Eng. Data.* **50** 11 – 15
[9] Sangeetha M, Mathammal R 2017. *J. Mol. Struct.* **1143** 192 – 203
[10] Qingxin M, Hong H, Chang L *Atmos Environ.* 2013. **69** 281 – 288
[11] Chadha R, Saini A, Arora P, Chanda S and Dharamvir S J. 2012. Int J Pharm Pharm Sci. **4** 244 – 250
[12] Stanton M K, Tufekcic S, Morgan C and Bak A 2009. *Cryst. Growth Des.* **9** 1344 – 1352
[13] Chegg 2003. Table of Acids with Ka and pKa Values. *Lab Manual for Zumdahl/Zumdahl's Chemistry*, 656
[14] Wouters J and Quere L 2012. *Pharmaceutical Salts and Co-Crystals*. (Royal Society of Chemistry).
[15] Williams D and Fleming I 2008. *Spectroscopic methods in organic chemistry.* (Berkshire: McGraw-Hill Higher Education)

Techno-economic Analysis of Acid Gas Removal and Liquefaction for Pressurized LNG

S H Lee[1], Y K Seo[2] and D J Chang[2]

[1] Graduate School of Ocean Systems Engineering, Department of Mechanical Engineering, Korea Advanced Institute of Science and Technology, Republic of Korea
[2] Department of Mechanical Engineering, Korea Advanced Institute of Science and Technology, Republic of Korea

computer10@kaist.ac.kr, djchang@kaist.ac.kr

Abstract. This study estimated the life cycle cost (LCC) of an acid gas removal and a liquefaction processes for Pressurized LNG (PLNG) production and compared the results with the cost of normal LNG production. PLNG is pressurized LNG that is liquefied at a higher pressure and temperature than normal LNG. Due to the high temperature, the energy for liquefaction is reduced. The allowable CO_2 concentration in PLNG is increased up to 3 mol% when the product pressure 25 bar. An amine process with 35 wt% of diethanolamine (DEA) aqueous solution and a nitrogen expansion cycle were selected for the acid gas removal and the liquefaction processes, respectively. Two types of CO_2 concentration in the feed gas were investigated to analyze their impacts on the acid gas removal unit. When the CO_2 concentration was 5 mol%, the acid gas removal unit was required for both LNG and PLNG production. However, the acid gas removal unit was not necessary in PLNG when the concentration was 0.5 mol% and the pressure was higher than 15 bar. The results showed that the LCC of PLNG was reduced by almost 35% relative to that of LNG when the PLNG pressure was higher than 15 bar.

1. Introduction

Small floating liquefied natural gas (FLNG) projects are under consideration to monetize small-scale or stranded gas fields located in near-shore areas [1]. If FLNG plants are installed at near-shore sites, the emission of greenhouse gases such as CO_2 from these plants will become a new environmental issue. Generally, a large amount of CO_2 is captured during the acid gas removal process at FLNG plants and vented to the atmosphere, as the CO_2 concentration in LNG should be maintained below 50 ppm to avoid the formation of solids. Pressurized Liquefied Natural Gas (PLNG) can provide a new economical solution for reducing the emissions of greenhouse gases in FLNG. PLNG is produced at a higher pressure and temperature than normal LNG. Due to its high temperature condition, CO_2 can be present in PLNG at rates as high as 1 mol% without the formation of solids when the pressure of PLNG is 15 bar; moreover, the level can reach 3 mol% when the pressure is 25 bar [2]. It is also possible to reduce the energy required and cost of liquefaction. The previous study have shown that when the product pressure is 10 bar compared to 1 bar, almost 35% of the cost for the liquefaction can be saved [3]. Several studies have suggested novel liquefaction processes for PLNG with a cryogenic CO_2 removal unit in which the CO_2 is removed by crystallization during the liquefaction process [2, 4]. However, a strategy for handling solid particles after the cryogenic removal unit in FLNG is still required. The purpose of this study is to estimate the lifecycle cost (LCC) of these separate two processes (acid gas removal and liquefaction) for PLNG and to investigate the economic feasibility of PLNG production by comparing it with the LCC of LNG production.

Content from this work may be used under the terms of the Creative Commons Attribution 3.0 licence. Any further distribution of this work must maintain attribution to the author(s) and the title of the work, journal citation and DOI.
Published under licence by IOP Publishing Ltd

IOP Conf. Series: Materials Science and Engineering **358** (2018) 012066 doi:10.1088/1757-899X/358/1/012066

2. Simulation methodology and economic evaluation

Before simulating the processes, the feed flow rate capacity is assumed to be 0.5 million tons per annum (MTPA) to satisfy the capacity of a small FLNG installation. Additionally, rich feed (with a CO_2 concentration in the feed gas of 5 mol%) and lean feed (with a concentration of 0.5 mol%) gases are considered for the feed gas. In the rich feed case, an acid gas removal process should be included for both LNG and PLNG liquefaction. In contrast, in the lean feed case, the acid gas removal process can be omitted for PLNG when the PLNG pressure is higher than 15 bar. Because the solubility of CO_2 in PLNG (higher than 15 bar) is higher than 0.5 mol%, no solidification occurs during the liquefaction process. The solubility of CO_2 depending on the product pressure is tabulated in Table 1.

Table 1. CO_2 solubility in LNG and PLNG [2].

Product pressure (bar)	CO_2 solubility (mol%)
1	50 ppm
5	0.1
15	1
25	3

2.1. Simulation of the Acid gas removal process

An amine process with chemical solvent is considered for the acid gas removal process, and 35 wt% of diethanolamine (DEA) in the form of an aqueous solution is used as a chemical solvent to meet the maximum concentration requirement of CO_2 for LNG production [5]. Figure 1 shows a process flow diagram (PFD) of this process. The amine process involves the use of an absorption unit and an amine regeneration unit. In the absorption tower, the feed gas enters the bottom of the tower and comes into contact with the solvent. The CO_2 in the feed gas is then dissolved into the solvent and the sweetened gas leaves the top of the tower. To maintain the pressure of the feed gas, the pressure at which the absorption tower operates is identical to that of the feed gas. After absorption, the rich amine solution which contains the CO_2 is flashed to 2 bar and temporarily stored in a flash tank to eliminate the dissolved hydrocarbons in the solution. In the regeneration column, the CO_2 in the rich amine solution is removed as vapor at the top of the column and the lean amine solution is generated at the bottom. After the regeneration step, the temperature of the lean amine solution is decreased and the pressure of the solution is increased to the operating condition of the absorption tower. The lean amine solution is then recycled to the tower. The simulation is performed using Aspen HYSYS (Aspen Technology, Inc., USA). The detailed parameters used in the simulation are given described in Table 2.

The circulation rate of the amine solution is determined by the CO_2 concentration in the sweet gas. Because PLNG can contain more CO_2 than LNG, the amount of CO_2 that should be removed in the tower for PLNG production is less than that for LNG production. Generally, the amount of the CO_2 removed in the absorption column is proportional to the circulation rate of the amine solution. The circulation rate is calculated by adjusting the flow rate of the lean amine solution to the absorption tower until the CO_2 concentration in the sweet gas reaches a retainable concentration. The retainable concentration of CO_2 in the sweet gas is assumed to be identical to the solubility of CO_2, as tabulated in Table 1.

2.2. Simulation of the Liquefaction process

A nitrogen expansion cycle, known as suitable for small-scale liquefaction, is considered for the liquefaction process [6]. Figure 2 illustrates the PFD of the liquefaction process. The nitrogen expansion cycle utilizes a cryogenic heat exchanger, a multi-stage compressor, and a turbo-expander. The nitrogen refrigerant is compressed at the multi-stage compressor up to 120 bar. The refrigerant is then pre-cooled in the heat exchanger. The pre-cooled refrigerant is expanded at the turbo–expander and become a cold refrigerant, which liquefies the feed gas. However, before the feed gas enters the liquefaction process, the water in the feed gas is eliminated by a dehydration process. In this study, the sweet gas which enters the liquefaction process is assumed to have been dehydrated earlier. The dehydrated sweet gas is liquefied by the cold refrigerant at the heat exchanger. After liquefaction, the sweet gas is expanded to the product pressure and sent to the storage tank. The simulation of the

IOP Conf. Series: Materials Science and Engineering **358** (2018) 012066 doi:10.1088/1757-899X/358/1/012066

liquefaction process is also performed using Aspen HYSYS, and the detailed parameters for this simulation are described in Table 3.

Table 2. Simulation parameters for the acid gas removal process.

Item	Value
Operating pressure in the absorption tower	70 bar
Inlet temperature of the feed gas and the lean amine	35 °C
Number of stages in the tower	10
Operating pressure in the regeneration column	2 bar
Number of stages in the column	10
Minimum temperature approach in the Heat exchanger	10 °C
Inlet temperature to the regeneration column	90 ~ 100 °C

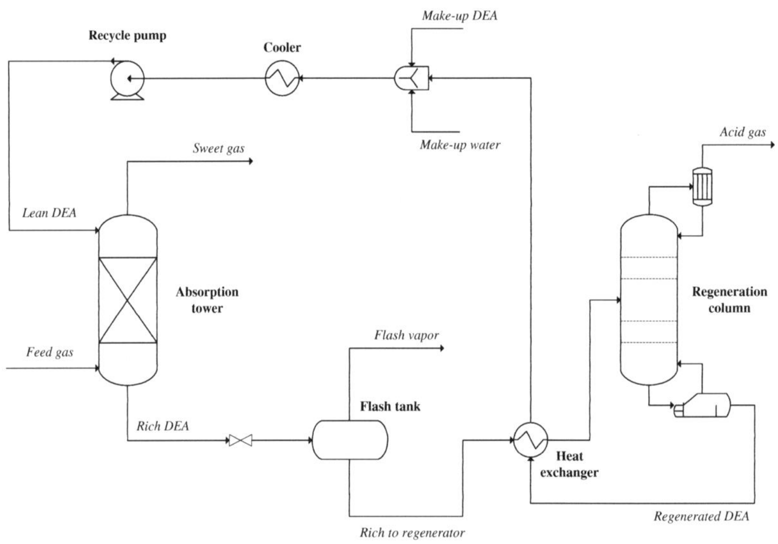

Figure 1. Process flow diagram of the acid gas removal process.

2.3. Methodology for the economic evaluation

To estimate the economic feasibility of PLNG, a LCC analysis is conducted. LCC is usually divided into capital expenditures (CAPEX) and operational expenditures (OPEX). CAPEX is calculated using what is known as the bare module cost method [8]. As shown in equation 1, this method estimates the capital investment cost of the process (C_{BM}) by multiplying the purchase cost of the equipment (C_p), a material factor (F_M), a pressure factor (F_p), and cost factors (B_1 and B_2). The purchase cost of the equipment is obtained from the Aspen process economic analyser (Aspen Technology, Inc. USA), and the material and pressure factors are estimated by considering the operating conditions of the equipment. Finally, the CAPEX is scaled to current prices using the chemical engineering plant cost index (CEPCI). OPEX is calculated by equation 2 [3]. OPEX usually expressed in a typical range of the utility ($C_{utility}$), labor (C_{OL}) and capital costs (C_{CAPEX}). The required utilities and the corresponding costs are estimated from the duty of the cooler, the compressor, the reboiler, and the condenser used in the processes. The labor cost is calculated by considering the number of pieces of equipment used in the processes. The lifecycle of the processes is assumed to be 15 years. After the estimation step, CAPEX and OPEX are expressed as normalized values.

$$\text{CAPEX}\ (USD) = \text{C}_{\text{BM}} = C_p F_{BM} = C_p[B_1 + B_2 F_p F_M] \quad (1)$$

$$\text{OPEX}\ (USD/yr) = 0.18\ C_{capex} + 2.73\ C_{OL} + 1.23\ C_{utility} \quad (2)$$

Table 3. Simulation parameters for the liquefaction process [7].

Item	Value
Number of stages in the multi-stage compressor	3
Inlet pressure at the compressor	6.3 bar
Outlet pressure at the compressor	120 bar
Minimum temperature approach in the heat exchanger	3 °C
Pressure drop in the heat exchanger	0 bar
Product pressure	1, 5, 15, 25 bar
Product temperature	- 167 °C~ -102 °C

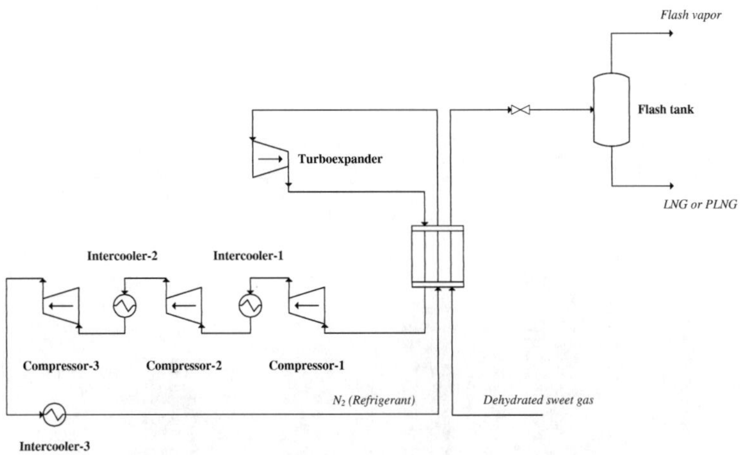

Figure 2. Process Flow Diagram of the liquefaction process.

3. Results and discussion

3.1 Simulation results of the processes

Figures 3 and 4 present the simulation results of the acid gas removal process. Figure 3 presents the circulation rate of the amine solution per unit of kgmole of the feed gas. For the rich feed gas, when the final product pressure is lower than 5 bar, a larger amount of solution is added to the absorption tower as compared to the feed gas to remove a large amount of CO_2 in the feed gas. However, when the final product pressure is higher than 15 bar, a smaller amount of solution than the feed gas is enough to remove the CO_2 in the feed gas. Figure 4 presents the total heating and cooling duty in the equipment per unit of kgmole for the feed gas. The duty in the equipment decreases considerably as the final product pressure increases. When the final product pressure is 15 bar, the equipment consumes only 65% of the energy relative to that at 1 bar. In the case of the lean feed gas, similar results are obtained.

Figure 5 presents the simulation results of the liquefaction process. As shown in this figure, when the product pressure is lower than 5 bar, the net power consumed in the compressor is identical in the two cases. However, when the product pressure exceeds 15 bar, the net power levels differ slightly because, in the rich feed case, the sweet gas, which enters the liquefaction process, contains the maximum amount of CO_2 in the stream. Hence, the compressor in the rich feed case consumes more energy than that in the lean feed case. However, the CO_2 containment in the sweet gas does not have much of an effect on the net power of the compressor.

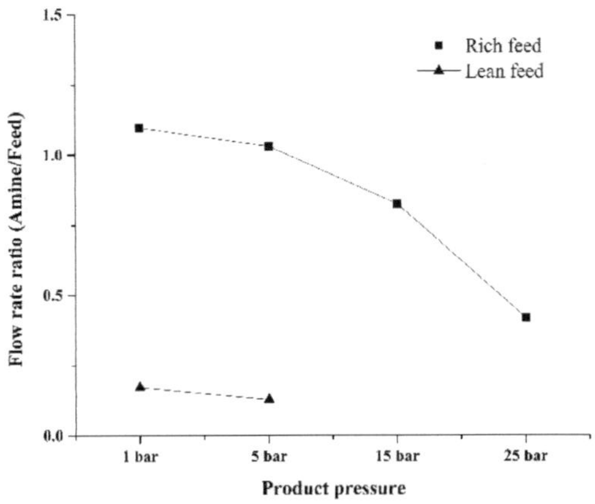

Figure 3. Circulation rate of the amine solution per unit of kgmole for the feed gas.

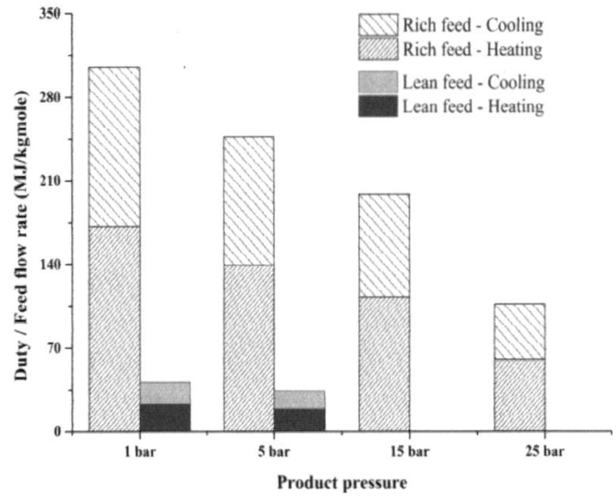

Figure 4. Heating and cooling duty per unit of kgmole for the feed gas.

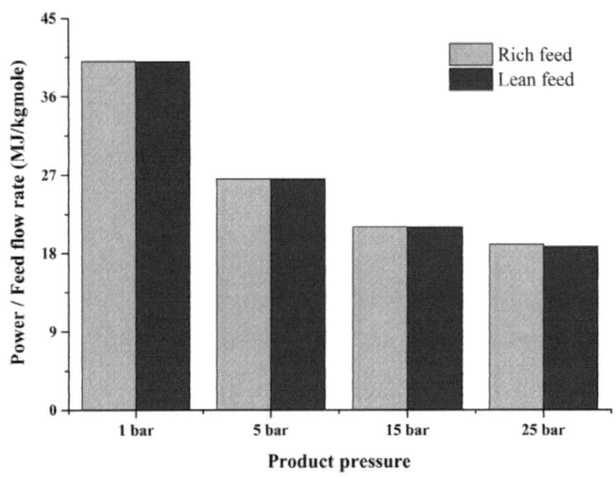

Figure 5. Net power in the compressor per unit of kgmole for sweet gas.

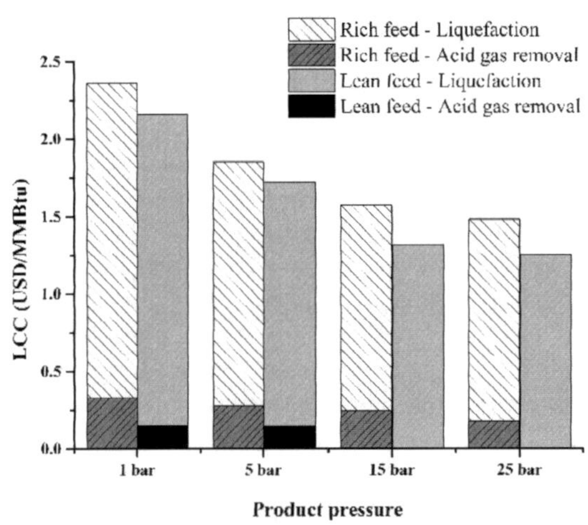

Figure 6. Total LCC for acid gas removal and liquefaction.

3.2 Results of the economic evaluation

Figure 6 presents the LCC for the acid gas removal and liquefaction processes with the product pressure. The estimation results show that the total LCC for the rich feed is decreased by nearly 35% when the liquefaction pressure exceeds 15 bar relative to that at 1 bar. Furthermore, 40% of the total LCC is saved when the acid gas removal unit is not required.

4. Conclusion

In this study, an economic analysis of acid gas removal and liquefaction processes for LNG and PLNG production were conducted and the results were compared to the cost for LNG production. When the CO_2 solubility in the feed gas exceeded the CO_2 concentration in the feed, the acid gas removal unit could be omitted and CO_2 venting to the atmosphere was not required. The evaluation results showed that the PLNG production was much more economical than LNG production. Consequently, PLNG can be an economical and eco-friendly technology for near-shore FLNG projects.

Acknowledgements
This research was supported by a grant from the LNG Plant R&D Centre funded by the Ministry of Land, Infrastructure and Transport of the Korean government.

References

[1] Jeanneau E, Chretien D, and Fuchs V 2013 17th Int. Conf. & exhibition on liquefied natural gas (Houston)
[2] Xiong X, Lin W, and Gu A 2015 *Energy* **93** 1-9
[3] Lee S, Seo Y, Lee J, and Chang D 2016 *J. of Natural Gas Science and Engineering* **33** 405-418
[4] Lee J, Jung J, and Kim K 2012 *Offshore Technology Conf.* (Houston)
[5] Kidnay A J, Parrish W R, and McCartney D G 2011 *Fundamentals of natural gas processing* Vol. 218 (Boca raton: CRC Press.)
[6] Barclay M A, Gongaware D F, Dalton K, and Skrzypkowski M P 2004 *Cryogenic Engineering Conf.* (Anchorage) Vol. 710 No. 1 pp 75-82 (Melville: AIP Publishing)
[7] Venkatarathnam G 2008 *Cryogenic mixed refrigerant processes* (New York: Springer) 149-220
[8] Turton R, Bailie R C, Whiting W B, and Shaeiwitz J A 2008 *Analysis, synthesis and design of chemical processes* (London: Pearson Education)

IOP Conf. Series: Materials Science and Engineering **358** (2018) 012067 doi:10.1088/1757-899X/358/1/012067

Rheological Behaviour of Water-in-Light Crude Oil Emulsion

H Husin[1,2], T S Taju Ariffin[1] and E Yahya[1]

[1]Department of Oil and Gas Engineering, Faculty of chemical Enginering, University Teknology MARA, 40450 Shah Alam, Selangor Darul Ehsan, Malaysia
[2]Core of Frontier Materials & Industry Applications, Universiti Teknology MARA(UiTM) 40450 Shah Alam, Selangor Darul Ehsan, Malaysia

hazlina858@salam.uitm.edu.my

Abstract. Basically, emulsions consist of two immiscible liquids which have different density. In petroleum industry, emulsions are undesirable due to their various costly problems in term of transportation difficulties and production loss. A study of the rheological behaviour of light crude oil and its mixture from Terengganu were carried out using Antoon Paar MCR 301 rheometer operated at pressure of 2.5 bar at temperature 30℃. Water in oil emulsions were prepared by mixing light crude oil with different water volume fractions (20%, 30% and 40%). The objectives of present paper are to study the rheological behaviour of emulsion as a fuction of shear rate and model analysis that fitted with the experimental data. The rheological models of Ostwald-De-Waele and Herschel-Bulkley were fitted to the experimental results. All models represented well the rheological data, with high values for the correlation coefficients. The result indicated that variation of water content influenced shear rate-shear stress rheogram of the prepared emulsions. In the case of 100% light crude oil, the study demonstrated non-Newtonian shear thickening behavior. However, for emulsion with different volume water ratios, the rheological behaviour could be well described by Herschel-Bulkley models due to the present of yield stress parameter (R^2 = 0.99807). As a conclusion, rheological studies showed that volume water ratio have a great impact on the shear stress and viscosity of water in oil emulsion and it is important to understand these factors to avoid various costly problems.

1. Introduction

Markets worldwide have shown an increasing demand for crude oil. Rheological behaviors of emulsion not only influence its visual and textural perception, but also affect its processing capabilities. For example, compared to Newtonian fluids, shear thinning fluids are more susceptible to applied stress. Thus, measuring shear thinning behavior is necessary as most fluids of interest come under this class of fluids. Crude oil is seldom produced alone and it generally is commingled with water. As the reservoir become older, the production of water can significantly increase. The water creates several problems and consequently, percentage of crude requiring treatment increases. The produced water must be separated from the oil and disposed properly. At the meantime, emulsions occur and might increase the cost of production. In order to correctly áddress the emulsion problem, it is vital to better understand the rheological properties of the emulsion. Rheological deals with the flow of fluid and deformation of matter under various kinds of strain and stress. The force can be applied in various ways: a compression, a tension, a shearing process, or some combination of three. Rheological properties of emulsion can be determined by measuring the relationship between stress, strain, temperature and time. Thorough knowledge of emulsion rheology is helpful in minimizing numerous difficulties during processes such as refining, transportation, separation and production. The rheological of emulsion commonly affected by temperature, volume fraction of dispersed phase, droplet size and interfacial tension. Emulsion can exhibits either Newtonian of non-Newtonian

Content from this work may be used under the terms of the Creative Commons Attribution 3.0 licence. Any further distribution of this work must maintain attribution to the author(s) and the title of the work, journal citation and DOI.
Published under licence by IOP Publishing Ltd

behavior depending upon its composition[1],2]. Newtonian liquid can be defined as constant viscosity independent of shear rate. Newtonian liquid also exhibits uniform resistance to flow independent of flow conditions. Meanwhile, non-Newtonian liquid is defined as viscosity which changes with shear rate and type of deformation. It exhibits different resistance to flow, at different shear and extensional rates. Typically, viscosity of emulsion changes from a few hundred mPa.s to about 100 000 mPa.s [3]. Figure 1, shows some rheological behavior of fluid.

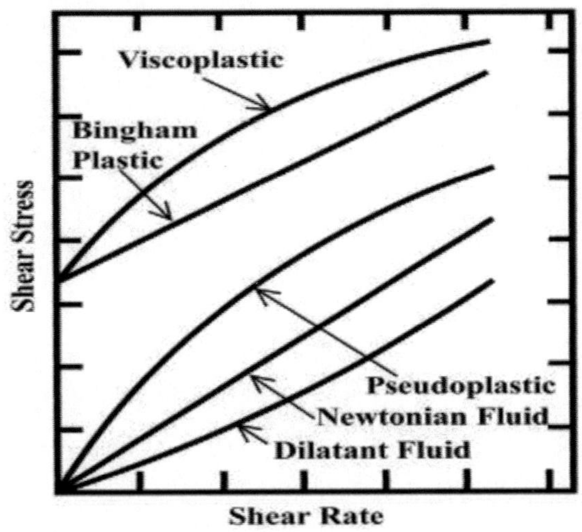

Figure 1. Rheological behavior of non-Newtonian fluid. [4]

The main objective of this investigation is to study the rheological properties of light crude oil and the characteristic of its emulsions. This investigation provide different rheological model to establish the viscosity rheogram as a function of volume water fraction (i.e., 20%, 30%, and 40%) at a temperature of 30 ℃.

2. Materials and methods

Light crude oil from Terengganu was used in this study for the preparation of emulsion. The emulsions were prepared with different volume fraction of water (i.e. 20%, 30%, and 40%). In order to prepare water in oil emulsions, water was added gradually to the oil in 250 ml beaker. Emulsions were homogenized with 10 000 rpm in 15 minutes. Emulsions were left in order to settle for a period of 24 hours in vials after homogenized. Rheological behaviors of the emulsions were analyzed by Antoon Paar MCR 301 rheometer which operates at pressure of 2.5 bar. The rheological test were conducted at temperature 30°C and shear rate varied from 0.1 – 1000 s^{-1}. Light crude oil used in the experiment has physical properties as given in Table 1[5]

Table 1. Properties of light crude oil[5].

Analysis	Light crude oil
Physical State	liquid
Specific Gravity	0.85
API gravity	34.87 °API
Pour Point °C	- 20

3. Results and discussions

3.1. Rheological behavior of light crude oil

Figure 2 shows the flow behavior of the light crude oil in terms of shear stress and shear rate. It was investigated over a wide range of shear rate at 30℃. The measurement test was conducted under the

IOP Conf. Series: Materials Science and Engineering **358** (2018) 012067 doi:10.1088/1757-899X/358/1/012067

controlled rate (CR) mode, and the values of shear stress and shear rate were obtained. From Figure 2, light crude oil exhibits non-Newtonian behavior, as shown by the non-linerity between the shear rate and shear stress plotted rheogram. Additionally, shear stress increases gradually with shear rate, evidence of the increase in viscosity with increase in shear rate, confirming the non-Newtonian dilatant behavior of light crude oil. On the other hand, in the majority of studies, light crude oil has been modeled under non-Newtonian Pseudoplastic behaviour [6][2][7]. This is due to the complex mixture and origin of the crude oil

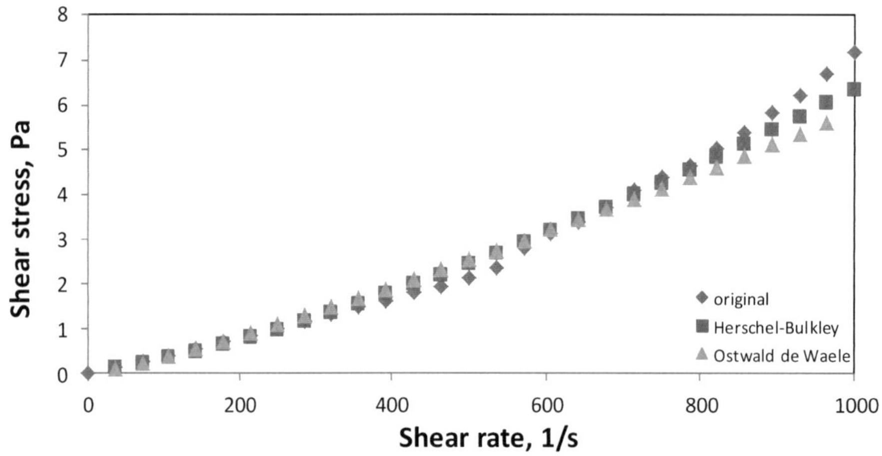

Figure 2. Relationship between shear stress and shear rate for light crude oil at 30℃

Figure 3 shows the relation between viscosity and shear rate. It is observed that viscosity values decrease with increasing shear rates at low region up to 100 s^{-1} and hence called as shear- thinning. However, when the shear rate is above 100 s^{-1}, viscosity of crude oil exhibits a strong non-Newtonion shear thinning behavior. Variation viscosity was influenced by several factors such as concentration and nature of oil and particle size[8]. As viscosity increased significantly with shear rate, the molecules are starting to expand and filled the void space. Hence, lead to much larger of shear stress that caused viscosity rapidly increasing[4].

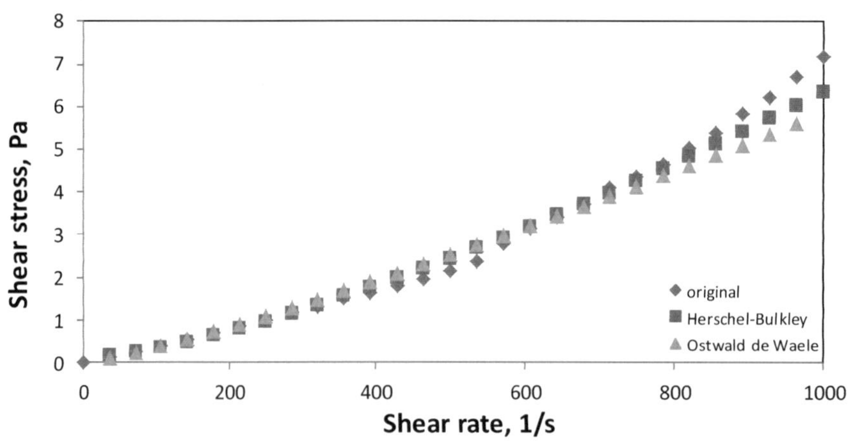

Figure 3. Relationship between viscosity and shear rate for light crude oil at 30℃.

3.2. Modeling the rheology of emulsion

The light crude oil was used in the study, shear stresses and shear rates were calculated using equations derived from (Ostwald de Waele and Herschel –Bulkley), Equations (1) and (2) respectively. These models were applied to transform emulsion to information on fluid rheological behavior. The Ostwald de Waele model (Eq 1), also known as the Power Law model. The equation has two parameters (k and n). Where n is the flow behaviour index and K gives the consistency index or viscosity (Pa.s)[9].

$$\tau = k(\gamma)^n \tag{1}$$

On the other hand, Herschel –Bulkley model consists of three adjustable parameters (τ_0, k, n) and it's considered as a precise model[10]. The Herschel –Bulkley model is expressed in equation 2, where k and n are consistency index and flow behaviour index respectively and τ_0 denotes the yield stress (Pa.s). The n-value in Equation (1) and (2) give fluid behavior information according to[11].

$$\tau = \tau_0 + k(\gamma)^n \tag{2}$$

$n < 1 \Rightarrow$ non-Newtonian Pseudoplastic behaviour
$n = 1 \Rightarrow$ Newtonian behaviour
$n > 1 \Rightarrow$ non-Newtonian Dilatant behavior

These two models are the best in describing the colloidal suspension of emulsion [12][13][14]. The calculated values were compared to the measured values to determine which of the model could predict better shear stresses and shear rate over the range of study. To assess the rheological behavior of crude oil emulsion at different water contents, the experimental data were fitted to the two proposed rheological models mentioned previously. The results of the experiments are shown in Table 2 and in Figures 4,5 and 6. It can be seen that both models satisfactorily represent the rheological behavior of crude oil emulsion. However, Herschel – Bulkley model provided the best fit of the rhelogical data. Regression coefficients (R^2) of Herschel – Bulkley model showed the highest value, which is 0.99807 for 40 % volume water fraction. This is because the present of parameter of yield stress. Meanwhile, Ostwald de Waele model consist only two parameters (does not consist parameter of yield tress). In industrial processes, yield stress is vital quality control parameter, particularly for comparing manufacturing products on different production lines[8]. The R^2 value shown in Table 1 indicates how well the regression analysis fits the shear stresses- shear rate data (see Figure 2,3,4). The closer the R^2 value is to 1.00, the better the linear regression; analysis has fit the data. Similar results have been reported by [12][7][5].

The decrease of water content in the three-different volume water fraction seemed to increase the consistency factor, (k). However, decrease the value of behavior index, (n). It shows that the values for the behavior index (n) are higher than 1 (except for 40% volume water fraction) indicating a non-Newtonian Dilatant behavior of light crude oil emulsion. This rheological parameter varied from 1.1172 to 0.9948 for the Ostwald de Waele model and from 1.3192 to 1.0046 for the Herschel-Bulkley model. The Herschel-Bulkley and Ostwald de Waele models both indicated a Dilatant behavior of 20% volume water fraction of emulsion since the value τ_0 was > 0 (0.13 Pa) and n > 1 (Table 2). Herschel-Bulkley model indicated that 30% volume water fraction of emulsion performed as a Dilatant behavior, since the yield stress value was > 0 (0.14 Pa) and a flow behavior index of 1.21 (Table 2). Results obtained by Ostwald de Waele confirmed a Dilatant behavior n > 1(1.0628) for 30% volume water fraction. However, 40% volume water fraction results obtained by the Ostwald de Waele models indicated a Pseudoplastic non-Newtonian behavior, since value n (0.99482) but Herschel - Bulkley model indicated a Newtonian behavior since n = 1.

Table 2. Modeling analysis of experimental data

Composition of the water fraction	Ostwald de Waele			Herschel - Bulkley			
	n	*k*	R^2	τ_0	*k*	n	R^2
0%	1.21000	0.00138	0.95497	0.09258	0.00037	1.4110	0.98340
20%	1.11720	0.00280	0.94889	0.13176	0.00073	1.3192	0.97819
30%	1.06280	0.00488	0.95931	0.14256	0.00182	1.2110	0.97870
40%	0.99482	0.01065	0.94740	0.01595	0.00997	1.0046	0.99807

IOP Conf. Series: Materials Science and Engineering **358** (2018) 012067 doi:10.1088/1757-899X/358/1/012067

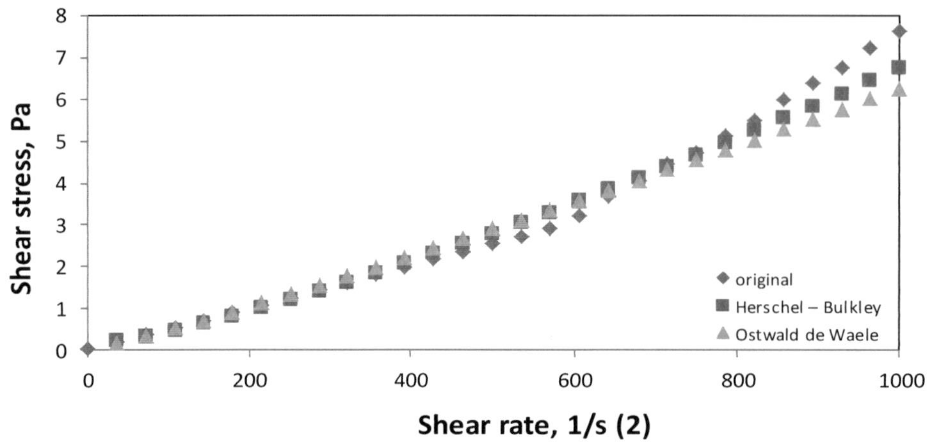

Figure. 4. Shear stress versus shear rate experimental data and fitting plots for 20% water volume fraction.

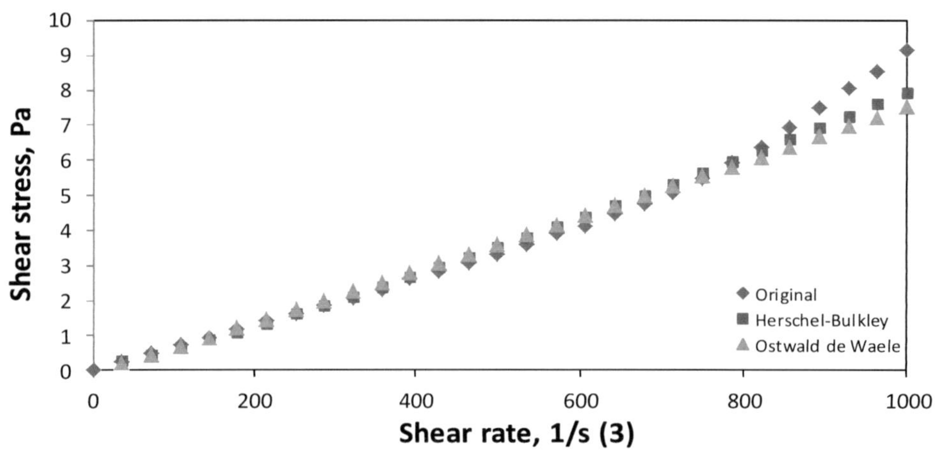

Figure. 5. Shear stress versus shear rate experimental data and fitting plots for 30% water volume fraction.

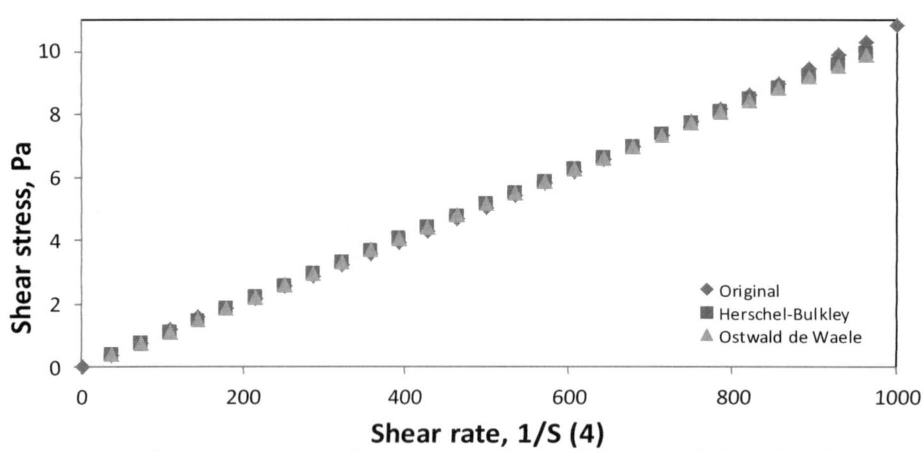

Figure 6. Shear stress versus shear rate experimental data and fitting plots for 40% water volume fraction.

IOP Conf. Series: Materials Science and Engineering **358** (2018) 012067 doi:10.1088/1757-899X/358/1/012067

4. Conclusion

Rheological behavior of light crude oil and its emulsion were investigated. Based on the results, the rheological behavior of light crude oil and its emulsion can be best presented by the Herschel-Bulkley model with highest regression correlation coefficient, R^2 value is 0.99807. The value of flow behavior index, n decrease (from 1.4110 to 1.0046) as volume water fraction increase, and characterizing non-Newtonian Dilatant behavior for light crude oil emulsion. In addition, experimental results showed the rheological behavior of studied emulsion were significantly influenced by their water content and the nature of the crude oil.

Acknowledgement

The authors would like to thank to Ministry of Higher Education (MOHE) and Universiti Teknolgi MARA (UiTM) for the financial support given through the Research Acculturation Grant Scheme (RAGS), 600-RMI/RAGS 5/3 (74/2013).

References

[1] Meriem-benziane M and Zahloul H 2013 Preparation *Int. J. Mech. Aerospace, Ind. Mechatronics Eng.***7** 3 223–227 2
[2] Ghannam M T, Hasan S W, Abu-Jdayil B and Esmail N 2012 *J. Pet. Sci. Eng.* **81** 122–128
[3] Fingas M and Fieldhouse B 2009 *Colloids Surfaces A Physicochem. Eng. Asp.* **333** 67–81
[4] Chhabra R P 2010 *Rheol. Complex Fluids*, pp. 3–34
[5] Ariffin T S T, Yahya E and Husin H 2016 *Procedia Eng.* **148** 1149–1155
[6] Liyana Nadirah M S 2014 *Int. J. Eng. Sci. Res. Technol.* **3** (1) 2277-9655
[7] Abdurahman N H and Georgiou G C 2013 *Int. J. Eng. Sci. Innov. Technol.* **2** 6 549–561
[8] Lopes A S, Mattietto R D A, De Menezes H C, Helena L and Pena S 2013 *Food Sci. Technol.* **33** 1 26–31
[9] Keleşoğlu S, Barrabino Ponce A, Humborstad Sørland G, Simon S, Paso K and Sjöblom J 2015 *J. Pet. Sci. Eng.* **126** 1–10
[10] Björn A, De La Monja P S, Karlsson A, Ejlertsson J and Svensson B H 2012 *Rheological characterization Biogas*. D. S. Kumar Ed. InTech pp 63-76
[11] Violeta Ghica M, Hirjau M, Lupuleasa D and Dinu-Pirrvu C E *Molecules* **21** 786 2016
[12] Meriem-benziane M, Abdul-Wahab S A, Benaicha M and Belhadri M 2012 *Fuel* **95** 97–107
[13] Hasan S W, Ghannam M T and Esmail N 2010 *Fuel* **89** 5 1095–1100
[14] Guzek A, Shufrin I, Pasternak E and Dyskin A V 2015 *J. Pet. Sci. Eng.* **135** 375–383

IOP Conf. Series: Materials Science and Engineering **358** (2018) 012068 doi:10.1088/1757-899X/358/1/012068

Relationship between Pipeline Wall Thickness (Gr. X60) and Water Depth towards Avoiding Failure during Installation

K Abdul Razak*, M I H Othman, S Mat Yusuf , M F I Ahmad Fuad and Effah Yahya

Faculty of Chemical Engineering, Universiti Teknologi MARA

*khalilabdulrazak@salam.uitm.edu.my,iqramhaikal11@gmail.com, suriatie0539@salam.uitm.edu.my,fazrilirfan@gmail.com, effahyahya@salam.uitm.edu.my

Abstract. Oil and gas today being developed at different water depth characterized as shallow, deep and ultra-deep waters. Among the major components involved during the offshore installation is pipelines. Pipelines are a transportation method of material through a pipe. In oil and gas industry, pipeline come from a bunch of line pipe that welded together to become a long pipeline and can be divided into two which is gas pipeline and oil pipeline. In order to perform pipeline installation, we need pipe laying barge or pipe laying vessel. However, pipe laying vessel can be divided into two types: S-lay vessel and J-lay vessel. The function of pipe lay vessel is not only to perform pipeline installation. It also performed installation of umbilical or electrical cables. In the simple words, pipe lay vessel is performing the installation of subsea in all the connecting infrastructures. Besides that, the installation processes of pipelines require special focus to make the installation succeed. For instance, the heavy pipelines may exceed the lay vessel's tension capacities in certain kind of water depth. Pipeline have their own characteristic and we can group it or differentiate it by certain parameters such as grade of material, type of material, size of diameter, size of wall thickness and the strength. For instances, wall thickness parameter studies indicate that if use the higher steel grade of the pipelines will have a significant contribution in pipeline wall thickness reduction. When running the process of pipe lay, water depth is the most critical thing that we need to monitor and concern about because of course we cannot control the water depth but we can control the characteristic of the pipe like apply line pipe that have wall thickness suitable with current water depth in order to avoid failure during the installation. This research will analyse whether the pipeline parameter meet the requirements limit and minimum yield stress. It will overlook to simulate pipe grade API 5L X60 which size from 8 to 20mm thickness with a water depth of 50 to 300m. Result shown that pipeline installation will fail from the wall thickness of 18mm onwards since it has been passed the critical yield percentage.

1. Introduction

Pipelines are major components of the oil and gas production as it transports either oil or gas from platform to the onshore for refinery and process to be the saleable product. Before the advance technologies develop, the first pipe was short and basic to get oil from drill holes and pass it or connect it to nearby tanks or refinery for continue the process. In recent years there has been an increased focus on oil and gas as a source of energy. From this thing, many technologies are invented and start to seek any field that have hydrocarbon reserves. Starting from onshore and bring to the offshore part. In order to overcome any challenges comes from both technical and economical for different level of water

Content from this work may be used under the terms of the Creative Commons Attribution 3.0 licence. Any further distribution of this work must maintain attribution to the author(s) and the title of the work, journal citation and DOI.
Published under licence by IOP Publishing Ltd

depth, considerations to pipeline and design installation must be made. For instance, design installation for deep water depth have large margin with ultra-deep water depth as the parameter of the characteristic is different from each other. To get the better result without any problem or in other word, to make pipelines installation and operations feasible in deep water depth, we need to concern on such many things. This is because significant challenges will appear regarding pipelines for oil and gas field developments in the area that have deep water depth. The large concerns that need to take action are the method of laying pipe, selection of pipeline concept and ability to do intervention. Besides that, limitation also needs to keep eye on as set the limitations to how deep pipeline can be installed or laid. Basically, pipeline installations are limited by laying vessel. But not only limitations to laying vessel tension capacity but also to technical design solution are important to consider. Usually the most common pipeline concepts are single steel pipelines. At this stage of pipeline concept, it is consider being the simplest engineering concept as has well known behaviour during the installation. Basically, this simplest engineering concept mostly suitable for more shallow fields and the cost are relatively low. Parallel to it, going to deeper waters has caused different concept and much complicated solution being considered and take action. Besides that, application in deep water we can use sandwich pipes and pipe-in-pipe as an alternation concepts to single steel pipelines. Besides that, in order to reduce the required wall thickness and pipeline weights of single steel pipelines, development of higher steel grades is explored and considered which may improve lay ability at such depths. Pipe laying vessel is one of the most important and it includes pipe conditioning apparatus located at its stern. The have several pipe conditioning apparatuses which is includes radius control apparatus, straightening apparatus and pipe guiding apparatus [1].

Every each of them have different role to play it as radius control apparatus basically for imparting a substantially uniform curve to the pipe after it is spooled from the reel. In addition, the important role for straightening apparatus is imparting a reverse bending force to the pipe opposite to the curvature imparted to the pipe by the reel. Moreover, the part of guiding pipe apparatus is to guiding the pipe into the water after it emerges from the straightening apparatus [1]. Besides that, the other apparatus is endless pipe conveyor apparatus. This apparatus doing their job to arranged around the radius control apparatus to convey the pipe over the radius control apparatus in the passage of the pipe from the reel into the pipe straightening apparatus. Basically, pipeline vessel function is laying offshore subsea pipelines for such as several uses as the gathering of oil and gas from offshore facilities or from offshore platform or from offshore subsea well. They have main method to lay the pipe. For instance, in the country like Mexico, they use "stovepiping" as their main method to lay the pipe. This method actually a pipeline is fabricated on the deck of a lay barge by welding it together individual lengths of pipe.The "stovepiping" method requires the skill from the welders and also their relatively bulking equipment accompany the pipelaying barge crew during the entire process to lay the pipe operations [1]. Generally, there have several steps that can follow in order to laying a subterranean pipeline [2]:

i. Suspending a support stinger pipe which is from vessel that having one ends removable that connected to the vessel and the second one end directed towards the subsea floor.
ii. Said stinger which including an elongated member which is having plurality of guide elements spaced longitudinally and to engage the inner walls of a pipe line segment.
iii. Adjusting the rigidity of said continuous support. This method is to regulate the curvature of the pipeline between the vessel and the subsea floor. This method settles by removable engaging additional support member.
iv. Sequentially passing pipeline segments about said stinger. This method is to regulate the pipe curvature during passage of the latter to the said floor.

In recent years the activity of the offshore drilling of oil and gas are has been increasingly. Results from this activity, the more well has created and need pipelines extending from the offshore sites to shore installations. In this case, the need for such pipeline that can stand with harsh conditions are been recognized. For instance, the pipelines that need to be laid in the underwater that is particularly exposed with saline water must be adequately protected such as coatings to against the corrosion from occur. The coating will create some problems during laying the pipeline. So, the machine for applying the coating material to the pipe and for compacting process was fully disclosed [3]. In addition, the pipe should be heavy enough to make sure the pipe sinks well to the bottom of the sea and stable against the action of disturbing influences [4]. The pipe must need something to holding them in stable position on or below the bottom of the body of water. There is some problem that has arisen when laying the pipeline along the floor of a deep body of water, i.e. kinking, excessive bending and difficult to control movement of length [5]. There have several systems that have been devised to control such a long length which is by using extensive guide structures and buoyant elements in order to support the pipeline [5]. Unfortunately, this system is expensive and the best way is by kept continuous controlling the tension of the pipeline that need to be laid, the excessive bending and kinking of the pipeline may be avoided without the extensive support structures or buoyant support means. It is so interesting to know much about the pipeline installation. This is because in the offshore operation, pipeline installation is one of the most challenging to handle it and requires a high level of engineering design to determine the criteria of the pipe, i.e. diameter, material and method. This criterion is important later on to make cost estimation and decide the installation vessel.

Two common methods to install pipeline which is S-lay and J-lay. *S-Lay* methods get their name from characterised by "S" curve during the laying of the pipeline to the seabed. The pipe will be stored and assembled on the vessel before it laying the pipeline to the bottom of the water. Basically, the pipe will leave the vessel at the stern which is at the back of the vessel and through a sloping ramp [6]. Upper curved part of the pipeline is known as the overbend or upper generator convex. In addition, certain angle the pipeline will lose contact with the stinger known as chosen angle. After the pipelines lose contact with the stinger, it will continue go deeper downward straight to the bottom of the sea and then gradually bends in the opposite direction which is known as the sagbend area. Starting from the sagbend area, the rest of the pipeline will continues to reach the bottom of the sea known as seabed at the touchdown point, Figure 1 [6].

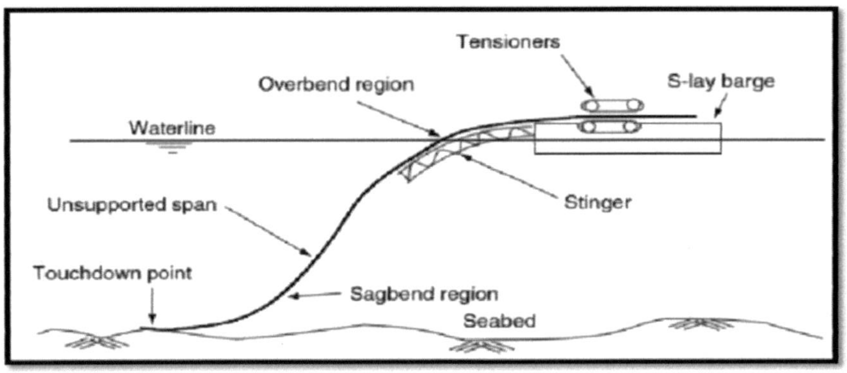

Figure 1. Pipe Laying Configuration Using S-Lay Method [6]

Besides that, in the sagbend area or lower generator concave, the combination must safely be sustained between bending and pressure loads. Besides that, in the sagbend area also, the tension has been applied at the top which is used basically to control the curvature in the respective area. Maximum strain and maximum bending movement can be controlled during the pipeline installation via the important parameters which is the stinger length, stinger radius, tension capacity and the longitudinal trim of the vessel [7]. J-lay method (Figure 2) suitable for bigger length pipe and required tension is bigger. This method is characterized by the pipeline which is leaving the vessel and nearly forms a vertical position and has J-shape. It can reduce the required over-bend curvature so shorter stinger should be enough. When further down into the water, the pressure will be increase and the tension will decrease progressively [6].

There has comparison between S-lay and J-lay, for instances the top tension category, S-lay configurations have higher required top tension compared to the J-lay configuration (Figure 3). Besides that, for the critical area, S-lay configuration, the most concern critical area is located at the overbend area whereas in the J-lay configuration, the most critical area is located opposite from the overbend which is at the sagbend of the pipeline. Usually, in the strain in the overbend region should satisfy the criteria that has stated in the DNV-OS-F101 (2007) [8].

IOP Conf. Series: Materials Science and Engineering **358** (2018) 012068 doi:10.1088/1757-899X/358/1/012068

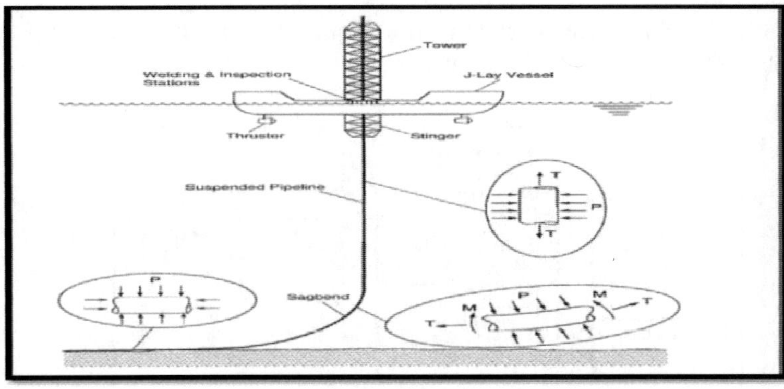

Figure 2. J-lay pipeline installation and associated pipeline loading [6].

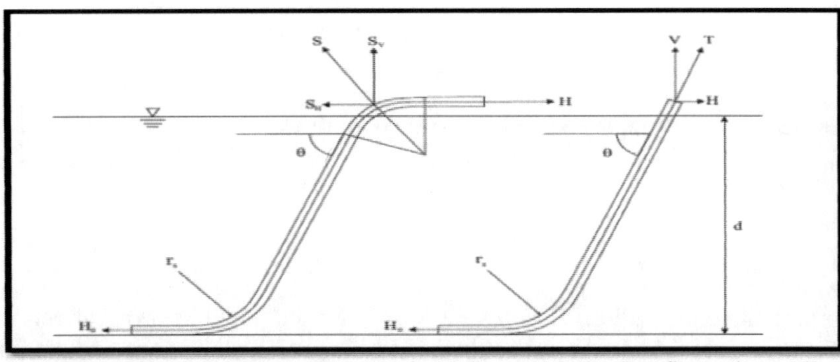

Figure 3. Comparison Tension for S-lay and J-lay Configurations [6]

Following material characteristic need to be consider, i.e. hardness, weldability, corrosion resistance, mechanical properties, fatigue resistance and fracture toughness [8]. Besides that, in order to identify the material process, following supplementary are required:

i. Supplementary requirement S, sour services which is pipeline that needs to transport fluid with existing of hydrogen sulphide (H2S) contents need to be evaluated for the "sour service" [8].
ii. Supplementary requirement F, fracture arrest properties that valid for gas pipelines which is carrying essentially pure methane up to 80% usage factor and up to a pressure of 15MPa with the wall thickness and the diameter of 30mm and 1120 mm respectively [8].
iii. Supplementary requirement P, Plastic Deformation which is applicable to the pipeline if the total nominal strain in any direction from a single event is exceeding 1.0%. Besides that, the tensile testing should be carried out in the longitudinal direction in order to satisfy the DNV requirements. This is for the pipes that require supplementary requirement (P).
iv. Supplementary requirement D, which is Dimensional Requirements which is should be consider the influences of dimensions for the requirements for tolerance that acting on the subsequent fabrication/installation activities and the welding facilities.
v. Supplementary requirement U, Utilization which is the purchaser may in retrospect upgrade a pipe delivery. For instance, in case of more than 50 test units, it must be demonstrated that the actual average yield stress is at least two (2.0) standard deviations above the SMYS. Besides that, the actual average yield stress shall as a minimum be 2.3, if the number of test units in between the range of 10 and 20. In addition, the actual yield stress of 2.1 will be held if the number of the test units in the range between 21 and 49 [8].

2. Methodology

OFFPIPE is a sophisticated, finite element method which is based on the computer program [9]. Basically, OFFPIPE has been developed in order to perform specifically for the modelling and structural analysis of the nonlinear problems. It also encountered in the installation and operation of the offshore pipeline. Static installation is summarized in Figure 4.

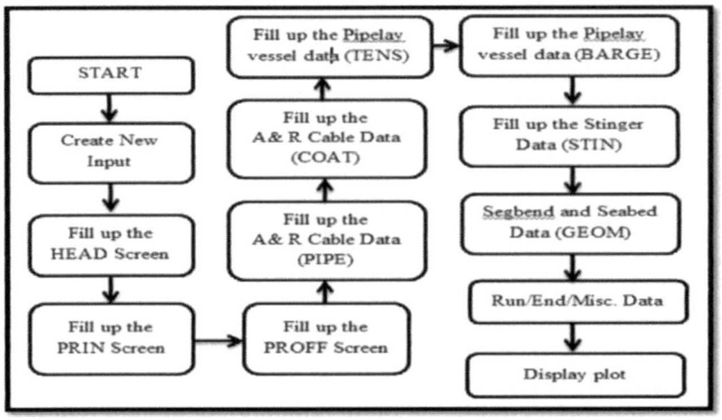

Figure 4. Static Installation using OFFPIPE software

2.1 Data Input

Samples used for these simulations were API 5L X60. Wall thicknesses (mm) used i.e. 12, 14, 16, 18 and 20. Water depth(m) use i.e. 50, 100, 150, 200, 250 and 300. Outside Diameter use i.e. 8", 12" and 16''

2.2 OFFPIPE Preparation

The most important thing in OFFPIPE software was the data input. It will affect the result of simulation as the result is so important to decide the data input is valid or not and the pipe laying is pass or not according the percent yield stress limitation of the pipe known as 85%. It also need make simplified criteria for overbend as reference as Table 1 [8].

Table 1. Simplified Criterion for Overbend

Criterion	*X70*	*X65*	*X60*	*X52*
I	0.270 %	0.250 %	0.230 %	0.205 %
II	0.325 %	0.305 %	0.290 %	0.260 %

2.3 OFFPIPE Manual

The first steps are creating a new input data which need to create the name of the input file, i.e. "Pipe Grade X65". After that, the Problem Heading Data, HEAD which is consist of problem title, job number, user name and problem input/ output units need to initiate. Then enter the Printed Output Control which will determine types of output that will appear in the result sheet. Next, pipe property table should be referred as to fill Property Table Index, Pipe Section Length, Elastic Modulus, Cross Section Area, Moment of Inertia, Weight in Air, Weight Submerged, Poisson's Ratio, Steel outside Diameter, Steel Wall Thickness, Yield Stress, Drag Coefficient, added mass coefficient and Coefficient Thermal Expansion. This data we refer what type of grade we use and here also we manipulated the wall thickness.

Then enter the Pipe Coating Properties such as steel density, corrosion density, corrosion thickness pipe joint length and density of contents. Next, enter Pipe Tension at Free End which is the input is static pipe tension as this simulation if for static installation and we can manipulate the input to get several results. Then enter the Barge Data and Geometry. This data comes from what type vessel we use. Every vessel has own data and coordinate of the stinger which is we need to fill up in Stinger Data and Geometry part. Then enter the Geometry Screen that here can manipulate the water depth. Lastly, run the simulation and get the result.

2.4 Wall Thickness Design Flowchart

Wall thickness design flowchart is illustrated in Figure 5.

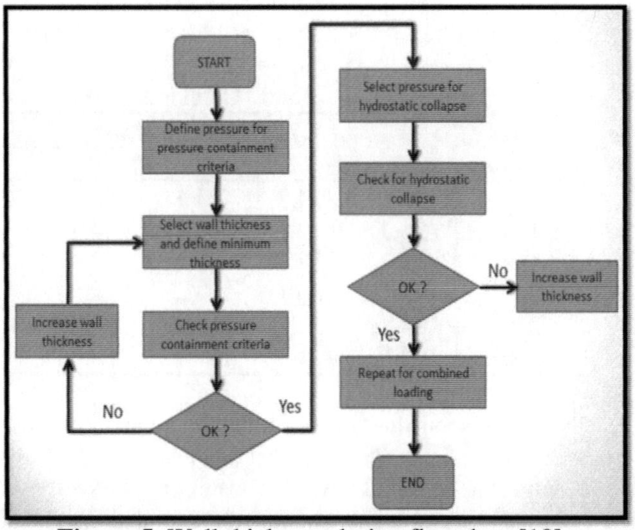

Figure 5. Wall thickness design flowchart [10].

3. Results and Discussion

3.1 Limitation Test

Limitation test were conducted by using OFFPIPE Software where we make reference for the sagbend area that Percent Yield Stress should be take place. Set up the limitation of the percent yield stress at 85%. From this limitation we can make assumption which pipeline that exceeds this kind of limitations, the pipeline will be consider fail because the pipeline will have damaged due to overstress. Opposite form this, which pipeline below this limitation consider pass for laying the pipeline.

3.2 OFFPIPE's Software Output.

Based on the software output, it can be seen clearly which criteria of the pipeline that does not meet the required and exceed the limitations of the yield stress. What Figure 6 shows that, the pipeline grade X60 and outside diameter of 8 inch was pass for the wall thickness of 12mm, 14mm and 16mm. The data is summarized in Table 2.

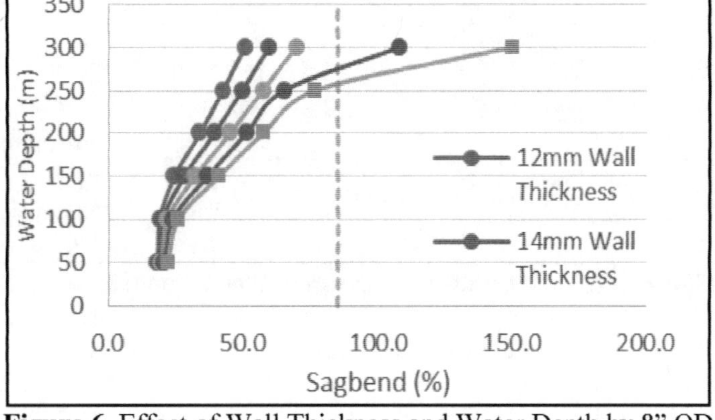

Figure 6. Effect of Wall Thickness and Water Depth by 8" OD

Unfortunately, the pipeline was failed at the wall thickness of 18mm and 20mm. From the graph it can see the critical yield percentage line was setup at 85%. Usually, the pipeline tends to fail because so many factor, in these experiments maybe wall thickness cannot stand enough to the such water depth and lead overpressure in the result of the stress

Table 2. Output for 8" Outside Diameter

Water	***Wall Thickness(mm)***				
Depth(m)	**12**	**14**	**16**	**18**	**20**
50	21.6	24.3	27.1	30.2	33.3
100	26.3	32.7	38.8	44.7	50.4
150	48.3	58.1	67.5	77.0	86.5
200	67.1	80.1	93.1	105.8	119.8
250	84.3	100.4	117.6	135.2	140.0
300	100.42	121.00	147.17	150.00	155.00

From the Figure 7, it is clearly seen that we run input for outside diameter of 12 inch. But, the main point here is to check either the listed wall thickness can stand such pressure or not. We can see that for wall thickness of 12 mm start to fail at 250 m of water depth and above 14 mm wall thickness starts to fail at 200 m and above (Table 3). The rest start to fail in range of 150 m to 200 m of water depth.

Table 3. Output for 12" Outside Diameter

Water	***Wall Thickness(mm)***				
Depth(m)	**12**	**14**	**16**	**18**	**20**
50	21.6	24.3	27.1	30.2	33.3
100	26.3	32.7	38.8	44.7	50.4
150	48.3	58.1	67.5	77.0	86.5
200	67.1	80.1	93.1	105.8	119.8
250	84.3	100.4	117.6	135.2	140.0
300	100.42	121.00	147.17	150.00	155.00

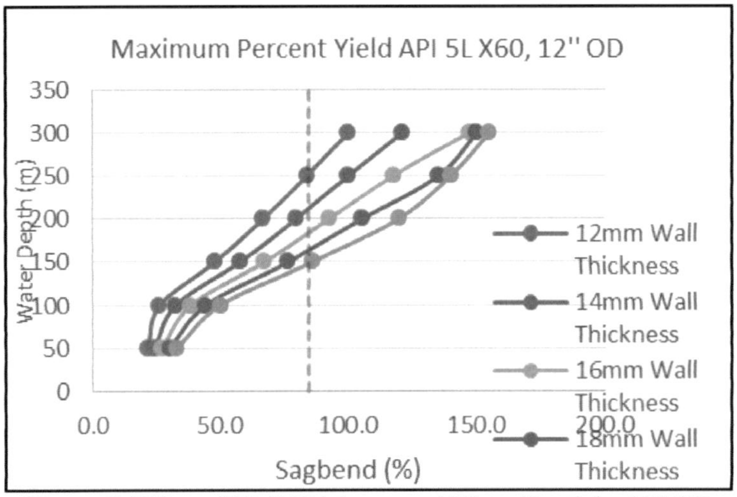

Figure 7. Effect of Wall Thickness and Water Depth by 12" OD

Figure 8, the graph shows that the 12mm wall thickness start to fail at water depth of 200m. The wall thickness of 14mm, 16mm, 18mm and 20mm was started to fail at water depth in range of 100m to 150m (Table 4). So, basically from this graph shoes that, for outside diameter of 16 inch is suitable for water depth of 200m and below for the respectively wall thickness. Above this range, the pipeline will tend to start fail.

Table 4. Output for 16" Outside Diameter

Water Depth(m)	*Wall Thickness(mm)*				
	12	**14**	**16**	**18**	**20**
50	22.6	27.8	33.2	38.2	43.1
100	30.6	42.7	52.7	62.1	70.9
150	59.3	75.9	91.0	105.1	119.4
200	83.0	103.8	124.3	144.9	165.9
250	103.8	130.2	156.5	184.1	190.0
300	123.75	155.64	188.67	200.00	250.00

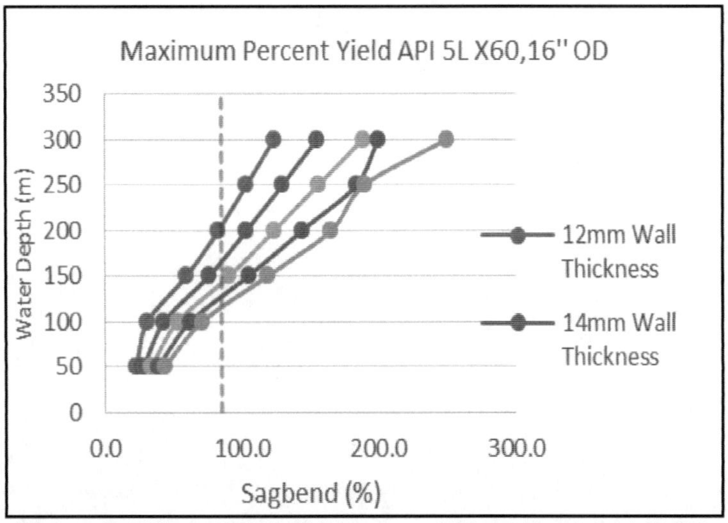

Figure 8. Effect of Wall Thickness and Water Depth by 16" OD

4. Conclusion

Generally, the research work was wholly based on the wall thickness stability in such water depth of 50m, 100m, 150m, 200m, 250m and 300m. From the graph, it was proving that the wall thickness is depends on the water depth. Besides that, every wall thickness has their own maximum water depth that they can stand with. That is why the pipeline will tend to be fail when wall thickness exceeds its own maximum limitation and this failure can be detect by such various way such as yield stress percentage. In this experiment, the yield stress percentage was setup as 85%. All samples that exceed the limitation will consider fail. This percentages were followed the standard and setup as reference to decide the yield stress is past or not. Besides that, the deeper the water depth the more pressure will act on the pipeline and the bigger wall thickness the heavier the pipeline. For future work and reference, this experiment need to be considered and run the OFFPIPE software properly with the correct input to get the best layable result and automatically it will reflect the economic part which is perfect layable will reduce the cost impact and can give better budget suggestion for the project.

Acknowledgement

This research is fully supported by LESTARI grant, 600-IRMI/DANA 5/3/LESTARI. The authors fully acknowledged Universiti Teknologi MARA for the approved fund which makes this important research viable and effective.

References

[1] Recalde C E 1996 *Pipelaying Vessel* U.S. Patent 5527134
[2] Hogan M G 1975 *Pipe laying Vessel with Stinger and Length* U.S. Patent 3911689
[3] Timothy P H and Shaw C W 1958 *Pipe Coating Method and Apparatus* U.S. Patent 2863204
[4] Timothy P H 1959 *Pipe laying apparatus and method* U.S. Patent 2910835
[5] Harold D C 1967 *Tension Pipe Laying Method* U.S. Patent 3331212
[6] Kyriakides S and Corona E 2007 *Mechanics of Offshore Pipelines* Vol 1, (Oxford: Elsevier) p...

IOP Conf. Series: Materials Science and Engineering **358** (2018) 012068 doi:10.1088/1757-899X/358/1/012068

[7] Iorio G, Bruschi R and Donati E 2000 *Challenges and Opportunities for Ultra Deep-Water Pipelines in Difficult Sea Bottoms: WPC 30319, 16th World Petroleum Congress (Calgary, Canada, 11 – 15 June 2000)* pp
[8] DNV-OS-F101 2007 *Submarine Pipeline Systems, Det Norske Veritas (DNV)* (Oslo: Norway)
[9] Malahy R C 1996 *Offpipe User's Guide, version 2.05 Available from:* http://www.offpipe.com/support.html Retrieved 28th November 2016
[10] Jee Ltd 2013 *Subsea Engineering and Training Expert* Subsea Pipeline Design IKM.

Kinetic Study on the Esterification of Palm Fatty Acid Distillate (PFAD) Using Heterogeneous Catalyst

U Rofiqah[1], R A Djalal[2], B Sutrisno[1] and A Hidayat[1]

[1] Chemical Engineering Department, Universitas Islam Indonesia, Indonesia
[2] Industrial Engineering Department, Universitas Islam Indonesia, Indonesia

arif.hidayat@uii.ac.id

Abstract. Esterification with heterogeneous catalysts is believed to have advantages compared to homogeneous catalysts. Palm Fatty Acid Distillate (PFAD) was esterified by ZrO_2-SO_4^{2-}/natural zeolite at temperature variation of 55°C, 60°C, and 65°C to produce biodiesel. Determination of reaction kinetics was done by experiment and modeling. Kinetic study was approached using pseudo-homogeneous model of first order. For experiment, reaction kinetics were 0.0031 s^{-1}, 0.0054 s^{-1}, and 0.00937 s^{-1} for a temperature of 55 °C, 60 °C and 65 °C, respectively. For modelling, reaction kinetics were 0.0030 s^{-1}, 0.0055 s^{-1}, and 0.0090 s^{-1} for a temperature of 55°C, 60°C and 65°C, respectively. Rate and conversion of reaction are getting increased by increasing temperature.

1. Introduction

Biodiesel is a renewable and environmentally friendly fuel for diesel engines. It was produced by transesterification or esterification of vegetable oils or animal fats with short chain alcohols in the presence of catalyst [1]. PFAD (Palm Fatty Acid Distillate) is a by-product of the palm oil industry and is considered as low grade oil with high free fatty acids [2]. Esterification is conventionally performed using homogeneous acid catalyst such as sulfuric acid or hydrochloric acid. However, it has a number of drawbacks such as need further processing in the separation of the catalyst from the product and may generate hazardous waste [3-5]. To overcome these problems, researchers have focused on using heterogeneous catalysts for biodiesel production. In this research, the esterification of free fatty acids (FFA) in PFAD with methanol, using ZrO_2-SO_4^{2-}/natural zeolite, was studied. The effects of the mass ratio of catalyst to oil (1 – 10%), the molar ratio of oil to methanol (1:6 – 1:12), and the reaction temperature (55 – 65°C) were studied for the conversion of FFA on PFAD to optimize the reaction conditions. The kinetics of FFA esterification (in PFAD) in the presence of ZrO_2-SO_4^{2-}/natural zeolite also has been investigated. Determination of kinetics of FFA esterification was conducted by using a first order pseudo-homogeneous kinetic model.

2. Materials and Methods

2.1 Materials

Raw materials used in this study was PFAD from Palm Oil Processing Industry Public Company Limited consisting of 93% (wt) of free fatty acids (45.6% of palmitic, oleic of 33.3%, 7.7% of linoleic, 3.8% of stearic , 1.0% of myristic, 0.6% of tetracosenoic, 0.3% of linolenic acid, 0.3% of ecosanoic, 0.2% of ecosenoic, and 0.2% of palmitoleic acid) and the rest were triglyceride (TG), diglycerides (DG), and monoglycerides (MG). Esterification was done using ZrO_2-SO_4^{2-}/natural zeolite. Other materials used were distilled water, $Ba(OH)_2$, nitrogen gas and oxygen gas.

Content from this work may be used under the terms of the Creative Commons Attribution 3.0 licence. Any further distribution of this work must maintain attribution to the author(s) and the title of the work, journal citation and DOI.
Published under licence by IOP Publishing Ltd

2.2 Method

2.2.1 Activation of natural zeolite

Natural zeolite of 200 g with the size of 200 mesh was dissolved in 500 ml of H_2SO_4 0.5 M. The solution was stirred for 4 hours at room temperature. Then it was cooled and filtered using Whatman 42 paper. The filtrate was discarded, while the solids were washed with distilled water to pH neutral. Washing is terminated if it were neutral to the test by using $Ba(OH)_2$. Then the solids were dried in an oven at 130°C for 24 hours.

2.2.2 The catalyst synthesis of Zirconia

Activated Zeolite soaked in a solution of Zirconium (IV) oxide chloride hexahydrate ($ZrOCl_2 \cdot 8H_2O$). It was refluxed at 90°C for 3 hours. Samples were filtered and washed with distilled water until the filtrate was clear then dried. Dried sample were calcined by flowing N_2 gas and oxidized by O_2 gas to make ZrO_2/natural zeolite catalyst. 10 g of ZrO_2/zeolite and 20 g of Ammonium Sulphate ($(NH_4)_2SO_4$) solids were mixed and crushed in a porcelain mortar until homogeneous. It calcined by flowing N_2 gas at 200°C for 4 hours. Heating rate is set at 5 °C/min. ZrO_2-SO_4^{2-}/natural zeolite was characterized by crystallinity measurements using X-Ray Diffraction (XRD). Determination of surface area and pore size distribution were done using BET Surface Area Analyzer, the analysis of mineral content by X-ray Fluorescence Spectroscopy (XRF), and analysis of functional groups with Fourier Transform Infrared (FT-IR) spectrometry.

2.2.3 Esterification of PFAD

PFAD was reacted with methanol in a stirred batch reactor. The catalyst used was ZrO_2-SO_4^{2-}/natural zeolite with concentration of 1%, 2.5%, 5% and 10%. Ratio of PFAD to methanol was 1: 6, 1: 8, 1:10, and 1:12. Esterification carried out at a temperature variation of 55°C, 60°C and 65°C.

2.2.4 Reaction kinetics of Pseudo-Homogeneous model

Kinetic of esterification of free fatty acid in PFAD with methanol can be appoximated by pseudo-homogeneous second order with assumption that the mass transfer of the liquid as a reactant into the solid as a catalyst is very rapid. Mathematically pseudo-homogeneous model of first order can be described in equation (1) as follows:

$$-r_A = -\frac{dC_A}{dt} = k\,{C_A}^{\alpha}{C_B}^{\beta} - k'{C_C}^{\gamma}{C_D}^{\lambda} \quad (1)$$

Which (-rA) is reaction rate of free fatty acids, C_A is concentration of free fatty acid; C_B is concentration of methanol; C_C and C_D are concentration of methyl ester and concentration of water formed during the reaction, respectively; k is reaction rate constant and k' is equilibrium constant. In the experiment, C_B was excess so that the concentration is assumed to be fixed and $k{C_B}^{\beta}$ can be symbolized as k_1 and the reaction rate is considered toward the product. The reaction rate can be expressed as pseudo first order.

$$(-r_A) = k_1\,C_A \quad (2)$$

Esterification of free fatty acids with the solid catalyst zeolite-sulfated zirconia run in batch.

$$(-r_A) = -\frac{dC_A}{dt} \quad (3)$$

Substitution of Equation (3) to Equation (2) thus obtained:

$$-\frac{dC_A}{dt} = k_1\,C_A$$

$$\frac{dC_A}{C_A} = -k_1\,dt$$

$$\ln C_A - \ln C_{A_0} = -k_1\, dt \qquad (4)$$

The parameters of the reaction kinetics (k_1) are through non-linear regression which gives the Sum of Square Quake Error (SSE) between the experimental data and the minimum calculation data using Matlab.

$$SSE\ (k_1) = \sum \left(X_{A_{experiment}} - X_{A_{calculation}} \right)^2 \qquad (5)$$

Which $X_{A_{experiment}}$ is conversion value of the experimental data and $X_{A_{calculation}}$ is a conversion value calculated from the kinetic equation proposed. Values of parameters obtained at each temperature is then plotted by the Arrhenius equation to obtain the activation energy data.

$$k_1 = A_r\ Exp\left(-\frac{E_a}{RT}\right) \qquad (6)$$

3. Result and Discussion

3.1 Effect of the amount of methanol to FFA conversion

Esterification is a reversible reaction. The equilibrium is shifted toward the product by making the number of moles of reactant becomes excessive. The influence of the amount of methanol to FFA conversion can be seen in Figure 1.

$$RCOOH + CH_3OH \Leftrightarrow RCOOCH_3 + H_2O \qquad (7)$$

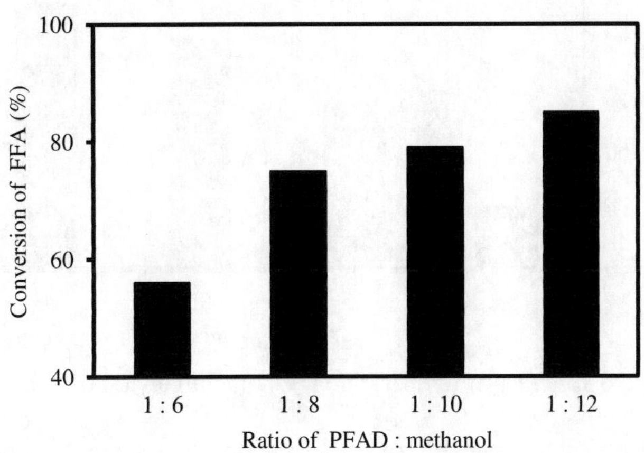

Figure 1. Effect ratio of PFAD to methanol on conversion of FFA (%).

Increasing the amount of methanol will increase FFA conversion. This happens because the larger reactants collide so that the conversion of the FFA even greater. The conversion occurs in processes that use ratio PFAD: methanol 1:12 was 83% of FFA conversion. From Figure 1 it is known that a significant increase in conversions that occur in reactions using the mole ratio PFAD with methanol = 1:10 and 1:12, but not as significant as ratio of 6 to 10. This is because the ratio of reactants 1:12 has begun to reach the maximum conversion.

3.2 Effect of catalyst concentration on the conversion of FFA

The influence of the amount of catalyst to the FFA is presented in Figure 2. The increasing amount of catalysts will increase FFA conversion. A significant rise in FFA conversion occurs when the amount of catalyst was increased from 1% to 2.5%. Its increasing is more significant than 5% to 10%. This happens the total number of active sites has begun to approach the maximum number of reactants needed. At a catalyst concentration of 10% it appears that the conversion slightly increased compared to the catalyst concentration of 2.5%.

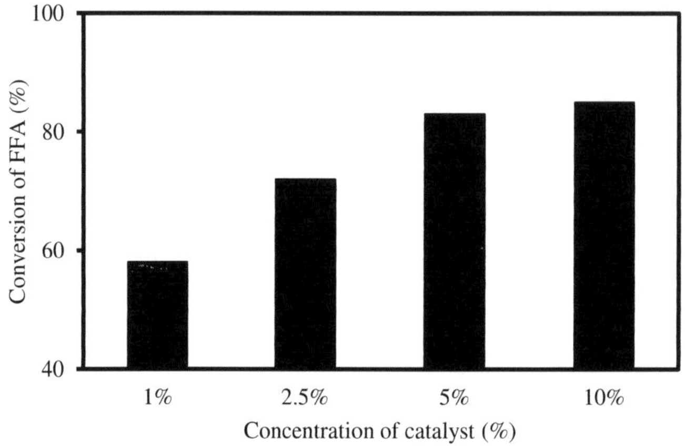

Figure 2. Effect of concentration of catalyst (%) on conversion of FFA (%).

3.3 Effect of reaction temperature on the conversion of FFA
With increasing temperature, FFA conversion will increase as illustrated in Figure 3.

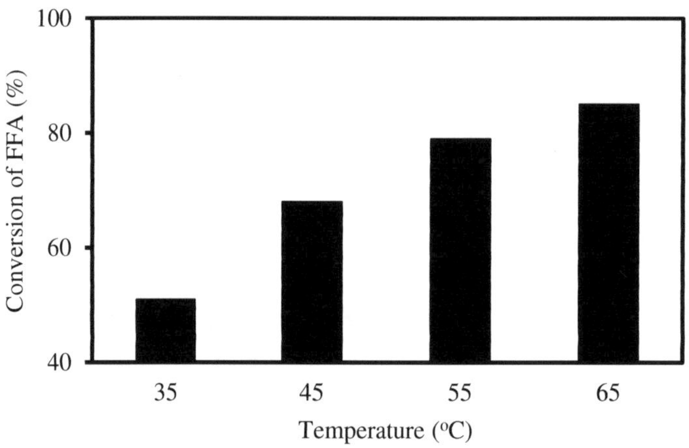

Figure 3. Effect of temperature (°C) on conversion of FFA (%).

The higher conversion was obtained from the reaction at temperature of 65°C. By raising the reaction temperature will raise FFA conversion. Rising temperature will cause the viscosity of the reaction down. That makes the mixing process will be more perfect and mass transfer of reactants to the catalyst surface will be better so the reaction runs perfectly.

3.4 Kinetics model
In this experiment it was seen that the conversion increased sharply after the catalyst was added to the reactor, and increased gradually over time. Something similar to the one delivered by Rattanaphra et al. [6], Alenezi et al. [7], and Park et al. [8]. This is due to several things: At the moment the catalyst is added the solution in the reactor is only the reactant (FFA and Methanol) so that diffusing to the active surface of the catalyst is only the reactant. Diffusion of the reactants to the active surface of the catalyst will accelerate the formation of FAME. After FAME formed and became a co-solvent in the solution, barriers to mass transfer of reactants to the catalytically active surface is greater so that the reaction speed becomes slower [8]. FFA conversion will increase the water that its existence can be deactivated catalyst [9] and the conversion of high FFA produce FAME make the formation

equilibrium of the reaction [7]. Spontaneous conversion occurred at the beginning of the reaction indicates that the rate of formation of FAME at the start of the reaction was controlled by the speed of chemical reactions. Yadav [10] said that the conversion increases with increasing temperature indicates that the reaction rate was controlled by the speed of chemical reactions. Effect of reaction temperature was used to determine the reaction kinetics. Reaction rate constant can be calculated by a pseudo-first-order reaction equation to generate minimum SSE. Data conversion and conversion count of each 10-minute interval is shown in Table 1 and Figure 4. The value of k at various reaction temperatures and SSE values are presented in Table 1.

Table 1. FFA conversion by experiment and model (mol PFAD: methanol = 1:10, catalyst concentration =10% of PFAD)

Time (minute)	T = 55°C		T = 60°C		T = 65°C	
	X_A experiment	X_A calculation	X_A experiment	X_A calculation	X_A experiment	X_A calculation
0	43,9554	43,9554	42,6150	44,8239	43,8298	43,8297
10	46,3783	45,3391	50,0305	47,2518	48,7786	48,4069
20	48,2605	46,6887	48,6616	49,5728	51,3678	52,6111
30	47,2241	48,0050	51,4416	51,7918	60,7201	56,4727
40	48,8007	49,2887	54,0710	53,9130	61,8794	60,0197
50	51,3480	50,5408	56,5581	55,9410	62,0061	63,2776
50	52,4297	51,7619	58,9105	57,8797	66,0904	66,2700
70	56,5467	52,9529	61,1355	59,7331	67,6226	69,0186
80	56,4099	54,1145	63,2400	61,5049	75,6839	71,5432
	k =0,0030		k =0,0055		k =0,0090	

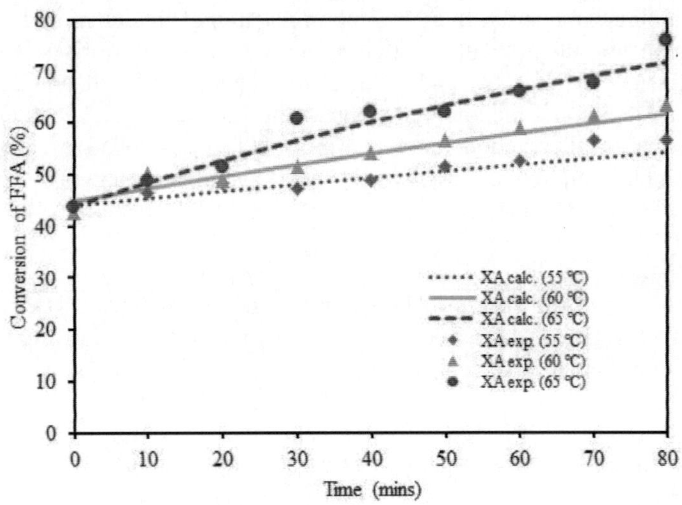

Figure 4. Calculation and experiment conversion of FFA (%) in variation of temperature (°C).

From Table 2 it can be seen that the value of SSE for each variation of the reaction temperature is quite small. The results show that the proposed kinetic model correcting ie homogeneous pseudo first order of the FFA is acceptable, this trend can also be seen from Figure 5.

Table 2. Value of k at different temperature

Temperature (°C)	**k_1(s^{-1})**	**k_1 Arhenius (s^{-1})**	**Rel.Err (%)**
55	0.0030	0.0031	4,0046
60	0.0060	0.0054	2,6281
65	0.0090	0.00937	4,4186

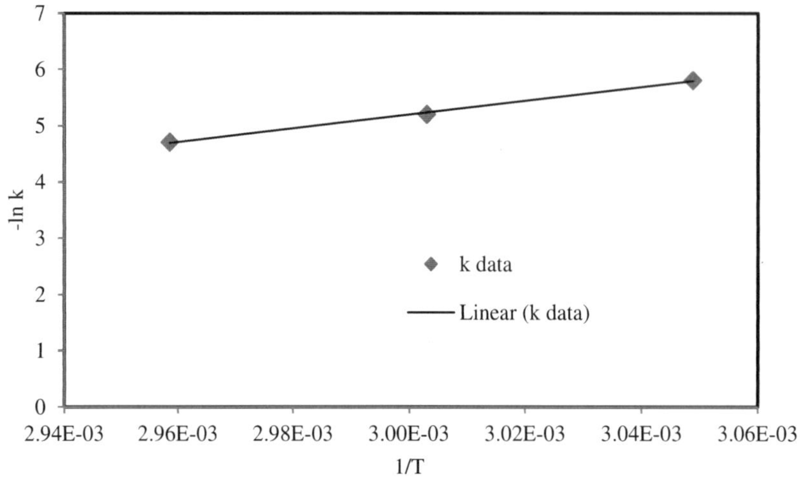

Figure 5. Arrhenius equation of k in difference temperature

From data of reaction rate constant (k) in Table 2 shows that increasing temperatures will increase the value of k. The value of k was used to calculate the value of the collision frequency factor (Ar) and the activation energy (Ea) with the Arrhenius equation. It was obtained by linear regression Ar of 4.3771 x 1013 s-1 and Ea of 999.99 kJ/mol.

4. Conclusion

ZrO_2-SO_4^{2-}/natural zeolite can be used to esterification reaction of free fatty acids. Increasing the mole ratio of PFAD to methanol, the amount of catalyst, and the temperature rises has been successful in increasing the conversion of free fatty acids. Correcting kinetic model of homogeneous pseudo first order can be used in determining the value of the reaction kinetics of PFAD esterification to become methyl ester. For experiment, reaction kinetics were 0.0031 s^{-1}, 0.0054 s^{-1}, and 0.00937 s^{-1} for a temperature of 55°C, 60°C and 65°C, respectively. For modelling, reaction kinetics were 0.0030 s^{-1}, 0.0055 s^{-1}, and 0.0090 s^{-1} for a temperature of 55°C, 60°C and 65°C, respectively.

Acknowledgments

The authors would like to acknowledge to Direktorat Jenderal Penguatan Riset dan Pengembangan, Direktorat Riset dan Pengabdian Masyarakat, Kemenristekdikti for financial support through research grant of Penelitian Fundamental 2016.

References

[1] Srivastava A., Prasad R., 2000 Triglycerides-based diesel fuels *Renew Sustainable Energy Rev.* **4**, 111–133

[2] Hidayat A, Rochmadi, Wijaya K and Budiman A 2016 Removal of free fatty acid in Palm Fatty Acid Distillate using sulfonated carbon catalyst derived from biomass wastefor biodiesel production IOP Conference Series: Materials Science and Engineering **105** (1) 012026.

[3] Vicente G, Martínez M, and Aracil J, 2007 Optimization of integrated biodiesel production part I a study of the biodiesel purity and yields *Bioresource Technology* **98** 1724–173.

[4] Hidayat A and Sutrisno B 2017 Esterification free fatty acid in sludge palm oil using ZrO_2/SO_4^{2-} - rice husk ash catalyst, AIP Conference Proceedings **1840**, 050001.

[5] Hidayat A, Rochmadi, Wijaya K and Budiman A 2015 Reaction kinetics of free fatty acids esterification in palm fatty acid distillate using coconut shell biochar sulfonated catalyst AIP Conference Proceedings **1699** 050005.

[6] Rattanaphra D, Harvey A P, Srinophakun P, 2010 Simultaneous conversion of triglyceride/free fatty acid mixture into biodiesel using sulfated zirconia *Top Catal.* **53** 773–782

IOP Conf. Series: Materials Science and Engineering **358** (2018) 012069 doi:10.1088/1757-899X/358/1/012069

[7] Alenezi R, Leeke G A, Winterbottom J M, Santos R C D and Khan A R (2010) Esterification kinetics of free fatty acids with supercritical methanol for biodiesel production *Energy Conversion and Management* 51 **5** 1055-59
[8] Park J Y, Kim D K and Lee J S 2010 Esterification of Free Fatty Acids using Water Tolerable Amberlyst as A Heterogeneous Catalyst *Bioresource Technol.* **101** S62-65
[9] Liu Y, Lotero E and Goodwin J 2006 Effect of Water on Sulfuric Acid Catalyzed Esterification *J. Mol. Catal.* **245** 132-140
[10] Yadav G D and Nair J J 1999 Sulfated Zirconia and Its Modified Versions as Promising Catalysts for Industrial Processes *Micropor. Mesopor Mat.* **33** 1-48.

IOP Conf. Series: Materials Science and Engineering **358** (2018) 012070 doi:10.1088/1757-899X/358/1/012070

Catalytic Efficiency of Titanium Dioxide (TiO_2) and Zeolite ZSM-5 Catalysts in the *in-situ* Epoxidation of Palm Olein

M Z Mohd Yunus, S K Jamaludin*, S F Abd. Karim, A Abd. Gani and A Sauki

[1,] Faculty of Chemical Engineering, Universiti Teknologi MARA, 40450 Shah Alam, Selangor, MALAYSIA

sitikhatijah@uitm.edu.my

Abstract. Titanium dioxide and zeolite ZSM-5 are the commonly used heterogeneous catalysts in many chemical reactions. They have several advantages such as low cost and environmental friendly. In this study, titanium dioxide and zeolite ZSM-5 act as catalyst in the *in-situ* epoxidation of palm olein. Epoxidation of palm olein was carried out by using *in-situ* generated performic acid to produce epoxidized palm olein in a semi-batch reactor at different temperatures (45°C and 60°C) and agitation speed of 400 rpm. The effects of both catalysts are studied to compare their efficiency in catalyzing the *in-situ* epoxidation. Epoxidized palm olein was analyzed by using percent of relative conversion to oxirane (RCO%) and fourier transform infrared spectroscopy (FTIR). Surface area of the catalysts used were then characterized by using BET. The results indicated that titanium dioxide is a better catalyst in the *in-situ* epoxidation of palm olein since it provides higher RCO% compared to Zeolite ZSM-5 at 45°C.

1. Introduction

Palm olein is a type of vegetable oil that is obtained from fractionation of palm oil. It consists of 53.5% unsaturated fatty acid and 46.5% saturated fatty acid, with iodine value ranging from 55.6 to 61.9 [1]. Since palm olein contains unsaturated fatty acid, it can be chemically adjusted to value added products by epoxidation process. Epoxidation can be defined as the conversion of unsaturated carbon-carbon double bond into an epoxide which is a cyclic ether with three atom ring [2]. The common process of epoxidation is by reacting performic acid with the unsaturated oil, also known as peracid epoxidation. According to Campanella et al. [3], this process can be conducted by reacting the unsaturated bond of the vegetable oils with formic acid, which acts as the oxygen carrier and hydrogen peroxide that acts as oxygen donor. Formic acid is more favored in most reactions due to its high reactivity. Figure 1 and figure 2 illustrate the reaction between formic acid and hydrogen peroxide to form performic acid and reaction of performic acid with unsaturation in fatty acid to form epoxide, respectively.

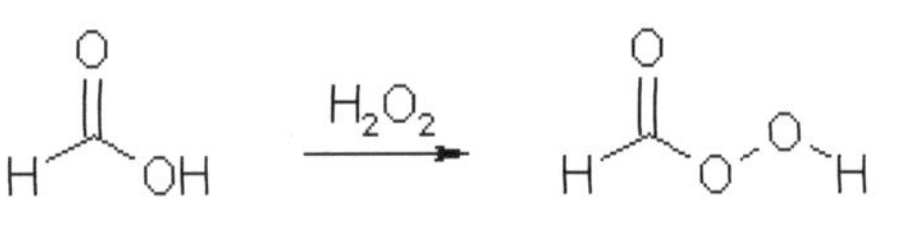

Figure 1. Reaction between formic acid and hydrogen peroxide to form performic acid.

Content from this work may be used under the terms of the Creative Commons Attribution 3.0 licence. Any further distribution of this work must maintain attribution to the author(s) and the title of the work, journal citation and DOI.
Published under licence by IOP Publishing Ltd

IOP Conf. Series: Materials Science and Engineering **358** (2018) 012070 doi:10.1088/1757-899X/358/1/012070

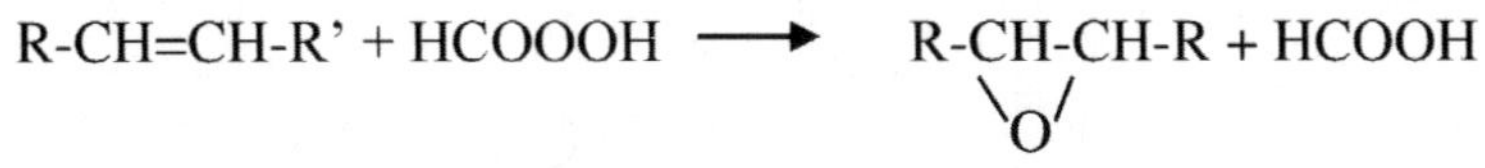

Figure 2. Reaction between performic acid and unsaturated fatty acid to form epoxide.

Epoxide has been used as an intermediate for organic synthesis of various chemicals such as alcohols, glycols, olefinic compounds, lubricants, plasticizer and stabilizer for polymers due to high oxirane reactivity [4]. Besides that, epoxides that obtained from the renewable raw materials and from their alkyl esters have many usages in daily life such as plasticizers and polymer stabilizers [5]. The catalysts used in this study are titanium dioxide (TiO_2) and zeolite ZSM-5. Both catalysts are heterogeneous which basically have many advantages since it is easy to separate by filtering the product and the catalyst can be recovered again in the manufacturing process. Titanium dioxide or TiO_2 basically is a pure white pigment that occurs naturally as crystal and has been applied widely since the surface properties of the rutile has many benefits and become ideal model system in the science of metal oxides [6]. Zeolite Socony Mobil-5 (ZSM-5) also has many applications in recent years. Zeolite ZSM-5 which is made up of various crystalline structures has large open pores in a very regular arrangement since it is composed by many pentansil units linked together to form a chain. Its ability to catalyze organic reactions under solvent-free conditions is one of its attractiveness, among many others [7].

In this study, the effects of temperatures and catalyst types on the epoxidation of palm olein were investigated. Characterization of both catalysts was also done with the intention to investigate the catalyst that performs better in catalyzing the reaction. Here, the characterization of catalyst basically refers to the surface area of the catalyst.

2. Methodology

2.1. Materials

Palm olein used in this study was obtained from a local refined palm olein manufacturing company in Malaysia. Formic acid (CHOOH, 98% purity) and hydrogen peroxide (H_2O_2, 30 w/w % solution) were used as oxygen carrier and oxygen donor respectively in epoxidation and were purchased from Merck Sdn Bhd. All materials were used as obtained without further purification. Hydrobromic acid (0.1N in acetic acid) was used for titration method to determine epoxy content. TiO_2 and Zeolite ZSM-5 catalysts were purchased from Merck Sdn Bhd.

2.2. Epoxidation of palm olein

The reaction was carried out at two different temperatures (45°C and 60°C) to investigate the effects of different temperatures on epoxidation. The molar ratio of hydrogen peroxide to formic acid to unsaturation in palm olein (H_2O_2: HCOOH: C = C) remained constant at 1:1:1. The first set of epoxidation was carried out at temperature 45°C and agitation speed of 400 rpm. The reaction was then carried out at three different conditions; no catalyst, which acts as control variable, 1 g of TiO_2 added and 1 g of zeolite ZSM-5 added. The reaction started with 100 g of palm olein and formic acid loaded into 250 mL reactor equipped with mechanical agitator that was used to ensure homogeneity of the reactor content. After about 30 minutes palm olein acidified with formic acid, hydrogen peroxide was added at the flow rate of 2 ml/s to accomplish the reaction.

2.3. Analytical procedure

The sample was retrieved every 3 minutes to determine the oxirane (epoxide) oxygen content which then was used for the calculation of relative conversion of double bond to oxirane, RCO%. The experimental oxygen oxirane content (OOC_{exp}) value in the epoxidized palm olein was determined by direct titration using hydrobromic acid solution in glacial acetic acid, according to AOCS Tentative Method Cd 9-57 [8], as in equation (1):

$$OOC_{exp}\,(\%) = 1.6 \times \quad \times (V-B) / W \qquad (1)$$

IOP Conf. Series: Materials Science and Engineering **358** (2018) 012070 doi:10.1088/1757-899X/358/1/012070

where N is the normality of HBr, V is the volume of HBr solution used for blank (in mL), B is the volume of HBr solution used for titration (in mL) and W is the weight of sample (in gram). The iodine value (IV) of palm olein used, which indicates the degree of unsaturation in the palm olein was calculated based on AOCS Standard Method Cd 1d-92, as in equation (2):

$$IV\,(g\ I_2\,/\,100g\ sample) = (B\text{-}S)\ x\ N\ x\ 12.69\,/\,mass\ of\ palm\ olein \tag{2}$$

where B is the volume of titrant for blank run (in mL), S is the volume of titrant for sample (in mL) and N is the normality of sodium thiosulfate. The iodine value then applied to calculate theoretical oxirane oxygen content (OOC_{theo}) [9], as in equation (3):

$$OOC_{theo} = (IV_{oA,}\,/\,2A_I)\,/(250 + \ (IV_{oA,}\,/\,2A_I)A_o\ )\ x\ A_o\ x\ 100 \tag{3}$$

where IV_o is initial iodine value, A_o is the molar mass of oxygen and A_I is the molar mass of iodine. Then, the percentage of relative conversion to oxirane is calculated using equation (4):

$$RCO\% = OOC_{exp}/OOC_{theo}\ x\ 100\% \tag{4}$$

Thereafter, the sample was analyzed using FTIR to determine the presence of oxirane ring in the epoxidized oil.

2.4. FTIR spectroscopy analysis

The sample was observed by using Fourier Transform Infrared Spectroscopy to examine the change in functional group before and after the reaction. The typical infrared spectrum of functional groups corresponding to their absorption range is shown in table 1 [10]:

Table 1. Characteristic absorption for functional group based on FTIR analysis.

Functional group	Characteristic Absorption (cm^{-1})
Alcohol	3200-3600
Carbon-carbon double bond	1620-1680
Carboxylic acid	1670-1820
Epoxy / oxirane ring	1210-1320

3. Results and discussion

To study epoxidation of palm olein, series of reactions were conducted by using *in situ* method which is reaction of the palm olein with the performic acid. This reaction samples were analyzed by several methods such as oxirane oxygen content (OOC), percent of relative conversion to oxirane (RCO%), and Fourier Transform IR (FTIR). Since the study was conducted to investigate the efficiency of the catalyst used (TiO_2 and Zeolite ZSM-5), the activities of the different catalyst were compared in the epoxidation of palm olein.

3.1. Effect of temperatures

Epoxidation was conducted at 45°C and 60°C, at 400 rpm agitation speed. Based on figure 3, the maximum yield of epoxidized palm olein obtained was 5.55% as indicated by the highest percentage of RCO% point at 10 min reaction time when the reaction was conducted at 45°C. On the other hand, the highest percentage of RCO% obtained was 2.22% at 60°C.

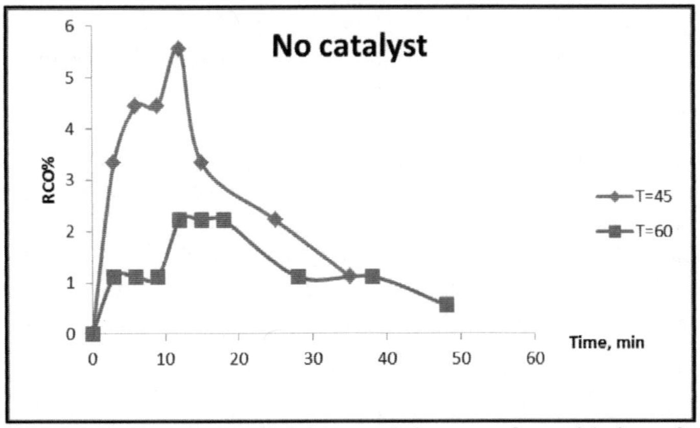

Figure 3. Conversion (RCO%) and the rate of epoxidation of palm olein for non-catalytic epoxidation.

The value of RCO% at 45°C was observed to be higher than at 60°C. This could be due to fact that at high temperatures, epoxidation occurred but was partially hydroxylized through oxirane cleavage, resulting in lower epoxy content [11]. Since the reaction has possibility for side reactions such as oxirane cleavage, epoxidation basically needs to be carried out at lower temperature. However, different temperatures lead to different results when different catalysts were used. This can be shown in figure 4 and figure 5 where the effect of temperatures on the reaction with different types of catalyst basically influences the RCO% of epoxidized palm olein.

As shown in figure 4, the value of RCO% in epoxidation at 45°C basically increase earlier as compared to 60°C but reach the maximum value only at 3.33%, less than if the reaction were carried out at 60°C. This shows that Zeolite ZSM-5 provides better stability at lower temperature but less conversion if compared to the epoxidation at higher temperature. This is because the content of aluminum in Zeolite ZSM-5 helps to lower the activation energy of the reaction, resulting in higher RCO% at 60°C compared to 45°C [12]. However, the RCO% decreased rapidly from peak value at 60°C due to the degradation of epoxy ring, as compared to the rate of epoxy ring degradation at 45°C which is much slower, leading to better ring stability as exhibited at around 7-18 minutes reaction time.

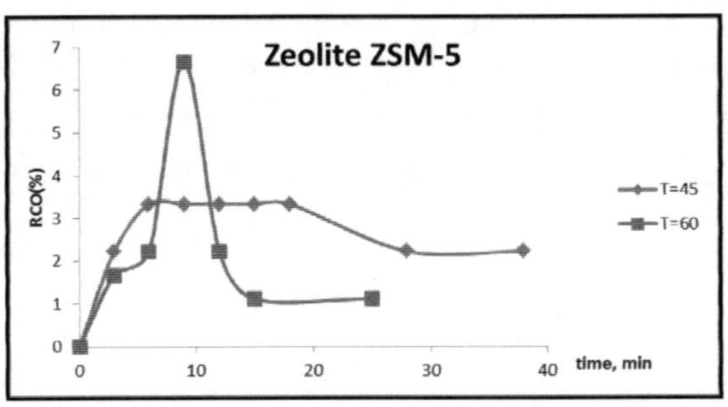

Figure 4. Conversion (RCO%) and the rate of epoxidation of palm olein with zeolite ZSM-5 as catalyst.

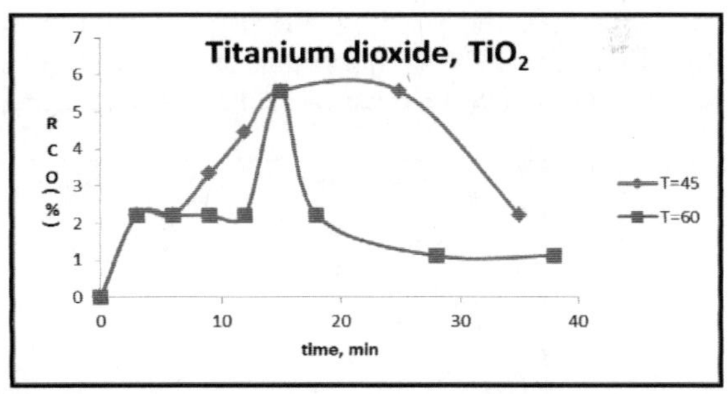

Figure 5. Conversion (RCO%) and the rate of epoxidation of palm olein with TiO_2 as catalyst.

In contrast to Zeolite ZSM-5, the value of RCO% for epoxidation using TiO_2 as catalyst is higher at 45°C which is about 5.55%, as illustrated in figure 5. Moreover, TiO_2 provides higher stability of oxirane ring at 45°C. According to Kim [13] who had carried out the epoxidation of methyl oleate with Ti-SiO_2, the conversion of the reaction increases when the temperature increased but then the selectivity decreases.

3.2. Effect of catalyst

To investigate the effect of heterogeneous catalyst on the epoxidation, different catalyst has been used and was conducted at different temperatures (45°C and 60°C). The types of catalyst varied are zeolite

ZSM-5 and titanium dioxide (TiO_2) while no catalyst acted as controlled variable. Figure 6 and figure 7 show the effect of different catalysts in the epoxidation at 45°C and 60°C.

Based on figure 6, both catalysts show the ability to achieve the epoxy ring stability at 45°C, as indicated by longer duration of peak conversion maintained before ring degradation started to take place (from 7 to 18 minutes reaction time for zeolite ZSM-5 and from 12 to 25 minutes reaction time for TiO_2). However, TiO_2 contributes to higher RCO% compared to Zeolite ZSM-5. This is because the content of titanium in TiO_2 influences the activity of the catalyst in epoxidation. This statement is supported by the research from Kim [13], which stated that the main factor that influences the catalytic activity is the catalyst framework or it structure. However, the result was different when the temperature was increased to 60°C, as illustrated in figure 7.

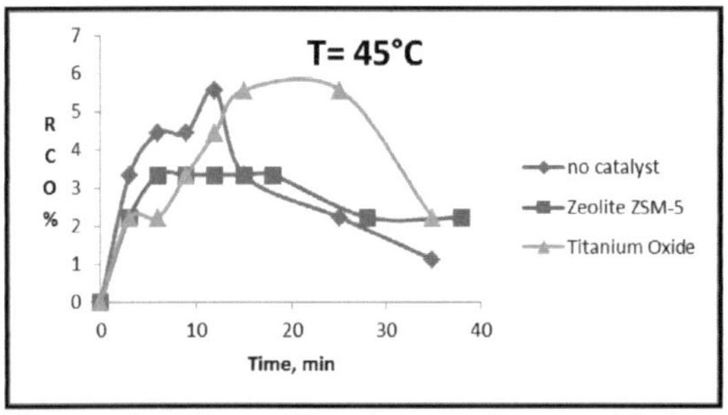

Figure 6. Effect of catalyst on the epoxidation at 45°C.

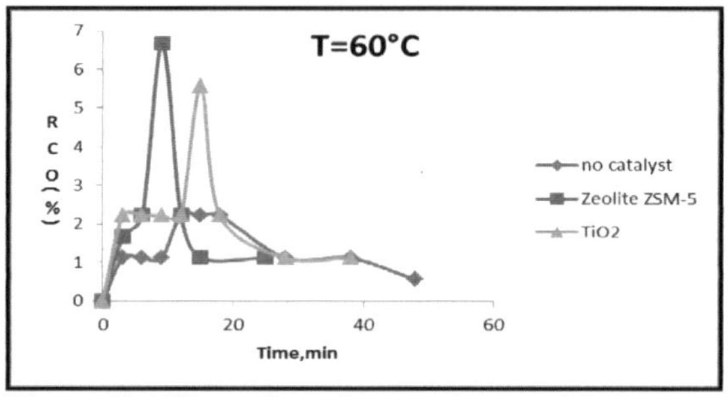

Figure 7. Effect of catalyst on the epoxidation at 60°C.

The effect of catalyst was different when the temperature increases. In figure 7, Zeolite ZSM-5 shows the highest percentage of RCO% compared to TiO_2 and the system without catalyst. Therefore, it can be conclude that Zeolite ZSM-5 catalyst is more efficient when the temperature is high. Besides that, the BET surface area of Zeolite ZSM-5 is larger than TiO_2 as shown in figure 9, thus providing more active sites for the reaction. However, the stability of epoxy ring decreases when the temperature increased, leading to ring opening. This was evidenced by the sharp drop of RCO% once the conversion reached the highest value.

3.3. FTIR of epoxidized palm olein

At the end of the reaction, the product sample was analyzed by using Fourier Transform Infrared Spectroscopy (FTIR) to examine the presence of oxirane group in the epoxidized palm olein. FTIR spectrum of epoxidized palm olein is shown in figure 8.

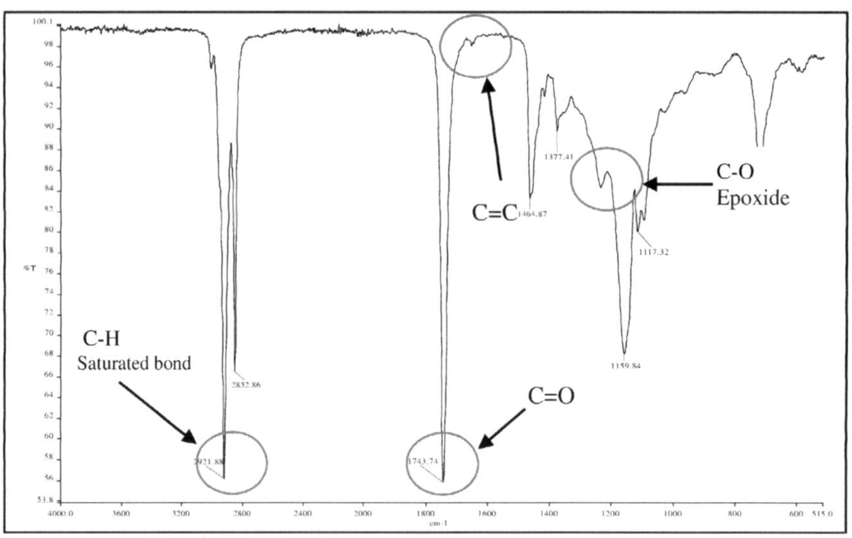

Figure 8. FTIR Spectrum of epoxidized palm olein.

The presence of epoxidized palm olein can be observed according to the absorption band signal as labeled in figure 8 (C-H saturated bond at 2921.88 cm^{-1}, C=O at 1743.74 cm^{-1} and epoxide bond C-O at 1210 cm^{-1}). The observation of epoxide signal band confirms the formation of oxirane ring in the epoxidized palm olein.

3.4. Characterization of catalyst

Zeolite ZSM-5 and TiO_2 were prepared into two different sample bottles and was characterized by using nitrogen adsorption analysis to determine the parameter such as pore volume (V_p) and the BET surface area (A_{BET}). BET surface area of the prepared catalyst was compared in figure 9, which indicates different nitrogen adsorption isotherm for the catalysts.

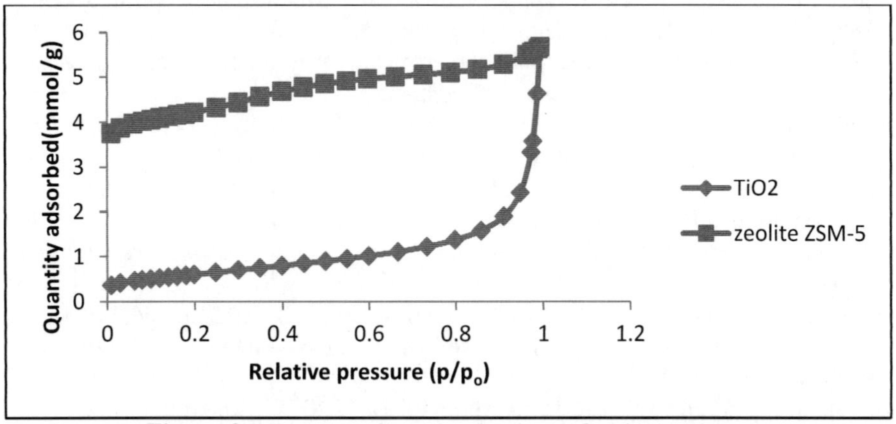

Figure 9. Nitrogen adsorption isotherm for the catalysts.

Table 2. Summary of BET surface area A_{BET} and pore volume (V_P) of the catalysts.

Sample	A_{BET} (m^2/g)	V_P (cm^3/g)
TiO_2	48.8972	0.12343
Zeolite ZSM-5	355.7150	0.15782

Based on table 2, it was observed that BET surface area and the pore volume of zeolite ZSM-5 was larger than TiO_2. This characteristic indicates that zeolite ZSM-5 catalyst provides more active sites for the epoxidation process which leads to higher conversion of the reaction. Wu et al. [14] provided similar result according to the BET surface area of TiO_2 and zeolite where TiO2 supported by zeolite provide large BET surface area (542.51 m^2/g) if compare to TiO_2(4.1952 m^2/g).

4. Conclusion

The efficiency of catalysts was determined based on the percentage of RCO in the epoxidation of palm olein. It can be concluded that TiO_2 has better performance in catalyzing the epoxidation since it is able provide higher oxirane content compared to zeolite ZSM-5 at 45°C. It shows the catalyst is preferable since the reaction should be carried out at lower temperature to avoid oxirane ring opening. Zeolite ZSM-5 has the best performance at higher temperature (60°C). However, the main factors that indicate the better catalyst is the catalyst's ability to provide high conversion and better epoxy ring stability in the epoxidation simultaneously at lower temperature. Therefore, TiO_2 was chosen as the best catalyst since it provides highest percentage of RCO (5.55%) and also epoxy ring stability at longer time when conducted at low temperature (45°C).

Acknowledgments

The authors would like to thank the Ministry of Education Malaysia for the financial support via Research Acculturation Grant Scheme (RAGS) (600-RMI/RAGS 5/3 (67/2015). The authors would like to thank the Faculty of Chemical Engineering, Universiti Teknologi MARA Shah Alam, Selangor for the use of services and facilities in conducting the research leading to this paper.

References

[1] Gunstone F D 2002 *Vegetable Oils in Food Technology : Composition, Properties and Uses* (Oxford: Blackwell Publishing)

[2] Padmasiri K G, Michael O and Laleen K 2009 Epoxidation of some vegetable oils and their hydrolysed products with peroxyformic acid - optimised to industrial scale *J. Natn. Sci. Foundation Sri Lanka* **37(4)** 229

[3] Campanella A, Fontanini C and Baltanas M 2008 High yield epoxidation of fatty acid methyl esters with performic acid generated in situ *Chem. Eng. J.* **144(3)** 466

[4] Flávia de A C, Felipe K S, Leandro S M M, Selma G F L, Rodrigo O M A D S and Ivana C R L 2012 Epoxidation of oleic acid catalyzed by PSCI-Amano lipase optimized by experimental design *Journal of Molecular Catalysis B: Enzymatic* **81** 7

[5] Luis R, David E and Fernando C 2013. Hydroxylation of vegetable oils using acidic resins as catalysts. *Industrial Crops and Products* **43** 183

[6] Diebold U 2003 Structure and properties of TiO2 surfaces: A brief review *Appl. Phys. A* **76** 681

[7] Massah A R, Kalbasi R J, Khalifesoltani M and Kordesofla F M 2013 ZSM-5-SO3H: An efficient catalyst for acylation of sulfonamides amines, alcohols, and phenols under solvent-free conditions *ISRN Organic Chemistry* **2013** Article ID 951749

[8] American Oil Chemists' Society 1963 *Tentative Method of Oxirane Oxygen: AOCS Cd 9-57* (Illinois)

[9] Petrovic Z S, Zlatanic A, Lava C C and Sinadinovic-Fiser S 2002 Epoxidation of soybean oil in toluene with peroxoacetic and peroxoformic acids - kinetics and side reactions *Eur. J. Lipid Sci. Technol.* **104** 293

[10] Noorfazlida M, Mohd J J, Siti K J and Ahmad R M D 2014 Formation of dihydroxystearic acid from hydrolysis of palm kernel oil-based epoxidized oleic acid. *J. Appl. Sci. & Agric.* **9(11)** 86

[11] Kim N 2014 Epoxidation and di-hydroxylation of camelina sativa oil *K-State Research Exchange (K-Rex)* Kansas State University Electronic Theses 1-74

[12] Choudary V R and Pataskar S G 1986 Thermal decomposition of TPA-ZSM-5 zeolites: effect of gas atmosphere and Si/Al ratio *Thermochimica Acta* **97** 1-10

[13] Kim K 2012 Epoxidation of vegetable oils using the heterogeneous catalysis, amorphous Ti-SiO2 *RWTH Publications* RWTH Aachen University Electronic Theses 1-113

[14] Wu L P, Li X J, Yuan Z H and Yong C 2009 The fabrication of TiO2-supported zeolite with core/shell heterostructure for ethanol dehydration to ethylene *Catalysis Communications* **11** 67-70

IOP Conf. Series: Materials Science and Engineering **358** (2018) 012071 doi:10.1088/1757-899X/358/1/012071

Pretreatment of Oil Palm Frond (OPF) with Ionic Liquid

I. S. Azmi, A. Azizan and R Mohd Salleh*

Faculty of Chemical Engineering, Universiti Teknologi MARA, Shah Alam, Malaysia

ruzitah@salam.uitm.edu.my

Abstract. Pretreatment is the key to unlock the recalcitrance of lignocellulose for cellulosic biofuel production. Increasing attention has been drawn to ionic liquids (ILs) for pretreatment of lignocellulosic biomass because this approach was considered as a green engineering method over other conventional methods. In this work, Oil palm frond (OPF) was pretreated by using the ionic liquid 1-ethyl-3-methylimidazolium acetate [EMIM] Ac at the temperature of 99°C for 3 hours. The characterization of the untreated and pretreated OPF was conducted by using different techniques which are Fourier Transform Infrared Spectroscopy (FTIR), X-Ray Diffraction (XRD) and Scanning Electron Microscopy (SEM). The pretreatment of OPF with [EMIM] Ac was demonstrated to be effective evidenced by the significant reduction of Lateral Order Index (LOI) from FTIR, reduction of Crystallinity Index (CI) based on XRD and the significant morphology changes indicated by SEM. The CI value for the pretreated OPF decreased from 0.47 (untreated sample) to 0.28 while the LOI value decreased from 1.10 to 0.24 after pretreatment with [EMIM]Ac and the SEM morphology showed that the pretreated OPF becomes distorted and disordered.

1. Introduction

Lignocellulosic materials can be represented as softwood, hardwood, agricultural and forest residues, grasses, food industry wastes and also municipal solid wastes. Cellulose, hemicellulose and lignin are the major components present in the lignocellulosic biomaterials [1]. In Malaysia, one of the most abundant resources available would be the cellulose content from all parts from palm tree (Elaeis guineensis). Palm residues (Elaeis guineensis) particularly comes from palm oil industry such as palm frond and palm trunk which contain high cellulose content of 50.78% and 41.88%, respectively [2]. In order to fully utilize the lignocellulosic biomass by making cellulose accessible for hydrolysis, there must be an efficient pretreatment method for deconstruction of biomass [3]. Nowadays, the most efficient and green engineering technology for pretreatment is by using ionic liquid. Among all of the ionic liquids, [EMIM] Ac has shown to be the most efficient ionic liquid in dissolving cellulose [3, 4, 5]. Ionic liquids have low toxicity, low hydrophobicity, high viscosity, low vapor pressure, thermal stability, enhanced electrochemical stability, and non-flammable properties, their use have low energy inputs and potentially minimal environmental impact. Therefore, in the last years, the use of ILs to dissolve many types of lignocellulosic biomasses has received more attention. In this study, the effect of [EMIM] Ac on OPF was discovered.

2. Material and Method

Oil Palm Frond (OPF) was provided by Forest Research Institute Malaysia (FRIM). It was first dried at 40°C in an oven before being ground by a horizontal mill. Next it was separated by a siever (Endecotts, USA) where the particle size falling between 250 and 75 microns. The moisture content of the sample was measured to be below than 10% before it was proceeded with pretreatment. 0.05 gram of OPF were mixed with 1 ml of [EMIM]Ac-water mixture in Eppendorf tube and shaken by using Bioshake iQ (Quantifoil Instrument GmbH, Germany) at 1800 rpm shaking frequency, temperature of

Content from this work may be used under the terms of the Creative Commons Attribution 3.0 licence. Any further distribution of this work must maintain attribution to the author(s) and the title of the work, journal citation and DOI.
Published under licence by IOP Publishing Ltd

99°C for 3 hours. After pretreatment, 1 ml of deionized water was added as an anti-solvent for cellulose precipitation. The pretreated mixture was centrifuged and the supernatant (liquid) was removed until the clear liquid was observed. Finally, the IL-pretreated biomass was collected and oven-dried at 60°C for 24 hours for next analysis.

2.1 Fourier Transform Infrared (FTIR) Analysis
The structural analysis by using Fourier Transform Infrared (FTIR) (Perkin Elmer, USA) where the sample is pressed into a disc and the scanning is in the range of 4000-515 cm^{-1} with a resolution of 4 cm^{-1}. The Lateral Order Index (LOI) for FTIR sample was calculated by using the absorbance (A) ratio A_{1430}/A_{898}. From the raw data of FTIR, the transmittance (T) value was converted to absorbance by using Equation 1 from Beer's Law.

$$A = 2 - \log_{10} (\%T) \quad (1)$$

2.2 X-Ray Diffraction (XRD) Analysis
The cellulose crystalline structure is analysed by using X-Ray Diffraction (XRD) (Multi Purpose X-Ray Diffractrometer, Ultima IV, Rigaku, USA). The scanning range, step size and step time was 10-30° (2θ), 0.02° and 1 second, respectively. The scans were conducted at 40 kV and 40 mA under ambient temperature. Crystallinity index (CI) was calculated by using peak height method developed by Segal and co-workers as in Equation 2.

$$CI (\%) = (I_{002} - I_{am}) / I_{002} \times 100 \quad (2)$$

I_{002} = intensity of crystalline 002 peak at 2θ around 22.5°
I_{am} = intensity of diffraction of the non-crystalline material which is taken at 2θ around 18.5°

2.3 Scanning Electron Microscopy (SEM) Analysis
The morphology of the biomass before and after pretreatment was observed by using Scanning Electron Microscope (SEM) (Hitachi TM3000, USA). The acceleration voltage was 5.0kV. The sample was mounted on aluminium sample stubs and before analysis, the sputter coated the sample with a thin layer of gold.

3. Result and Discussion

3.1 FTIR Analysis
FTIR spectroscopy was frequently used to investigate the structure of constituents and its chemical changes occurring during the pretreatment of lignocellulosic biomass [10, 11]. Table 1 show the characteristic and variations of wavenumber in FT-IR spectra possible for cellulose. The FTIR spectra of cellulose samples irradiated with different doses are shown in Figure 1. In general, the patterns of the spectrum of the pretreated OPF were similar with the untreated OPF but the transmittance percentages of some bands were different. From Figure 1 at peaks 10, the obvious changes of the peaks are at the wavelength of 3100-3600 cm-1 which correspond to alcohols (O-H) stretching vibration which give considerable information concerning the hydrogen bonds [6]. The untreated OPF has a broad band at this wavelength and the transmittance percentage increases causing flattening curve to be observed after pretreatment with [EMIM]Ac. The flattening peaks may represent the possible breakage of hydrogen bonds between O-H links and also possible disruption of crystalline cellulose. This result was supported by [7] which found that the peaks around 3100-3600 has a lower intensity compared to untreated samples and concluded that this was the effect of the scission of the intramolecular and intermolecular hydrogen bonds.

Table 1. LOI Values.

Samples	LOI
Raw OPF	1.10
Pretreated OPF	0.24

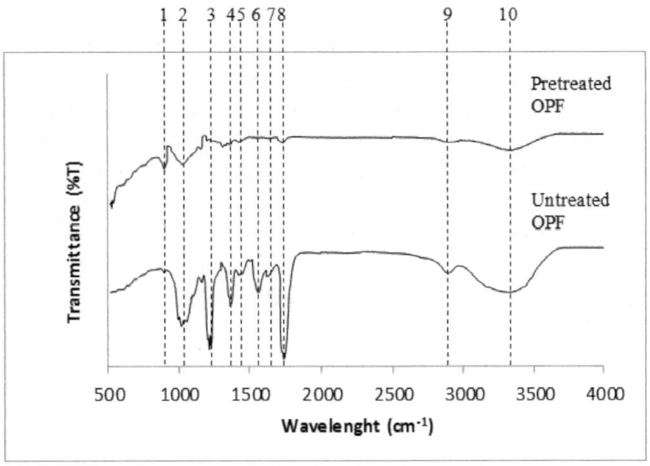

Figure 1. FTIR Spectra of Untreated OPF and Pretreated OPF.

The next wavelength is the peaks around 2900 cm-[1] at peak 9 in Figure 1 which indicates an alkane functional group which is C-H stretching vibration. After pretreatment with [EMIM]Ac, the peak of the FTIR spectrum again flatten and the transmittance percentage value increases. This result shows that the methyl (CH_3) group in OPF was disrupted. The decrease of the band at wavelength 2900 cm^{-1} was also found by [8] and [9] which concluded that the sample was more amorphous. The FTIR peak at the wavelength of 898 cm^{-1} as shown as in Figure 1 at peak 1 which characterizes the C-O-C stretching at β-1,4-glycosidic bond and also known as "amorphous absorption band" becomes more intense and strong with increasing of [EMIM]Ac concentration. It can be concluded that the cellulose becomes more amorphous after pretreatment with [EMIM]Ac. The finding of current study is consistent with [10] who found that the peaks at 897 cm^{-1} is becoming strong and sharp after pretreatment with various ionic liquids and concluded that the sample becomes more amorphous.
The peak at 1430 cm^{-1} in Figure 1 as indicated by peak 5 which is assigned to CH_2 bending vibration and also known as a "crystallinity band" becomes decrease after OPF was pretreated with [EMIM]Ac. This band is strong in crystalline cellulose and weak in amorphous cellulose [11]. The decrease of the peak indicates reduction in the degree of crystallinity in OPF. A low cellulose crystallinity index also indicates that a part of crystalline structure of cellulose was transformed into amorphous form. With the amorphous structure presence, higher possibilities of enzymatic hydrolysis could occur in comparison to the previous ordered arrangement of crystalline structures. The wavelength of ~1732 cm^{-1} as shown as a very sharp descending peak indicated by peak 8 in Figure 1 is assigned to the C=O stretching of carboxylic acids [10] and also corresponds to aldehydes in hemicellulose [12]. This particular peak was still retained in the pretreated OPF as shown as in Figure 1. However, it is very significant that the peak nearly disappeared after the pretreatment which may indicate a significant removal of hemicellulose.

Lignin main functional group is an aromatic compound (phenolic hydroxyl group) which has a structure of benzene ring. In FTIR spectroscopy, the wavelength which corresponds to the aromatic compound is ~1500 – 1600 cm^{-1} [7, 13]. From Figure 1, it can be observed that the peaks of ~1550 and ~1600 cm^{-1} at peaks 6 and 7, respectively, were almost disappeared after the pretreatment step which may indicate some removal of lignin. Other significant sharp peaks observed from the FTIR spectrum as shown as in Figure 1 are at the wavelength of ~1031 and ~1230 cm^{-1} as indicated at the peaks 2 and 3, respectively, corresponds to C-O-C stretching. While the observed peak of wavelength of ~1365 at peak 4 shown is assigned to C-H and C-O stretching in polysaccharides. All of the peaks was decreased after pretreatment with [EMIM]Ac. This also signifies the destruction of the ordered structure and dissolution of polysaccharides which are cellulose, hemicellulose and lignin.

3.2 XRD Analysis

XRD analysis was conducted in order to further examine the crystallinity of cellulose since the determination of the crystallinity index by FTIR spectroscopy only gives the relative values from both crystalline and amorphous regions. The XRD spectra and CI values for the effect of [EMIM]Ac pretreatment on OPF is as shown in Figure 2 and Table 2, respectively. In this study, two typical diffraction peaks were observed at 2θ = 15° and 22.5° which correspond to (101) and (002) peaks, respectively. The CI value for the pretreated OPF decreased from 0.47 (untreated sample) to 0.28. It might be due to the swelling of the cellulose matrix by [EMIM]Ac. From Figure 2, it can be observed that the XRD spectra for the pretreated OPF become decrease and flattened compared to the sharp peaks of raw OPF. Therefore, the CI result from XRD supported the FTIR result by showing a significance decrease of CI. The lower CI may thus provide a larger surface area accessible by cellulose enzyme which which may then enhanced the reducing sugar yield after enzymatic hydrolysis of the IL-pretreated biomass. These results may suggest that [EMIM]Ac pretreatment can reduce the cellulose crystallinity in OPF.

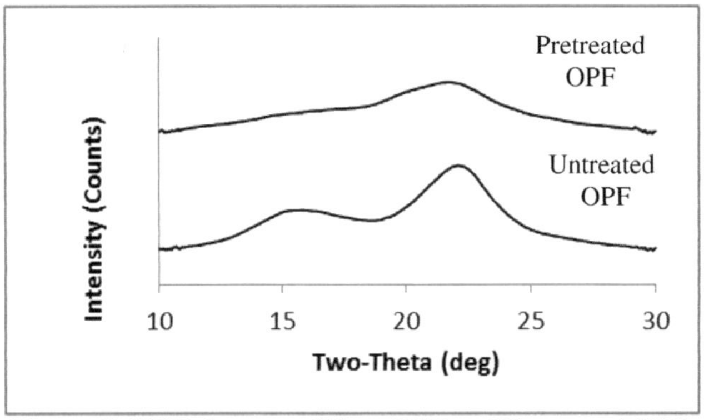

Figure 2. XRD Spectra of Raw OPF and Pretreated OPF.

Table 2. Values of Crystallinity Index (CI).

Samples	CI
Untreated OPF	0.47
Pretreated OPF	0.28

3.3 Scanning Electron Microscopy (SEM)

Scanning Electron Microscopy (SEM) which was further used to investigate the morphological structures of OPF after pretreatment with [EMIM]Ac can be seen as shown as in Figure 3(a) and Figure 3 (b).

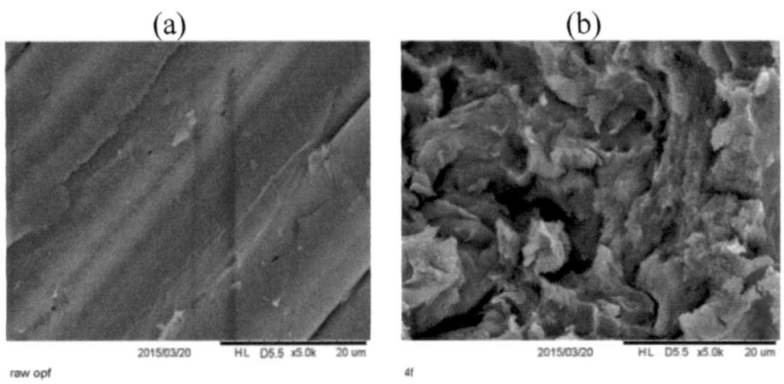

Figure 3: (a) Untreated OPF and (b) Pretreated OPF

It can be clearly seen that the raw OPF has a rigid, compact and ordered structure. After the pretreatment with [EMIM]Ac, the structure becomes disordered, loose and had some cracks or distorted surfaces as compared to the earlier image representing the untreated OPF. These changes of morphology probably happened because of the swelling of cellulose crystalline structure and disruption of hydrogen bonds in cellulose after pretreatment with [EMIM]Ac. SEM analysis supported the results from FTIR and XRD where [EMIM]Ac pretreatment is effective in reducing the crystallinity of cellulose.

4. Conclusion

In summary, from the findings found from the investigations of the pretreatment of OPF with [EMIM]Ac can be said effective because of the significant decrease in Lateral Order Index (LOI) and the Crystallinity Index (CI) as well as the observed significant changes on the OPF morphology. The decrease of cellulose crystallinity will enhance the digestibility during enzymatic hydrolysis which may increase sugar yield for bioethanol production. Thus, it can be concluded that [EMIM]Ac is a promising green solvent to reduce cellulose crystallinity during pretreatment step.

Acknowledgments

The work was financially supported by Fundamental Research Grant Scheme (FGRS) (600-RMI/FRGS 5/3 (99/2014)) & Ministry of Higher Learning. Appreciation is also given to the Faculty of Chemical Engineering, University Teknologi MARA, Malaysia.

References

[1] Lynd L R, Weimer P J, van Zyl W H and Pretorius I S 2002 Microbial Cellulose Utilization: Fundamentals and Biotechnology *Microbiology and Molecular Biology Reviews*. **66(3)** 506–577

[2] Lai L W and Idris A 2013 Disruption of oil palm trunks and fronds by microwave-alkali pretreatment *BioResources*. **8(2)**, 2792–2804

[3] da Costa Lopes A M, João K G, Morais, A R C, Bogel-Łukasik E and Bogel-Łukasik R 2013 Ionic liquids as a tool for biomass fractionation *Sustainable Chemical Processes*. **1(1)** 1-31

[4] Zavrel M, Bross D, Funke M, Büchs J and Spiess A C 2009 High-throughput screening for ionic liquids dissolving (ligno-)cellulose. *Bioresource Technology*. **100(9)** 2580–87

[5] Qiu Z and Aita G M 2013 Pretreatment of energy cane bagasse with recycled ionic liquid for enzymatic hydrolysis *Bioresource Technology*. **129** 532–537.

[6] Azizan A, Shafaei N S M, Sidek N S, Hanafi F, Mokti N and Zaharudin S 2016 Fourier Transform Infrared Spectroscopy interpretation on pretreated Acacia Auriculiformis, Melastoma Malabathricum and Leucaeana Leucocephala. *International Journal of Applied Engineering Research*. **11(20)** 10048-51

[7] Li C, Knierim B, Manisseri C, Arora R, Scheller H V, Auer M and Singh S 2010 Comparison of dilute acid and ionic liquid pretreatment of switchgrass: Biomass recalcitrance, delignification and enzymatic saccharification *Bioresource Technology*. **101(13)** 4900–06

[8] Ciolacu D, Ciolacu F and Popa V I 2011 Amorphous Cellulose–Structure and Characterization *Cellulose Chemistry and Technology*. **45(1–2)** 13–21

[9] Kumar, P., Barrett D M, Delwiche M J and Stroeve P 2009 Methods for pretreatment of lignocellulosic biomass for efficient hydrolysis and biofuel production *Industrial & Engineering Chemistry Research*. **48(8)** 3713-29

[10] Mood S H, Golfeshan A H, Tabatabaei M, Abbasalizadeh S, Ardjmand M and Jouzani G S 2014 Comparison of different ionic liquids pretreatment for corn stover enzymatic saccharification *Preparative Biochemistry & Biotechnology*. **44** 451–63

[11] Orozco R S, Hernández P B, Morales G R, Núñez F U, Villafuerte J O, Lugo V L and Vázquez P C 2014 Characterization of lignocellulosic fruit waste as an alternative feedstock for bioethanol production. *BioResources*. **9(2)** 1873–85.

[12] Zhao X, Zhou Y, Zheng G and Liu D 2010 Microwave pretreatment of substrates for Cellulase production by solid-state fermentation *Applied Biochemistry and Biotechnology*. **160(5)** 1557–71.

[13] Hsu T-C, Guo G-L, Chen W-H and Hwang W-S 2010 Effect of dilute acid pretreatment of rice straw on structural properties and enzymatic hydrolysis *Bioresource Technology*. **101(13)** 4907–13

[14] Qiu Z, Aita G M and Walker M S 2012 Effect of ionic liquid pretreatment on the chemical composition, structure and enzymatic hydrolysis of energy cane bagasse. *Bioresource Technology*. **117** 251–256

IOP Conf. Series: Materials Science and Engineering **358** (2018) 012072 doi:10.1088/1757-899X/358/1/012072

Encapsulation of Volatile Citronella Essential Oil by Coacervation: Efficiency and Release Study

M A Manaf[1,a], I Subuki[1,b], J Jai[1,c], R Raslan[1,d] and A N Mustapa[1,e]
[1] Faculty of Chemical Engineering, Universiti Teknologi MARA, 40450 Shah Alam, Selangor, Malaysia

[a]mastura.abdmanaf@gmail.com, [b]istikamah@salam.uitm.edu.my, [c]junejai@salam.uitm.edu.my,[d]rafeqah@salam.uitm.edu.my, [e]anajwa@salam.uitm.edu.my

Abstract. The volatile citronella essential oil was encapsulated by simple coacervation and complex coacervation using Arabic gum and gelatin as wall material. Glutaraldehyde was used in the methodology as crosslinking agent. The citronella standard calibration graph obtained with R2 of 0.9523 was used for the accurate determination of encapsulation efficiency and release study. The release kinetic was analysed based on Fick's law of diffusion for polymeric system and linear graph of Log fraction release over Log time was constructed to determine the release rate constant, k and diffusion coefficient, n. Both coacervation methods in the present study produce encapsulation efficiency around 94%. The produced capsules for both coacervation processes were discussed based on the capsules morphology and release kinetic mechanisms.

1. Introduction

These guidelines, written in the style of a submission to *IOP Conf. Ser.: Mater. Sci. Eng.*, discuss how Few reported works on the formulation of Citronella oil (CO) claimed that the active ingredient has only 3 minutes to 6 hours retained in any formulation of solution [1-3]. Encapsulation is one of the recent methods to preserve or retain the high value volatile compounds in several chemical solutions. The basic form of encapsulation is by emulsifying the target compounds in an aqueous solution by entrapping the oil drops in a carrier material [4]. Common goals in the development of encapsulated essential oil are to protect the compounds from degradation or from losses by evaporation, to achieve a controlled release, and to facilitate handling [5].

Encapsulation by coacervation is a phase separation of one or many hydrocolloids from the initial solution. It involves the deposition of the newly-formed coacervate phase around the active ingredient suspended or emulsified in the same reaction media under specific pH, temperature or composition of the solution [6, 7]. Chemical agent or cross-linking may require strengthening the capsules. Simple coacervation involves the use of single polymer such as gelatin or ethyl cellulose. While, complex coacervation involves neutralization of two oppositely charge polymer in aqueous solution. The widely used method is neutralization between negatively charge Arabic gum and positively charge gelatin [8]. The release behavior of the core material from the capsules was studied by using a generalized equation that is based on Fick's law of diffusion for a polymeric system as Eq. 1. Where ($Mt / M\infty$) is the fractional release at time t. Mt is percent release at specific time while $M\infty$ is total material release, k is a constant characteristic of the polymer encapsulated and n is diffusion coefficient.

$$Mt / M\infty = ktn \tag{1}$$

Content from this work may be used under the terms of the Creative Commons Attribution 3.0 licence. Any further distribution of this work must maintain attribution to the author(s) and the title of the work, journal citation and DOI.
Published under licence by IOP Publishing Ltd

2. Procedure

2.1 Preparation of Citronella Standard Graph.
A stock solution of 2 mg/ml of citronella standard solution was prepared in chloroform. Working solutions were prepared by diluting the stock solution with the same solvent to contain 2, 5, 10, 20, 40, 60, 80 and 100 μg/mL. Each solution was injected into GC-MS system and the concentration of citronella against peak area was obtained for each sample.

2.2 Encapsulation of Citronella oil by Simple Coacervation
For simple coacervation method, 60 mL Arabic gum solution was prepared at 12.5 % w/v concentration and mixed using magnetic stirrer at 250 rpm and temperature 27°C. The Arabic gum solution was first mixed with 4 mL citronella essential oil and 10 mL deionized water. To enhance coacervation stage, the mixture was added with 90 mL deionized water and mixed at speed 500 rpm, temperature 50°C for 20 minutes. pH of the mixture was adjusted to 4.5 by dropwise addition of 10% acetic acid solution. Then, 100ml distilled water was added and emulsified at stirring speed of 250 rpm, temperature below 10°C in 60 minutes. Twenty ml 25% of glutaraldehyde solution was added as crosslinking agent. pH was altered to value of the mixture of 9.7 by 10% sodium hydroxide solution addition the stirring was kept at the speed of 250 rpm, in 30 minutes, below 10°C to stabilize the microcapsules. The emulsion was put in separation funnel for 24 hours.

2.3 Encapsulation of Citronella oil by Complex Coacervation
In the present complex coacervation method, two wall materials used which are gelatin and Arabic gum. In the study, on to one ratio of gelatin and Arabic gum was practiced. It was reported produced spherical, scattering, maximum encapsulation efficiency with relatively equal size capsules. The gum Arabic solution was prepared at 12.5 % w/v concentration and mixed using magnetic stirrer at 250 rpm and room temperature. The gelatin solution was prepared at 12.5 % w/v concentration and mixed using magnetic stirrer at 250 rpm and temperature 40°C. The 30 mL 12.5 % gum Arabic solution was first mixed with 4 mL citronella essential oil and 5 mL deionized water at speed 500 rpm, temperature 50°C for 20 minutes. Then, the 30 mL gelatin solution was mixed together with the emulsion at room temperature, stirring speed of 500 rpm for another 20 minutes. After that, 195 ml distilled water was added to the emulsion, then continued agitating in 30 minutes, at temperature 50°C and same stirring speed 500 rpm. To enhanced coacervation, pH of the mixture was adjusted to 4.5 by dropwise addition of 10% acetic acid solution. Then, 200 ml distilled water was added and emulsified at stirring speed of 250 rpm, temperature below 10°C in 60 minutes. 20 ml 25% of glutaraldehyde solution was added as crosslinking agent. The stirring was kept at the speed of 250 rpm, in 30 minutes, below 10°C to stabilize the microcapsules. The emulsion was put in separation funnel for 8 hours.

2.4 Determination of Encapsulation Efficiency (EE)
The encapsulation process efficiency was calculated as Eq. 2. In this study, the amount of un-capsulated CO was determined by analyzing the amount of excess citronella oil presented at each layer for both process after the separation.

$$EE\ (\%) = [(\text{Initial CO} - \text{un-capsulated CO}) / \text{Initial CO in emulsion}] \times 100 \quad (2)$$

2.5 GC-MS Analysis
The CO presence was determined by using inert GC-MS Agilent 7890A Series of GC system, USA. The analysis methods were referred to Xiao et al. [12]. For the upper layer, before analysis, 1 g of the capsules layer were suspended in a flask containing 10 ml n-hexane. The mixture was shaked gently and allowed to settle. The hexane layer was collected and added with 5 g of anhydrous sodium sulfate to remove water. After mixed well the mixture, the supernatant was collected and filter to be used for GC-MS analysis. The same procedure was applied for the middle and bottom layer with 20 ml sample, 20 ml n-hexane and 10 g of anhydrous sodium sulfate.

2.6 Determination of Release Kinetic

Generalized equation developed from Fick's law of diffusion was used to analyze the release behavior of the citronella from capsules. By using this equation, the representative plot of 'Log (fractional release)' as a function of 'Log (release time)' was developed by using the data to determine the release kinetic characterization. The constants n and k, both are best determined by means of a Log-Log plot of Mt/M∞ versus time as Eq.3. The slope of this plot is the dimensionless exponent, n and the intercept on the Y-axis gives the release rate constant, k [9].

$$\text{Log}\,(Mt/M\infty) = n\text{Log}\,t + \text{Log}\,k \quad (3)$$

To determine Mt, the procedure was conducted according to Siow and Ong [11] and Dong et al. [6]. 0.5 g of wet capsules were placed into six separated sealed containers which each of it contained 10 ml of n-hexane. The suspensions were put at room condition at stirring speed of 50 rpm. Two ml samples were taken out once from one suspension at 30, 60, 120, 180, 240, and 300 minutes by using syringe. For determination of M∞, the 0.5 g of wet capsules were suspended in same solvent and extracted in ultrasonic for 10 minutes. Then, the collected samples were filtrated before used for GC-MS analysis.

2.7 Capsules Morphology Study

Morphology of the capsules were examined under light microscope, (MEIJI Techno, RZ, Jepan) which incorporated with a monitor display. The wet capsules were put on a microscope slide and observed under objective magnification of 10x. Image was captured and determined the size using the Video Toolbox Pro software.

3. Result and Discussion

3.1. Encapsulation Efficiency

Preliminary calibration experiments were carried to determine the citronella compound presence in the polymeric solution. The standard graph was constructed based on GC-MS peak area versus the citronella oil concentration from 2 µg/ml to 100 µg/ml as in Fig. 1. The standard calibration graph is a linear graph with R2 is 0.9523. The same citronella standard graph was used for release kinetic study in order to determine the emitted citronella to the medium at specific time. Encapsulation efficiency data for both simple and complex coacervations are shows at Table 1. The present study shows that both process able to provide good encapsulation process with encapsulation efficiency around 94%. According to Yeo et al.[15] encapsulation efficiency increases with increasing polymer concentration. High concentration of polymeric solution, provide high viscosity media and faster the solidification of the dispersed phase contributed to reducing porosity of the microparticles as well. In the both coacervation process, equal polymeric solution concentration was applied which is 12.5 %w/v Arabic gum solution and 12.5 %w/v gelatin solution. Compare to other method, this concentration is high enough to produce good encapsulation efficiency.

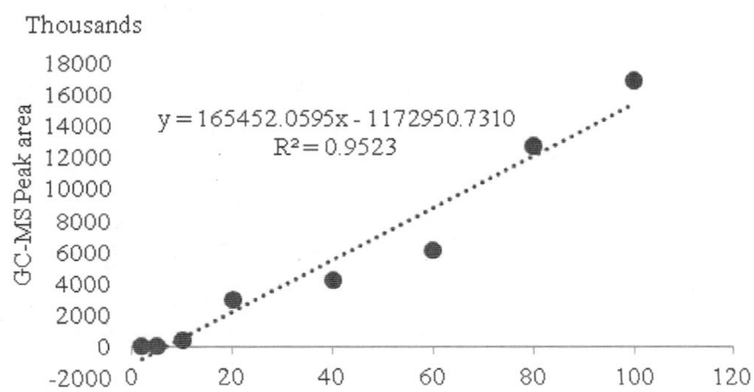

Figure 1. Citronella standard graph of GC-MS peak area over varies concentration of citronella oil

Table 1. Encapsulation Efficiency of Simple and Complex Coacervation

Efficiency	Simple coacervation	Complex coacervation
CO presence in bottom layer (g)	0.0929	0.0042
CO presence in middle layer (g)	-	0.1059
CO presence in upper layer (g)	0.1271	0.0911
Encapsulation Efficiency (%)	93.9116	94.4218

*CO refer to Citronella oil.

3.2. Oil Release Study

Figure 2 shows the plots of percentage of citronella release from polymeric wall system over time for simple coacervation and complex coacervation. Simple coacervation shows higher percent release at specific time compare to complex coacervation. This is partly because the low molecular weight polymer is more soluble in the organic solvent and undergoes slow solidification to produce more porous microparticles. On the other hand, it can also be attributed to the smaller size of the microparticles, which provides more surface area for drug diffusion. With increase in wall material cross-linking, the release was slowed down which happen in complex coacervation where gelatin and Arabic gum were combined to develop coacervates polymeric wall. The release kinetic mechanism is study further by plotting Log-log graph of fraction release verses time as shows in Fig. 3.

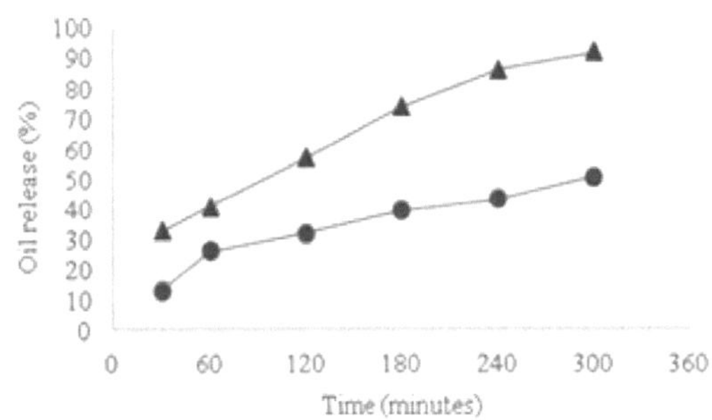

Figure 2. Percent release curve of encapsulants at specific times. (▲): simple coacervation. (●): complex coacervation.

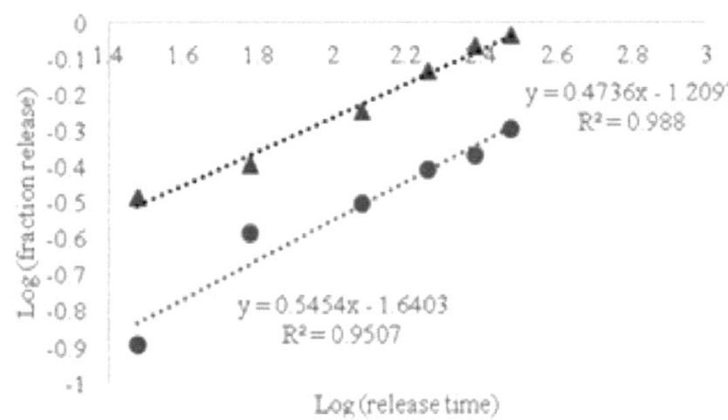

Figure 3. Percent release curve of encapsulants at specific times. (▲): simple coacervation. (●): complex coacervation.

The graph represents the plot of Log fraction release as function of Log release time for microcapsules from simple coacervation and complex coacervation process. The calculated values for k and n in the present study are given in Table 2. The release constant rate, k affecting by the characteristic of the polymeric wall of the microcapsules while the dimensionless exponent, n, represents the type of release mechanism of the active ingredient through the capsules wall.

Table 2. Release Rate Constant and Diffusional Exponent of Coacervates

Process	k ($hour^{-1}$)	n
Simple coacervation	0.0617	0.4736
Complex coacervation	0.0229	0.5454

*k is constant release in unit $hour^{-1}$, while n is dimensionless exponent

Theoretically, the k constant decrease with increasing degree of cross-linking of the polymeric structure [9]. For dimensionless exponent, n, mathematical solutions of the model have been described in three cases to elaborate the types of release mechanism of the coacervates. First case is where $n=0.5$ indicates Fickian encapsulant diffusion. Second is non-Fickian or deviating from the regular form (anomalous) when $0.5<n<1.0$. The third case with $n<0.5$ which is 'Less Fickian' diffusion. Zero-order release kinetics is a constant rate of release for the encapsulant when $n=1$ [9, 13]. In this study, both coacervation methods reported dimensionless coefficient, n=0.5 which is Fickian diffusion mechanism where the release of core material is significant with time.

3.3Capsule Morphology

The morphology of the capsules produced by simple coacervation and complex coacervation are presented in Fig. 4. Coacervates made from both process is spherical in shape. The capsules morphology gives further descriptions and correlation for the release mechanism of the encapsulant for both process. Capsules from simple coacervation have more porous surface compare to capsules made by complex coacervation which has smoother surface. Capsules with porous surface because higher initial burst effect resulted high release rate of the core material. Within 24 hours at 25°C temperature, the capsules already melt indicates fully oil release from the capsules. Capsules made by complex coacervation with lower release rate have more rigid shape and can sustained longer in room condition. The capsules still in the shape after 7 days. The initial burst of the capsules is directly affected by the molecular weight of the polymer. In general, low molecular weight polymer result in high burst release of encapsulate.

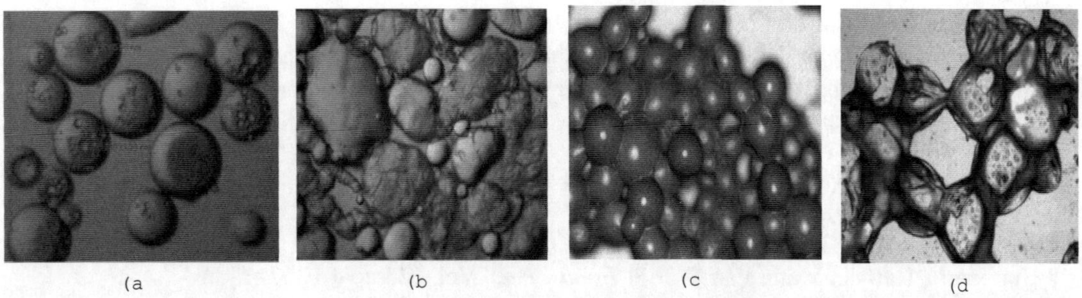

Figure 4. Capsules morphology under light microscope with 10x magnification objective. (a): Simple coacervation freshly after capsules harvesting. (b): Simple coacervation after 24 hours in room condition. (c): Complex coacervation freshly after capsules harvesting. (d): Complex coacervation after 7 days in room condition.

This is because the low molecular weight polymer is more soluble in the organic solvent and undergoes slow solidification to produce more porous microparticles thus provide higher surface area for core material release [15]. The capsules from simple coacervates made by Arabic gum only, have lower molecular weight compared to capsules made from complex coacervation with combination of gelatin and gum Arabic. The release mechanism of capsules from each process can be determined from Fig. 4b and Fig. 4d. Generally, different release mechanisms of encapsulated materials provide controlled, sustained, or target release of core material. From the figures, the capsules made by present simple coacervation possess dissolution or melting of capsules wall of release mechanism where water is absorbed into the microcapsule core through the pore and swelling the capsule's wall. Subsequent swelling was ruptured the microcapsule shell thus change the capsule's shape and release the core material to environment. The core materials release by diffusion through the wall for the capsules made by complex coacervation. Diffusion is wherein the core material leak out through the interstitial channels of capsule's wall. The overall release depends on the rate at which the core material leaks out and disperse from the surface [16].

4. Conclusion

The simple coacervation and complex coacervation method in present study able to provide good encapsulation process with encapsulation efficiency around 94%. For release kinetic study, release constant of simple coacervation higher than complex coacervation indicates that the complex coacervation produce higher cross-linking of polymeric structure thus lower the release rate of the core material. Both coacervation methods reported dimensionless coefficient, n=0.5 which is Fickian diffusion mechanism. Coacervates made from both process is spherical in shape. Simple coacervation possess dissolution or melting of capsules wall of release mechanism while the core materials release by diffusion through the wall for the capsules made by complex coacervation.

Acknowledgement
Deepest gratitude goes to Malaysia Ministry of Education and Universiti Teknologi MARA for the financial support for this project from the research grants of 600-RMI/RAGS 5/3(76/2013).

References

[1] Arie M, Linder C, and David S 2010 "Formulations containing microencapsulated essential oils," 2010.
[2] Martín Á, Varona S, Navarrete A and Cocero M J 2010 *Open Chem. Eng. J.* Vol. **04**, 3
[3] Solomon B, Sahle F F, Gebre-Mariam T, Asres K and Neubert R H H 2012 *Eur. J. Pharm. Biopharm* Vol. **80,** no. 1, 61–66
[4] Umer H, Nigam H, Tamboli A M and Nainar M S M 2011 *Int. J. Res. Pharm. Biomed. Sci.* Vol. 2 (2) 474–481
[5] Bansode S S, Banarjee S K, Gaikwad D D, Jadhav S L and Thorat R M 2010 *Int. J. Pharm. Sci.* Vol. **1**, no. 2 38–44
[6] Dong Z, Ma Y, Hayat K, Jia C, Xia S and Zhang X 2011 *J. Food Eng.* Vol. **104** no. 3 455–460
[7] Park J K and Chang H N 2000 *Biotechnol. Adv.* Vol. 18 no. 4 303–19
[8] Karsa D R and Stephenson R A 1993 *Encapsulation and Controlled Release*. (United Kingdom: Royal Society of Chemistry)
[9] Vahabzadeh F, Zivdar M and Najafi A 2004 *IJE Trans.* Vol. **17** 333–343
[10] Nguyen H X and Nguyen C N 2012 *Pharm. J.* Vol. **52** no 437 6–10
[11] Siow L and Ong C 2013 Vol. **4** no 1 1–5
[12] Jun-xia X, Hai-yan Y and Jian Y 2011 *Food Chem.* Vol. **125** no 4 1267–1272
[13] H. B. Bohidar, 2008 Vol. **24** no. 3 105–124
[14] Jyothi1 N V N, Prasanna P M, Sakarkar S N, Prabha K S, Ramaiah P S and Srawanl G Y 2010, *J. Microencapsul.* Vol. **2010** 27 3,187–197
[15] Yeo Y and Park K 2004 *Arch. Pharm. Res.* Vol. **27** no. 1 1–12
[16] Singh M N, Hemant K S Y, Ram M and Shivakumar H G 2010 Vol. **5** no. 2 65–77.

IOP Conf. Series: Materials Science and Engineering **358** (2018) 012073 doi:10.1088/1757-899X/358/1/012073

Study of Vapour Cloud Explosion Impact from Pressure Changes in the Liquefied Petroleum Gas Sphere Tank Storage Leakage

Z A Rashid, A F Mohd Suhaimi Yeong, A B Alias, M A Ahmad and S AbdulBari Ali

Faculty of Chemical Engineering, Universiti Teknologi MARA, Malaysia.

zulmas06@yahoo.com.my[1]; anisfarhanahmsy@gmail.com; azilbahari@salam.uitm.edu.my.

Abstract. This research was carried out to determine the risk impact of Liquefied Petroleum Gas (LPG) storage facilities, especially in the event of LPG tank explosion. In order to prevent the LPG tank explosion from occurring, it is important to decide the most suitable operating condition for the LPG tank itself, as the explosion of LPG tank could affect and cause extensive damage to the surrounding. The explosion of LPG tank usually occurs due to the rise of pressure in the tank. Thus, in this research, a method called Planas-Cuchi was applied to determine the Peak Side-On Overpressure (P^o) of the LPG tank during the occurrence of explosion. Thermodynamic properties of saturated propane, (C_3H_8) have been chosen as a reference and basis of calculation to determine the parameters such as Explosion Energy (E), Equivalent Mass of TNT (W_{TNT}), and Scaled Overpressure (P_S). A cylindrical LPG tank in Feyzin Refinery, France was selected as a case study in this research and at the end of this research, the most suitable operating pressure of the LPG tank was determined.

1. Introduction

LPG comprises flammable combination of hydrocarbon gases such as propane (C_3H_8) and is used to power many of the equipment we have today. However, LPG can be extremely dangerous if mistake occurs. Some of the accidents happening at these petrochemical plants can be studied through historical record. For instance, the Flixborough (Nypro UK) Explosion in 1974 resulted in 30 tonnes of cyclohexane leakage from a vertical crack of one of the onsite reactors, causing the death of 28 workers and injured 36 more. Another major LPG accident was the PEMEX LPG Terminal tragedy in Mexico City in 1984, which claimed over 500 lives and some 4,000 people injured. It was caused by a rupture at an 8-inch pipe that was connecting a sphere and a series of cylinders and the operators' inability to identify the source of the pressure drop, causing the LPG to leak continuously. The first BLEVE (Boiling Liquid Expanding Vapour Explosion) occurred 15 minutes after the initial release and a series of BLEVEs ensued. The explosions were so catastrophic that they caused severe damage to the immediate surroundings with window/glass thrown away up to 4km from the source. There are also nearly similar incidents such as BP Texas City Refinery Explosion in 2005 and one that is in close proximity to Malaysia, Shell's Pulau Bukom (Singapore) Plant Fire Breakout in 2011. These incidents were usually caused by vapour cloud explosion (VCE) coming from pressure changes in the LPG sphere tank leakage. To ensure these incidents are not repeated, assessments will be made using models related to VCE.

Content from this work may be used under the terms of the Creative Commons Attribution 3.0 licence. Any further distribution of this work must maintain attribution to the author(s) and the title of the work, journal citation and DOI.
Published under licence by IOP Publishing Ltd

2. Method of Analysis

2.1 Study Area: LPG Storage Facility at Feyzin Refinery, France

On the morning of January 4th, 1966, a series of explosion went down at a typical LPG storage facility in Feyzin (Rhône), France. Due to a human error caused by an operator handling the valves, there came to be a leak coming from T61-443 propane sphere that brought about a disastrous BLEVE incident. According to the French Ministry of the Environment (1), the Feyzin refinery has a total capacity of 13,000 m^3 and is located 22.5 m from the a highway called the A7 highway. The fireball created a destructive blast wave through Rhône valley, shattering windows up to a distance of 8 km. Forty-five minutes after the first BLEVE occurred at tank T61-443, a second BLEVE ensued at tank T61-442. The explosion resulted in a death toll of 18 people, 89 injured and damaging 6 fire trucks, and 1475 shelters with its explosions. Twelve storage vessels were also destroyed; 6 spheres, 2 cylinders and 4 floating cap tanks while tonnes of flammable materials were burned; 1012 t of propane, 2027 t of butane and 1500 t of LPG products (2). Based on Figure 1, the propane gas started to escape from the 2-inch bottom-venting pipe of sphere tank T61-443 for approximately 10 minutes when the operator failed to close the first valve. During that time, Török et al. (2011) estimated that the initial mass flow rate of propane released into the atmosphere was 11.5 kg/s.

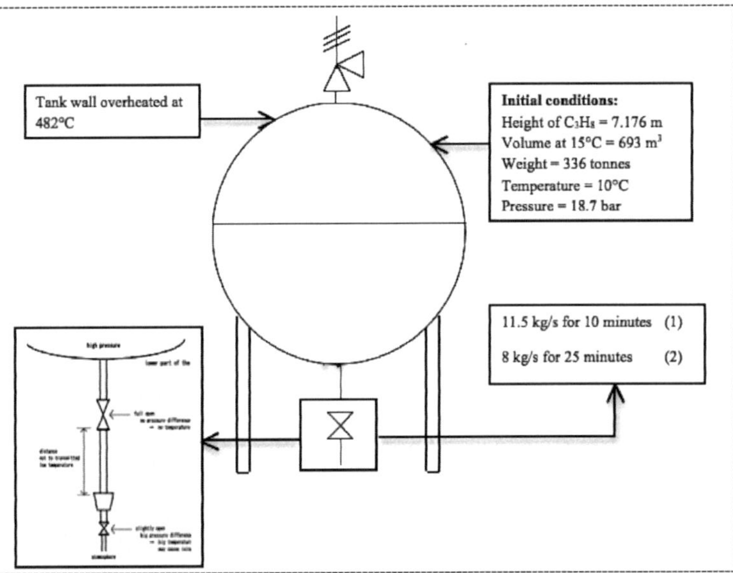

Figure 1. Simplification of fire explosion occurrence at sphere tank T61.443

Circumstances above lead to the released propane at estimated 6.9 tonnes (11.5kg/s x 600s). In the first 10 minutes after the leakage started, no fire or explosion had occurred in the Feyzin refinery plant. Witnesses stated that the fire incident at tank T61-443 only happened 25 minutes after the dispersion of propane cloud around the plant. The cause of fire came from a car that had moved into the propane cloud, resulting the ignition of the cloud, producing a flash fire that propagated back to tank T61-443 (1). Based on Davenport et al. (3) findings, the delayed ignition occurred only 60 minutes after release time of propane gas at sphere T61-443 and the drop of flow rate to 8kg/s after 10 minutes and the quantity of released propane for the next 25 minutes was estimated at 12 tonnes (8kg/s x 1500s). This makes the total amount of liquid propane released was 18.9 tonnes (6.9+12 tonnes). Although there was an uncontrolled release (leakage) from the valve opening to minimize the pressure changes, the gas released was insufficient to relieve the pressure rise in the sphere. Table 1 provides vital information on the chemical and physical properties of LPG (propane) that was stored in tank T61-443. The data are needed to determine the impacts from the incident.

IOP Conf. Series: Materials Science and Engineering **358** (2018) 012073 doi:10.1088/1757-899X/358/1/012073

Table 1. Physical and Chemical Properties of Propane, C_3H_8

PHYSICAL PROPERTIES
• Appearance: Colourless gas. (Cold vapour cloud may be white but the lack of visible gas cloud does not indicate the absence of gas). A colourless liquid when pressurized. • Vapour density: 1.87 kg/m^3 at 15°C and 1 atm • Liquid density: 580 kg/m^3
CHEMICAL PROPERTIES
• Molecular weight: 44.09 • Melting point/freezing point: -189.7°C (-309.4°F) • Boiling point: -42.11°C (-43.8°F) • Flash point: -104°C (-155.2°F) • Auto-ignition temperature: 450°C (842°F)

2.2 Prediction Methods

In this paper, selected models used in the analysis of VCE will be discussed. Thus, the result obtained from the aforementioned accidents will be carefully analyzed to estimate the suitability of these models.

2.2.1TNT Equivalent and Baker-Strehlow Methods versus Planas-Cuchi + TNT Equivalent Coupling Method

Previously, the TNT and Baker-Strehlow (BS) methods were often used to predict VCE blast load impacts. TNT model is considered to be less accurate as it does not take into account the obstacles that may affect the pressure contours. On the other hand, Baker-Strehlow includes obstacle into its calculation, making it more accurate than TNT. Nevertheless, all 3 models do not take into account for what happens to the flammable material (LPG) contained in the sphere tank with respect to its thermodynamics aspect, thus an inaccurate result is produced. The Planas-Cuchi and TNT Equivalent Coupling method considers the moment the flammable material is within the operating conditions until just before it explodes and forms VCE. What happens inside the sphere is already justified by the experimental work done using the Peng-Robinson Equation of State simulation prepared using the MATLAB (Source: BiTP Vol. 30 Issue 2, 2013, pp. 31-39). This reduces the inaccuracies gap of the calculation result. Taking the operating pressure as a criterion for analyzing the magnitude of consequences impact, a pressure of 10 bars to design pressure of sphere tank of 60 bar was constructed at an interval of 10 bar. The worst pressure consequences can be determined from these stages of processes. The safety valve was set to lift open at 20 bar (corresponding to propane temperature of 60℃) to prevent the internal pressure of tank from reaching its rupture pressure. Therefore, it is safe to assume that the pressure inside the vessel had remained at 20 bar whilst boiling off the liquid propane into vapour. There was a sudden physical process related to the disintegration of tank and rapid transition in the state of the LPG present in the tank that created a wave of overpressure that propagated through the atmosphere, causing some serious damages from its immense energy. When the wall of the tank begun to fracture, a rapid pressure drop occurred down to p_{atm}, at which the boiling temperature for liquefied gases is significantly lower than the ambient temperature. The liquid was released in which part of it evaporated and rapidly created a boiling pool, or its vapour burnt when ignited. Then, propane will rapidly change its state from that of liquid to gas. This liquid-to-gas transition will result in a tremendous increase in volume taken up by the LPG in the tank, causing it to exceed the critical parameters; a change from liquid to 'overcritical liquid' state. This will inevitably result in an explosion of the 'overcritical liquid' contained in the tank.

2.2.2Planas-Cuchi + TNT Equivalent Coupling Method

This coupling method is used to determine the Peak Side-On Overpressure, P°, by considering the thermodynamic properties of propane (C_3H_8) at various pressure differences. To get the Explosion

Energy, E value at the respective pressure difference, ΔP, the thermodynamic properties of propane was identified. Table 2 below provides the thermodynamic properties of saturated propane by view of the pressure (bar).

Table 2. Thermodynamic properties of saturated propane (C_3H_8) – pressure table

Pressure,	Temperature		Specific Volume, m^3/kg		Internal Energy, kJ/kg	
bar	°C	K	v_f	v_g	u_f	u_g
17.00	49.65	322.8	2.227	0.02606	228.3	472.7
18.00	52.30	325.45	2.253	0.02441	236.2	474.9
19.00	54.83	327.98	2.280	0.02292	243.8	476.9
20.00	57.27	330.42	2.308	0.02157	251.3	478.7
22.00	61.90	335.05	2.364	0.01921	265.8	481.7
24.00	66.21	339.36	2.424	0.01721	279.7	484.3
26.00	70.27	343.42	2.487	0.1549	293.1	486.2
28.00	74.10	347.25	2.555	0.01398	306.2	487.5
30.00	77.72	350.87	2.630	0.01263	319.2	488.1
35.00	86.01	359.16	2.862	0.009771	351.4	486.3

The steps in the Planas-Cuchi methodology below show the sequence on how to determine the Peak Side-On Overpressure. As a VCE occurs at an LPG storage facility, the vapour fraction with respect to total mass involved in the incident is to be calculated using Equation (1).

$$x = \frac{m_T P_{v_{LO}} - V_T P + m_T u_{LO} - U}{[(u_{LO} - u_{VO}) - (v_{VO} - v_{LO})P] m_T} \quad (1)$$

Once the value of U is obtained, it is then substituted into Equation (2) to calculate the overall variation of the internal energy of the tank's content, −ΔU.

$$-\Delta U = (u_{LO} - u_{VO}) m_T x - m_T u_{LO} + U \quad (2)$$

This methodology considers the real expansion work done by taking into account the real gas behaviour and adiabatic irreversible expansion from when the whole content of the tank changes from the state of explosion until it reaches its final state. Therefore, this work must be equivalent to the change in internal energy of the tank's content [refer to Equation (3)].

$$E = -P\Delta V = \Delta U \quad (3)$$

In order to get the TNT equivalent mass, W_{TNT} from the explosion energy, E, Equation (4) may be utilized.

$$W_{TNT} = \frac{\beta E}{\Delta H_{TNT}} = 1.068 \text{ x } 10^{-7} E \quad (4)$$

Once the TNT equivalent mass is calculated, the scaled distance, Z_e can be known using Equation (5).

$$Z_e = \frac{R}{(W_{TNT})^{1/3}} \quad (5)$$

Scaled Pressure, P_S can be calculated using Equation (6).

$$P_S(\text{kPa}) = \frac{1616\left[1+\left(\frac{Ze}{4.5}\right)^2\right]}{\sqrt{1+\left(\frac{Ze}{0.048}\right)^2}\sqrt{1+\left(\frac{Ze}{0.32}\right)^2}\sqrt{1+\left(\frac{Ze}{1.35}\right)^2}} \quad (6)$$

where, P^o is calculated by using an Equation (7) and (8).

$$P_S = \frac{P^0}{P_a} \quad (7)$$

$$\text{P}^0(\text{kPa}) = \text{P}_\text{a} \cdot \frac{1616\left[1+\left(\frac{Ze}{4.5}\right)^2\right]}{\sqrt{1+\left(\frac{Ze}{0.048}\right)^2}\sqrt{1+\left(\frac{Ze}{0.32}\right)^2}\sqrt{1+\left(\frac{Ze}{1.35}\right)^2}} \quad (8)$$

whereby, P^o is the overpressure, and P_a is the atmospheric pressure.

2.3 *Prediction of Burst Pressure over Sphere Wall T61-443*

The rupture pressure can be estimated from knowledge of the membrane stress in a spherical vessel.

Tensile strength of structural steel $\sigma_{TS} = 620 \text{ MN/m}^2$,

$$\text{Rupture pressure, } P_R = \frac{4t\sigma_{TS}}{D} \quad (9)$$

2.4 *Prediction of Impact from Sphere Wall T61-443 Fire Explosion*

Table 3 and 4 are referred to estimate the consequences of overpressure on human and structures.

Table 3. Consequences of Overpressure on Human and Structures (4)

Overpressure (kPa)	Effect on Structures	Effect on the Human Body
6.9	Window glass shutters	Light injuries from fragments occur
13.8	Moderate damage to houses (windows and doors blown out and severe damage to roofs)	People injured by flying grass or debris
20.7	Residential structures collapse	Serious injuries are common, fatalities may occur

34.5	Most buildings collapse	Injuries are universal, fatalities are widespread
69.0	Reinforced concrete buildings are severely damaged or demolished	Most people are killed
137.9	Heavily built concrete buildings are severely damaged or demolished	Fatalities approach 100%

Table 4. Consequences of Overpressure on Building and Structures (4)

Overpressure (kPa)	Damage
0.21(E_1)	Occasional breaking of large glass windows already under strain
0.69(E_2)	Breakage of small windows under strain
1.03(E_3)	Typical pressure for glass breakage
2.07(E_4)	"Safe distance" (probability 0.95 of no serious damage below this value); projectile limit; some damage to house ceilings; 10% window glass broken
2.76(E_5)	Limited minor structural damage
3.4 – 6.9(E_6)	Large and small windows usually shatter; occasional damage to window frames
4.8(E_7)	Minor damage to house structures
6.9(E_8)	Partial demolition of houses, inhabitable, corrugated asbestos shatters; corrugated steel/aluminium panels, fastenings fail, followed by buckling; wood panels (standard housing), panels blow in
6.9 – 13.8(E_9)	Partial collapse of walls and roofs of houses, concrete or cinder block walls, not reinforced, shatter
13.8(E_{10})	50% destruction of brickwork of houses
13.8 – 20.7(E_{11})	Frameless, self-framing steel panel buildings demolished; rupture of oil storage tanks
17.2(E_{12})	Cladding of light industrial buildings ruptures
20.7 – 27.6(E_{13})	Wooden utility poles snap; tall hydraulic presses (40,000 lb) in buildings slightly damaged
34.5 – 48.2(E_{14})	Nearly complete destruction of houses
68.9(E_{15})	Probable total destruction of buildings; heavy machine tools (7,000 lb), moved and badly damaged, very heavy machine tools (12,000 lb) survive

3. Results and Discussion

In this analysis, the amount of propane involved was 336,000 kg with a total volume of 1,218 m^3. To compare the Planas-Cuchi and TNT Equivalent Coupling method with TNT equivalent and Baker-Strehlow methods, the same radius of 50 m, 100 m, 150 m, 160 m, 300 m, 2.2 km, 4.2 km, 8 km, and 16 km have been used as fixed variable in the comparison of peak overpressure between the 3 methods.

IOP Conf. Series: Materials Science and Engineering **358** (2018) 012073 doi:10.1088/1757-899X/358/1/012073

Table 5: Comparison prediction of peak overpressure between Planas-Cuchi and TNT Equivalent Coupling, TNT equivalent and Baker-Strehlow methods.

Incident Case	Radius	Planas-Cuchi + TNT equivalent method	TNT equivalent method	Baker-Strehlow method	ARIA Report for Ministry of Environment French (witness' observation)
336 t of propane at T61-443	r (m)	P^0 (kPa)	P^0 (kPa)	P^0 (kPa)	Diagnose
	50	91.78 (E_{15})	1219.18 (E_{15})	4.36 (E_6)	Fatalities approach 100% (E_{15})
	100	27.63 (E_{13})	255.52 (E_{15})	4.36 (E_6)	Injuries are universal, fatalities are widespread (E_{13})
	150	16.05 (E_{11})	100.25 (E_{15})	4.36 (E_6)	Serious injuries, fatality may occur (E_{11})
	160	14.83 (E_{11})	86.79 (E_{15})	4.36 (E_6)	Serious injuries, fatality may occur (E_{11})
	300	7.31 (E_9)	24.73 (E_{13})	2.94 (E_5)	Serious injuries, fatality may occur (E_{11})
	2,200	0.96 (E_3)	4.22 (E_7)	0.42 (E_1)	Roofs damaged (E_4)
	4,200	0.50 (E_2)	2.18 (E_4)	0.22 (E_1)	Walls moved (E_4)
	8,000	0.26 (E_1)	1.146 (E_3)	0.13 (E_1)	Blast from explosion was felt, doors opened (E_2)
	16,000	0.13 (E_1)	0.284 (E_1)	0.10 (E_1)	Blast from explosion was felt, doors opened (E_1)

Based on Equation [9], the blast from sphere tank's wall will occur when the pressure inside the sphere is close or over 80 bar ($P_R = \frac{4t\sigma_{TS}}{D} = \frac{4\ x\ 0.045\ x\ 620x10^6}{14} = 79.71$ bar ≈ 80 bar). Taking $\Delta P = 800$ kPa to be constant at each radius, Table 5 shows the comparison between the 3 models. The overpressures at different radii were calculated using the 3 methods above and were compared to one another whilst using the French Ministry of Environment report as guideline in determining the accuracy of each model. Table 4 shows the Planas-Cuchi + TNT Equivalent Coupling method was the most accurate or most similar diagnose as the ones done in ARIA Report. On the other hand, TNT equivalent method alone shows some accuracy in for the near-range distances but soon begins to differ from ARIA's report. Meanwhile, Baker-Strehlow shows quite different values from the witness' observation. Between 50 – 160 m, the overpressure value generated from the explosion at tank T61-443 had dropped quite drastically for the Coupling method and TNT equivalent method, particularly in the latter while Baker-Strehlow model shows consistency. The overpressure effects had gone south of Rhône valley, causing damage to ceilings and room at 2.2 km away. At a distance of 4.2 km, it was observed that some walls were moved and damaged while inflicting minor structure damage and breaking windows at 8 km away. In addition, some villagers at Vienne, which was located at 16 km upstream from the refinery, had claimed that they felt the blast from the explosion. Although the damage impact done on building structures was interpreted through the means of calculating the overpressure from the Coupling and Baker-Strehlow models, this was found to have obviously deviated from the actual structural building analyzed in the report compared to the TNT equivalent model impact analysis results at 4.2 km, 8 km and 16 km. At 2.2 km, Planas-Cuchi + TNT equivalent model gave more correct value to the said report.

4. Conclusion

Current studies have shown that chemical process industries involving propane storages are risky and bear a high potential for the occurrence of incidents such as VCE and BLEVEs. All 3 models demonstrate that there was a decrease in overpressure value as the distance from the source of explosion became greater. This is practically understood as the energy of explosion reduces by time and distance during energy dissipation and dispersion. Nevertheless, both TNT Equivalent and Baker-Strehlow models show a great deviation in the value of overpressure produced as compared to that of Planas-Cuchi and TNT Equivalent Coupling model. This is particularly seen at distance 50 - 2200 m in which the Coupling model displays higher precision in the results produced when compared with what witness(es) had observed and analyzed in the French Ministry of Environment report. Table 5 proves the accuracy of the Planas-Cuchi and TNT Equivalent Coupling method's result on the overall distance; from 50 m to 16 km. Although the TNT Equivalent method shows more precise impact properties in accordance to the ARIA report from 4.2 km to 16 km, the result deviation from the Planas-Cuchi and TNT Equivalent Coupling method is not obvious for the distances said. Thus, for future fire explosion analysis that concerns the condition of pressure changes in a vessel, the Planas-Cuchi and TNT Equivalent Coupling model would be most recommended.

Acknowledgment

We, the authors, would like to extend our sincerest gratitude to the Ministry of Higher Education (MoHE) for administering the provision to us in a form of research grant with the grant number (reference code) of FRGS/1/2016/TK02/UITM/02/13. We would also like to thank the Research Management Institute (RMI) of Universiti Teknologi MARA (UiTM) (600-RMI/ FRGS 5/3(0094/2016)) for managing our affairs in attending this conference. We appreciate all the hard work and dedication provided by the staff and all those involved.

References

[1] Environment FM of the BLEVE in an LPG storage Facility at a rafinery Feyzin (Rhône) 2008 *ARIA*.**1** p1–20

[2] Török Z, Ajtai N, Turcu AT, Ozunu A 2011 Comparative consequence analysis of the BLEVE phenomena in the context on Land Use Planning; Case study: The Feyzin accident. *Process Saf Environ Prot*. **89(**1) p1–7

[3] Lenoir EM, Davenport JA 1993 A survey of vapor cloud explosions: Second update *Process Saf Prog [Internet]* Jan [cited 2017 May 31];**12**(1):12–33. Available from: http://doi.wiley.com/10.1002/prs.680120104

[4] Ferradás EG, Alonso FD, Miñarro MD, Aznar AM, Gimeno JR, Pérez JFS 2008 Consequence analysis by means of characteristic curves to determine the damage to buildings from bursting spherical vessels *Process Saf Environ Prot*. **86**(3) p175–81.